Antenna and Array Technologies for Future Wireless Ecosystems

IEEE Press
445 Hoes Lane
Piscataway, NJ 08854

Antenna and Array Technologies for Future Wireless Ecosystems

Edited By

Y. Jay Guo
University of Technology Sydney (UTS), Ultimo, Australia

Richard W. Ziolkowski
University of Technology Sydney (UTS), Ultimo, Australia

IEEE PRESS
WILEY

Published by John Wiley & Sons, Inc., Hoboken, New Jersey.
Published simultaneously in Canada.

Library of Congress Cataloging-in-Publication Data applied for:

Hardback ISBN: 9781119813880

Cover Design: Wiley
Cover Images: © Chi Hou Chan, Christopher Holloway, Inigo Liberal, Stefano Maci, Y.P. Zhang, and Richard W. Ziolkowski

Set in 9.5/12.5pt STIXTwoText by Straive, Chennai, India

We bring to you this book with the hope that our collection of contributions to the science and engineering of electromagnetics from eminent authors from around the world will provide useful knowledge for current and future generations of researchers and will help stimulate creative ideas to achieve advanced wireless technologies for the benefit of humankind.

Contents

Author Biographies *xv*
List of Contributors *xvii*
Preface *xxi*

1 **Surface-Wave Based Metasurface Antennas** *1*
Enrica Martini, Marco Faenzi, David González-Ovejero, and Stefano Maci
1.1 Introduction *1*
1.2 Typologies of Pixels *2*
1.3 Flat Optics Analysis *5*
1.4 Multiscale Analysis and Synthesis *7*
1.4.1 Continuous BC Synthesis *9*
1.4.2 Element Synthesis *9*
1.4.3 Virtual Prototyping *10*
1.5 Dual Polarization *10*
1.5.1 Double-Mode Excitation by a Single Feed *10*
1.5.2 Aperture Sharing and Double-Feed *11*
1.5.3 Central and Peripheral Excitations *12*
1.6 Beam Shaping *15*
1.7 Limit of Aperture Efficiency and High-Gain Examples *15*
1.7.1 Various Types of Efficiencies *15*
1.7.2 Bounds on the Maximum Aperture Efficiency *17*
1.7.3 Feed Efficiency *19*
1.7.4 Example of MTS Antenna with High Aperture Efficiency *20*
1.7.5 Example of a Very High-Gain Antenna *21*
1.8 Wideband and Limit of Bandwidth-Gain Product *22*
1.8.1 Gain and Bandwidth for Uniform Phase Modulation *22*
1.8.2 Gain and Bandwidth for Non-uniform Phase Modulation *22*
1.8.2.1 Exponential Modulation *24*
1.8.2.2 General Form of the Modulation Period *24*
1.8.2.3 Range of Applicability *28*
1.8.3 Phase Center Stability *28*
1.9 Multibeam and Multi-frequency *29*
1.9.1 Single Point Source and Dual Beam *29*
1.9.2 Multi-point Sources and Multibeam, with Total Aperture Sharing *30*
1.9.3 Radar Monopulse Application *31*

1.9.4 Multi-frequency Metasurface Antennas *34*
1.10 Conclusions *37*
Acknowledgments *37*
References *37*

2 Techniques for Designing High Gain and Two-Dimensional Beam Scanning Antennas for 5G *43*
Raj Mittra, Abdelkhalek Nasri, Ravi K. Arya, Prashant Chaudhary, James R. Kelly, Shaker Alkaraki, and Alejandro L. Borja
2.1 Introduction *43*
2.2 Luneburg Lens Designs *44*
2.2.1 Introduction *44*
2.2.2 Fabrication of Luneburg Lenses *45*
2.2.2.1 Use of Dielectric Sheets with Holes *45*
2.2.2.2 Use of 3D Printing *46*
2.2.2.3 Use of Specialized Materials *47*
2.2.3 Luneburg Lens with a Flat-Base and a Waveguide Array Excitation *47*
2.2.4 Validation of Luneburg Lens Design *48*
2.2.5 Dual-Polarized Luneburg Lens Antenna Feed Systems *49*
2.2.5.1 Dual-Linear Feed Design *51*
2.2.6 Size-Reduction of Luneburg Lens *53*
2.3 Gain Enhancement Approaches for Antennas and Arrays *58*
2.3.1 High Gain Wideband Microstrip Patch Antenna and Arrays Using Wings *58*
2.3.2 High Gain Slotted Waveguide Antenna Arrays *61*
2.3.2.1 Gain Enhancement Using Grooves and Tilted Wings *62*
2.3.2.2 Gain Enhancement Using V-Shaped Wings *63*
2.3.2.3 Gain Enhancement by Using Two Arrays, Each with Wings *63*
2.3.3 Two-Dimensional Beam Scanning with Slotted Array Antennas *66*
2.3.3.1 Longitudinal Plane Beam Scanning Using U-Shaped Phase Shifter *66*
2.3.3.2 Transversal Beam Scanning Using Phase Tapers in the Feed Ports *68*
2.3.3.3 Two-Dimensional Beam Scanning Using Three SCAAs Augmented with Wings *68*
2.4 Reconfigurable Liquid Metal-Based SIW Phase Shifter *70*
2.4.1 Concept and Structure of the SIW Phase Shifter *71*
2.4.1.1 Switched-Line SIW Phase Shifter for Coarse Phase Steps *71*
2.4.1.2 Reactively Loaded SIW Phase Shifter for Fine Phase Steps *73*
2.4.2 Practical Considerations of Fabrication *73*
2.4.3 Results and Discussion *74*
2.5 Summary *76*
References *79*

3 Low-Cost Beam-Reconfigurable Directional Antennas for Advanced Communications *83*
Qi Luo, Steven Gao, Xue-xia Yang, and Josaphat T. Sri Sumantyo
3.1 Introduction *83*
3.2 Beam-Reconfigurable Antenna Using Active Frequency Selective Surfaces *83*
3.3 Beam-Reconfigurable Antenna Using Parasitic Elements *88*
3.4 1-Bit Reflectarray and Transmitarray *91*

3.4.1 1-Bit Reflectarray *92*
3.4.2 1-Bit Transmitarray *96*
3.5 Beam-Switching and Multi-Beam Lens *98*
3.5.1 Extended Hemispherical Lens *98*
3.5.2 Luneburg Lens *100*
3.6 Beam-Reconfigurable Array Using Tunable Dielectrics *103*
3.7 Other Techniques to Realize a Beam-Reconfigurable High-Directivity Antenna *105*
3.7.1 Fixed-Frequency Beam-Scanning Leaky-Wave Antennas *105*
3.7.2 Mechanical 2D Beam-Steering *106*
3.7.3 Beam-Steering Fabry–Perot Resonator Antennas *108*
3.7.4 Beam-Reconfigurable Arrays Using Low-Cost Beamforming Networks *110*
3.8 Summary *112*
References *113*

4 Smart Leaky-Wave Antennas for Iridescent IoT Wireless Networks *119*
José L. Gómez-Tornero
4.1 Leaky-Wave Antennas for Efficient Wireless Systems *119*
4.2 Low-Cost Printed-Circuit Frequency-Scanning LWAs *123*
4.3 LWAs for Efficient Communications in Iridescent Wireless Networks *131*
4.4 LWAs Applied for Localization in Practical Wireless Networks *140*
4.4.1 FS-LWAs for DoA in Wi-Fi WLANs *144*
4.4.2 FS-LWAs for DoA in BLE WPANs *153*
4.4.3 Two-Dimensional Localization *157*
4.4.4 Near-Field Effects *158*
4.4.5 Use of FS-LWA in WSNs *162*
4.5 LWAs for Efficient Wireless Power Transfer *163*
4.6 Conclusion *168*
References *168*

5 Antenna-in-Package Design for Wireless System on a Chip *183*
Y. P. Zhang
5.1 Introduction *183*
5.2 High-Volume Manufacturing of an AiP Module *185*
5.2.1 LTCC *185*
5.2.2 HDI *186*
5.2.3 FOWLP *186*
5.3 Design Consideration *187*
5.3.1 Radiating Element *187*
5.3.2 Ground Plane *189*
5.3.2.1 Meshed Ground Plane *190*
5.3.2.2 Defected Ground Plane *190*
5.3.2.3 Patterned Ground Plane *191*
5.3.3 Feed Network *191*
5.3.3.1 Wire *192*
5.3.3.2 Bump *194*
5.3.3.3 Via *196*
5.3.4 Metal Fill *196*

5.3.5 Shielding Structure *197*
5.3.5.1 Coupling Mechanism *197*
5.3.5.2 CFS and CPS *198*
5.3.6 Cooling Method *199*
5.3.6.1 Passive Cooling *199*
5.3.6.2 Active Cooling *200*
5.4 Testing AiP *200*
5.4.1 Test System *200*
5.4.2 Testing Strategy *201*
5.5 Three AiP Examples *202*
5.5.1 A Wire-Bond AiP for a Receiver at 60 GHz *202*
5.5.2 A Flip-Chip AiP for a 5G NR at 28 GHz *204*
5.5.3 A Fan-Out AiP for a Drone Radar at 122 GHz *208*
5.6 Concluding Remarks *209*
References *209*

6 Terahertz Lens Antennas *215*
Geng-Bo Wu, Kwun Wing Cheung, Ka Fai Chan, and Chi Hou Chan
6.1 Introduction *215*
6.2 Printing with 3D Printers from Formlabs *217*
6.3 Measurement Platforms for Lens Antennas *218*
6.4 Pixel Design of the Discrete Dielectric Lenses *220*
6.5 Integration of a Dielectric Polarizer with a Lens for CP Radiation *222*
6.6 Design, Fabrication, and Testing of THz Lens Antennas *224*
6.6.1 Far-Field and Near-Field Focusing LP Lenses *225*
6.6.2 Far-Field and Near-Field Focusing CP Lenses *228*
6.6.3 Bessel Beam LP Lenses *232*
6.6.4 Bessel Beam LP Lenses Carrying OAM *235*
6.7 Summary *244*
Acknowledgment *247*
References *247*

7 Photonics-Based Millimeter-Wave Band Remote Beamforming of Antenna Arrays Integrated with Photodiodes *253*
Shigeyuki Akiba and Jiro Hirokawa
7.1 Introduction *253*
7.2 Configuration of Photonics-Based Antenna Beamforming Utilizing the RoF Technique *254*
7.2.1 Basic Concept and Configuration *254*
7.2.2 Generation of RoF Signals with Different RF Phases to be Fed to Antenna Element *255*
7.2.3 Transmission of Multiple RoF Signals with Different RF Phases *256*
7.3 Two-Dimensional (4 × 2) Antenna Array Integrated with Photodiodes for 60 GHz Band Applications *257*
7.3.1 Design and Fabrication *257*
7.3.2 Beamforming Utilizing WDM RoF Transmission *261*
7.3.3 Fading of an Amplitude Modulated RoF Signal Due to Chromatic Dispersion *263*

7.3.4 Digital Signal Transmission at 3.5-Gbit/s *264*
7.3.5 Wireless Communication Range *266*
7.4 Compact Antenna Module Integrated with Photodiodes to Achieve a Flexible Array Pattern *267*
7.4.1 Design and Fabrication of an Integrated Photonic Antenna Module *267*
7.4.2 Beamforming by an Array of the Antenna Array Modules *269*
7.5 Direct Delay Control for Beamforming by Variable Optical Delay Devices with 10 Gbit/s Class Data Transmission *271*
7.5.1 Direct Delay Control by Optical Variable Line *271*
7.5.2 Demonstration of Beamforming of a Digital Signal *273*
7.6 Antenna Array Beamforming Using a 1.3-μm Band EML as the Light Source *276*
7.7 Perspectives *279*
7.8 Estimation of the Direction of a User Terminal *279*
7.9 Summary *281*
Abbreviations *282*
References *283*

8 Contemporary Array Analysis Using Embedded Element Patterns *285*
David B. Davidson and Karl F. Warnick
8.1 Introduction *285*
8.2 Design Methods: Classical Array Factor *286*
8.2.1 Array Theory: The Uniform Linear Array *286*
8.3 Embedded Element Patterns *288*
8.4 Approximate Analysis Methods: The Lossless, Resonant, Minimum Scattering Approximation *290*
8.4.1 LRMSA Examples *291*
8.4.2 Isotropic Radiators *292*
8.5 Exact Design Methods: Full-Wave Computational Electromagnetics Modeling *292*
8.5.1 MoM Modeling of Thin Wires *293*
8.5.2 Beamforming and Beam Steering *294*
8.5.2.1 Limitations of the Array Factor Method *294*
8.5.2.2 The Maximum Directivity Beamformer *295*
8.5.2.3 An Example: A Base Station Array *295*
8.5.3 Summarizing Array Analysis Options *297*
8.6 Receive-Only Systems *297*
8.6.1 Array Receivers: Network Model *297*
8.6.2 Array Receivers: Reciprocity and EEPs *298*
8.6.3 Array Receivers: Figures of Merit *299*
8.6.4 An Example of Receiver System Temperature Calculation *300*
8.7 An Open Question Around EEPs *301*
8.8 Conclusions *302*
References *302*

9 Angle-of-Arrival Estimation in Large-Scale Hybrid Antenna Arrays *305*
Kai Wu and Y. Jay Guo
9.1 Introduction *305*
9.1.1 Satellite Communications *306*

9.1.2 Simultaneous Wireless Information and Power Transfer (SWIPT) *306*
9.1.3 High-Speed Railway (HSR) Communications *307*
9.1.4 Angle-of-Arrival (AoA) Estimation *308*
9.2 Popular Hybrid Structures of Large-Scale Antenna Arrays *308*
9.3 AoA Estimation – State of the Art *310*
9.3.1 Narrowband Hybrid Array of Phased Subarrays *310*
9.3.2 Wideband Hybrid Array of Phased Subarrays *311*
9.3.3 Lens Antenna Arrays *312*
9.4 Fast and Accurate AoA Estimation Techniques for Large-Scale Arrays of Phased Subarrays *314*
9.4.1 Narrowband AoA Estimation *314*
9.4.1.1 Phase Shift Calculation *315*
9.4.1.2 Estimation of *Nu* *316*
9.4.1.3 Estimation of *u* *318*
9.4.2 Wideband AoA Estimation *318*
9.5 Fast and Accurate AoA Estimation *323*
9.5.1 Wide-Beam Synthesis *323*
9.5.2 DFT Beam Difference *326*
9.6 Conclusions *327*
References *328*

10 Electrically Small Antenna Advances for Current 5G and Evolving 6G and Beyond Wireless Systems *335*
Richard W. Ziolkowski
10.1 Introduction *335*
10.2 ESA Figures of Merit *336*
10.2.1 Bandwidth *336*
10.2.2 Directivity and Efficiency *338*
10.3 Overcoming Conventional ESA Stigmas and Trade-Offs *339*
10.3.1 Metamaterial-Based ESAs *339*
10.3.2 Metamaterial-Inspired ESAs *341*
10.3.3 Non-Foster Circuits *344*
10.4 More Complex Electrically Small NFRP Antennas *348*
10.4.1 Multifunctional Designs *348*
10.4.2 Reconfigurable Designs *349*
10.5 Directive Electrically Small NFRP Antennas *351*
10.5.1 Quasi-Yagi *351*
10.5.2 Huygens Dipole Antennas *353*
10.5.3 Additional Reported ESAs *356*
10.6 Forward Looking ESA Applications *357*
10.6.1 WPT and Rectennas *357*
10.6.2 Wirelessly Powered Sensors *359*
10.6.3 Pregtronics *360*
10.6.4 Millimeter-Wave Designs *362*
10.6.5 Beam-Steerable Huygens Dipole Antenna Arrays *363*
10.6.5.1 HDA Theory *364*
10.6.5.2 Array of Huygens Dipole Sources *365*

10.6.5.3 Proof-of-Concept Prototype *369*
10.7 Summary *372*
Acknowledgments *373*
References *373*

11 Overview of Rydberg Atom-Based Sensors/Receivers for the Measurement of Electric Fields, Power, Voltage, and Modulated Signals *393*
Christopher L. Holloway, Matthew T. Simons, Alexandra B. Artusio-Glimpse, and Joshua A. Gordon
11.1 Introduction *393*
11.2 Electric-Field Strength: EIT (On Resonant and Stark Shift) *394*
11.3 Uncertainties *400*
11.4 Detection of AM and FM Modulated Signals *400*
11.5 Phase Detection and Phase Modulated Signal Detection *401*
11.6 RF Power Measurements *405*
11.7 Voltage Measurements *406*
11.8 Other Applications: RF Camera, Angle-of-Arrival, Waveform Analyzer, Plasma Measurements, and Thermometry *407*
11.9 Conclusion and Discussion *412*
References *414*

12 Quantum Antenna Arrays *419*
Iñigo Liberal and Richard W. Ziolkowski
12.1 The Rise of Quantum Technologies *419*
12.2 Quantum Antenna Array Theory *420*
12.2.1 Macroscopic Quantum Electrodynamics *420*
12.2.2 The Quantum Field in the Far-Zone *422*
12.2.3 Decay Dynamics in the Noninteracting Markovian Regime *424*
12.3 Photon Statistics *426*
12.4 Linear Array of Quantum Emitters *427*
12.4.1 First-Order Correlations *429*
12.4.2 Second-Order Correlations *431*
12.4.3 *N*th-Order Correlations *432*
12.5 Isotropic Single-Photon Sources *434*
12.6 Quantum Superdirectivity? *437*
12.7 Quantum Antenna Array Technologies *438*
12.8 Forward-Looking Directions *440*
References *441*

Index *447*

Author Biographies

Y. Jay Guo is a Distinguished Professor and the founding Director of Global Big Data Technologies Centre (GBDTC) at the University of Technology Sydney (UTS), Australia. He is the founding Technical Director of the New South Wales Connectivity Innovation Network. Prior to joining UTS in 2014, he served as a Director in CSIRO for over nine years. Before joining CSIRO, he held various senior technology leadership positions in Fujitsu, Siemens and NEC in the United Kingdom. His research interests include antennas, mm-wave and THz communications, and sensing systems, as well as big data technologies. He has published five books and over 600 research papers including over 280 IEEE Transactions papers, and he holds 26 international patents. His main technical contributions are in the fields of Fresnel antennas, reconfigurable antennas, hybrid antenna arrays, and, most recently, analogue multi-beam antennas and joint communications and sensing (JCAS) systems for 6G.

Prof Guo is a Fellow of the Australian Academy of Engineering and Technology and a Fellow of IEEE. He was a member of the College of Experts of Australian Research Council (ARC, 2016–2018). He has won a number of the most prestigious Australian national awards including the Engineering Excellence Awards (2007, 2012) and CSIRO Chairman's Medal (2007, 2012). He was named one of the most influential engineers in Australia in 2014 and 2015, and one of the top researchers across all fields in Australia in 2020 and 2021, respectively. He and his students have won numerous best paper awards at international conferences.

Prof. Guo has chaired many international conferences and served as a guest editor for a number of IEEE publications. He was the Chair of International Steering Committee, International Symposium on Antennas and Propagation (2019–2021). He was the International Advisory Committee Chair of IEEE VTC2017, General Chair of ISAP2022, ISAP2015, iWAT2014 and WPMC'2014, and TPC Chair of 2010 IEEE WCNC, and 2012 and 2007 IEEE ISCIT. He served as Guest Editor of special issues on "Low-Cost Wide-Angle Beam Scanning Antennas," "*Antennas for Satellite Communications*," and "*Antennas and Propagation Aspects of 60–90 GHz Wireless Communications*," all in IEEE Transactions on Antennas and Propagation, Special Issue on "*Communications Challenges and Dynamics for Unmanned Autonomous Vehicles*," IEEE Journal on Selected Areas in Communications (JSAC), and Special Issue on "*5G for Mission Critical Machine Communications*," IEEE Network Magazine.

Richard W. Ziolkowski received the B.Sc. (*magna cum laude*) degree (Hons.) in physics from Brown University, Providence, RI, USA, in 1974; the M.S. and Ph.D. degrees in physics from the University of Illinois at Urbana-Champaign, Urbana, IL, USA, in 1975 and 1980, respectively; and an Honorary Doctorate degree from the Technical University of Denmark, Kongens Lyngby, Denmark in 2012.

Prof. Ziolkowski was the recipient of the 2019 IEEE Electromagnetics Award (IEEE Technical Field Award). He is a Life Fellow of the Institute of Electrical and Electronics Engineers (IEEE, 1994). He became a Fellow of OPTICA (previously the Optical Society of America, OSA) in 2006 and the American Physical Society (APS) in 2016. He was the 2014–2015 Australian DSTO Fulbright Distinguished Chair in Advanced Science and Technology. He served as the President of the IEEE Antennas and Propagation Society (AP-S) in 2005 and has had many other AP-S leadership roles. He is also actively involved with the International Union of Radio Science (URSI), the European Association on Antennas and Propagation (EurAAP), and the International Society for Optics and Photonics (SPIE) professional societies.

He is currently a Distinguished Professor in the Global Big Data Technologies Centre in the Faculty of Engineering and Information Technologies (FEIT) at the University of Technology Sydney, Ultimo NSW Australia. He became a Professor Emeritus at the University of Arizona in 2018, where he was a Litton Industries John M. Leonis Distinguished Professor in the Department of Electrical and Computer Engineering in the College of Engineering and was also a Professor in the College of Optical Sciences. He was the Computational Electronics and Electromagnetics Thrust Area Leader with the Engineering Research Division of the Lawrence Livermore National Laboratory before joining The University of Arizona, Tucson, AZ, USA, in 1990. His current research interests include the application of new mathematical and numerical methods to linear and nonlinear problems dealing with the interaction of electromagnetic and acoustic waves with complex linear and nonlinear media, as well as metamaterials, metamaterial-inspired structures, nano-structures, and other classical and quantum applications-specific configurations.

List of Contributors

Shigeyuki Akiba
Tokyo Institute of Technology
Department of Electrical and
Electronics Engineering
Tokyo
Japan

Shaker Alkaraki
Queen Mary University of London
School of Electronic Engineering and
Computer Science
London
UK

Alexandra B. Artusio-Glimpse
U.S. Department of Commerce, Boulder
Laboratories, National Institute of Standards
and Technology (NIST)
Boulder, CO
USA

Ravi K. Arya
XSL
Zhongshan Institute of CUST
China

Alejandro L. Borja
Universidad de Castilla-La Mancha
Departamento de Ingeniería Eléctrica,
Electrónica, Automática y Comunicaciones
Escuela de Ingenieros Industriales de Albacete
Albacete
Spain

Chi Hou Chan
State Key Laboratory of Terahertz
and Millimeter Waves (City University
of Hong Kong)
Hong Kong SAR
China

and

City University of Hong Kong
Department of Electrical Engineering
Hong Kong SAR
China

Ka Fai Chan
State Key Laboratory of Terahertz and
Millimeter Waves (City University of
Hong Kong)
Hong Kong SAR
China

Prashant Chaudhary
University of Delhi South Campus
Department of Electronic Science
New Delhi
India

Kwun Wing Cheung
State Key Laboratory of Terahertz and
Millimeter Waves (City University of
Hong Kong)
Hong Kong SAR
China

David B. Davidson
Curtin University
International Centre for Radio Astronomy Research
Perth, WA
Australia

Marco Faenzi
University of Sienna
Department of Information Engineering and Mathematics
Sienna
Italy

Steven Gao
University of Kent
School of Engineering and Digital Arts
Canterbury
UK

José L. Gómez-Tornero
Technical University of Cartagena
Department of Information and Communication Technologies
Telecommunication Faculty
Cartagena
Spain

David González-Ovejero
University of Rennes
CNRS
Institut d'Électronique et de Télécommunications de Rennes (IETR)
Rennes
France

Joshua A. Gordon
National Institute of Standards and Technology (NIST)
U.S. Department of Commerce
Boulder Laboratories
Boulder, CO
USA

Jiro Hirokawa
Tokyo Institute of Technology
Department of Electrical and Electronics Engineering
Tokyo
Japan

Christopher L. Holloway
National Institute of Standards and Technology (NIST)
U.S. Department of Commerce
Boulder Laboratories
Boulder, CO
USA

Y. Jay Guo
University of Technology Sydney
Global Big Data Technologies Centre
New South Wales, Sydney
Australia

James R. Kelly
Queen Mary University of London
School of Electronic Engineering and Computer Science
London
UK

Iñigo Liberal
Institute of Smart Cities (ISC), Public University of Navarre (UPNA)
Department of Electrical, Electronic and Communications Engineering
Pamplona
Spain

Qi Luo
University of Hertfordshire, School of Physics Engineering and Computer Science
Hatfield
UK

Stefano Maci
University of Sienna
Department of Information Engineering and Mathematics
Sienna
Italy

Enrica Martini
University of Sienna
Department of Information Engineering and Mathematics
Sienna
Italy

Raj Mittra
University of Central Florida
Electrical and Computer Engineering
Orlando, FL
USA

and

King Abdulaziz University
Electrical and Computer Engineering Department, Faculty of Engineering
Jeddah
Saudi Arabia

Abdelkhalek Nasri
University of Central Florida
Electrical and Computer Engineering
Orlando, FL
USA

Matthew T. Simons
National Institute of Standards and Technology (NIST)
U.S. Department of Commerce
Boulder Laboratories
Boulder, CO
USA

Josaphat T. Sri Sumantyo
Chiba University
Center for Environmental Remote Sensing
Chiba
Japan

Karl F. Warnick
Brigham Young University
Department of Electrical and Computer Engineering
Provo, UT
USA

Geng-Bo Wu
State Key Laboratory of Terahertz and Millimeter Waves (City University of Hong Kong)
Hong Kong SAR
China

Kai Wu
University of Technology Sydney
Global Big Data Technologies Centre
New South Wales, Sydney
Australia

Xue-xia Yang
Shanghai University
School of Communication and Information Engineering
Shanghai
China

Y. P. Zhang
Nanyang Technological University
School of Electrical and Electronic Engineering
Singapore
Singapore

Richard W. Ziolkowski
Global Big Data Technologies Centre
School of Electrical and Data Engineering
University of Technology Sydney
Ultimo, New South Wales
Australia

Preface

We are living in an age of technology acceleration. Over the last two decades, wireless technologies in particular have fundamentally changed the way we live and work. Now, for the majority of us, a life without Wi-Fi and mobile devices would be unimaginable. We are constantly demanding better mobile connectivity wherever we are. To advance wireless technologies further, research on the sixth generation (6G) wireless communications networks has started in earnest. For antenna researchers and engineers, one big question remains: What will the antenna technologies of the future look like? Instead of consulting a crystal ball, we have invited a number of world-class antenna researchers to join us in presenting their latest research findings and perspectives. This edited book is the result. It aims to cover some of the hottest and most promising antenna research topics and technologies from the astronomical to the quantum levels. We sincerely hope that the book will stimulate many creative ideas to advance wireless technologies for the future benefit of humankind.

Metamaterials have undoubtedly been one of the hottest antenna research topics over the last two decades. Consequently, we start the book with the latest achievements on beamforming using modulated metasurfaces by Prof. Stefano Maci's group (Italy). Metasurfaces (MTS) are thin metamaterials composed of a dielectric layer loaded with subwavelength inclusions whose geometrical features are properly varied in space to control the resulting macroscopic electromagnetic properties. The inclusions in the microwave range are represented, for example, by printed metallic elements deposited as a regular lattice on top of a dielectric layer. The elements in the terahertz (THz) range include metallic pins protruding from a ground plane. Chapter 1 is particularly focused on a special class of modulated MTS antennas based on the interactions between a cylindrical surface wave (SW) launched from a vertical monopole, and a modulated metasurface. The SW generated by the monopole is perturbed by the modulated boundary conditions (BCs) imposed by the metasurface and subsequently is transformed into a general curvilinear wavefront leaky-wave (LW) mode. The authors demonstrate how the SW-MTS interference can be tailored to generate single or multiple radiated beams with a variety of attractive performance characteristics including high gain and shaped patterns for single and dual frequency operations.

Owing to the demand for multi-beam antennas for 5G and beyond, research interest in Luneburg lens antennas has re-emerged recently. Chapter 2 presents a number of new concepts for designing practical Luneburg lens antennas and antenna arrays by a group of international authors led by Prof. Raj Mittra (USA). The focus of this chapter is on achieving high antenna gain (>30 dBi) together with a wide-angle beam scanning capability. It begins with new designs of Luneburg lens antennas supporting beam scanning over a very wide angular range. The feed array design for the lens is then discussed to facilitate beam scanning and the generation of multiple beams. The design of a hemispherical Luneburg lens antenna to reduce the lens volume and height follows. Other topics covered include the gain enhancement of a slotted waveguide antenna array for scanning in both the longitudinal and transverse planes.

Reconfigurable antennas have served as a major research topic over the last two decades. Current endeavors include the development of high performance reconfigurable beamforming antennas and their applications to practical systems in order to reduce system costs and power consumption. Chapter 3, which is contributed by an international team led by Dr. Qi Luo and Prof. Steven Gao (UK), focuses on low-cost solutions to beam-reconfigurable directional antennas. In principle, the cost of beamforming antennas can be reduced in two ways, namely, by limiting the number of active antenna elements or by avoiding the use of expensive RF components. Following this premise, a comprehensive overview of a wide range of techniques is presented. These include using low-bit phase quantization for reflectarray and transmitarray designs, employing active frequency selective surfaces (AFSS) or parasitic elements, introducing mechanically rotated metasurfaces, using lenses with multiple feeds, integrating tunable materials into the antenna design, combining tunable high impedance surface with Fabry-Perot cavity antennas, using low-cost beamforming networks, designing fixed-frequency beam-scanning leaky wave antennas, and exploiting SIW-based technologies.

Leaky wave antenna (LWA) innovations have risen steadily in the last decade because of their advantageous low cost and simplicity. Nevertheless, they have not appeared in many practical applications. The frequency scanning feature of conventional LWAs has been largely targeted to radar applications. In Chapter 4, Prof. José Luis Gómez-Tornero (Spain) presents a fresh investigation into how frequency-scanning LWAs can be used as low-cost smart antennas for access points in wireless networks such as those associated with the Internet of Things (IoT). The underlining premise is that one can exploit the inherent frequency dependence of the LWA beam directions to cover different areas using different frequency channels. LWA designs are presented that generate frequency-scanning directive beams which provide coverage in a desired angular region and which use the frequency channels available in the physical (PHY) layer protocol to facilitate radio access and communications.

The development of 5G millimeter wave (mm-wave) systems and modern mm-wave automotive radars are changing the ways we design mm-wave transceivers. To reduce costs and increase compactness, reliability and performance, a high degree of integration of a transceiver's antennas, and mm-wave circuits becomes essential. This approach has stimulated the growth and adoption of antenna in package (AiP) technologies. The AiP technology provides a good balance among system performance, size, and cost. Prof. YP Zhang (Singapore) presents a comprehensive overview and future outlook on AiP technologies in Chapter 5. It covers high-volume manufacturing of AiP modules in low temperature co-fired ceramic, high-density interconnects, and fan-out wafer level packaging technologies. Some important design considerations for AiP radiating elements, feed networks, ground planes, metal fills, shielding structures, and cooling methods are discussed. The issue of over-the-air testing of AiP modules in a production line is elaborated. Finally, three examples, i.e. wire-bond, flip-chip, and fan-out AiPs, are given with each representing one of the main AiP fabrication approaches.

The sixth generation (6G) of wireless communications networks are expected to occupy part of the terahertz (THz) spectrum to deliver the promised terabits per second data rates. To this end, there has been accelerated research on THz antennas. Challenges in THz antenna designs include significant metal and dielectric material losses and limited fabrication tolerances. In Chapter 6, the group led by Prof. Chi-Hou Chan (Hong Kong) presents their latest research on THz lens antennas. They have addressed these practical challenges by developing phase modulation techniques that are integrated into the THz lens antenna designs. In particular, they focus on discrete dielectric lenses (DDLs) as high-gain THz antenna candidates. DDLs offer a number of advantages such as low loss, simple feeding networks, arbitrary aperture phase control, and ease of fabrication. The authors demonstrate how to utilize inexpensive 3D printing technology to fabricate novel DDL antennas that can generate high-gain linearly-polarized (LP) beams. By integrating the THz LP lens with a 3D printed dielectric grating, circularly-polarized (CP) performance is also demonstrated. Modifications to the phase modulation that occurs at the lens aperture facilitates achieving focused LP and CP beams for THz imaging applications. A THz Bessel beam launcher is realized to obtain a high intensity beam focused over a considerable distance for near-field ultrahigh-speed data exchange.

Beamforming is a technology critical to 5G, 6G, and beyond communications networks. To increase user data rates, 5G and beyond metropolitan networks are adopting the concept of densely distributed small cells operating at mm-wave frequencies. Feeding signals to these small cells whilst supporting cost-effective beamforming and power efficiency is a major challenge. Profs. Shigeyuki Akiba and Jiro Hirokawa (Japan) present a new photonic approach to address this problem in Chapter 7. They employ a radio-over-fiber (RoF) transmission technique and photodiodes integrated with the array antenna elements. The RF signals in such a system are delivered to antennas at 60 GHz through low-loss optical fibers without changing the signal format. Direct delay control is achieved with variable optical delay devices. Compared with conventional systems using RF cables, their innovative approach avoids the power losses in the cables as well as in the phase shifters. Hence, it circumvents the conventional use of amplifiers. A method for estimating the direction of user equipment (UE), which is necessary for effective analogue beamforming, is also given.

Many applications, from radio astronomy and defense systems to 5G and beyond networks, require massive antenna arrays. These arrays can have hundreds and even thousands of antenna elements. Accurate analysis of these arrays is not only essential for saving development time and cost, but it is also an engineering necessity particularly since it is often impractical, if not impossible, to conduct all the testing required for a successful design. Chapter 8, written by Prof David B. Davidson (Australia) and Karl F. Warnick (USA), addresses this issue. The chapter outlines contemporary analysis methods for array designs. The key concept is to specifically include mutual coupling either approximately or exactly in the antenna design process. They describe how this is done by using embedded element patterns (EEPs). The theory of EEPs is reviewed along with a suitable framework for modelling mutual coupling in arrays. Properly including the electronics connected to the array, such as low noise amplifiers and beamforming networks, is also discussed. A particular beamformer which employs EEPs, the maximum directivity beamformer, is presented. Furthermore, an intermediate approach between the classical array factor analysis and the rigorous full-wave simulation methodology that requires obtaining the EEPs is introduced and characterized. The utility of this lossless, resonant, minimum scattering approximation (LRMSA) is demonstrated.

Massive antenna arrays not only pose challenges to array analysis, but also are costly and power hungry. These issues are particularly true if multi-beams and their associated beamformers are required. They have led to the birth of hybrid antenna arrays. In a hybrid antenna array, the antenna elements are grouped into analogue subarrays and the output signals from the subarrays are processed jointly by digital signal processing techniques. For massive multiple-input-multiple-output (MIMO) communications in the millimeter-wave and terahertz frequency bands, the main challenge in signal processing is the accurate estimation of the angle-of-arrival (AoA). In Chapter 9, Dr. Kai Wu and Prof. Y. Jay Guo (Australia) present an overview of their contributions to this new field. They begin by reviewing several important applications of large-scale antenna arrays in 5G and beyond 5G (B5G) communications systems. A number of popular hybrid structures associated with large-scale arrays are outlined. The importance and challenges of the AoA estimation in large-scale hybrid arrays and their applications are discussed. Some cutting edge AoA estimation methods for large scale hybrid antenna arrays are described. Finally, their recent progress on fast and accurate AoA estimation is reviewed.

A major feature of 5G and B5G systems is the support of the IoT. Many IoT devices are physically small and, hence, they require physically small antennas to enable their wireless functions. As it stands, mm-wave and THz transceivers are still too costly for low-cost IoT devices. As a consequence, the connectivity of IoT devices remains provided mostly at microwave frequencies. This leads to the demand of highly efficient electrically small antennas (ESAs). Moreover, advanced IoT devices and their applications demand multifunctional performance, particularly when their platforms are very space limited. Prof. Richard W. Ziolkowski (Australia, USA) provides an overview of past and current research on ESAs and several of their innovative applications in Chapter 10. The discussion is focused particularly on near field resonant parasitic (NFRP) antennas, and highlights the issues and realizations of multifunctional, reconfigurable, passive, and active ESA systems. Their Huygens dipole antenna versions and advantages are stressed. The chapter concludes with discussions on future research and development directions of ESAs including wireless power transfer (WPT) rectennas, wirelessly power sensors, structurally integrated systems (Pretronics), millimeter wave examples, and beam-steerable, wide angular coverage Huygens dipole antenna arrays for WPT applications.

While analysis and simulations are integral aspects in the design of antenna and antenna arrays, measurements provide the necessary validation of their efficacy. In fact, one of the keys to developing new science and technologies is to have sound metrological tools and techniques. Whenever possible, we would like them to make absolute measurements of the physical quantity of interest. A goal of the international metrology organizations is to make all measurements directly traceable to the International System of Units (SI). Measurements based on atoms provide one direct SI traceability path and enable absolute measurements of physical quantities. Written by Dr. Christopher L. Holloway and his colleagues at NIST-Boulder (USA), Chapter 11 presents an overview of the fundamentally new Rydberg atom-based measurement approach. Rydberg atoms are atoms with one or more electrons excited to a very high principal quantum number n; and, hence, measurements with them are SI traceable. They have several useful properties including the fact that their dipole moments are exceptionally large, scaling with n^2. These large dipole moments make Rydberg atoms very sensitive to electric fields and, hence, they are very useful for measuring them. Rydberg atom-based sensor and receiver systems now include power-sensors, spectrum analyzers, angle-of-arrival sensors and receivers for modulated communication signals. Rydberg atom-based systems will be beneficial for 5G and beyond measurements since they will facilitate traceable calibrations of both electric field strength and power at frequencies above 100 GHz, which are currently not available.

The book concludes with a futuristic chapter on quantum antenna arrays written by Dr. Iñigo Liberal (Spain) and Prof. Richard W. Ziolkowski (Australia, USA). While arrays of electrically small classical antennas are described in Chapter 10, the quantum antennas are physically extremely small, obeying the laws of quantum mechanics instead of those of classical electromagnetics. Its main focus is on how photon statistics can be engineered by designing the geometry of a quantum antenna array. The basic theory behind quantum antenna arrays is presented in a style that will provide antenna designers with the tools to explore how their knowhow in conventional antenna designs could be applied to the exciting field of quantum technologies. To this end, the authors explore intriguing non-classical light sources with no classical counterparts, including directionally entangled photon bunches and perfectly isotropic single-photon sources. A number of possible future research directions of quantum antenna arrays are provided to conclude the chapter and our book.

It has been a great privilege for us to work with all the contributing authors, many of whom are world-leading researchers in their fields. We thank them wholeheartedly for their generous time and efforts as well as their insights into current and future antenna and antenna array research topics. We are also very grateful for the friendships we have enjoyed with many of them. Despite the current Covid-19 pandemic across the globe, we are optimistic and look forward to being with all of our contributing authors in person, as well as you the readers, in the near future!

1

Surface-Wave Based Metasurface Antennas

Enrica Martini[1], Marco Faenzi[1], David González-Ovejero[2], and Stefano Maci[1]

[1] *Department of Information Engineering and Mathematics, University of Sienna, Sienna, Italy*
[2] *University of Rennes, CNRS, Institut d'Électronique et de Télécommunications de Rennes (IETR), UMR 6164, 35000 Rennes, France*

1.1 Introduction

Metasurfaces (MTSs) are thin metamaterials composed of a dielectric layer loaded with subwavelength inclusions, whose geometrical features are properly varied in space for controlling the macroscopic electromagnetic properties [1]. Several kinds of devices have been recently developed for the control of the reflection or transmission of space waves [2–6] or the dispersion of surface waves (SWs) [7–13]. The inclusions in the microwave range are represented by printed metallic elements deposited on a regular Cartesian or hexagonal lattice [14–25] that constitute a film on top of the dielectric layer (the same technology as a printed circuit board [PCB]). The elements in the terahertz (THz) range can be formed by protruding metallic pins on a ground plane [26–28].

In this chapter, we treat a special class of modulated MTS antennas based on an interaction between a cylindrical SW launched from a vertical monopole, typically in the center of a circular aperture, and a modulated MTS. The SW generated by the monopole is perturbed by the modulated boundary conditions (BCs) imposed by the MTS, and it is transformed into a general curvilinear wavefront leaky-wave (LW) mode [29, 30]. The SW–MTS interference can generate a large variety of radiation patterns. Figure 1.1 shows examples of MTS antennas in PCB technology operating at microwaves.

Several benefits are obtained with such smart modulated apertures, i.e. low profiles, light weights, and small losses. Their fabrication is low cost, relying on standard printing processes. Their in-plane feeding structures, comprised of a low-profile monopole connected to a circular patch, render the antennas impressively thin and simple. Furthermore, one of their most interesting aspects is that the analysis of the currents flowing in the MTS film can be carried out with excellent accuracy in quasi-analytical form by Flat Optics (FO), a method based on the adiabatic Floquet wave (FW) expansion of currents and fields. Both topics will be reviewed herein. This method relies on the homogenized approximation of the sheet ("transparent") impedance of the metallic cladding. The printed elements are indeed characterized by sub-wavelength dimensions (on the order of $\lambda/10$ up to $\lambda/5$, where λ is the free-space wavelength at the working frequency). Consequently, the SW feels continuous BCs. The printed elements behave therefore as pixels of an image and will be denoted as "pixels" hereinafter. The MTS equivalent homogenized surface impedance is "non-uniform," which means that it is constituted by elements whose geometry varies in space.

Antenna and Array Technologies for Future Wireless Ecosystems, First Edition.
Edited by Y. Jay Guo and Richard W. Ziolkowski.

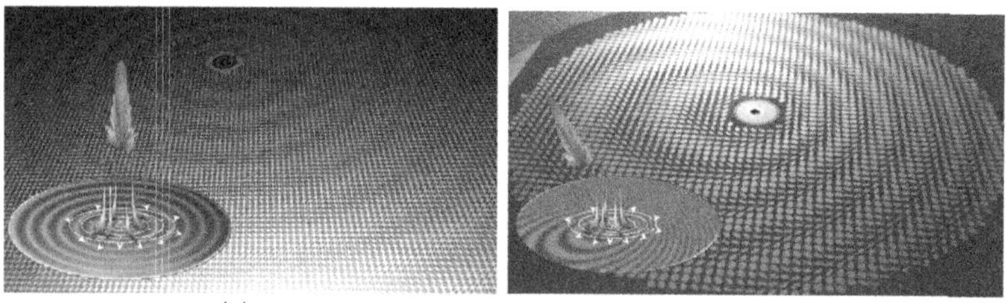

(a) (b)

Figure 1.1 Examples of MTS surface-wave based antennas. (a) Broadside radiation. The periodicity of the radial modulation matches the wavelength of the surface wave supported by the average impedance. (b) Elliptical modulation for tilted beam radiation.

The many MTS antennas models, design methods, bounds and high-performance realizations recently introduced by the authors are reviewed. Different examples of designs and implementations are shown that exhibit high gain, shaped patterns, multiple beams, and broadband or dual frequency operation. The different types of sub-wavelength pixels are described in Section 1.2 and the performance of each of them is discussed. The FO analysis method and the associated adiabatic FW expansion are presented in Section 1.3. They constitute the basis for the electromagnetic description of the homogenized MTS. A global picture of a multiscale design process is given in Section 1.4 with a focus on the synthesis methods for the design of the impedance BC at the mesoscopic scale. Possible solutions for achieving dual-polarization capabilities are shared in Section 1.5. Examples of beam shaping are shown in Section 1.6. Sections 1.7 and 1.8 are dedicated to the limits of the aperture efficiency and the bandwidth-gain product. Examples of multi-beam and dual-frequency systems are presented in Sections 1.9.3 and 1.9.4, respectively. Conclusions are drawn in Section 1.10.

1.2 Typologies of Pixels

The basic structure of a modulated MTS antenna operating in the microwave range includes a grounded dielectric layer top loaded by a thin film formed with metallic pixels. These pixels are positioned at the nodes of a regular sub-wavelength lattice of Cartesian or hexagonal shape [31] (see Figure 1.1). The geometrical parameters characterizing the shape and dimensions of the pixels are chosen to reproduce the local value of the reactance profile defined by a certain level of homogenization. This way, a global aperture, continuous reactance pattern results, that is discretized sampled and reconstructed by several thousands of unit elements.

A simple pixel, i.e. an element realized with PCB technology, consists of a circular or square metallic patch whose dimension and orientation can vary. The layer constructed with such pixels leads to an almost isotropic surface. It is shown in [15] that these simple elements can create a beam with circular polarization (CP), but with poor performance in terms of polarization purity. This aspect can be improved by using anisotropic MTSs. As an example, Figure 1.2 shows a comparison between the radiation patterns of two CP MTS antennas composed of either isotropic (circular) or anisotropic (circular with a cut) patches, but with their other design parameters, i.e. slab thickness and relative permittivity, average impedance, modulation index, and radius of the antenna, the same. It is seen that the anisotropy permits a much better control of the polarization and a consequent increase of the antenna gain. It is therefore evident that anisotropic pixels offer much better performance. Some geometries of anisotropic pixels are presented in Figure 1.3,

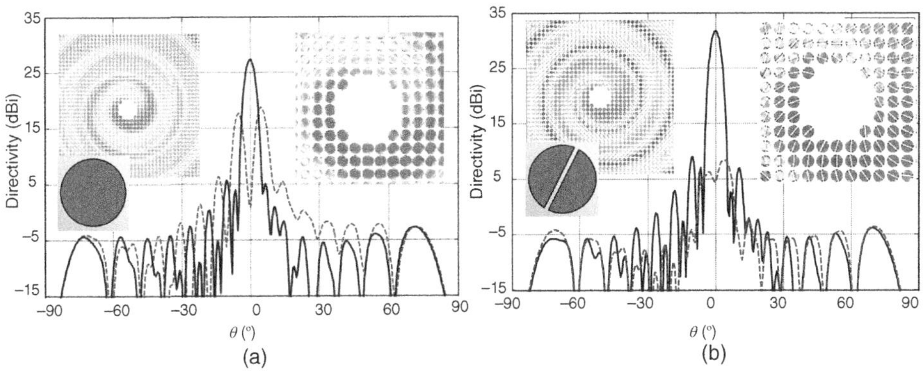

Figure 1.2 Comparison between the radiation patterns of (a) isotropic and (b) anisotropic pixel-based MTS antennas.

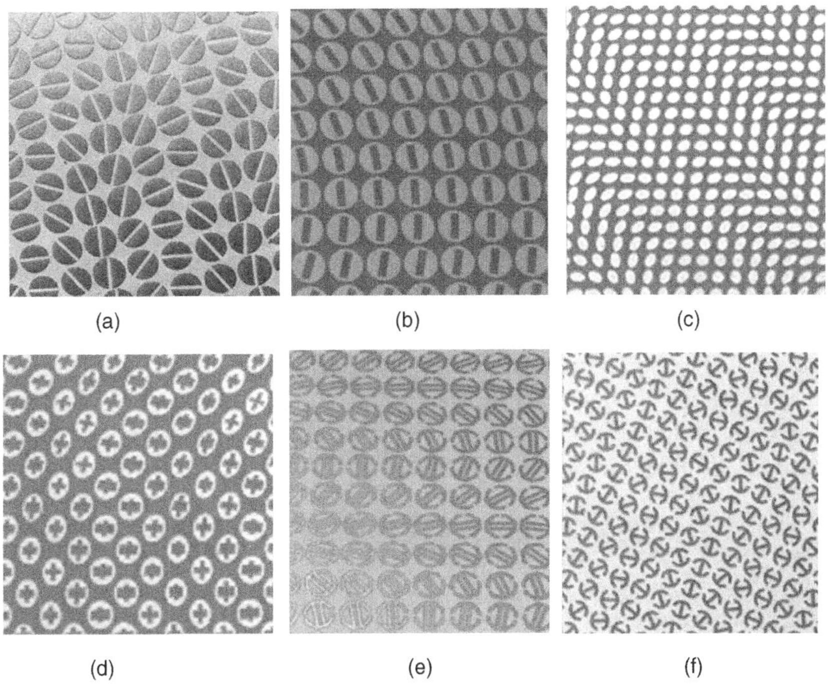

Figure 1.3 Details of the MTS pixels (a) Coffee bean element. (b) Circular patch with slot. (c) Grain of rice element. (d) Elliptical patch with cross-slot. (e) Double π element. (f) Double anchor element.

and their performance is summarized in Figure 1.4. All of them show two orthogonal symmetry axes, a property that when combined with the hypothesis of the absence of losses yields purely real cross-diagonal entries in the reactance tensor. The elements in Figure 1.4 show at least two geometrical parameters relevant to the pixel size, a, and rotation angle, ψ. Some of them show also an additional parameter b that accounts for the aspect ratio.

The table in Figure 1.4 indicates each pixel's performance in terms of some features of interest, including the realizable anisotropy level, the range of reactance, the losses, the element bandwidth, and the sensitivity to fabrication tolerances. In particular, the element bandwidth is strictly

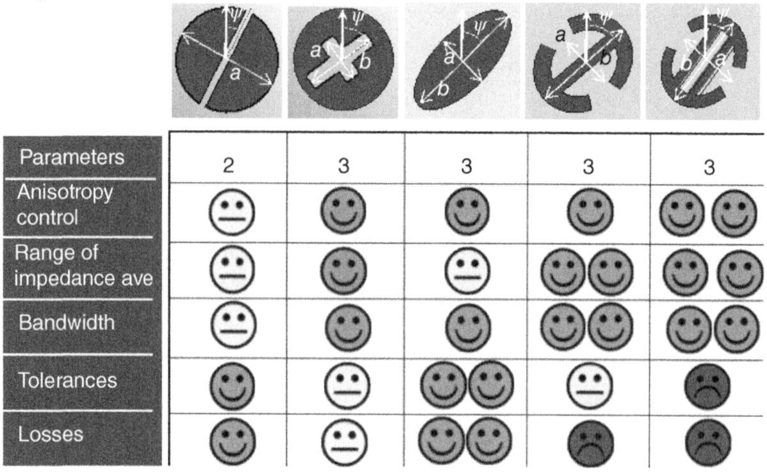

Figure 1.4 Performance of different shaped pixels for the synthesis of MTS antennas.

related to its capability of providing a highly capacitive load, i.e. the higher the capacitive loading is in a fixed lattice, the lower the relative effective dielectric constant of the slab is. The improvement in terms of bandwidth results from an increase of the group velocity associated with a corresponding propagation constant and depends on the reduction of the effective dielectric constant of the slab.

The PCB technology can be used up to 30–40 GHz. For increasing frequencies, the losses impose the use of different types of pixels that are created with micro-structured pillars. Figure 1.5 shows examples of all-metal antennas operating at 300 GHz, where the MTS is formed by micro-pillars having a (i) cylindrical or (ii) elliptical cross section. The corresponding radiation diagrams are shown in Figure 1.6. It is seen that the control of the polarization is much better when using elliptical cross-section micro-pillars, analogous to the PCB type pixels. This control feature arises from the extra degree of freedom the elliptical structure offers to the MTS design.

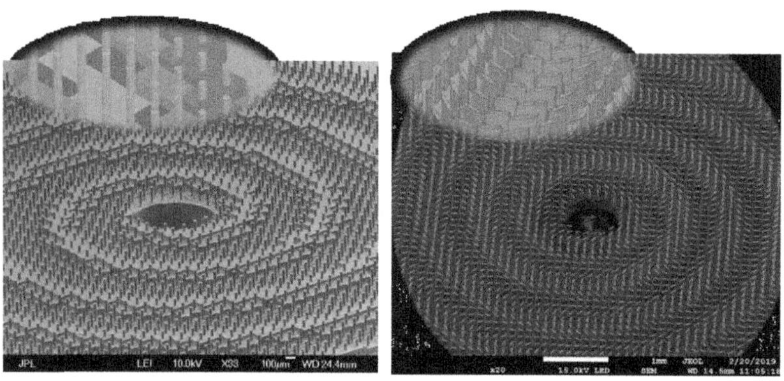

(a) (b)

Figure 1.5 Examples of surface-wave based MTS antenna for terahertz applications. (a) Isotropic pin-type MTS. Source: Courtesy of G. Chattopadhyay. (b) MTS consisting of elliptical cross-section micro-pillars (Courtesy of O. de Sagazan, University of Rennes 1, France).

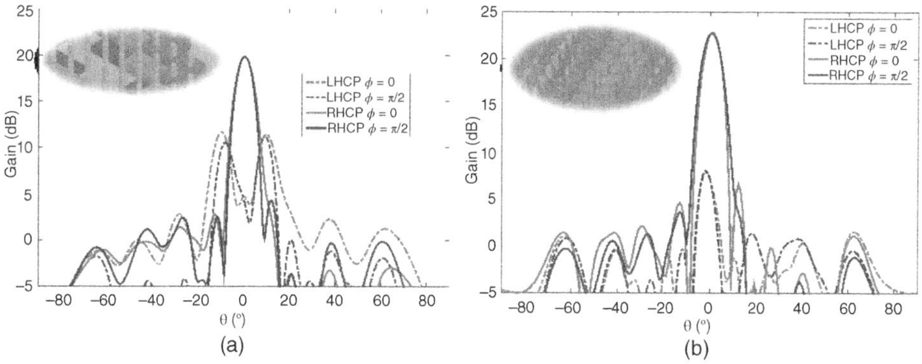

Figure 1.6 Comparison between the radiation patterns of generated by the (a) isotropic and (b) anisotropic micro-pillar-based MTSs.

1.3 Flat Optics Analysis

The analysis of MTS antennas modeled with homogeneous BCs can be performed through the FO method. It was applied to SW-based MTS antennas for the first time in [29]. It is formulated as a sort of Physical Optics approach and uses an adiabatic extension of the rigorous FW expansion for 1D periodic problems that locally match the 2D homogenized BCs of the MTS (Figure 1.7). This method provides a good description of the local currents flowing on the MTS when the local periodicity is defined by using the gradient of the phase of the modulation function, even if the global modulated reactance is not strictly periodic. This description has been successfully applied to define a general design procedure for MTS antennas [30]. The key points of the method are summarized in the following.

We assume that there are no losses in the grounded dielectric slab, which is characterized by a constant relative permittivity ε_r and a constant thickness h. The metallic elements are also assumed to be lossless, and the radius of the radiating aperture is a. The top film of the MTS, composed of many sub-wavelength metallic elements, is modeled with a penetrable BC that is locally periodic as shown in Figure 1.7. It is represented by a shunt capacitive reactance in the corresponding transverse transmission line model.

It is convenient to adopt a cylindrical coordinate reference system (ρ, ϕ, z) with unit vectors $(\hat{\boldsymbol{\rho}}, \hat{\boldsymbol{\varphi}}, \hat{\mathbf{z}})$. Its origin is located at the feeder position and usually coincides with the geometrical center

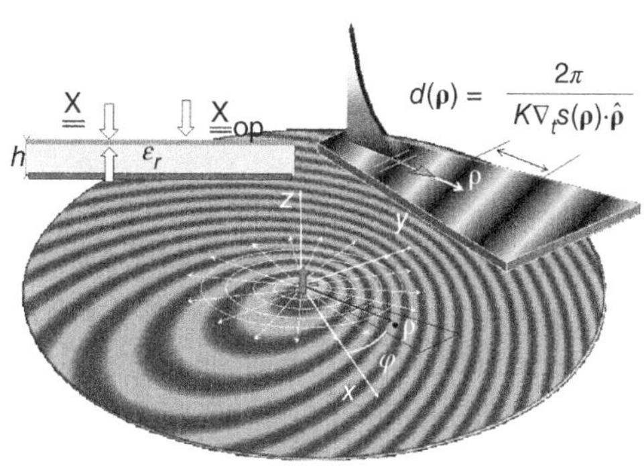

Figure 1.7 Geometry of the MTS antenna and relevant model. The top film is composed of many subwavelength patches. It is modeled as a tensorial sheet reactance that defines a penetrable BC on a grounded slab as depicted in the inset in the upper left corner. The inset in the upper right corner shows the local periodic problem for the definition of the adiabatic Floquet modes on the reactance surface.

of the aperture on the top surface of the grounded slab. In the following, bold characters will denote vectors and bold characters underlined by double bars will indicate tensors.

Since it is the metallic film top covering of the grounded dielectric slab, the MTS imposes the following BC on the tangential electric and magnetic fields established in its immediate vicinity:

$$\mathbf{E}_t = j\underline{\underline{\mathbf{X}}} \cdot \hat{\mathbf{z}} \times (\mathbf{H}_t|_{0^+} - \mathbf{H}_t|_{0^-}) = j\underline{\underline{\mathbf{X}}} \cdot \mathbf{J} \tag{1.1}$$

where $\underline{\underline{\mathbf{X}}}$ is the tensor reactance accounting for the metallic cladding and $\mathbf{J}$ is the electric current flowing in it. The tensor $\underline{\underline{\mathbf{X}}}$ is expressed as:

$$\underline{\underline{\mathbf{X}}} = \hat{\boldsymbol{\rho}}\hat{\boldsymbol{\rho}}X_{\rho\rho} + (\hat{\boldsymbol{\rho}}\hat{\boldsymbol{\varphi}} + \hat{\boldsymbol{\varphi}}\hat{\boldsymbol{\rho}})X_{\rho\varphi} + \hat{\boldsymbol{\varphi}}\hat{\boldsymbol{\varphi}}X_{\varphi\varphi} \tag{1.2}$$

$$X_{\rho\rho} = \overline{X}_\rho[1 + m_\rho(\boldsymbol{\rho})\cos(Ks(\boldsymbol{\rho}) + \Phi_\rho(\boldsymbol{\rho}))] \tag{1.3}$$

$$X_{\rho\varphi} = \overline{X}_\rho m_\varphi(\boldsymbol{\rho})\cos(Ks(\boldsymbol{\rho}) + \Phi_\varphi(\boldsymbol{\rho})) \tag{1.4}$$

$$X_{\varphi\varphi} = \overline{X}_\varphi[1 - m_\rho(\boldsymbol{\rho})\cos(Ks(\boldsymbol{\rho}) + \Phi_\rho(\boldsymbol{\rho}))] \tag{1.5}$$

with its components featuring a constant average value across the aperture. The average is non-zero (zero) for the diagonal (off-diagonal) components. The averages for the diagonal entries are represented by the term in front of the brackets and are capacitive in nature. The terms s and Φ in (1.3)–(1.5) are functions that are slowly varying with respect to the wavelength. The term K is a large constant. Consequently, the following condition involving the transverse gradients of the phase terms is then true: $K|\nabla_t s| \gg |\nabla_t \Phi_{\rho,\varphi}|$.

The different modulation parameters in (1.3)–(1.5) are tailored to control the SW interaction with the MTS. In particular, the local modulation period is related to the gradient of the s function. It is chosen to have the -1 Floquet mode fall in the visible region of the wavenumber spectrum and, thus, is identifiable with the radiating aperture field. The factor Ks thus represents a quickly varying phase contribution that determines the phase of the radiating field. On the other hand, the terms $\Phi_{\rho,\varphi}$ exhibit a much slower variation; they control the polarization of the radiated fields. Finally, the modulation indexes $m_{\varphi,\rho}$ determine the leakage rate of the mode and have a major impact on the amplitude distribution on the aperture and, hence, the illumination efficiency. It is important to note that all of these parameters can be varied independently since they enable the full control of the field distribution over the radiating aperture.

As already mentioned, the tensor reactance in (1.1) is characterized by a constant, diagonal average tensor $\overline{\underline{\underline{\mathbf{X}}}}$. This property ensures that the average current flowing in the metallic coating can be treated as a perturbation of the TM-polarized (transverse magnetic-polarized) SW current that would be excited by the source in a non-modulated MTS characterized by the tensor $\overline{\underline{\underline{\mathbf{X}}}}$. This current source can be expressed as:

$$\mathbf{J}_0 = J_0 H_1^{(2)}(\beta_{sw}\rho)\hat{\boldsymbol{\rho}} \tag{1.6}$$

where $H_1^{(2)}$ is the Hankel function of second kind and first order. In (1.6), β_{sw} is the wavenumber of the SW mode. It is determined by solving a dispersion equation relevant to the "penetrable" capacitive reactance $\overline{\underline{\underline{\mathbf{X}}}}$ and can also be expressed as $\beta_{sw} = k\sqrt{1 + \overline{X}_{op}^2/\zeta^2}$, where k and ζ are the free space wavenumber and impedance, respectively. The term $\overline{X}_{op}$ is the TM "impenetrable" inductive reactance; it accounts for the contributions of both the metallic cladding and the grounded slab. It is equal to the ratio of the total tangential electric and magnetic fields on the top interface. A closed-form approximation of β_{sw} is available in [32].

When a modulation is added to the average reactance, there is a progressive SW-to-LW conversion. The average current representing the dominant 0-indexed mode of the adiabatic Floquet expansion can be derived in this case from (1.6) by a *local* transformation of the unperturbed β_{SW} into a complex wavenumber that reads $k^{(0)} = \beta_{\mathrm{SW}} + \beta_{\Delta} - j\alpha$. The wave attenuation parameter $\alpha = \alpha(\boldsymbol{\rho})$ accounts for the radial power leakage associated with the progressive radiation incurred while the SW travels toward the structure's edges. The term $\beta_{\Delta} = \beta_{\Delta}(\boldsymbol{\rho})$ accounts for the small deviation of the propagation constant from the value associated with the average impedance; it arises from the local modulation [33]. Notice that both of these terms directly depend on the *local* modulation tensor $\underline{\underline{\mathbf{X}}}$. It is also important to notice that $k^{(0)}$ is a spatially varying complex wavenumber and that the *global* phase of the dominant mode can then be retrieved by integrating $k^{(0)}$ along the propagation path.

The adiabatic Floquet expansion of the current flowing in the coating can thus be obtained using the following local modes

$$\mathbf{J}^{(n)} = \left(J_{\varphi}^{(n)}\hat{\boldsymbol{\rho}} + J_{\rho}^{(n)}\hat{\boldsymbol{\phi}}\right) e^{-jnKs(\boldsymbol{\rho})} H_1^{(2)}\left(\int_0^{\rho} k^{(0)} d\rho\right) \tag{1.7}$$

By considering the large argument asymptotic form of the Hankel function, it is clear that the n-th mode in (1.7) propagates with the *local* n-indexed Floquet mode wavevector

$$\boldsymbol{\beta}^{(n)} = \mathrm{Re}\nabla_t\left[\int_0^{\rho} k^{(0)} d\rho + nKs\right] = (\beta_{\mathrm{SW}} + \beta_{\Delta})\hat{\boldsymbol{\rho}} + nK\nabla_t s \tag{1.8}$$

Furthermore, all of the modes leak energy with the same rate, $\alpha(\boldsymbol{\rho})$.

The adiabatic FW expansion of the electric field can then be obtained by introducing the spectral Green's function $\underline{\underline{\mathbf{Z}}}_{\mathrm{GF}}^{(n)}$ of the grounded slab, evaluated with $\beta^{(n)}$ (see Appendix in [29]), which leads to

$$\mathbf{E}_t(\boldsymbol{\rho}) = \sum_n \mathbf{E}^{(n)} \approx \sum_n \underline{\underline{\mathbf{Z}}}_{\mathrm{GF}}^{(n)} \cdot \mathbf{J}^{(n)} \tag{1.9}$$

Although the basis used for the modal expansion is based on an adiabatic asymptotic approximation that holds far from the excitation and it does not rigorously satisfy Maxwell's equations, this model provides a good accuracy in the representation of the currents and fields. This representation can be used to obtain an adiabatic solution of the local problem (e.g. the local wavevector $k^{(0)}$) by imposing the following BC, obtained by inserting (1.9) in (1.1)

$$\sum_n \underline{\underline{\mathbf{Z}}}_{\mathrm{GF}}^{(n)} \cdot \mathbf{J}^{(n)} = \sum_n j\underline{\underline{\mathbf{X}}} \cdot \mathbf{J}^{(n)} \tag{1.10}$$

From this condition, a set of equations is obtained by equating the terms that show an analogous rapid phase variation. This set of equations constitutes a system of the kind $j\underline{\underline{\chi}} \cdot \mathbf{J}^{(0)} = 0$ that has a nontrivial solution only for vanishing values of the determinant of the system matrix, i.e. for $\det[\underline{\underline{\chi}}] = 0$. This condition identifies a *local dispersion equation* that depends on $\boldsymbol{\rho}$, having the *local* wavenumber of the dominant mode, $k^{(0)}$ as its solution. Therefore, this approach allows for finding $\beta_{\Delta} - j\alpha$, i.e. the local complex deviation of $k^{(0)}$ from the unperturbed propagation constant of the SW β_{SW}. Furthermore, the eigenvector associated with the zero eigenvalue provides the local field structure in terms of the FW expansion.

1.4 Multiscale Analysis and Synthesis

The electromagnetic modeling of these antennas is a complex computational task. The MTS aperture can consist of several thousands of elements and several millions of method of moments

(MoMs) unknowns. This computational size prevents the use of a general purpose full-wave simulator for design and optimization. The complexity arises from the presence of multiple scales in the same problem: (i) the "macroscale" of the overall antenna, which can have a diameter up to 50λ; (ii) the "mesoscale," which is relevant to the periodicity of the homogenized impedance modulation and is comparable with λ; (iii) the "microscale" of the elementary patches (pixels) composing the MTS, whose size is around $\lambda/6$–$\lambda/8$; (iv) the "nanoscale" of the fine details of the elements, which can be on the order of $\lambda/50$ in size. This latter scale dictates the choice of the basis functions to be used for modeling the currents on each element. They can be either entire domain basis functions with edge-singularities or sub-domain basis functions on the order of $\lambda/50$ in size to properly capture the wave physics. However, it is well known that an antenna of 50λ cannot be studied with basis functions of size $\lambda/50$ due to the excessive computational burden and the associated ill-conditioning problems. Therefore, it is not possible to perform the design based on the brute force application of a numerical method like, for instance, the MoM.

In order to define an effective analysis approach, each scale needs to be properly addressed [34]. In particular, we will see that any synthesis at the macroscale level is performed with the assumption of homogenized BCs. This choice allows one to get rid of the fine details of the constituent elements [35, 36]. On the other hand, the nano-scale subdomain functions are used only in combination with the assumption of the local microscale periodicity. This choice allows one to reduce the computational domain to a single unit cell. Only the final analysis is carried out by a Fast-Multiple MoM, but using synthetic functions on the pixels.

Generally speaking, the design of MTS antennas can be conveniently divided into distinct tasks related to the different scales of the MTS. Three main steps can be identified. The first step is concerned with the definition of the global impedance modulation that is suitable to generate a certain desired far field when illuminated by the feeder. The second step is relevant to the sampling and reconstruction of the continuous global modulation through sub-wavelength elements. Finally, the last design step consists of the global antenna analysis by means of a dedicated full-wave solver. These three steps are represented in Figure 1.8 and discussed in the following sections.

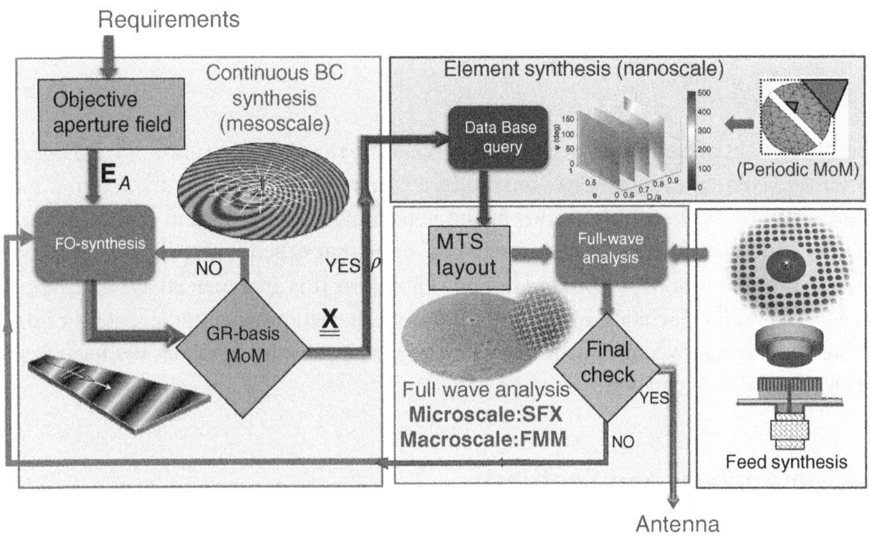

Figure 1.8 Schematic of the design process for an MTS antenna.

1.4.1 Continuous BC Synthesis

In the first step of the described design procedure (left side gray box in Figure 1.8), the MTS is specified by a continuous spatially variable tensor $\underline{\underline{\mathbf{X}}}$ that represents the equivalent penetrable BC. The FO approach is applied to obtain the MTS reactance tensor $\underline{\underline{\mathbf{X}}}$. It is assumed that the radiating components of the target aperture field are known [29, 30].

An alternative to the FO approach for the synthesis of the continuous impedance profile starting from the desired radiation pattern is presented in [37–39]. It is based on a decomposition of the unknown impedance profile into entire-domain basis functions whose coefficients are determined though a quasi-direct inversion of the electric field integral equation after deriving the desired surface current from the objective pattern. In contrast to the FO approach, it does not rely on the locally sinusoidal modulation assumptions, and it therefore allows for the exploration of alternative solutions. However, it is at the price of an increased computational burden. Nevertheless, this increase is limited thanks to the homogenization of the BCs and the use of the entire-domain basis functions.

Once the continuous impedance tensor profile has been derived, it is input into a customized and efficient MoM tool based on Gaussian-Ring (GR) basis functions [35]. The GR basis functions have a Gaussian shape along the radial direction and a linear phase variation $\exp(-jn\,\phi)$ along the azimuth direction. With these basis functions, the evaluation of the MoM matrix is extremely fast, even for very large apertures. This feature is thanks to the availability of closed form expressions for the reaction integrals. Alternative entire-domain basis functions have been proposed in [40, 41]. The aforementioned approaches render the synthesis and analysis at the macro- and meso-scales extremely fast and accurate.

1.4.2 Element Synthesis

After the "mesoscale" design step, the synthesis of the pixels of the MTS is considered (the "microscale" design step). The single pixels are analyzed by using a dedicated periodic spectral MoM [42]. Periodic BCs are applied at the unit cell edges and are modeled in terms of their equivalent anisotropic impedance. This approach is based on a "local micro-periodicity" hypothesis that assumes that any element can be considered as immersed in a homogeneous environment because of the gradual variation of the pixel's geometry across the MTS. This allows for the use of a periodic Green's function in the integral equation formulation and for reducing the number of unknowns to the ones relevant to a single unit cell.

The validity of the micro-periodicity assumption has been verified through rigorous numerical simulations in [43]. The periodic spectral MoM generally makes use of subdomain basis functions whose dimensions are on the order of $\lambda/50$–$\lambda/100$. The pixel analysis is repeated for a set of characterizing geometrical parameters in order to construct a database of pixel equivalent impedances. The database construction process is usually performed at the design frequency of the antenna. Additional frequency samples can be useful in the cases where a broader operational bandwidth is required. To this end, the pole-zero matching method [44] or the analytical models in [32, 45] can also be adopted. Interpolation of the pixel geometry samples can be applied to a have a finer description without increasing the computational burden.

When the pixel impedance database is available, the continuous BC provided at the previous step is discretized according to the lattice structure. The target impedance pattern is then reconstructed cell by cell by using the best fitting pixel through a least square error (LSE) minimization. The outcome of this process is the discretized reconstruction of the MTS through a "pixelated layout," to be further analyzed for performance checks.

1.4.3 Virtual Prototyping

The final design step consists of an accurate analysis of the pixelated layout with a global full-wave solver. As noted, simulating an entire MTS aperture could result in a huge computational effort (up to several millions of unknowns) due to the high number of pixels and the number of unknowns potentially needed to model each unit cell. For this reason, a dedicated optimized tool should be adopted for the global antenna analysis. More specifically, our analysis is done by combining a fast multipole method (FMM) [46, 47] with entire-domain basis functions. In such an MoM analysis, the Galerkin method is applied. A standard FMM is used because it speeds up the evaluation of the matrix-vector products relevant to the far-interaction contributions [48, 49]. On the other hand, the entries relevant to the near-field interactions are evaluated by a spectral MoM using the method in [50] and closed-form basis functions. This choice leads to a drastic reduction of the number of unknowns per unit cell [51, 52]. A considerably agile, rigorously accurate tool for the analysis of MTS apertures is thus attained.

The closed form spectra of the entire-domain basis functions are directly available for the "grain of rice" elements [11]. For a generic element shape for which analytical entire-domain basis functions are not directly available, the closed form spectra can be obtained by an aggregation of the transformable Rao–Wilton–Glisson (RWG) basis functions. We remark that the "grain of rice" elements are versatile enough to be applied to several antenna synthesis examples.

In the next sections, different prototype examples will be introduced on the basis of the described design methodology. These prototypes have been recently realized and measured in the framework of different projects mainly supported by the European Space Agency (ESA). The performance characteristics exhibited by these examples are indicative of the level of technological maturity that has been reached in the development of modulated MTS antennas.

1.5 Dual Polarization

In the next subsections, we present three alternative ways to realize dual polarization with MTS antennas. They are based on single, double, and central/peripheral feeds.

1.5.1 Double-Mode Excitation by a Single Feed

The excitation of an MTS antenna is normally provided by a TM SW mode, which is the dominant mode for a grounded slab loaded with an MTS. In order to obtain dual linear polarization (LP) performance, one possible approach is based on the excitation of both a rotationally symmetric transverse electric (TE) and TM SW mode [25, 53]. A properly designed centered circular waveguide-feed can be used to excite the two orthogonal (hence, decoupled) SW modes and to guarantee the balance of the power associated with each of them. By a proper selection of the anisotropy and modulation of the MTS pattern, the phase velocities of the two modes are synchronized and leakage is generated in the broadside direction with two orthogonal linear polarized fields.

An example of an X-band dual polarized antenna realized with this concept is shown in Figure 1.9. The antenna design has been presented in [25]; and its practical implementation, which includes the description of the dual mode-feed, is given in [53]. The TE and TM modes launched by the dual-mode feed are decoupled in power and possess degenerate dispersion

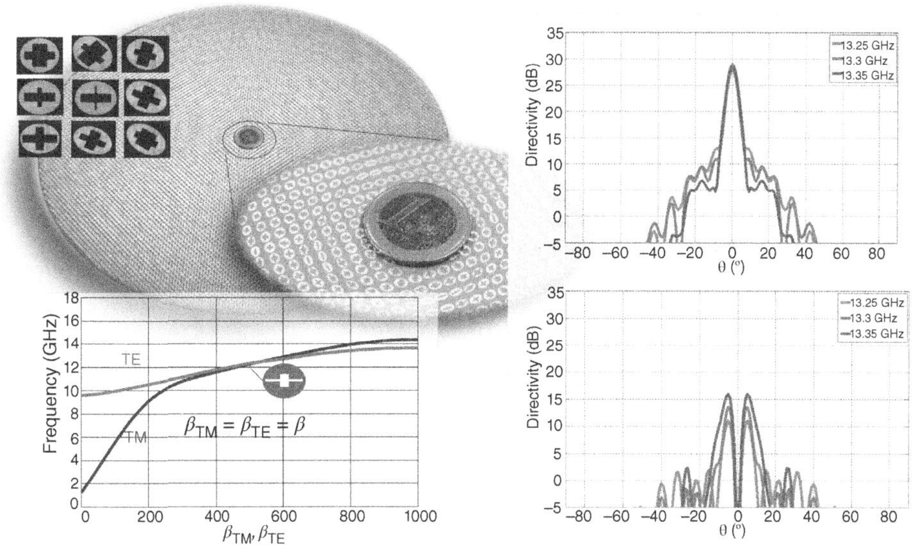

Figure 1.9 X-band dual polarized MTS antenna excited by two modes in the Ku band. Top left, pixels; bottom left, dispersion diagram of the TE and TM modes for a certain periodic element. Right hand side: Directivity patterns of the LHCP (top) and RHCP (bottom) fields for three X-band frequencies.

diagrams in a certain frequency band (see the left-bottom inset of Figure 1.9). Namely, they propagate in that frequency range with the same phase velocity. By employing the dual-linear circular waveguide, their amplitudes can also be equalized. Both SWs interact with the rotationally symmetric, anisotropic MTS. Modulating the MTS impedance with a period that matches the common SW wavelength leads a field leakage at broadside with individual control of the two polarizations. Adjusting the phase shift between the two orthogonal TE_{11} modes of the circular waveguide facilitates simultaneous RHCP (right hand circular polarization) and LHCP (left hand circular polarization) from a single MTS aperture.

The matching of the dispersion curves of the two modes on a single anisotropic MTS impedance corresponds to the balanced condition: $X_{TE}\, X_{TM} = -\zeta^2$, where ζ is the free-space impedance and X_{TE} and X_{TM} are the average values of the azimuthal and radial entries of the impenetrable reactance tensor (see inset on the bottom left of Figure 1.9), respectively. The pixel elements chosen for this purpose are ellipses with a cross-slot (top-left inset of Figure 1.9). Different lengths of the arms of their slots allow for individual control of the local propagation constants of the two modes. The radiation patterns obtained for this design are shown in Figure 1.9 on the right-hand side. The aperture efficiency, including the feed efficiency, is 24%. This method has a reduced aperture efficiency that is limited to about 33% for both polarizations. The polarization purity has been found to be very good, an outcome that substantially depends on the technological realization of the feed [53].

1.5.2 Aperture Sharing and Double-Feed

An alternative, yet simple way to obtain a dual mode excitation consists of superimposing the analytical expressions of two counter-rotating aperture spiral modulations that individually correspond to RHCP and LHCP responses. The two spirals have shifted centers, and the two feeds are located at their centers. When feed 1 is turned-on, the relevant SW launched on the surface couples coherently with the modulation of spiral 1 and it radiates the dominant CP field. On the other hand, the coupling of the SW excited by the feed 2 with the modulation of spiral 1 produces spurious radiation, mainly cross polarized. This method is quite similar to the one used for dual beam and/or dual frequency operations described in Sections 1.9 and 1.10, respectively. It can also be used to realize LP.

A double-feed example is shown in Figure 1.10. The antenna has the radius $R = 10.0$ cm, central frequency $f_0 = 26.4$ GHz, and it is realized on a substrate with relative permittivity $\varepsilon_r = 6.15$ and thickness $h = 0.635$ mm. Two counter-rotating spiral-type modulations with modulation-index $m = 0.3$ and average transparent capacitive impedance $X_\rho = -249\,\Omega$ have been superimposed. The resulting interference pattern defines a transparent reactance tensor, which is eventually implemented with the double anchor elements. The radial entry of this tensor is shown in Figure 1.10a (top). The feeds, placed at the center of their respective impedance modulation layouts, are located at a distance of 1.7λ from each other (see the bottom of Figure 1.10a). The radiation patterns shown in Figure 1.10b (u–v plane) and 1.10c (principal cuts) have been obtained with a full-wave analysis based on a homogenized impedance [35]. An aperture efficiency of about 29% was found for both polarizations. It is worth noting that dual polarization with similar performance has been obtained by the direct inversion method described in [37].

1.5.3 Central and Peripheral Excitations

An alternative method to obtain CP fields consists of exciting an SW from the center (outward SW) and an SW from the periphery (inward SW) on the same modulated MTS. In such a case, an RH spiral modulation produces an RHCP beam when excited from the center and an LHCP beam when excited from the periphery (Figure 1.11).

The arrangement of the feeds is shown in Figure 1.11a. It consists of a double coaxial line in which the internal coaxial line excites the center probe, and an external coaxial line that excites a cylindrical-wavefront transverse electromagnetic (TEM) mode in a parallel plate waveguide on the back of the antenna. A peripheral corner reflector transmits the mode to the top of the MTS producing the second polarization. The main difficulty of this solution is concerned with finding the appropriate amplitude modulation that is able to produce similar gain and beamwidth for both polarizations. The aperture distribution of the -1 indexed field contribution is not the same for both wave directions. This complication is essentially due to the spreading factors of the two SWs in relation to their attenuation constants. Indeed, the asymptotic behavior of the amplitude of the zero-indexed outward mode field is approximately $\exp(-\alpha\rho)/\sqrt{\rho}$, where α is the ρ-dependent leakage parameter. On the other hand, the inward mode has a $\exp(\alpha\rho)/\sqrt{\rho}$ like behavior. Therefore, the amplitude modulation should be defined to equalize the different distributions of the exciting zero-mode amplitudes. An additional impairment is due to the fact that the inward wave exhibits a caustic at the central point and, therefore, it produces an undesired outward mode when crossing the center, which radiates a cross-polar component.

An example of a design in the Ka band (29.75 GHz) is shown in Figure 1.11b. The antenna has a radius of 20λ, and it is printed on a substrate with relevant permittivity $\varepsilon_r = 0.762$ and thickness $h = 0.635$ mm. The MTS has been implemented using double anchor elements. The aperture efficiency is 60% for the outward case and 70% for the inward case. However, the inward case exhibits higher side lobe levels and cross-polar lobes, as expected.

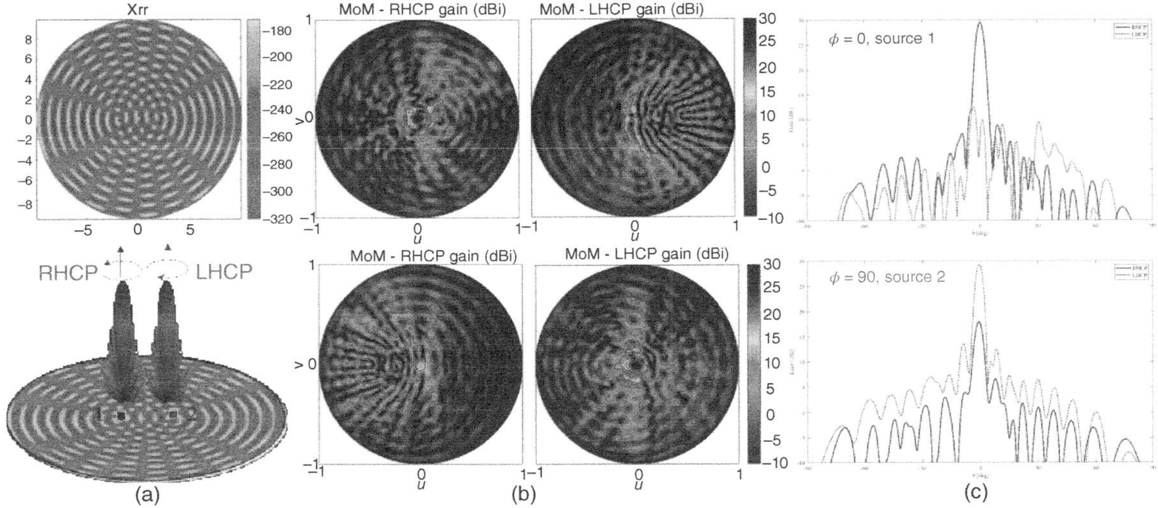

Figure 1.10 Dual polarized MTS antenna obtained by metasurface aperture sharing. The antenna has a radius of 10.0 cm and is printed on a substrate with $\varepsilon_r = 6.15$ and thickness $h = 0.635$ mm. (a) Top: radial component of the transparent reactance tensor obtained by superimposing counter-rotating spirals; bottom: feed associated with each spiral and relevant polarizations for feeds 1 and 2. (b) Gain patterns in the u–v plane for excitation of source 1 (top) and source 2 (bottom). (c) Gain patterns in the two orthogonal planes: top, source 1, $\phi = 0°$; bottom: source 2, $\phi = 90°$. These numerical results were obtained with the homogeneous impedance MoM.

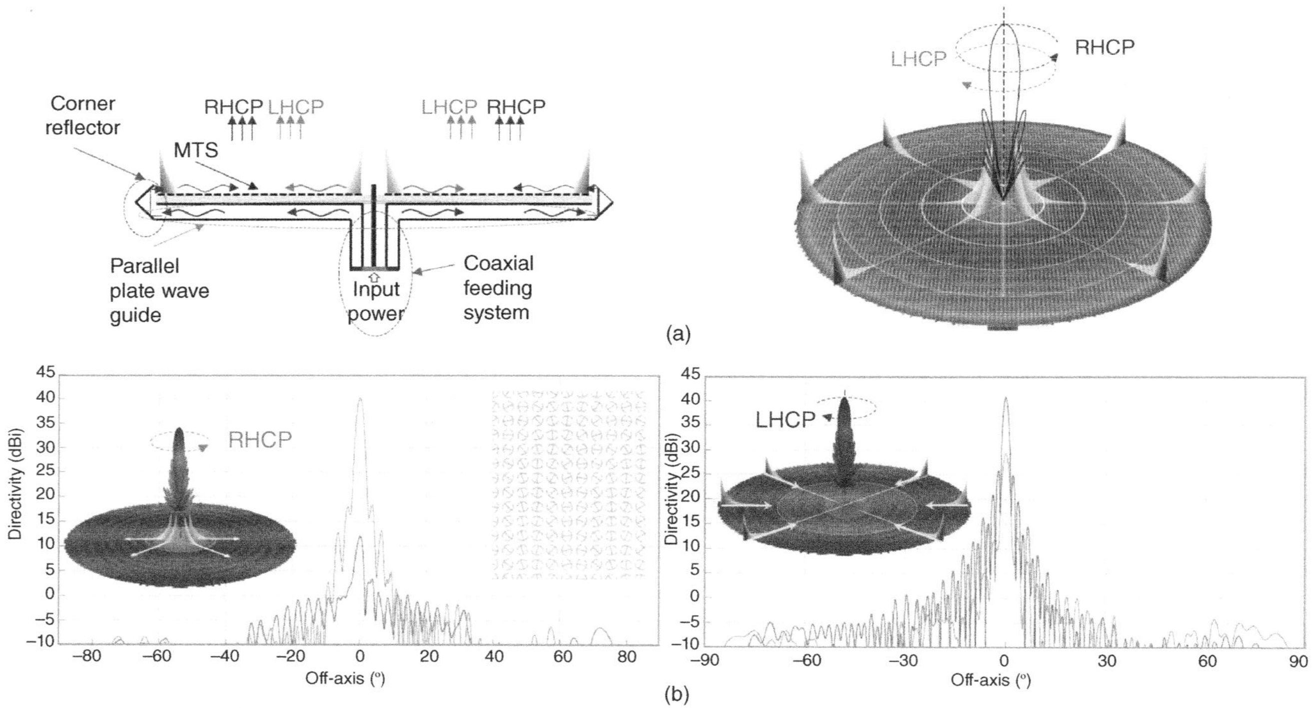

Figure 1.11 Dual polarized MTS antenna obtained with central and peripheral excitations. The antenna has a radius of 20 free space wavelengths at 29.75 GHz and it is printed on a substrate with $\varepsilon_r = 0.762$ and thickness $h = 0.635$ mm. The MTS has been implemented using double anchor elements. (a) Left: feed mechanism; Right: picture of the outward (light gray) and inward (dark gray) SW excitation. (b) Directivity patterns for outward (left) and inward (right) wave excitation. Numerical results obtained by a full-wave fast multipole MoM analysis. Source: Design by G. Minatti (Wave Up).

1.6 Beam Shaping

Beam shaping can be obtained with the FO approach. An example of this application is the MTS antenna with the sectorial isoflux patterns shown in Figure 1.12. This kind of pattern is useful mainly for antennas onboard low Earth orbit (LEO) satellites for Earth observation missions. The purpose of the pattern shaping in elevation is to ensure that a uniform power-flux density reaches the Earth's surface for different observation angles [18].

In order to increase the communication data rate, higher antenna gains are required. Thus, it is useful to consider sectorial isoflux beams that concentrate the radiated power in a specific azimuthal direction, while ensuring the path loss compensation with respect to the observation angle. The radiating aperture is then mechanically rotated while the satellite orbital motion progresses to maintain beam alignment with the ground station.

The homogenized impedance modulation required to obtain a sectorial isoflux pattern at Ka-band (central frequency 26.7 GHz) was obtained through the FO approach. The antenna features RHCP fields, has a radius of 9λ and is based on a grounded slab characterized by a relative dielectric constant $\varepsilon_r = 9.8$ and thickness 0.5 mm. The requested radial entry of the target reactance function is displayed in the inset of Figure 1.12a. The obtained continuous impedance profile is then analyzed by using the MoM code described in Section 1.4.1. The calculated directivity pattern is successfully compared in Figure 1.12a with the one predicted with the FO analysis.

A prototype has been built based on this design. A photo of it is shown in Figure 1.12d. One can see the feed structure at the center of the aperture and details of the implementation of the pixels. The radiation performance of this prototype was measured, and the comparisons between the simulated and measured directivity patterns in the u–v plane are shown in Figure 1.12b,c. We remark that the feed point is located at the geometrical center of the aperture. The feeder is essentially formed by the inner conductor of a coaxial waveguide with a metal hat added to increase the coupling between the guided mode in the coaxial cable and the SW.

1.7 Limit of Aperture Efficiency and High-Gain Examples

1.7.1 Various Types of Efficiencies

In addition to their ability to control the radiation pattern shape, another advantage of primary importance of MTS antennas is their capability to tailor the aperture efficiency through an extremely fine control of the phase and amplitude of the radiating field supported by the MTS. Aperture field tapering is only one of many factors that play a role in determining the overall efficiency of the antenna. Other factors are the power coupling between the feeder and the SW (namely, the feed-efficiency $\varepsilon_{\text{feed}}$), the power losses in the metal and in the dielectric (ohmic efficiency ε_Ω), and ultimately the power diffracted at the antenna edge (spill-over efficiency $\varepsilon_{\text{spill}}$). As a result, the antenna gain is given by the product of several factors and can be expressed as

$$G = \varepsilon_{\text{feed}}\,\varepsilon_\Omega\,\varepsilon_{\text{spill}}\,\varepsilon_{\text{tap}}\,(ka)^2 \tag{1.11}$$

where k the free space wavenumber and a is the antenna's radius. The term ε_{tap} is the conventional tapering efficiency, a parameter that directly depends on the modulation index on the surface.

The wave mechanisms of the MTS antenna can be visualized as illustrated in Figure 1.13 [54]. The input power P_{in} is in part directly radiated by the feeder into free space, and in part delivered to the SW. The interaction with the modulated MTS transforms the SW into an LW, which radiates

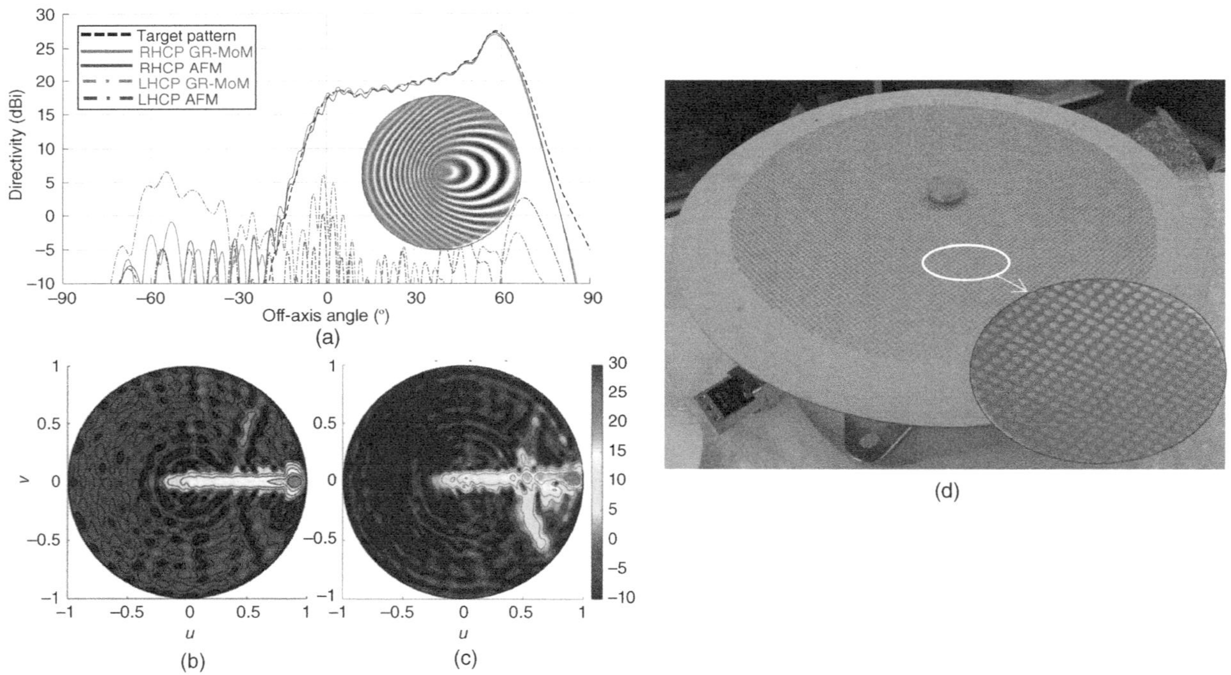

Figure 1.12 Sectorial isoflux MTS antenna for Earth observation missions from LEO satellites. (a) Simulated directivity pattern. Inset: sketch of the aperture impedance pattern (radial component of the reactance tensor). Simulated (b) and measured (c) RHCP component of the directivity in the u–v plane. (d) Photo of the antenna realized with a pixelated surface. The feed is on the center of the aperture and covered by a metallic cap to improve the antenna matching and the SW launching efficiency.

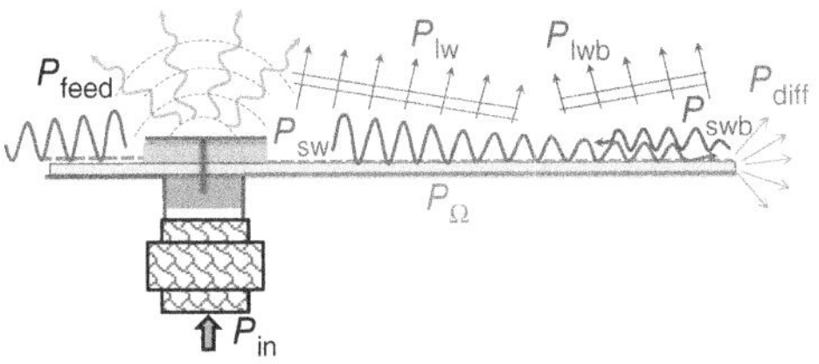

Figure 1.13 Visualization of the time averaged power contributions in an MTS antenna. P_{in}: input power. P_{sw}: power delivered by the feed to the SW. P_{feed}: power directly radiated by the feed as a space wave. P_{diff}: power diffracted at the edge of the aperture. P_{swb}: power transported by the SW excited at the edge. P_{lw}: power transferred from the SW and higher order FWs to the LW and radiated in free space. P_{lwb}: power transferred from the SW and higher order FWs excited at the edge to the LW. P_{Ω}: power dissipated by the ohmic losses. Source: From [54] / with permission from IEEE.

the desired field, with a consequent attenuation of the SW. The attenuated wave that reaches the edge of the aperture is partially diffracted, producing a space-wave, and partially reflected, producing an inward-propagating SW. The latter SW interacts with the MTS and transfers its power to a spurious LW.

We define the "feed-efficiency" as the quantity

$$\varepsilon_{feed} = \frac{P_{sw}}{P_{in}} = 1 - \frac{P_{feed}}{P_{in}} \tag{1.12}$$

where P_{feed} is the power directly radiated by the feed as a space wave. It can be also represented as $\varepsilon_{sw} = 1 - l_{feed}$, where $l_{feed} = P_{feed}/P_{in}$ is the loss associated with the feed. We note that the ratio between the field radiated into free-space and the input power in conventional patch antennas denotes the efficiency and not the loss. In the absence of ohmic losses, namely when $P_{\Omega} = 0$, we define the "*conversion efficiency*" or "*spill-over*" efficiency the quantity

$$\varepsilon_{spill} = \frac{P_{lw}|_{P_{\Omega}=0}}{P_{sw}} \tag{1.13}$$

which relates the power radiated by the LW to that transported by the SW. For a broadside beam, the aperture tapering efficiency is given by

$$\varepsilon_{tap} = \frac{\lambda^2}{4\pi A} \frac{S_{max} 4\pi r^2}{P_{lw}} \tag{1.14}$$

where S_{max} is the maximum amplitude of the Poynting vector at a distance r in the far zone of the antenna, and A is the area of the antenna's aperture. Methods for an efficient numerical computation of these efficiencies are discussed in [55].

1.7.2 Bounds on the Maximum Aperture Efficiency

It was demonstrated in [54] that the maximum product between the tapering efficiency and conversion efficiency (namely the "aperture efficiency") when the modulation index across the antenna is uniform is always limited by 58%. This value is obtained with the leakage constant equal to 0.9 times the reciprocal of the radius. However, this limit can be overcome if one applies an amplitude modulation across the aperture. This additional feature leads to the bound

$$\varepsilon_{tap}\varepsilon_{spill} \leq \frac{a}{\lambda} \Big/ \left(\frac{a}{\lambda} + 2\right) \tag{1.15}$$

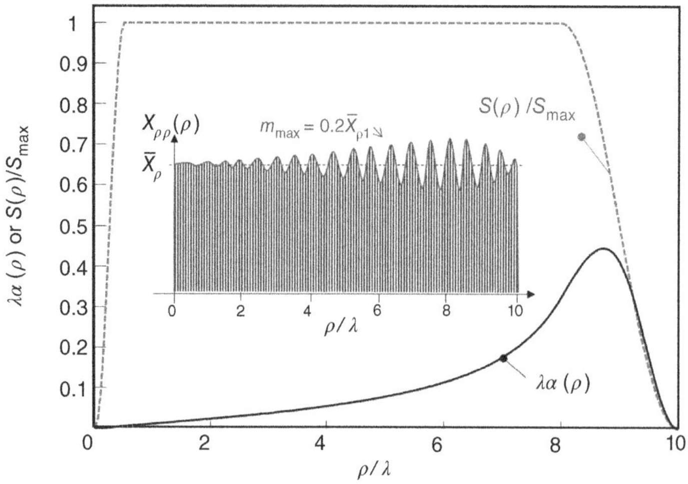

Figure 1.14 Behavior of $\lambda\alpha(\rho)$ as a function of ρ/λ for an MTS antenna of radius 10λ designed for the maximum aperture efficiency and the corresponding behavior of the power density of the −1 indexed (leaky) mode on the aperture. In the inset: schematic behavior of the radial entry of the reactance tensor.

The limit value can be reached by a specific behavior of the leakage constant. An example of this outcome is given in Figure 1.14.

According to the FO analysis, the dependence of α on the parameters of the MTS modulation can be obtained by solving the canonical problem with a 1D modulation that adiabatically matches the local periodicity. This leads to a functional dependence of the following type

$$\alpha = f(m, \overline{X}_{\text{op}}/\zeta, kh, \varepsilon_r) \tag{1.16}$$

For moderate values of the modulation index ($m < 0.1$), the dependence of α on m becomes quadratic, namely, $\alpha \approx m^2 \text{A}(\overline{X}_{\text{op}}/\zeta, kh, \varepsilon_r)$. Examples are shown in Figure 1.15a,b, where $\alpha\lambda$ as

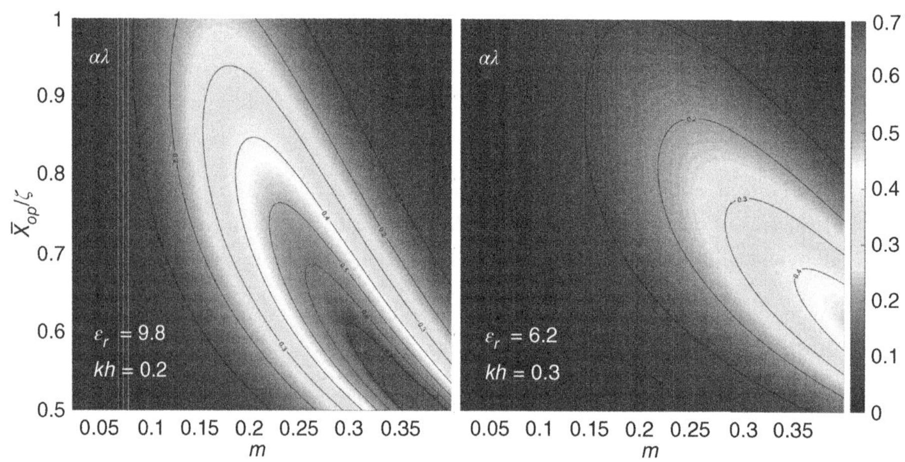

Figure 1.15 Normalized attenuation constant $\alpha\lambda$ as a function of the normalized average opaque reactance $\overline{X}_{\text{op}}/\zeta$ and of the local modulation index m for an MTS antenna whose grounded slab has a relative dielectric permittivity $\varepsilon_r = 9.8$ and two different values of its normalized slab thickness: (a) $kh = 0.3$ and $\varepsilon_r = 9.8$; (b) $kh = 0.2$ and $\varepsilon_r = 6.2$.

a function of m and $\overline{X}_{op}/\zeta$ for $\varepsilon_r = 9.8$ and two different values of kh is presented as 2D contour plots. Note that because of the adopted normalization, the maps in Figure 1.15 do not depend on the frequency. By comparing Figure 1.15a,b, it can be seen that larger values of $\alpha\lambda$ can be obtained for lower values of the substrate thickness and values of $\overline{X}_{op}/\zeta$ that range from 0.5 to 1; and, therefore, $\beta_{sw}/k \in (1.16, 1.41)$. Since it is difficult in practice to get values of m larger than 0.3, the maximum practical values of $\alpha_{max}\lambda$ are in the range $\alpha_{max}\lambda \in (0.45, 0.6)$, depending on the choice of the substrate.

1.7.3 Feed Efficiency

Another important element of an MTS antenna is its SW launcher. As noted, it is essentially formed with a vertical monopole fed by a coaxial waveguide. Additional structures such as metallic rings, circular patches, or metallic hats can be added to this basic structure for matching purposes or to optimize the SW launching efficiency. Note that ideally, the amount of space wave directly launched by the feeder should be negligible. Also notice that these additional structures do not alter the cylindrical phase front of the radial SW.

The problem of the optimum efficiency of SW launchers has been considered in [54]. There, in order to estimate the feeder efficiency, it was assumed that the source radiates over a non-modulated MTS characterized by an average reactance. The feed-efficiency will change depending on $\overline{X}_\rho$ and on the feeder type. Practical TM feeders can be realized by: (i) a coaxially-fed monopole on a circular metallic disc (Figure 1.16a); (ii) a circular waveguide fed by a TM_{01} (coax-type) mode, coupled to the MTS through a circular slot in the ground plane (Figure 1.16b); and (iii) a probe with a corrugated top cover (Figure 1.16c). These three types of feeders have been modeled with the simplified equivalent current arrangements shown in the same figure.

The efficiency is then calculated starting from these simplified models through a spectral domain analysis. The results shown in Figure 1.17 were obtained by adjusting the geometrical parameters of the source in such a way that the maximum of the modal current spectrum outside of the visible

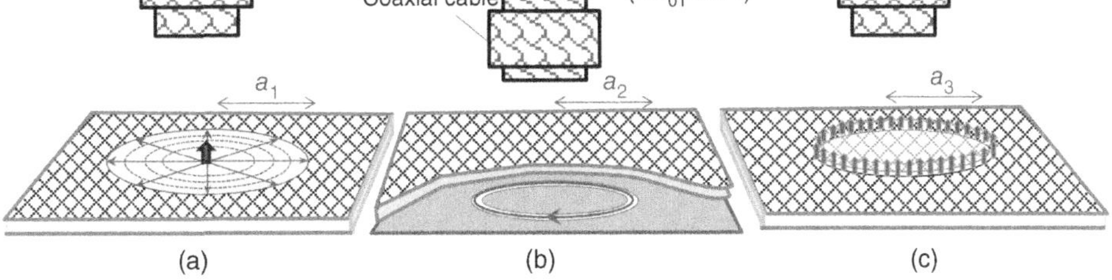

Figure 1.16 Different approaches to exciting the MTS antenna. Upper figures: Types of TM feeders: (a) Probe with a patch. (b) Slot ring on the ground plane. (c) Corrugated hat above a vertical probe. Lower figures: corresponding simplified models for the analysis of the efficiency of these three feeders.

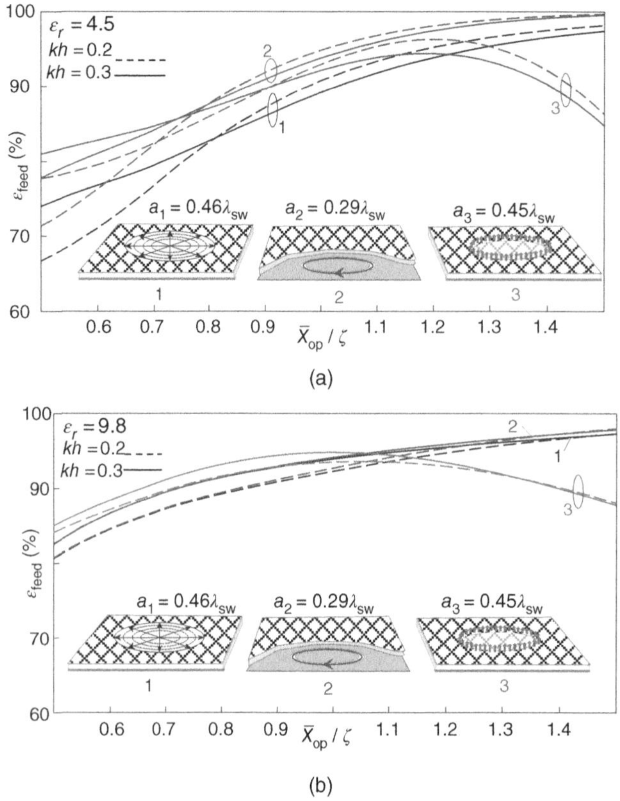

Figure 1.17 Feeder efficiency of the optimized canonical sources as a function of the normalized opaque impedance. Dashed lines, $kh = 0.2$, and continuous lines $kh = 0.3$. (a) $\varepsilon_r = 9.8$, (b) $\varepsilon_r = 4.5$.

region occurs at the location of the SW pole of Green's function for the grounded slab having a transparent average reactance on its top surface.

1.7.4 Example of MTS Antenna with High Aperture Efficiency

An MTS antenna with maximum aperture efficiency can be designed by implementing the modulation profile introduced in Section 1.7.2. A Ka-band prototype antenna has been designed and fabricated to verify this approach.

The antenna features a diameter of 27λ at the center frequency of 29.75 GHz. It was designed to radiate a boresight beam with a product between the tapering and spill-over efficiencies of 85%. The prototype was realized by using a Rogers RO3003 laminate characterized by a thickness of 0.762 mm and a relative dielectric constant equal to 3.0, with a loss factor equal to $\tan\delta = 0.001$.The measured peak gain was equal to 37 dBi corresponding to an overall efficiency $\varepsilon_{tot} = \varepsilon_{feed}\,\varepsilon_{\Omega}\,\varepsilon_{spill}\,\varepsilon_{tap} \approx 0.70$ and an effective product $\varepsilon_{feed}\,\varepsilon_{\Omega} \approx 0.80$. This antenna currently constitutes a benchmark in terms of the maximum achievable efficiency attained by MTS apertures. The simulated and measured co-polar and cross-polar gain patterns are shown in Figure 1.18a,b. The measured patterns are compared with those produced by the analysis based on the adiabatic Floquet modes (AFMs); by the GRs MoM code based on continuous BCs; and by the FMM analysis introduced in Section 1.4.3, which accounts for the actual pixelated layout. Figure 1.18a,b also show excellent agreement of the measured and simulated results predicted by the mentioned analysis methods.

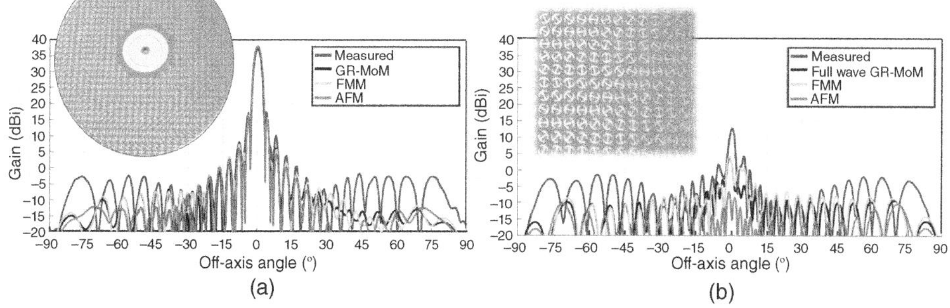

Figure 1.18 Layouts and performance of two highly efficient MTS antennas. (a) Co-polar and (b) Cross-polar gain patterns. The curves provide a comparison of the measurements (dark gray lines) with three different numerical analysis results (light gray lines: FO method (labelled AFM); gray lines: GR-basis function continuous impedance BCs MoM [35]; and black line: full wave FMM for the textured layout). (b) MTS antenna and details of the feed before insertion of the monopole.

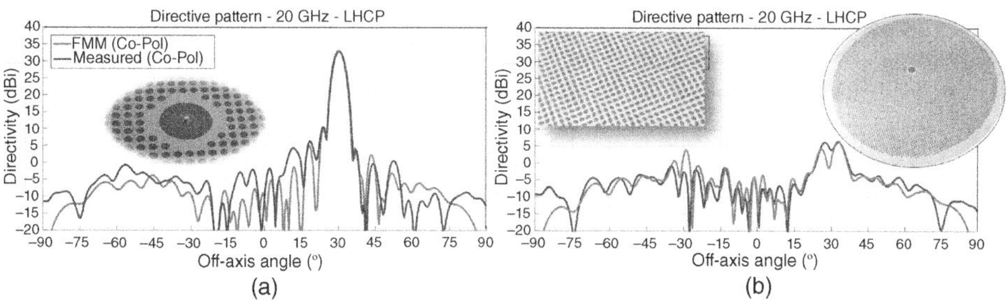

Figure 1.19 Comparisons between the measured and simulated gain patterns of a tilted beam high efficiency MTS antenna realized by the "grain of rice" elements. The antenna has been designed to radiate a tilted RHCP beam pointed at 30° at 20 GHz. (a) Co-polar component. (b) Cross-polar component. Insets: photos of the overall antenna that include details of the aperture implementation and of the feeder.

The performance characteristics and pixels layouts for another MTS antenna prototype example are illustrated in Figure 1.19. A comparison between the simulated and measured gain patterns at 20 GHz of a modulated MTS antenna designed to radiate an RHCP beam that is tilted 30° off from boresight is presented. The antenna has a radius of 10λ at the working frequency. One notes that excellent agreement between the measured gain patterns and the calculated ones was obtained, corroborating the effectiveness of the design procedure. The achieved peak gain is equal to 33 dBi and corresponds to an aperture efficiency of 0.58.The polarization purity was extremely good. Figure 1.19 also shows the overall antenna layout and some details of the feeder, which was composed essentially of a coaxial guide coupled to some patches of annular shape for matching purposes.

1.7.5 Example of a Very High-Gain Antenna

An example of a very high gain MTS antenna prototype is shown in Figure 1.20. It attained the largest gain that has been measured until now with this type of antenna. The prototype was designed to operate at 30 GHz and was fabricated using a Rogers RO3003 substrate of 0.762 mm thickness. It had a 46.0 cm diameter. The feeder consisted of a coaxial square-flanged connector

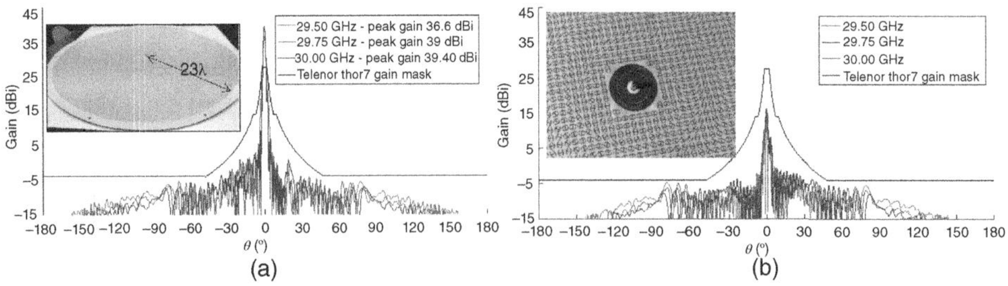

Figure 1.20 Measured gain pattern of the very high gain MTS antenna prototype. (a) Co-polar. (b) Cross polar. Inset: details of the feeder. Source: Design and measurement done by Wave Up.

placed at the center of the antenna that was surrounded by metallic rings on the MTS side. This configuration achieved both a low input reflection coefficient and a high SW launching efficiency.

The measured gain patterns of the antenna are shown in Figure 1.20. The peak value reached 39.5 dBi at 30 GHz. The corresponding peak directivity was 41.2 dBi. The losses of this antenna did not exceed 1.7 dB across the operational bandwidth of 500 MHz. It is an unprecedented value for an MTS antenna. This value also includes the losses in the SMA connector, which was not the optimal choice for this frequency range. These results indicate that a 40 dBi peak gain could be achieved by simply adopting a connector optimized for Ka band operation.

1.8 Wideband and Limit of Bandwidth-Gain Product

1.8.1 Gain and Bandwidth for Uniform Phase Modulation

The usable bandwidth of modulated MTS antennas that feature a radially constant modulation period (uniform phase of the reactance) depends mainly on the aperture dimensions and on the dispersion properties of the SW. As shown in [56], the gain-bandwidth product for an antenna whose radius is larger than 3λ can be estimated as

$$GB \approx 22\frac{v_g}{c}a_\lambda < 22\,a_\lambda \text{ Uniform amplitude modulation} \tag{1.17}$$

$$GB \approx 47\left[\frac{a_\lambda}{a_\lambda+2}\right]\frac{v_g}{c}a_\lambda < 47a_\lambda \text{ Non-uniform amplitude modulation} \tag{1.18}$$

where G is the aperture gain, $B = \Delta f/f_0$ is the bilateral 3 dB fractional bandwidth, a_λ is the antenna radius normalized to the free-space wavelength at the center frequency f_0, c is the speed of light, and v_g is the group velocity of the SW supported by the average reactance at f_0. The relations (1.17) and (1.18) represent an upper bound obtained by maximizing the product $\varepsilon_{\text{spill}}\,\varepsilon_{\text{tap}}$ for uniform and non-uniform modulation indexes, respectively. In fact, the bandwidth for apertures having a uniform modulation period can be estimated as $B \approx 0.95v_g/(ca_\lambda) < 0.95/a_\lambda$ for a uniform modulation index and as $B \approx 1, 2v_g/(ca_\lambda) < 1.2/a_\lambda$ for a non-uniform modulation index.

1.8.2 Gain and Bandwidth for Non-uniform Phase Modulation

A technique to enlarge the operative bandwidth of an MTS antenna consists in radially modifying the modulation periodicity, i.e. the periodicity becomes a function of the radial coordinate. Its values are chosen according to the SW frequency dispersion. The antenna can then be conceived as an active region device where different annular rings are characterized by their own local

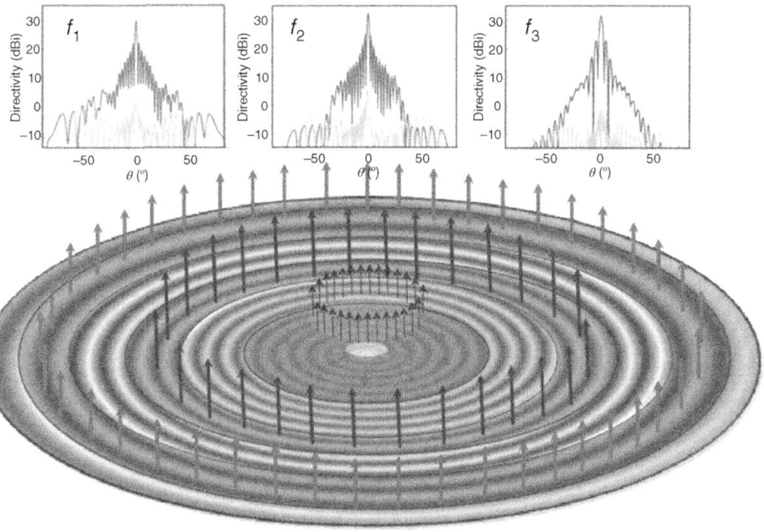

Figure 1.21 Active regions at three different frequencies and examples of the resulting radiation patterns ($f_1 < f_2 < f_3$).

period and are activated alternately as the frequency is changed to radiate a beam in a desired direction. Figure 1.21 illustrates these mechanisms and the annular active regions of a wideband MTS antenna. The directivity patterns are color-coded to the active regions that produce them. Note that the active region slides inward on the aperture away from its edge as the working frequency increases. This phenomenon is due to a local resonance that couples the SW power to the LW power. Conversely, the aperture area outside of an active region barely interacts with the SW at certain frequencies and, hence, it radiates very weakly at them.

A functional wideband design, for instance, for a broadside radiating aperture is ensured by matching the modulation phase $Ks(\rho)$ in (1.3)–(1.5) with the frequency dependent SW phase through radially shaping the periodicity of the modulation function. Note that the vector position of the modulation phase will be assumed to be independent of the azimuthal angle throughout this section. As the SW wavenumber $\beta_{sw}(\omega, \rho)$ depends dominantly on the "opaque" reactance $\overline{X}_{op}(\omega)$ of the MTS antenna, any wideband design needs to take into consideration both the dispersion of the metallic coating and the contribution from the grounded slab over the design bandwidth. Because the dispersion of the capacitive reactance of the metallic coating is quite weak when compared with the one of the grounded slab, it can be evaluated at a single frequency [42] (usually the central one). The in-band frequency behavior can then be extrapolated with a linear approximation of the capacitive admittance. When needed, a more refined estimation of the texture reactance can be obtained by using the method illustrated in [32]. We note that β_{sw} depends at second order on the amplitude and phase of the impedance modulation, which changes point by point over the surface.

For the broadside apertures treated in this section, the design relation that holds for an active region links the SW phase and the modulation phase as $\beta_{sw}(\omega, \rho) = Ks(\rho) \doteq 2\pi\rho/\gamma_p(\rho)$, where $\gamma_p(\rho)$ is a "periodicity transition function." This relation is imposed for a set of frequencies along the radius of the antenna. Note that the choice of the function $\gamma_p(\rho)$ is not univocal, but can be the object of an optimization that aims to increase the pattern stability. On the other hand, the local period is linked to the function $\gamma_\rho(\rho)$ as $d(\rho) = [\partial(\rho/\gamma_\rho)/\partial\rho]^{-1}$. This period function ranges between two bounds, d_1 and d_2. The two bounds are chosen to match the SW wavelength at the two extremes of the design band, i.e. $2\pi/d_{1,2} = \beta_{sw}(f_{1,2})$, where $d_1 = d(\rho)|_{\rho=0}$, $d_2 = d(\rho)|_{\rho=a}$, and $f_{1,2}$ is its lowest and highest working frequencies, respectively.

1.8.2.1 Exponential Modulation

The authors treated the case $m_\rho = m_\varphi$ in [57] with an exponential variation of the radial period of the form

$$d(\rho) = \frac{d_1 e^\delta - d_2}{e^\delta - 1} + \frac{d_2 - d_1}{e^\delta - 1} e^{\delta\rho/a} \tag{1.19}$$

where δ is a non-dimensional constant that defines the speed of the exponential growth. The period function bounds take the values that the local period of the modulation assumes at $\rho = 0$ and at $\rho = a$; i.e. $d_1 = d(0)$, $d_2 = d(a)$. The optimal choice of δ is obtained in the Appendix of [57] as $\delta \doteq \delta_{opt} = \ln(d_2/d_1 - 0.984) + 4.13$. It was also shown that this choice for δ leads to the average antenna gain ([57], Eq. 28)

$$G_{ave} \approx \frac{a(4\pi)^2}{d_1\left[1 - \frac{2d_2}{a}\right]} \frac{35\,\sigma_{min}^2}{\delta_{opt}^3} \tag{1.20}$$

where σ_{min} is the ratio of the SW wavelength to the free-space wavelength at the minimum frequency of operation, which occurs approximately for $\lambda_{sw,min} \approx 1.1 d_1$. The antenna bandwidth is bounded by the two frequencies for which the SW wavelength assumes its minimum $\lambda_{sw,min} = \gamma_1 d_1$ and maximum $\lambda_{sw,max} = \gamma_2 d_2$ values. The factors $\gamma_{1,2}$ are larger than one and depend on the acceptable gain with respect to the average gain used to define the bandwidth. For instance, the factors $\gamma_{1,2}$ approach unity for a −6 dB gain bandwidth.

Figure 1.22 shows the results for an aperture designed to work at Ka-band. Its substrate has the thickness $h = 0.635$ mm and relative permittivity $\varepsilon_r = 6.15$. The modulation is characterized by an exponential profile of the radial periodicity that is bounded at the center and at the rim by the values $d_1 = 7.0$ mm and $d_2 = 13.7$ mm, respectively. The antenna has a 3 dB bandwidth ranging from 23.0 to 29.1 GHz and exhibits a ±2 dB in-band gain oscillation. Even though the present design is quite effective in extending the operability of MTS antennas in terms of their pattern bandwidths, these gain oscillations with respect to its average are still non-negligible.

Figure 1.23 shows the directivity versus the frequency of the fabricated MTS antenna designed to work between $f_1 = 24$ GHz and $f_2 = 30$ GHz. It presents a comparison of measured directivity with the ones computed with the continuous impedance boundary condition (IBC) model developed into the IBC-MoM [35], and with the commercial software CST Microwave Studio (MWS). The inset shows the elliptical patches that were used to synthesize the impedance tensor. The measured average gain of this prototype is $G_{ave} = 27.5$ dBi for an aperture radius $a = 11.11$ cm. The in-band gain oscillation was ±1.5 dB.

1.8.2.2 General Form of the Modulation Period

The modulated tensor for a broadband aperture can be designed with different procedures. In particular, an alternative method of construction of the periodicity function $d(\rho)$ was outlined in [58]. The method is based on an optimization procedure that aims at determining an optimal radial profile of the function $d(\rho)$that provides a maximally flat gain response over the designed operational frequency band.

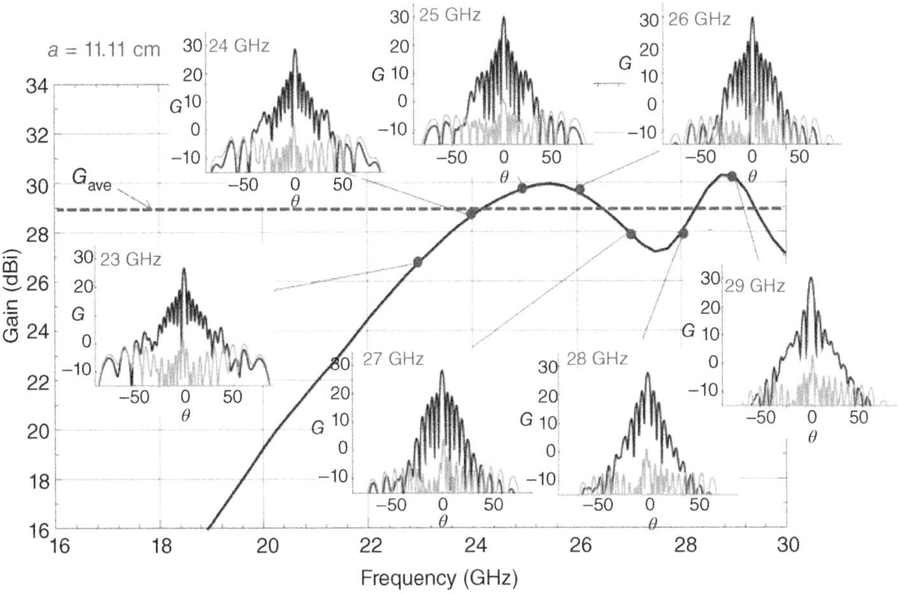

Figure 1.22 Gain as a function of the source frequency for an MTS antenna with exponential tapering and $\delta = \delta_{opt}$, $a = 11.11$ cm, $d_1 = 7.0$ mm and $d_2 = 13.7$ mm. The peak gains and the gain patterns in the insets have been obtained with a full wave MoM solution for the homogenized impedance. The average gain is $G_{ave} = 29.0$ dBi with ± 1.0 dB in-band gain oscillation.

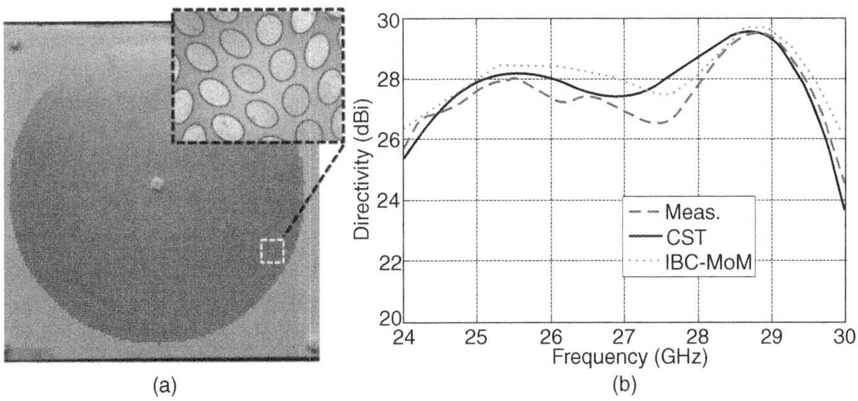

Figure 1.23 Ka-band MTS antenna. (a) Photo of the fabricated prototype and an inset showing the elliptical patches used to synthesize its impedance tensor. (b) Directivity versus frequency response: measured (dashed line) and simulated directivities. The simulated values were computed with the IBC-MoM in [35] (dotted line) taking into consideration the solution for the ideal homogenized reactance and with the commercial software CST.

As a first step, one simply needs to set the bandwidth of the design, i.e. f_{min} and f_{max}. Then using [58] one has

$$G_{ave} \frac{\Delta f}{f_0} \approx 28.7 \frac{v_{g,ave}}{c} \frac{a}{\lambda_0} \tag{1.21}$$

where f_0 represents the central frequency and $v_{g,ave}$ is computed as the integral average of the group velocity associated with the average reactance $\overline{X}_\rho$ over the design bandwidth. One can then determine the aperture radius a that can facilitate the desired maximum gain.

The average reactance is usually determined starting from the choice of the texture, and its dependence on the frequency within the operating band is found by analyzing the reference pixel, immersed in a homogeneous environment, with the full-wave solver described in Section 1.4.1. A dispersion database β_{sw} is built up in this manner. Then, a set of "*phase control frequencies*," equally spaced in the band of interest, are selected and associated with the corresponding SW wavelengths $\lambda_{sw,n} = 2\pi/\beta_{sw,n}$ through the created database. Any value $\lambda_{sw,n}$ is thus linked to a radial control point ρ_n that corresponds to the center of the n-th active region at the frequency f_n. The other key parameter is the "*amplitude control frequency*," f_p, at which the power density of the −1 Floquet mode is imposed as a constant along ρ.

The optimizer, therefore, comes into play at each iteration by adaptively choosing the frequency f_p and the radial positions of the control points ρ_n. Their increasing order is preserved in order to maintain $d(\rho)$ as a monotonically increasing function. The algorithm evaluates the aperture frequency response in terms of the calculated gain by means of a cost function

$$C = \sum_i \frac{|G(\omega_i) - G_{ave}|}{G_{ave}} \tag{1.22}$$

where the antenna gain G is evaluated by means of the FO-based formula [29]:

$$G(\omega) = \frac{8\pi^2}{\lambda^2} \frac{\left| \int_0^a \sqrt{S(\rho,\omega)} e^{j\Psi(\rho,\omega)} \rho d\rho \right|^2}{\int_0^a S(\rho,\omega) \rho d\rho} \tag{1.23}$$

where the −1 Floquet mode phase $\Psi(\rho, \omega)$ depends on the modulation phase $Ks(\rho)$ and on the cumulative wavenumber deviation due to $\Delta\beta_{sw}(\rho, \omega)$. The aperture gain G depends on the angular frequency ω by means of both the power distribution profile S and the phase Ψ. These quantities can be determined once the curve $S(\rho, \omega_p)$ is imposed at ω_p and the related profile of the attenuation parameter $\alpha(\rho, \omega_p)$ is derived according to [58]. Furthermore, it is possible to associate the modulation index that locally best fits the desired $\alpha(\rho, \omega_p)$ curve to completely characterize the reactance tensor $\underline{\underline{X}}$ at ω_p by solving the local canonical problem that locally matches the modulation features in terms of $\overline{X}_\rho$ and $d(\rho)$. At this point, the aperture can be simply synthesized using subwavelength patch elements. A reactance database must be constructed that associates to each characteristic pixel geometry, the related anisotropic reactance. This process is performed using the full-wave periodic spectral MoM solver described in Section 1.4.2. By solving the canonical problem that locally matches the modulation features at each frequency ω_i, one can then retrieve the SW wavenumber deviation $\Delta\beta_{sw}(\rho, \omega_i)$. The phase $\Psi(\rho, \omega_i)$ is obtained with it and $S(\rho, \omega_i)$ is calculated using the inverse formula developed in [54].

At this stage of the iterative process the gain $G(\omega_i)$ can be evaluated at any frequency sample ω_i. The computation time of a single iteration is quite fast with this procedure. For instance, the evaluation of the cost function C employs around one minute to evaluate 25 frequency samples in the Ka-band for an aperture of 11.11 cm radius on an Intel I7 core machine. The optimization workflow summarized earlier has been used to design two broadband MTS antennas with $a = 11.11$ cm and with the following target operating bands and average gains:

- Case 1: Operating band: 23.5–27.5 GHz, average gain 29.3 dB.
- Case 2: Operating band: 20.5–29.5 GHz, average gain 25.4 dB.

The in-band response computed from the optimization was obtained after around 1000 iterations of the algorithm. The results are shown in Figure 1.24. They compare a full wave analysis obtained

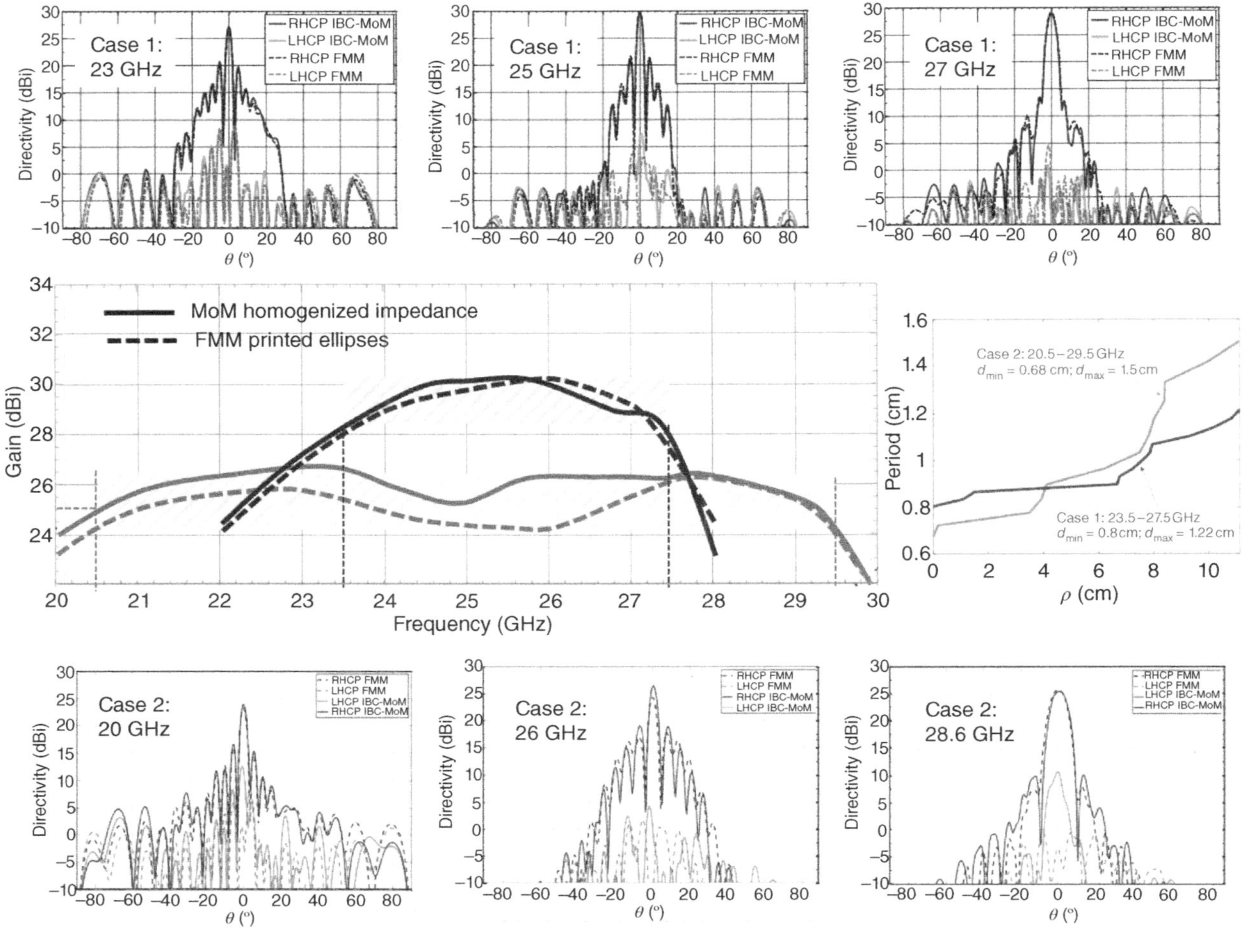

Figure 1.24 The simulated performance characteristics of two optimized antennas whose radius is 11.11 cm: Case 1 (dark gray lines) and Case 2 (light gray lines). Top row: Comparisons of the directivity patterns at 23, 25, and 27 GHz obtained for the broadband MTS (Case 1) with the homogenized IBC-MoM and the FMM analysis with its implementation of the elliptical patches. Central row left hand side: Gain versus frequency responses. The solid lines correspond to the full-wave analysis of the homogenized impedance MoM. The dashed lines correspond to the FMM simulation of the MTS synthesized by elliptical patches. Central row, right hand side: Final distribution of the period as a function of the radial distance. Bottom row: same comparison as in the first row, but for Case 2.

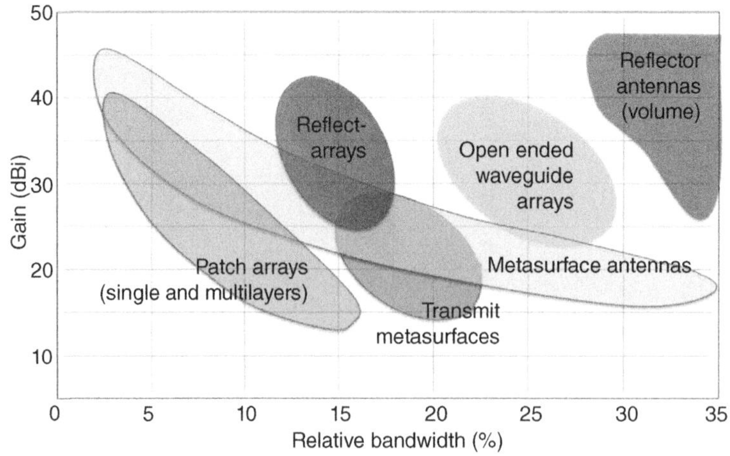

Figure 1.25 Gain versus relative bandwidth diagram for several flat antenna typologies.

with the homogenized impedance MoM with those obtained with a FMM analysis obtained from the implementation of the final results with elliptical patches printed on a Rogers RO3006 substrate with relative permittivity $\varepsilon_r = 6.15$ and thickness $h = 0.635$ mm. In both configurations, the basic square unit cell has a 1.6 mm side length.

1.8.2.3 Range of Applicability

The range of operability of MTS antennas in terms of their main performance characteristics is compared with those of other flat microwave antennas in Figure 1.25. The diagram represents the obtained gain versus the relative bandwidth. The MTS antennas cover a large region of this diagram. They are able to have not only up to a 40 dB peak gain with few percent bandwidth, but also a large bandwidth with medium gain.

1.8.3 Phase Center Stability

The frequency stability of the phase center is an important property in some applications. It can be shown that the phase center (PHC) of an MTS antenna is quite stable with frequency [58]. This desirable feature occurs because their layout is approximately symmetric and usually fed by a single monopole at its center. To illustrate this property, we have calculated the PHC for the antenna fabricated and measured in [57]. The PHC calculation follows the method described in [59] and it is summarized here for convenience. First, the pattern phase is obtained in the half power angular range of the main beam, i.e. within the −3 dB level. Next, the angular direction in elevation angle of the null of the phase gradient is determined numerically. Finally, the PHC is found as the radii of the osculating spheres that match the phase gradient up to the second order derivative around the direction of the null. The measured and calculated (using both MoM and CST) PHC results are displayed in Figure 1.26.

It is worth noting that this procedure differs from those typically implemented in commercial software. Indeed, in the latter, there is no search of the phase gradient null. Moreover, the osculating sphere approximation is performed directly on the data of the main beam in a range symmetrically positioned with respect to the maximum amplitude value. This commercial software process has been found to provide significant errors in the phase center estimate even for small deviations between the maximum of the beam and the null of the phase gradient.

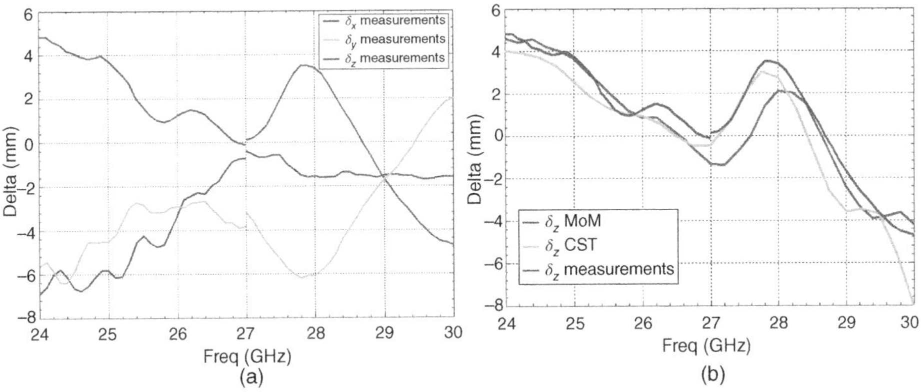

Figure 1.26 Measured and calculated phase centers as functions of the source frequency. (a) PHC coordinates calculated from measurements using the method in [59]. (b) Comparison between the z components of the phase centers calculated using the measured patterns and using the IBC-MoM and CST software.

1.9 Multibeam and Multi-frequency

A major practical goal for 5G and beyond is to obtain multiple directive beams from a single aperture. Multiple beams at the same frequency will be considered first. Multi-frequency operability from an individual antenna by means of the so-called aperture sharing technique will be considered next. These aperture sharing concepts foresee, in essence, the superposition of two or more modulation functions point by point on the entire aperture with the purpose to obtain additional functionalities or specific pattern shapes with respect to a single harmonic modulation. An example of such MTS antenna was developed in [60]. In these specific cases, the reactance modulation tensor assumes a more general form that reads

$$\underline{\underline{\mathbf{X}}} = \frac{1}{4}\sum_{n=1}^{4}\left[\hat{\boldsymbol{\rho}}\hat{\boldsymbol{\rho}}\, X_{\rho\rho}^{(n)} + (\hat{\boldsymbol{\rho}}\hat{\boldsymbol{\varphi}} + \hat{\boldsymbol{\varphi}}\hat{\boldsymbol{\rho}})X_{\rho\varphi}^{(n)} + \hat{\boldsymbol{\varphi}}\hat{\boldsymbol{\varphi}}\, X_{\varphi\varphi}^{(n)}\right] U_A \tag{1.24}$$

where $U_A = u(a - \rho)$ and $u(x)$ is the unit step function. The superscript n here denotes the index of each modulation that is superimposed on the aperture.

1.9.1 Single Point Source and Dual Beam

The antenna we present here has a radius of 6λ and is designed at 17 GHz to radiate two tilted beams at the zenith angles of $\theta = 30°$ and $\theta = -30°$. The antenna is fed by a single point source that is located at the geometrical center of the aperture. The feeder is formed by a coaxial guide inner conductor soldered to a circular patch and coupled to an outer annular metal ring as depicted in the inset of Figure 1.27b. Elliptical metal elements were selected to implement the reactance pattern. The Rogers RO3010 substrate was employed. This copper-clad laminate is characterized by the dielectric constant $\varepsilon_r = 10.2$ and thickness $h = 0.635$ mm.

A comparison between the simulated and measured results for the developed prototype is presented in Figure 1.27a. It displays the directivity patterns at 17 GHz on the principal plane with azimuthal angle $\phi = 0$. The numerical evaluation of the directivity pattern was performed using the GR basis-functions based MoM with the ideal continuous BCs [35]. The pixelated layout behavior was instead evaluated using the commercial software ANSYS HFSS.

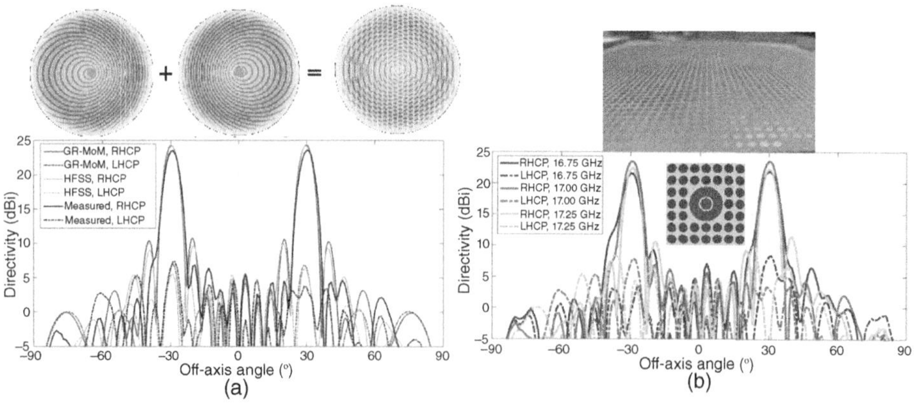

Figure 1.27 Multibeam MTS antenna directivity performance. (a) Simulated results (dark gray line and light gray line for the IBC-MoM and full-wave analyses, respectively). The measured (black line) directivity pattern for the multibeam shared aperture. All patterns were obtained at 17 GHz. Two beams point at the desired ±30° directions from broadside. The cross polar discrimination and beam symmetry is quite good. (b) Measured directivity patterns at 16.75, 17.0, and 17.25 GHz.

The agreement between the simulated and measured patterns is very good. Two beams appear at the expected zenith angles. They have the measured co-polar peak directivity of 23.5 dBi and the cross-polar level of −16.5 dBi. The quasi-orthogonality of the beams translates into quasi-orthogonal reactance patterns, which allow this design approach. Its accuracy and consistency is demonstrated by the measurements given in Figure 1.27. The stability of the directivity patterns with frequency is demonstrated in Figure 1.27b. The frequencies at which the directivity patterns are shown are 16.75, 17.0, and 17.25 GHz. It is observed that the symmetry of the directivity patterns is preserved, and a quite stable frequency response was attained.

1.9.2 Multi-point Sources and Multibeam, with Total Aperture Sharing

The previous two-beam example considered a transmitting antenna that was excited by a vertical electric dipole (VED) and that radiated two beams at the same frequency. However, if it was used as a receiver, the waves incoming from the directions of its two principal lobes will add up in the backend. Thus, it is not possible to process the received signals individually. This situation can be avoided by assigning one feed point per beam instead of a single, unique feed. One could consider dividing the aperture in distinct angular sectors, each excited by a different source. This approach requires a feed network scheme in which each source illuminates only its corresponding sector. Consequently, the overall system is more complex in comparison to the single feed version. Nevertheless, one can still apply the superposition of the modulation patterns as was exploited in Section 1.9.1 to obtain one independent beam per source. The main difference is that each $X^{(n)}$ in (1.24) will have to be referenced to a local coordinate system whose origin is located at the corresponding feed point location.

Figure 1.28 shows an example of the multibeam MTS antenna that has four feed points to synthesize four independent RHCP beams. The aperture radius is now $a = 12\lambda$, and the substrate dielectric constant and thickness are now, respectively, $\varepsilon_r = 9.8$ and $h = 0.508$ mm. The location of the four phase centers of this antenna are given by $\rho_1 = 1.7\,\lambda\,\hat{\mathbf{x}}$, $\rho_2 = -1.7\,\lambda\,\hat{\mathbf{x}}$, $\rho_3 = 1.7\,\lambda\,\hat{\mathbf{y}}$, and

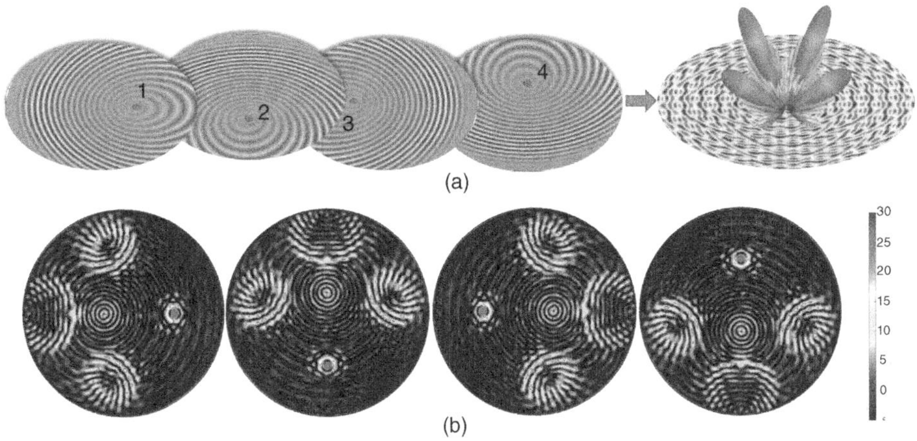

Figure 1.28 Four-beam MTS antenna. (a) Aperture sharing to generate the four beams using four VED sources and the superposition of their modulated impedance patterns. The colors on the aperture represent the amplitude of the impedance modulation. (b) Far-field directivity patterns in the *u*–*v* plane that were obtained with the GRBF-MoM simulations.

$\rho_4 = -1.7\,\lambda\,\hat{\mathbf{y}}$. Figure 1.28b shows the simulated u–v plane RHCP and LHCP far-field directivity patterns that were obtained by exciting a single VED located at the phase center position $\boldsymbol{\rho}_1$. Note that the u–v plane directivity patterns for each one of the other three possible excitation locations: $\boldsymbol{\rho}_2$, $\boldsymbol{\rho}_3$, and $\boldsymbol{\rho}_4$, are quite similar to the one obtained from the VED at $\boldsymbol{\rho}_1$. In all the cases, a peak gain of 27.5 dBi was obtained for each beam. This result implies that the aperture efficiency was approximately $\varepsilon_{ap} = 46\%$.

1.9.3 Radar Monopulse Application

The flexibility of the modulated MTS antenna designs can be exploited to generate single and multiple monopulse beams for radar applications. A specific design case is illustrated in Figure 1.29. The aperture consists of a circular domain of radius $a = 7.5$ cm that is divided into four identical angular sectors, each one illuminated by a phased monopole.

Being individually defined on a sector, the modulation over each of these four sub-apertures is capable of radiating an LP broadside beam. The overall reactance tensor then assumes the form:

$$\underline{\underline{X}} = \sum_{n=1}^{4} \underline{\underline{X}}_n U_n = \sum_{n=1}^{4} [\hat{\boldsymbol{\kappa}}_n \hat{\boldsymbol{\kappa}}_n X_{\kappa\kappa,n} + (\hat{\boldsymbol{\kappa}}_n \hat{\boldsymbol{\gamma}}_n + \hat{\boldsymbol{\gamma}}_n \hat{\boldsymbol{\kappa}}_n) X_{\kappa\gamma,n} + \hat{\boldsymbol{\gamma}}_n \hat{\boldsymbol{\gamma}}_n X_{\gamma\gamma,n}] U_n \tag{1.25}$$

where U_n is an angular unit step function defined on the n-th angular region as: $U_n = 1$ on the angular interval $-\pi/4 + (n-1)\pi/2 < \phi \leq \pi/4 + (n-1)\pi/2$ for $n = 1,\ldots,4$ and on the radial interval $0 < \rho < a$, and $U_n = 0$ elsewhere. The unit vectors $\hat{\boldsymbol{\kappa}}_n$, $\hat{\boldsymbol{\gamma}}_n$ are defined locally for each angular region of the aperture. Their origin is located at the corresponding feed position ρ_n. Consequently, they are defined as $\hat{\boldsymbol{\kappa}}_n = (\rho - \rho_n)/R_n$ and $\hat{\boldsymbol{\gamma}}_n = \hat{\mathbf{z}} \times \hat{\boldsymbol{\kappa}}_n$, where the relative distance $R_n = |\rho - \rho_n|$ for ρ being a generic observation point on the aperture with respect to the global coordinate system whose origin is located at its geometric center. Finally, each of the local modulation functions $X_{\chi\chi,n}$, where $\chi = \kappa, \gamma$, assume a sinusoidal shape and are able to radiate an LP field along the azimuthal cuts $\phi = \pm 45°$.

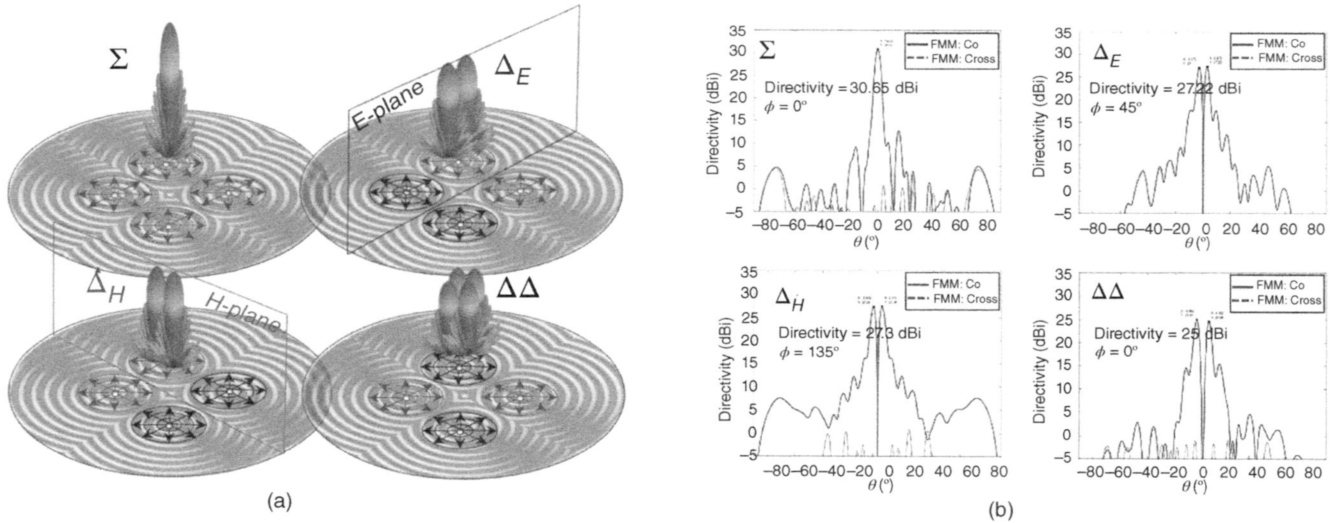

Figure 1.29 Monopulse MTS antenna. (a) Illustration of the conceptual scheme. The aperture is partitioned into four identical sub-apertures, each fed by an independent cylindrical surface wave source. The colors on the aperture represent the impedance modulation. Four distinct beams Σ, Δ_E, Δ_H, $\Delta\Delta$ are obtained by the phased recombination of contributions from each element to the radiation in the far-field zone. The phasing scheme of these simple sources is represented by a colored coding of the four beams. The sources colored in light gray and dark gray feature a zero and π phase, respectively. (b) Σ, Δ_E, Δ_H, and $\Delta\Delta$ directivity patterns calculated with the FMM described in Section 1.4. The prototype antenna was realized with printed ellipses. The directivity patterns were calculated along the cuts containing their maxima, at the azimuthal angles written on the top left corner of each frame. The values of their peak-directivities are also indicated there. (c) Fabricated antenna and CAD model of one feed. (d) Measured monopulse normalized intensity beams.

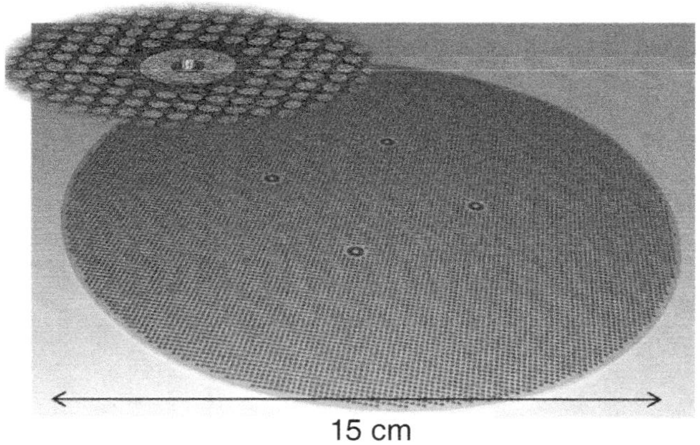

(c)

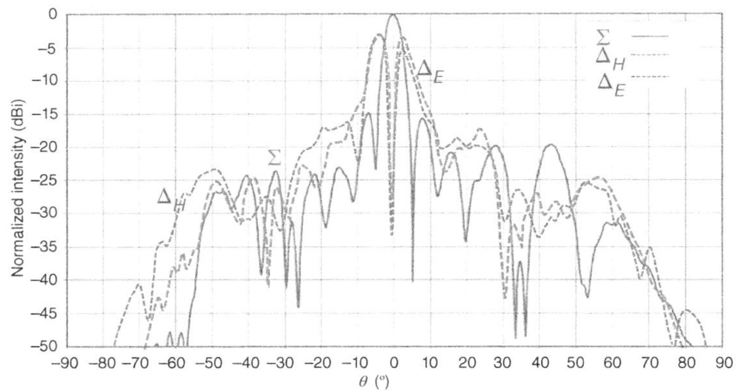

(d)

Figure 1.29 (*Continued*)

An LP single-fed antenna based on an aperture partition was also obtained in [20]. In contrast to that design, we use the aperture partitioning in this design to obtain a set of distinct LP beams. The individual partitions in the design here directly yield the LP fields. With a very simple source phasing scheme, the phased recombination of the four radiated beams, one from each sector of the partition, allows one to generate four distinct monopulse type beams, i.e. a sum pattern (Σ), a delta–azimuth pattern (Δ_E), a delta–elevation pattern (Δ_H), and a delta–delta pattern ($\Delta\Delta$). They are displayed in Figure 1.29. The Δ_E and Δ_H patterns have their maxima aligned in azimuth along $\phi = 45°$ and $\phi = 135°$, respectively. The $\Delta\Delta$ pattern features four maxima aligned along $\phi = 0°$ and $\phi = 90°$.

The four sub-apertures are virtually separated by properly designing the MTS modulation over each of them. To further this end, the LW attenuation constant is calibrated to sufficiently damp each individual SW and thus prevent interactions with any adjacent region. This outcome ensures a low cross interaction between the SW launched by the feed i and the modulation relevant to sub-region j, where $i, j = 1, 2, 3, 4$ and $i \neq j$. Therefore, the printed structure is not delimited by any physical separation, but only by a continuous change of the equivalent BCs.

1.9.4 Multi-frequency Metasurface Antennas

The same aperture sharing concept can also be applied to the design of multi-frequency modulated MTS apertures. This technique was presented in [61]. Dual band designs were achieved based on aperture sharing with two distinct modulations being defined. Different from the multi-beam applications, these two modulation functions were superimposed point-by-point on the whole aperture. They featured periodicities that are tuned to radiate a fixed beam at two separate frequencies. Without any loss of generality, we treat here an aperture sharing design that produces broadside pointing CP beams at two distinct frequencies. The two modulations can share the same phase center and, hence, they are illuminated by the same source. On the other hand, they can be illuminated by two distinct points by decoupling the feed positions. This variation is called a self-diplexed aperture. The modulation functions relevant to the decoupled feeds or the aperture centered-single source cases can be defined following formula (1.24), where the index $n = 1, 2$. The associated reactance functions assume the shape detailed with [61]. We present two design examples that feature the dual band single point aperture and the self-diplexed aperture with decoupled sources. The aperture for both examples has the radius $a = 10.0$ cm.

The single point source aperture is implemented by using an elliptically shaped texture embedded into a Cartesian lattice of periodicity 1.4 mm. It is printed on a Rogers RO3006 laminate that has the dielectric constant $\varepsilon_r = 6.15$ and thickness $h = 0.635$ mm. The two overlapping modulations feature periodicities tuned to radiate two broadside beams in the Ka-band at $f_1 = 26.25$ GHz and $f_2 = 32.0$ GHz. The modulation indexes were designed by imposing a quasi-uniform power distribution profile $S(\rho, \omega_0)$ at the middle angular frequency ω_0, i.e. at $\omega_0 = 2\pi \times (26.25 + 32)/2$. The related leakage parameter is obtained using [54]. The local modulation index is retrieved by resolving a generalized version of the local canonical problem in which the modulation front originates from the composition of the two frequencies. The aperture radiation at f_1 and f_2 and over the related sub-bands has been calculated using the homogenized reactance model developed into Gaussian rings basis functions (GRBFs) based MoM framework and the FMM-MoM solver. These simulation results are shown in Figure 1.30.

A dual band prototype that is a self-diplexed configuration, i.e. the SW illumination, is provided alternatively or simultaneously from two decoupled feeds, has also been fabricated and tested.

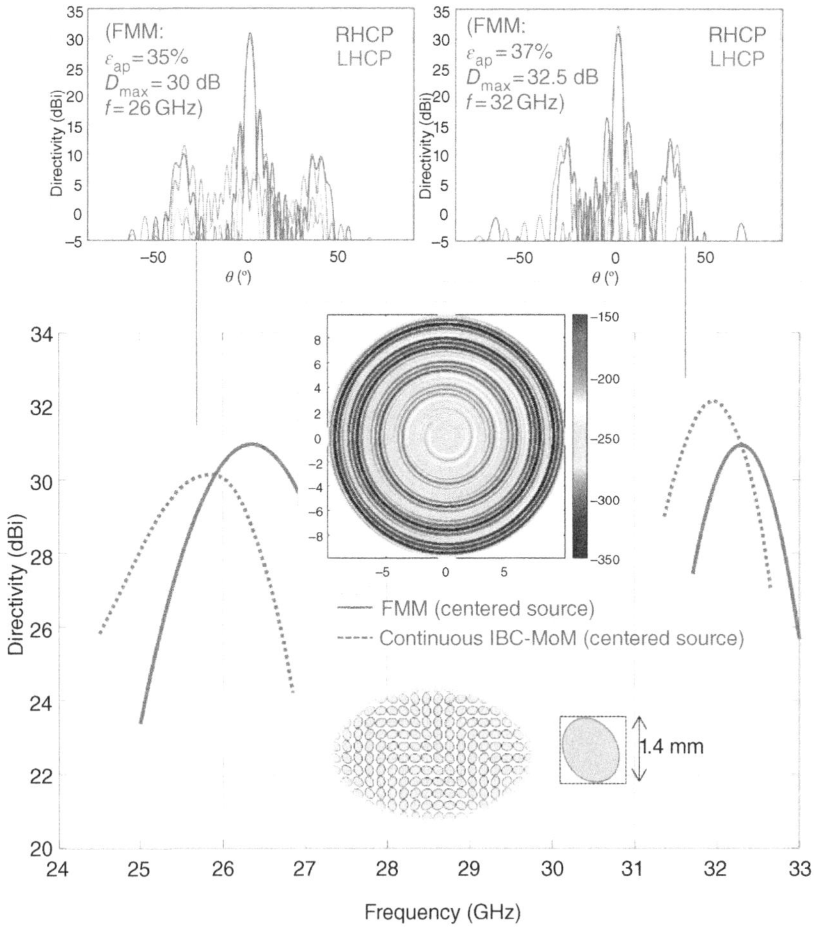

Figure 1.30 Dual-band MTS antenna with a single point source aperture. The top left and right subplots are the directivity patterns at f_1 and f_2, respectively. The light gray curves (solid and dashed) represent the cross-polar component evaluated using the FMM and the GRBF-MoM analyses. The dark gray curves (solid and dashed) represent the co-polar field components coded with the same line style. The bottom plot displays the corresponding simulated directivities as functions of the source frequency. Its bottom inset shows the adopted texture printed on the substrate and a sketch of the sampling lattice. Its top inset shows the radial component of the tensor impedance.

The prototype features the same 10.0 cm aperture radius. It was designed to work at $f_1 = 25$ and $f_2 = 30$ GHz. Its two feed points were separated by a distance of 4.1 cm.

A photo of the prototype and details of its feeds are given in Figure 1.31. The measured S-parameter values as functions of the source frequency and the normalized measured directivity patterns at $f_1 = 25$ and $f_2 = 30$ GHz are also presented. These patterns were obtained on the plane orthogonal to the line connecting the two feeds. The directivity patterns reasonably met expectations. The input parameters indicate that good impedance matching for both ports was achieved at the design frequencies. The average isolation level between the two ports was around 27 dB over a wide band of frequencies.

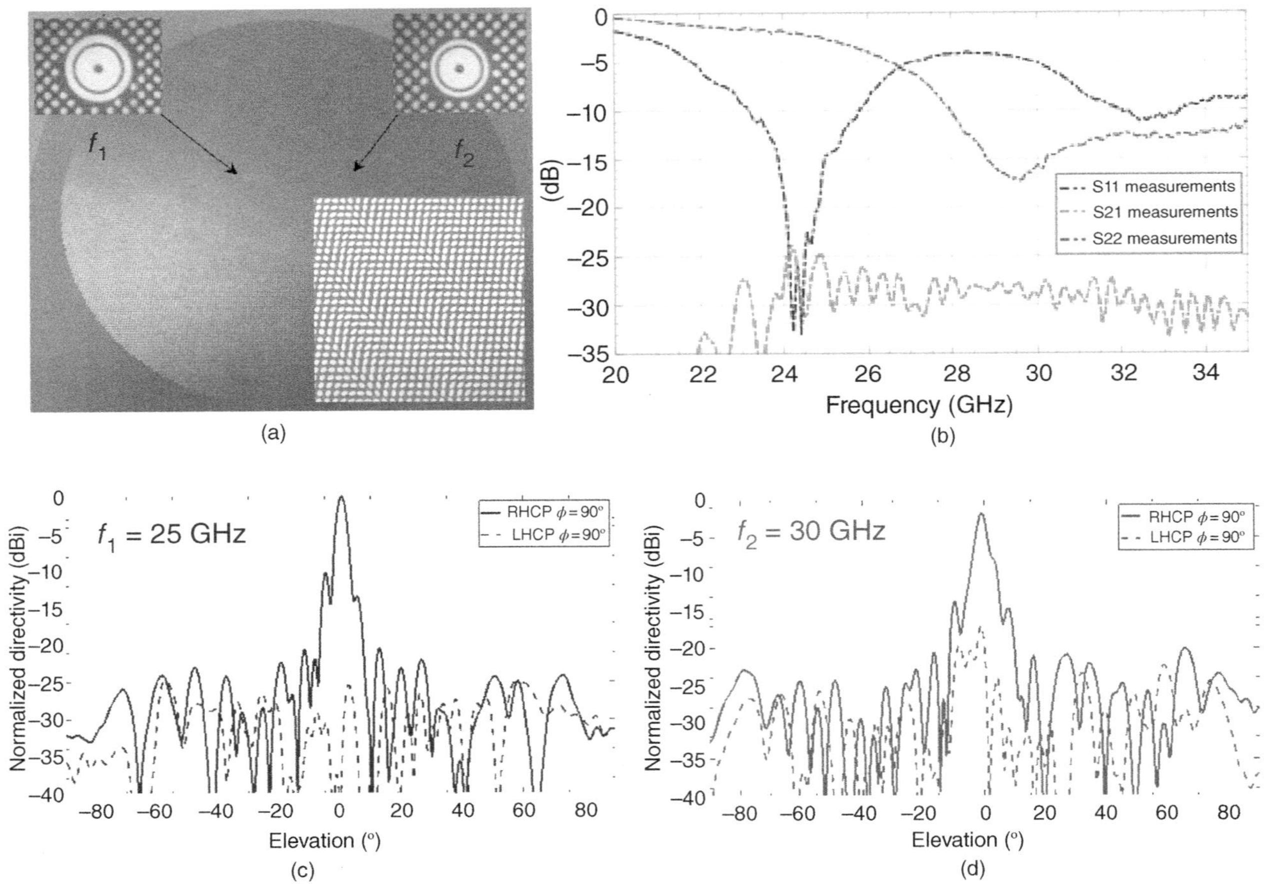

Figure 1.31 MTS antenna with a self-diplexed configuration for the generation of broadside pointing bi-chromatic beams. (a) Photo of the fabricated prototype. It had the radius $a = 10.0$ cm. Details of its two different feeders are shown in the insets. The MTS is composed of small elliptical patches embedded in a square lattice of 1.0 mm periodicity. They were printed on a grounded substrate whose thickness was 640 μm and relative permittivity was 10.2. The two feeders were separated by a distance of 4.1 cm and were formed by vertical monopoles with a circular slot-loaded patch printed on the top of the same MTS substrate. (b) Measured S-parameters. The normalized measured co-polar and cross-polar directivity patterns on the plane orthogonal to the line segment connecting its two feeds are given at (c) $f_1 = 25$ GHz and (d) $f_2 = 30$ GHz.

1.10 Conclusions

MTS antenna designs have emerged and substantially evolved during the last ten years. They have become a significant innovation in the field of single and multi-beam antenna systems. Concisely, they are a class of LW antennas in which the energy carried by an SW propagating on an impedance BC plane is gradually radiated. Owing to its interaction with this surface impedance, the SW is transformed into a radiated LW whose amplitude and polarization can be finely controlled. The MTS basically serves to impose an artificially tailored impedance BC that is seen by the SW. Rather than shaping conductors, the BCs are designed and implemented to obtain the desired radiation characteristics. These include tilted or shaped single or multiple directive beams. All of these very challenging properties occur while preserving the light weight and extremely low profile of these transmitting or receiving systems. These features constitute the key concepts behind MTS antennas.

We have briefly summarized the design process that we have developed. We have provided several known and new examples of MTS antenna designs. We have demonstrated that by properly designing the impedance BCs over the entire aperture or parts of it, one can obtain single or multiple shaped beams and attain measured aperture efficiencies up to 70%. We have also proven that a proper design of the impedance BC allows one to control the antenna bandwidth as well as to realize an antenna operating in two different frequency bands.

Several of our MTS antennas have been integrated into terrestrial communication networks and have been qualified for use on space platforms. While these successful systems are contributing to practical applications, many challenging scientific and engineering issues associated with the designs, analysis, simulations, and implementations of MTS antennas still exist. We believe that their continuing evolution will have a positive and influential impact on current 5G and future 6G and beyond terrestrial and non-terrestrial wireless ecosystems.

Acknowledgments

Some of the work presented in this Chapter has been carried out within the Joint Metasurface Laboratory between the University of Siena and Wave Up S.r.l, Via Roma, 77 – 53100 Siena – Italy. The authors would like in particular to acknowledge Francesco Caminita, Gabriele Minatti, and Cristian Della Giovampaola of Wave Up for contributing to the results presented in Sections 5.3, 7.4, and 7.5, respectively.

References

1 C. L. Holloway *et al.*, "An overview of the theory and applications of metasurfaces: the two-dimensional equivalents of metamaterials," *IEEE Antennas Propag. Mag.*, vol. 54, no. 2, pp. 10–35, 2012.
2 C. Pfeiffer and A. Grbic, "Metamaterial Huygens' surfaces: tailoring wave fronts with reflectionless sheets," *Phys. Rev. Lett.*, vol. 110, no. 19, p. 197401, 2013.
3 M. Selvanayagam and G. V. Eleftheriades, "Discontinuous electromagnetic fields using orthogonal electric and magnetic currents for wavefront manipulation," *Opt. Express*, vol. 21, no. 12, pp. 14409–14429, 2013.
4 N. Yu *et al.*, "Flat optics: controlling wavefronts with optical antenna metasurfaces," *IEEE J. Sel. Top. Quantum Electron.*, vol. 19, no. 3, pp. 4700423, 2013.

5 F. Monticone, N. M. Estakhri, and A. Alù, "Full control of nanoscale optical transmission with a composite metascreen," *Phys. Rev. Lett.*, vol. 110, no. 20, p. 203903, 2013.

6 N. Yu and F. Capasso, "Flat optics with designer metasurfaces," *Nat. Mater.*, vol. 13, pp. 139–150, 2014.

7 S. Maci, G. Minatti, M. Casaletti, and M. Bosiljevac, "Metasurfing: addressing waves on impenetrable metasurfaces," *IEEE Antennas Wireless Propag. Lett.*, vol. 10, pp. 1499–1502, 2011.

8 A. M. Patel and A. Grbic, "Transformation electromagnetics devices based on printed-circuit tensor impedance surfaces," *IEEE Trans. Microw. Theory Tech.*, vol. 62, no. 5, pp. 1102–1111, 2014.

9 A. M. Patel and A. Grbic, "Effective surface impedance of a printed-circuit tensor impedance surface (PCTIS)," *IEEE Trans. Microw. Theory Tech.*, vol. 61, no. 4, pp. 1403–1413, 2013.

10 A. M. Patel and A. Grbic, "The effects of spatial dispersion on power flow along a printed-circuit tensor impedance surface," *IEEE Trans. Antennas Propag.*, vol. 62, no. 3, pp. 1464–1469, 2014.

11 M. Mencagli, E. Martini, and S. Maci, "Surface wave dispersion for anisotropic metasurfaces constituted by elliptical patches," *IEEE Trans. Antennas Propag.*, vol. 63, no. 7, pp. 2992–3003, 2015.

12 K. Achouri *et al.*, "Synthesis of electromagnetic metasurfaces: principles and illustrations," arXiv preprint arXiv:1510.05997, 2015.

13 S. B. Glybovski, S. A. Tretyakov, P. A. Belov, Y. S. Kivshar, and C. R. Simovski, "Metasurfaces: from microwaves to visible," *Phys. Rep.*, vol. 634, pp. 1–72, 2016.

14 B. H. Fong, J. S. Colburn, J. J. Ottusch, J. L. Visher, and D. F. Sievenpiper, "Scalar and tensor holographic artificial impedance surfaces," *IEEE Trans. Antennas Propag.*, vol. 58, no. 10, pp. 3212–3221, 2010.

15 G. Minatti, F. Caminita, M. Casaletti, and S. Maci, "Spiral leaky-wave antennas based on modulated surface impedance," *IEEE Trans. Antennas Propag.*, vol. 59, no. 12, pp. 4436–4444, 2011.

16 A. M. Patel and A. Grbic, "A printed leaky-wave antenna based on a sinusoidally-modulated reactance surface," *IEEE Trans. Antennas Propag.*, vol. 59, no. 6, pp. 2087–2096, 2011.

17 G. Minatti *et al.*, "Modulated metasurface antennas for space: synthesis, analysis and realizations," *IEEE Trans. Antennas Propag.*, vol. 63, no. 4, pp. 1288–1300, 2015.

18 G. Minatti, S. Maci, P. De Vita, A. Freni, and M. Sabbadini, "A circularly-polarized isoflux antenna based on anisotropic metasurface," *IEEE Trans. Antennas Propag.*, vol. 60, no. 11, pp. 4998–5009, 2012.

19 M. Faenzi *et al.*, "Realization and measurement of broadside beam modulated metasurface antennas," *IEEE Antennas Wireless Propag. Lett.*, vol. 15, pp. 610–613, 2016.

20 S. Pandi, C. A. Balanis, and C. R. Birtcher, "Design of scalar impedance holographic metasurfaces for antenna beam formation with desired polarization," *IEEE Trans. Antennas Propag.*, vol. 63, no. 7, pp. 3016–3024, 2015.

21 D. Sievenpiper, J. Schaffner, J. J. Lee, and S. Livingston, "A steerable leaky-wave antenna using a tunable impedance ground plane," *IEEE Antennas Wireless Propag. Lett.*, vol. 1, no. 1, pp. 179–182, 2002.

22 D. F. Sievenpiper, J. H. Schaffner, H. J. Song, R. Y. Loo, and G. Tangonan, "Two-dimensional beam steering using an electrically tunable impedance surface," *IEEE Trans. Antennas Propag.*, vol. 51, no. 10, pp. 2713–2722, 2003.

23 D. F. Sievenpiper, "Forward and backward leaky wave radiation with large effective aperture from an electronically tunable textured surface," *IEEE Trans. Antennas Propag.*, vol. 53, no. 1, pp. 236–247, 2005.

24 J. L. Gómez-Tornero, F. D. Quesada-Pereira, and A. Álvarez-Melcón, "Analysis and design of periodic leaky-wave antennas for the millimeter waveband in hybrid waveguide-planar technology," *IEEE Trans. Antennas Propag.*, vol. 53, no. 9, pp. 2834–2842, 2005.

25 A. T. Pereda *et al.*, "Dual circularly polarized broadside beam metasurface antenna," *IEEE Trans. Antennas Propag.*, vol. 64, no. 7, pp. 2944–2953, 2016.

26 M. G. Silveirinha, C. A. Fernandes, and J. R. Costa, "Electromagnetic characterization of textured surfaces formed by metallic pins," *IEEE Trans. Antennas Propag.*, vol. 56, no. 2, pp. 405–415, 2008.

27 G. Chattopadhyay, C. D. Jung-Kubiak, T. J. Reck, D. Gonzalez-Ovejero, and M. A. del Pino. Low-Profile and High-Gain Modulated Metasurface Antennas from Gigahertz to Terahertz Range Frequencies, U.S. Patent Application No. 15/473 485, 2017.

28 D. González-Ovejero, T. J. Reck, C. D. Jung-Kubiak, M. Alonso-DelPino, and G. Chattopadhyay, "A class of silicon micromachined metasurface for the design of high-gain terahertz antennas," in *2016 IEEE International Symposium on Antennas and Propagation (APSURSI)*, June 2016, pp. 1191–1192.

29 G. Minatti, F. Caminita, E. Martini, and S. Maci, "Flat optics for leaky-waves on modulated metasurfaces: adiabatic Floquet-wave analysis," *IEEE Trans. Antennas Propag.*, vol. 64, no. 9, pp. 3896–3906, 2016.

30 G. Minatti, F. Caminita, E. Martini, M. Sabbadini, and S. Maci, "Synthesis of modulated-metasurface antennas with amplitude, phase, and polarization control," *IEEE Trans. Antennas Propag.*, vol. 64, no. 9, pp. 3907–3919, 2016.

31 M. Li, S. Q. Xiao, and D. F. Sievenpiper, "Polarization-insensitive holographic surfaces with broadside radiation," *IEEE Trans. Antennas Propag.*, vol. 64, no. 12, pp. 5272–5280, 2016.

32 M. Mencagli, E. Martini, and S. Maci, "Transition function for closed-form representation of metasurface reactance," *IEEE Trans. Antennas Propag.*, vol. 64, no. 1, pp. 136–145, 2016.

33 A. Oliner and A. Hessel, "Guided waves on sinusoidally-modulated reactance surfaces," *IRE Trans. Antennas Propag.*, vol. 7, no. 5, pp. 201–208, 1959.

34 M. Faenzi, G. Minatti, D. González-Ovejero *et al.*, "Metasurface antennas: new models, applications and realizations," *Sci. Rep.*, vol. 9, p. 10178, 2019.

35 D. González-Ovejero and S. Maci, "Gaussian ring basis functions for the analysis of modulated metasurface antennas," *IEEE Trans. Antennas Propag.*, vol. 63, no. 9, pp. 3982–3993, 2015.

36 M. A. Francavilla, E. Martini, S. Maci, and G. Vecchi, "On the numerical simulation of metasurfaces with impedance boundary condition integral equations," *IEEE Trans. Antennas Propag.*, vol. 63, no. 5, pp. 2153–2161, 2015.

37 N. Bodehou, "Modulated metasurface antennas: analysis and direct numerical synthesis," PhD dissertation, UCL Louvain, 2020.

38 M. Bodehou, C. Craeye, E. Martini, and I. Huynen, "A quasi-direct method for the surface impedance design of modulated metasurface antennas," *IEEE Trans. Antennas Propag.*, vol. 67, no. 1, pp. 24–36, Jan 2019.

39 M. Bodehou, E. Martini, S. Maci, I. Huynen, and C. Craeye, "Multibeam and beam scanning with modulated metasurfaces," *IEEE Trans. Antennas Propag.*, vol. 68, no. 3, pp. 1273–1281, Mar. 2020.

40 M. Bodehou, D. González-Ovejero, C. Craeye, and I. Huynen, "Method of moments simulation of modulated metasurface antennas with a set of orthogonal entire-domain basis functions," *IEEE Trans. Antennas Propag.*, vol. 67, no. 2, pp. 1119–1130, Feb. 2019.

41 F. Vernì, M. Righero, and G. Vecchi, "On the use of entire-domain basis functions and fast factorizations for the design of modulated metasurface," *IEEE Trans. Antennas Propag.*, vol. 68, no. 5, pp. 3824–3833, May 2020.

42 N. Engheta and R. W. Ziolkowski, Eds., *Metamaterials: Physics and Engineering Explorations.* Wiley, 2006.

43 E. Martini, F. Caminita, and S. Maci, "Double-scale homogenized impedance models for periodically modulated metasurfaces," *EPJ Appl. Metamater.*, vol. 7, no. 12, 2020.

44 S. Maci, M. Caiazzo, A. Cucini, and M. Casaletti, "A pole-zero matching method for EBG surfaces composed of a dipole FSS printed on a grounded dielectric slab," *IEEE Trans. Antennas Propag.*, vol. 53, no. 1, pp. 70–81, 2005.

45 M. Mencagli, C. Della Giovampaola, and S. Maci, "A closed-form representation of isofrequency dispersion curve and group velocity for surface waves supported by anisotropic and spatially dispersive metasurfaces," *IEEE Trans. Antennas Propag.*, vol. 64, no. 6, pp. 2319–2327, 2016.

46 V. Rokhlin, "Rapid solution of integral equations of scattering theory in two dimensions," *J. Comput. Phys.*, vol. 86, no. 2, pp. 414–439, 1990.

47 C. R. Anderson, "An implementation of the fast multipole method without multipoles," *SIAM J. Sci. Stat. Comput.*, vol. 13, no. 4, pp. 923–947, 1992.

48 J. S. Zhao, W. C. Chew, C. C. Lu, E. Michielssen, and J. Song, "Thin-stratified medium fast-multipole algorithm for solving microstrip structures," *IEEE Trans. Microwave Theory Tech.*, vol. 46, no. 4, pp. 395–403, 1998.

49 L. Tsang, C. H. Chan, and K. Pak, "Monte Carlo simulation of a two-dimensional random rough surface using the sparse-matrix flat-surface iterative approach," *Electron. Lett.*, vol. 29, no. 13, pp. 1153–1154, 1993.

50 M. Albani, A. Mazzinghi, and A. Freni, "Asymptotic approximation of mutual admittance involved in MoM analysis of RLSA antennas," *IEEE Trans. Antennas Propag.*, vol. 57, no. 4, pp. 1057–1063, 2009.

51 L. Matekovits, V. A. Laza, and G. Vecchi, "Analysis of large complex structures with the synthetic-functions approach," *IEEE Trans. Antennas Propag.*, vol. 55, no. 9, pp. 2509–2521, 2007.

52 V. A. Prakash and R. Mittra, "Characteristic basis function method: a new technique for efficient solution of method of moments matrix equations," *Microwave Opt. Technol. Lett.*, vol. 36, no. 2, pp. 95–100, 2003.

53 A. T. Pereda, F. Caminita, E. Martini, I. Ederra, J. C. Iriarte, R. Gonzalo, and S. Maci, "Experimental validation of a Ku-band dual-circularly polarized metasurface antenna," *IEEE Trans. Antennas Propag.*, vol. 66, no. 3, pp. 1153-1159, Mar. 2018.

54 G. Minatti, E. Martini, and S. Maci, "Efficiency of metasurface antennas," *IEEE Trans. Antennas Propag.*, vol. 65, no. 4, pp. 1532–1541, Apr. 2017.

55 M. Bodehou, D. González-Ovejero, C. Craeye *et al.*, "Power balance and efficiency of metasurface antennas," *Sci. Rep.*, vol. 10, p. 17508, 2020.

56 G. Minatti, M. Faenzi, M. Sabbadini, and S. Maci, "Bandwidth of gain in metasurface antennas," *IEEE Trans. Antennas Propag.*, vol. 65, no. 6, pp. 2836–2842, Jun. 2017.

57 M. Faenzi, D. González-Ovejero, and S. Maci, "Wideband active region metasurface antennas," *IEEE Trans. Antennas Propag.*, vol. 68, no. 3, pp. 1261–1272, Mar. 2019.

58 M. Faenzi, D. González-Ovejero, and S. Maci, "Flat gain broadband metasurface antennas," *IEEE Trans. Antennas Propag.*, vol. 69, no. 4, pp. 1942–1951, Apr. 2021.

59 M. Albani and S. C. Pavone, "On the (local) phase center of an antenna and its explicit derivation," in *Proceedings of International Conference on Electromagnetics in Advanced Applications (ICEAA)*, Granada, Spain, Sep 2019, p. 1303.

60 D. González-Ovejero, G. Minatti, G. Chattopadhyay, and S. Maci, "Multibeam by metasurface antennas," *IEEE Trans. Antennas Propag.*, vol. 65, no. 6, pp. 2923–2930, Jun. 2017.

61 M. Faenzi, D. González-Ovejero, and S. Maci, "Overlapped and sequential metasurface modulations for Bi-Chromatic beams generation," *Appl. Phys. Lett.*, vol. 118, no. 18, p. 181902, 2021.

2

Techniques for Designing High Gain and Two-Dimensional Beam Scanning Antennas for 5G

Raj Mittra[1,4], Abdelkhalek Nasri[1], Ravi K. Arya[2], Prashant Chaudhary[3], James R. Kelly[5], Shaker Alkaraki[5], and Alejandro L. Borja[6]

[1] *University of Central Florida, Electrical and Computer Engineering, Orlando, FL 32816, USA*
[2] *XSL, Zhongshan Institute of CUST, China*
[3] *University of Delhi South Campus, Department of Electronic Science, New Delhi 110021, India*
[4] *King Abdulaziz University, Electrical and Computer Engineering Department, Faculty of Engineering, Jeddah 21589, Saudi Arabia*
[5] *Queen Mary University of London, School of Electronic Engineering and Computer Science, London, E1 4NS, UK*
[6] *Universidad de Castilla-La Mancha, Departamento de Ingeniería eléctrica, electrónica, automática y comunicaciones, Escuela de Ingenieros Industriales de Albacete, Campus Universitario, 02071 Albacete, Spain*

2.1 Introduction

Typically, an antenna with a gain level of 30 dBi or higher is frequently desired for a wide variety of communication applications including those associated with 5G and beyond 5G wireless technologies. The gain level of an antenna is determined by its aperture size, and this level can be predicted theoretically under the assumption that both the amplitude and phase distributions in this aperture are uniform. How to excite this aperture in order to realize the theoretical gain is not obvious, and various strategies for meeting this goal have been proposed in the literature. The simplest strategy is to use an array configuration of antenna elements that are fed in unison via a power dividing network. If beam scanning is required, then phase shifters can be inserted into the feed network of the antenna array. The array configuration is determined by whether the beam should scan in one dimension or two. Phase shifters are also essential components of several microwave and millimeter-wave (mm-wave) systems, such as phased arrays, smart antennas, modulators, and microwave instrumentation. However, the majority of passive phase shifters suffer from high insertion loss (IL) at microwave and mm-wave frequencies. For example, the average IL of a phase shifter operating at X-band frequencies (e.g. at 10 GHz) is 4.69 dB with a variance of 5.24 dB [1–20]. Moreover, it is considerably higher in the millimeter wave range, as is the cost of these higher frequency phase shifters. The IL problem can be mitigated by employing active phase shifters, which incorporate integrated amplifiers. However, active phase shifters consume considerably more direct current (DC) power than passive ones. For this reason, one could argue that active phase shifters are ill-suited for power-constrained applications.

Of the available X-band phase shifters, mechanical reconfiguration or the use of an Microelectromechanical systems (MEMS)-switched capacitor offers the lowest IL. The use of mechanical moving parts increases the need for maintenance and repair. MEMS suffer from a range of disadvantages, including poor reliability, e.g. due to contact welding effects. These issues have prevented their widespread use in practice.

Antenna and Array Technologies for Future Wireless Ecosystems, First Edition.
Edited by Y. Jay Guo and Richard W. Ziolkowski.

Given this background, the main focus of this chapter is to explore ways of designing antennas and arrays, primarily for current and future mm-wave applications. Those antennas and arrays exhibit realized gain levels of 30 dBi and offer the capability of both one- and two-dimensional (2D) beam scanning, but without the use of conventional phase shifters. Section 2.2 begins with the Luneburg lens configuration which is spherical in shape and whose refractive index varies radially from the center to the edge of the sphere. The issues of fabrication and two different techniques for practically realizing the Luneburg lens are presented. It next goes on to discuss the feed design for dual linear polarization with a high isolation between the two polarizations. Additionally, the section discusses the issues of two-dimensional beam scanning by using feed networks which utilize switches as opposed to conventional phase shifters. Finally, Section 2.2 concludes by presenting a novel technique for wide angle beam scanning by using reconfigurable frequency selective surfaces (FSSs) or tilted metal panels.

We then describe in Section 2.3 a number of innovative array designs that achieve the desired gain level of 30 dBi and are also capable of scanning the beam either in one or two dimensions. The scanning again occurs without using conventional phase shifters. Alternative designs are used instead which either employ electronic switches or drill holes filled with liquid metal. Section 2.4 is entirely devoted to the topic of phase shifters based around SIWs. These phase shifters are reconfigured using liquid metal. It presents not only detailed designs, but also simulation and measurement results, which validate the designs. Finally, Section 2.5 includes a summary of the three main sections which make up the body of the chapter, closing it out with some concluding remarks.

2.2 Luneburg Lens Designs

In this section, we describe beam scanning techniques based on the Luneburg lens, which is well known for its wide-angle scanning capability with virtually no scan loss. We also present the results of a study which investigates a hemispherical Luneburg lens with a circular perfect electric conductor (PEC) base. The lens exhibits a wide-angle scan capability over a wide frequency band and can be mounted on conformal surfaces. The hemispherical lens has a smaller footprint and is easy to fabricate.

2.2.1 Introduction

The Luneburg lens [21] is a spherically symmetric gradient-index lens whose refractive index decreases radially from the center to its outer surface. The traditional Luneburg lens has a radially varying permittivity profile, ε_r, defined as:

$$\varepsilon_r(r) = 2 - \left(\frac{r}{R}\right)^2, (0 < r < R) \tag{2.1}$$

where r is the distance from the center of the lens, and R is the radius of the lens. Figure 2.1 shows the ray picture which explains how the lens works when operating in the receive mode.

Theoretically, we need a large number of concentric shells with varying refractive indices to attain a smooth bending of the rays inside the lens. However, in practice, only a limited number of shells are used to fabricate the lens. As an example, the spherical lens shown in Figure 2.2a has six shells as the sectional view in Figure 2.2b shows. Its indices were optimized [22] via numerical simulation to achieve the best performance. Table 2.1 shows the optimum values of the refractive indices of the six-shell design.

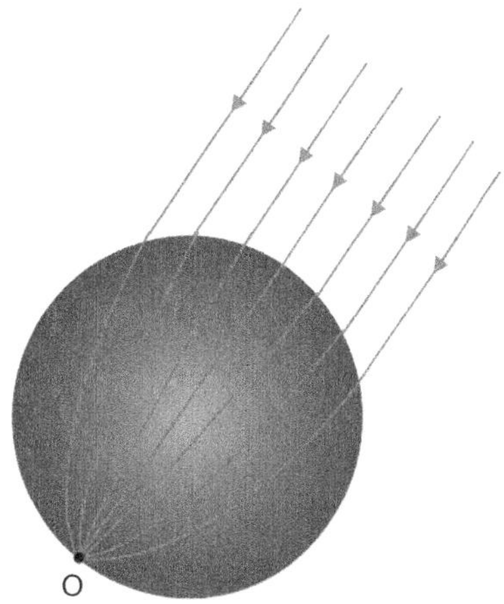

Figure 2.1 Luneburg lens – principle of operation.

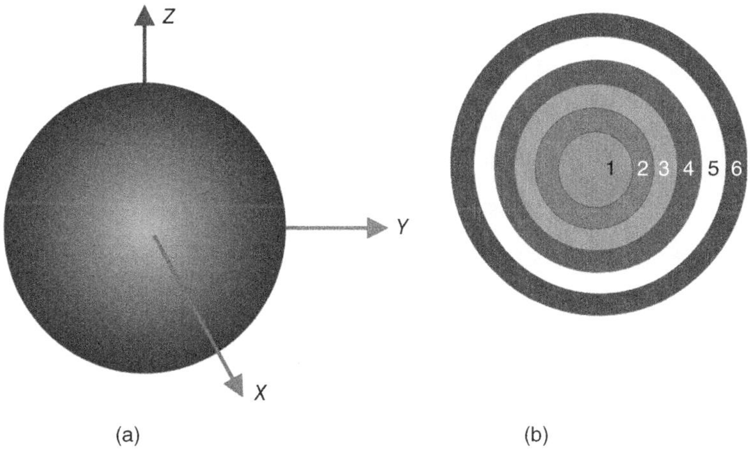

Figure 2.2 (a) Luneburg lens geometry (b) sectional view.

2.2.2 Fabrication of Luneburg Lenses

In a real-world scenario, it is very difficult to find the required dielectric materials with the permittivities specified in Table 2.1 since they are not commercially available off-the-shelf. Therefore, it becomes necessary to synthesize these materials artificially by using techniques [23], which we describe below.

2.2.2.1 Use of Dielectric Sheets with Holes

To synthesize a material with the desired permittivity, we begin with a base material that is commercially available in a sheet form. Next, we drill holes in the dielectric sheet which can be achieved using either a laser or a milling machine. We determine the required sizes of the holes by carrying out numerical simulation of "holey" plates to drive their equivalent dielectric constants [24]. We determine the equivalent dielectric constants by using a numerical simulation method that we have developed since we found that the available mixing formulas were not accurate. The hole pattern of the disk-shaped sheets is determined systematically for each disk. These sheets are then assembled to form the Luneburg lens. The number of disks depends upon the diameter of the lens and the thickness of the dielectric sheet used to form it. The total number of disks can run upward of 100

Table 2.1 Luneburg lens dimensions.

Layer	Outer shell radius (mm)	ε_r
1	29.6	1.92
2	41.9	1.77
3	51.3	1.62
4	59.2	1.46
5	66.2	1.31
6	72.5	1.15

Figure 2.3 Different stages of assembly of a Luneburg lens constructed with holey dielectric sheets.

or more. The lens formed by using the "holey disk" approach is shown in Figure 2.3, and further details may be found in [23].

2.2.2.2 Use of 3D Printing

An alternate approach to synthesizing the desired artificial dielectric for the lens is to use 3D printing. More details about this technique are discussed in [24–26]. Figure 2.4 shows the 3D-printed prototype lens. The inner layers (1–5) were 3D-printed. The sixth layer was fabricated by using a foam material because it was not practical to realize it by using 3D printing. Once the lens has been 3D-printed, it is enveloped in a foam layer (shell) as shown in Figure 2.4.

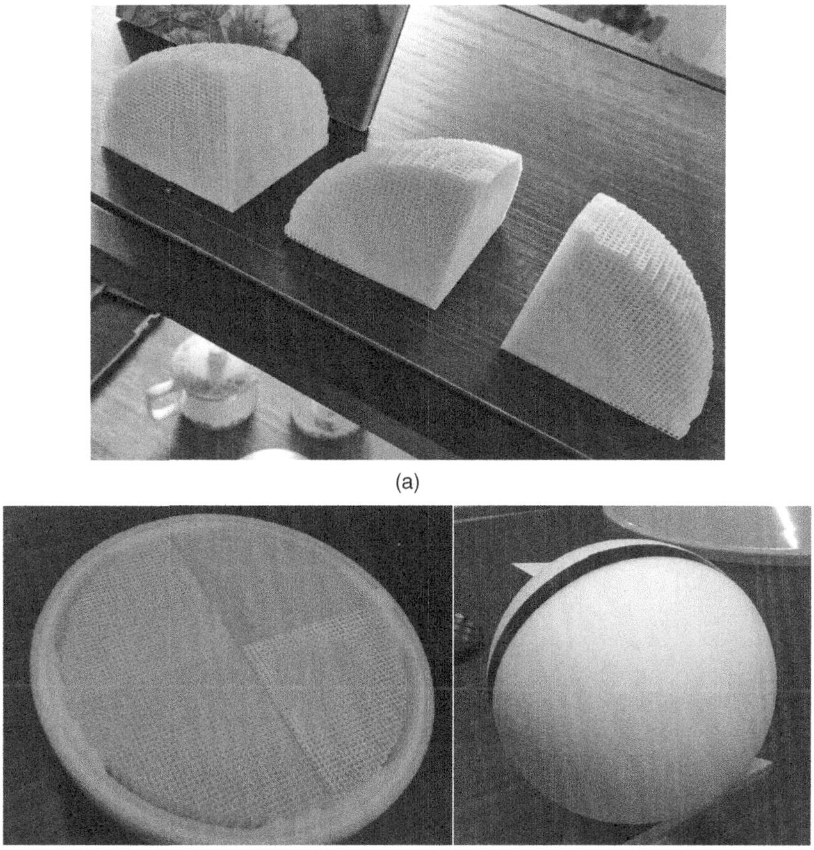

(a)

(b) (c)

Figure 2.4 3D-printed Luneburg lens (different stages of assembly). (a) One eighth of lens. (b) Half lens with all six layers. (c) Fully assembled lens.

2.2.2.3 Use of Specialized Materials

Yet another way to fabricate the Luneburg lenses is to use synthetic materials realized by varying the volume fractions of nano-sized particles with the permittivity values that a Luneburg lens demands. It is important to note that this method can be expensive as materials with the desired permittivity must be specially designed [27] and may require the use of sophisticated tools. A Luneburg lens fabricated by this method is shown in Figure 2.5.

2.2.3 Luneburg Lens with a Flat-Base and a Waveguide Array Excitation

The excitation of the Luneburg lens can be somewhat problematic because of the curved nature of the lens surface. In this section, we describe a novel flat-base waveguide type of feed system which is very convenient to use. We begin with a Luneburg lens designed by using 11 layers to achieve a smooth transition of the permittivity from one layer to the next as described in [28]. An array of 6×6 square waveguides was used to excite the lens. The antenna was able to scan its beam by up to 72° from boresight in both azimuth and elevation with only a slight degradation of the radiation pattern and with only a relatively small scan loss. Similarly, a Luneburg lens excited by an array of

Figure 2.5 Luneburg lens fabricated by using specialized materials.

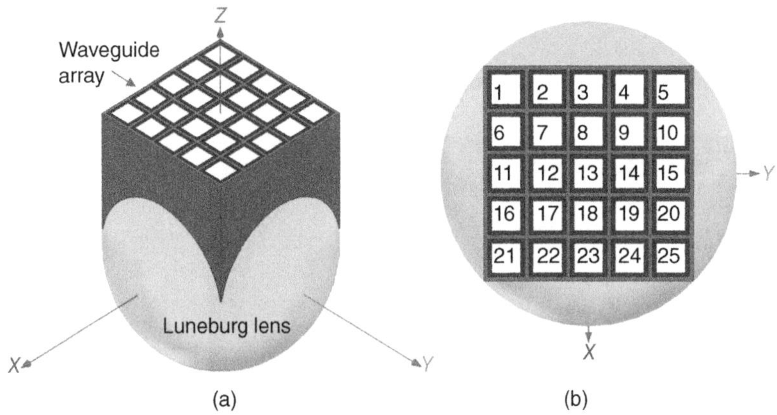

Figure 2.6 Luneburg Lens with a flat-base waveguide array feed. (a) Isometric view. (b) Top view.

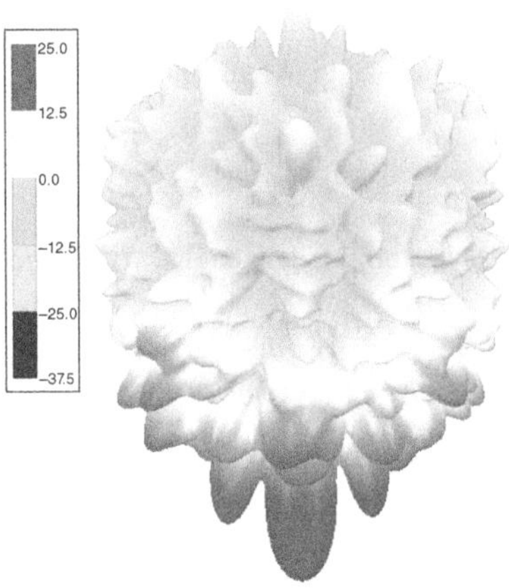

Figure 2.7 Realized gain pattern of the Luneburg lens when waveguide-13 is excited on the top of the lens.

5 × 5 waveguides is shown in Figure 2.6. A typical radiation pattern realized by exciting one of the waveguides of the array located on top of the lens is shown in Figure 2.7.

2.2.4 Validation of Luneburg Lens Design

To test the validity of method "*c*", mentioned in Section 2.2, a Luneburg lens was fabricated using the lens parameters listed in Table 2.1. The diameter of the lens was 14.5 cm. Figure 2.8a shows the

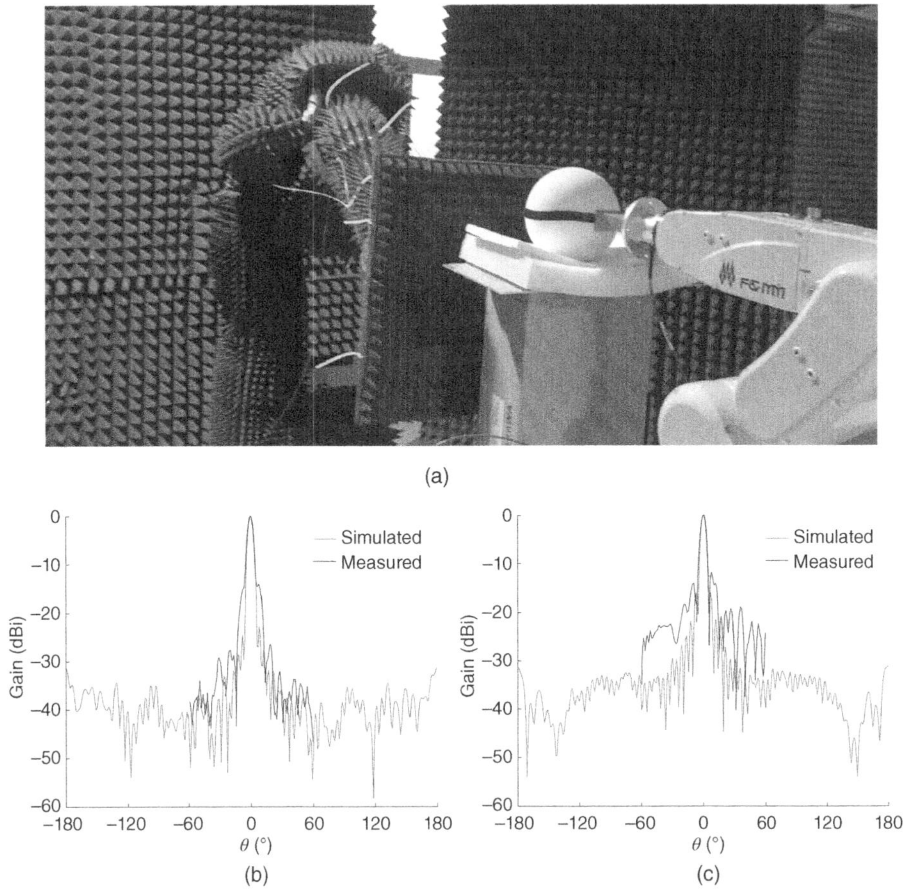

Figure 2.8 Luneburg lens antenna. (a) Measurement setup. Measured and simulated radiation patterns in the (b) $\phi = 0°$, and (c) $\phi = 90°$ principal planes.

measurement setup for the Luneburg lens with an open-ended waveguide feed. Figure 2.8b,c plot the radiation patterns for the lens at 26 GHz, for the $\phi = 0°$ and 90° cuts, respectively. The simulated peak realized gain of the lens is 30 dB, while the measured value was found to be 29.2 dB at 26 GHz. We conjecture that the slight loss in the realized gain is attributable to manufacturing tolerances that become significant when the operating frequency is in the millimeter wave range. Though not explicitly shown here, it is obvious that the symmetry of the spherical lens guarantees that it can scan to wide angles, with virtually no scan loss.

2.2.5 Dual-Polarized Luneburg Lens Antenna Feed Systems

In order to scan a beam when using the Luneburg lens, multiple feeds are employed such that an individual feed can be switched ON to point the beam in a desired direction. It is evident from the ray picture in Figure 2.1 that the feed should be located diametrically opposite to the desired beam direction. A prototype system to achieve such a wide beam scanning is shown in Figure 2.9. In this design, the first level of switching action is accomplished by the printed circuit board PCB1, which determines which PCB2 board is to be activated. Once a particular PCB2 board is selected,

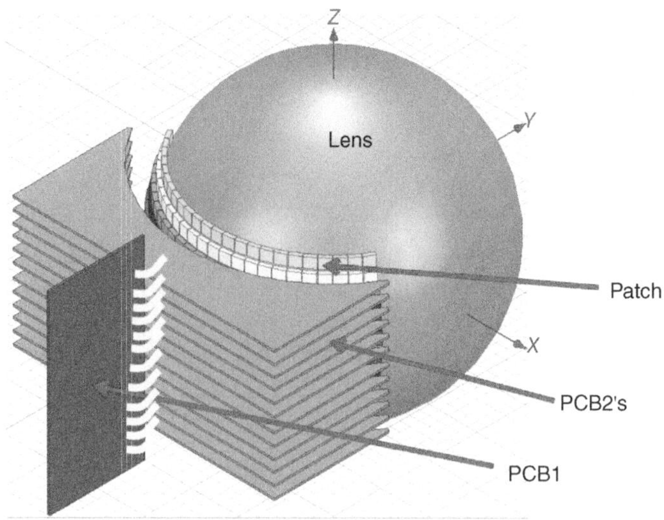

Figure 2.9 Luneburg lens with the scanning array feed consisting of microstrip patches and switches.

the switch on it activates the desired microstrip antenna feed. In the following paragraph we give details about the whole system. This is work in progress so only partial results are shown.

The physical and performance requirements for this antenna system are:

1) Frequency range: 24.75–27.5 GHz
2) Patch feed with $|S_{12}|$ less than −18 dB and $|S_{11}|/|S_{22}|$ should be less than −10 dB
3) Volume of the antenna: 15 cm × 15 cm × 15 cm
4) Desired antenna realized gain: 30 dBi
5) Elevation scanning range of −60° to +60° and azimuthal scanning range of −40° to +40°

Twelve PCB2 boards are used to feed 576 output ports which excite the 288 dual-polarized patches, each of which has two ports for the two polarizations. A single input port is located on the PCB1 board, which connects to 12 PCB2 boards where the output ports are located. Once the PCB1 and PCB2 boards are designed, all the switching and control circuits and connections need to be installed on the boards. Figure 2.10a shows the layout of the PCB1 board with all the

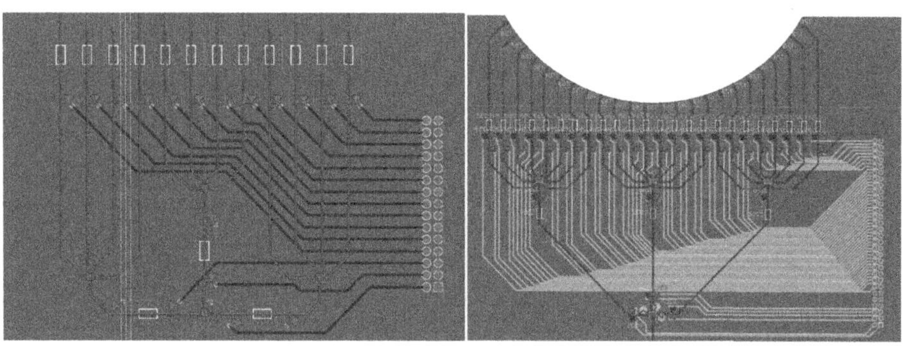

(a) (b)

Figure 2.10 Snapshot of the designed layout of the boards (a) PCB1 and (b) PCB2.

(a) (b)

Figure 2.11 Fabricated structures. (a) Fabricated test board for SP4T and SP8T switches. (b) Test fixtures for evaluating the insertion loss of the microstrip line and coaxial-to-microstrip line transition.

microstrip feed lines and control wires, while Figure 2.10b illustrates the snapshot of the detailed layout of a PCB2 board.

Based on the requirements of the feeding network, single pole 4 throw (SP4T) and SP8T switches were identified and acquired, keeping in mind that the IL of the switches must be relatively small. This choice circumvented the need for radio frequency (RF) amplifiers to compensate for the loss. The next step was to add the associated DC block and biasing circuits, leading to the fabricated board with both SP4T and SP8T switches, as shown in Figure 2.11a. For the purpose of the initial testing, the boards were terminated in K-connectors. Since the K-connectors also contribute to losses in the system, they too were chosen with care and tested. Figure 2.11b shows a typical test fixture for characterizing the IL of these lines and transitions.

2.2.5.1 Dual-Linear Feed Design

Most of the research on Luneburg lens feed design such as that reported in current literature, e.g. [29, 30], concentrates on linearly polarized (LP) radiation. It is relatively straightforward to achieve. However, some recent applications call for a feed design that is dual-linear and exhibits a high level of isolation (better than 18 dB) between the two polarizations. In contrast to the LP designs, these requirements are highly challenging. Microstrip patches are typically used as radiating elements when linear, dual-linear, or circular polarization is required.

Figure 2.12 presents the design of a microstrip patch antenna with dual-linear polarization. While achieving it is relatively straightforward, the same cannot be said about realizing a design with low mutual coupling between the two polarizations. Considerable optimization is required to achieve a design which would meet the specification that the coupling be less than −18 dB between the two polarizations.

The simulated S-parameters of the designed dual-LP microstrip patch antenna are shown in Figure 2.13. It shows that reflection coefficients at both ports are better than 10 dB while mutual coupling between both ports is better than 18 dB within the band 24.75–27.75 GHz.

Next, a test array was designed to check the response of the lens. The antenna element was replicated to form a 3 × 3 conformal antenna array, as shown in Figure 2.14. The edge-to-edge distance between two adjacent antenna elements is 1.0 mm. The composite conformal array plus lens system is shown in Figure 2.15a. The simulated realized gain patterns of the array and the lens are presented in Figure 2.16. The setup used for the antenna measurement is shown in Figure 2.15b. The measured total electric field patterns are also plotted in Figure 2.16. The peak realized gain of

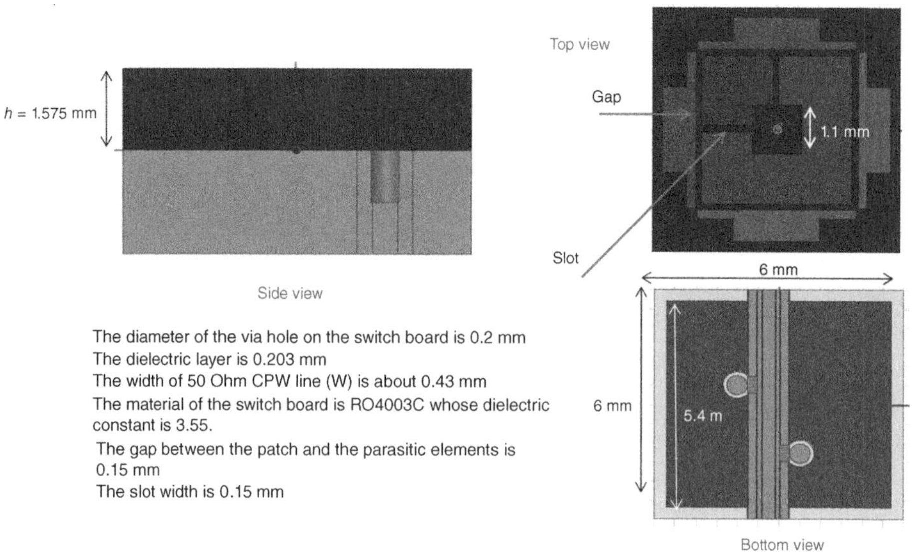

Figure 2.12 Microstrip patch antenna element with a dual-polarization capability.

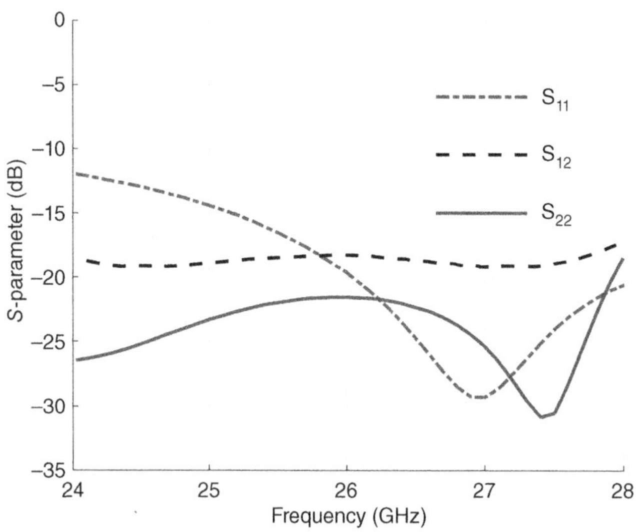

Figure 2.13 S-parameters of the dual-LP microstrip patch antenna element.

the array plus Luneburg lens combination (see Figure 2.8), was found to be 17.3 dBi when only the central element was excited and other ports were left open-circuit. The isolation between ports 1 and 2 was 18.3 dB. Owing to high losses in the measurement system, the peak realized gain was found reduced by approximately 10 dB as compared with the level predicted by the simulations.

The simulated results for this antenna show that this array is able to scan the beam from −5° to +5°, with a step of 5° and an almost constant realized gain of 28 dBi. A larger array would be needed to scan the beam over a wider angle range, but the Luneburg lens itself is well capable of handling wider scan angles.

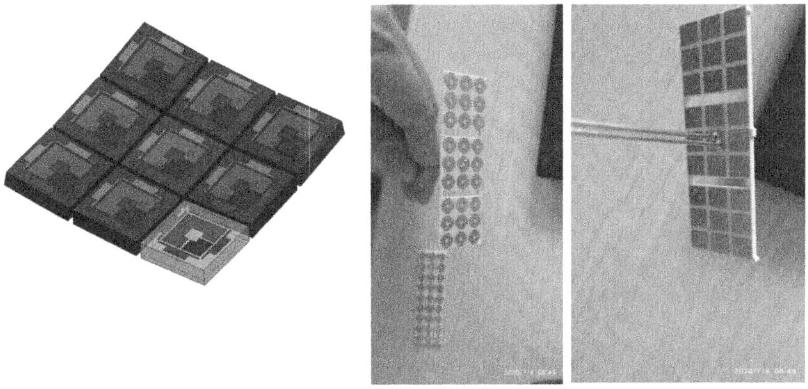

(a) (b)

Figure 2.14 3 × 3 conformal dual-LP microstrip patch array. (a) Software model. (b) Fabricated prototype.

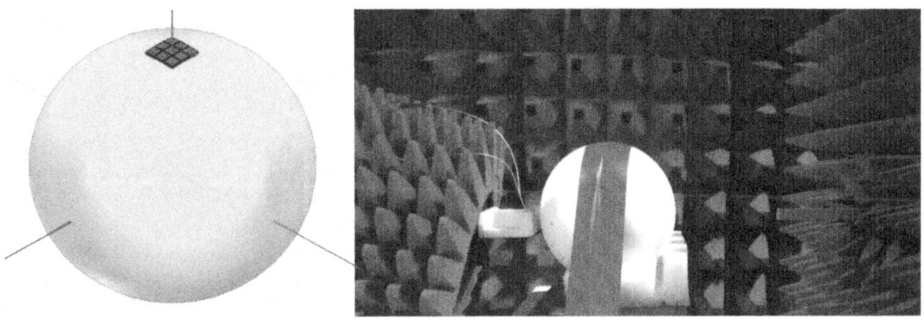

(a) (b)

Figure 2.15 Luneburg lens antenna. (a) 3 × 3 conformal array exciting a. (b) Measurement setup.

2.2.6 Size-Reduction of Luneburg Lens

The Luneburg lens is obviously bulkier than a planar array. Consequently, it is desirable to reduce its volume to lower its profile without sacrificing its beam-scanning capability. A straightforward approach to reducing the profile of the lens is to make it smaller in size. However, as shown in Figure 2.17, the realized gain of the lens, not unexpectedly, is reduced when the lens diameter decreases. Compromising the gain of the lens to reduce its size is not an acceptable solution for many applications. It is therefore desirable to seek alternative solutions to this problem, and a number of different strategies have been proposed to handle this issue.

One such technique is to flatten the lens [27] by using Transformation Optics (TO) or its variations. While the volume of the lens can be decreased by this method, the drawback of this strategy is that it calls for materials whose permittivity and/or permeability characteristics may be difficult to realize in practice. Also, one may have to accept that the performance of the lens would be compromised when realistic substitute materials are used to fabricate the lens instead of those called for by the TO algorithm. Another issue with this technique is that it only provides a limited range of beam steering, up to 30° or so, which is insufficient for most of the recent applications of the lens, 5G being one of them. By changing the shape of the lens [31], the scan range can be enhanced up to 50°; but it still needs relative permittivity (ε_r) values up to 20. Low loss materials with such high ε_r values are not readily available. Moreover, requiring permittivity values less than 1.0 is typically

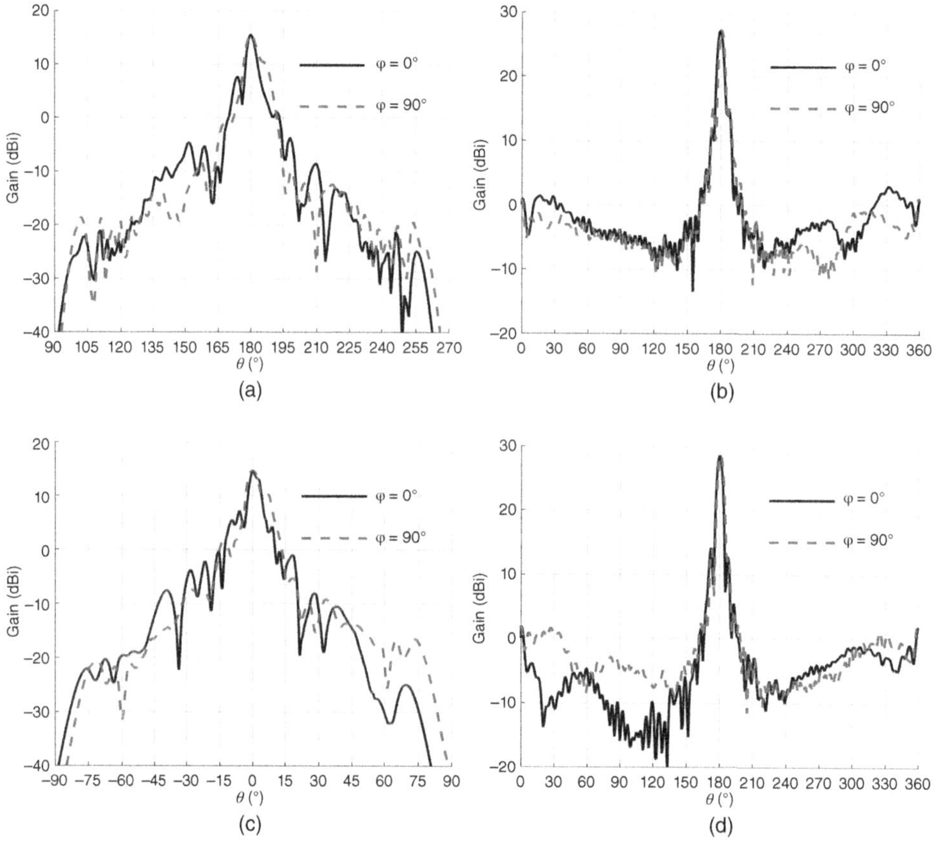

Figure 2.16 Radiation performance. (a) Measured total electric field patterns when port-1 is excited. (b) Simulated realized gain patterns when port-1 is excited. (c) Measured total electric field patterns when port-2 is excited. (b) Simulated realized gain patterns when port-2 is excited.

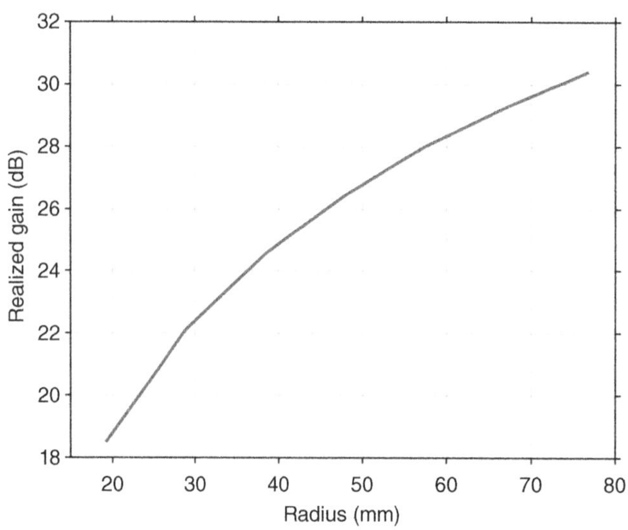

Figure 2.17 Realized gain when outer radius of the lens is varied.

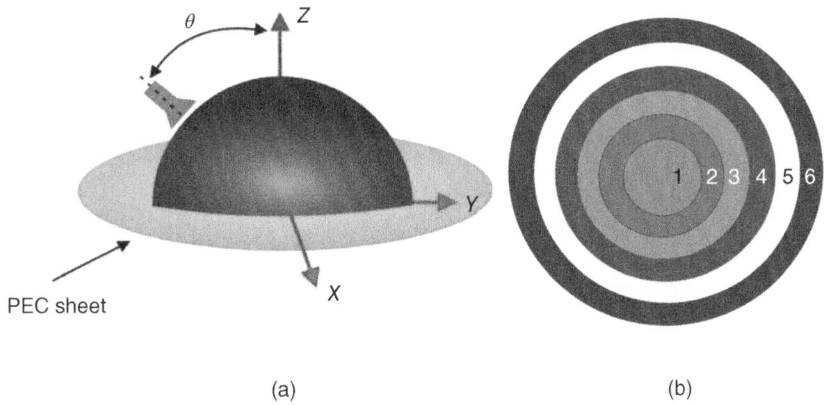

Figure 2.18 Hemispherical lens. (a) Lens with circular PEC sheet backing. (b) Different layers of the lens.

Table 2.2 Lens dimensions and material parameters.

Layer	Outer shell radius (mm)	Permittivity
1	8.20	1.92
2	11.60	1.77
3	14.20	1.62
4	16.40	1.46
5	18.33	1.31
6	20.00	1.15

associated with having to deal with permeability values which can vary over a wide range. This unfortunate reality simply exacerbates the problem in the TO approach of designing the flat lens.

In the following text, we present an alternative strategy for reducing the size of the lens. Toward this end, we first split the conventional lens in half, as shown in Figure 2.18. Next, we place a PEC sheet, of radius R, below the hemispherical lens. This radius is slightly larger than that of the sphere. Introducing such an image plane serves to restore the performance of the hemispherical lens so that its gain level becomes comparable to that of the full-size spherical lens.

To demonstrate the efficacy of this design strategy for size reduction, we consider a hemispherical Luneburg lens whose diameter is 40 mm. The proposed lens structure consists of six optimized layers [22] of dielectric materials with radii ranging from $R1$ to $R6$, as shown in Figure 2.18 and Table 2.2. This grounded hemispherical lens is well suited for applications in which the lens is to be conformally mounted on a metallic surface. In accordance with the guidelines of geometric optics, the dimension of the lens should be a multiple of the free-space operating wavelength λ_0. We choose the lens diameter to be approximately 4 λ_0, where λ_o corresponds to the frequency $f_o = 26$ GHz. The lens is designed to operate in the frequency range 24–28 GHz with a center frequency of 26 GHz and fed by an open-ended waveguide.

The working principle of the hemispherical Luneburg lenses is based on the concept of reflection from the image plane below the lens. In common with the full-size lens, the input feed is chosen

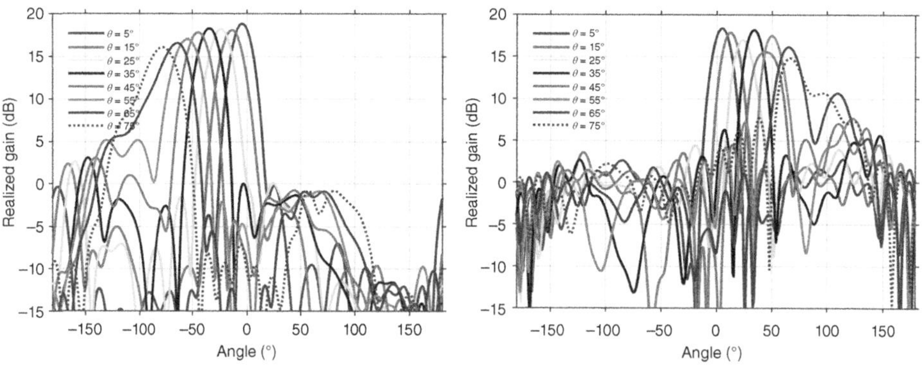

Figure 2.19 Realized gain patterns of the lens with different feed locations with the diameter of PEC sheet being 40 mm (same as the lens diameter). (a) TE polarization. (b) TM polarization.

to be an open-ended waveguide placed at an angle θ with respect to the zenith. Beam scanning is achieved by changing either the azimuth or the elevation angle of the feed to scan in the azimuth or elevation planes. The plane wave excitation from the feed traverses through the lens and reflects off the PEC backing to exit the lens. In turn, this specular reflection defines the pointing direction of the beam.

Figure 2.19 shows the realized gain patterns of the hemispherical lens with a circular PEC backing equal to the diameter of the lens. The patterns for both the transverse electric (TE) and transverse magnetic (TM) polarizations are displayed. The maximum gain value is seen to be 18 dB for both the TE and TM polarizations at low angles. The scan loss is as high as 2.7 dB for the TE polarization and 4.0 dB for the TM polarization. These levels of scan loss are unacceptable for most communication systems. Furthermore, the sidelobe levels (SLLs) become high for wider scan angles and the beamwidth broadens, which are also undesirable. In contrast, the spherical Luneburg lens maintains its gain level of 19.5 dB regardless of the scan angle. To mitigate the scan-loss problem of the hemispherical lens, we now modify the original design to restore its performance. This is achieved by increasing the size of the PEC sheet underneath the hemispherical lens. To determine the optimized size of the PEC sheet, we track the maximum scan loss as we increase the diameter of the image plane.

Figure 2.20 shows that the reflected field strength depends upon the radius of the PEC sheet, R_S. The realized gain values change as it is varied from 20 to 50 mm. These figures indicate that the choice of 80 mm for the diameter of the PEC sheet provides a good overall performance in terms of the gain and SLL with only a moderate increase in its footprint.

Figure 2.21 shows the realized gain patterns for both the TE and TM polarizations when the feed is placed at different θ locations along the diameter. The diameter of the PEC sheet was chosen to be 80 mm instead of the original 40 mm. We note from Figures 2.19 and 2.21 that the scan loss reduces when the diameter of the PEC sheet is increased, but the improvement is relatively small if we further increase the diameter beyond 80 mm. The maximum scan loss for both the TE and TM polarizations was found to be 2 dB when the diameter of the PEC sheet was chosen to be 80 mm. This level of scan loss is quite acceptable. On the other hand, the highest gain level was close to that of the full spherical lens until the scan angles became large and the scan loss begins to appear. The response of TE polarization is found to be more stable than that of the TM, perhaps because the polarization of the electric field parallel to the PEC plane and, hence, the diffraction from the edges of the plate is less. It is also evident that the hemispherical lens has low SLLs.

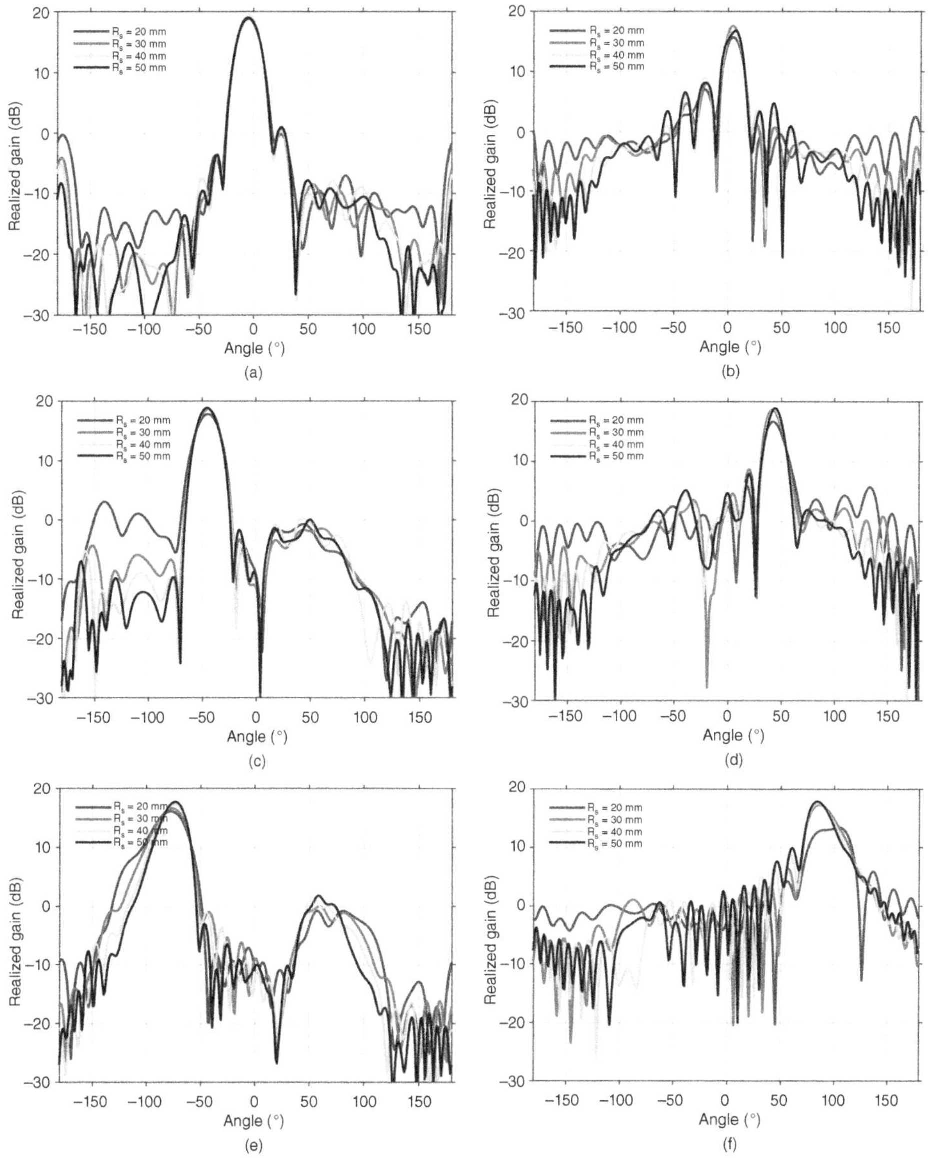

Figure 2.20 Realized gain patterns of the lens for different angles of incidence as the radius of the PEC sheet varies from 20 to 50 mm. (a) TE Mode, 5° incidence. (b) TM Mode, 5° incidence. (c) TE Mode, 45° incidence. (d) TM Mode, 45° incidence. (e) TE Mode, 75° incidence. (f) TM Mode, 85° incidence.

In summary, the hemispherical lens is a cost-effective alternative to the full spherical lens. It uses half the material compared with that of the full spherical lens. Moreover, it is half the height profile of the conventional Luneburg lens.

In Section 2.3, we first explore a variety of ways for enhancing the gain of a single element of an array to achieve our goal of a gain of level of 30 dBi. Following this review we go on to discuss several approaches to beam scanning, in both the transverse and longitudinal planes.

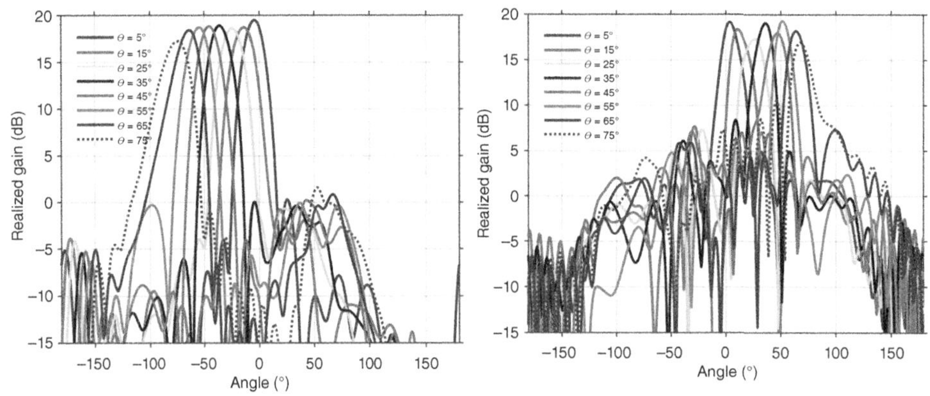

Figure 2.21 Realized gain patterns of the lens with diameter of PEC sheet being 80 mm for different angles of incidence. (a) TE Mode. (b) TM Mode.

2.3 Gain Enhancement Approaches for Antennas and Arrays

While the Luneburg lens offers the unique capability of very wide angle scans in both elevation and azimuth planes, it is not well suited for applications where a relatively low profile antenna is desired. With this in mind, we explore in this section a number of different ways by which we can enhance the gain of microstrip antennas and of slotted waveguide arrays to achieve the desired level of 30 dBi, while retaining the capability of scanning in the longitudinal or transverse planes, or both. We begin by presenting several strategies for gain enhancement to achieve this gain level goal. We then consider various techniques for beam scanning of the high-gain antennas, which is also desired in many communication applications.

2.3.1 High Gain Wideband Microstrip Patch Antenna and Arrays Using Wings

We propose a novel technique for enhancing the gain of a single radiating element and boosting it from approximately 6 to 30 dBi. We begin by employing a conventional microstrip patch antenna as the basic radiating element as shown in Figure 2.22. The antenna consists of a square patch, which is printed on a dielectric substrate and excited from the bottom by a microstrip line. The top, bottom, and side views of the element are shown in the figure. The proposed design has a moderately large bandwidth, almost 20% as shown in Figure 2.23, and a gain of 6.6 dBi as displayed in Figure 2.24.

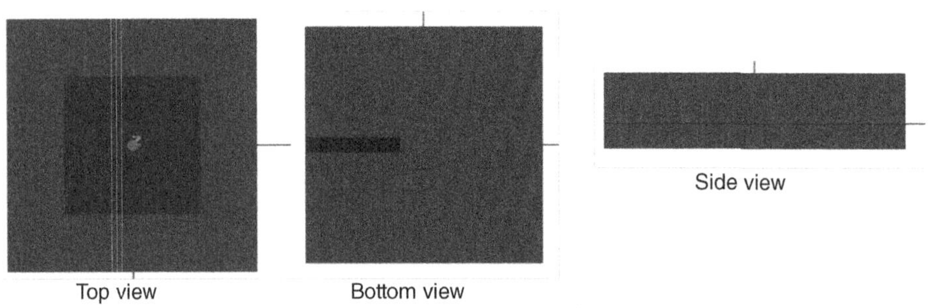

Figure 2.22 Proposed microstrip patch antenna.

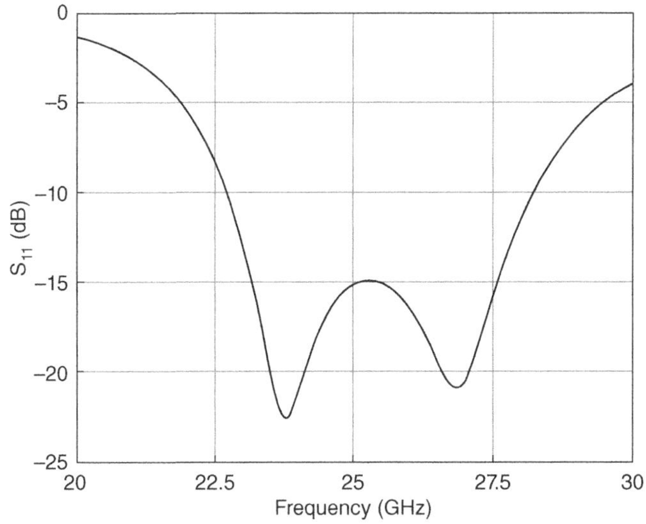

Figure 2.23 Simulated reflection coefficient of the proposed microstrip patch.

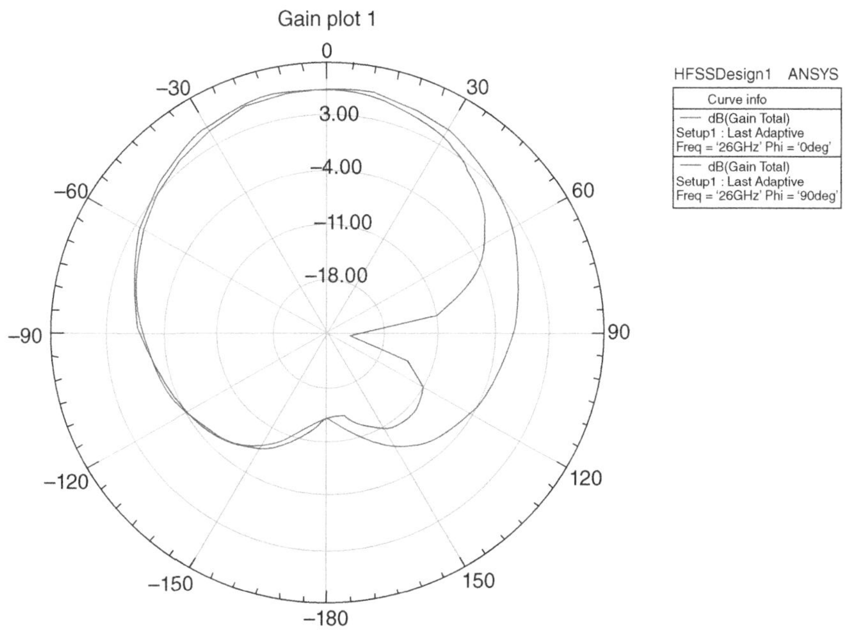

Figure 2.24 Simulated radiation patterns of the proposed microstrip patch.

Our next step is to add four metallic wings along the four sides of the antenna element as shown in Figure 2.25. Figure 2.26 presents the gain patterns of the patch antenna with wings. It shows that the proposed design can enhance the gain from 6.6 to 24.7 dBi.

Our next step is to further enhance the gain to achieve our objective, the 30 dB level. To this end, we design a 3 × 3 patch array, with each element having its own set of wings, as shown in Figure 2.27. Next, we add a pigeon-hole type of superstrate above the array with the objective of controlling the SLL. Figure 2.28 presents the simulation results for the gain patterns for this antenna

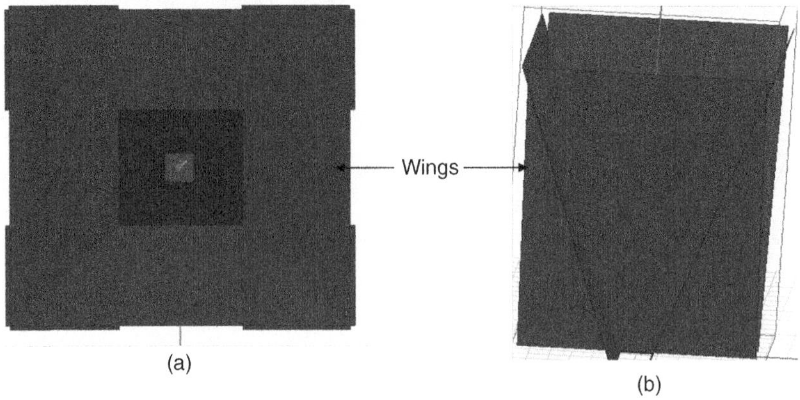

Figure 2.25 Top and side views of proposed microstrip patch antenna incorporating four wings. (a) Top view. (b) Side view.

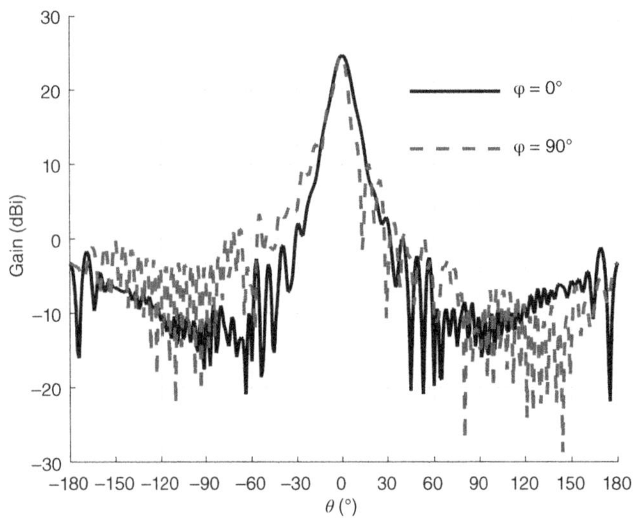

Figure 2.26 Simulated gain patterns of the proposed microstrip patch with four wings.

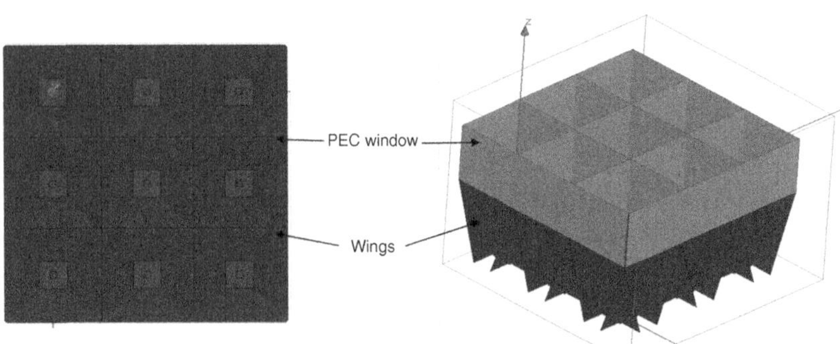

Figure 2.27 Proposed 3 × 3 patch array with wings and the pigeon-hole superstrate.

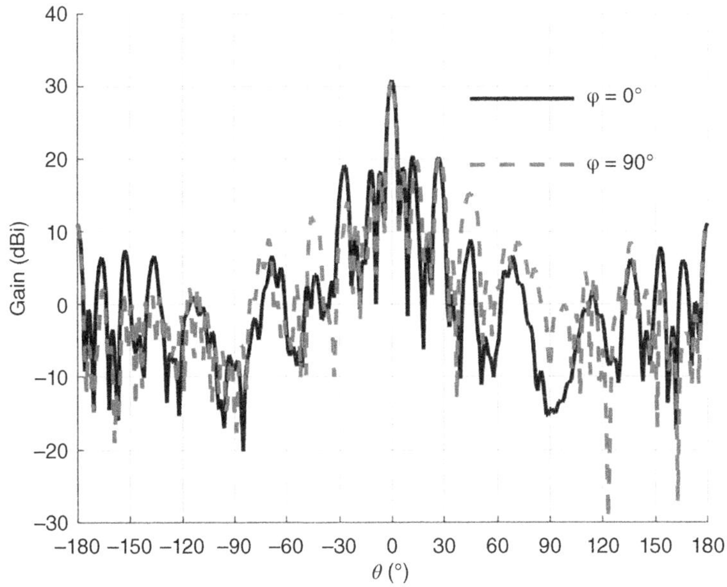

Figure 2.28 Simulated gain patterns of the proposed 3 × 3 patch array with wings.

Table 2.3 Performance comparison of microstrip patch antennas using four wings.

Design	Number of radiating elements	Aperture size (mm × mm)	Height (mm)	Gain (dB)	SLL (dB)
Microstrip patch antenna (MPA)	1	—	—	6.5	—
PA + wings	1	82 × 82	127	24.71	—
PA + wings	2	82 × 164	127	27.72	5
PA + wings	4	164 × 16	127	30.65	5
PA + wings + PEC windows	9	150 × 150	120	30.8	10.4

array configuration. Table 2.3 compares the performance characteristics of the different proposed designs obtained by progressively increasing the number of radiating elements in the array. As one sees, the gain has been enhanced from 6.6 to 30.8 dBi. The final size of the radiating aperture is 13 × 13 wavelengths squared, which is consistent with the gain level achieved.

2.3.2 High Gain Slotted Waveguide Antenna Arrays

We now present some examples of gain enhancement of antenna arrays that are based on three different strategies to those described earlier. As before, our goal remains unchanged, namely, to achieve a gain level of 30 dBi. We begin by describing the SWAA shown in Figure 2.29. This design has been detailed in [32]. It has 41 slots and the separation distance between two consecutive slots is equal to 0.75 λ_0 at 26 GHz.

Figure 2.29 Slotted waveguide array antenna.

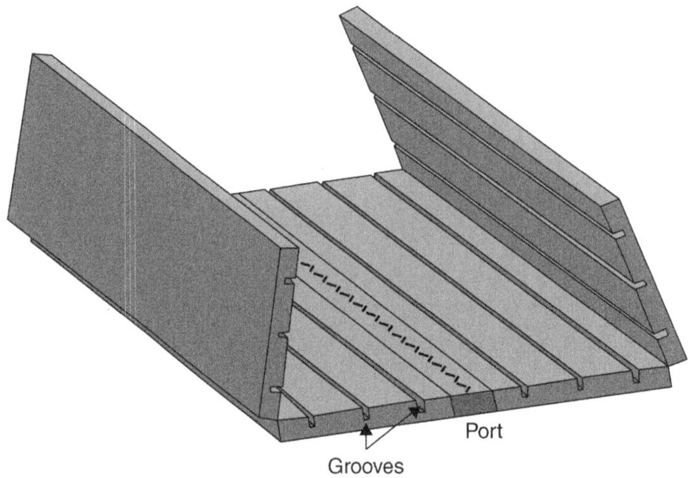

Figure 2.30 Slotted waveguide antenna array with tilted wings and grooves.

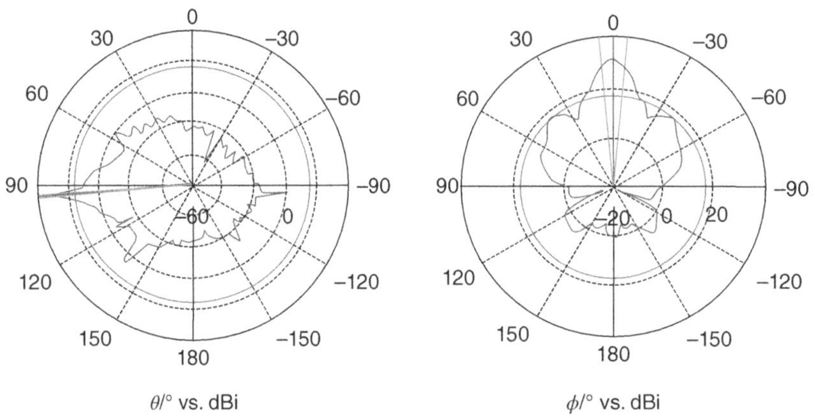

Figure 2.31 Simulated realized gain patterns of the slotted waveguide array antenna with tilted wings and grooves. (a) x–z plane (b) y–z plane.

2.3.2.1 Gain Enhancement Using Grooves and Tilted Wings

The first technique to increase the gain is to use a single slotted waveguide array with extensions of the top wall of the guide in which grooves have been introduced as shown in Figure 2.30. The groove design is symmetric. Each side-extension has two segments that are also shown. These segments are comprised of horizontal and tilted sections. The tilt angle is 75° between them. Figure 2.31 clearly demonstrates that a 30 dB gain level has been realized by using this tilted grooves technique.

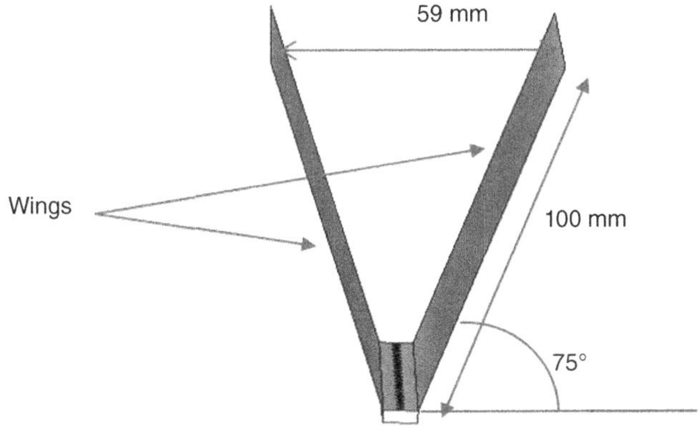

Figure 2.32 V-shaped slotted waveguide array antenna.

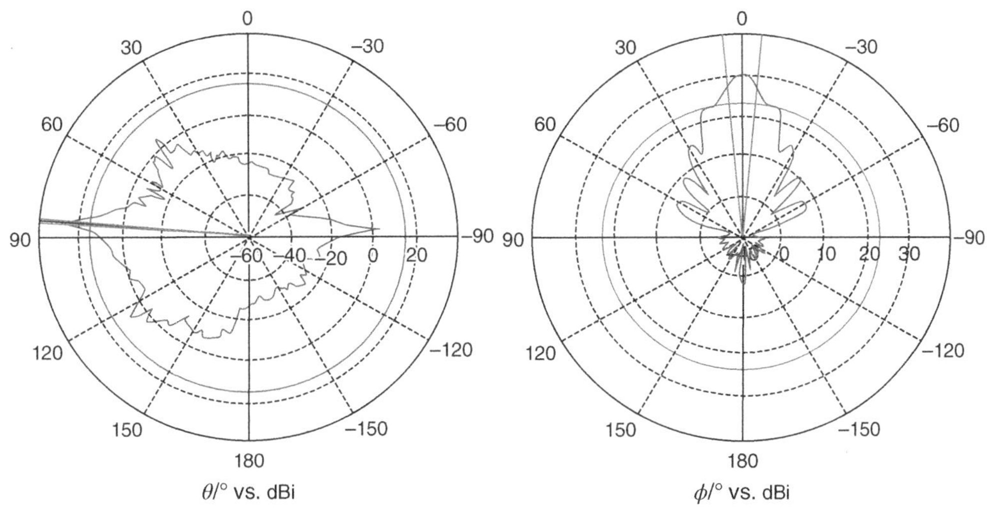

Figure 2.33 Realized gain patterns of the V-shaped slotted waveguide array antenna. (a) x–z plane (b) y–z plane.

2.3.2.2 Gain Enhancement Using V-Shaped Wings

The second technique proposed for gain enhancement is based on removing the grooves and simply adding two planar metallic wings to the slotted waveguide array to form a V-shaped system. Figure 2.32 presents the dimensions and parameters of the proposed wing structure. The simulation results, shown in Figure 2.33, demonstrate a gain level of 30.1 dB at 26 GHz, but with a SLL of 7.1 dB in the transverse plane. This SLL is higher than what is typically acceptable.

Next, we insert a dielectric lens in the aperture as illustrated in Figure 2.34 to mitigate the SLL problem. The realized gain patterns shown in Figure 2.35 demonstrate that the lens has corrects the phase distribution over the aperture. The SLL has been improved from 7.1 to 10.4 dB.

2.3.2.3 Gain Enhancement by Using Two Arrays, Each with Wings

Finally, we further embellish the winged array design by adding the wings to two slotted waveguide array antennas to enhance the overall performance. The principal difference between the previous enhancement technique and the present one is that the height profile of the new configuration is

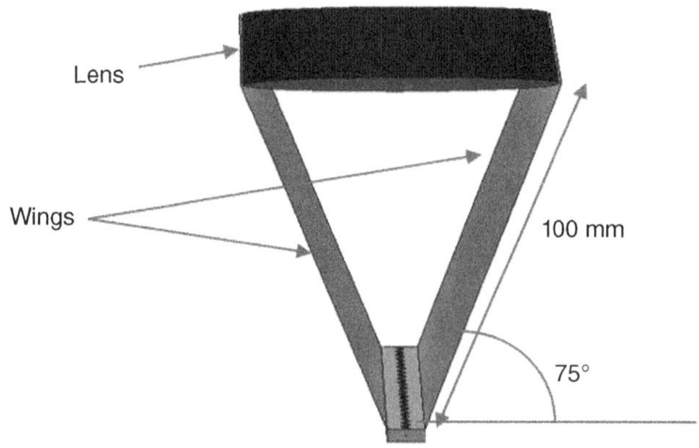

Figure 2.34 Slotted waveguide array antenna with two tilted wings giving its V-shaped appearance and with a lens covering its aperture.

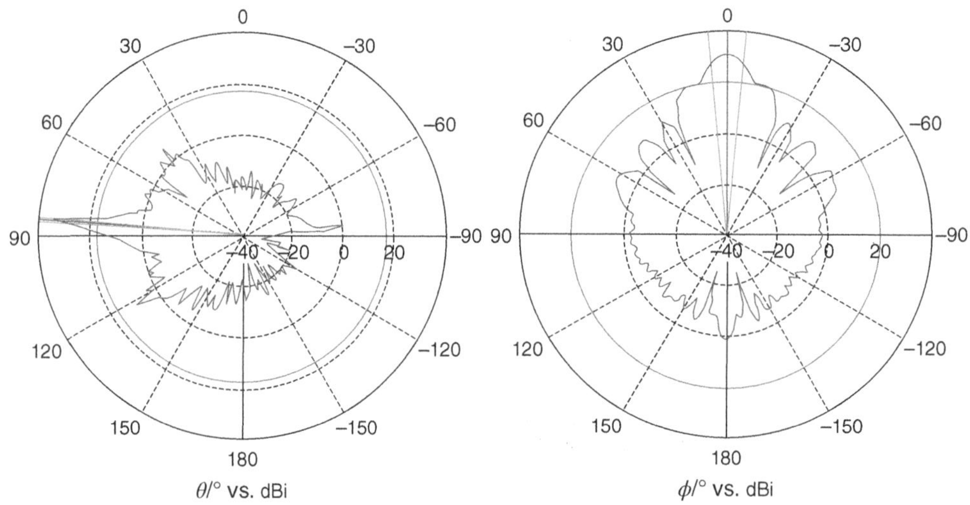

Figure 2.35 Realized gain patterns of the slotted waveguide array antenna with two wings and a lens. (a) x–z plane (b) y–z plane.

one-half that of the earlier one shown in Figure 2.32 at the cost of it being twice as wide. Note that the profile reduction has been achieved in this approach without compromising the gain level. The design of two slotted-waveguides with wings is illustrated in Figure 2.36. Its realized gain patterns are presented in Figure 2.37. We note that a peak realized gain level of ~30 dB and a SLL larger than 13.6 dB has been achieved in both the E- and H-planes.

All of the techniques proposed earlier are simple, low cost, low loss, and relatively easy to fabricate. They are suitable for different mm-wave applications which require a high gain that is on the order of 30 dBi. Table 2.4 provides a performance comparison of the different presented configurations.

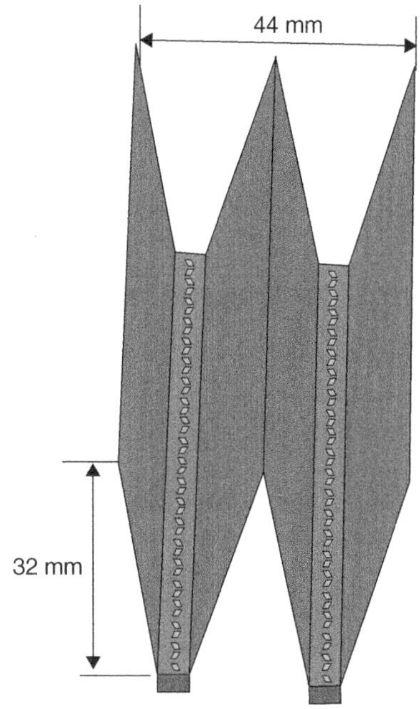

Figure 2.36 Antenna formed with two V-shaped slotted waveguide antenna arrays.

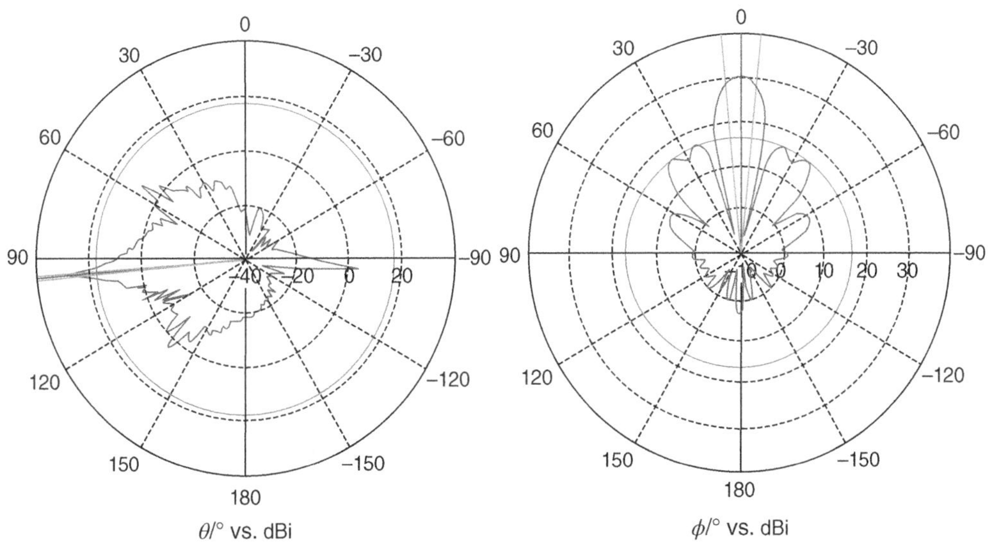

Figure 2.37 Simulated radiation patterns of the two slotted waveguide antenna arrays with wings. (a) x–z plane (b) y–z plane.

Table 2.4 Performance comparison of the presented high gain designs.

Design	Number of arrays	Design aperture (mm × mm)	Design height (mm)	Gain (dB)
SWAA + tilt grooves	1	74 × 368.4	50	30.2
SWAA + wings + lens	1	59 × 368.4	100	30.8
SWAA + wings	2	44 × 368.4	32	30.2

2.3.3 Two-Dimensional Beam Scanning with Slotted Array Antennas

We discuss two techniques for two-dimensional (2D) beam scanning of an antenna array. The proposed designs are low cost, simple to fabricate and are well suited for many recent millimeter wave applications.

2.3.3.1 Longitudinal Plane Beam Scanning Using U-Shaped Phase Shifter

As earlier, we begin with the SWAA shown in Figure 2.38. It is the baseline design, i.e. the building block, for the scanning array. Next, we insert U-shaped waveguide sections between two adjacent slots of the array. They function as phase shifters. These phase shifters are utilized to scan the beam in the longitudinal plane. Figure 2.39 presents two examples of the proposed U-shaped design for the phase shifters. They consist of empty tubes that are partially filled to achieve the desired changes in the effective path length through the U-shaped tubes to realize the desired phase taper needed to point the beam at a specified angle. The filler material may either be a dielectric or a liquid metal, whichever is best suited for the application in mind.

Our next step is to modify the baseline design to achieve the 30 dB gain level goal. Toward this end we increase the number of SWAAs with U-shaped phase shifters to eight as depicted in Figure 2.40.

Figure 2.38 Baseline slotted waveguide array antenna.

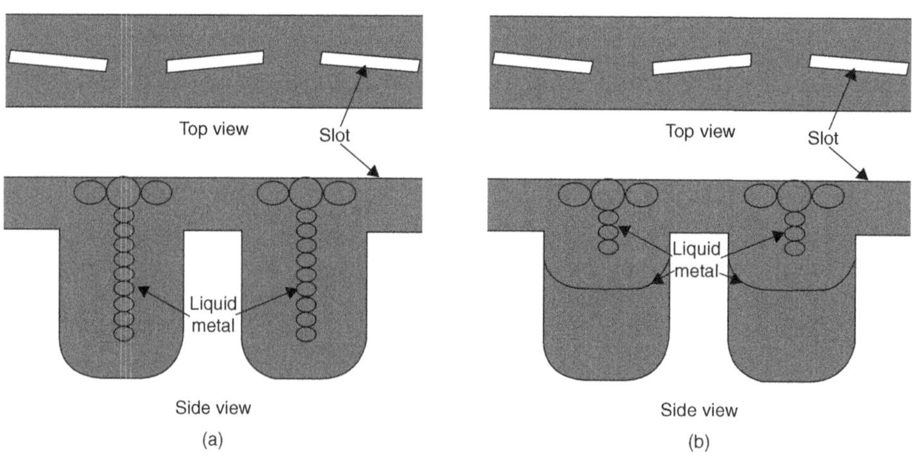

Figure 2.39 Examples of the proposed U-shaped phase shifter designs. (a) Design-1. (b) Design-2.

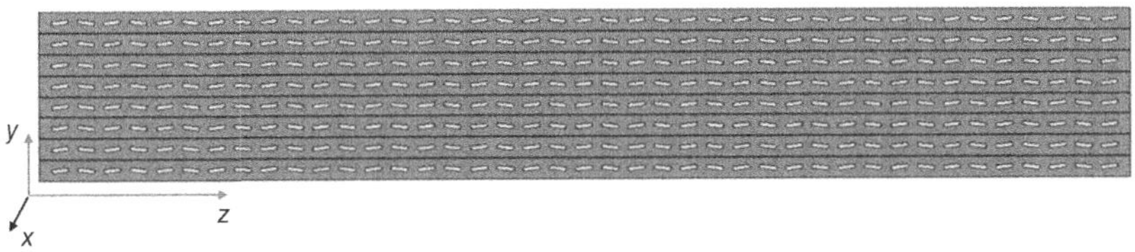

Figure 2.40 Combination of eight slotted waveguide antenna arrays for transverse scanning.

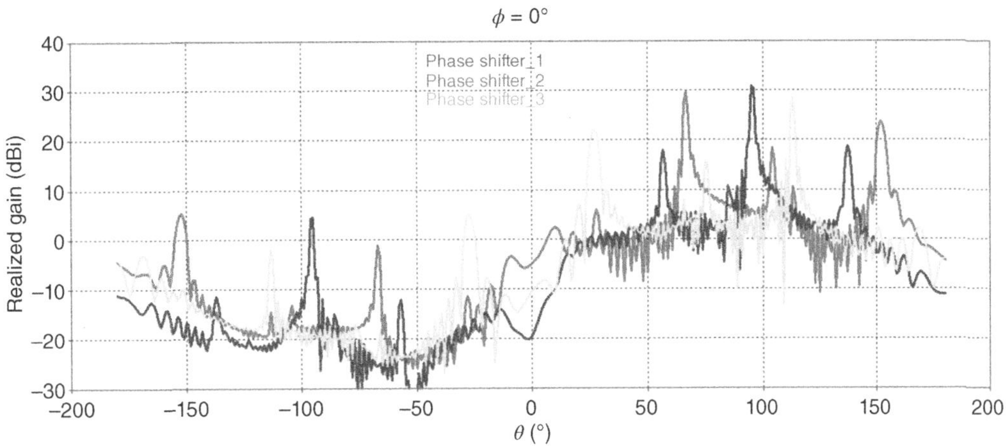

Figure 2.41 Realized gain patterns of the three U-shaped phase shifter-based SWAAs in the x–z plane.

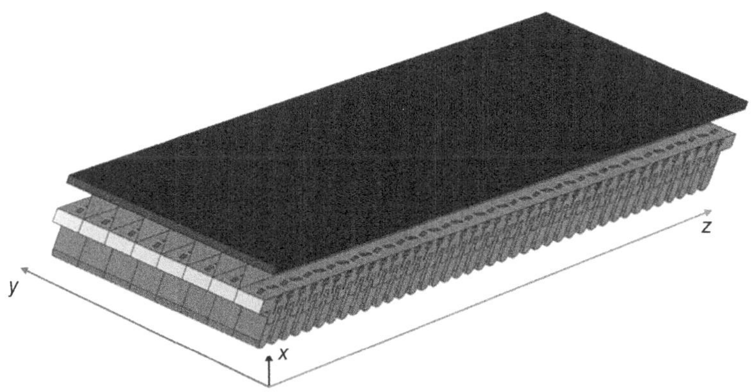

Figure 2.42 Combination of eight slotted waveguides antenna arrays with a dielectric superstrate.

However, we find that for certain scan angles, the SLLs are worse than 10 dB. This is ascertained from the realized gain patterns, for three different phase shifters, given in Figure 2.41.

Next, to address the SLL issue, we add a dielectric superstrate [33] above the slot arrays as depicted in Figure 2.42. The simulated realized gain patterns given in Figure 2.43 demonstrate that inserting the superstrate above the arrays improves the SLL to better than 10.5 dB. Moreover, the realized gain of this superstrate augmented scanning array with U-shaped phase shifters also improves to better than 30 dBi. Again, our goal has been achieved.

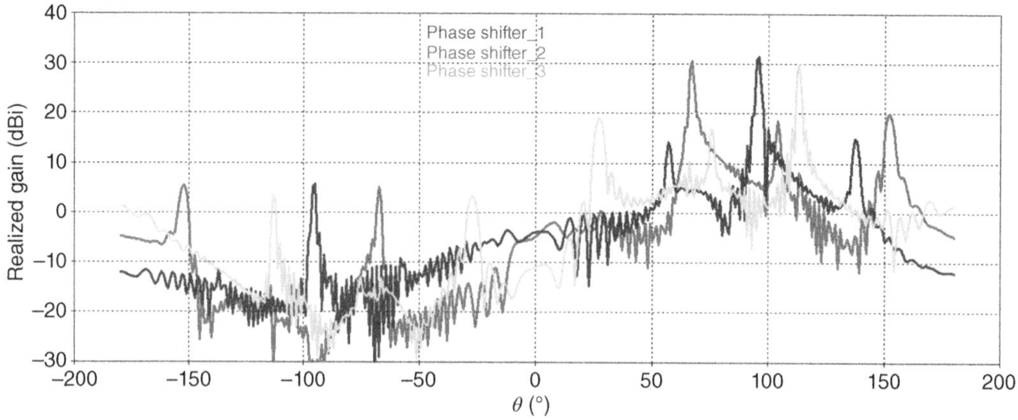

Figure 2.43 Realized gain patterns in the x–z plane of the superstrate augmented U-shaped phase shifter-based SWAAs.

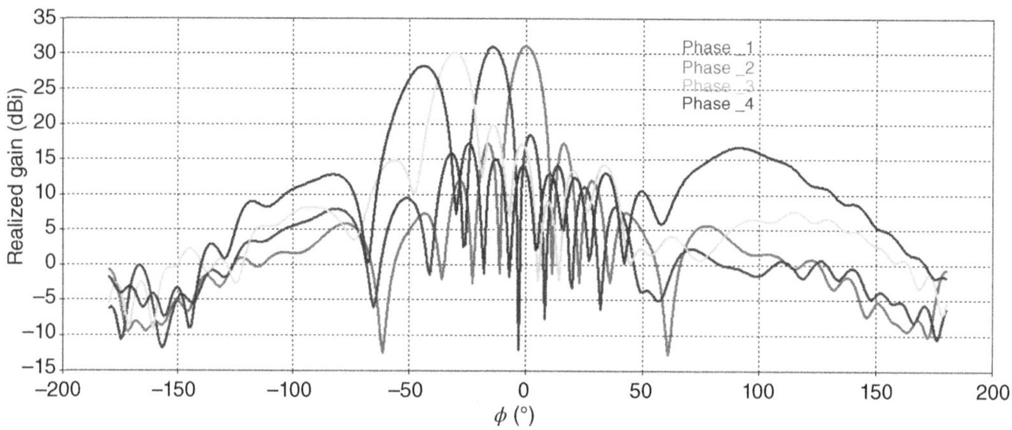

Figure 2.44 Beam scanning characteristics of the combined eight U-shaped phased shifter-based SWAAs in the transverse (x–y) plane with a phase taper applied to the excitation ports.

2.3.3.2 Transversal Beam Scanning Using Phase Tapers in the Feed Ports

Next, we introduce a phase taper between the waveguide feed ports of the combined eight U-shaped phased shifter-based SWAAs augmented with the dielectric superstrate to address the problem of scanning the beam in the transverse plane. The realized gain patterns displayed in Figure 2.44 demonstrate that this approach successfully scans the beam in the range of −44° to +44° in the transverse plane. The maximum scan loss is less than 2.5 dB.

2.3.3.3 Two-Dimensional Beam Scanning Using Three SCAAs Augmented with Wings

Finally, we move to the issue of two-dimensional beam scanning, which is required for some applications. A straightforward way to extend the beam scanning in both the longitudinal and transverse directions to a SWAA ensemble such as the one shown in Figure 2.40 is to introduce phase tapers

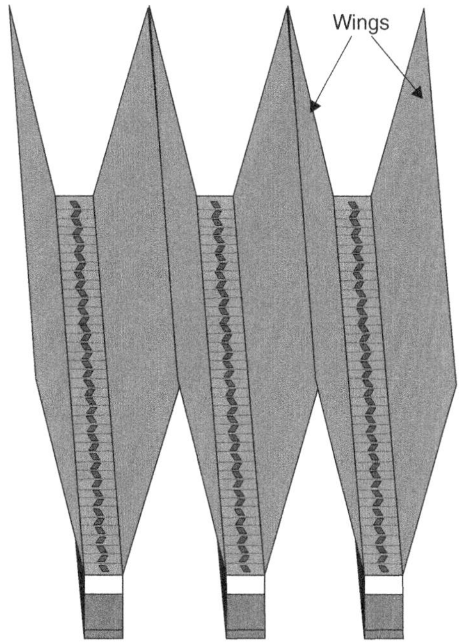

Figure 2.45 Three element U-shaped phase shifter-based SWAA ensemble in which each SWAA is augmented with wings.

in those waveguides to scan the beam in the transverse plane. One would continue to utilize the U-shaped waveguide inserts as the phase shifters to accomplish the already demonstrated longitudinal scan capability.

The alternate approach we propose for the two-dimensional beam scanning differs from the earlier approach in three ways. First, it utilizes three SWAAs instead of eight. Second, it adds metallic wings to the three arrays as shown in Figure 2.45. This augmentation facilitates the ensemble realizing the required gain level of 30 dBi. This gain level could not be achieved by using the three waveguide arrays alone. Finally, as shown in Figure 2.46, a reflecting panel with a tilt angle whose value can be changed is introduced to achieve the desired transverse beam scan.

The 3D realized gain patterns of the three element U-shaped phase shifter-based SWAA ensemble with wings, and without and with it being augmented with the reflecting panel are presented in Figure 2.47 for the sake of comparisons. The ensemble without the panel generates a beam pointed in the broadside direction. The one with it is designed to generate a beam pointed in the direction ($\theta = 60°$, $\phi = 0°$). It is evident from Figure 2.47 that the tilted panel approach has the capability of wide-angle scanning with virtually no scan loss. In fact, the peak realized gain without (with) the panel is 30.13 (30.1) dBi.

The panel itself may be realized either by using liquid metal technology, as discussed in Section 2.4, or by using reconfigurable FSSs or even by a mechanical means. The choice depends upon the desired speed of the beam scanning, which is determined by the application. The reconfigurable FSS panel is likely to offer a faster switching speed in comparison to the one provided by either the liquid metal or mechanical version. Nevertheless, the FSS panel is considerably more expensive to fabricate and is more power hungry in comparison to its liquid metal counterpart.

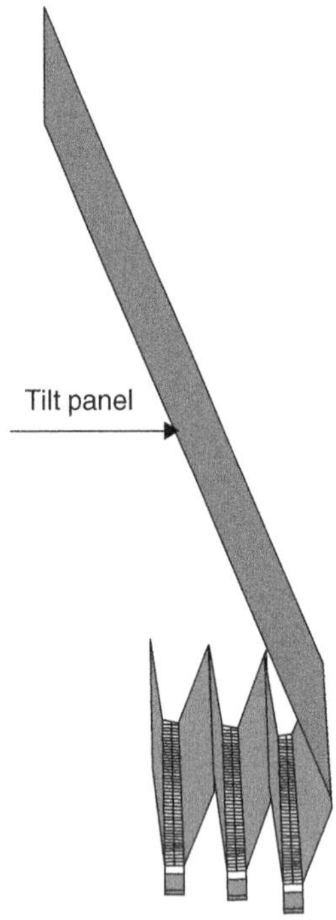

Figure 2.46 Three element U-shaped phase shifter-based SWAA ensemble with wings augmented with a tiltable reflecting panel.

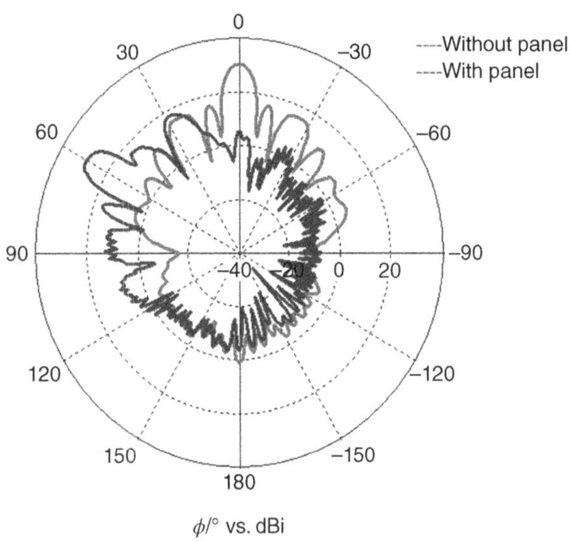

Figure 2.47 Comparison of the three element U-shaped phase shifter-based SWAA ensemble with wings, and without and with it being augmented with the reflecting panel.

2.4 Reconfigurable Liquid Metal-Based SIW Phase Shifter

We focus on two topics in this section, namely: phase shifters and their practical realizations. Phase shifters play an important role in the design of electronically scanned arrays. Of course, the art of phase shifter design is very mature at microwave frequencies, where ferrite-based phase shifters placed in conventional waveguides, are typically used. However, such phase shifters become highly lossy and costly when the operating frequency is in the millimeter wave range. Replacing the conventional passive phase shifters with active ones is not always a viable alternative either. The power dissipation and cost of active phase shifters can become prohibitively high. This makes them unattractive for many applications. This background on phase shifters motivates us to explore alternative strategies for their design. One is to use liquid metals rather than ferrite-filled waveguides.

Earlier we used phase shifters based on a U-shaped waveguide inserted between two radiating elements and remarked that phase shifting could be realized either by filling drill holes with liquid metal or by using electronic switches/tuning elements. The latter are typically pin or varactor diodes. Next, we present a phase shifter design that is a slight modification of the U-shaped guide one and employs liquid metal for reconfiguration. We begin by providing a brief background on liquid metal and follow this up by outlining the design of the phase shifter. Finally, we present the measured and simulated performance of the phase shifter.

Liquid metal, based on alloys of gallium (Ga), have already been used to design several reconfigurable microwave devices, e.g. filters [34–38], switches [39, 40], and antennas [41–44]. A phase shifter that can be reconfigured using liquid metal was presented in [45]. The design was based around a defected ground plane. The phase shifter operated at 5.6 GHz and provided a total phase shifting range of 67.2° in uneven steps.

For the first time we propose an SIW phase shifter that can be reconfigured by using liquid metal technology. The SIW technology combines some of the advantages of conventional hollow waveguides, such as low loss and high power-handling capability, together with the advantages of planar technology, such as ease of fabrication. The phase shifter exhibits a low IL. The phase state is altered by filling and emptying the vias with liquid metal. The proposed phase shifter performance has been validated numerically and experimentally. Relative to competing devices, the proposed phase shifter is expected to offer significantly improved linearity [41]. The proposed phase shifter is suitable for applications for which high power-handling capability and low loss are priorities. However, the reconfiguration time is expected to range from seconds to milli-seconds [12, 46]. This is slower than that for pin diodes or varactors.

2.4.1 Concept and Structure of the SIW Phase Shifter

Figure 2.48 identifies the dimensional parameters of the proposed phase shifter. The numerical values of those dimensions are given in Table 2.5. The proposed phase shifter employs two different stages of reconfigurable phase shift in concert to achieve fine-grained phase control over a 180° phase-shifting range. Specifically, the concept relies on achieving coarse changes in the phase shift by means of a switched-line phase shifter, along with fine changes in the phase shift by means of reactive loading. By combining these two different forms of reconfiguration, the proposed phase shifter is able to achieve a phase shifting range of 0° to 180°, in steps of 10°.

The proposed SIW phase shifter has been designed to operate at 10 GHz. Nonetheless, the design can be readily adapted to operate at higher frequencies, going into the millimeter wave range. It is designed around a Rogers RO4003C substrate, which has the dielectric constant $\varepsilon_r = 3.55$, a loss tangent ($\tan\delta$) of 0.0027, and a thickness (h) of 0.813 mm.

2.4.1.1 Switched-Line SIW Phase Shifter for Coarse Phase Steps

Coarse changes in the phase shift are realized, in the proposed phase shifter, by guiding the traveling electromagnetic (EM) waves through paths that have different electrical lengths, as illustrated in Figure 2.49. The switched-line phase shifter was designed to produce three different phase-shift values, namely, −60°, −120°, and −180°. These values were achieved by adjusting the physical length of the SIW line. Depending on that length, the phase shifter was configured either in a straight

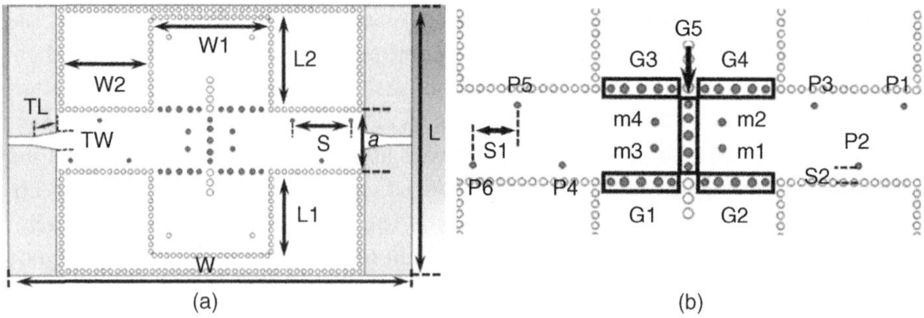

Figure 2.48 Proposed liquid metal-based phase shifter. (a) Top view showing the dimensional parameters. (b) Zoomed-in top view incorporating labels for the walls and vias.

Table 2.5 Dimensions of the proposed liquid metal phase shifter (Units: mm).

Length of phase shifter (L) = 56.2	Width of phase shifter (W) = 87.2
Length of the microstrip to SIW transition (TL) = 5.1	Width of the microstrip to SIW transition (TW) = 3.5
d = distance between tuning vias and center of the SIW = 4.17	Length of the SIW (a) = 13.0
Width of SIW tilted path (W1) = 26.23	Width of the SIW straight path (W2) = 19.24
Length of SIW long paths (L1) = 17.43	Length of SIW short paths (L2) = 19.24
Diameter of liquid metal vias used into RF blocking walls (LD) = 1.2	Distance between tuning vias (S) = 12.6
LD = diameter of liquid metal tuning vias = 1.2	Distance between tuning vias (S1) = 6.3
LD1 = diameter of the matching vias = 1.1	LD2 = diameter of liquid metal tuning vias = 1.0

(Figure 2.49a) or folded path (Figure 2.49b,c). Reconfiguration between these different paths is achieved by using switches formed from liquid metal.

Each switch consists of a group of holes drilled into the substrate as shown in Figure 2.48. When a particular set of holes is filled with liquid metal, a blocking wall is formed, which prevents the propagation of traveling EM waves. Blocking walls can be added along either the E-plane or H-plane of the waveguide. An individual via is said to be ON or OFF depending on whether it is filled with or emptied of liquid metal. When all of the holes associated with a particular wall are filled with liquid metal, it is said to be turned ON. The individual paths within the switched-line phase shifter can be opened or closed independently by adding or removing the relevant blocking walls.

The blocking walls are labeled G1 to G5 in Figure 2.48. A third set of vias – termed matching vias – are used to improve the impedance matching associated with the different paths. Those vias are labelled m1 to m4. Table 2.6 details the configurations of the various walls and matching vias required to realize each operating state. For example, consider that vias G1, G2, G3, and G4 are switched ON while vias G5, m1 to m4, and P1 to P6 are switched OFF. Under this condition the EM waves are confined in the central section of the SIW line and a phase shift of −60° is realized. This state is illustrated in Figure 2.49a,d. The blocking walls are used in this manner to reconfigure the phase shifter between three operating states corresponding to paths having different electrical lengths, namely: state A: 60°; state B: 120°; and state C: 180°.

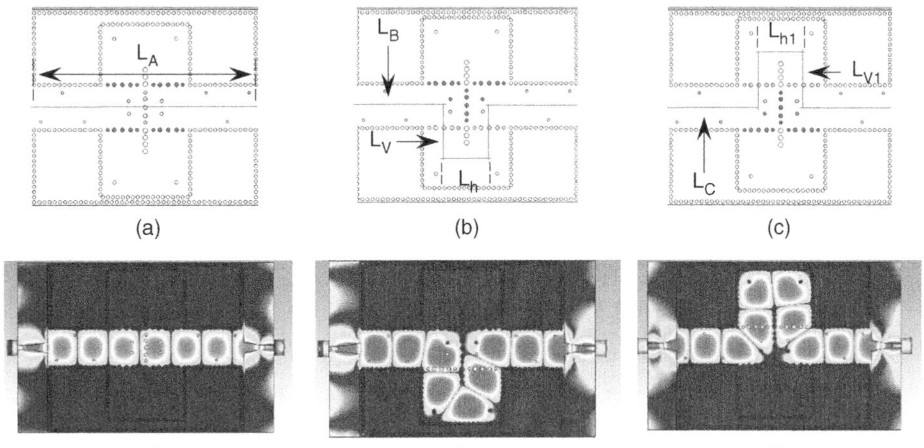

Figure 2.49 Schematics illustrating the three main operating states of the proposed phase shifter. (a) State A = −60°. (b) State B = −120°. (c) State C = −180°. The electric field (E-field) distributions corresponding to states A, B and C, respectively.

Table 2.6 The active via configurations of the main three paths of the proposed phase shifter.

State	Vias ON	Phase Shift (°)
State A	G1, G2, G3, G4	−60
State B	G4, G3, G5, m2 and m4	−120
State C	G1, G2, G5, m1 and m3	−180

2.4.1.2 Reactively Loaded SIW Phase Shifter for Fine Phase Steps

Fine changes in the phase shift are realized in the proposed phase shifter by introducing reactive loading into the waveguide. The reactive loading is introduced by filling dedicated drill holes (termed tuning vias) with liquid metal. The proposed phase shifter incorporates a total of six tuning vias, labelled P1 to P6, in Figure 2.48b. The tuning vias behave as inductive posts that modify the propagation of EM waves inside the SIW. As shown in Figure 2.49b, P1 to P3 are added at the input of the phase shifter, while P4 to P6 are added at its output. The amount of phase shift provided by each tuning via can be controlled by the diameter and position of each tuning via. The tuning vias in the proposed phase shifter enable the phase shift to be varied from 0° to −60° in steps of 10°. Fine tuning of the phase is enabled in this manner within each of the coarse steps of the phase control.

2.4.2 Practical Considerations of Fabrication

The fabrication and actuation of the hardware prototype are depicted in Figure 2.50. The channel structure, used to contain and guide the liquid metal, consists of two layers of clear Perspex. These layers are labelled layer-1 (#L1) and layer-2 (#L2) in Figure 2.48. Each layer incorporates several liquid metal reservoirs (LMRs). The layers of Perspex were held in place by using 12 metallic screws which were located exterior to the SIW.

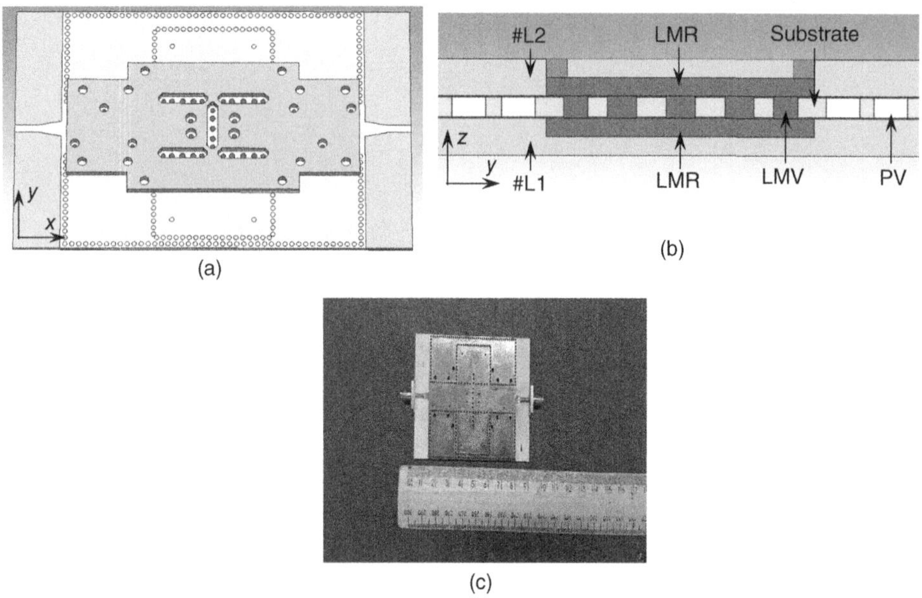

Figure 2.50 The proposed phase shifter with Perspex cover. (a) Perspective view. (b) Cross section in the z–y plane. (c) Fabricated prototype. Terms: #L1: Perspex layer-1, #L2: Perspex layer-2, LMR: Liquid metal reservoir, LMV: Liquid metal via, PV: copper plated via.

A syringe was used to inject or remove liquid metal from each via. This approach for actuation is commonly employed in the literature [42, 43, 47–55]. Alternative methods of actuation have also been reported including electrochemically controlled capillary action [44, 56] and the use of a micropump [12, 34, 47, 57, 58]. One of these existing techniques could be employed in this application. However, liquid metal actuation is not the main focus of this discussion. Moreover, altering the method of actuation would have minimal effect on the RF performance of the proposed phase shifter because the actuation circuits would be located beneath the ground plane. The electric and magnetic field strengths are minimal beneath the ground plane. In our experiments, no residues of liquid metal were observed inside the vias after the liquid metal was removed. We actuated the proposed switches several times in order to test their repeatability, and found that the performance was largely unchanged with the passage of time. We typically reused the same batch of liquid metal in our experiments. Once again, this reuse was found to have negligible effect on the RF performance.

2.4.3 Results and Discussion

Figure 2.51 shows the simulated electric field (E-field) distribution associated with the proposed phase shifter in different operating states when achieving a phase shift of −60° (straight path). Figure 2.51a shows the structure with no fine-tuning vias activated. Figure 2.51b shows the structure with one fine-tuning via activated (namely P2). Figure 2.51c shows the structure with three fine-tuning vias (namely: P1 to P3) activated, while Figure 2.51d shows the structure with five fine-tuning vias (P1 to P6) activated.

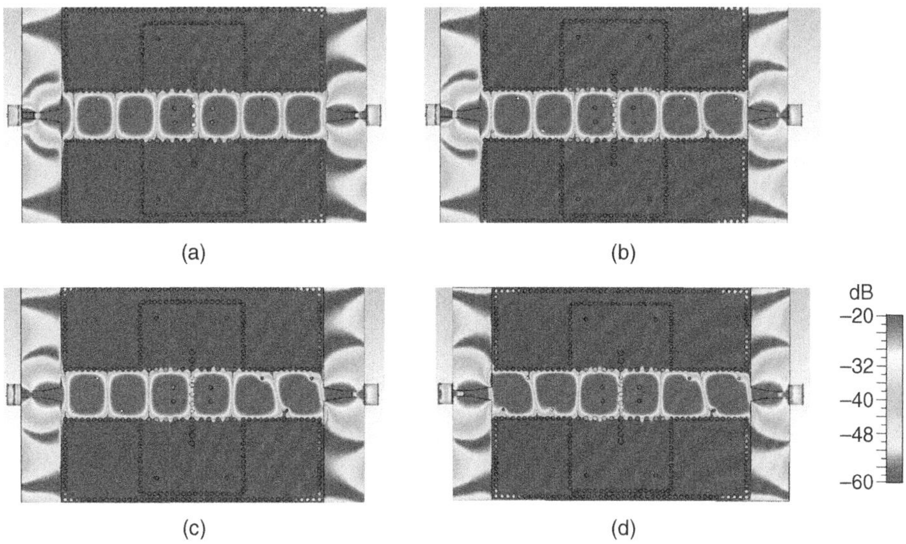

Figure 2.51 The effect of the tuning vias on the magnitude of the E-field distribution of the proposed phase shifter at 10 GHz. (a) P1 to P6 are OFF. (b) P2 only is ON. (c) P1 to P3 are ON. (d) P1 to P6 are ON.

It can be observed that the tuning vias perturb the propagating EM waves, and therefore their associated phase shifts. The total phase shift that can be achieved by combining the coarse and small phase shifts varies from 0° to 180° in steps of 10°. Table 2.7 details the configuration of the various walls and matching vias that are required to realize each operating state, along with the associated phase shift. It is important to mention that any combination of the six different vias (viz. P1 to P6) could be switched ON to realize any particular 10° increment in the phase shift. In other words, the phase shift is determined by the number of vias that are turned ON, rather than by the positions of those vias. Table 2.7 describes just one possible solution. Note that there are several other permutations also possible.

The measured phase difference between states A and B is 62.0° at 10 GHz. The difference, between states B and C, is 62.1°. The measured and simulation results are seen to be in very good agreement. They attain a phase difference of 60° between each of the three main states. Moreover, the smallest available phase shift within each of the three coarse phase steps is 10°. The measurement results are summarized in Table 2.8. They show that the measured phase step for the proposed phase shifter ranges from 8.4° to 11.3°, which is in good agreement with the simulated steps of 10°. This effectively confirms that we can tune the phase from 0° to 180° in steps of 10°.

Figures 2.52 and 2.53 show the measured S-parameters for the proposed phase shifter. Figure 2.52 shows the magnitude of the reflection coefficients, $|S_{11}|$ and $|S_{22}|$, associated with the three main states of the phase shifter. It can be observed that the phase shifter has a measured −10 dB bandwidth of at least 5 GHz ranging from 7.5 to 12.5 GHz. Figure 2.53 shows $|S_{12}|$ and $|S_{21}|$ for the three main operating states of the phase shifter. The measured IL for the three operating states ranges from 1.2 to 2.0 dB at 10 GHz. The IL losses are only due to the dissipation and matching losses.

Table 2.7 Phase shifter scheme.

Phase shift (°)		Coarse vias ON	Matching vias ON	Small vias ON
0	State A	G1 to G4	None	P1 to P6
−10		G1 to G4	None	P1 to P5
−20		G1 to G4	None	P1 to P4
−30		G1 to G4	None	P1 to P3
−40		G1 to G4	None	P1 to P2
−50		G1 to G4	None	P1
−60		G1 to G4	None	None
−60	State B	G3 to G5	m2 and m4	P1 to P6
−70		G3 to G5	m2 and m4	P1 to P5
−80		G3 to G5	m2 and m4	P1 to P4
−90		G3 to G5	m2 and m4	P1 to P3
−100		G3 to G5	m2 and m4	P1 to P2
−110		G3 to G5	m2 and m4	P1
−120		G3 to G5	m2 and m4	None
−120	State C	G1, G2 and G5	m1 and m2	P1 to P6
−130		G1, G2 and G5	m1 and m2	P1 to P5
−140		G1, G2 and G5	m1 and m2	P1 to P4
−150		G1, G2 and G5	m1 and m2	P1 to P3
−160		G1, G2 and G5	m1 and m2	P1 to P2
−170		G1, G2 and G5	m1 and m2	P1
−180		G1, G2 and G5	m1 and m2	None

2.5 Summary

This chapter began with the Luneburg lens antenna, which supports beam scanning over a very wide-angle range, both in the elevation and azimuthal planes. The concepts associated with it have been discussed in Section 2.2, where the classical spherical version was detailed, and a novel hemispherical version of the Luneburg lens was introduced. The design of a feed system, comprising of a two-dimensional array of specially designed microstrip patch antenna elements which generate dual-linear polarized beams, was presented. Moreover, two different approaches for fabricating the lens were described. Next, in Section 2.3, we have presented a number of techniques for enhancing the gain of an SWAA, as well as for scanning the array in both the longitudinal and transverse planes. Beam scanning has been achieved without the use of conventional phase shifters that are both lossy and costly. The proposed U-shaped phase shifters are based on the use of either electronic switches or vias of liquid metal partially filling drilled holes. Both approaches facilitate modifying the electrical length of the U-shaped waveguides that are inserted between the adjacent radiating

Table 2.8 Measured phases of the proposed phase shifter.

Simulated phase shift (°)	Measured phase shift	Step difference measured
0	0°	0
−10	−8.4°	8.4
−20	−18.6°	10.2
−30	−27.7°	9.1
−40	−37.1°	9.4
−50	46.1°	9
−60	55.3°	9.2
−60	62.0°	—
−70	68.7°	6.7
−80	77.2°	8.5
−90	88.5°	11.3
−100	96.5°	8.0
−110	105.8°	9.3
−120	115.4°	9.6
−120	124.1°	—
−130	133.9°	9.8
−140	143.5°	9.6
−150	152.9°	9.4
−160	161.4°	8.5
−170	171.0°	9.6
−180	181.1°	10.1

elements to provide the requisite phase taper. Additionally, the enhanced performance characteristics of both high-gain and scanning array antennas were reviewed in this section.

Section 2.4, has presented an alternate design of a phase shifter based around SIW technology, which can be reconfigured by using liquid metal. The designed phase shifter operates at 10 GHz and exhibits a relatively low IL. Phase shifting is achieved by using a combination of coarse- and fine-phase tuning which enable the proposed phase shifter to achieve a phase shift ranging between 0° and 180°, in steps of 10°. Coarse phase shifts are realized by reconfiguring the device between three different SIW paths, having different electrical lengths. In this way it is possible to achieve phase shifts of 60°, 120°, and 180°. The reconfiguration between these paths was achieved by adding/removing RF blocking walls comprising of a group of liquid metal vias. Furthermore, six liquid metal tuning vias were used to achieve fine control over the phase shift. These tuning vias control the phase change in increments of 10° using the mechanism of reactive loading.

The presented measured and simulated results serve to demonstrate that the performance of the proposed phase shifter performance is satisfactory. It was shown to achieve a measured phase shift of up to 180° with approximately 10° phase steps with a low IL. The proposed phase shifter has several advantages in comparison with competing technologies. Those advantages include: (i) bidirectional and reciprocal operation; (ii) very low IL considering the 10° resolution of the available

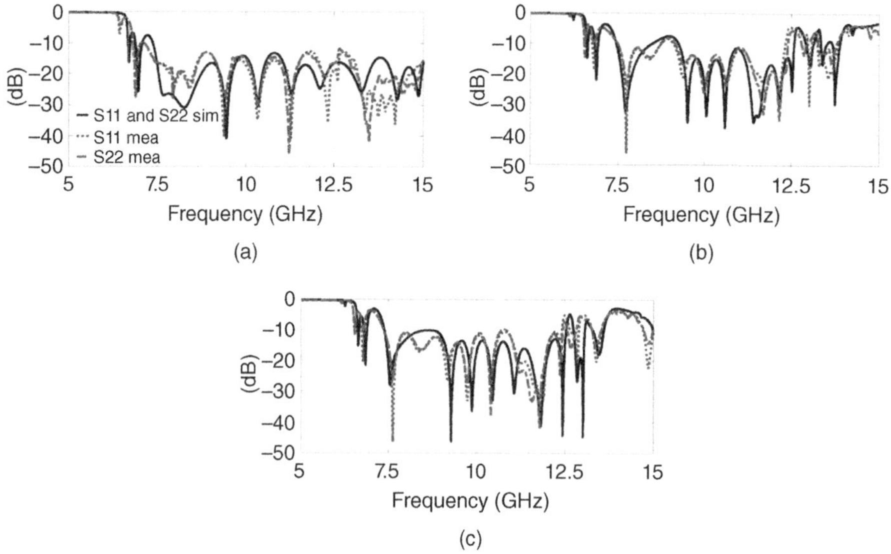

Figure 2.52 Measured and simulated $|S_{11}|$ and $|S_{22}|$ of the main three states of the proposed phase shifter. (a) State A (−60°). (b) State B (−120°). (c) State C (−180°).

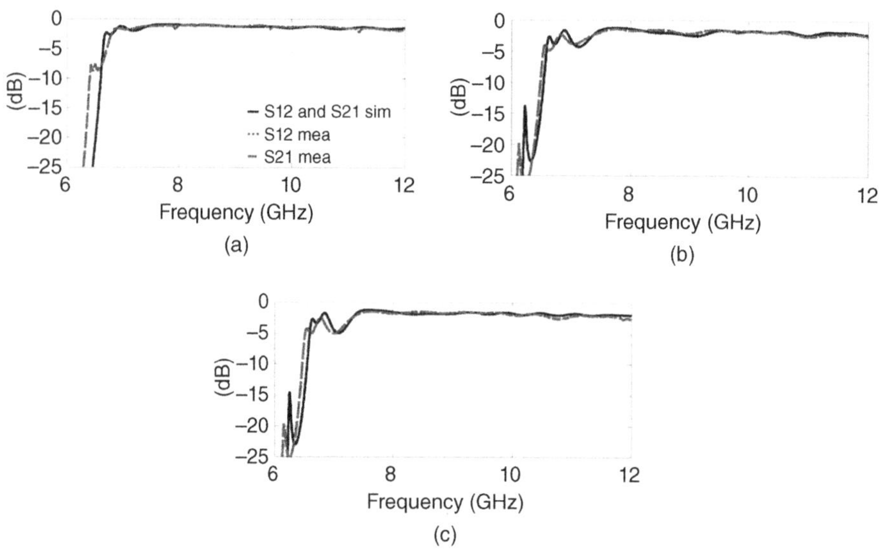

Figure 2.53 Measured and simulated $|S_{21}|$ and $|S_{12}|$ of the main three states of the proposed phase shifter. (a) State A (−60°). (b) State B (−120°). (c) State C (−180°).

phase control; (iii) low root mean square (RMS) phase error; and (iv) the potential for improved power handling. The proposed phase shifter will be applicable within a wide range of different reconfigurable microwave devices.

It is anticipated that a number of concepts for the design of scanning arrays presented in this chapter will be implemented in the near future in 5G systems in the millimeter wave regime, where the design challenges abound at present, and where novel concepts are welcomed.

References

1 S. Hayden and G. M. Rebeiz, "2-bit MEMS distributed X-band phase shifters," *IEEE Microwave Guided Wave Lett.*, vol. 10, no. 12, pp. 540–542, Dec 2000.

2 A. E. Martynyuk, A. G. Martinez-Lopez, and J. I. Martinez Lopez, "2-bit X-band reflective waveguide phase shifter with BCB-based bias circuits," *IEEE Trans. Microwave Theory Tech.*, vol. 54, no. 12, pp. 4056–4061, Dec 2006.

3 B. T. W. Gillatt, M. D'Auria, W. J. Otter, N. M. Ridler, and S. Lucyszyn, "3-D printed variable phase shifter," *IEEE Microwave Wireless Compon. Lett.*, vol. 26, no. 10, pp. 822–824, Oct 2016.

4 J. J. P. Venter, T. Stander, and P. Ferrari, "X-band reflection-type phase shifters using coupled-line couplers on single-layer RF PCB," *IEEE Microwave Wireless Compon. Lett.*, vol. 28, no. 9, pp. 807–809, Sep 2018.

5 M. A. Morton, J. P. Comeau, J. D. Cressler, M. Mitchell, and J. Papapolymerou, "5 bit, silicon-based, X-band phase shifter using a hybrid pi/t high-pass/low-pass topology," *IET Microwaves Antennas Propag.*, vol. 2, no. 1, pp. 19–22, Feb 2008.

6 Y. Du, J. Bao, and X. Zhao, "5-bit MEMS distributed phase shifter," *Electron. Lett.*, vol. 46, no. 21, pp. 1452–1453, Oct 2010.

7 G.-L. Tan, R. E. Mihailovich, J. B. Hacker, J. F. DeNatale, and G. M. Rebeiz, "A 2-bit miniature X-band MEMS phase shifter," in *IEEE Microwave Wireless Compon. Lett.*, vol. 13, no. 4, pp. 146–148, April 2003.

8 K. Gharibdoust, N. Mousavi, M. Kalantari, M. Moezzi, and A. Medi, "A fully integrated 0.18 CMOS transceiver chip for X-band phased-array systems," *IEEE Trans. Microwave Theory Tech.*, vol. 60, no. 7, pp. 2192–2202, July 2012.

9 M. Cho, D. Baek, and J. Kim, "An X-Band 5 bit phase shifter with low insertion loss in 0.18 μm SOI technology," *IEEE Microwave Wireless Compon. Lett.*, vol. 22, no. 12, pp. 648–650, Dec 2012.

10 T. N. Ross, K. Hettak, G. Cormier, and J. S. Wight, "Design of X-band GaN phase shifters," *IEEE Trans. Microwave Theory Tech.*, vol. 63, no. 1, pp. 244–255, Jan 2015.

11 W. Luo, H. Liu, Z. Zhang, P. Sun, and X. Liu, "High-power X-band 5b GaN phase shifter with monolithic integrated E/D HEMTs control logic," *IEEE Trans. Electron Devices*, vol. 64, no. 9, pp. 3627–3633, Sep 2017.

12 A. Qaroot and G. Mumcu, "Microfluidically reconfigurable reflection phase shifter," *IEEE Microwave Wireless Compon. Lett.*, vol. 28, no. 8, pp. 684–686, Aug 2018.

13 S. Yeo, J. Chun, and Y. Kwon, "A 3-D X-Band T/R module package with an anodized aluminum multilayer substrate for phased array radar applications," *IEEE Trans. Adv. Packag.*, vol. 33, no. 4, pp. 883–891, Nov 2010.

14 S. Sim, L. Jeon, and J. Kim, "A compact X-band bi-directional phased-array T/R chipset in 0.13 μm CMOS technology," *IEEE Trans. Microwave Theory Tech.*, vol. 61, no. 1, pp. 562–569, Jan 2013.

15 C. Chang *et al.*, "A new compact high-power microwave phase shifter," *IEEE Trans. Microwave Theory Tech.*, vol. 63, no. 6, pp. 1875–1882, Jun 2015.

16 B. Acikel, T. R. Taylor, P. J. Hansen, J. S. Speck, and R. A. York, "A new high performance phase shifter using $Ba_xSr_{1-x}TiO_3$ thin films," *IEEE Microwave Wireless Compon. Lett.*, vol. 12, no. 7, pp. 237–239, Jul 2002.

17 S. Kagita, A. Basu, and S. K. Koul, "Characterization of LTCC-based ferrite tape in X-band and its application to electrically tunable phase shifter and notch filter," *IEEE Trans. Magn.*, vol. 53, no. 1, pp. 1–8, Jan 2017.

18 J. Zhang, S. Ortiz, and A. Mortazawi, "Design and performance of a new digital phase shifter at X-band," *IEEE Microwave Wireless Compon. Lett.*, vol. 14, no. 9, pp. 428–430, Sep 2004.

19 D. J. Chung, R. G. Polcawich, J. S. Pulskamp, and J. Papapolymerou, "Reduced-size low-voltage RF MEMS X-band phase shifter integrated on multilayer organic package," *IEEE Trans. Compon. Packag. Manuf. Technol.*, vol. 2, no. 10, pp. 1617–1622, Oct 2012.

20 R. Ramadoss, A. Sundaram, and L. M. Feldner, "RF MEMS phase shifters based on PCB MEMS technology," *Electron. Lett.*, vol. 41, no. 11, pp. 654–656, May 2005.

21 R. K. Luneburg, *Mathematical Theory of Optics*. Providence, Rhode Island: Brown University, 1944, pp. 189–213.

22 B. Fuchs, L. Le Coq, O. Lafond, S. Rondineau, and M. Himdi, "Design optimization of multi-shell Luneburg lenses," *IEEE Trans. Antennas Propag.*, vol.55, no. 2, pp. 283–289, Feb 2007.

23 R. Mittra, D. Oueslati, A. Nasri, R. K. Arya, and A. Ghalib, "Fixed and scanned-beam antenna arrays for 5G applications," in *Wideband, Multiband, and Smart Antenna Systems*, M. A. Matin, Springer, 2022 145-208.

24 R. K. Arya, S. Pandey, and R. Mittra, "Flat lens design using artificially engineered materials," *Progr. Electromagn. Res. C*, vol. 64, pp. 71–78, 2016. DOI: https://doi.org/10.2528/PIERC16021302

25 S. Zhang, R. K. Arya, W. G. Whittow, D. Cadman, R. Mittra, and J. C. Vardaxoglou, "Ultra-wideband flat metamaterial GRIN lenses assisted with additive manufacturing technique," *IEEE Trans. Antennas Propag.*, vol. 69, no. 7, pp. 3788–3799, Jul 2021

26 S. Zhang, R. K. Arya, S. Pandey, J. C. Vardaxoglou, W. G. Whittow, and R. Mittra, "3D-printed planar graded index lenses," *IET Microwaves Antennas Propag.*, vol. 10, no. 13, pp. 1411–1419, 2016.

27 C. Mateo-Segura, A. Dyke, H. Dyke, S. Haq, and Y. Hao, "Flat Luneburg lens via transformation optics for directive antenna applications," *IEEE Trans. Antennas Propag.*, vol. 62, no. 4, pp. 1945–1953, April 2014.

28 S. Jain, R. Mittra, and S. Pandey, "Flat-base broadband multibeam Luneburg lens for wide-angle scan," *J. Electromagn. Waves Appl.*, vol. 29, no. 10, pp. 1329–1341, Jul 2015.

29 C. Pfeiffer and A. Grbic, "A printed, broadband Luneburg lens antenna," *IEEE Trans. Antennas Propag.*, vol. 58, no. 9, pp. 3055–305, Sep 2010.

30 Q. Cheng, H. F. Ma, and T. J. Cui, "Broadband planar Luneburg lens based on complementary metamaterials," *Appl. Phys. Lett.*, vol. 95, no. 18, 181901, 2009.

31 J. Gao, C. Wang, K. Zhang, Y. Hao, and Q. Wu, "Beam steering performance of compressed Luneburg lens based on transformation optics," *Results Phys.*, vol. 9, pp. 570–575, Jun 2018.

32 E. C. V. Boas, R. Mittra, and A. C. Sodre, "A low-profile high-gain slotted waveguide antenna array with grooved structures," *IEEE Antennas Wireless Propag. Lett.*, vol. 19, no. 12, pp. 2107–2111, Dec 2020.

33 A. Maifuz, R. K. Arya, A. K. Yerrola, L. Murmu, and A. Kumar," Bandwidth and gain enhancement with cross-polarization suppression in microstrip antenna using superstrate," in *2021 XXXIV General Assembly and Scientific Symposium of the International Union of Radio Science (URSI GASS)*, 28th August- 4th September, 2021, Rome, Italy: Sapienza Faculty of Engineering.

34 S. N. McClung, S. Saeedi, and H. H. Sigmarsson, "Band-reconfigurable filter with liquid metal actuation," *IEEE Trans. Microwave Theory Tech.*, vol. 66, no. 6, pp. 3073–3080, Jun 2018.

35 J. H. Dang, R. C. Gough, A. M. Morishita, A. T. Ohta, and W. A. Shiroma, "A tunable X-band substrate integrated waveguide cavity filter using reconfigurable liquid-metal perturbing posts," in *IEEE MTT-S International Microwave Symposium (IMS)*, Phoenix AZ, 17–22 May 2015, DOI: https://doi.org/10.1109/MWSYM.2015.7167060.

36 S. N. McClung, S. Saeedi, and H. H. Sigmarsson, "Single-mode-dual-band to dual-mode-single-band bandpass filter with liquid metal," in *IEEE 18th Wireless and Microwave Technology Conference (WAMICON)*, Cocoa Beach FL, 24–25 April 2017, DOI: https://doi.org/10.1109/WAMICON.2017.7930270.

37 A. H. Pham, S. Saeedi, and H. H. Sigmarsson, "Continuously-tunable substrate integrated waveguide bandpass filter actuated by liquid metal," in *IEEE MTT-S International Microwave Symposium (IMS)*, Boston MA, 2–7 June 2019, DOI: https://doi.org/10.1109/MWSYM.2019.8700809.

38 T. B. Chaieb, A. Nasri, and H. Zairi, "Liquid filled method for substrate integrated waveguide reconfigurable filter," in *International Conference on Advanced Systems and Electric Technologies (IC_ASET)*, Hammamet, Tunisia, 22–25 March 2018, DOI: https://doi.org/10.1109/ASET.2018.8379842.

39 S. Khan, N. Vahabisani, and M. Daneshmand, "A fully 3-D printed waveguide and its application as microfluidically controlled waveguide switch," *IEEE Trans. Compon. Packag. Manuf. Technol.*, vol. 7, no. 1, pp. 70–80, Jan 2017.

40 N. Vahabisani, S. Khan, and M. Daneshmand, "A K-band reflective waveguide switch using liquid metal," *IEEE Antennas Wireless Propag. Lett.*, vol. 16, pp. 1788–1791, Mar 2017.

41 M. Wang, I. M. Kilgore, M. B. Steer, and J. J. Adams, "Characterization of intermodulation distortion in reconfigurable liquid metal antennas," *IEEE Antennas Wireless Propag. Lett.*, vol. 17, no. 2, pp. 279–282, Feb 2018.

42 A. M. Morishita, C. K. Y. Kitamura, A. T. Ohta, and W. A. Shiroma, "A liquid-metal monopole array with tunable frequency, gain, and beam steering," *IEEE Antennas Wireless Propag. Lett.*, vol. 12, pp. 1388–91, Oct 2013.

43 C. Wang, J. C. Yeo, H. Chu, C. T. Lim, and Y.-X. Guo, "Design of a reconfigurable patch antenna using the movement of liquid metal," *IEEE Antennas Wireless Propag. Lett.*, vol. 17, no. 6, pp. 974–977, Apr 2018.

44 M. Wang, C. Trlica, M. R. Khan, M. D. Dickey, and J. J. Adams, "A reconfigurable liquid metal antenna driven by electrochemically controlled capillarity," *J. Appl. Phys.*, vol. 117, no. 19, 194901, May 2015.

45 J. H. Dang, R. C. Gough, A. M. Morishita, A. T. Ohta, and W. A. Shiroma, "Liquid-metal-based phase shifter with reconfigurable EBG filling factor," in *2015 IEEE MTT-S International Microwave Symposium*, Phoenix, AZ, 2015, pp. 1–4.

46 A. Dey and G. Mumcu, "Microfluidically controlled frequency tunable monopole antenna for high-power applications," *IEEE Antennas Wireless Propag. Lett.*, vol. 15, pp. 226–229, Jun 2016.

47 W. Chen, Y. Li, R. Li, A. V. Thean, and Y. Guo, "Bendable and stretchable microfluidic liquid metal-based filter," *IEEE Microwave Wireless Compon. Lett.*, vol. 28, no. 3, pp. 203–205, Mar 2018.

48 C. Koo, B. E. LeBlanc, M. Kelley, H. E. Fitzgerald, G. H. Huff, and A. Han, "Manipulating liquid metal droplets in microfluidic channels with minimized skin residues toward tunable RF applications," *J. Microelectromech. Syst.*, vol. 24, no. 4, pp. 1069–1076, Aug 2015.

49 M. A. Rafi, B. D. Wiltshire, and M. H. Zarifi, "Wideband tunable modified split ring resonator structure using liquid metal and 3-D printing," *IEEE Microwave Wireless Compon. Lett.*, vol. 30, no. 5, pp. 469–472, May 2020.

50 A. Ha and K. Kim, "Frequency tunable liquid metal planar inverted-F antenna," *Electron. Lett.*, vol. 52, no. 2, pp. 100–102, 2016.

51 A. M. Watson *et al.*, "Physically reconfigurable RF liquid electronics via Laplace barriers," *IEEE Trans. Microwave Theory Tech.*, vol. 67, no. 12, pp. 4881–4889, Dec 2019.

52 L. Song, W. Gao, C. O. Chui, and Y. Rahmat-Samii, "Wideband frequency reconfigurable patch antenna with switchable slots based on liquid metal and 3-D printed microfluidics," *IEEE Trans. Antennas Propag.*, vol. 67, no. 5, pp. 2886–2895, May 2019.

53 T. Kim, K. Kim, S. Kim, J. Lee, and W. Kim, "Micropatterning of liquid metal by dewetting," *J. Microelectromech. Syst.*, vol. 26, no. 6, pp. 1244–1247, Dec 2017.

54 J. Low, P. Chee, and E. Lim, "Deformable liquid metal patch antenna for air pressure detection," *IEEE Sens. J.*, vol. 20, no. 8, pp. 3963–3970, April 2020.

55 V. T. Bharambe, J. Ma, M. D. Dickey, and J. J. Adams, "Planar, multifunctional 3D printed antennas using liquid metal parasitics," *IEEE Access*, vol. 7, pp. 134245–134255, 2019.

56 M. Wang, M. R. Khan, C. Trlica, M. D. Dickey, and J. J. Adams, "Pump free feedback control of a frequency reconfigurable liquid metal monopole," in *Proceedings of the 2018 IEEE International Symposium on Antennas and Propagation & USNC/URSI National Radio Science Meeting*, Vancouver, Canada, 19–24 July 2015, pp. 2223–2224.

57 D. Rodrigo, L. Jofre, and B. A. Cetiner, "Circular beam-steering reconfigurable antenna with liquid metal parasitics," *IEEE Trans. Antennas Propag.*, vol. 60, no. 4, pp. 1796–1802, April 2012.

58 S. Singh, J. Taylor, H. Zhou, A. Pal, A. Mehta, H. Nakano, and P. Howland, "A pattern and polarization reconfigurable liquid metal helical antenna," in *Proceedings of the 2018 IEEE International Symposium on Antennas and Propagation & USNC/URSI National Radio Science Meeting*, Boston, MA, 2018, pp. 857–858.

3

Low-Cost Beam-Reconfigurable Directional Antennas for Advanced Communications

Qi Luo[1], Steven Gao[2], Xue-xia Yang[3], and Josaphat T. Sri Sumantyo[4]

[1] *University of Hertfordshire, School of Physics, Engineering and Computer Science, Hatfield, UK*
[2] *University of Kent, School of Engineering and Digital Arts, Canterbury, UK*
[3] *Shanghai University, School of Communication and Information Engineering, Shanghai, China*
[4] *Chiba University, Center for Environmental Remote Sensing, Chiba, Japan*

3.1 Introduction

The conventional technique for realizing a beam-reconfigurable directional antenna is to use a phased array, where the radiating elements are placed in an aperture to form a linear or planar array. To reconfigure the beam, the phase of each array element needs to be adjusted, by using either analogue beamforming, digital beamforming, or hybrid beamforming. Although the phased array shows high-directivity and good beam-steering performance, generally speaking, the beamforming networks (BFNs) have high-costs and bulky configurations. Even if the array antenna is designed to use corporate or series-fed networks, the BFN is still complex especially when there is a large number of array elements.

How does one reduce the cost of a beam-reconfigurable directional antenna? This question can be tackled from different angles. The first approach is to reduce the number of array elements without changing the aperture size. In this way, the number of RF components required to reconfigure the beam can be reduced. The second approach is to avoid the use of expensive RF components to reconfigure each array element. For example, positive intrinsic negative (PIN) diodes or varactors can be used to replace the phase shifters to provide lower bit phase quantizations. Alternatively, the phase shift can be obtained by using tunable materials. This method requires that the array element be compatible with this phase control method and some phase synthesis needs to be performed to optimize the far-field radiation pattern of the array. The third approach is to reduce the complexity of the BFN. The complexity of a BFN increases significantly with an increasing number of radiating elements in the array. Thus, it is highly desirable to develop novel techniques to realize low-complexity BFNs. The ultimate goal of these approaches is to achieve beam-steering performance comparable to a conventional phased array. In this chapter, the associated design techniques are discussed with examples.

3.2 Beam-Reconfigurable Antenna Using Active Frequency Selective Surfaces

Frequency selective surfaces (FSSs) belong to the family of metamaterials. An FSS consists of set of two-dimensional (2D) planar periodic structures and operates as a bandstop or bandpass filter to

Antenna and Array Technologies for Future Wireless Ecosystems, First Edition.
Edited by Y. Jay Guo and Richard W. Ziolkowski.

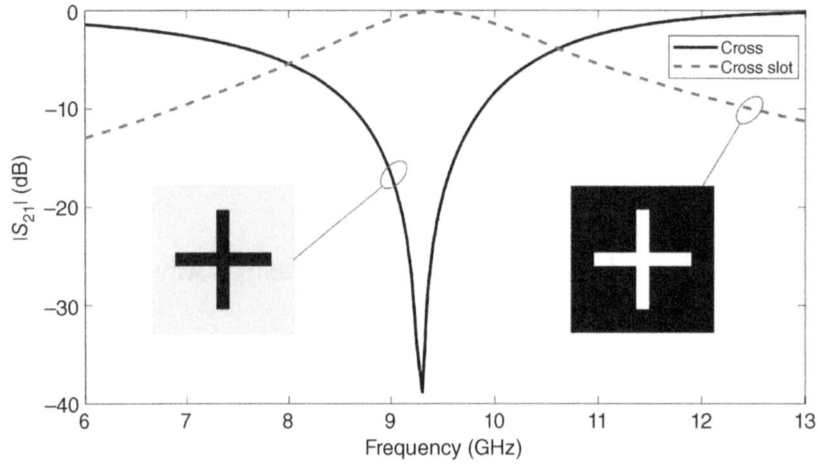

Figure 3.1 The amplitude of the transmission coefficient, $|S_{21}|$, of two complementary FSS unit cells.

the incident waves. Equivalent circuit models consisting of series and shunt connections of capacitors and inductors are usually used to analyze the performance of an FSS and synthesize its design [1]. As demonstrated in Figure 3.1, the cross type FSS shows bandstop performance, while its complementary version, the cross-slot type FSS, shows bandpass filter performance. In this example, both FSS unit cells have the same values of the design parameters and are printed on the same dielectric substrate.

FSSs have found many applications in the design of antennas, filters, polarizers, radomes, and electromagnetic shielding. Through the embedding of active devices into the unit cells, the FSS becomes an active frequency selective surface (AFSS) and its responses are reconfigurable. This characteristic can be used to realize low-cost beam-reconfigurable antennas that do not require any RF phase shifters as traditional phased arrays do. Thus, a significant reduction in the weight, power consumption, complexity, and cost of an antenna can be achieved. Figure 3.2 demonstrates the concept of using PIN diodes to switch the operating modes of an AFSS in a frequency range of interest. The unit cell of this AFSS example consists of a simple rectangular slot printed on a dielectric substrate with a PIN diode placed in the middle of the slot. Please note that the biasing circuit for the diode is not included in the EM simulation of this example. Moreover, the PIN diode is assumed to be lossless. As shown in Figure 3.2b, the FSS shows a typical bandpass response with center frequency at 9.1 GHz when the PIN is OFF. The equivalent circuit model of the unit cell of the FSS in this case consists of shunt connections of one capacitor and two inductors. When the PIN is switched ON, the rectangular slot is shorted. The response of the FSS then becomes a high-pass filter that has bandstop characteristics at the frequency of interest. The equivalent circuit model in this case consists of shunt connections of two capacitors and three inductors.

Inspired by the concept of an AFSS, a low-cost beam-reconfigurable antenna that switches its beam in the azimuth plane can be developed. The concept and operating principle of this type of antenna is shown in Figure 3.3. The AFSS is formed as a cylinder and an omnidirectional antenna is placed in the center as its feed. RF switches are used to electronically configure the AFSS to operate at different modes, bandpass or bandstop, to the incident waves. The AFSS is linearly polarized and the feed must have its polarization match the polarization of the AFSS. When the AFSS is in the bandpass mode at its operating frequency, it is equivalent to being electromagnetically (EM)

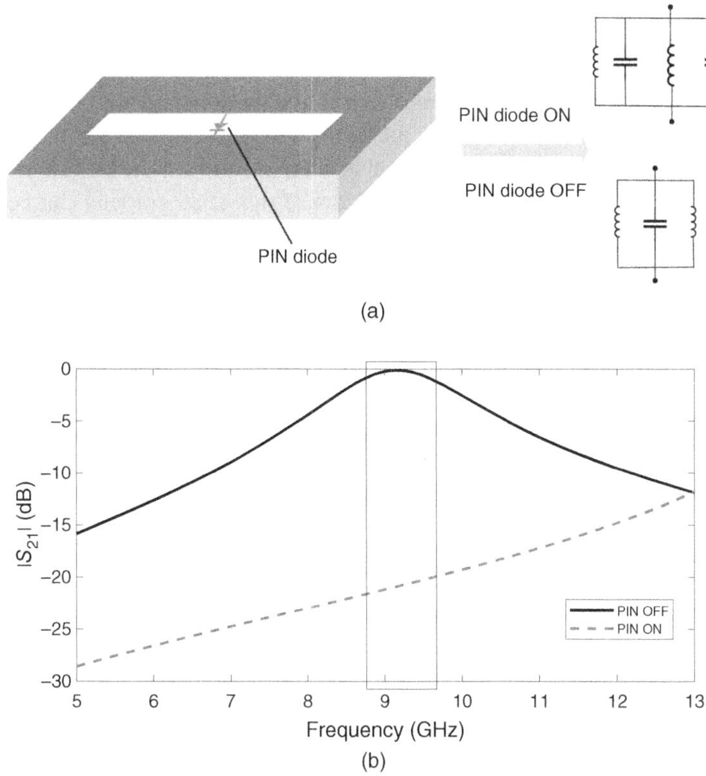

Figure 3.2 An active rectangular slot-based FSS. (a) The configuration of the unit cell of this AFSS and the equivalent circuits of its ON and OFF states. (b) The magnitude of its simulated transmission coefficient, $|S_{21}|$. The incident wave is polarized orthogonal to the slot.

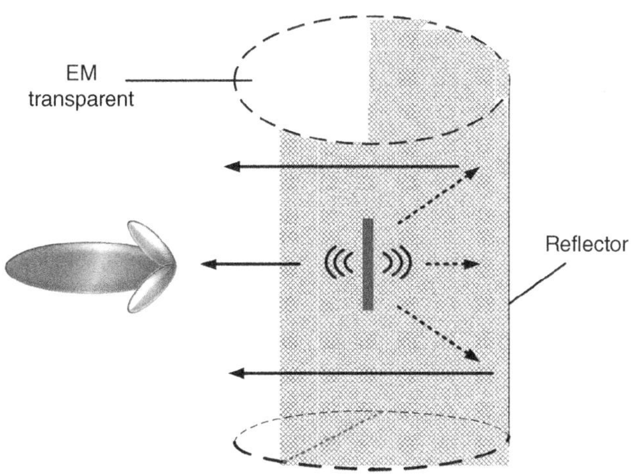

Figure 3.3 The concept and operating principle of an AFSS beam-switching antenna. In this example, both the AFSS and its feed are vertically polarized.

transparent. On the other hand, when the AFSS is operating in its bandstop mode, it is equivalent to a reflector.

The cylindrical AFSS is divided into several sectors. When a section of the AFSS is configured as EM transparent and the rest of it is configured to work as a reflector, at least half of the radiated waves from the feed would, in theory, be reflected back by the equivalent reflector. As a result, the AFSS antenna would show a more directional radiation pattern in comparison with the feed alone. By controlling the operating modes of each AFSS sector, the beam of the AFSS antenna can be switched within a full 360° angle range over the entire azimuth plane.

Figure 3.4 shows some examples of active microstrip FSS designs that only use one RF switch – a PIN diode – in each FSS unit cell. Reconfigurability is realized by changing the resonance behavior of the FSS by controlling the states of its PIN diodes. The AFSS unit cell shown in Figure 3.4a is a two layer design in which an H-shaped slot is printed on one layer and a strip with one PIN diode placed at its center is printed on the other layer. The total length of the slot is approximately one half-wavelength at the center frequency of the system. When the PIN is switched ON, the microstrips operate as a polarizer that effectively reflects back the incident waves whose polarization is parallel to the microstrips. The substrate used in this design is 0.005 mm thick Mylar polyester film, which has $\epsilon_r = 3.55$ and a loss tangent of 0.0023. The AFSS unit cell shown in Figure 3.4b is a single layer design and the unit cell consists of two types of elements, a strip and

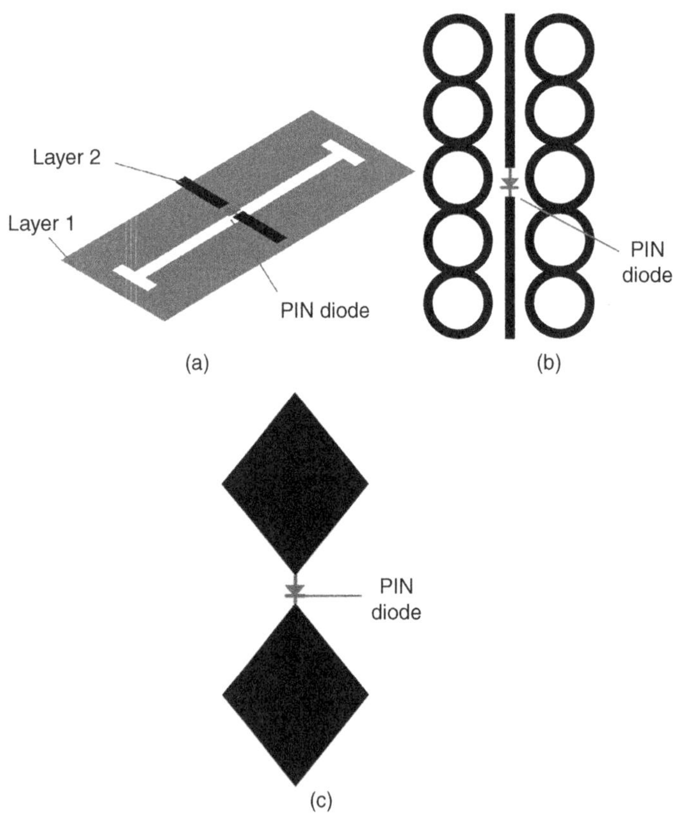

Figure 3.4 Examples of active microstrip FSS unit cell designs. (a) Slot type AFSS. Source: Adapted from [2]. (b) Single layer hybrid element. Source: Adapted from [3]. (c) Single layer design using diamond-shape patches. Source: Adapted from [4].

a set of ring loops. By changing the states of the PIN diode, this FSS unit cell produces a passband response at two different frequencies. The FSS surface is fabricated by using a thin flexible substrate, Rogers RO3003, with $\epsilon_r = 3.0$ and a loss tangent of 0.0013. The AFSS unit cell shown in Figure 3.4c is also a single-layer design. This AFSS element consists of two diamond-shaped patches connected by one PIN diode. When the PIN diode is ON or OFF, the FSS exhibits a reflection or a transmission characteristic, respectively, at the frequency of interest. Instead of conforming the AFSS to a cylinder, the AFSS is printed as three panels that are arranged in an equilateral triangle shape. Thus, by configuring one FSS panel in the transmission mode and the other two panels in the reflection mode, the AFSS antenna can switch the beam in 120° steps in the azimuth plane.

The AFSS antennas realized by using the unit cells shown in Figure 3.4 can switch their beams in the azimuth plane with peak directivities up to 10 dBi. In many applications, it would be desirable to further improve the directivity of the antenna and to optimize its beamwidth. Several methods have been reported to achieve this goal. The AFSS screen is conformed into a cylindrical shape that is divided into six equal sectors by metallic sheets in [5]. Two metallic cones are placed on the top and the bottom of the cylinder. The concept of this design is shown in Figure 3.5. It was found in this study that the directivity and backlobe level of the antenna could be improved by optimizing the diameter of the cone and the height of metallic sheets. The fabricated antenna prototype of this design had a bandwidth from 2.3 to 3 GHz with a 3 dB beamwidth of 60°. A similar technique was used in [6] and a maximum gain improvement of 7 dB was reported.

Another technique to improve the gain of a traditional cylindrical AFSS antenna is presented in [7]. Figure 3.6 shows the concept of this design. Eight AFSS panels are used to form an octagon shape and a printed monopole is placed in the center as the feed antenna. Different from the conventional design, the top and bottom faces of the cylinder are conductors. During its operation, only two screens (one quarter) of the AFSS are configured to operate in their EM transparent mode and the rest of the screens are configured to work in their reflector mode. Thus, the design is equivalent to placing a feed antenna in a reconfigurable cavity. It leads to the increased directivity of the antenna. Figure 3.7 illustrates the simulated beam-switching performance of the antenna. The simulation results show that this design achieves a peak gain of 14 dBi and switches the beam in steps of 45° in the azimuth plane with low sidelobes and backlobes.

The AFSS antennas presented earlier perform their beam-switching in the azimuth plane. A recent work [2] extends the beam-switching AFSS antenna to perform limited beam-switching in the elevation plane in addition to switching its beams in the azimuth plane. The beam-switching

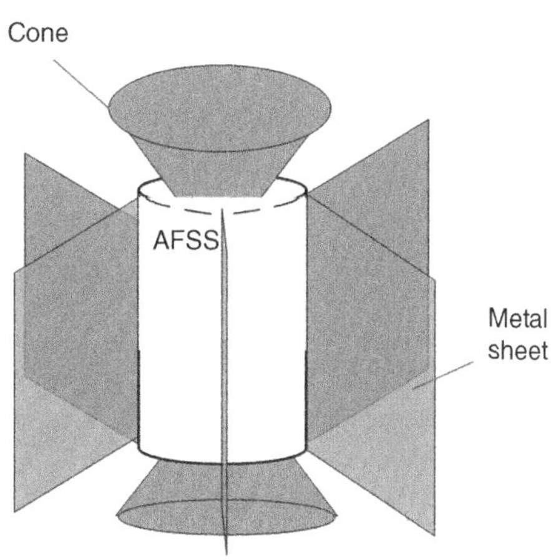

Figure 3.5 Configuration of the high gain AFSS antenna presented in [5]. Source: Adapted from [5].

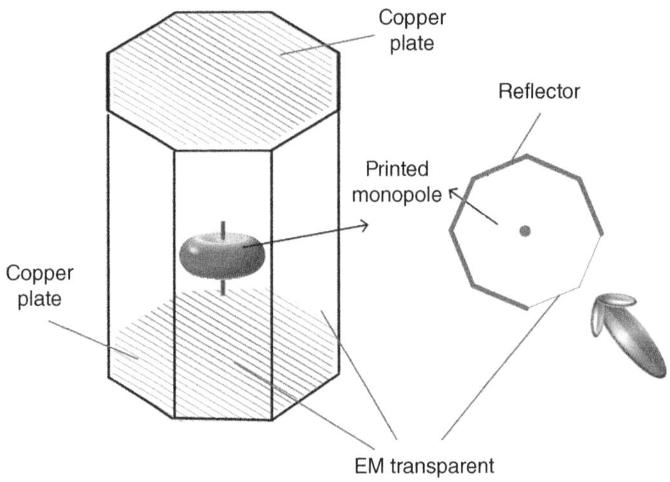

Figure 3.6 Concept of the AFSS antenna with improved gain presented in [7]. Source: Adapted from [7].

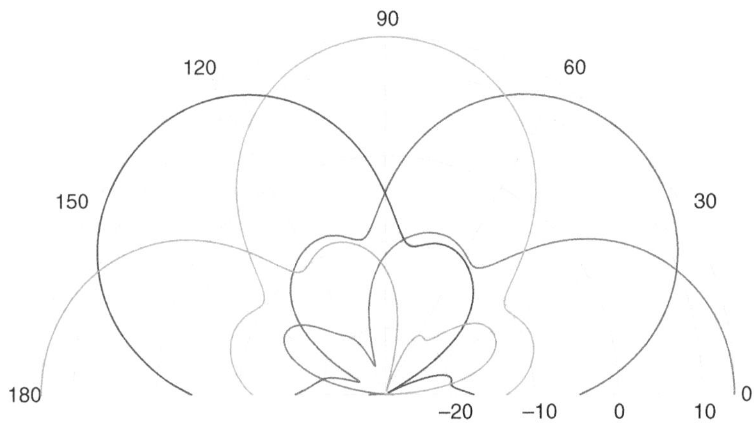

Figure 3.7 Simulated beam-switching performance of the AFSS antenna presented in [7]. Source: Adapted from [7].

in the elevation plane was achieved by using a two-element linear array as the feed and by adjusting their excitation phases. The AFSS antenna was able to scan its beam from −15° to 16° in the elevation plane with a peak gain that was higher than 9 dBi.

3.3 Beam-Reconfigurable Antenna Using Parasitic Elements

Another method to realize beam-steering in the azimuth plane is through the use of a parasitic array antenna, which consists of a driven element and several parasitic elements. The driven antenna is the only active radiating element. Thus, the RF energy is distributed from the driven element to the parasitic elements by electromagnetic coupling.

One well-known parasitic array antenna is the electronically steerable parasitic array radiator (ESPAR). It obtains reconfigurable radiation patterns by applying tunable components to its

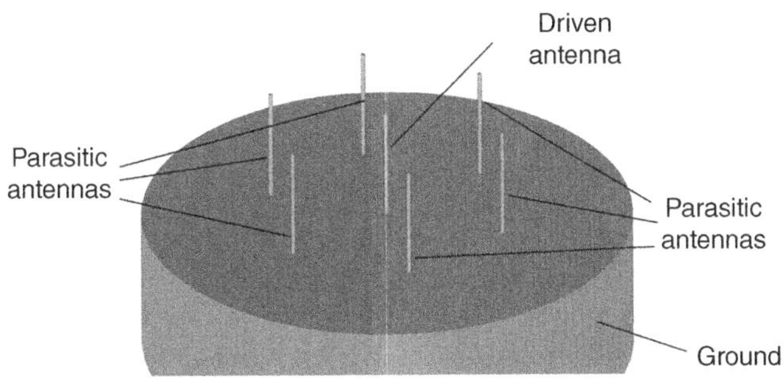

Figure 3.8 The configuration of the classic seven-element ESPAR antenna.

parasitic antenna elements. Figure 3.8 shows the configuration of a classic seven-element ESPAR antenna. The element located at the center is the driven element. It is the only element connected to the RF front end. This driven element is surrounded by six parasitic elements. Each of parasitic elements is loaded by an electronically tunable component such as sliding loads, PIN diodes, varactors or microelectromechanical systems (MEMS) switches. The tunable loads can be adaptively controlled by a digital signal processor (DSP). Beamforming is performed by adjusting the reactance loads of the parasitic antennas to yield a weighted signal synthesis. For example, each parasitic element can be configured to operate as a director or as a reflector. Since the RF energy is distributed from the driven element to the parasitic elements by electromagnetic coupling, the ESPAR antenna does not require any phase shifters or feed network. Thus, it is a low-cost solution to realize a beam-reconfigurable antenna. The theory and operating principles of EPSAR systems are detailed in [8].

There are many reported research works in the field of ESPAR design. They include using folded monopoles, embedding the ESPAR in a homogeneous dielectric material, and using a planar configuration [9–14]. These designs show promising beam-switching performance over the entire azimuth plane from 0° to 360°. Nevertheless, their directivities are not high, being in the range from 4 to 8 dBi. A simple and effective method to improve the gain of the traditional ESPAR was presented in [15]. The design placed a reconfigurable director array near its parasitic elements as illustrated in Figure 3.9. Six director arrays were placed in a "Star Topology" adjacent to the ESPAR. Each director consists of two hollow cylinders connected by a PIN diode. When the PIN diodes are forward biased, the radiated power of the system is increased in the forward direction. When the PIN diodes are reverse biased, the induced currents along the reconfigurable small directors are attenuated. Thus, they have little effect on the radiated power in the forward direction. The measurement results presented in [15] show that compared to the traditional seven-element ESPAR, the gain of this modified ESPAR is increased by 3 dB with an improved front-to-back ratio (FBR).

It has been shown in [16] that by introducing some reflectors on the top of a planar parasitic antenna, its peak gain can be improved by more than 3 dB. The concept of this approach is presented in Figure 3.10. The radiating antenna consists of a circular patch and six parasitic elements. Both the parasitic elements and the circular patch have a group of shorting vias integrated in them that are reconfigured with PIN diodes. The reflectors are passive and are positioned symmetrically around the patch. The height of the reflectors is a key design parameter that needs to be optimized in order to increase the gain of the antenna. The results given in [16] show that a 10 dBi gain is obtained at 5.3 GHz when the heights of the reflectors are increased to 14 mm. Beam-steering is achieved by

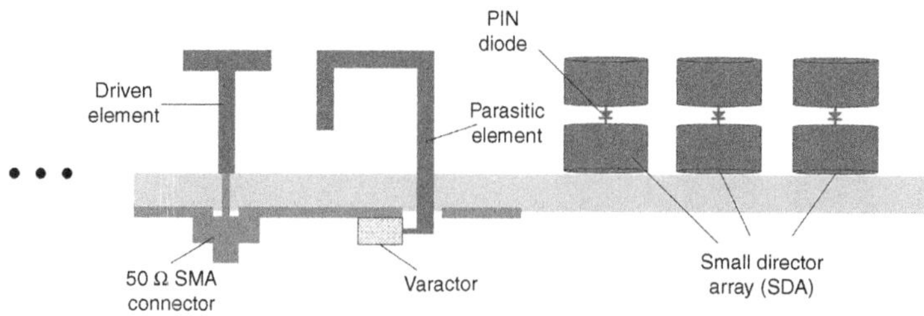

Figure 3.9 The concept of an ESPAR with improved gain presented in [15]. Source: Adapted from [15].

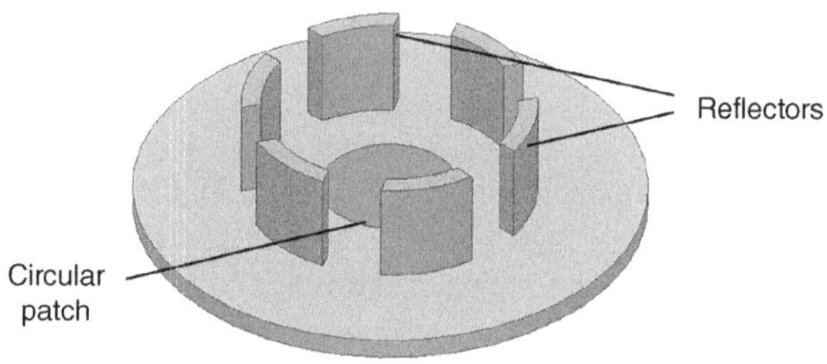

Figure 3.10 The parasitic antenna with improved gain developed in [16]. The parasitic elements and the vias are not shown in this figure. Source: Adapted from [16].

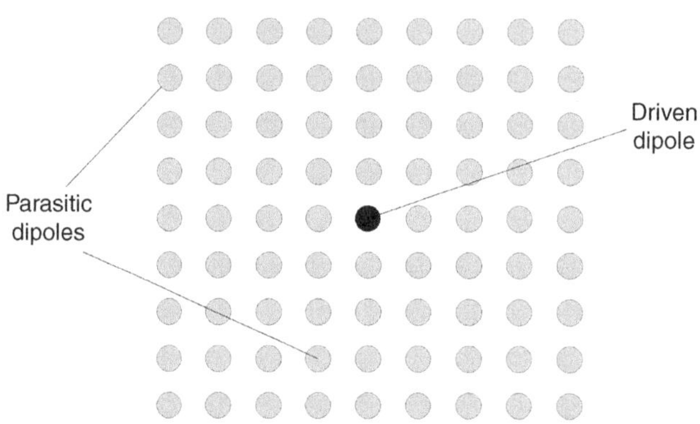

Figure 3.11 The concept of the RECAP presented in [17]. Source: Adapted from [17].

rotationally shorting two adjacent parasitic patches to the ground and by shorting the vias of the circular patch in the desired direction.

There are also some theoretical studies on using the concept of a reconfigurable aperture (RECAP) to realize a parasitic array antenna [17, 18]. Figure 3.11 illustrates the general concept of a parasitic RECAP array. The driven element is positioned in the center and the rest of the elements are parasitic and are terminated with reconfigurable reactive loads. The radiation pattern realized by this array is synthesized by superposing the complex vectorial far-field radiation pattern of each parasitic element [17]. The beamforming of an $N \times N$ planar parasitic array antenna was explored by studying the number of reconfigurable elements (REs) per wavelength and the number of reconfiguration bits (RBs) per reconfigurable element.

$$E(\theta, \phi) = \sum_{j=1}^{N^2} v_j E_j(\theta, \phi) \tag{3.1}$$

where $E_j(\theta, \phi)$ is the radiation pattern of the parasitic element N_j and v_j is the voltage applied to it. The voltage on each of the parasitic elements satisfies

$$\begin{bmatrix} i_1 \\ i_2 \end{bmatrix} = \begin{bmatrix} y_{11} & y_{12} \\ y_{21} & y_{22} \end{bmatrix} \begin{bmatrix} v_1 \\ v_2 \end{bmatrix} \tag{3.2}$$

where y is the admittance matrix of the complete array; i_1 and v_1 are the scalar current and voltage on the feed, respectively; and i_2 and v_2 are the currents and voltages on the ports of the reconfigurable parasitic elements, respectively. By optimizing the number of parasitic elements and their reconfigurable states, it was shown in [17] that the radiation pattern of the array can be reconfigured to different angles with high directivity and low sidelobe levels (SLLs). The only condition is that the states of each parasitic element must be reconfigurable and there must be at least four of them.

3.4 1-Bit Reflectarray and Transmitarray

Reflectarrays (RAs) and transmitarrays (TAs) have received a significant amount of research interest recently. RAs combine the advantages of reflector antennas and phased arrays. A conventional microstrip RA consists of an array of reflecting elements and a feed antenna(s). Each reflecting element provides a pre-adjusted phase to form a focused beam in the desired direction. An RA antenna can be a substitute over its operating frequencies for the traditional parabolic reflector, but with a lightweight and planar structure. These features make it an attractive solution for satellite on-the-move (SOTM) applications [19, 20]. A TA also consists of a feed antenna(s) and an array of unit cells. Different from an RA, it is an array of unit cells that transmit the waves instead of reflecting them back toward the feed antenna(s). Thus, a TA is referred to as a "planar lens" in some literature. The TA places the feed antenna(s) directly next to the aperture without incurring any blockage loss. To produce a collimated beam in a given direction, the required phase distribution over the aperture is calculated by using the formula given in [21]:

$$\phi_R = k_0(d_i - (x_i \cos\varphi_b + y_i \sin\varphi_b)\sin\theta_b) \tag{3.3}$$

where ϕ_R is the phase of the reflection (for an RA design) or transmission (for a TA design) coefficient of the antenna element i; k_0 is the phase constant in vacuum; (x_i, y_i) are the coordinates of the array element i; d_i is the distance from the phase center of the feed to the radiating element; and (θ_b, φ_b) is the targeted scan angles of the beam in the spherical coordinate system.

To realize continuous beam-steering with low-cost, a 1-bit phase shift can be applied to the design of the RA and TA unit cells. Using this approach, there is no need for phase shifters because a 1-bit phase shift can be obtained by incorporating RF switches into the unit cells. These switches can be PIN diodes or MEMS. The RF PIN diodes exhibit fast switching time with reasonable insertion loss and isolation. RF MEMS can work at a higher frequency than a PIN diode with lower insertion loss, but with a higher cost. Since the phase has only two states, they can be quantized after calculating the desired phase delays ϕ_R (using Eq. (3.3)) for each element as

$$\phi_q = +\frac{\pi}{2}, \quad \text{when} \quad 0 \leq \phi_R(modulo\ 2\pi) < \pi \tag{3.4}$$

$$\phi_q = -\frac{\pi}{2}, \quad \text{when} \quad \pi \leq \phi_R(modulo\ 2\pi) < 2\pi \tag{3.5}$$

where ϕ_q is quantized phase value assigned to one of the PIN states. Figure 3.12 shows the calculated phase distribution using continuous phase resolution and 1-bit phase resolution. For a low-cost high gain array antenna design, 1-bit RAs and TAs are quite appealing for the following reasons. First, they employ spatial feeding instead of using traditional feed networks. This is an important advantage as it reduces the system complexity and its cost. This is particularly true in the case that there is a need for using a large number of antenna elements to achieve a high directivity. Second, there is no need to use phase shifters and the control of the RF switches is binary (either ON or OFF). These features significantly reduce the cost and complexity of the overall antenna system. Third, they can be manufactured using low-cost printed circuit board (PCB) technologies and can be designed to have planar structures as well as ones that are conformal to the bodies of satellites, airplanes, or vehicles.

There are quantization errors associated with any 1-bit phase quantization. They lead to a decrease of the gain and aperture efficiency of the array antenna, as well as to an increase of the SLLs. The gain reduction of a phased array due to the phase quantization error is discussed in [22]

$$\Delta G = \frac{\sin^2(\delta_{\max})}{\delta_{\max}^2} \tag{3.6}$$

where ΔG is the gain reduction and $\delta_{\max}$ is the maximum phase quantization error in radians. In the case of single bit phase resolution, $\delta_{\max} = \pi/2$ and the calculated gain reduction of the array is about 3.9 dB when compared to a continuous phase distribution. This is a trade-off between having low cost and high efficiency. By optimizing the phase distribution, it is feasible to obtain promising steering performance of RAs and TAs with 1-bit phase resolution. A recent study [23] demonstrates that the scan loss of an RA with 1-bit phase shift can be reduced by optimizing the phase distribution. Moreover, the SLLs can be reduced as well when the size of the aperture is increased.

3.4.1 1-Bit Reflectarray

Figure 3.13 shows the typical configurations of an RA. Similar to a reflector antenna, the RA can be either center-fed or offset-fed. Each reflecting element is designed to provide a pre-adjusted phase to form a collimated beam in the desired direction. There are several approaches to adjust the phase of each reflecting element. These include varying the sizes of the reflecting elements (e.g. patches), loading the radiating elements with open-circuited microstrip stubs of varied length, rotating the radiating elements, etc.

A straightforward approach to realize a 1-bit phase shift is to use a microstrip patch loaded with an open-circuited stub as the reflecting element and to introduce one RF switch on it. Figure 3.14 illustrates this concept. To obtain a 180° reflection phase difference, the electrical length of the second part of the stub should be 90° at the frequency of interest. The electrical length of the phase delay line can then be changed by controlling the states of the RF switches. The outcome is equivalent to a 1-bit phase shifter. A beam-switching 1-bit RA using this approach was presented in [24].

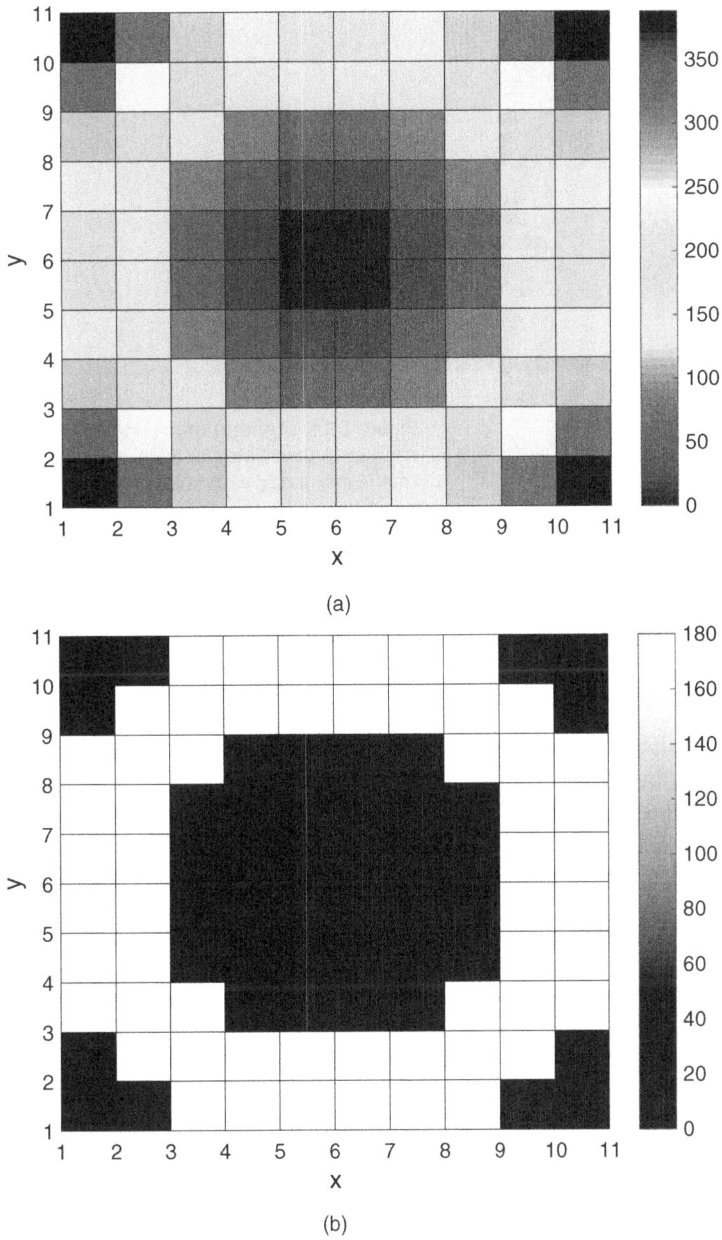

Figure 3.12 The phase distribution of an aperture. (a) Continuous phase distribution. (b) Discrete 1-bit phase resolution.

The unit cell of the RA in this design was an aperture coupled microstrip patch antenna. The patch and the open-ended microstrip phase delay lines were printed on different layers separated by a ground plane. The MEMS switches were monolithically integrated on the open-ended microstrip phase delay lines. By changing the states of these MEMS, the equivalent electrical lengths of the open-ended transmission lines were adjusted. This RA was designed to operate at 26.5 GHz with

Figure 3.13 Configurations of a classic reflectarray. (a) Center-fed. (b) Offset-fed.

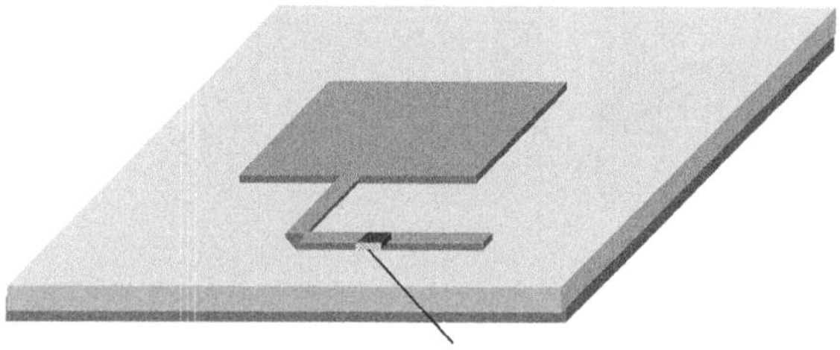

Figure 3.14 Concept of loading a microstrip patch with an open-ended stub that is integrated with one PIN diode.

100 reflecting elements. The obtained peak directivity was higher than 22 dBi, and the main beam of the RA was switched between the broadside and 40° off broadside directions.

Figure 3.15 shows other methods that can be used to realize a 1-bit phase shift reconfigurable reflecting element using only one RF switch. The RA element in [25] is a patch with a short-circuited stub loaded with a PIN diode. Changing the ON/OFF state of the diode alters the resonant behavior of the element from a microstrip patch to an inverted-F antenna (PIFA). This state switch produces a 180° difference in the reflected phase. The RA reported in [25] consists of 160×160 microstrip square patches and is designed to operate at the 60 GHz band. The control of the reconfigurable RA is realized using a Field Programmable Gate Array (FPGA). Its main beams can be scanned in both the azimuth and elevation planes within ±25° with a 3-dB beamwidth from 0.55° to 0.63° with an average SLL of −25 dB. The RA represented in [26] has 12×12 elements and the states of the PIN diodes are controlled by an 8-bit microcontroller unit (MCU). The RA element is a slotted metallic square patch with a PIN diode placed in the center of the slot. Switching the ON/OFF state of the PIN changes the current distribution on the patch, which leads to a 180° phase shift. Beam scanning within a ±50° angle range was achieved in both the *xoz* and *yoz* planes at 5 GHz. The RA unit cell in [27] consists of four squared patches with a PIN diode used to connect two diagonal patches that are shorted to the ground. The required 180° phase shift is obtained by changing the polarization of the reflected wave. The RA designed with this unit cell has 20×20 elements and is able to scan the beam within a ±45° angle range with a maximum gain of 23.5 dBi. The sub-array concept can be used to save the number of RF switches and further reduce the system complexity. The design presented in [28] grouped the phase delay lines of a pair of patches with a common microstrip line. The microstrip lines are EM coupled to the rectangular patches through coupling slots. Because the phase delay line is shared, the pair of patches are configured to have the same

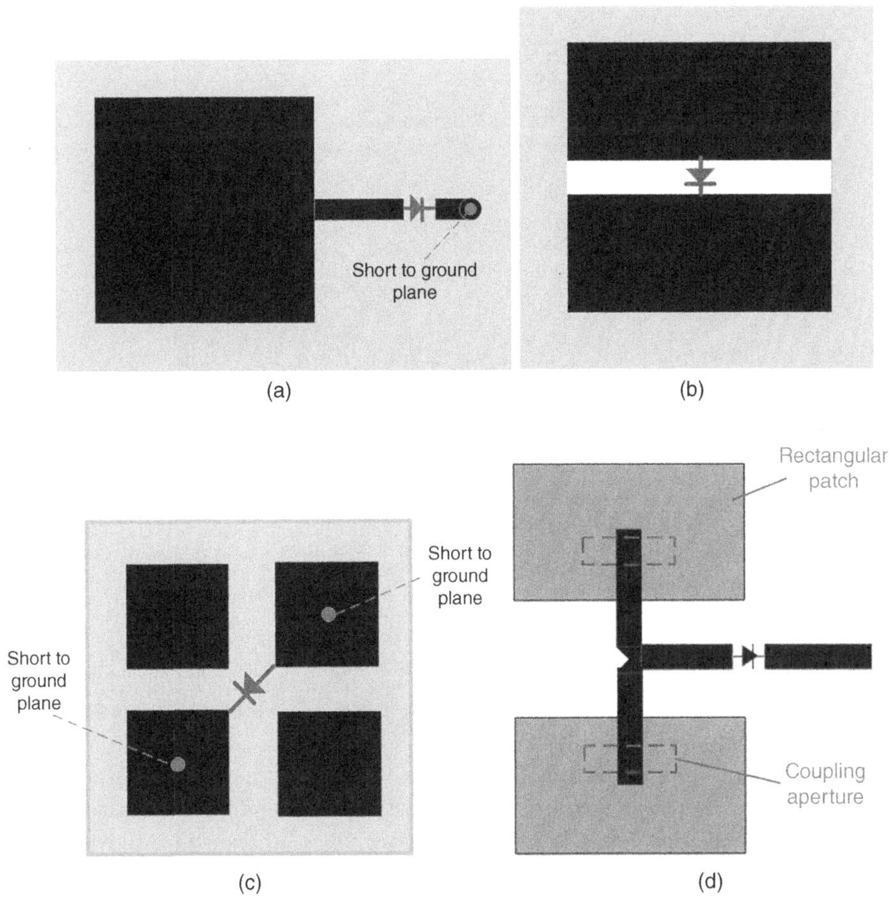

Figure 3.15 Examples of reported 1-bit RA unit cells that use only one RF switch. (a) Microstrip patch with a short circuited stub loaded with a PIN diode. Source: Adapted from [25]. (b) Square patch with a slot in the center. Source: Adapted from [26]. (c) Four squared patches with two diagonal patches shorted. Source: Adapted from [27]. (d) Two rectangular patches shares one RF switch. Source: Adapted from [28].

phase delay. Thus, the beam-steering angle of the RA is limited in order to avoid the appearance of grating lobes. The latter are caused by the fact that the period of the sub-array becomes much greater than half a wavelength. This RA has 244 elements and was designed to operate at 10.4 GHz. Its main beam can be switched between $-5°$, $0°$, and $+5°$ with a peak directivity higher than 26 dBi.

Since the reflecting element of these 1-bit RAs only has two phase states, the phase distribution over the RA aperture can be optimized to improve its gain as well as its beam-steering performance. By adding one constant reference phase, $\Delta\varphi$, in Eq. (3.3), it has been shown that the gain of a 10×10 1-bit RA is improved by more than 2 dB [29]. This reported RA steers the beam within a $\pm 50°$ angle range, as shown in Figure 3.16, with a maximum aperture efficiency of 17.9%.

While the design techniques and examples discussed earlier use only one RF switch per unit cell, there are studies that use multiple RF switches on one unit cell to realize the 1-bit phase shift. For example, the reported RA unit cell in [30] uses four PIN diodes and the 180° phase shift is obtained by reversing the polarization of the reflected waves.

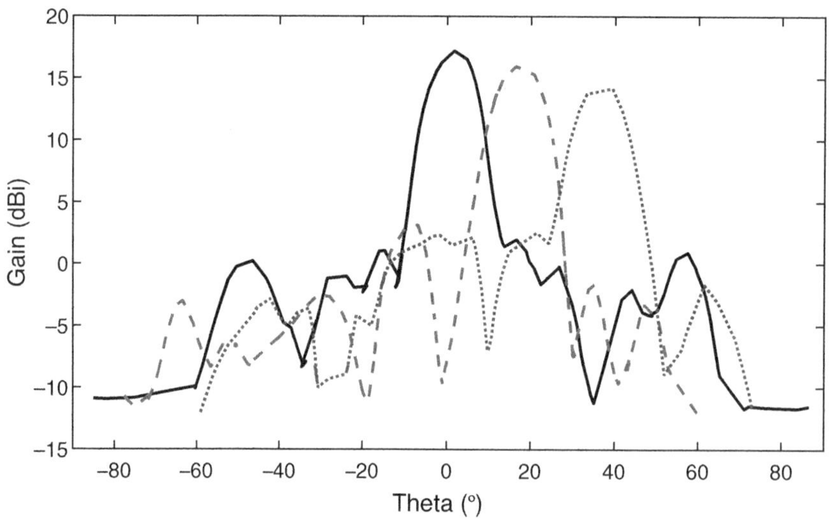

Figure 3.16 The measured beam-scanning performance at 13.25 GHz of the 1-bit RA reported in [29]. Source: Adapted from [29].

3.4.2 1-Bit Transmitarray

The TA configuration is very similar to that of the RA. It consists of an illuminating feed source and a transmitting surface, which is typically a planar array. Different from RA, the TA transforms the incident wave into outgoing waves instead of reflected ones. The TA is designated a planar lens antenna in some papers in the literature. Figure 3.17 shows a typical transmitarray configuration.

A preferred approach to realize a 1-bit TA is to place the RF switches either on the top or/and bottom layer of the TA panel. The TA panel can then be fabricated using conventional multilayer PCB

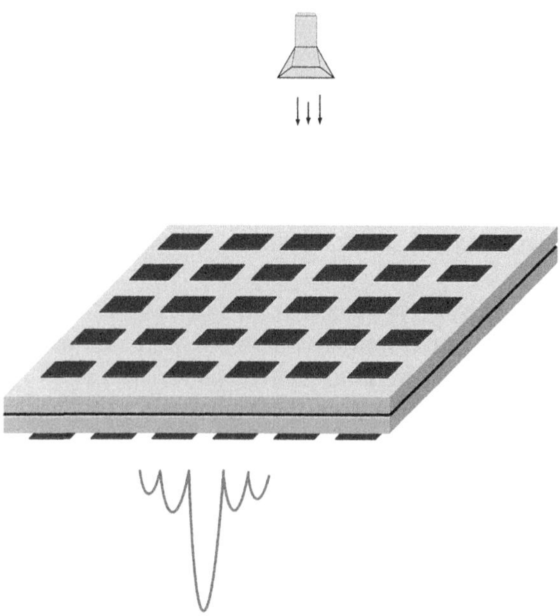

Figure 3.17 Typical transmitarray configuration.

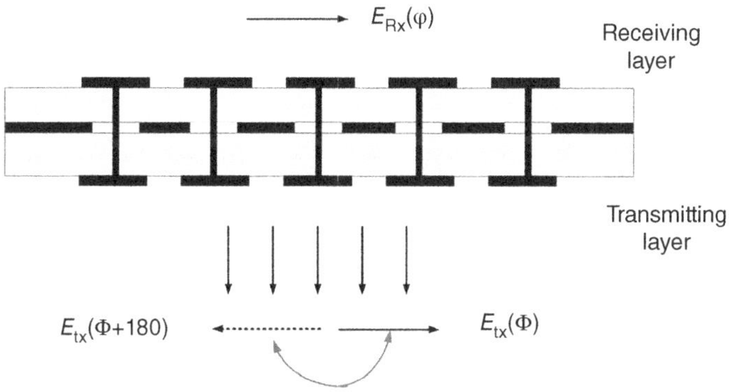

Figure 3.18 The concept of reversing the polarization of the transmitted field to obtain the desired 180° phase shift.

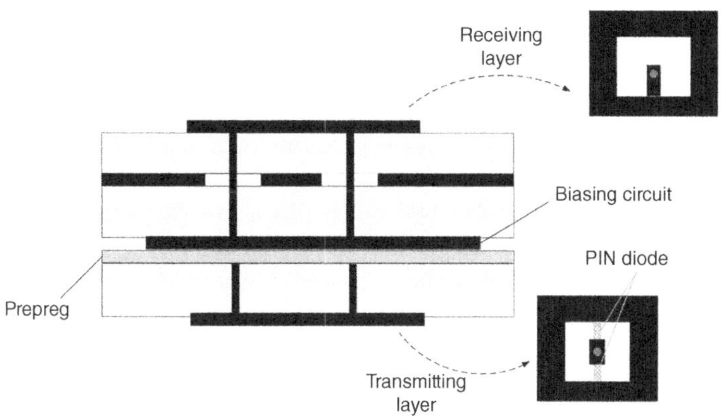

Figure 3.19 The configuration of the 1-bit Ka-band TA unit cell developed in [31]. Source: Adapted from [31].

technology. The phase shift of 180° associated with the 1-bit approach can be obtained by reversing the polarization of the received RF signal on the transmit side of the TA panel. This technique is depicted in Figure 3.18. It has been used in many 1-bit reconfigurable TA designs. As an example, Figure 3.19 shows the configuration of the 1-bit Ka-band TA unit cell reported in [31]. The receiving layer is a rectangular patch with a U-shaped slot, and the transmitting layer is a rectangular patch loaded with an O-shaped slot. Two PIN diodes are placed on the O-shaped slot. The current on the transmitting patch is reversed by switching ON only one PIN diode at a time. This current reversal provides the 180° phase change. The insertion loss of this TA unit cell is less than 1.3 dB for both phase states at the center frequency, 29 GHz. The 3-dB transmission bandwidth is about 11.2%.

A 1-bit Ku-band TA unit cell using an equivalent magnetic dipole element was presented in [32]. Its TA unit cell configuration is shown in Figure 3.20. The receiving layer is a rectangular patch and the transmitting layer is a side-shorted patch that is equivalent to a magnetic dipole. This magnetic dipole has a wide half-power beamwidth in the *E*-plane, which contributes to the wide-angle scanning of the TA. Two PIN diodes are placed at the middle of the side-shorted patch and are biased in opposite states. The shorted patch resonates when the diode is OFF. The 180° phase difference of the transmitted signal is generated by switching the ON/OFF state of the PIN diode. The 1-bit

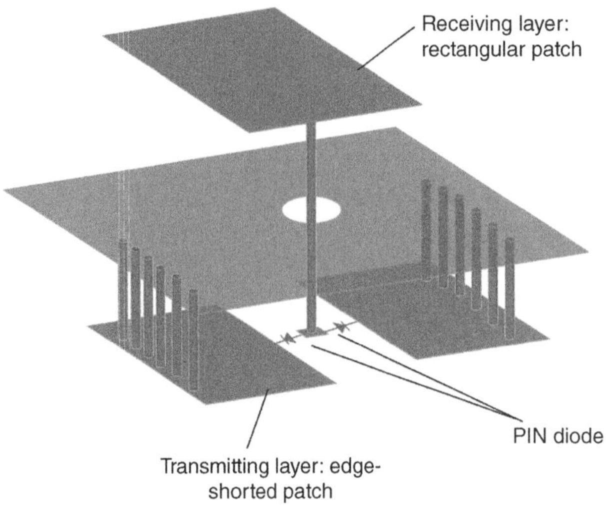

Figure 3.20 The configuration of the 1-bit Ku-band TA unit cell developed in [32] that employs an equivalent magnetic dipole element. Source: Adapted from [32].

TA using this unit cell was reported to be able to scan within ±60° in the *E*-plane with a scan loss of 3.6 dB.

Figure 3.21 shows another two techniques that can be applied to realize a 1-bit TA element. The TA element shown in Figure 3.21a is an FSS-based design. The FSS works as a free-space phase shifter whose phase delays are obtained by adjusting its filter response. A single layered FSS normally provides a limited phase shift range; multiple layered designs would be used to improve the phase shift range. The TA element in this design consists of two substrates separated by an air gap; the PIN diodes are placed in the middle of these two layers. The TA element shown in Figure 3.21b is a receive-transmit unit cell; its 180° phase shift is realized by reversing the polarization of the transmitted signal. Instead of placing the PIN diodes on the same layer as the radiating element, a reconfigurable feed line is used to reverse the current so that the transmitting patch is excited with a 180° phase difference.

3.5 Beam-Switching and Multi-Beam Lens

The section presents some techniques to realize a low-cost beam-switching or multi-beam lens antenna. This type of antenna normally consists of a lens and several passive feed antennas. If all of the feed elements are simultaneously excited, multi-beam radiation is obtained. A beam-switching network that can switch between different RF ports is required in order to switch the beam direction. Note that this beam-switching network does not need to perform any beamforming, i.e. it does not need to provide different phase distributions, e.g. as a Butler Matrix [35, 36] does. The beam-switching network can be designed by using single-pole n-throw (SPnT) switches to realize a multi-level switching. This approach is illustrated in Figure 3.22.

3.5.1 Extended Hemispherical Lens

A simple but very effective approach to obtain a high directivity antenna is to use a lens. A beam in the boresight direction can be generated by placing a feed antenna at the center of the lens. On

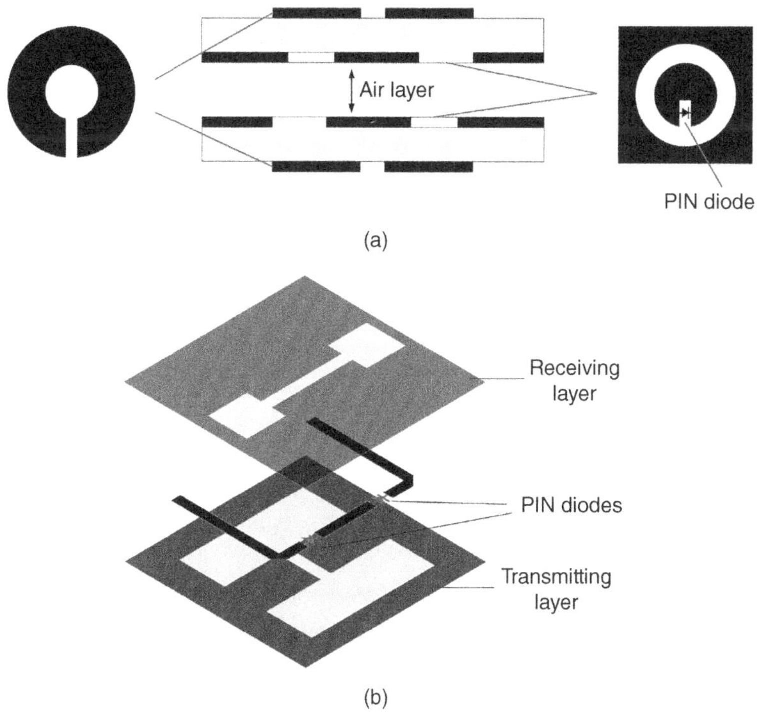

Figure 3.21 Examples of two reported 1-bit TA unit cell designs. (a) An FSS-based TA element. Source: Adapted from [33]. (b) A receive-transmit unit cell with a reconfigurable feed line. Source: Adapted from [34].

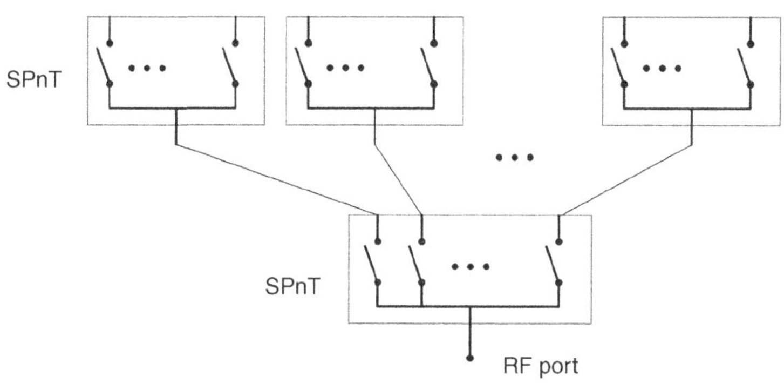

Figure 3.22 Illustration of using multiple SPnT switches to realize a beam-switching network.

the other hand, a tilted beam can be produced by placing the feed antenna off that center position. The angle of the tilted beam is determined by the offset position of the feed. This approach can be extended to achieve beams directed at different angles by introducing multiple offset antennas to feed the dielectric lens and by exciting the RF port that will produce the beam pointed in the desired direction. To obtain a low-cost design, it is desirable to use only one type of dielectric material to fabricate the lens rather than employing a stepped-index arrangement.

A good design example was presented in [37]. A beam-steerable lens antenna was developed for an E-band antenna system for 5G applications. The lens in this design was fed by a 64-element

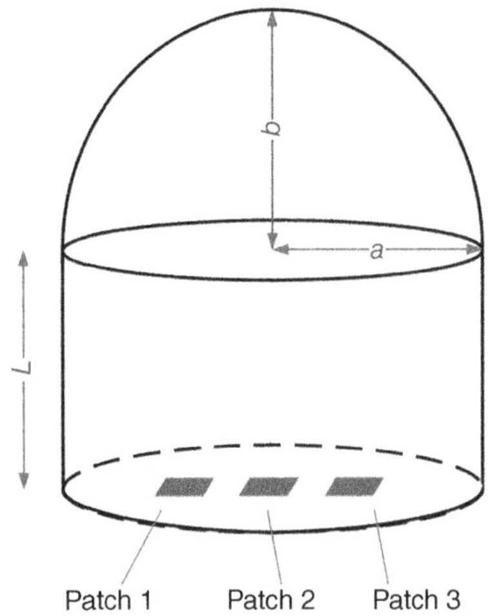

Figure 3.23 Configuration of an extended hemispherical lens antenna using multiple microstrip patches as the feed elements.

array consisting of aperture coupled microstrip-line fed patches. The system produced 64 beams within ±4° in the *E*-plane and ±17° in the *H*-plane.

Higher directivity can be attained by using an extended hemispherical lens that consists of half of an ellipse and a cylindrical extension. Figure 3.23 shows such a configuration that is feed with a linear microstrip patch array antenna. The directivity of a lens antenna is proportional to the dimensions of its lens. The dimensions of an extended hemispherical lens are calculated by using the formulas given in [38]:

$$a = b \times \sqrt{\frac{\varepsilon_r - 1}{\varepsilon_r}} \tag{3.7}$$

where ε_r is the relative permittivity of the material and *a* and *b* are the minor and major axes of the ellipsoidal lens, respectively. The radius of the cylindrical extension is equal to the minor axes of the ellipse and its height is given by the expression:

$$L = \frac{b}{\sqrt{\varepsilon_r}} \tag{3.8}$$

Figure 3.23 depicts an extended hemispherical lens antenna that is fed with a microstrip patch array. The lens was designed using Rexolite in its simulation model. It is a low-cost substrate and has a stable relative permittivity (ε_r) of 2.54 from low frequencies up to mm-wave frequencies. The diameter of the lens was 28 mm. The patch was designed to resonate at the center frequency of 12.5 GHz. The distance between the radiating elements was $0.32\lambda_{12.5\ \mathrm{GHz}}$. Figure 3.24 shows the simulated beam-switching radiation patterns produced by this configuration. The beam generated by this system can be is switched from broadside to 36° in the *E*-plane with a peak gain that varies from 13.3 to 12.2 dBi simply by activating the corresponding element(s).

3.5.2 Luneburg Lens

The Luneburg lens is one of the most well-known lenses used in antenna designs. It has a radially symmetric configuration and it requires a continuous refractive index profile given by the relation [39]

$$n = \sqrt{\epsilon_r} = \sqrt{2 - r^2} \tag{3.9}$$

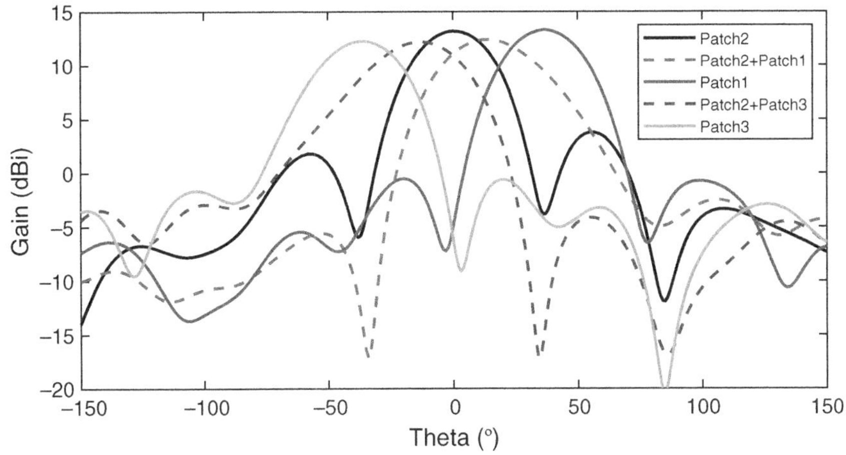

Figure 3.24 Simulated beam-switching radiation patterns of the lens antenna shown in Figure 3.23. The lens is designed to operate at the center frequency of 12.5 GHz. The patches are excited separately to steer its beam.

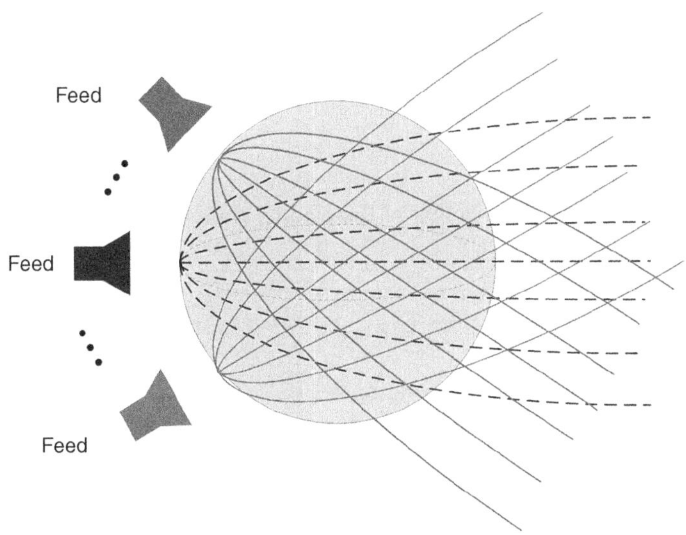

Figure 3.25 The operation of a Luneburg lens.

where n is the refractive index, ϵ_r is the permittivity of the dielectric, and r is the distance normalized against the radius of the lens. The dielectric permittivity ϵ_r of the material at the center of the lens is two. It reduces to one at the surface of the lens to provide a perfect impedance match to free-space. Because of its symmetrical configuration, the Luneburg lens is an ideal lens for high gain and wide-angle scanning applications. A group of feed elements can be placed at the foci of the lens, as demonstrated in Figure 3.25, to steer the output beam.

Because the refractive index of an ideal Luneburg lens is a function of its radius, its construction is complex. To overcome this difficulty in practice, the lens is developed with a stepped-index profile, i.e. the lens is divided into several regions and the permittivity within each region is constant [40]. Figure 3.26 illustrates an example of a stepped-index microwave Luneburg lens that corresponds

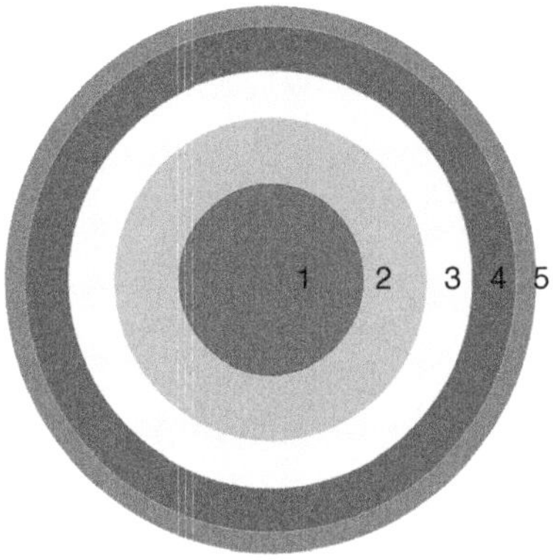

Figure 3.26 Stepped index Luneburg lens.

to the one developed in [41]. It was constructed with five different dielectrics and a half-spherical array that consisted of 46 microstrip antennas was used to feed it. The simulation results show that this Luneburg lens can switch the beam over 360° in the azimuth plane and 74° in the elevation plane.

There are several approaches to realize a Luneburg lens-based beam-steerable antenna. One is to use a dielectric-filled parallel-plate waveguide. The effective index of refraction in a parallel-plate waveguide is given by the relation [42]:

$$n = \sqrt{\epsilon_r - \left(\frac{\lambda_0}{2h}\right)^2} \tag{3.10}$$

where λ_0 is the free-space wavelength and h is the height of the parallel-plate waveguide. If the parallel-plate guide is air-filled, it simplifies to:

$$n = \sqrt{1 - \left(\frac{\lambda_0}{2h}\right)^2} \tag{3.11}$$

An air-filled parallel-plate cylindrical Luneburg lens operating at 30 GHz band was presented in [43]. Because it was a cylindrical shape, the Luneburg lens produced fan-shaped beams, as illustrated in Figure 3.27. The design used 21 antennas as the feeds of the lens, which had a diameter of 100 mm. Its measured 3 dB beamwidths were 8.6° and 68° in the E- and H-plane, respectively. It achieved a measured ±80° scanning coverage with less than 3-dB gain variations.

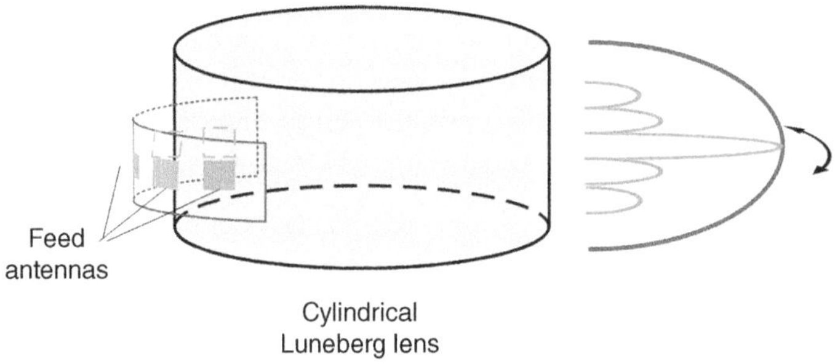

Figure 3.27 The cylindrical Luneburg lens concept reported in [43]. Source: Adapted from [43].

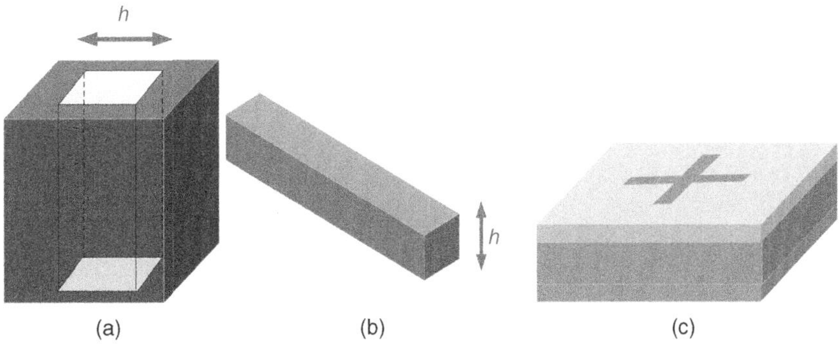

Figure 3.28 Example of the unit cells that have been used to realize a Luneburg lens. Source: Adapted from [45, 46]. (a) Drill a hole in dielectric material and vary its width. (b) Using dielectrics of different heights. (c) Using printed circuit.

A glide-symmetric metallic Luneburg lens was presented in [44]. Vertical metallic inclusions were introduced to gain more flexibility in choosing the distance between its parallel plates. Moreover, the design avoided the use of small air gaps.

Another method to obtain the gradient of the dielectric permittivity required to achieve the Luneburg lens is to use periodic structures. Studies have shown that such a design can be realized by using periodic unit cells either in the form of a PCB structure or a dielectric only unit cell [45, 46]. The dielectric-only design provides the possibility of fabricating the lens with 3D printing technologies. Its effective dielectric permittivity can be extracted from its S-parameters. Figure 3.28 shows some examples of the unit cells that have been used to realize a Luneburg lens.

3.6 Beam-Reconfigurable Array Using Tunable Dielectrics

Recall that the RAs and TAs in Section 3.4 were based on a 1-bit phase quantization which used RF switch(es) to act as equivalent low-bit phase shifters. This design approach led to low-cost beam-steering array antennas. An alternative solution is considered in this section that uses tunable dielectrics to realize a phased array.

The tunable dielectrics have the characteristic that their dielectric permittivity varies with the applied external voltage. Ferroelectric materials and liquid crystals are two types of tunable dielectrics that have been successfully employed in antenna applications. The tunability of a dielectric material is defined as [47]:

$$\text{Tunability} = \frac{|\epsilon_0 - \epsilon_v|}{\epsilon_0} \tag{3.12}$$

where ϵ_0 and ϵ_v are the dielectric permittivity of the material with no bias voltage and with a biased voltage, respectively. It was shown in [48] that the dielectric permittivity of a 400 nm thick thin-film of barium strontium titanate ($Ba_{0.5}Sr_{0.5}TiO_3$) changes from 500 to 275 at 31.6 GHz when the E-field varies from 0 to 35 V/μm and has a maximum loss tangent of 0.12. Ferroelectric materials have a very high dielectric permittivity. Consequently, a multi-layered structure should be used when they are combined with a microstrip transmission line to realize a phase shifter. Figure 3.29 shows two examples of using ferroelectric materials to design a phase shifter. One uses a CPW line and

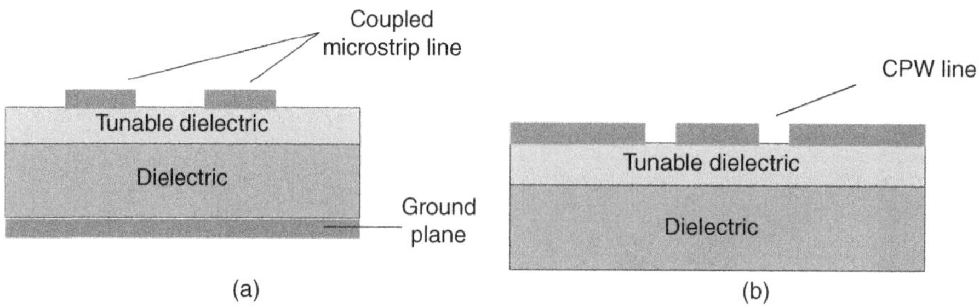

Figure 3.29 Examples of using tunable materials to realize a phase shifter. (a) Phase shifter using a coupled microstrip line. Source: Adapted from [49]. (b) Phase shifter using a CPW line. Source: Adapted from [50].

the other one uses coupled microstrip lines. As shown, a relatively thick dielectric material is sandwiched between the ferroelectric material and the ground plane. This feature decreases the effective dielectric permittivity of the combination.

A figure of merit (FoM) was defined in [51] to characterize the insertion loss of a tunable material-based phase shifter

$$\text{FOM} = \frac{\Delta\text{Phase}}{IL} \tag{3.13}$$

where ΔPhase is the relative phase shift in degrees and *IL* is the maximum insertion loss in dB. The phase shifter presented in [50] had an FoM of 14°/dB at 20 GHz. A linear three-element continuous transverse stub (CTS) array operating at 10 GHz was designed using it. It was demonstrated that the array could scan its beam within a ±20° angle range with a HPBW of about 47°. When the beam was scanned to 20°, the peak gain decreased by 2.2 dB. The scan loss was mainly from the loss of the phase shifter when the bias voltage was applied to the ferroelectric material.

A monolithically BST-integrated Ka-Band RA was presented in [48]. The BST thin film in this design was integrated with a capacitively-loaded patch as depicted in Figure 3.30. The thin-film BST layer was deposited in the gap of the patch. By varying the bias voltage, the dielectric permittivity of the BST thin film and, hence, the effective capacitance of the patch changed. In turn, the resonance frequency of the patch is affected by this change in the capacitance. A phase shift of 250° was obtained with a maximum loss of about 8 dB.

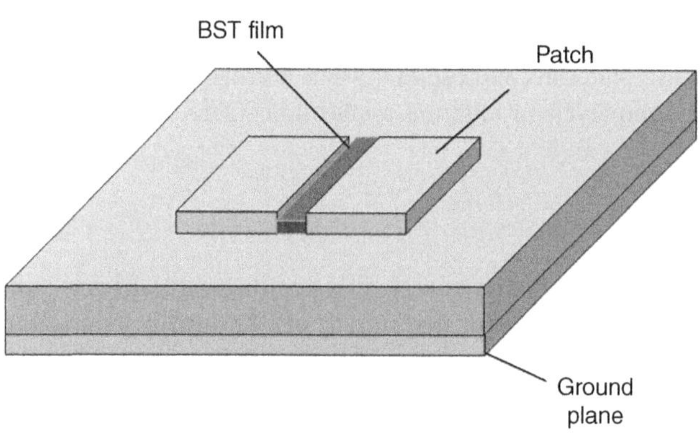

Figure 3.30 The configuration of the patch integrated with a voltage tunable BST thin film developed in [48]. Source: Adapted from [48].

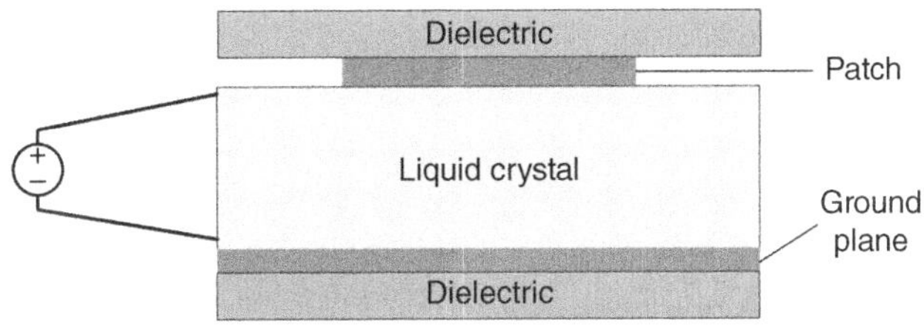

Figure 3.31 The microstrip patch on a liquid crystal substrate reported in [53]. Source: Adapted from [53].

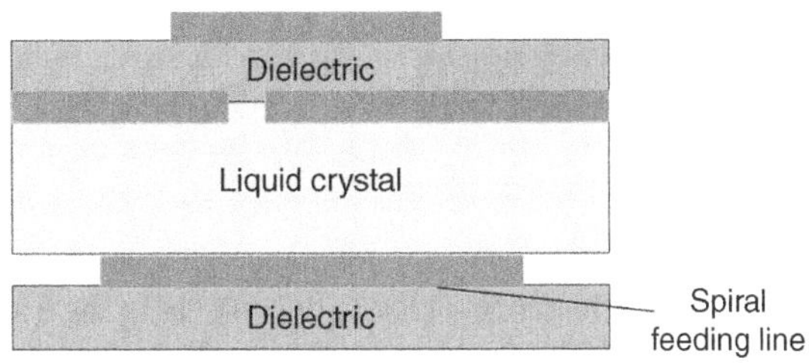

Figure 3.32 The concept of the patch element developed in [54] that was fed by phase delay lines with variable phase delays. Source: Adapted from [54].

Liquid crystals have a much lower permittivity and smaller tunability range than ferroelectrics, but they have a lower loss tangent. The commercial liquid crystal used in [52] had a dielectric permittivity in the range from 2.39 to 3.27 at 19 GHz with a maximum loss tangent of 0.007. By using a liquid crystal whose dielectric constant varied from 2.47 to 3.16 as the substrate of a rectangular patch, a relative phase shift range of 305° at 77 GHz was obtained in [53]. This patch antenna configuration is shown in Figure 3.31. A folded RA was designed using it as its reflecting elements. The same voltage in the range of 0–20 V was applied to the elements in one row of the RA to verify its beam-steering performance. The beam was steered to 17° with a maximum gain of 25.1 dBi.

Liquid crystals can also be used to design phase delay lines with variable phase delays. A liquid crystal was used in [54] as the substrate of the spiral delay line whose total length was $5\lambda_0$. This spiral delay line had an FoM of 105°/dB. It was used to design a feed network of a 2×2 microstrip patch array operating at 17.5 GHz. The results showed that by varying the DC biasing voltage in the range of 0–15 V, the beam of the array could be steered within a 2D ±25° angle range. The concept of this design is illustrated in Figure 3.32.

3.7 Other Techniques to Realize a Beam-Reconfigurable High-Directivity Antenna

3.7.1 Fixed-Frequency Beam-Scanning Leaky-Wave Antennas

Leaky-wave antennas are known for radiating frequency-scanning pencil beams with simple configurations and low-profiles [55]. Recently, there have been developments in the fixed-frequency beam-scanning leaky-wave antennas, where the leaky-wave antennas can be controlled by means of using components such as varactors to electronically steer the beam at a fixed frequency. The principle to design such a leaky-wave antenna is as follows. For a traditional leaky-wave antenna, the angle of the beam at frequency f_n is determined by [56]

$$\theta(f_n) = \sin^{-1}[\beta(f_n)/k_0(f_n)] \tag{3.14}$$

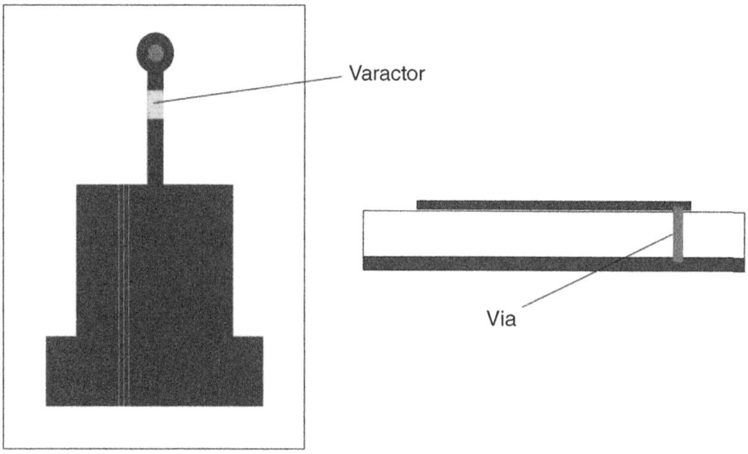

Figure 3.33 The configuration of the unit cell of the fixed-frequency beam-scanning leaky-wave antenna developed by Wang et al. [57]. Source: Adapted from [57].

where $\theta(f_n)$ is the angle of the beam, $\beta(f_n)$ is the phase constant of the leaky-wave mode, and $k_0(f_n)$ is the corresponding propagation constant in the air for frequency f_n. From this equation, it can be seen that if the value of $\beta(f_n)$ can be varied, the leaky-wave antenna then can scan the beam at a fixed frequency.

One method to control $\beta(f_n)$ is to reconfigure the surface impedance of the leaky-wave antenna. In [57], varactors were used to load a corrugated microstrip line and the design concept of the unit cell is shown in Figure 3.33. By changing the DC bias voltage from 0 to 20 V, at 5.8 GHz the normalized value of $\beta(f)/k_0(f)$ varied from −0.3 to 0.32. As a result, the beam scanned in the angle range of 45°. Another approach to realize fixed-frequency scanning of a leaky-wave antenna is through the use of composite right-/left-handed (CRLH) structures, as demonstrated in [58]. The phase constant β of the CRLH structures can be controlled by varying the values of the capacitances and inductances in their corresponding equivalent circuit models. The configuration of the unit cell developed in [58] and its equivalent circuit model are shown in Figure 3.34. On each of the unit cells, there were two varactors placed on the central rectangular patch. By varying the capacitances of these two varactors, both values of L_L and C_L were changed. The measurement results showed that this leaky-wave antenna scanned the beam from −37° to 32° at 5 GHz.

The fixed beam-scanning leaky-wave antenna can also be realized by incorporating reconfigurable phase shifting structures to the leaky-wave antenna. For example, in their design, Lou et al. [59] placed reconfigurable phase-shifting structure between the adjacent radiating elements. The radiating element was interdigital slots and the reconfigurable phase-shifting structure was designed by using fan-shaped open stubs with different sizes and PIN diodes. This antenna achieved beam-scanning range of 25° with a maximum gain of 12.4 dBi at 5.8 GHz.

3.7.2 Mechanical 2D Beam-Steering

RAs have also been used to realized 2D beam-steering designs. For example, a circularly polarized RA was demonstrated in [60]. It was shown that when the incident wave:

$$\vec{E}_{\text{incident}} = (\hat{u}_x + j\hat{u}_y)ae^{-jkz}e^{-j\omega t} \tag{3.15}$$

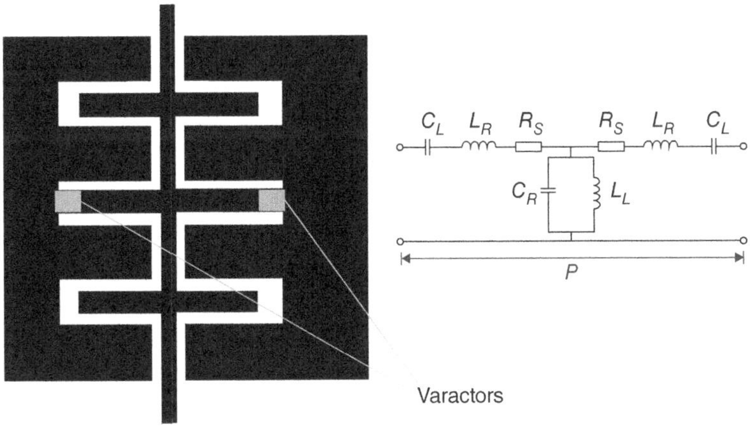

Figure 3.34 The configuration of the CRLH unit cell of the fixed-frequency beam-scanning leaky-wave antenna developed by Chen et al. [58] and its equivalent circuit model. Source: Adapted from [58].

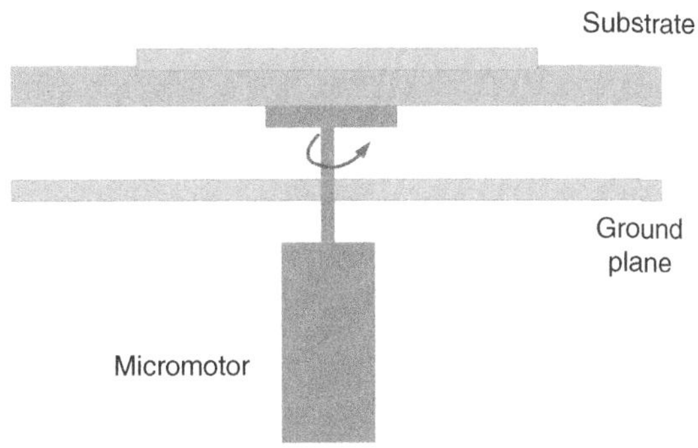

Figure 3.35 The configuration of the mechanically rotated reflectarray unit cell developed in [61]. Source: Adapted from [61].

impinges on a circularly polarized element that is counter-clockwise rotated by φ rad, the phase of the reflected wave is delayed by 2φ rad, i.e.

$$\vec{E}_{\text{reflected}} = (\hat{u}_x - j\hat{u}_y)ae^{jkz}e^{j\omega t}e^{2j\varphi} \tag{3.16}$$

With the recent development of mechanical controlling techniques, it has become feasible to incorporate a miniaturized micro-motor or actuator to control the rotation of a RA unit cell. One advantage of using angularly rotated elements is that their loss is low. A beam-reconfigurable RA that used mechanically rotated elements was reported in [61]. It was designed to operate with a center frequency of 8.3 GHz. The RA unit cell consisted of two concentric dual split rings printed on one substrate layer. The substrate is placed above a ground plane and a micromotor is mounted beneath it. The concept of this mechanically rotated element is shown in Figure 3.35. A continuous 360° phase shift was obtained by physically rotating the antenna element. Its loss was less than 0.2 dB. It was demonstrated that the RA could continuously steer its beam within the ±60° angular range with a maximum scan loss of 3.7 dB. The control of the micromotors was realized by using FPGAs, each of which controlled 30 micromotor drivers. Note that there is a trade-off when using a mechanically rotating system to achieve beam steering rather than an electronic one, i.e. its response time is larger. The reported overall response time of the tested prototype was 0.55 seconds, including a 0.35 seconds system latency and a 0.2 seconds maximum rotation time.

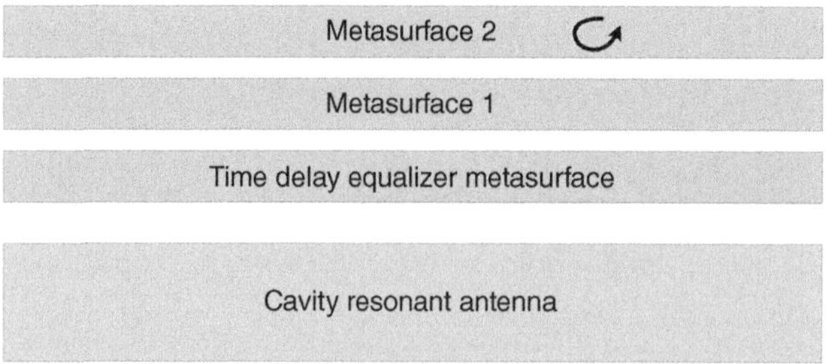

Figure 3.36 The concept of the beam-steering antenna realized by rotating metasurfaces over a cavity resonant antenna reported in [62]. Source: Adapted from [62].

A resonant cavity antenna generally exhibits a medium-to-high peak gain. By placing metasurface layers above and in the near field of a resonant cavity antenna, studies [62, 63] have demonstrated that 2D beam-steering can be achieved by mechanically rotating the metasurface layers while keeping the resonant cavity antenna in a fixed position. Figure 3.36 depicts the concept of this antenna. It has multiple layers. The top two layers are identical passive metasurfaces that can be rotated. They are placed within the near field of a resonant cavity antenna. Between the metasurface layer and the resonant cavity antenna, there is a time-delay equalizer metasurface. It produces a near field having a uniform phase distribution. The two metasurfaces can be rotated independently around the antenna axis. The metasurfaces were designed to provide a spatial sawtooth time delay which transforms the phase distribution of the near-field and leads to a tilted beam. According to the measurement results provided in [62], the beam of the antenna can be steered within 0°–360° in the azimuth plane and 0°–51° in the elevation plane with a peak realized gain of 19.4 dBi when the top metasurface layer is rotated. The gain variation during the beam-scanning is only 1.9 dB when the beam is scanned within the angle range of 0°–46° in the elevation plane.

Two layers of phase-shifting surfaces were placed in the near-field region of a dual-polarized continuous CTS array in [64]. Each phase-shifting surface consisted of four layers of periodic unit cells. The CTS array radiated a quasi-plane wave in its near-field region. By simultaneously rotating the two phase-shifting surfaces, the antenna scanned the beam in the angle range ±40° in the elevation plane and over 360° in the azimuth plane with high directivity.

3.7.3 Beam-Steering Fabry–Perot Resonator Antennas

A Fabry–Perot (FP) cavity antenna consists of a feed antenna located in a resonant cavity formed between a reflector and a partially reflective surface (PRS). As a resonant cavity antenna, an FP antenna shows medium to high directivity. Figure 3.37 shows the configuration of a typical FP cavity antenna.

Ray optics theory can be used to explain the operating principles of a FP cavity antenna. The directivity of the antenna is significantly increased through multiple reflections between the ground plane and the PRS. Resonance is attained if the cavity height of the FP cavity antenna satisfies the condition [65]:

$$h = \frac{\lambda}{4\pi}(\phi_{\text{PRS}} - \pi) + \frac{\lambda}{2} \times N, \quad N = 0,1,2,\ldots \tag{3.17}$$

where h is the cavity height, λ is the wavelength of the operating frequency, and ϕ_{PRS} is the reflection phase of the PRS.

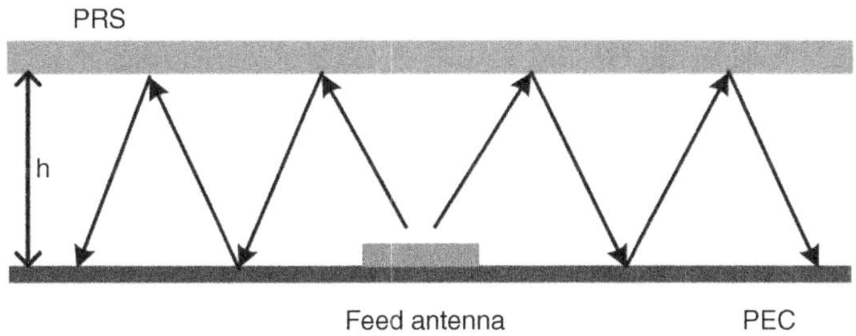

Figure 3.37 Configuration of a classic Fabry–Perot cavity antenna.

There have been some studies on the use of a tunable high impedance surface (HIS) to realize an electronically beam-steering FP cavity antenna. For instance, a fixed frequency beam-steering one-dimensional FP leaky-wave antenna was presented in [66]. This antenna consists of a parallel-plate waveguide loaded with a planar PRS and a tunable HIS loaded with varactors. This arrangement is shown in Figure 3.38. The operating principle of this tunable design is that the effective resonant length of the HIS is changed by varying the capacitance of the varactors. The leaky-mode phase constant at a fixed frequency increases when the capacitance of the varactors increases. Another critical parameter is the height of the cavity; it determines the scanning range of the antenna. As calculated in [66], the optimum height of the cavity should be chosen in the way that the leaky-mode cut-off condition matches with the minimum capacitance of the varactors. The resulting antenna was designed to operate at 5.6 GHz and the radiating aperture had a length of $5\lambda_0$. Continuous beam scanning from 9° to 30° with a maximum gain of 12.95 dBi was obtained when the bias voltage of the varactors was varied from 18 to 5 V.

This 1D design concept was extended to a 2D FP antenna that can simultaneously scan a pencil beam in both the elevation and azimuth planes [67]. Figure 3.39 depicts the configuration of this design. It consists of a planar passive PRS and a tunable HIS layer separated by a distance H. A stacked patch is placed at the center as the feed source. The tunable HIS is divided into four sectors. Each sector can be independently configured to three different operating states described as a scanning region, a cut-off region, and a splitting condition. The antenna can then perform TM scanning, TE scanning, hybrid scanning, and no scanning. As was demonstrated, the developed 2D FP antenna can radiate in the broadside direction and can perform continuous elevation scanning within the angle range ±15° around the broadside direction at eight selectable azimuthal directions. The directivity varied from 11.6 to 16.9 dBi with a maximum aperture efficiency of 50%.

Figure 3.38 The configuration of the 1D beam-steering FP antenna reported in [66]. Source: Adapted from [66].

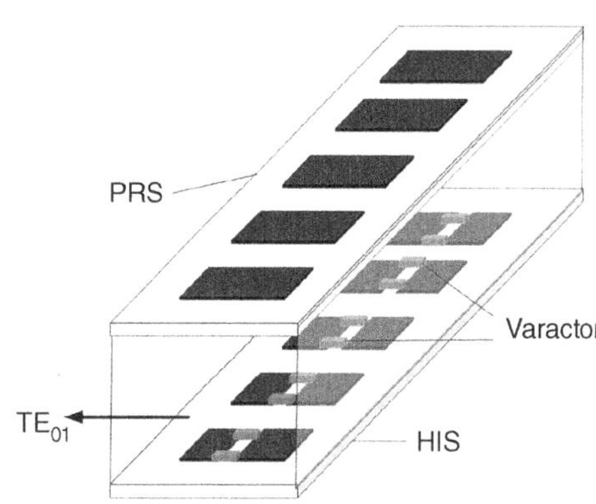

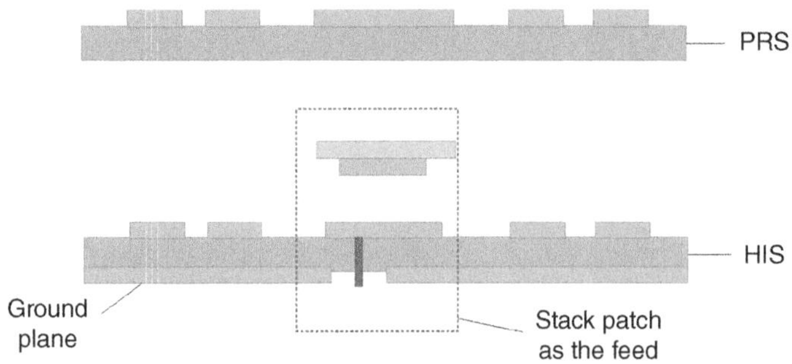

Figure 3.39 The configuration of the 2D beam-steering FP antenna reported in [67]. Source: Adapted from [67].

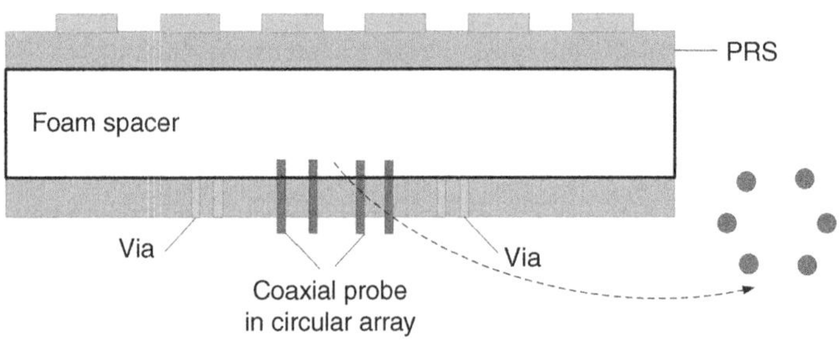

Figure 3.40 The configuration of the 2D beam-steering FP antenna reported in [68]. Source: Adapted from [68].

Another method to realize a 2D beam steering FP cavity antenna is to combine frequency scanning and fixed frequency beam-steering as demonstrated in [68]. The reported antenna consists of a PRS layer placed above a ground plane and a circular array of six vertical coaxial probes. This arrangement is illustrated in Figure 3.40. The PRS has a period much smaller than the wavelength of its operating frequency. The radial distance between the feeds is $0.5\lambda_0$. Via rings are introduced to improve the impedance matching to the feed. Cylindrical leaky waves are excited by the feeds, which are fed with equal phases. The resulting far field has a conical pattern and achieves frequency scanning from 18 to 23 GHz in the elevation plane. When the feeds are excited with different phases, the beam of the FP antenna is steered in the azimuth plane.

3.7.4 Beam-Reconfigurable Arrays Using Low-Cost Beamforming Networks

The Butler matrix and the Rotman lens are two types of well-known BFNs. They have N inputs and M outputs. Each of the input ports is isolated from the other input ports. The outputs of the BFN are connected to an array antenna. By exciting one input port, the array is excited with a corresponding phase distribution and the beam of the array antenna is steered to a specific direction. If multiple beams are required, multiple input ports need to be simultaneously excited. There are several reported works on the integration of the BFN with the array antenna using substrate-integrated waveguide (SIW) technology [35, 69, 70].

The pillbox antenna is a cylindrical reflector sandwiched between two parallel plates. It has been used for communication with base stations. It has a high gain with a cosecant-squared pattern

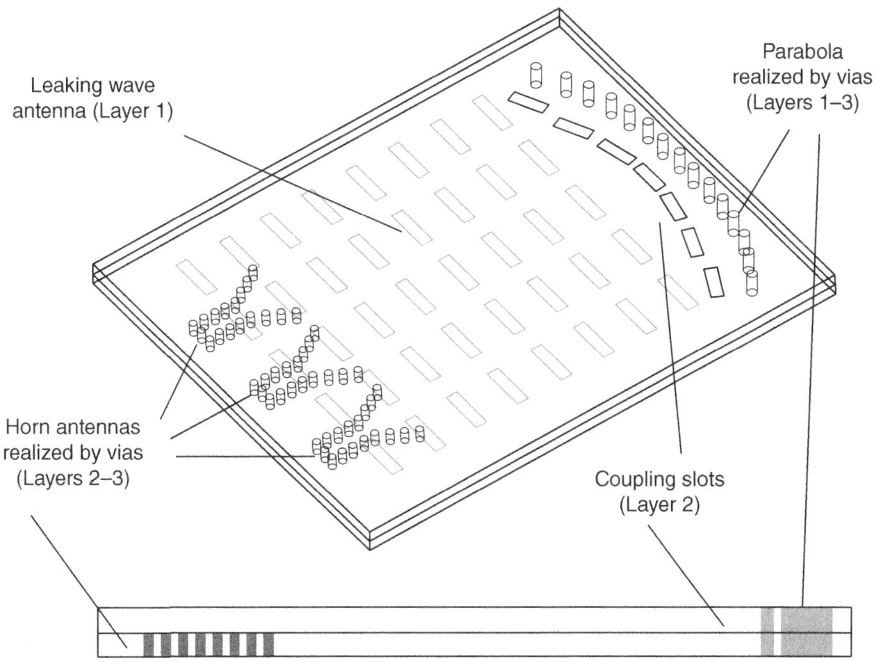

Figure 3.41 The configuration of the SIW pillbox leaky-wave antenna presented in [72]. Source: Adapted from [72].

in the elevation plane and a broad beam in the azimuth plane [71]. The concept of the pillbox antenna has been applied recently to the design of the feed network of SIW-based leaky wave and slot antennas [72–74]. Since it can be fabricated by using conventional PCB techniques, the overall manufacturing cost is not high. This aspect makes the pillbox antenna an attractive solution for mm-wave beam-switching or multi-beam arrays.

Figure 3.41 shows the concept of the SIW pillbox leaky-wave antenna reported in [72, 74]. The antenna has two substrates with three copper layers. The first layer is the leaky-wave antenna, which is a planar array of rectangular slots. A horn antenna is used as the feed antenna and is placed on the bottom layer. The horn antenna can be realized by using vias, i.e. with SIW technology. Another array of vias is used to form a parabola, which realizes the cylindrical reflector of the pillbox antenna. The incident wave from the feed is a quasi-TEM mode. It is reflected by the parabola and then transferred through aperture coupling to the substrate where the leaky-wave antenna is printed. This arrangement forms a quasi-optical system that can also transform cylindrical waves into plane waves. Since the antenna is a leaky-wave system, it has a frequency-scanning characteristic in the elevation plane. When the feed is placed at an offset position and the leaky-wave array is fed with a different phase distribution, the output beam can be pointed at different angles. Note that spillover must be considered when choosing the feed position taking into account the profile of the parabola. If it is not, the SLLs will increase. By combining the frequency-scanning and multiple feeds, the SIW pillbox leaky-wave antenna can be designed to have a high directivity beam pointing at different elevation angles. The SIW pillbox leaky-wave antenna developed in [74] consisted of seven integrated horn antennas as the feeds. It operated from 23.6 to 26.0 GHz and had a peak gain higher than 20 dBi. The measurement results showed that this antenna could scan within ±30° in the azimuth plane and 20° in the elevation plane.

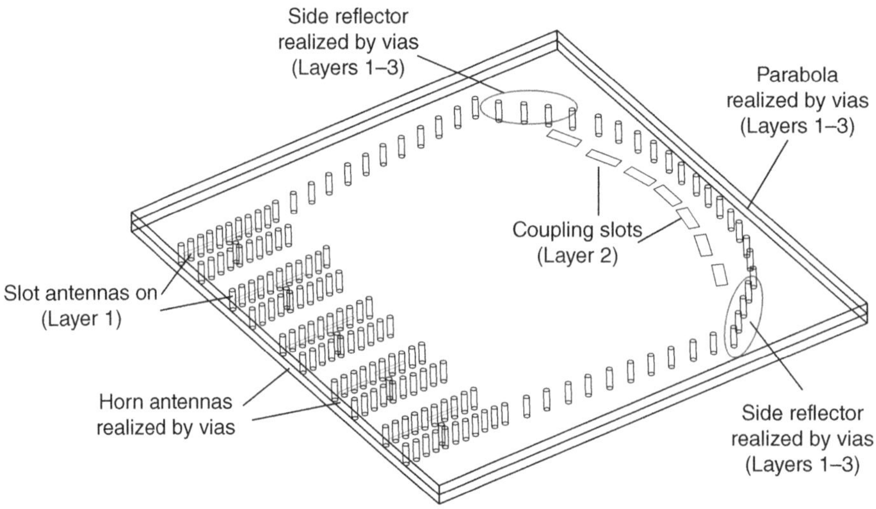

Figure 3.42 The configuration of the SIW pillbox slot array antenna presented in [73]. Source: Adapted from [73].

Figure 3.42 shows the concept of the SIW pillbox slot array antenna developed in [73]. Its configuration is similar to the one shown in Figure 3.41. The main difference is that the radiating antenna is an array of slot antennas. The design reported in [73] operates in the 28 GHz band. It consists of a 2×7 SIW slot antenna array and five feeds that are formed by vias. Two side reflectors are introduced to reduce the SLLs. Five beams within the angle range of $\pm 45°$ in the elevation plane were obtained with SLLs lower than -12 dB. The measured efficiency was 80% when the beam was pointing at boresight and 50% for the $\pm 45°$ beams.

Instead of using a pillbox antenna to transform cylindrical waves into plane waves, an SIW flat lens can be used to convert the cylindrical wave-front from an SIW horn into a planar wave-front as demonstrated in [75]. The plane waves are then used to excite a CTS array. By placing SIW horns at offset locations with respect to the focal point of the reflector, the beam generated by the CTS array was scanned in the elevation plane.

3.8 Summary

Recent progress on the techniques that can be applied to design low-cost beam-reconfigurable and multi-beam directional antennas were discussed in this chapter. They included using low-bit phase quantization for RA and TA designs, employing AFSS or parasitic elements, introducing mechanically rotated metasurfaces, using lenses of multiple feeds, integrating tunable materials to the antenna design, combining tunable HIS with the FP cavity antenna, using low-cost BFN, designing fixed-frequency beam-scanning leaky-wave antenna, and exploiting SIW-based technologies. In summary, the cost of the beam-configurable high-directivity antenna can be reduced from two aspects: reducing the number of active antenna elements and avoiding the use of expensive RF components. This is a research area full of possibilities and needs to be further studied in order to overcome the shortcomings of different techniques. The goal of the research is to obtain a beam-steering performance that is comparable to a conventional phased array with low-complexity, low-cost, and promising directivity and efficiency.

References

1 B. A. Munk, *Frequency Selective Surfaces: Theory and Design*. John Wiley and Sons, 2000.
2 C. Gu, S. Gao, B. Sanz-Izquierdo, E. A. Parker, F. Qin, H. Xu, J. C. Batchelor, X. Yang, and Z. Cheng, "3-D coverage beam-scanning antenna using feed array and active frequency-selective surface," *IEEE Trans. Antennas Propag.*, vol. 65, no. 11, pp. 5862–5870, 2017.
3 M. Bouslama, M. Traii, T. A. Denidni, and A. Gharsallah, "Beam-switching antenna with a new reconfigurable frequency selective surface," *IEEE Antennas Wirel. Propag. Lett.*, vol. 15, pp. 1159–1162, 2016.
4 G. H. Elzwawi, H. H. Elzwawi, M. M. Tahseen, and T. A. Denidni, "Frequency selective surface-based switched-beamforming antenna," *IEEE Access*, vol. 6, pp. 48042–48050, 2018.
5 A. Edalati and T. A. Denidni, "Frequency selective surfaces for beam-switching applications," *IEEE Trans. Antennas Propag.*, vol. 61, no. 1, pp. 195–200, 2013.
6 J. Li, Q. Zeng, R. Liu, and T. A. Denidni, "A gain enhancement and flexible control of beam numbers antenna based on frequency selective surfaces," *IEEE Access*, vol. 6, pp. 6082–6091, 2018.
7 Q. Luo, S. Gao, B. S. Izquierdo, X. Yang, X. Ren, and J. Wu, "Low-cost smart antenna using active frequency selective surfaces," in *2019 International Symposium on Antennas and Propagation (ISAP)*, pp. 1–3, 2019.
8 Q. Luo, S. (Shichang) Gao, W. Liu, and C. Gu, *Low-cost Smart Antennas*. John Wiley and Sons, Ltd, 2019.
9 H. T. Liu, S. Gao, and T. H. Loh, "Electrically small and low cost smart antenna for wireless communication," *IEEE Trans. Antennas Propag.*, vol. 60, no. 3, pp. 1540–1549, 2012.
10 J. Lu, D. Ireland, and R. Schlub, "Dielectric embedded ESPAR (DE-ESPAR) antenna array for wireless communications," *IEEE Trans. Antennas Propag.*, vol. 53, no. 8, pp. 2437–2443, 2005.
11 S. Zhang, G. H. Huff, J. Feng, and J. T. Bernhard, "A pattern reconfigurable microstrip parasitic array," *IEEE Trans. Antennas Propag.*, vol. 52, no. 10, pp. 2773–2776, 2004.
12 X. S. Yang, B. Z. Wang, S. H. Yeung, Q. Xue, and K. F. Man, "Circularly polarized reconfigurable crossed-Yagi patch antenna," *IEEE Antennas Propag. Mag.*, vol. 53, no. 5, pp. 65–80, 2011.
13 S. Jeong and W. J. Chappell, "A city-wide smart wireless sewer sensor network using parasitic slot array antennas," *IEEE Antennas Wirel. Propag. Lett.*, vol. 9, pp. 760–763, 2010.
14 M. Tang, Y. Duan, Z. Wu, X. Chen, M. Li, and R. W. Ziolkowski, "Pattern reconfigurable, vertically polarized, low-profile, compact, near-field resonant parasitic antenna," *IEEE Trans. Antennas Propag.*, vol. 67, no. 3, pp. 1467–1475, 2019.
15 H. Liu, S. Gao, and T. Loh, "Small director array for low-profile smart antennas achieving higher gain," *IEEE Trans. Antennas Propag.*, vol. 61, no. 1, pp. 162–168, 2013.
16 Y. Yang and X. Zhu, "A wideband reconfigurable antenna with 360° beam steering for 802.11ac WLAN applications," *IEEE Trans. Antennas Propag.*, vol. 66, no. 2, pp. 600–608, 2018.
17 R. Mehmood and J. W. Wallace, "Diminishing returns with increasing complexity in reconfigurable aperture antennas," *IEEE Antennas Wirel. Propag. Lett.*, vol. 9, pp. 299–302, 2010.
18 R. Mehmood and J. W. Wallace, "Interference suppression using parasitic reconfigurable aperture (RECAP) antennas," in *2011 International Workshop on Antenna Technology (iWAT)*, pp. 82–85, 2011.
19 Q. Luo, S. Gao, C. Zhang, D. Zhou, T. Chaloun, W. Menzel, V. Ziegler, and M. Sobhy, "Design and analysis of a reflectarray using slot antenna elements for Ka-band SatCom," *IEEE Trans. Antennas Propag.*, vol. 63, no. 4, pp. 1365–1374, 2015.

20 Q. Luo, S. Gao, W. Li, M. Sobhy, I. Bakaimi, C. H. K. de Groot, B. Hayden, I. Reaney, and X. Yang, "Multibeam dual-circularly polarized reflectarray for connected and autonomous vehicles," *IEEE Trans. Veh. Technol.*, vol. 68, no. 4, pp. 3574–3585, 2019.

21 J. Huang and J. A. Encinar, *Reflectarray Antennas*. Wiley-Blackwell, 2007.

22 O. Mohammed Bakr and M. Johnson, "Impact of phase and amplitude errors on array performance," Technical Report UCB/EECS-2009-1, EECS Department, University of California, Berkeley, Jan 2009.

23 H. Yang, F. Yang, S. Xu, M. Li, X. Cao, J. Gao, and Y. Zheng, "A study of phase quantization effects for reconfigurable reflectarray antennas," *IEEE Antennas Wirel. Propag. Lett.*, vol. 16, pp. 302–305, 2017.

24 O. Bayraktar, O. A. Civi, and T. Akin, "Beam switching reflectarray monolithically integrated with RF MEMS switches," *IEEE Trans. Antennas Propag.*, vol. 60, no. 2, pp. 854–862, 2012.

25 H. Kamoda, T. Iwasaki, J. Tsumochi, T. Kuki, and O. Hashimoto, "60-GHz electronically reconfigurable large reflectarray using single-bit phase shifters," *IEEE Trans. Antennas Propag.*, vol. 59, no. 7, pp. 2524–2531, 2011.

26 J. Han, L. Li, G. Liu, Z. Wu, and Y. Shi, "A wideband 1 bit 12 × 12 reconfigurable beam-scanning reflectarray: Design, fabrication, and measurement," *IEEE Antennas Wirel. Propag. Lett.*, vol. 18, no. 6, pp. 1268–1272, 2019.

27 Y. Zhong, H. Luyen, and N. Behdad, "A 1-bit, low-complexity, 20×20-element electronically reconfigurable reflectarray antenna," in *2019 13th European Conference on Antennas and Propagation (EuCAP)*, pp. 1–3, 2019.

28 E. Carrasco, M. Barba, and J. A. Encinar, "X-band reflectarray antenna with switching-beam using pin diodes and gathered elements," *IEEE Trans. Antennas Propag.*, vol. 60, no. 12, pp. 5700–5708, 2012.

29 H. Yang, F. Yang, S. Xu, Y. Mao, M. Li, X. Cao, and J. Gao, "A 1-bit 10x10 reconfigurable reflectarray antenna: Design, optimization, and experiment," *IEEE Trans. Antennas Propag.*, vol. 64, no. 6, pp. 2246–2254, 2016.

30 M. Zhang, S. Gao, Y. Jiao, J. Wan, B. Tian, C. Wu, and A. Farrall, "Design of novel reconfigurable reflectarrays with single-bit phase resolution for Ku-band satellite antenna applications," *IEEE Trans. Antennas Propag.*, vol. 64, no. 5, pp. 1634–1641, 2016.

31 L. Di Palma, A. Clemente, L. Dussopt, R. Sauleau, P. Potier, and P. Pouliguen, "1-Bit reconfigurable unit cell for Ka-band transmitarrays," *IEEE Antennas Wirel. Propag. Lett.*, vol. 15, pp. 560–563, 2016.

32 Y. Wang, S. Xu, F. Yang, and M. Li, "A novel 1 bit wide-angle beam scanning reconfigurable transmitarray antenna using an equivalent magnetic dipole element," *IEEE Trans. Antennas Propag.*, vol. 68, no. 7, pp. 5691–5695, 2020.

33 B. D. Nguyen and C. Pichot, "Unit-cell loaded with pin diodes for 1-bit linearly polarized reconfigurable transmitarrays," *IEEE Antennas Wirel. Propag. Lett.*, vol. 18, no. 1, pp. 98–102, 2019.

34 M. Wang, S. Xu, F. Yang, and M. Li, "Design and measurement of a 1-bit reconfigurable transmitarray with subwavelength H-shaped coupling slot elements," *IEEE Trans. Antennas Propag.*, vol. 67, no. 5, pp. 3500–3504, 2019.

35 Q. Yang, S. Gao, Q. Luo, L. Wen, X. Ren, J. Wu, Y. Ban, and X. Yang, "A low complexity 16x16 butler matrix design using eight-port hybrids," *IEEE Access*, vol. 7, pp. 177864–177873, 2019.

36 Q. Yang, S. Gao, Q. Luo, L. Wen, Y. L. Ban, X. Ren, J. Wu, X. Yang, and Y. Liu, "Millimeter-wave dual-polarized differentially fed 2-D multibeam patch antenna array," *IEEE Trans. Antennas Propag.*, vol. 68, no. 10, pp. 7007–7016, 2020.

37 J. Ala-Laurinaho, J. Aurinsalo, A. Karttunen, M. Kaunisto, A. Lamminen, J. Nurmiharju, A. V. Räisänen, J. Säily, and P. Wainio, “2-D beam-steerable integrated lens antenna system for 5G *e* -band access and backhaul,” *IEEE Trans. Microw. Theory Tech.*, vol. 64, no. 7, pp. 2244–2255, 2016.

38 X. Wu, G. V. Eleftheriades, and T. E. van Deventer-Perkins, “Design and characterization of single- and multiple-beam MM-wave circularly polarized substrate lens antennas for wireless communications,” *IEEE Trans. Microw. Theory Tech.*, vol. 49, no. 3, pp. 431–441, 2001.

39 G. Peeler and H. Coleman, “Microwave stepped-index luneberg lenses,” *IRE Trans. Antennas Propag.*, vol. 6, no. 2, pp. 202–207, 1958.

40 H. Mathis, “Checking design of stepped luneberg lens,” *IRE Trans. Antennas Propag.*, vol. 8, no. 3, pp. 342–343, 1960.

41 C. Zhao, S. Yang, K. Liu, Y. Chen, and S. Qu, “A circularly polarized luneberg lens antenna for half-space beam coverage,” in *2019 IEEE 2nd International Conference on Electronics Technology (ICET)*, pp. 427–430, 2019.

42 X. Wu and J. Laurin, “Fan-beam millimeter-wave antenna design based on the cylindrical luneberg lens,” *IEEE Trans. Antennas Propag.*, vol. 55, no. 8, pp. 2147–2156, 2007.

43 C. Hua, X. Wu, N. Yang, and W. Wu, “Air-filled parallel-plate cylindrical modified luneberg lens antenna for multiple-beam scanning at millimeter-wave frequencies,” *IEEE Trans. Microw. Theory Tech.*, vol. 61, no. 1, pp. 436–443, 2013.

44 O. Quevedo-Teruel, J. Miao, M. Mattsson, A. Algaba-Brazalez, M. Johansson, and L. Manholm, “Glide-symmetric fully metallic luneburg lens for 5G communications at Ka-band,” *IEEE Antennas Wirel. Propag. Lett.*, vol. 17, no. 9, pp. 1588–1592, 2018.

45 Y. Li, L. Ge, M. Chen, Z. Zhang, Z. Li, and J. Wang, “Multibeam 3-D-printed luneburg lens fed by magnetoelectric dipole antennas for millimeter-wave mimo applications,” *IEEE Trans. Antennas Propag.*, vol. 67, no. 5, pp. 2923–2933, 2019.

46 C. Pfeiffer and A. Grbic, “A printed, broadband luneburg lens antenna,” *IEEE Trans. Antennas Propag.*, vol. 58, no. 9, pp. 3055–3059, 2010.

47 M. F. Iskander, Z. Yun, Z. Zhang, R. Jensen, and S. Redd, “Design of a low-cost 2-D beam-steering antenna using ferroelectric material and CTS technology,” *IEEE Trans. Microw. Theory Tech.*, vol. 49, no. 5, pp. 1000–1003, 2001.

48 K. K. Karnati, M. E. Trampler, and X. Gong, “A monolithically bst-integrated K_a -band beam-steerable reflectarray antenna,” *IEEE Trans. Antennas Propag.*, vol. 65, no. 1, pp. 159–166, 2017.

49 R. R. Romanofsky, “Advances in scanning reflectarray antennas based on ferroelectric thin-film phase shifters for deep-space communications,” *Proc. IEEE*, vol. 95, no. 10, pp. 1968–1975, 2007.

50 W. Kim, M. F. Iskander, and W. D. Palmer, “An integrated phased array antenna design using ferroelectric materials and the continuous transverse stub technology,” *IEEE Trans. Antennas Propag.*, vol. 54, no. 11, pp. 3095–3105, 2006.

51 J. F. White, “Figure of merit for varactor reflection type phase shifters,” in *IEEE Northeast Electron. Eng. Meeting Rec*, pp. 206–207, Nov 1965.

52 H. Tesmer, R. Reese, E. Polat, M. Nickel, R. Jakoby, and H. Maune, “Liquid-crystal-based fully dielectric lateral wave beam-steering antenna,” *IEEE Antennas Wirel. Propag. Lett.*, vol. 18, no. 12, pp. 2577–2581, 2019.

53 S. Bildik, S. Dieter, C. Fritzsch, W. Menzel, and R. Jakoby, “Reconfigurable folded reflectarray antenna based upon liquid crystal technology,” *IEEE Trans. Antennas Propag.*, vol. 63, no. 1, pp. 122–132, 2015.

54 O. H. Karabey, A. Gaebler, S. Strunck, and R. Jakoby, “A 2-D electronically steered phased-array antenna with 2x2 elements in LC display technology,” *IEEE Trans. Microw. Theory Tech.*, vol. 60, no. 5, pp. 1297–1306, 2012.

55 D. R. Jackson and A. A. Oliner, *Leaky-Wave Antennas*, Chapter 7, John Wiley and Sons, Ltd, 2008, pp. 325–367.

56 S.-L. Chen, D. K. Karmokar, Z. Li, P.-Y. Qin, R. W. Ziolkowski, and Y. J. Guo, “Circular-polarized substrate-integrated-waveguide leaky-wave antenna with wide-angle and consistent-gain continuous beam scanning,” *IEEE Trans. Antennas Propag.*, vol. 67, no. 7, pp. 4418–4428, 2019.

57 M. Wang, H. F. Ma, H. C. Zhang, W. X. Tang, X. R. Zhang, and T. J. Cui, “Frequency-fixed beam-scanning leaky-wave antenna using electronically controllable corrugated microstrip line,” *IEEE Trans. Antennas Propag.*, vol. 66, no. 9, pp. 4449–4457, 2018.

58 S.-L. Chen, D. K. Karmokar, Z. Li, P.-Y. Qin, R. W. Ziolkowski, and Y. J. Guo, “Continuous beam scanning at a fixed frequency with a composite right-/left-handed leaky-wave antenna operating over a wide frequency band,” *IEEE Trans. Antennas Propag.*, vol. 67, no. 12, pp. 7272–7284, 2019.

59 T. Lou, X.-X. Yang, H. Qiu, Q. Luo, and S. Gao, “Low-cost electrical beam-scanning leaky-wave antenna based on bent corrugated substrate integrated waveguide,” *IEEE Antennas Wirel. Propag. Lett.*, vol. 18, no. 2, pp. 353–357, 2019.

60 J. Huang and R. J. Pogorzelski, “A Ka-band microstrip reflectarray with elements having variable rotation angles,” *IEEE Trans. Antennas Propag.*, vol. 46, no. 5, pp. 650–656, 1998.

61 X. Yang, S. Xu, F. Yang, M. Li, Y. Hou, S. Jiang, and L. Liu, “A broadband high-efficiency reconfigurable reflectarray antenna using mechanically rotational elements,” *IEEE Trans. Antennas Propag.*, vol. 65, no. 8, pp. 3959–3966, 2017.

62 M. U. Afzal and K. P. Esselle, “Steering the beam of medium-to-high gain antennas using near-field phase transformation,” *IEEE Trans. Antennas Propag.*, vol. 65, no. 4, pp. 1680–1690, 2017.

63 A. A. Baba, R. M. Hashmi, K. P. Esselle, M. Attygalle, and D. Borg, “A millimeter-wave antenna system for wideband 2-D beam steering,” *IEEE Trans. Antennas Propag.*, vol. 68, no. 5, pp. 3453–3464, 2020.

64 T. Lou, X. Yang, H. Qiu, Z. Yin, and S. Gao, “Compact dual-polarized continuous transverse stub array with 2-D beam scanning,” *IEEE Trans. Antennas Propag.*, vol. 67, no. 5, pp. 3000–3010, 2019.

65 G. V. Trentini, “Partially reflecting sheet arrays,” *IRE Trans. Antennas Propag.*, vol. 4, no. 4, pp. 666–671, 1956.

66 R. Guzman-Quiros, J. L. Gomez-Tornero, A. R. Weily, and Y. J. Guo, “Electronically steerable 1-D Fabry-Perot leaky-wave antenna employing a tunable high impedance surface,” *IEEE Trans. Antennas Propag.*, vol. 60, no. 11, pp. 5046–5055, 2012.

67 R. Guzmán-Quirós, A. R. Weily, J. L. Gómez-Tornero, and Y. J. Guo, “A Fabry-Pérot antenna with two-dimensional electronic beam scanning,” *IEEE Trans. Antennas Propag.*, vol. 64, no. 4, pp. 1536–1541, 2016.

68 D. Comite, S. K. Podilchak, M. Kuznetcov, V. G. G. Buendía, P. Burghignoli, P. Baccarelli, and A. Galli, “Wideband array-fed Fabry-Perot cavity antenna for 2-D beam steering,” *IEEE Trans. Antennas Propag.*, vol. 69, no. 2, pp. 784–794, 2021.

69 Y. J. Cheng, W. Hong, K. Wu, Z. Q. Kuai, C. Yu, J. X. Chen, J. Y. Zhou, and H. J. Tang, “Substrate integrated waveguide (SIW) Rotman lens and its Ka-band multibeam array antenna applications,” *IEEE Trans. Antennas Propag.*, vol. 56, no. 8, pp. 2504–2513, 2008.

70 J.-W. Lian, Y.-L. Ban, H. Zhu, and Y. J. Guo, "Reduced-sidelobe multibeam array antenna based on SIW Rotman lens," *IEEE Antennas Wirel. Propag. Lett.*, vol. 19, no. 1, pp. 188–192, 2020.

71 E. L. Holzman, "Pillbox antenna design for millimeter-wave base-station applications," *IEEE Antennas Propag. Mag.*, vol. 45, no. 1, pp. 27–37, 2003.

72 M. Ettorre, R. Sauleau, L. Le Coq, and F. Bodereau, "Single-folded leaky-wave antennas for automotive radars at 77 GHz," *IEEE Antennas Wirel. Propag. Lett.*, vol. 9, pp. 859–862, 2010.

73 S. Yan, M. Zhao, Y. Ban, J. Lian, and Z. Nie, "Dual-layer siw multibeam pillbox antenna with reduced sidelobe level," *IEEE Antennas Wirel. Propag. Lett.*, vol. 18, no. 3, pp. 541–545, 2019.

74 M. Ettorre, R. Sauleau, and L. Le Coq, "Multi-beam multi-layer leaky-wave siw pillbox antenna for millimeter-wave applications," *IEEE Trans. Antennas Propag.*, vol. 59, no. 4, pp. 1093–1100, 2011.

75 H. Qiu, X. Yang, Y. Yu, T. Lou, Z. Yin, and S. Gao, "Compact beam-scanning flat array based on substrate-integrated waveguide," *IEEE Trans. Antennas Propag.*, vol. 68, no. 2, pp. 882–890, 2020.

4

Smart Leaky-Wave Antennas for Iridescent IoT Wireless Networks

José L. Gómez-Tornero

Technical University of Cartagena, Department of Information and Communication Technologies, Telecommunication Faculty, Cartagena, Spain

4.1 Leaky-Wave Antennas for Efficient Wireless Systems

Most advanced wireless mobile communications systems depend on adaptive "smart" antennas to provide maximum range and data rate [1–3]. Unfortunately, those smart antennas tend to be complex, costly, and power hungry compared with passive antennas. Therefore, smart adaptive antennas have not been widely used in mobile terminals and Internet of things (IoT) devices.

Leaky-wave antennas (LWAs) are a type of traveling-wave antenna that can produce directive scanning beams using a single feed [4]. Compared with phased-array antennas, they avoid the use of complex feeding networks and electronically tunable components to provide high-gain scannable radiation patterns. The directive beam of an LWA can be steered by simply changing the operation frequency. Since LWAs are strongly dispersive, they have the inherent property of being *directive frequency-beam scanning*. This is illustrated in Figure 4.1a, where the configuration of an LWA of length L and fed from a single port is depicted. As shown in Figure 4.1a, a directive beam is created, pointing at a scan angle θ_R measured from its perpendicular direction, and with a given half-power beamwidth (HPBW) $\Delta\theta$. The radiated-beam characteristics depend on the dispersion of the leaky wave propagating through the LWA.

A leaky wave can be characterized by a complex propagation constant, whose real part β accounts for the phase variations (measured in RAD/m), and whose imaginary part α accounts for the leakage rate (measured in NEP/m):

$$k = \beta - j\alpha \,(\text{m}^{-1}) \tag{4.1}$$

Since leaky modes are strongly dispersive, the complex propagation constant depends on the frequency, $k = k(f)$. The phase constant β determines the beam scan angle θ_R according to the following equation [4]:

$$\sin\,\theta_R(f) \approx \frac{\beta(f)}{k_0} \tag{4.2}$$

where k_0 is the free-space wavenumber. As indicated by Eq. (4.2), the scan angle θ_R depends on the frequency as a result of the dispersion of the leaky-mode phase constant $\beta(f)$. Figure 4.1b shows a general frequency beam-scanning dispersion diagram. It can be seen that the scanning band of the LWA is limited between the frequencies f_{min} and f_{max}, which, respectively, determine the angular scanning range from θ_{min} to θ_{max}. Depending on the type of LWA, the angular scanning

Antenna and Array Technologies for Future Wireless Ecosystems, First Edition.
Edited by Y. Jay Guo and Richard W. Ziolkowski.

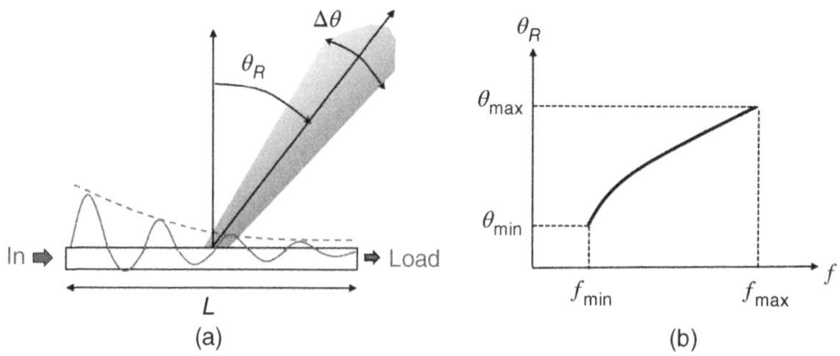

Figure 4.1 Leaky-wave antenna. (a) Basic configuration. (b) Frequency beam-scanning response.

range can be wider or narrower. Uniform and quasi-uniform LWAs can only radiate in the forward quadrant ($\theta_{min} > 0°$ and $\theta_{min} < +90°$), while periodic and metamaterial-based LWAs can provide backward-to-forward frequency beam scanning ($\theta_{min} > -90°$ and $\theta_{max} < +90°$) [4].

The HPBW is determined by both the LWA length and the scan angle [4]:

$$\Delta\theta \approx \frac{57°}{\frac{L}{\lambda_0} \cos \theta_R} \tag{4.3}$$

where λ_0 is the free-space wavelength. It is important to notice that the LWA aperture length L must be efficiently illuminated for Eq. (4.3) to be valid. As sketched in Figure 4.1a, the leaky wave shows an exponentially-decaying aperture-field distribution. Thus, the amount of energy reaching the output port of the LWA depends on the leakage rate and determines the radiation efficiency, i.e.

$$\eta_R = 1 - e^{-2\alpha L} = 1 - e^{-4\pi \frac{\alpha}{k_0} \frac{L}{\lambda_0}} \tag{4.4}$$

Therefore, the normalized leakage rate α/k_0 of an LWA must be properly adjusted according to the normalized antenna length L/λ_0 in order to provide high radiation efficiency. The amount of energy reaching the output port of the LWA is normally absorbed by a matched load to avoid undesired standing waves and reflected lobes [4].

In this chapter, we show how frequency-scanning leaky-wave antennas (FS-LWAs) can be used as low-cost smart antennas connected to access points (APs) in IoT wireless networks. The underlining idea is to synthesize frequency-scanned directive beams which provide coverage in the desired angular region and use the frequency channels available in the physical layer (PHY) protocol. Figure 4.2 illustrates this basic idea by representing an AP connected to an LWA for a wireless protocol with three available channels having the respective central frequencies f_1, f_2, and f_3, and channel bandwidth Δf. In practice, the scanning frequency band will be determined by the operating bandwidth of the PHY protocol. Again, the angular scanning range is limited by the minimum and maximum scanning angles, θ_{min} and θ_{max}.

The specified number of beams and the angular scanning range determine the needed HPBW, i.e. each adjacent scanned beam should intersect at their half-power points in order to properly scan the angular zone [θ_{min}, θ_{max}]. As an example, we fix $\theta_{min} = +10°$ and $\theta_{max} = +50°$. Thus the optimum HPBW would be $\Delta\theta = 20°$ to create $N = 3$ beams covering this angular zone, as illustrated in Figure 4.2a. In general, with N beams and a given scanning angular range from θ_{min} to θ_{max}, the needed HPBW is given by:

$$\Delta\theta \approx \frac{\theta_{max} - \theta_{min}}{N - 1} \tag{4.5}$$

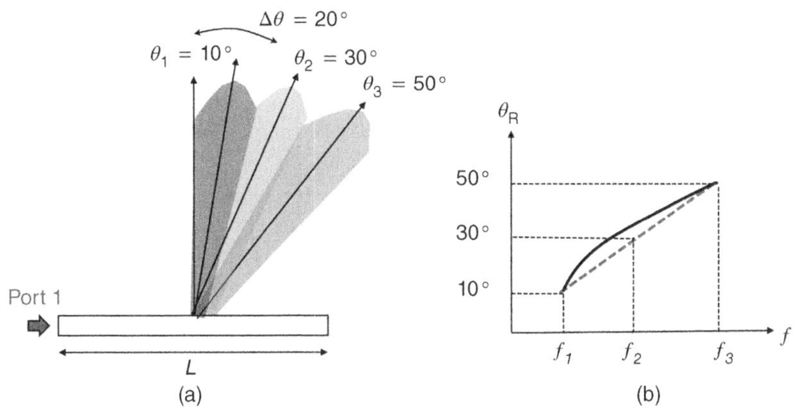

Figure 4.2 Example of an LWA with an angular scanning range between 10° and 50° using three frequency channels and a single antenna port.

According to Eq. (4.3), the LWA length should be approximately $L = 3\lambda_0$ to create directive beams with a HPBW of $\Delta\theta = 20°$. The design of an LWA that meets these specifications involves the careful *dispersion engineering* of its leaky-wave phase and leakage rates, i.e. $\frac{\beta}{k_0}$ and $\frac{\alpha}{k_0}$, to provide the desired scanning angles at the desired frequencies according to Eq. (4.2), and with good radiation efficiency according to Eq. (4.4). This task is generally complicated; it cannot be precisely satisfied for all desired frequencies. For instance, an ideal linear frequency beam-scanning dispersion curve is shown with the black dashed line in Figure 4.2b, while a real dispersion one is not linear as depicted with the black continuous line. Note that Figure 4.2 represents a forward FS-LWA example, i.e. only positive scanning angles are obtained. As will be explained later, homogeneous LWAs provide simpler designs compared with periodic or metamaterial-based LWAs, but at the cost of providing only forward scanning.

Nevertheless, the simplicity of forward-scanning homogeneous LWAs can be combined with the fact that the two opposite ports of a symmetric homogeneous LWA can be simultaneously used to provide mirror-symmetric beams. This feature is illustrated in Figure 4.3a, where the same antenna as in Figure 4.2 is now simultaneously fed from ports 1 and 2. As a consequence of also using port 2, the scanning range of this LWA is extended to negative angles, from $-\theta_1 = -10°$ at f_1 to $-\theta_3 = -50°$ at f_3. In this manner, a simple homogeneous symmetric LWA can perform both backward to forward scanning when its two feeding ports are both used. As will be illustrated in the following sections, modern APs provide MIMO (Multiple Input Multiple Output) and antenna diversity functionalities, which can be realized by connecting a single AP to the two opposite ports of an LWA. This configuration provides simultaneous operation of 2*N beams, where N is the number of operating frequency channels. The overall coverage of this LWA with a two-port MIMO AP is illustrated in Figure 4.3b. It shows the angular coverage has been extended from negative to positive angles. At this point it is important to note that a frequency-scanning configuration is only possible if the PHY allows for *dynamic channel assignation schemes*. From the example shown in Figure 4.3b, it is evident that each mobile terminal will request a given frequency channel for an optimum radio link depending on its location with respect to the AP. Therefore, the AP must somehow dynamically assign the most appropriate channel to each individual mobile terminal at any particular moment.

This is typically performed by using *channel hopping techniques* to sweep all the available channels, and then choose the optimum channel for each mobile terminal. If this dynamic channel

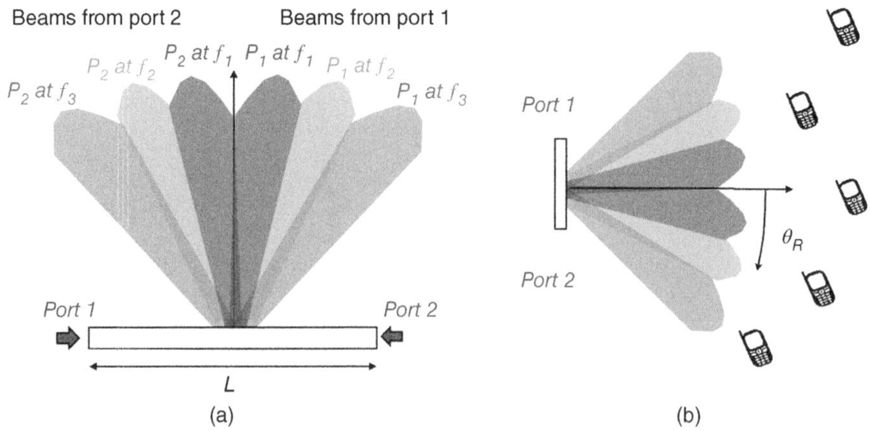

Figure 4.3 Example of an LWA with an angular scanning range between 10° and 50° using three frequency channels and a two antenna ports.

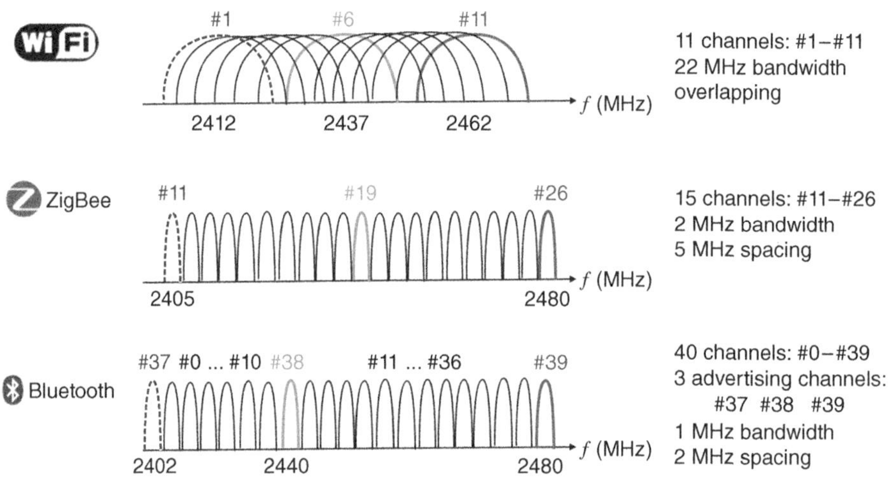

Figure 4.4 Wi-Fi, Zigbee, and Bluetooth channels in the ISM 2.4 GHz band.

technique is affordable, this topology is extremely simple and can support a smart adaptive antenna system. The latter provides high gain in a wide angular region without the need of expensive hardware and power-consuming signal processing associated with phased-array techniques.

As an example, Figure 4.4 shows the channels employed in the 2.45 GHz ISM band in some of the most common wireless networks. Wi-Fi Wireless Local Area Networks (WLANs) follow the IEEE 802.11 standard and typically use 11 channels with central frequencies extending from 2.412 GHz (channel #1) to 2.462 GHz (channel #11). These channels typically have a 22 MHz bandwidth, and they overlap as shown in Figure 4.4. On the other hand, Wireless Sensor Networks (WSNs) using the Zigbee protocol defined by the IEEE 802.15.4 standard commonly use 15 channels from channel #11 centered at 2.405 GHz to channel #26 at 2.48 GHz, with 2 MHz bandwidth and 5.0 MHz spacing. Finally, another example is the Bluetooth protocol used in Wireless Personal Area Networks (WPANs), which defines 40 channels from 2.402 to 2.48 GHz, with typical 1.0 MHz bandwidth and 2.0 MHz spacing. In all cases in Figure 4.4, the lowest channels are represented in dotted line, the mid channels in light gray, and the highest channels in dark gray. These choices are simply consistent with the frequency assignations applied to the scanned beams shown in Figure 4.3.

Naturally, one needs to design the LWA to fit the available channels of each wireless protocol in the PHY radio layer. In Section 4.2, we describe a low-cost printed-circuit LWA technology which allows one to adapt its frequency-beam scanning response to fit with the desired bandwidth and number of channels. Later, practical examples and applications of the use of smart LWAs applied to modern wireless systems are described.

4.2 Low-Cost Printed-Circuit Frequency-Scanning LWAs

Among the vast variety of LWAs [4], the one based on printed-circuit microstrip leaky lines is probably the simplest topology [5]. Certainly, in contrast to waveguide-based LWAs, their guiding structures can be fabricated using conventional printed-circuit processes.

The full-width microstrip leaky-wave antenna (MLWA) was first proposed by Deschamps and Menzel and studied in detail by Oliner [5]. It is basically formed by a metallic patch of width W and length L, which is printed on a conductor-backed dielectric substrate of thickness H and relative dielectric permittivity ε_r, as sketched in Figure 4.5a. Its radiation is based on the excitation of an EH_1-type leaky mode whose electric field distribution inside the dielectric substrate is also sketched in the cross-section view given in Figure 4.5a. The electric field under the microstrip patch induces magnetic currents along its two long lateral sides which radiate out into free space. Since the EH_1 leaky-mode is a higher-order mode, a full-width MLWA must be carefully fed to avoid the excitation of the fundamental EH_0 non-radiative mode of the microstrip.

A half-width version of the MLWA was proposed by Zelinski *et al.* [6]. As shown in Figure 4.5b, one lateral side of the microstrip patch in this design is short circuited by a perfect electric conductor (PEC) wall, which can be effectively realized in practice by using metallic vias that short the top patch to ground. As a result, the fundamental EH_0 microstrip mode is eliminated, and the antenna, now with a half width $W/2$, can be operated in its first $EH_{1/2}$ leaky mode. The resulting half-width microstrip leaky-wave antenna (HWM LWA) radiates only from its one long open lateral side as indicated in Figure 4.5b. It is more compact than the original full-width counterpart and can be fed

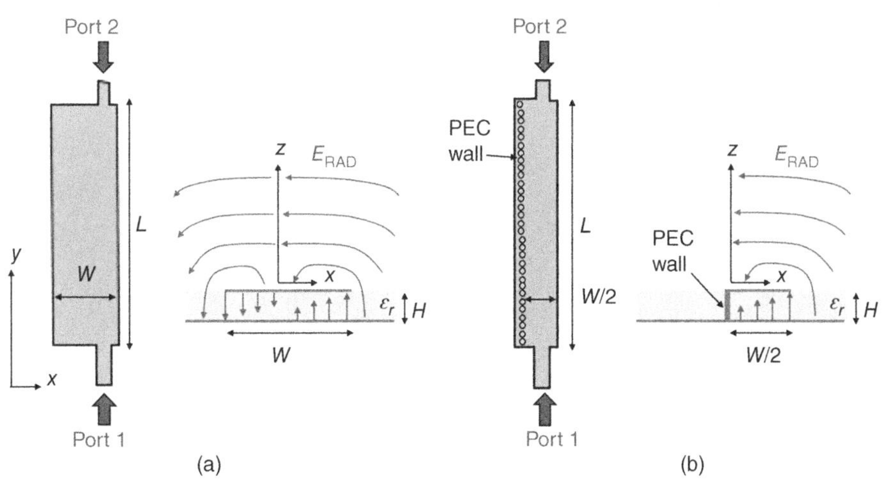

Figure 4.5 Microstrip leaky-wave antennas. (a) Full-width scheme. (b) Its half-width counterpart.

with a simpler circuit. As a result, the HWM LWA is a very cheap and simple antenna topology to create frequency-scanned directive beams. Thus, it has been appealing for applications in low-cost smart wireless systems. It must be noticed that this type of antenna can be directly fed by simple microstrip lines at its two symmetric ports, i.e. ports 1 and 2 in Figure 4.5, to obtain beam scanning through both positive and negative angles as previously commented. Moreover, an SIW (substrate integrated waveguide) version of this type of planar antenna was proposed in [7, 8] and was shown to have a high design flexibility at the cost of a more expensive manufacturing process due to the use of the SIW technology.

Figure 4.6a shows a picture of a fabricated HWM LWA. Its metallic patch of length L and width W is printed on a thin dielectric substrate of height H and relative dielectric permittivity ε_r. The PEC wall that short-circuits one side of the patch is achieved using metallic posts of diameter d located at close distances along the waveguide with period $P \ll \lambda$. A detailed zoom of the microstrip feeding circuit is shown in Figure 4.6b; it illustrates how an SubMiniature version A (SMA) connector can be used to directly feed the LWA through either one of the two opposite ports. As indicated in Figure 4.6c, a narrow piece of metallic patch of safeguard width W_0 separates the vias from the neighboring edge of the patch. A matching circuit is needed to optimize the return losses when feeding the radiating leaky patch with a simple microstrip line. It is formed by a width-tapered microstrip and an inset at the patch input. One sees that the HWM LWA is planar, structurally very simple and totally passive. It does not require any active circuits or a complex feeding network to create directive scanned beams. This simplicity is important in the paradigm of low-cost smart antenna designs for commercial wireless networks and IoT applications.

Furthermore, this type of antenna has an important advantage related to its electromagnetic design. In general, an LWA can be directly designed by obtaining the dispersion relation of the corresponding leaky mode. Certainly, as explained with Eqs. (4.1)–(4.4), the leaky-mode complex phase constant $\beta(f)$ can be determined for a given LWA geometry. The main beam direction of the radiation pattern of the antenna can then be obtained directly. Compared with other

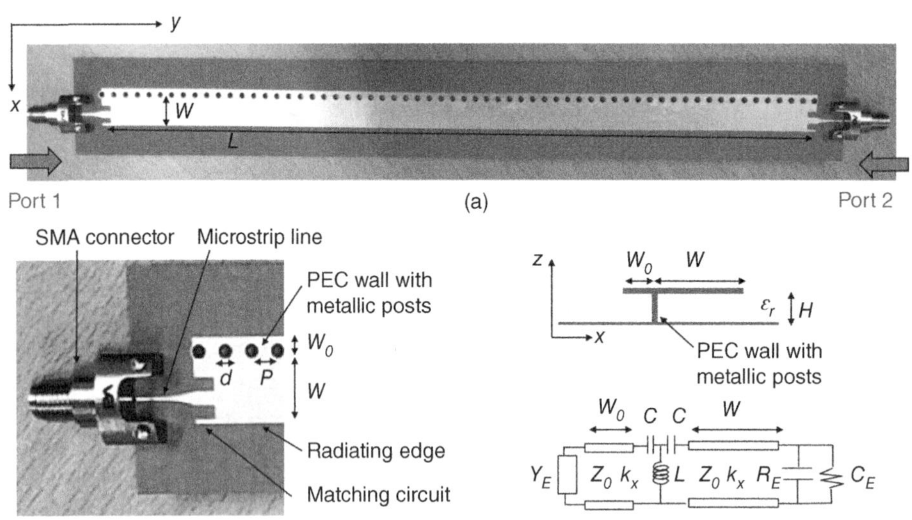

Figure 4.6 HWM LWA. (a) Photo of fabricated prototype. (b) Feeding circuit details. (c) Transverse equivalent network (TEN).

directive antenna technologies which require time consuming electromagnetic numerical analysis techniques, an LWA can be analyzed in a very efficient way using its transverse equivalent network (TEN), with which the leaky-mode dispersion relation is obtained numerically [4]. The TEN along the transverse x-direction for the HWM LWA shown in Figure 4.6a is depicted in Figure 4.6c. It can be seen that it is basically formed by two sections of the transmission lines which correspond to the patch sections of widths W and W_0, connected by the PEC metallic posts. They are modeled by two equivalent series capacitors and one shunt inductor whose properties depend on the diameter d of the vias and their periodicity P [8]. The radiating edges of the patch are modelled with an equivalent admittance, Y_E, formed by the radiating resistance R_E and fringe capacitance C_E. This TEN yields a transverse resonance equation (TRE) that can be solved numerically to obtain the complex transverse leaky wavenumber k_x. Finally, the longitudinal complex propagation constant of the leaky wave, k_y, can then be derived from the transverse one, k_x, with the relation:

$$k_y(f) = \sqrt{k_0^2\varepsilon_r - k_x^2(f)} = \beta(f) - j\alpha(f) \tag{4.6}$$

More details about the TEN and TRE can be found in [8]. Once β and α are obtained as functions of the frequency and the main antenna dimensions, one can design the HWM LWA to operate in the desired frequency band. To illustrate this design process, it will be shown how the main dimensions of the HWM LWA must be chosen to obtain a desired frequency-scanning response. As was described in Eqs. (4.1)–(4.5), the HWM LWA can synthesize N directive beams that cover a given angular zone from θ_{min} and θ_{max}, and facilitate a given set of N channels with respective central frequencies from $f_{min} = f_1$ to $f_{max} = f_N$. Moreover, Eq. (4.5) provided a specified scanned-beam HPBW $\Delta\theta$ to cover this angular zone with optimal beam overlap. The patch length L is directly obtained from Eq. (4.3) once $\Delta\theta$ is known.

Nevertheless, the most difficult task is obtaining the dispersion of the leaky-mode phase constant in Eq. (4.2) which satisfies the desired scanning angles versus frequency, $\theta(f)$, as well as the normalized leakage rate dispersion, $\alpha/k_0(f)$, which provides the optimum radiation efficiency (Eq. (4.4)) in the specified scanning band. Therefore, this *dispersion engineering* design process requires a technique that facilitates the realization of these desired LWA properties. The dispersion engineering synthesis is not always affordable since it sometimes relies on the numerical optimization of metamaterial-based antennas. The latter are based on complicated unit cells with many coupled tuning parameters [9–11]. Design simplicity is of paramount importance in the case of low-cost smart antennas. It is for that reason we have chosen a simple passive HWM LWA. Since we have only three structural parameters in our case, they can be easily tuned to obtain the desired frequency dispersion characteristics. They are the substrate height H, its relative dielectric permittivity ε_r, and the printed patch width W. As commented previously, the printed patch length L is directly obtained from the desired HPBW $\Delta\theta$ using Eq. (4.3). A numerical optimization technique has been developed in [11] to obtain the optimum HWM LWA parameters H, ε_r, and W, which provide a desired frequency scanning response. Here we present numerical dispersion results to describe how each parameter directly affects them [12].

As a design example, let us assume that the LWA must operate in the 2.45 GHz ISM band used in commercial IoT wireless protocols such as Wi-Fi, Zigbee, and Bluetooth. As was sketched in Figure 4.4, this band is defined between f_{min} = 2.40 GHz and f_{max} = 2.48 GHz. We use the previous example with $N = 3$ beams that cover the angular zone from $\theta_{min} = +10°$ to $\theta_{max} = +50°$, with an optimum HPBW $\Delta\theta = 20°$. Again, according to Eq. (4.3), this HPBW requires an antenna length of approximately three free-space wavelengths. Therefore, the leaky patch length must be designed to this value: $L = 3\lambda_0 = 36$ cm. The second design step involves the selection of the patch width W. First we must fix the diameter of the metallic posts, their periodicity, and the safeguard

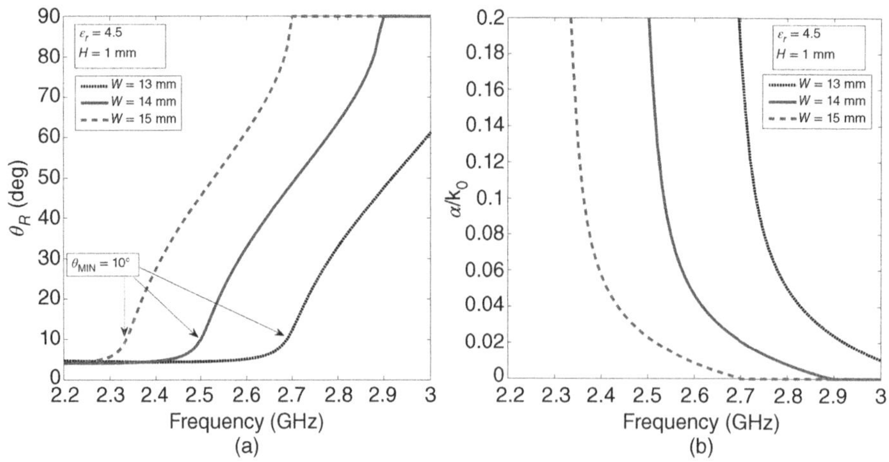

Figure 4.7 Dispersion curves of the HWM LWA when the patch width W is varied. (a) Scanning angle. (b) Leakage rate.

width W_0. It is then necessary to choose a dielectric substrate with a fixed height H and relative permittivity ε_r. In our case we chose an flame retardant 4 (FR4) substrate with $\varepsilon_r = 4.5$ and $H = 1.0$ mm. A good PEC wall condition can then be obtained in the assumed frequency band with $d = 2.0$ mm, $P = 4.0$ mm, and $W_0 = 3.0$ mm [11]. Once all of these design parameters have been determined, we can obtain the leaky-wave frequency dispersion when the patch width W is varied. These results are shown in Figure 4.7. Figure 4.7a shows the variation of the leaky scan angle $\theta_R(f)$ obtained from the leaky-mode phase constant (Eq. (4.2)), while Figure 4.7b represents the dispersion of the normalized leakage rate $\alpha/k_0(f)$. As can be seen in Figure 4.7a, the strip width W allows one to shift the frequency-scanning response. Wider leaky strips produce a lower cut-off frequency f_{min} that satisfies $\theta_R(f_{min}) = \theta_{min} = 10°$. Conversely, narrower strips increase this cut-off frequency.

As shown in Figure 4.7a, the minimum scan angle $\theta_{min} = 10°$ is produced at $f_{min} = 2.7$ GHz for $W = 13$ mm, $f_{min} = 2.5$ GHz for $W = 14$ mm, and $f_{min} = 2.34$ GHz for $W = 15$ mm. In our design example we needed $f_{min} = 2.40$ GHz. Therefore, the leaky strip width should be a value between $W = 14$ mm and $W = 15$ mm. The final design value is $W = 14.6$ mm, which, as shown in Figure 4.8a, provides the desired condition $\theta_{MIN}(2.4\text{ GHz}) = 10°$ for the selected substrate with $\varepsilon_r = 4.5$. However, this design with $\varepsilon_r = 4.5$ and $W = 14.6$ mm (light gray line in Figure 4.8a) has dispersion properties which do not yield the condition $\theta_R(f_{max}) = \theta_{max} = 50°$ for $f_{max} = 2.48$ GHz. It scans to a lower angle $\theta_R = 33°$ for f_{max}. In fact, the angle $\theta_{max} = 50°$ is obtained at the higher frequency $f = 2.6$ GHz rather than the desired one. This outcome is related to the scanning ratio of the LWA, which is defined as the sensitivity of the scan angle to the frequency, i.e. $\partial\theta_R(f)/\partial f$. In this sense, the scanning ratio of a HWM LWA can be controlled by varying the relative dielectric permittivity of the substrate ε_r [11, 12]. This feature is illustrated in Figure 4.8.

If the dielectric permittivity is reduced, the scanning ratio of the HWM LWA is reduced. This behavior is shown in Figure 4.8a for the design with $\varepsilon_r = 2.2$ (dashed line). In this case, once the strip width is tuned to $W = 20.7$ mm, the condition $\theta_{min}(2.4\text{ GHz}) = 10°$ is satisfied. However, now the angle $\theta_{max} = 50°$ is obtained at an even higher frequency, $f = 2.8$ GHz. Consequently, a lower permittivity substrate, e.g. $\varepsilon_r = 2.2$, scans to an even lower maximum angle $\theta_R = 22°$ for the maximum frequency of 2.48 GHz. On the other hand, a substrate with a higher permittivity, e.g. $\varepsilon_r = 10$, provides a sufficient scanning ratio that facilitates the beam being steered from $\theta_{min} = +10°$ to $\theta_{max} = +50°$ in the desired scanning band from $f_{min} = 2.40$ GHz to $f_{max} = 2.48$ GHz. This result

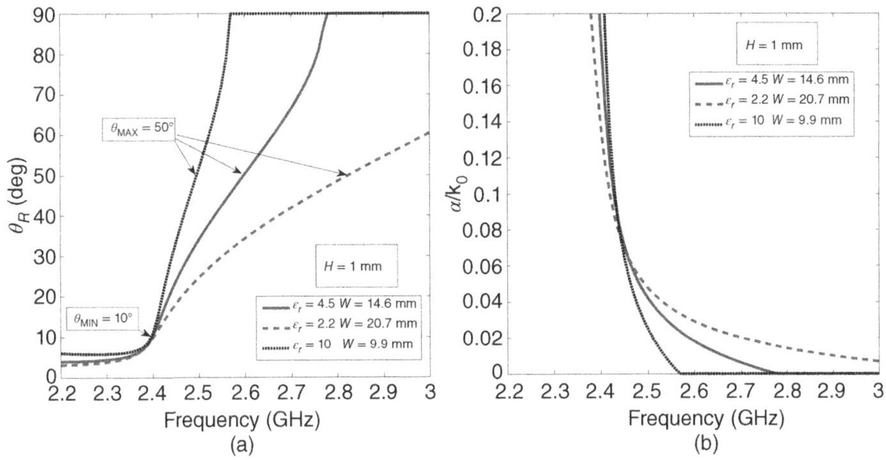

Figure 4.8 Dispersion curves of the HWM LWA when the relative dielectric permittivity ε_r of the substrate is varied. (a) Scanning angle. (b) Leakage rate.

is illustrated with the black dotted line in Figure 4.8a for a HWM LWA design with $\varepsilon_r = 10$ and $W = 9.9$ mm. Therefore, the dielectric permittivity ε_r of the substrate and the strip width W must be optimized simultaneously to cover the desired angular range in the prescribed scanning frequency band.

Once the desired frequency-beam scanning response $\theta_R(f)$ has been obtained, the last step is to tune the substrate thickness H to ensure that the leakage rate $\alpha/k_0(f)$ provides both a good radiation efficiency and a good aperture efficiency. The radiation efficiency is given by the value of the leakage rate and the LWA length (Eq. (4.4)). As explained in detail in [13] and [14], a too low leakage rate produces a low radiation efficiency since the leaky mode will not have radiated most of its guided energy when it reaches the far end of the LWA and, hence, too much guided power reach the far end of the antenna. On the other hand, a too high leakage rate also creates a low aperture efficiency. Most of the power will have been radiated from the first longitudinal section of the LWA, and the far end will not have been illuminated, thus reducing the effective radiating length. A good trade-off between the radiation and aperture efficiencies usually involves a selection of radiation efficiency, e.g. $\eta_{rad} = 90\%$ [13, 14]. Nonetheless, it must be considered that the leakage rate values $\alpha/k_0(f)$ reduce as the frequency is increased and the beam is scanned from the boresight $\theta_R = 0°$ to the endfire $\theta_R = 90°$ direction. This effect can be seen in all of the dispersion curves for the leakage rate presented in both Figures 4.7b and 4.8b.

It is important to be able to control the quality of the leakage rate curve $\alpha/k_0(f)$, while not affecting the frequency-beam scanning curve $\theta_R(f)$. This can be accomplished by controlling the substrate thickness H and, again, by adjusting the radiating strip width W to correct for any deviation in the cut-off frequency of $\theta_R(f)$.

Thicker substrates generally will provide higher leakage values, while thinner substrates will deliver lower leakage [11, 12]. This is illustrated in Figure 4.9 using a dielectric with $\varepsilon_r = 10$ for three different values of the substrate thickness H. Figure 4.9a shows that for each value of H, the strip width W can be readjusted to keep the desired frequency-scanning response $\theta_R(f)$, i.e. $\theta_{min} = +10°$ and $\theta_{max} = +50°$ for $f_{min} = 2.40$ GHz and $f_{max} = 2.48$ GHz. The leakage rate curves $\alpha/k_0(f)$ are plotted in Figure 4.9b. They clearly show how the leakage level increases for thicker substrates. Finally, the antenna radiation efficiency as a function of frequency is represented in Figure 4.9c. It shows the aforementioned antenna efficiency drop as the frequency and scan angle are increased.

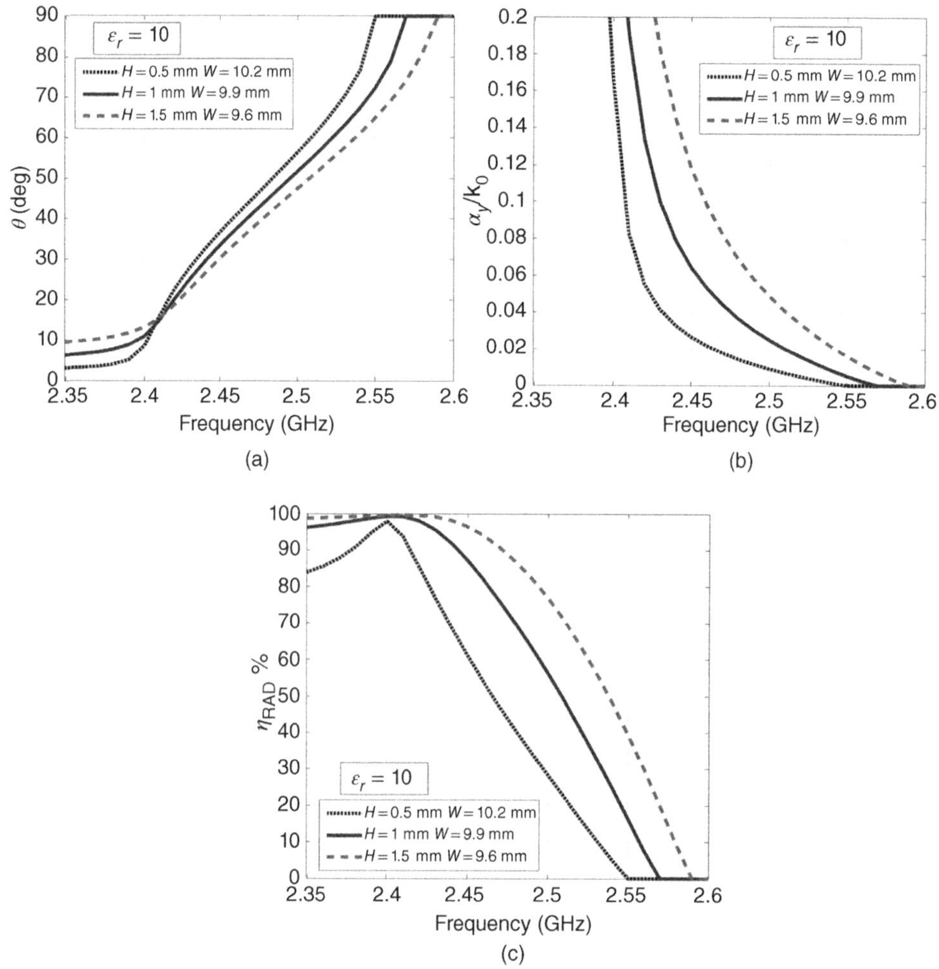

Figure 4.9 Dispersion curves of the HWM LWA when the substrate thickness H is varied. (a) Scanning angle. (b) Leakage rate. (c) Radiation efficiency.

A thin substrate with $H = 0.5$ mm provides the maximum radiation efficiency $\eta_{\text{rad}} = 90\%$ for the low 2.4 GHz band but only $\eta_{\text{rad}} = 30\%$ for the high 2.5 GHz band. A substrate with triple thickness, $H = 1.5$ mm, shows a radiation efficiency above $\eta_{\text{rad}} = 80\%$ over the whole scanning band from 2.4 to 2.5 GHz at the cost of low antenna aperture efficiency. A tradeoff between these characteristics is to select $H = 1.0$ mm. It yields a reasonable balance with a radiation efficiency above 60% over the entire scanning band, which assures high aperture efficiency as well.

Since the LWA is typically several wavelengths long in its longitudinal dimension (length L in Figure 4.6) and narrow in its transverse direction (width W in Figure 4.6), a magnetic line source along the radiating edge of the long metal strip serves as a good radiation model. As a result, the radiated beam is narrow in its H plane (YZ plane) and wide in its E plane (ZX plane). Thus, a *scanned fan beam* is formed. It must be recalled that the scanning plane is the YZ plane, which corresponds to the longitudinal propagation direction of the leaky wave. As expressed in Eq. (4.3), the HPBW of the directive beam in the longitudinal YZ plane is dictated by the LWA length L. On the

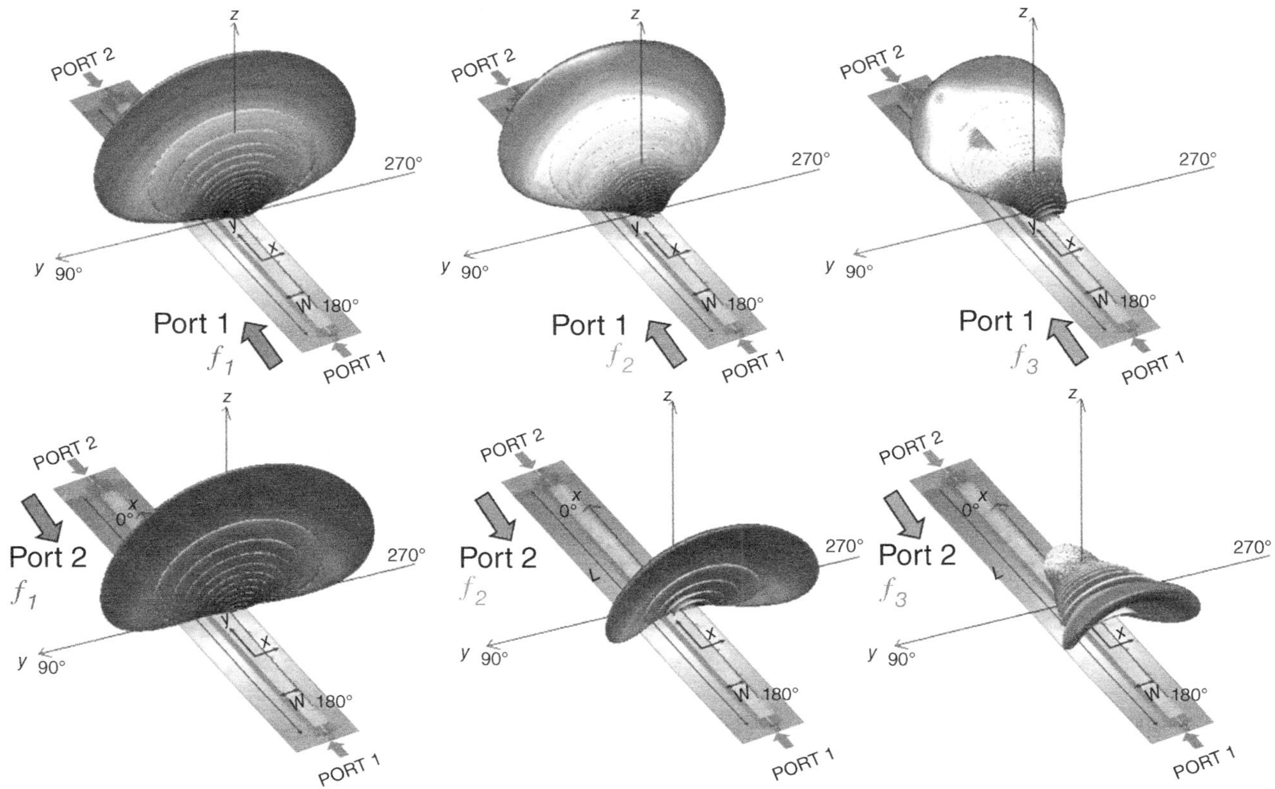

Figure 4.10 Fan-beam scanning with a single-element HWM LWA.

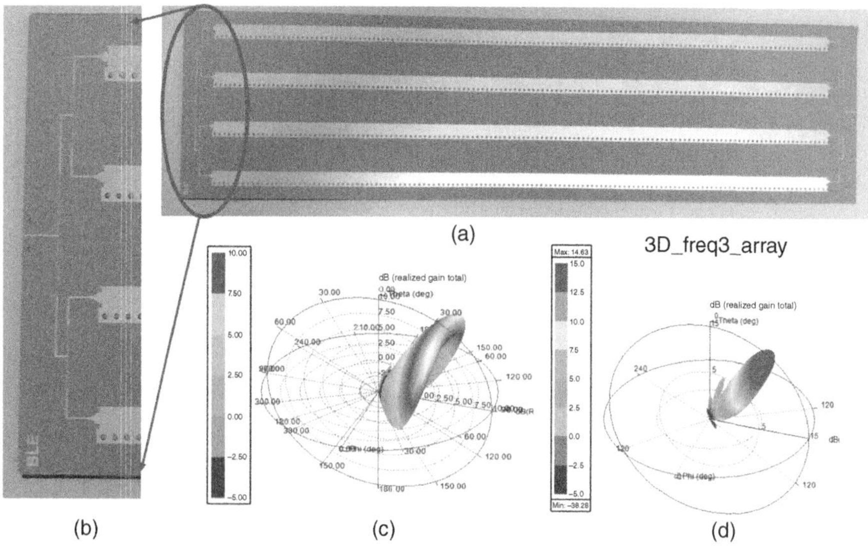

Figure 4.11 Array of four HWM LWAs to increase the directivity in the transverse plane. (a) Photo of the antenna. (b) Feeding circuit details. Realized gain pattern of (c) the individual HWM LWA, and (d) the HWM LWA array.

other hand, the beam width in the transverse *ZX* plane is large, covering the range between 60° and 130° depending on the scan direction [15]. As a result, the type of *frequency-scanned fan-beam radiation patterns* obtained by this type of planar antenna is shown in Figure 4.10. The example in Figure 4.3, which used three different frequencies and the two opposite ports, is again employed here to illustrate the generation of a total of six scanned fan beams covering the angular space.

The transverse beam width can be reduced if several HWM LWAs are arranged in parallel as an array. A practical design of this configuration involving four HWM LWAs is shown in Figure 4.11a and its matched microstrip feeding circuit is detailed in Figure 4.11b. The feed network provides an in-phase excitation with equal amplitude to the four individual elements of this LWA array. When the radiation pattern of an individual HWM LWA element in Figure 4.11c is compared with the one obtained by the array in Figure 4.11d, it is clearly demonstrated that the transverse beam width can be reduced from 120° to 40°. Thus, the directivity is increased by more than 5 dBi [16]. Therefore, an array of HWM LWAs can be used for *pencil-beam frequency scanning* to increase the directivity and gain of the antenna by reducing the beam width in the transverse plane. Nevertheless, it must be taken into account that the frequency-scanning of the beam is only performed along the longitudinal plane. Therefore, the transverse beam width must be adjusted to give sufficient signal coverage in this transverse plane. A too narrow beam in the transverse plane might not cover angles far from the direction perpendicular to the LWA array.

In order to scan the pencil beam in the transverse plane, active switched-beam or phase-scanning can be combined together with frequency scanning to obtain two-dimensional steering [17]. However, this is at the cost of higher complexity since active phase-shifters and radio frequency (RF) switches are needed to produce such a scan. However, active systems are beyond the scope of this chapter. We aim here to emphasize a *totally passive smart scanning antenna*. In the following sections, some real applications of the HWM LWA for efficient practical wireless networks will be described.

4.3 LWAs for Efficient Communications in Iridescent Wireless Networks

The first example application of LWAs for efficient wireless systems is to serve as adaptive smart antennas which can scan their directive beams to the locations of the mobile devices. This functionality is desired in order to increase the signal level when compared with the use of conventional isotropic antennas, as was illustrated in Figure 4.3b. For instance, modern Wi-Fi multi-user Multiple Input Multiple Output (MU-MIMO) APs use beam-forming techniques in WLANs, to focus the AP signal to the desired user location and thus increase the signal-to-noise ratio (SNR) [18]. This arrangement eventually facilitates an optimum modulation and coding scheme (MCS) which provides a higher data throughput. Normally, beam-forming practices rely of phased-array techniques [18]. As depicted in Figure 4.12a, several antennas (typically three) are used in an array configuration. A beam-forming algorithm must then control the proper amplitude and phase of the RF signal exciting each antenna in order to steer the beam toward the desired direction in space as illustrated in Figure 4.12b. Conventionally, an array of monopole antennas is used on the AP as shown in Figure 4.12a.

Among the new features of the Wi-Fi WLAN IEEE 802.11ax standard (a.k.a. Wi-Fi 6 [19]), advanced MU-MIMO schemes based on simultaneous (concurrent) multiple-channel operation have been proposed [20]. Multi-user beam-forming (MU-BF) techniques dedicate a Wi-Fi channel to a single user (or several users) located in a similar region in space. As sketched in Figure 4.13a

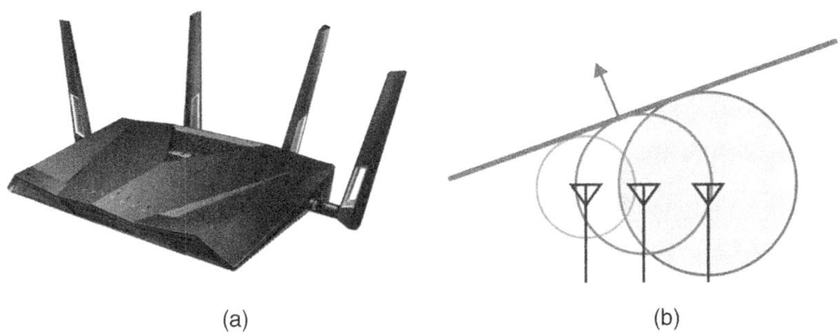

Figure 4.12 MU-MIMO Wi-Fi AP example. (a) Illustration of an actual AP. (b) Phased-array beam forming and scanning depiction. Source: (a) ASUSTeK Computer Inc, (b) From [18] / with permission from O'Reilly Media, Inc.

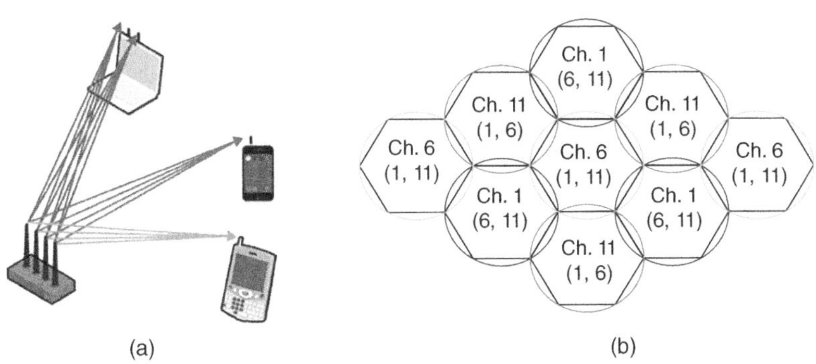

Figure 4.13 MU-BF "multi user beam forming." (a) Application illustration. Source: From [18] / with permission from O'Reilly Media, Inc. (b) Spatial sectorization with the channel slicing proposed in [21]. Source: From [21] / Edgewater Wireless System Inc.

they focus an electromagnetic beam toward the proper location. Three simultaneous and different Wi-Fi channels (namely, #1, #6, and #11) are represented in different colors (black, light gray, and dark gray), each one being focused to a different user location.

The use of beamforming techniques together with multi-user (multi-channel) techniques can thus produce a high quality of service in dense WLAN scenarios. The WLAN users perceive high bandwidth and high signal quality with lower time waits, lower latency, by sharing the Wi-Fi channels [20]. As shown in Figure 4.13a, the AP must allocate a spatial beam with a dedicated Wi-Fi channel to the particular location of each user. This configuration delivers both a high signal level and a high instantaneous bandwidth and, hence, an optimum performance from the PHY. The monopole array must be properly configured for each channel to achieve this outcome, and this requires complex and power-consuming signal processing techniques. As an example, the new Quantenna QSR10GU-AX PLUS Wi-Fi chipset [22] incorporates 8×8 dual-band MIMO, 4×4 in the 2.4 GHz band plus 4×4 in the 5 GHz band, with a total of 4 and 8 spatial streams in each band, respectively. The support of several concurrent Wi-Fi channels simultaneously can provide a wider instantaneous bandwidth for each user connected to the MU-MIMO AP since each channel is directed to each user location.

Some innovative developers have proposed a "Multi-Channel Single Radio" (MCSR) spectrum slicing scheme [21], which supports three simultaneous Wi-Fi concurrent channels (namely, #1, #6, and #11) in the 2.45 GHz band. This phased-array MU-BF technique is useful for dense WLAN deployments. As sketched in Figure 4.13b, a single MU-MIMO AP must create this channel-space multiplexing so that each color (black for channel #1, light gray for channel #6, and dark gray for channel #11) is focused to a different region in space. As mentioned, a directive focused beam using a dedicated Wi-Fi channel in each sectorized zone provides a high signal quality and the full bandwidth of the corresponding channel. In technical reports [18, 23], it is stated that an increase from 2 to 5 dB in the signal level can be obtained using this MU-BF technique with a MU-MIMO AP with an array of three monopole antennas for each frequency channel. Moreover, it is mentioned that a higher number of antenna elements would be needed to further increase the array directivity and gain and, thus, the beamed signal level. However, this approach (using an AP with more than three antennas operating simultaneously) would notably increase the hardware cost and the signal processing complexity. Thus it would eventually lead to a higher power consumption and processing delays.

In this context, smart LWAs are able to produce MU-BF in a simple manner, dispensing with any active circuits and a higher number of antenna terminals. Effectively, LWAs are totally passive antennas with only two RF ports. Consequently, they do not need any active reconfiguration circuitry to *multiplex different frequency-channels into different sectorized angular beams*. The LWA needs only two connectors and uses its inherent frequency-scanning response to simultaneously focus each one of its three Wi-Fi channels toward a different sector in space with high directivity and gain. Thus, it avoids the conventional power-consuming and expensive phased-array techniques to create the beam forming.

This feature was illustrated in Figure 4.3, which showed an LWA with two RF ports, each port splitting three Wi-Fi channels in the ISM 2.45 GHz band and radiating them into different directions. The lower channels (2.40 GHz, black arrow) are radiated close to the perpendicular direction of the antenna; the intermediate channels (2.45 GHz, light gray arrow) are beamed to intermediate angles; and the high frequency channels (2.50 GHz, dark gray arrow) are scanned toward endfire direction. Since the LWA has two symmetrical ports, the same antenna can be used to generate symmetrical beams radiating toward either positive or negative angles depending on the chosen port. It is important to observe that the electromagnetic beams can be very narrow, thus focusing

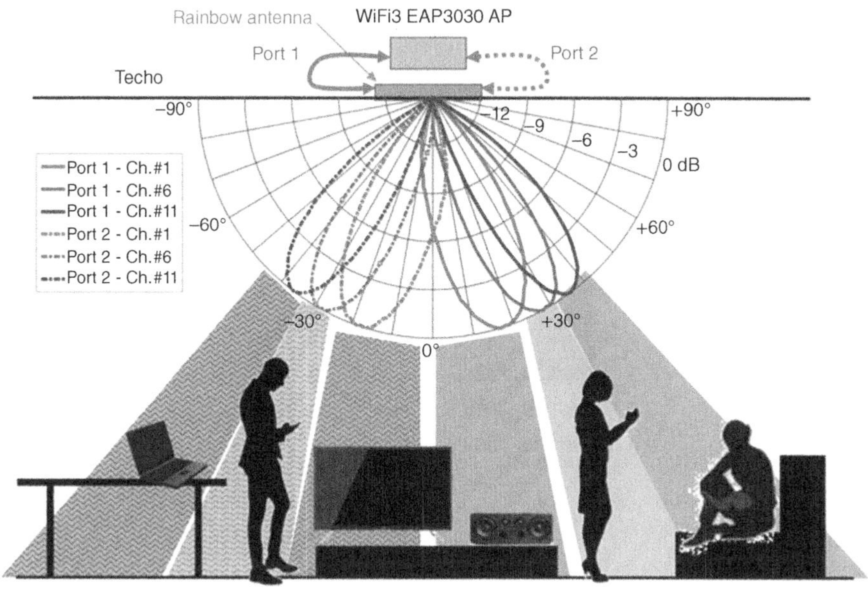

Figure 4.14 LWA connected to a Wi-Fi MU-MIMO AP using spatial sectorization with channel slicing. MAC, medium access control protocol; LLC, logical link control.

the radiated signal into a narrow portion of space. This is important so that directive beams produce focused regions where the signal level is high. This creates spatial sectorization with frequency assignments and high signal levels for dense WLAN deployments as proposed in [20, 21, 23].

Figure 4.14 illustrates the use of the HWM LWA designed in [16] (see Figure 4.11) in this context. The two ports of the HWM LWA array are connected to a MU-MIMO Wi-Fi3 AP, which implements the multi-channel scheme. The antenna is mounted on the ceiling to enable a dense WLAN scenario. As shown in the radiation patterns in Figure 4.14, the light gray color corresponds to the beam radiated at Wi-Fi channel #1 (2.412 GHz), which is beamed almost perpendicular to the AP site. Two symmetrical light gray beams are generated, one to the right when using port 1 (left side of antenna) and one to the left side when using port 2 (right side of the antenna). Similarly, the two beams for channel #6 (2.437 GHz) are represented in dark gray color and radiate at the intermediate directions, while the black beams correspond to channel #11 (2.462 GHz) and radiate at the larger angles. As one can see, the six beams thus produce a full-space scanning which covers complementary zones, following the recommendation of [21] as shown in Figure 4.13b.

To illustrate the benefit of using LWAs together with Wi-Fi MU-MIMO APs and channel slicing schemes, simulated radio coverage maps are represented in Figure 4.15. The Wi-Fi3 MU-MIMO AP is connected to the HWM LWA at a height of 3.5 m. An office with dimensions 10 × 5 m and with four tables is sketched for reference. The relative signal level created at the tables height (90 cm) for each one of the six colored beam is represented with contour curves. Using the afore-described MU-MIMO beamforming technique, a total of six sectorized, complementary, and concurrent beamed zones can be created, each one for each concurrent Wi-Fi channel. Channel #1 (black beams) covers the central zone, which is almost perpendicular to the LWA. Light gray beams (channel #6) continue scanning toward more distant positions, and finally dark gray beams (channel #11) illuminate the far zones of the room. Any Wi-Fi device located in this office will receive high Wi-Fi signal level from this smart AP powered with Wi-Fi6 AP together with *smart FS-LWAs*.

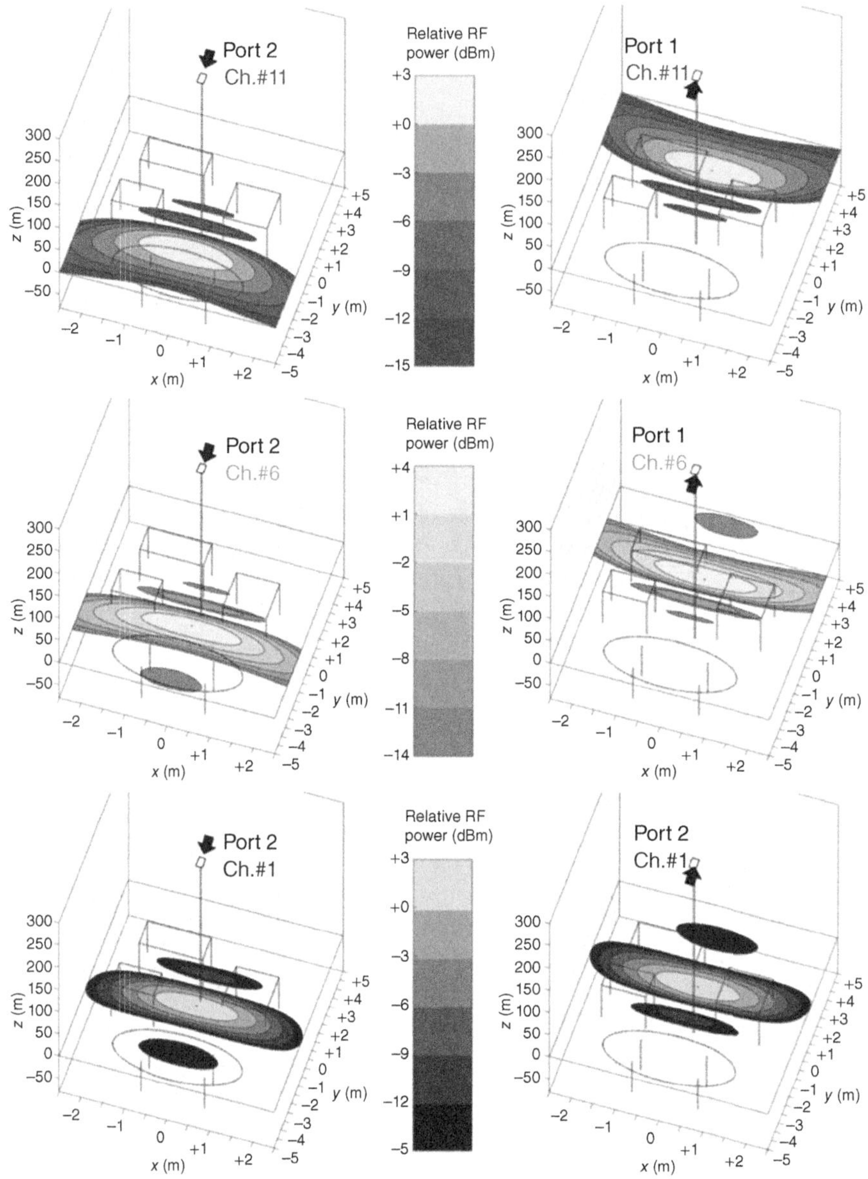

Figure 4.15 Signal-level radio maps for six scanned beams (channel #1, #6, and #11 and two ports) using an FS-LWA connected to a MU-MIMO AP with MU-BF channel-slicing scheme (iridescent Wi-Fi network).

To visualize the signal boost due to the described beam forming technique, Figure 4.16 shows the radio coverage map obtained with a conventional monopole antenna (4 dBi gain) located in the same office scenario as in Figure 4.15. No spatial frequency diversity is obtained with the simple monopole. The antenna produces an almost isotropic radiation pattern for all frequencies and, hence, provides poor directivity and gain for all Wi-Fi channels. As a result, a −7 dBm signal level is obtained below the antenna location (at the office center). Moreover, the signal level decays to

−12 dBm at a distance of 3.5 m. The low directivity characteristic of such conventional antennas makes the signal be spread equally to all directions. The signal level thus decays rapidly far from the antenna location. This is a waste of signal when compared with beamforming techniques which can focus the radiated signal to the user location.

If we compare the Wi-Fi signal distribution in the office using the MU-MIMO beamforming technique together with the FS-LWA as shown in Figure 4.15, we observe that the relative signal level at the room center is +3 and 0 dBm at a similar distance of 3.5 m. This result represents a relative Wi-Fi signal boost of more than 10 dB when compared with the conventional AP antenna. This signal boost value is even greater at the further zones of the office. It is also very important to remark that the proposed MU-BF technique is extremely simple. The passive LWA creates the corresponding focused zones for each allocated Wi-Fi channel, dispensing with any active beamforming technique which consumes power and time, as well as requiring specific expensive hardware.

One can find in the literature electronically reconfigurable antennas suited for smart wireless APs, e.g. see [24–28]. These designs normally rely on expensive RF control circuits with tunable RF components (RF switches, phase-shifters, etc.) in order to perform the electronic beam scanning. Similarly, one can find electronically reconfigurable LWAs that have been proposed to improve the radio-link quality in modern wireless networks [29–32]. However, again electronically reconfigurable LWAs rely on tunable RF components and electronic control signals to scan the directive beams. On the other hand, the HWM LWA is a totally passive scanning antenna which performs the directive beam steering by simply sweeping the channel frequency. Therefore, passive FS-LWAs can be used as low-cost smart antenna designs. They are very well suited for wireless protocols which allow for agile frequency-channel assignation schemes, as recently in proposed in [16, 33–37].

This type of wireless network, which uses a smart FS-LWA, can be defined as an *Iridescent Wireless Network*. Iridescence (also known as *goniochromism*) is the phenomenon associated with certain surfaces that appear to gradually change color as the viewing angle changes. Examples of iridescence include soap bubbles, feathers, butterfly wings, and seashell nacres (as shown in Figure 4.16b), as well as certain minerals. The word iridescence is derived in part from the Greek word îris, meaning rainbow. The goddess Iris of Greek mythology is the personification of the rainbow and acted as a messenger of the gods. Certainly, as shown in the 2D scheme in Figure 4.15 and in the 3D representation in Figure 4.16, the described wireless network topology using FS-LWAs provides an *iridescent frequency-to-space mapping* or a *colored distribution of the radio signal.*

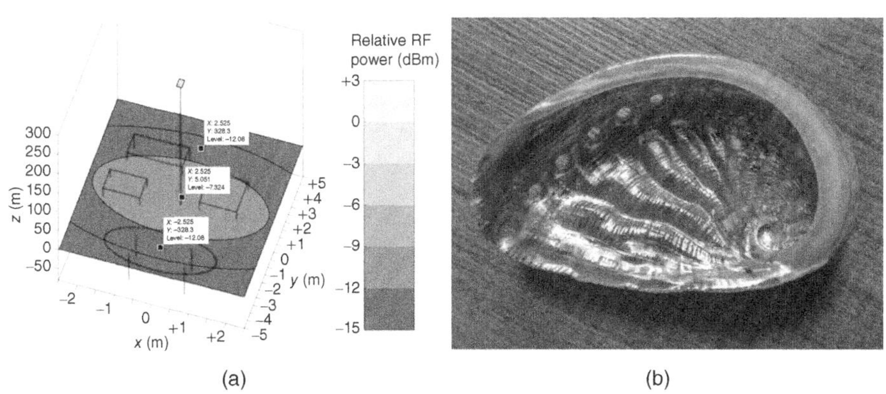

Figure 4.16 Conventional Wi-Fi AP antenna in the same scenario as in Figure 4.15. (a) Radio map. (b) An iridescent sea shell directing different colors toward different directions (goniochronism).

Although the example discussed in this section has been dedicated to the IEEE 802.11ax (Wi-Fi6) standard with MU-MIMO channel-slicing capacity [19, 21], other wireless protocols allow for flexible and *dynamic channel assignation schemes*. As was summarized in Figure 4.4, we can find a multitude of wireless protocols for different applications if we focus on the 2.45 GHz ISM band. Wi-Fi (IEEE 802.11) is intended for high-rate WLANs, and it offers 11 channels with 22 MHz bandwidth along with central frequencies extending from 2.412 GHz (channel #1) to 2.462 GHz (channel #11). As commented, these channels could not be dynamically assigned until the latest standard (IEEE 802.11ax [19]). This newest Wi-Fi6 standard allows the MU-MIMO technique to provide concurrent and simultaneous multi-channel operations. Similarly, as noted earlier, Zigbee [38] is a commercial protocol based on the IEEE 802.15.4 standard [39] that is intended for low-rate WSNs. Zigbee allows for *dynamic channel assignation and channel hopping schemes*. Similarly, the Bluetooth protocol [40] used in low-to-medium data-rate WPANs is also based on IEEE 802.15.4 standard. It assigns 40 channels with typical 1 MHz bandwidth, ranging from 2.402 to 2.48 GHz.

The Wi-Fi 802.11 standard also offers several channels in the 5 GHz band. The central frequency ranges from 5.160 to 5.885 GHz, and the channel bandwidths are tunable between 20, 40, 80, or 160 MHz using the channel bonding method [41]. This standard also allows one to use either the 2.4 or 5.0 GHz band. The 802.11ad/ay Wi-Fi standard, also known as WiGig, operates in the 60 GHz V band and also parts of the ISM bands [42]. The bandwidth allocation, number of channels, and associated channel bandwidths, depend on the regulations in each country. Similarly, in the lower-part of the spectrum, we can find a multitude of wireless systems operating in the 900 MHz ultra high frequency (UHF) band. Wi-Fi 802.11ah (also known as Wi-Fi HaLow) operates in the sub-gigahertz unlicensed band to offer an extended range [43, 44]. Again, each country has different regulations. Europe limits the bandwidth between 863 and 868 MHz for the UHF ISM band, while the bandwidth in the United States ranges from 902 to 928 MHz. The basic channel bandwidth is 1.0 MHz. However, one can use the channel bonding method to increase the bandwidth up to 16 MHz.

The UHF ISM band is also used in other types of wireless protocols apart from the UHF Wi-Fi one. For instance, the UHF band is especially well suited for low-rate long-range communications, particularly in comparison with higher frequency bands which suffer from increased propagation losses over those longer ranges. For this reason, this part of the spectrum is widely used for WSN systems, such as LoRa WAN active WSN or passive RFID (Radio Frequency IDentification) WSN. The LoRa WAN is a long range, low power, wide area networking protocol designed to wirelessly connect battery operated devices to the internet [45]. It allows use of 64 uplink channels from the end-node device to the gateway (channels 0–63), with central frequencies ranging from 902.3 to 914.9 MHz, a 125 KHz channel bandwidth and a 200 KHz channel spacing. For the gateway to end-node device communications, there are 8 downlink channels (channels 0–7) with central frequencies from 923.3 up to 927.5 MHz, a 500 kHz channel bandwidth and a 600 KHz channel spacing [46]. On the other hand, passive RFIDs in the United States use the UHF ISM band from 902 to 928 MHz with a total of 25 channels having a 250–500 KHz bandwidth [47]. Passive RFIDs use backscattering communications. This means that the end-node devices (called tags) are totally passive. The tags passively modulate and retransmit the received signals to identify objects without the need of batteries.

For any wireless protocol, the LWA must be designed to adapt its frequency-scanning response to the available bandwidth and number of channels in the PHY. The 40° scan and 80 MHz frequency ranges in the 2.45 GHz ISM band demonstrated in Figure 4.8a correspond to an approximate maximum scanning ratio of $\Delta\theta R(f)/\Delta f \approx 5°/10\,\text{MHz}$. A higher frequency-scanning bandwidth allows a higher scanning range for a given limited LWA scanning ratio. Moreover, it is important to note that this scanning ratio depends on the operating frequency band. The aforementioned

Table 4.1 Comparison of the available scanning bandwidths and potential angular scanning ranges for a HWM LWA with $\varepsilon_r = 10$ that provides a typical scanning ratio of 50° over a FBW = 3.28%.

Wireless protocol	f_{min} (GHz)	f_{max} (GHz)	Scanning bandwidth (MHz)	FBW (%)	Scanning range θ_{max}
BLE/Zigbee	2.402	2.48	78	3.2	+50°
	2.405		75	3.1	+48°
Wi-Fi	2.412	2.462	50	2.0	+32°
RFID	0.902	0.928	26	2.8	+44°
LoRa	0.9023	0.9149	12.6	1.4	+22°

scanning ratio, 5°/10 MHz, applies only to the 2.45 GHz ISM band. It is better to define the maximum available scanning ratio in terms of the relative or fractional bandwidth (FBW). For the aforementioned case the FBW = 3.28% (80 MHz over 2.44 GHz). If we compare the FBWs that are available in the different wireless protocols for the different ISM frequency bands described earlier, we can estimate the maximum angular scanning range that can be obtained with a fast-scanning HWM LWA with the same material and performance constraints as used for Figure 4.8. These results are summarized in Table 4.1.

The angular scanning range of a fast-scanning HWM LWA depends on the available fractional scanning bandwidth. In the case of the Bluetooth low energy (BLE) standard with a 3.28% FBW, i.e. the 80 MHz bandwidth in the 2.45 GHz ISM band that was shown in Figure 4.8a, the LWA scans up to $\theta_{max} = +50°$. A similar angular range can be attained with the Zigbee standard due to the almost identical scanning bandwidth (see Figure 4.4). However, this angular range is reduced to $\theta_{max} = +32°$ for the Wi-Fi standard due to the reduced scanning bandwidth, i.e. 50 MHz in the same 2.45 GHz ISM band, giving FBW = 2.04%. The passive RFID frequency bandwidth of 26 MHz in the UHF 900 MHz ISM band allows for a 2.89% relative scanning bandwidth, which translates to a $\theta_{max} = +44°$. This angular range is strongly reduced to $\theta_{max} = +22°$ for LoRa due to the more restricted available fractional scanning bandwidth of 1.4%, i.e. only 12.6 MHz in the 900 MHz band.

As can be understood from the comparison Table 4.1, there is a strong link between the PHY specifications for each wireless protocol and the scanning performance of the smart-antenna used to improve the efficiency of the overall communication system. Moreover, it is of paramount importance not only to design the FS-LWA to match the available channels, but also to adapt the electromagnetic response of the LWA to the PHY and the DATA-link layer. This strong link between the different layers of the open systems interconnection (OSI) communications model gives rise to the concept of "*cross-layer optimization*" [48–53], which breaks the conventional strict and independent boundaries between the OSI layers. The cross-layer approach facilitates dynamic exchanges between the different layers via the layer boundaries to optimize the efficiency of the wireless network. In our case, it is evident that:

(a) The smart-antenna frequency-scanning response must be optimized to the frequency band and number of channels provided by the PHY and must provide a given angular scanning range and number of sectorized beams depending on each wireless protocol's available spectrum (as it was exampled in Table 4.1). The EM (electromagnetic) layer of the smart LWA must furnish these properties to the PHY.

(b) The PHY and DATA-link layer connections must enable direct interactions to facilitate the dynamic interchange of information. In particular, once the LWA has been optimized for a given channel allocation in the PHY, the PHY must then scan the available channels in such a

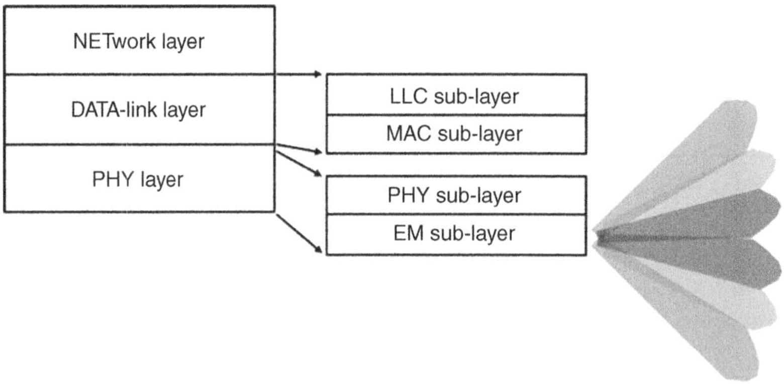

Figure 4.17 OSI communications model with a new EM layer for cross-layer optimization.

way that the smart LWA scans the specified angular space. This functionality expedites the optimization of the parameters of the DATA-link layers such as the signal level, SNR, bite-error-rate (BER), throughput, etc.

The smart-LWA design plays a key role in the optimization of the overall wireless protocol. Its electromagnetic response, i.e. its frequency-scanning radiation patterns, must not only be designed to match a given PHY channel allocation scheme, but also the PHY must take into account the electromagnetic response of the LWA to optimize the performance of the overall wireless network. Clearly the dynamic channel allocation algorithm and associated channel-hopping schemes used by the PHY and requested by the upper layers are of key importance in optimizing tradeoffs and achieving throughput maximization. This is illustrated in Figure 4.17, which shows the OSI communications model with an added sublayer called the *smart-antenna electromagnetic layer*. It is below the PHY. Normally, wireless network engineers have treated the antenna as a black box that simply allows radio transmission and reception. In the proposed *iridescent network topology*, the electromagnetic frequency-scanning response of the smart LWA is a key parameter in the design and optimization of the performance of the overall wireless communications.

As an example, Figure 4.18 shows the variation of the received signal strength indicator (RSSI) measured in a Zigbee iridescent network when a HWM LWA with a frequency-scanning response such as the one introduced in Figure 4.11 is used. The smart LWA has been optimized to scan the 16 available channels of the Zigbee PHY, from channel #11 to #26, which cover the central frequencies from 2.405 to 2.48 GHz. Depending on the position of the Zigbee end-node (called the mote) with respect to the Zigbee AP (called the coordinator), there will be an optimum channel which maximizes the signal level. The signal levels can be dynamically monitored by performing a channel-hopping algorithm so that the RSSI measured at each mote can be obtained as a function of the 16 available channels.

This *cognitive radio* technique [49, 54] is illustrated in Figure 4.18 with the dotted bars exemplifying that the optimum channel that maximized the digital Zigbee signal strength for this example is channel #14. For other frequency channels, the RSSI decays as a result of the mote location with respect to the smart Zigbee coordinator. This coordinator generates an *iridescent signal distribution* in the WSN area as was shown in Figure 4.15. This RSSI distribution, i.e. the digital power level acquired by the PHY of the Zigbee protocol, as a function of the used Zigbee channel, is consistent with the analog power level produced by the smart antenna as a function of the frequency. This feature is indicated with the dotted line in Figure 4.18.

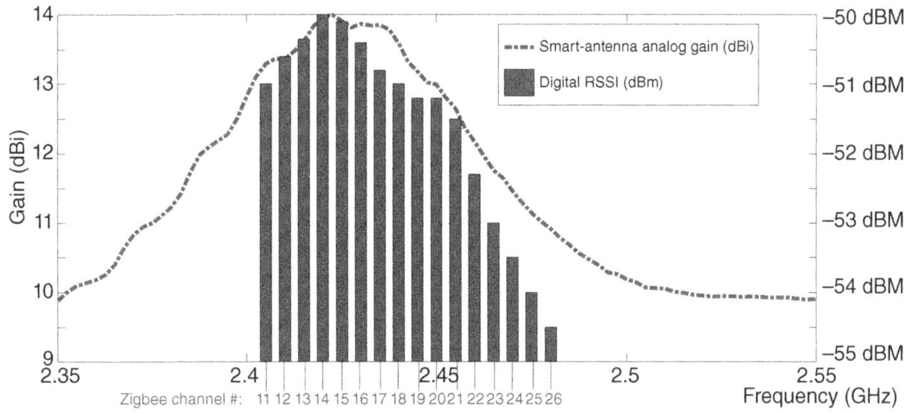

Figure 4.18 Example of RSSI monitorization using an LWA in an iridescent Zigbee network with a channel-hopping algorithm.

This *channel-to-space mapping* information, which arises from the electromagnetic sub-layer corresponding to the optimized smart LWA design, allows the PHY and the DATA-link layers to use efficient algorithms to optimize the wireless network performance. For instance, the iridescent Zigbee coordinator can dictate a sequential sweep over the 16 available channels. The node will then follow this channel-sweeping and will acknowledge it with the RSSI shown in Figure 4.18. A similar check can be performed over the perceived BER or the throughput as a function of the usable frequency channels. In this way, the coordinator can monitor the most adequate channel for each node in the WSN and can then assign this node-channel matching to a look-up table that can be dynamically refreshed. Thus, a smart WSN can adapt to changes in the radio channel, movements of the nodes, or an appearance of any obstacle or radio interference.

Extensive work has been done to increase the efficiency of WSNs; it has given rise to Smart WSNs [55, 56]. Cross-layer optimization using smart antennas has been developed to extend the range of a WSN, increase its signal quality, and optimize its power efficiency [57, 58]. Normally, smart WSNs are based on active smart-antenna technologies such as phased arrays or switched-beam sectorized antennas [59–76]. Smart Zigbee coordinators based on LWAs provide a much simpler solution for smart WSNs to produce beam scanning by simply using channel-hopping schemes. They avoid the power-consuming and expensive active reconfigurable smart antennas. As commented before, it must also be highlighted that a WSN based on the IEEE 802.15.4 standard [38] such as Zigbee, *directly implements time-slotted channel hopping as part of its PHY*.

Normally this channel diversity is used to optimize the use of the available spectrum and to reduce interferences [77–83]. In the case of an *iridescent WSN*, the channel-hopping is used as part of the cross-layer design together with the electromagnetic layer of the smart-antenna coordinator to create a dynamic assignation of the channel to each end-node mote. This arrangement can drastically increase the signal level, and thus extend the range of the WSN or increase its energy efficiency. It must be taken into account that the power constraints of the sensor in a WSN are of key importance. This aspect is in contrast to Wi-Fi WLANs where the end-terminals are not so power dependent. Interestingly, this frequency-space mapping information can be used to localize the mobile terminals using an FS-LWA together with an iridescent wireless network, as will be explained in Section 4.4.

4.4 LWAs Applied for Localization in Practical Wireless Networks

Indoor real-time localization systems (RTLSs) are one of the key enabling technologies for efficient IoT applications and services [84, 85]. The Global Navigation Satellite System (GNSS) cannot be precisely employed in many indoor scenarios. This leads to the necessity of using deployed commercial wireless networks to localize mobile devices. In this context, we can find a multitude of indoor RTLS solutions to localize IoT devices using Wi-Fi WLANs [86] and WPANs based on BLE beacons [87]. They are also found in WSNs using active Zigbee sensors [88] or passive RFID sensors [89]. In general, RTLS approaches can be divided into lateration, fingerprinting (FP), time of arrival (ToA), and direction of arrival (DoA) techniques [90]. They are represented in Figure 4.19 in which three fixed APs of a wireless network (represented with the triangles) are used to estimate the position of a mobile device (represented as a smartphone).

Lateration is based on ranging estimation using the RSSI and propagation models. As shown in Figure 4.19a, once the ranging distance R_i is estimated to each one of the fixed APs, three circumferences with radii R_i can be described. From the intersection of these three circles, the location of the mobile device can be estimated. Despite being the simplest approach, lateration is used in BLE networks. However, it provides inaccurate distance estimations due to multipath fading [91, 92]. In the fingerprinting approach sketched in Figure 4.19b, the RSSI is acquired by a multitude of APs in the calibration phase. It facilitates the construction of a reference radio map that describes the variation of the RSSI as a function of the position. Later in the estimation phase, the set of RSSI levels coming from the mobile device with unknown location are measured at each AP. The ith set is labeled $RSSI_i$. They are compared with the recorder radio map to estimate the position. This estimation is normally performed by some short of trained machine learning (ML) decision tool, such as a neural network (NN) or other type of artificial intelligence (AI) technique. The BLE fingerprinting approach provides more accurate predictions than the lateration technique does, although it relies on cumbersome on-site surveys to construct the radio map. The map might require periodic recalibrations due to variations over the time in a large building [93, 94]. Similarly, fingerprinting is a quite extended RTLS technique for Wi-Fi WLANs [95] and Zigbee WSN [96]. Again, the main drawbacks of fingerprinting are related to the extensive and time-consuming off-line calibration phase [97, 98], and the need of ML techniques to estimate the location with accuracy. On the other hand, the main advantage of fingerprinting is that it relies on the RSSI data, which can be acquired directly from any BLE beacon, Wi-Fi card adapter, or Zigbee coordinator.

In contrast, ToA methods rely on complex synchronization to accurately measure the time of flight (ToF) of the radio signal to each AP, t_i being the time to the ith AP. As represented in

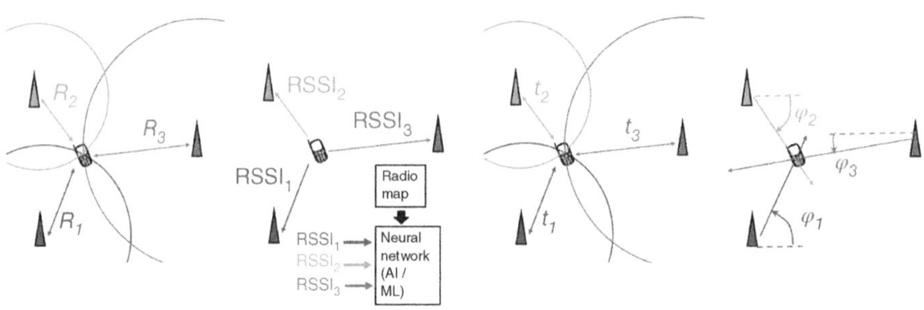

Figure 4.19 Techniques used for indoor localization using wireless networks. (a) Lateration. (b) Fingerprinting. (c) Time of Arrival (ToA). (d) Direction of arrival (DoA).

Figure 4.19c, at least three APs are needed again to perform triangulation using the ToA or ToF. Due to the high speed of light, a small error in the estimation of the ToF causes a significant error in the computed distance. For this reason, Wi-Fi ToA RTLSs require acquisition of complex in-phase and quadrature (IQ) (quadrature signal) data, which involves not only a power measurement as with the RSSI, but also phase information with accurate synchronization [99]. In the case of a Wi-Fi network, the ToF can be extracted from channel state information (CSI) data. It involves a complex calibration which is not in the Wi-Fi standard nor available in commercial off-the-shelf (COTS) APs [100, 101]. More recently, the 802.11mc (Wi-Fi Fine Time Measurement, FTM) standard [102] has provided fine ToF ranging capabilities to Wi-Fi employing the round-trip-time (RTT) procedure. According to the Wi-Fi Alliance and some recent papers [103–106], an accuracy down to a meter level can be reached using Wi-Fi RTT for ToA localization. The fourth RTLS approach is based on the estimation of the DoA of the radio signal propagating between the mobile device and each AP. The corresponding subtending or bearing angles are represented in Figure 4.19d, each φ_i defining the line of sight from the ith AP to the mobile device. A minimum of two APs and normally three APs, are required to intersect three lines for accurate estimation of the position of the mobile device.

Radio direction finding usually requires an antenna-array system to scan all possible angles φ to obtain the DoA [107]. This is illustrated in Figure 4.20 for the two basic beam-scanning techniques for DoA estimation. The first one is based on extracting phase information from the arriving radio signal. As sketched in Figure 4.20a, an incident plane wave front coming from an angle φ creates a phase-shift between the elements of an array of antennas equal to $\Delta\phi = kd \sin \varphi$, where d is the array inter-element distance and $k = 2\pi/f$ is the free-space propagation constant. Therefore, one can estimate the incoming direction φ by measuring the phase shift of the radio signal $\Delta\phi$ created between the elements of the array [108]. This DoA estimation is done by using signal-processing algorithms such as MUSIC [109] or ESPRIT [110]. Because this technique requires phase information of the radio signal received at each element of the array, it can only be obtained using IQ sampling and good synchronization between the array elements. In practice, these issues limit the applicability of this phased-based DoA estimation approach for practical applications since accurate phase information is not accessible in COTS wireless cards.

In contrast, amplitude information can be extracted from the RSSI which is directly accessible in all wireless standards. For this reason, RSSI-based DoA estimation is of much interest. Amplitude-based DoA estimation techniques rely on the use of a smart antenna which can synthesize a reconfigurable radiation pattern to sequentially scan the angular space in the form of a steered directive beam. This technique is illustrated in Figure 4.20b. Some short of control signal is needed to scan the directive beam toward the desired direction/sector. The smart antenna acts in this case as an angular filter which can measure the power received from each sectorized

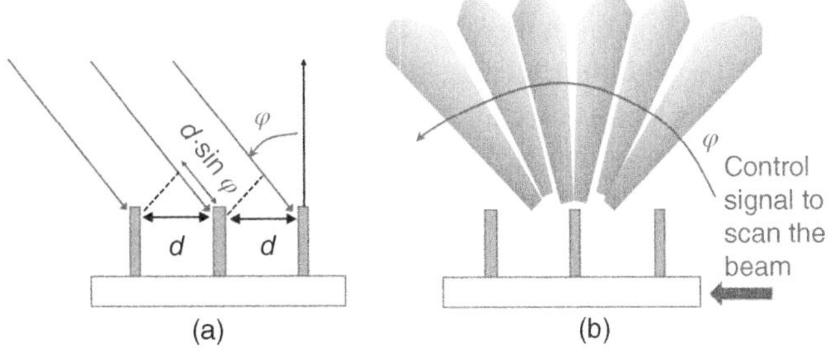

Figure 4.20 DoA estimation techniques. (a) Phase-based estimation. (b) Amplitude-based estimation.

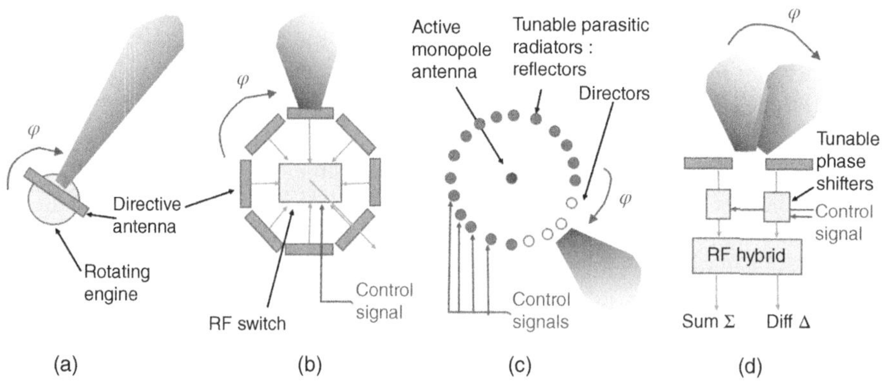

Figure 4.21 Beam-scanning smart antennas developed for RSSI-based DoA estimation. (a) Mechanically revolving directive antenna. (b) Switched beam antenna. (c) ESPAR antenna. (d) Scanned monopulse antenna.

beam. This received power is then processed to estimate the DoA without the need of any phase information of the received radio signal [111]. Clearly, RSSI-based DoA estimation techniques can be easily extended for low-cost efficient localization in practical wireless networks, thus avoiding the need of complex synchronization hardware for accurate phase information. On the other hand, the design of a beam-scanned smart antenna which synthesizes the *scanning power beams* in Figure 4.20b can also increase the cost and complexity of the overall system depending on the selected technology. Figure 4.21 shows the most common beam-scanning antennas used for localization in wireless networks.

One approach for RSSI-based DoA estimation is based on *mechanically scanned antennas* [112–115]. They use an engine to rotate the antenna around one or more axes as sketched in Figure 4.21a. Clearly, this rotating or revolving smart antenna is not an efficient solution compared with electronically-scanned antennas since the scanning is performed slowly and the mechanical components are prone to failures.

One type of electronically-scanned antenna is the *switched beam antenna* (SBA) shown in Figure 4.21b [81, 116–119]. Several directive antennas are arranged so that each one covers an adjacent azimuthal sector, forming a sectorized antenna array. Using an RF switching network connected to the multitude of antenna ports, the control signal commutes between the sectorized beams to govern the beam scanning. Another solution is the ESPAR (electronically steerable parasitic array radiator) antenna [120–122] which uses a single active monopole antenna surrounded by a circular array of parasitic radiators that are controlled with tunable RF components. As sketched in Figure 4.21c, the control signal adjusts the electromagnetic response of these parasitic radiators, which behave either as reflectors or as directors. In this way, a directive beam can be sequentially focused and steered to the desired azimuthal sector. Finally, *electronically-scanned amplitude-monopulse antennas* have also been developed for RSSI-based DoA estimation in wireless networks [123–125]. The control signal in this case tunes the RF phase-shifters response to produce the scanning of the monopulse pattern as sketched in Figure 4.21d.

In summary, all of these electronic beam-scanning smart antennas require multiple antenna elements that can be switched or tuned by an active control circuit, i.e. by RF switches or RF tunable components controlled by external electronic circuitry. These topologies increase the cost, power needs, and the hardware and software complexity. In this context, fixed-frequency electronically-scanning leaky-wave antennas (ES-LWAs) offer a simpler solution to synthesize directive beam scanning with a single-aperture and single-feed antenna. As shown in Figure 4.22a, the ES-LWAs are formed by a single radiating leaky transmission line loaded by tunable

RF components (normally varactor diodes or RF switches). The external electronic control signal tunes the impedance of the RF components, which in turn produces a variation in the leaky-mode propagation constant and thus in the beam scanning direction. Using periodic or metamaterial-based ES-LWAs, several solutions have been developed for DoA estimation [126–131]. Although ES-LWAs offer a simpler solution due to their single-feed topology, they still require expensive RF tunable components and the associated external control circuitry to produce the necessary electronic beam scanning.

FS-LWAs can produce beam-scanning by simply modifying the frequency of the radiating RF signal, as explained in preceding sections. This is illustrated in Figure 4.22b, which shows each scanned beam with a different color to stress that each beam corresponds to a different frequency. As can be seen, there is no need of tunable components and associated electronic control circuits. This is a very interesting advantage which can strongly reduce the cost of the DoA estimation system, assuming one can employ agile channel assignation schemes. Several examples of FS-LWAs proposed for DoA estimation can be found in [132–138]. In contrast, ES-LWAs produce fixed-frequency beam-scanning, i.e. all of the beams illustrated in Figure 4.22a correspond to the same frequency, and for this reason they are plotted with the same color.

The FS-LWA represented in Figure 4.22b is a backward-to-forward scanning antenna with a single RF feed port. It can be used to create the beams scanning from negative to positive angles. As commented in previous sections, forward-scanning LWAs are structurally simpler than periodic/metamaterial-based backward-to-forward LWAs. The former utilize the two opposite RF feeding ports of the LWA to scan both the positive and negative quadrants. This feature minimizes the angular scanning range for a limited scanning bandwidth since each port is dedicated to scan a reduced angular range. This can be an advantageous issue since many wireless commercial protocols are narrowband with operational fractional bandwidths below 3% as it was summarized in Table 4.1.

In the following, the HWM LWA will be used as a smart passive FS-LWA for RSSI-based DoA estimation in a practical iridescent wireless network. As commented in Section 4.3, a synonym for *iridescence* is *goniochromism*, which is derived from the Greek words gonia, meaning "angle," and chroma, meaning "color." The *frequency-to-angular mapping* obtained when a smart FS-LWA is employed in a wireless network can be used to efficiently estimate the angular location of a mobile device using simple RSSI information. It has no need for active electronically-reconfigurable smart antennas.

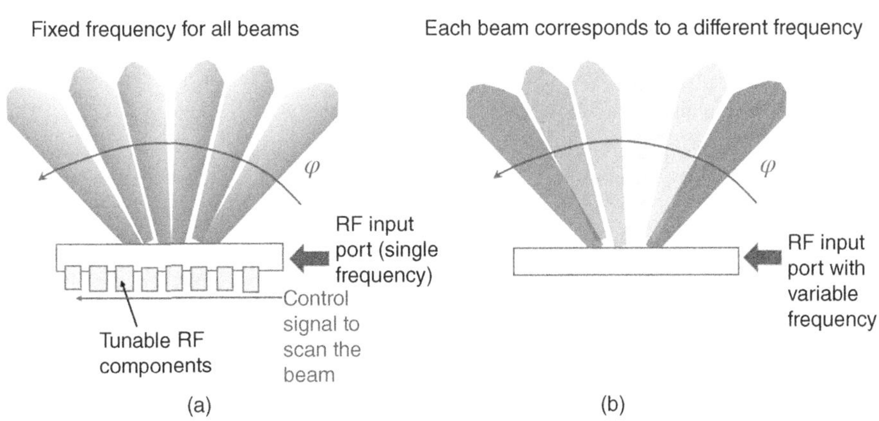

Figure 4.22 Beam-scanning leaky-wave antennas (LWAs) developed for RSSI-based DoA estimation. (a) Fixed-frequency electronically-scanned LWA. (b) Frequency-scanned LWA.

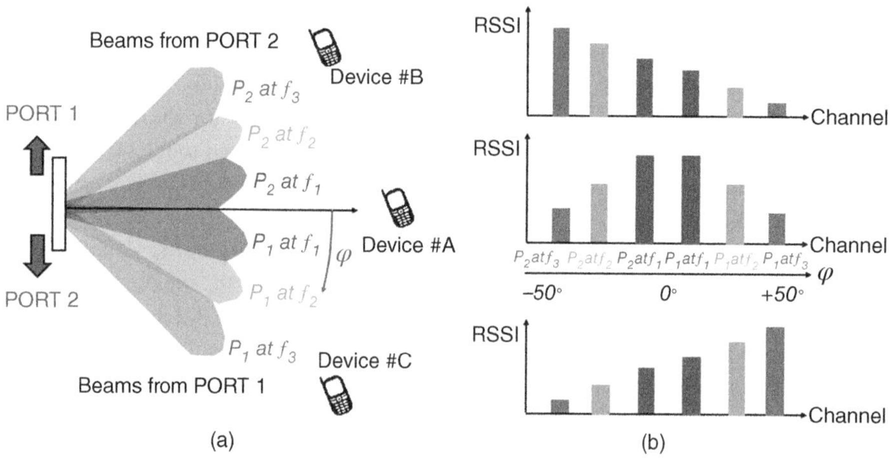

Figure 4.23 RSSI-based DoA estimation using an FS-LWA. (a) FS-LWA based iridescent wireless network illustration. (b) RSSI versus channel spectrums acquired from each one of the three mobile devices.

The basic idea for the DoA estimation using FS-LWAs is illustrated in Figure 4.23. In Figure 4.23a, a two-port FS-LWA as the HWM LWA designed in previous sections is located in an iridescent wireless network that services three frequency channels (f_1, f_2, and f_3). Its two RF ports are designated P_1 and P_2. The six associated beams produced for each frequency channel and port are represented by colors. The RSSI data from three different mobile devices are acquired at each port of the FS-LWA and for the three frequency channels. As represented in Figure 4.23b, a RSSI versus channel power spectrum is obtained from each mobile device. The channels are ordered from P_2 at f_3 to P_2 at f_1 and then from P_1 at f_1 to P_1 at f_3 in order to have a continuous sweep from the negative to positive azimuthal angles φ. Depending on the angular location of each mobile device, this power spectrum will have a unique distribution. For instance, the device #A located in the direction perpendicular to the FS-LWA will have a centered symmetric RSSI spectrum, i.e. the RSSI received by the centered beams P_2 at f_1 and P_1 at f_1, will be higher than the powers received from the rest of the beams scanning the side directions. The RSSI will be lower for the beams P_2 at f_3 and P_1 at f_3 pointing toward the outermost side directions. On the other hand, device #B is located at a negative angle φ. Because beam P_2 at f_3 is pointing toward device #B, it produces a RSSI spectrum with its peak shifted toward its angular direction. On the other side, device #C is located in the positive φ quadrant in the direction of beam P_2 at f_3.

Thus, it creates a RSSI power spectrum that is shifted toward its angular location as shown in Figure 4.23b. As it is qualitatively illustrated, a unique and distinguishable RSSI profile can be obtained that depends on the angular position of each mobile device inside the iridescent network. Consequently, the DoA can be estimated using only the RSSI information. The general principles of RSSI-based DoA estimation using FS-LWAs will be demonstrated with real examples in Wi-Fi and Bluetooth wireless networks in the following sections. Furthermore, two-dimensional localization and near-field effect issues will be briefly described.

4.4.1 FS-LWAs for DoA in Wi-Fi WLANs

The DoA in IEEE 802.11 Wi-Fi WLANs finds a multitude of applications for indoor localization, user location tracking, context-aware services, and security [86, 139]. In the paradigm of a Smart

Home and the IoT, it is desired that a single AP performs the Wi-Fi sensing and direction finding of mobile devices [139]. In this context, several single-AP Wi-Fi direction finding systems have been developed. Most of them estimate the DoA with MIMO antennas and using CSI [140–145]. As noted earlier, CSI yields complex data of the received signal, including phase information from which phased-array techniques can be applied to estimate the DoA [140, 142]. However, CSI data is not directly accessible on commodity Wi-Fi APs. Ad-hoc hardware and software modifications to attain dedicated synchronization are needed to calibrate the system [141, 143–145]. Since RSSI data is directly accessible from commercial Wi-Fi APs, received signal strength (RSS)-based DoA techniques for single-Wi-Fi AP localization have also been developed because they do not rely on complex time and/or phase synchronization.

As explained, RSSI-based DoA techniques require smart beam-scanning antennas. In the case of Wi-Fi, mechanically-scanning solutions including very high frequency omni-directional range base station (VORBA) [121], SBAs [81, 116, 119, 146], and amplitude-monopulse antennas [106, 147, 148] can be found. All of these smart antennas rely on power-consuming active scanning mechanisms to mechanically revolve the VORBA [121]; to switch between the multiplied scanned beams in the case of the SBA [81, 116–119, 146]; or to electronically produce the phase-scanning in monopulse antennas [124, 125].

Note that the static Wi-Fi monopulse solutions presented in [106, 147, 148] attain only a limited field of view (FoV) of 60°. Table 4.2 summarizes these Wi-Fi DoA solutions, highlighting the type of DoA estimation technique and the obtained FoV and DoA estimation mean error along with a brief description of the main drawbacks that can be overcome using FS-LWAs operating an iridescent Wi-Fi WLANs. The smart FS-LWA is a novel RSSI-based direction finding technique for Wi-Fi WLANs. It avoids the use of CSI data or the need of complex active-scanning beam antennas. This novel technique is based on the use of a simple passive frequency-scanning antenna which can be directly connected to a single MIMO Wi-Fi AP to perform the beam scanning required for the RSSI-based DoA estimation without any external control hardware. As highlighted in Table 4.2, the beam scanning is obtained by a sequential channel-hopping scheme which can be directly performed by commercial IEEE 802.11 APs.

Table 4.2 Comparisons of different Wi-Fi DoA estimation techniques.

System	DoA estimation technique	FoV	DoA mean error	Drawbacks
ArrayTrack [140] Phaser [142] OpArray [145]	Phased-array techniques	180°	9° [145]	Phase information from dedicated and calibrated hardware
CUPID [141] SpotFi [143]	CSI to derive phase-delay between MIMO AP antennas	180°	20° [141] 10° [143]	Synchronization to obtain phase from CSI
VORBA [121]	Mechanically revolving antenna	360°	22° [121]	Mechanic steering
SBA [81, 116, 119]	Electronically-controlled switched beam antenna	360°	7.1° [81]	Active RF electronics control circuit to switch
Monopulse [106, 125, 147, 148]	RSSI-based monopulse with two panel antennas	60°	4.3° [148]	Active RF electronics for phase-steering
FS-LWAs [149]	Frequency-scanned LWA	180°	5.4°	Channel hopping is requested

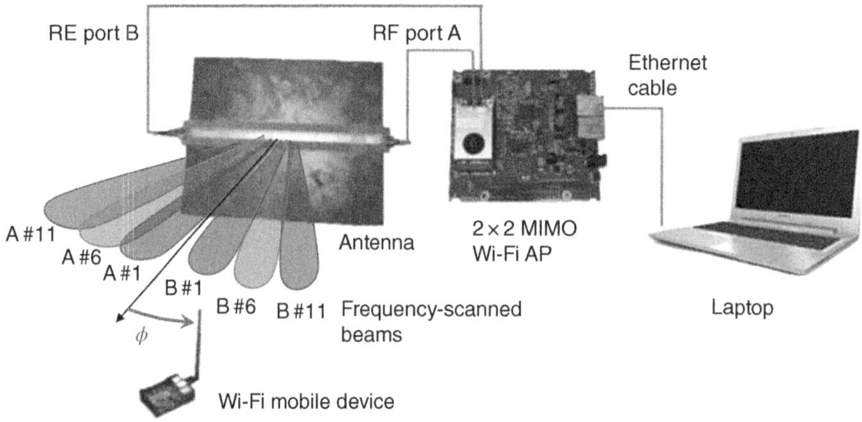

Figure 4.24 Illustration of the smart FS-LWA demonstrator and its connection to a Wi-Fi MIMO card.

Figure 4.24 illustrates the smart FS-LWA Wi-Fi DoA system demonstrator.

An embedded board with a mini-PCIe (peripheral component interconnect express) Atheros AR9380 Wireless 3 × 3 MIMO card is connected to an FS-LWA as shown. The two RF ports (port A and port B) of the LWA are connected to the two RF interfaces of the MIMO Wi-Fi card. The LWA is tuned to operate in the Wi-Fi 2.4 GHz band as described previously. The main idea is that different Wi-Fi channels can be focused to different angular sectors with high spatial directivity. This approach can perform directive beam scanning using a simple passive antenna.

To demonstrate this, the system was characterized in an anechoic chamber as shown in Figure 4.25a. The LWA connected to the Wi-Fi AP was mechanically rotated using a turntable. A mobile device transmitted frames using a fixed Wi-Fi frequency channel, and the MIMO AP card measured the RSSI levels for different azimuthal angles ϕ as it was being rotated. The radiation patterns were obtained for the different Wi-Fi channels, #1–#11, and for the two RF ports, A and B. They are shown in Figure 4.25b and demonstrate the successful generation of directive beams that were digitally steered simply by sweeping the selected Wi-Fi channel. As depicted in Figure 4.25b, Wi-Fi channel #1 with central frequency 2.412 GHz was received at port B with the directive sectorial beam directed at $\phi \approx +20°$. As the Wi-Fi channel was increased, the beam direction was steered to higher scanning angles, i.e. $\phi \approx +30°$ for channel #6 centered at 2.437 GHz, and $\phi \approx +40°$ for channel #11 centered at 2.462 GHz. The rest of the beams associated with all 11 Wi-Fi channels are not shown for the shake of clarity. They demonstrate a smooth continuous scanning response of their peak gain direction from $\phi \approx +20°$ to $\phi \approx +40°$.

Similarly, channel-scanning to negative angles ϕ was performed when the Wi-Fi frames from port A of the MIMO Wi-Fi card were acquired. Recall that the electromagnetic response of the frequency scanning antenna (FSA) is mirror-symmetric. When the two ports of the MIMO Wi-Fi AP and the 11 Wi-Fi channels were combined together, a total of 22 directive scanned beams were obtained. The resulting overall digital Wi-Fi channel-hopping scanning response (i.e. the variation the of scan angle as the Wi-Fi channel is swept) is plotted in Figure 4.26a with 22 black circles, each one corresponding to the direction of the maximum Wi-Fi RSSI level, ϕ_M, produced by each one of the 22 scanned beams. This digital Wi-Fi response is compared with the analog frequency-scanning curve of the LWA in Figure 4.26a. The analog results are shown with the dashed black line. The analog scanning response was obtained by using a microwave vector network analyzer (VNA) connected to the FSA to measure the associated analog radiation patterns.

The measured Wi-Fi digital scanning response is consistent with the expected analog microwave response of the FSA plotted in Figure 4.26a. The small discrepancies observed between them are

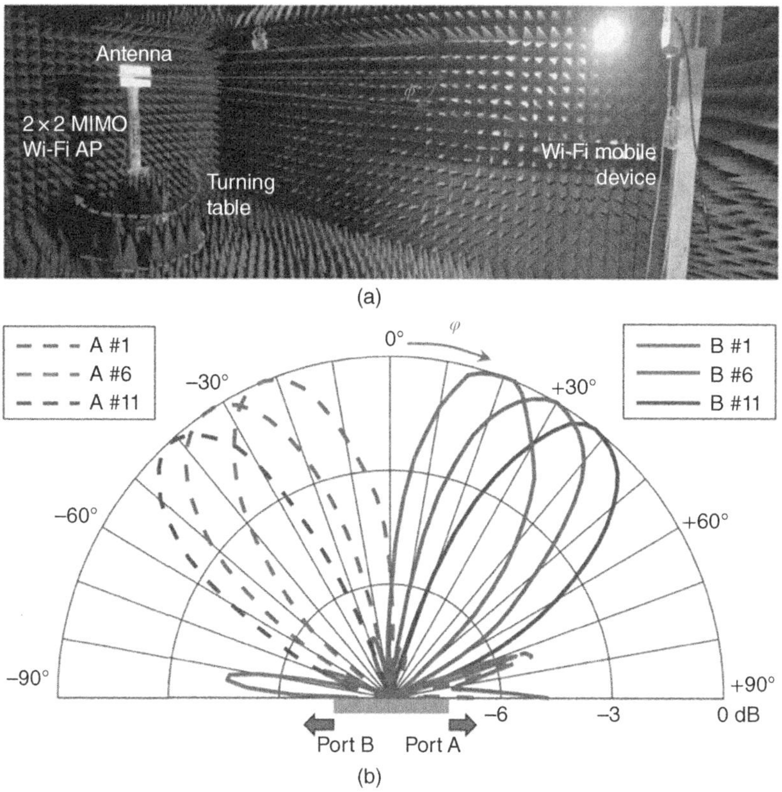

Figure 4.25 Smart FS-LWA antenna system. (a) Experimental setup in an anechoic chamber. (b) Measured digital beam patterns in Wi-Fi channels #1, #6, and #11 using port A and port B.

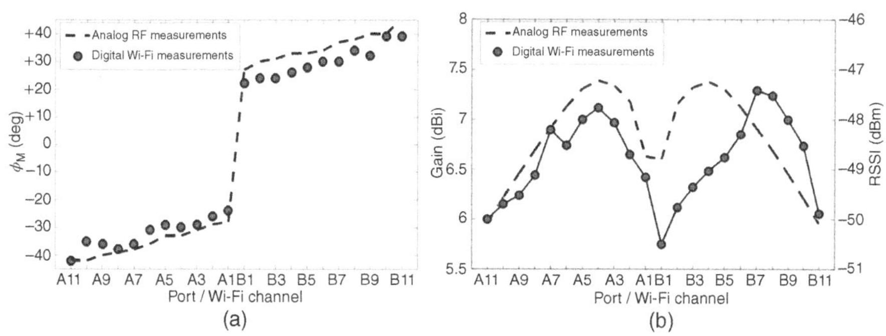

Figure 4.26 Characterization of the FS-LWA operating in the Wi-Fi channels. (a) Maximum radiation angle. (b) Peak gain of the beams.

due to the fact that the Wi-Fi patterns were measured using the RSSI of the Wi-Fi digital frames modulated with a 20 MHz bandwidth. The analog results were obtained using pure RF tones generated by a microwave VNA and were acquired at analog power levels. These differences between the analog and the digital Wi-Fi measurements produced a small shift in the beam scanning responses.

Similarly, the peak gain of the beams as they are frequency-scanned in space must be characterized to calibrate the Wi-Fi system response in the digital domain. Certainly, the 22 obtained beams can show different peak power levels for the different scan directions. The measured Wi-Fi peak RSSI levels (dBm) that were obtained in the anechoic chamber for each beam are shown in Figure 4.26b with black circles. It is corroborated that the peak power response is not uniform. A maximum RSSI level of -47.5 dBm was measured for beam B7 (port B and channel #7, pointing at $\phi \approx +30°$) while the minimum RSSI level of −50.5 dBm was measured for beam B1 directed at $\phi \approx +20°$.

This non-uniform digital power response can be attributed to several factors. First, it arises from the fact that the FS-LWA has an analog RF peak gain response which is not constant as the beams are scanned. This is demonstrated by observing the analog gain response (in dBi) plotted in Figure 4.26b with the dashed black line. The central beams A4 and B4 had the highest analog peak gain, 7.4 dBi. The analog peak gain decreases near to the perpendicular scan angles, i.e. a 6.6 dBi reduction for beams A1 and B1. Decreases also occurred for the beams scanned toward the largest side angles, i.e. a 6 dBi reduction for beams A11 and B11. This fluctuating behavior of the analog peak gain also occurs in the digital gain response. Therefore, it is the variation of the analog gain of the LWA antenna as it scans with frequency that is the main cause of the digital RSSI non-uniform peak gain response as the Wi-Fi channel is changed. Also, note that the digital peak gains obtained in port A (negative scan angles) and port B (positive scan angles) are not completely symmetrical. For instance, beam B7 shows the maximum RSSI level of −47.5 dBm, while its mirrored counterpart, beam A7, shows a lower RSSI of −47.8 dBm. This difference is due to the electronics of the Wi-Fi MIMO card when measuring digital RSSI levels from port A or from port B. Consequently, the digital Wi-Fi RSSI gain response for all 22 beams must be calibrated to construct the MUSIC (MUltiple Signal Classification) steering vectors [109, 128], as described in the next paragraphs.

The MUSIC algorithm [109] has been applied extensively to estimate the DoA in Wi-Fi systems using MIMO APs (normally with three or more antennas as it was shown in Figure 4.13), and using phase-shift information obtained from CSI data [140–143]. Mathematically, the MUSIC angular pseudo-spectrum $P(\phi)$ is constructed with the following equation:

$$P_{\mathrm{MU}}(\phi) = \frac{1}{a(\phi)^H \mathbf{E_N} \boldsymbol{E}_{\boldsymbol{N}}^{\boldsymbol{H}} a(\phi)} \tag{4.7}$$

where $a(\phi)$ represents the so-called MUSIC steering vectors, H denotes the Hermitian, and $\mathbf{E_N}$ is a matrix whose columns are the eigenvectors of the MUSIC matrix. Recall that single-AP Wi-Fi DoA estimation systems use phase-shift information between the different antennas of the MIMO AP in order to construct the steering vectors for different bearing angles ϕ. In our case, we use amplitude information (RSSI data) from each one of the 22 available scanned beams to construct the RSSI-based steering vectors:

$$a(\phi) = [\mathrm{RSSI}_1(\phi), \mathrm{RSSI}_2(\phi), \ldots, \mathrm{RSSI}_n(\phi)] \tag{4.8}$$

with $n = 1 \ldots N = 22$.

As an example, the steering vectors for three illustrative directions: $\phi = -60°$, $\phi = 0°$, and $\phi = +60°$, are represented in Figure 4.27. They have 22 elements corresponding to the RSSI levels acquired at each of the 22 scanned beam directions. Once the RSSI levels are normalized using the digital peak gains in Figure 4.26b, these RSSI levels are arranged from channel A #11 to channel B #11. The shapes of the calibrated steering vectors have a geometrical interpretation which can be connected directly to the DoA. This feature was qualitatively anticipated in Figure 4.23b.

When a Wi-Fi signal arrives perpendicular to the FSA ($\phi = 0°$), the associated steering vector presents a RSSI profile with a power distribution showing the maximum of the received signal

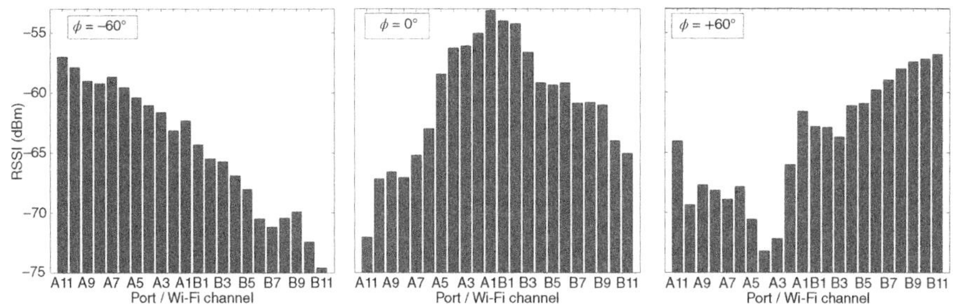

Figure 4.27 MUSIC steering vectors at three illustrative bearing angles: $\phi = -60°$, $\phi = 0°$, and $\phi = +60°$.

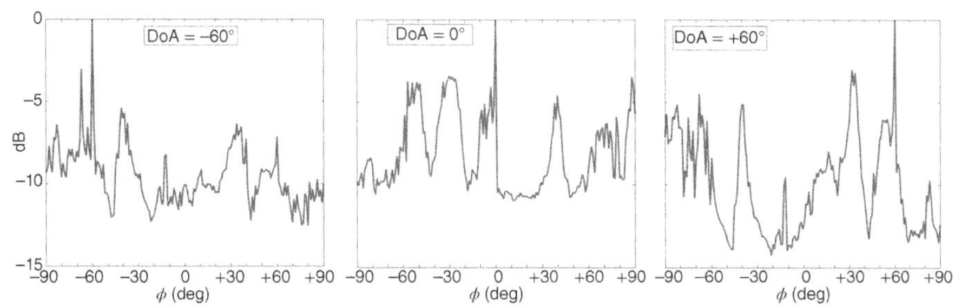

Figure 4.28 MUSIC angular pseudo-spectrums for three different DoA directions: $\phi = -60°$, $\phi = 0°$, and $\phi = +60°$.

for the central channels, i.e. A1 and B1. This result is shown in Figure 4.27. As the DoA of the Wi-Fi signal moves to the left direction (negative angles), the RSSI peak shifts to the left channels as illustrated with the Wi-Fi signal coming from $\phi = -60°$. The maximum received power level corresponds to channel A11, whose associated beam points at $\phi = -40°$, i.e. it receives a higher RSSI than the rest of the beams. Conversely, if the Wi-Fi signals move to right (positive) angles, the RSSI peak of the associated steering vector moves to the right channels. This behavior is also illustrated in Figure 4.27 with the Wi-Fi signal coming from $\phi = +60°$. Similarly, the power profile distribution has the maximum RSSI at channel B11, corresponding to the beam pointing at $\phi = +40°$. Therefore, the RSSI profile associated to the calibrated steering vectors can be geometrically connected to the angular quadrant from which the signal arrives.

This rough qualitative interpretation is given for the sake of simplicity. Obviously the quantitative accurate estimation of the DoA is performed by applying the MUSIC mathematical formulation, which uses the calibrated steering vectors (Eq. (4.8)), to compute the angular pseudo-spectrum (Eq. (4.7)). The MUSIC calculated angular pseudo-spectra for the aforementioned studied DoA angles: $\phi = -60°$, $\phi = 0°$, and $\phi = +60°$, are shown in Figure 4.28. The angular pseudo-spectrum must present a peak at the actual DoA of any arriving Wi-Fi signal. This is demonstrated for the example directions. For instance, the spectrum displayed for a Wi-Fi signal coming from the direction $\phi = -60°$ shows a clear unique peak at the real DoA. The same happens for the other arrival directions. Similar successful DoA estimations were achieved for all of the directions in the wide angular region of 180°. It must be highlighted that the results in Figure 4.28 were obtained for ideal static conditions inside the anechoic chamber in which the same Wi-Fi direction-finding system was calibrated.

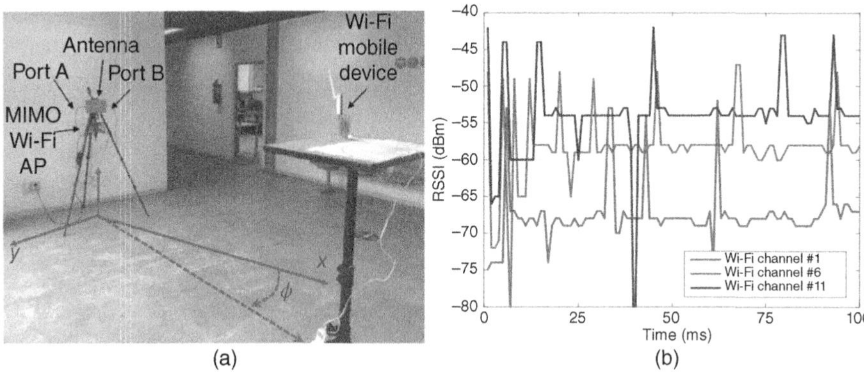

Figure 4.29 Wi-Fi DoA system evaluation. (a) Experimental set-up. (b) RSSI acquired from the Wi-Fi signals.

Once the digital Wi-Fi MUSIC steering vectors had been calibrated in the anechoic chamber in static conditions, the single-AP DoA Wi-Fi system was evaluated in a room scenario. The configuration of this experimental setup is presented in Figure 4.29a. The FS-LWA connected to the MIMO Wi-Fi AP was mounted on a tripod in front of a wall. A Wi-Fi compatible mobile device was connected to the AP Service Set IDentifier (SSID), which dictated the channel frequency. The mobile device was located at different angular points on a semi-circle with a radius of 3 m from the AP. The DoA was estimated for each angular position by acquiring the RSSI levels as the AP performed its channel-hopping scheme. This step was followed by the mobile terminal refreshing the active Wi-Fi channel. The estimated DoA was then compared with the real bearing angle.

The dynamic nature of the Wi-Fi channel hopping plays an important role in the DoA estimation performance. For real-time applications the DoA system must be able to scan the full 180° angular space as fast as possible in order to determine the DoA with a minimum delay. Consequently, the iridescent Wi-Fi AP must perform the channel-hopping dynamically and acquire the RSSI levels of the incoming Wi-Fi signals for all frequency channels. The channel-hopping technique has been used in other single-AP Wi-Fi DoA systems such as Chronos [150] or S-Phaser [151]. As an example of this dynamic channel-hopping RSSI acquisition, Figure 4.29b shows the RSSI data recorded when the system switched between three different Wi-Fi channels: #1, #6, and #11. The Wi-Fi AP configures its operating channel every 100 ms and then hops to the next channel. The Wi-Fi frames are sampled each millisecond so that 100 RSSI data are recorded per channel.

As one can see in Figure 4.29b, the RSSI levels are not stable in time. Notably, the RSSI varies during the first 20 ms after the AP has switched from one channel to the next one. After initial time period, the RSSI level shows a more stable mean value with sudden peaks which might be due to interferences or multipath effects. These strong variations of the acquired RSSI levels observed in the first 20 ms might be due to the mobile device; it needs this time to follow the channel hop dictated by the Wi-Fi AP. Nevertheless, the data show that the system needed to skip the first 20 RSSI samples to attain a more stable RSSI signal.

Under these dynamic channel-hopping conditions, the single-AP Wi-Fi direction finding performance was evaluated by estimating the DoA for different angles in a wide 180° FoV. The mobile device was again connected to the AP at a distance of 3 m following the setup in Figure 4.29. The AP executed a sequential channel-hopping scheme over a set of N channels. The mobile device followed the connection at the channel sequentially dictated by the MIMO AP, which measured the RSSI levels acquired by each port of the FSA. The connection was held in each Wi-Fi channel for

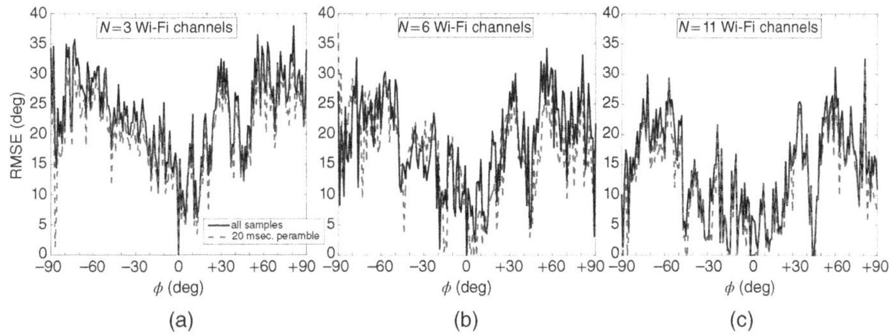

Figure 4.30 Angular distribution of the RMSE for the DoA estimation using different numbers of Wi-Fi channels N. (a) $N = 3$. (b) $N = 6$. (c) All $N = 11$ channels.

0.1 s, and a total of 100 RSSI samples per channel was acquired during this time. The RSSI values for each channel were averaged. Once the RSSI from the N Wi-Fi channels were acquired, the MUSIC technique was applied to estimate the DoA. The process was repeated for all testing angles from $\phi = -90°$ to $\phi = +90°$, and the root mean square error (RMSE) between the real bearing angle and the estimated DoA was evaluated. Figure 4.30 shows the obtained distribution of the RMSE in the DoA estimation for all testing angles as a function of the number of Wi-Fi channels N used in the channel-hopping technique. The DoA estimation error when $N = 3$ Wi-Fi channels were used is lower when the Wi-Fi signal comes from angles close to the direction perpendicular to the antenna ($\phi = 0°$), showing a RMSE below 20° in this central zone.

The error increases up to 30° RMSE as the bearing angle approaches $\phi = \pm 60°$. The better performance for the near perpendicular DoA directions is due to the richest angular diversity provided by the LWA directive beams at these directions. It leads to more angle-sensitive steering vectors for $\phi = 0°$ than for higher angles, e.g. $\phi = 60°$, as shown in Figure 4.27. Note that the results are plotted skipping the first 20 samples to avoid the large variations in the preamble of each channel switch. Thus, the angular error slightly decreases as expected. Similar angular distributions of the DoA RMSE were obtained when the number of Wi-Fi channels employed was increased to $N = 6$, namely, #1, #3, #5, #7, #9, and #11. The RMSE in the central angular zone decreased below 15°, while the RMSE kept higher, around 30°, for angles in the $\phi = \pm 60°$ zones. When all $N = 11$ Wi-Fi channels were used, a reduction of the RMSE below 10° in the perpendicular direction and below 25° in the side directions was realized. Therefore, despite the number of Wi-Fi channels used, the DoA error was lower for angles near the perpendicular direction and the 20 ms preamble was removed to decrease the angular estimation error.

It is apparent from Figure 4.30 that there is a positive effect when the number of Wi-Fi channels N increases, i.e. the RMSE decreases for all observation angles in the wide 180° FoV. Clearly, the MUSIC algorithm benefits from a larger number of independent elements in the steering vectors. It can then resolve the angular space with a higher accuracy [17]. As N is increased, the length of the steering vectors (Eq. (4.8)) increases to ($2*N$) corresponding to the number of associated scanned beams. Therefore, a higher number of employed channels improves the angular sensitivity and thus the ability to distinguish between different DoAs. Simply, a system with more directive beams sniffing the angular space yields more accurate DoAs. To better illustrate this improvement, the DoA RMSE as a function of the evaluated FoV is presented in Figure 4.31a when $N = 3$, 6, and 11 channels are used. The results obtained using and not using 20 ms preamble are compared. The FoV

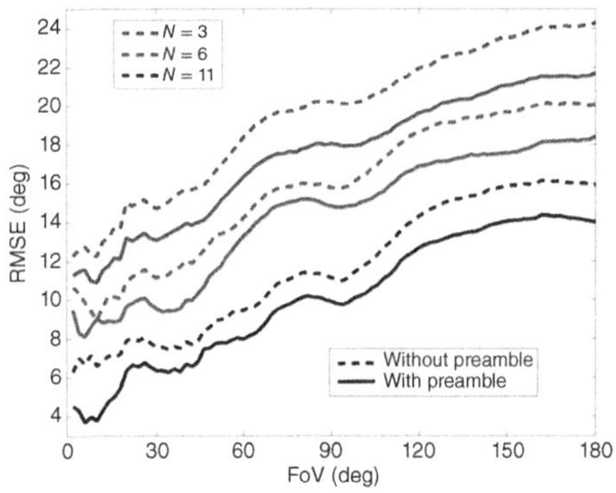

(a)

FoV	$N = 3$	$N = 6$	$N = 11$
45°	14.1°	10.5°	7.1°
60°	16.3°	13.3°	7.9°
90°	18.0°	14.8°	9.9°
120°	19.6°	16.9°	12.7°
180°	21.7°	18.4°	14.0°
Proc. time	0.3 sec	0.6 sec	1.1 sec

(b)

Figure 4.31 DoA estimation as a function of the field of view (FoV) for different numbers (N) of Wi-Fi channels employed. (a) RMSE. (b) Detailed summary.

is defined as the angular region symmetrically centered on the $\phi = 0°$ direction. For instance, a FoV = 30° means that the RMSE is from $\phi = -15°$ to $\phi = +15°$. As the FoV widens, the angular RMSE in Figure 4.31a increases despite the number N of Wi-Fi channels being used. Nevertheless, the preamble improves the DoA estimation performance by reducing the RMSE by approximately 2°, independent of N and the evaluated FoV. Similarly, when N Wi-Fi is doubled, the RMSE in the DoA estimation is reduced approximately by 4° for any FoV evaluated.

These results are summarized in the table in Figure 4.31b. Note, however, that when the number N of Wi-Fi channels used in the channel-hopping scheme is increased, the RSSI acquisition time also rises. Each extra frequency channel used in our example needs 0.1 s to acquire 80 RSSI samples per channel, which are averaged to eliminate the effects of the RSSI peaks as observed in Figure 4.29b. Thus, 1.1 s is needed to perform the channel hopping through all $N = 11$ Wi-Fi channels to obtain the best DoA estimation performance. If less channels are used, this acquisition time can be reduced as also summarized in the table. Moreover, it is clear from Figure 4.31 that the overall DoA estimation performance degrades, showing a higher RMSE, as the operational FoV is increased. The FoV determines the total angular range for the DoA estimation. It actually depends on the application in the context of Wi-Fi sensing in Smart Homes [139]. Two illustrative cases are highlighted in the table of Figure 4.31b. If the Wi-Fi AP is located at the corner of a rectangular room, a FoV of 90° is sufficient for the proposed system to report a DoA estimation RMSE below 10° (with $N = 11$ channels). Conversely, if the AP is located on the ceiling or a side wall, it might be necessary to cover a complete semi-circular FoV of 180° which yields a higher RMSE of 14° in the DoA estimation. Therefore, depending on the FoV (which depends on the topology and location of the Wi-Fi AP) and the requested angular accuracy and delay response, one might choose a channel-hopping scheme using a different number N of Wi-Fi channels.

In summary, the DoA RMSE of the single-AP RSSI-based DoA system using an FS-LWA and an iridescent Wi-Fi network with channel-hopping is 14° for the wide 180° FoV. The corresponding mean error is only 5.4°. Its performance is competitive with current state-of-the-art single-AP Wi-Fi DoA systems, as indicated in Table 4.2. The system time response for data acquisition and angular estimation is 1.1 s. This DoA solution has two main advantages. First, it is based on RSSI data which is directly accessible from commercial APs. It does not need any precise phase information nor time synchronized data. Second, the antenna is completely passive and performs the beam scanning for RSSI-based DoA estimation without the need of electronically switched-beams or mechanically rotating antennas. Consequently, the possibility of using FS-LWAs together with iridescent Wi-Fi

WLANs for low-cost single-AP Wi-Fi RSSI-based DoA estimation is demonstrated. This RSSI-based angle-of-arrival estimation approach can be combined with RTT-based ranging techniques [104] in order to achieve a two-dimensional radio detection and ranging (RADAR) single-AP Wi-Fi location system [106].

4.4.2 FS-LWAs for DoA in BLE WPANs

Another widespread wireless network used for RTLS is Bluetooth [84, 85]. A BLE system uses beacons, which are simple IoT devices that transmit radio beacon (continuous wave or periodic) signals in the 2.45 GHz ISM band [87]. BLE systems are used extensively for proximity, guiding, tracking [152], and localization [153] in indoor scenarios. As noted in Figure 4.4, a BLE system employs 40 physical narrowband channels in the 2.4 GHz ISM band, each separated by 2 MHz. Two transmissions types employ Bluetooth: data and advertising. The data physical channel uses up to 37 RF channels for communications between connected devices. The advertising physical channel uses three RF channels (2402 MHz with index #37, 2426 MHz with index #38, and 2480 MHz with index #39) for discovering devices, initiating a connection, and broadcasting data [154]. Sequential channel hopping between these three advertising channels is directly implemented and performed by BLE beacons. Therefore, in contrast to the Wi-Fi APs, there is no need to program any channel hopping algorithm. There are currently five BLE standards: 4.0, 4.1, 4.2, 5.0, and the recent 5.1 [155]. Note that RSSI information is directly accessible from BLE beacon frames as it is with Wi-Fi WLANs and most radio PHYs.

As noted, current BLE localization services are mainly based on the acquisition of the RSSI from advertising messages transmitted by a beacon [87]. As was illustrated in Figure 4.20, three different techniques are used for such RSSI-based localization: lateration, fingerprinting, and angulation [84]. The BLE lateration is based on ranging estimation. Although it is the simplest approach, it provides inaccurate distance estimations due to multipath fading [91, 92]. The BLE fingerprinting provides more accurate predictions, but it relies on cumbersome on-site surveys to construct the reference radio map [93, 94]. More recently, angular-based location techniques have been developed for BLE positioning. They exhibit performance superior to ranging techniques [156, 157], while dispensing with the acquisition of any time-consuming fingerprinting radio-maps. The BLE angulation is based on the estimation of the direction of departure (DoD) or the DoA between the mobile BLE devices to be located and a set of reference BLE beacons. A BLE angulation system must incorporate some type of smart antenna and signal processing.

As explained for Wi-Fi phase-based angulation (e.g. the Phaser system [142]), the BLE phased-array DoA estimation is performed using IQ data which provides inter-element array phase information [158]. However, phase-based angular estimation requires special hardware which is not yet compatible with COTS BLE 4.0 devices. The new BLE 5.1 standard [155] incorporates phase information at the cost of extra hardware for accurate time synchronization. It is not backward compatible with current BLE COTS products. For all these reasons, BLE RSSI-based DoA is still a topic of much interest for low-cost indoor-positioning applications. As noted, RSSI-based DoA requires smart beam-scanning antennas. This is normally obtained using SBAs [81, 116–119], ESPAR antennas [120–122], or monopulse-antenna topologies [123–125]. In particular, switched scanning monopulse antennas have been proposed for RSSI-based DoA in BLE [157, 159, 160]. In all of these cases, the directive scanned beams are steered and/or switched using RF control circuits.

As a BLE beacon, the FS-LWA must be tuned to scan the respective three central frequencies (2402, 2426, and 2480 MHz) to the desired scan angles because the sequential channel hopping is directly performed among the three available advertising channels (#37, #38, and #39). Since only $N = 3$ frequency channels can then be used, only $2*N = 6$ scanned beams can obtained with a single FS-LWA using its two symmetrical ports. This is a handicap compared with Wi-Fi where a total of $N = 11$ channels and therefore $2*N = 22$ beams could be synthesized. To overcome this issue, an array of two HWM FS-LWAs can be used [161] as shown in Figure 4.32a. To cover a wide FoV using this FS-LWA array configuration, the frequency-beam scanning responses of the two HWM LWA forming the array must be shifted. As shown in Figure 4.32b, the LWA using port A scans from approximately $\phi = 10°$ for channel #37 to $\phi = 30°$ for channel #39, while the LWA using port B scans from $\phi = 40°$ (#37) to $\phi = 55°$ (#39). The measured analog and the BLE digital radiation patterns (obtained from the measured RSSI) are plotted in Figures 4.32c and d, respectively. These results confirm the generation of the aforementioned $4*N = 12$ scanned beams covering a wide 120° FoV.

It must be emphasized that the BLE HWM LWAs were designed using the low-cost FR4 substrate ($\varepsilon_r = 4.55$, $H = 1.0$ mm, and $\tan\delta = 0.016$). They produced the desired scanning response covering 30° over the BLE bandwidth. This scanning response is less sensitive (lower slope or lower scanning velocity), than the one obtained with a higher-permittivity dielectric as the one used for the Wi-Fi HWM LWA (i.e. $\varepsilon_r = 10.2$), which required an angular scanning range of 50° for an even lower scanning bandwidth (see Figure 4.26a). As was explained previously (see Figures 4.7–4.9), the substrate thickness, dielectric permittivity, and strip width must be properly designed to tune the HWM LWA to achieve the desired frequency-scanning response [11, 12]. The BLE design with a lower-permittivity FR4 substrate has a slower scanning response than the Wi-Fi design with its higher angular scanning range and lower scanning bandwidth characteristics. Nevertheless, fiber-based low-cost FR4 substrate was chosen instead of ceramic higher-permittivity substrate to reduce the associated expenses.

A photo of the HWM FS-LWA array connected to four BLE beacons is presented in Figure 4.32a. Each BLE beacon transmits advertising frames using a channel-hopping scheme over the three channels while using a unique identifier. A BLE sniffer acquires the RSSI levels from ach advertising frame and identifies the beacon ID and the channel used for each sampled RSSI. As was necessary for the Wi-Fi DoA system shown in Figure 4.25b, the digital radiation patterns in Figure 4.32d must be first calibrated in an anechoic chamber to characterize the reference steering vectors. Then, the MUSIC algorithm can be applied to estimate the DoD from the RSSI levels obtained from the four BLE beacons signals and three frequency channels.

An example of the steering vectors obtained for a DoD to $\phi = 0°$ is plotted in Figure 4.33a. It is evident that the steering vectors are now composed of $4*N = 12$ elements. This configuration corresponds to the 12 scanned beams shown in Figure 4.32d. As expected, the steering vector for this DoD $\phi = 0°$ example presents higher RSSI levels for the beams A37 and A'37 which are the ones pointing near to the $\phi = 0°$ direction (see Figure 4.32d). The rest of the RSSI levels decrease as the beams point further toward the lateral directions, acquiring a minimum value for beams B39 and B'39, which point close to the endfire angles.

This RSSI distribution follows the expected behavior qualitatively described in Figure 4.23b and as quantitatively observed for the particular case of the Wi-Fi steering vectors in Figure 4.27. The MUSIC angular pseudospectrum obtained for this case is plotted in Figure 4.33b. It has a peak very close to the real DoD and, thus, it estimated the DoD with a low error. An outdoor experimental campaign was performed. It covered a wide 120° FoV and a maximum distance of 12 m. An illustration of this measurement arrangement is shown Figure 4.34a. A mobile BLE device was positioned at different distances: 1, 2, 5, 8, and 12 m, for different azimuthal angles ϕ in 10° steps from $\phi = -60°$ to $\phi = +60°$. The DoD was estimated for each test location. A photo of this experimental set-up is shared in Figure 4.34b. A comparison of the real DoD and its estimated azimuthal angle is plotted in Figure 4.34c. As seen, the angular estimation error is below 10° for all of the test points.

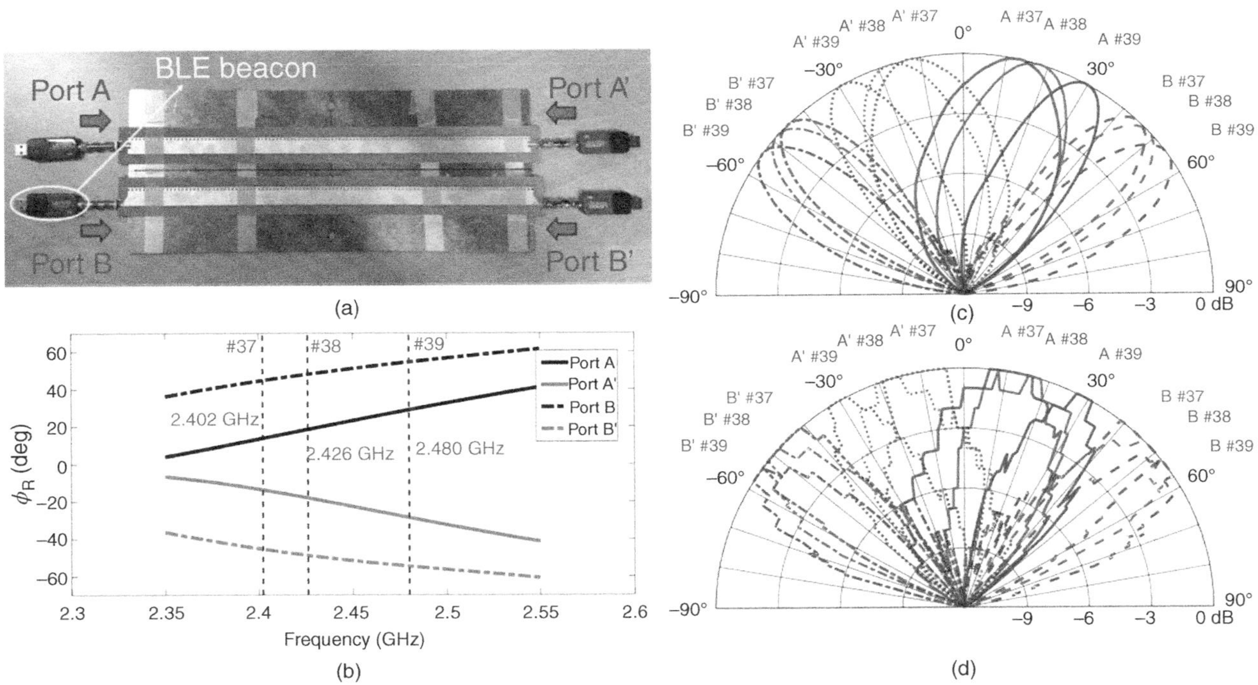

Figure 4.32 Array of two HWM LWAs for BLE DoA estimation. (a) Photo of the array. (b) Frequency-scanning array response. (c) Analog radiation patterns. (d) BLE digital radiation patterns.

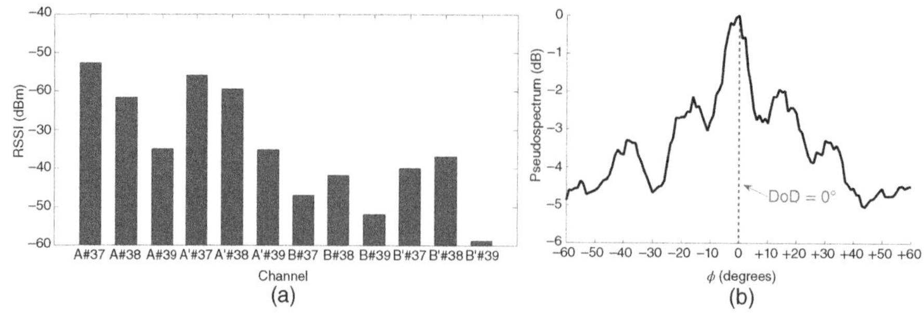

Figure 4.33 BLE DoD estimation results. (a) RSSI level-based MUSIC steering vectors. (b) Angular pseudospectrum for the DoD = 0° example.

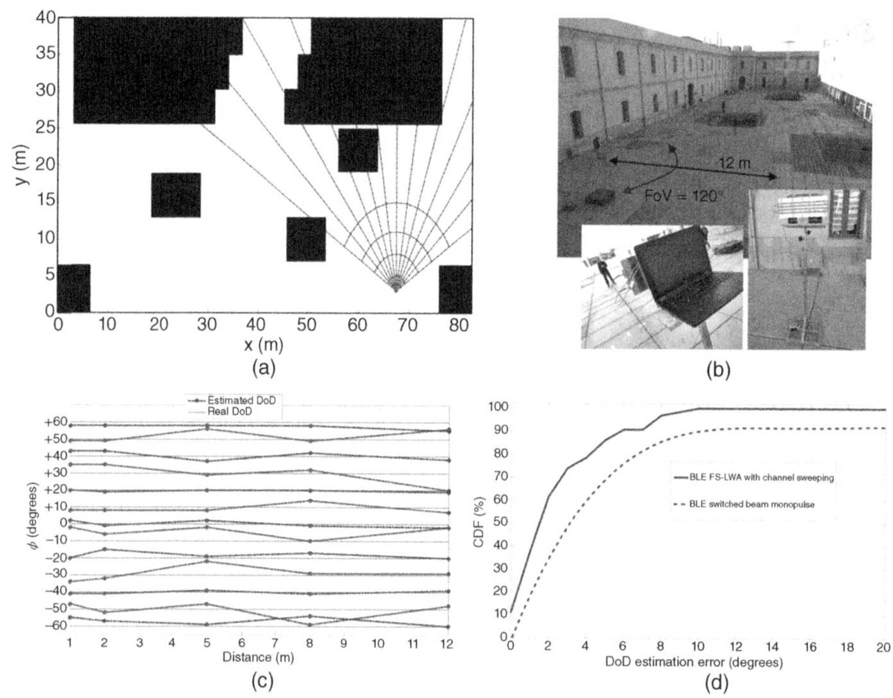

Figure 4.34 BLE DoD estimation experiment. (a) Outdoor test arrangement. (b) Picture of the experimental set-up. (c) DoD estimation as a function of the distance and azimuthal angle. (d) CDF (angular error) of the BLE DoD estimation using the FS-LWA and monopulse results from [160]. Source: (a, c, d) Adapted from [160], (b) From [160] / with permission from IEEE.

These results are compared with those obtained in [160] using switched-beam monopulse antennas for the RSSI-based DoD estimation in BLE wireless networks. The cumulative distribution function (CDF) of the angular estimation error is plotted in Figure 4.34d. The channel-hopping FS-LWA BLE DoD smart antenna system achieves a 50 and 90 percentile error of 1.5° and 5.8°, respectively, while a 50 and 90 percentile error of 3.9° and 8.9°, respectively, was reported in [160] (see Figure 19 in [160]). Moreover, the FoV reported in [160] is at most 90° since the BLE beacons were located at the corners of a room. The FS-LWA smart array for the BLE DoD estimation provides a wider 120° FoV and a higher angular estimation accuracy. Furthermore, it must be highlighted that the FS-LWA DoD estimation technique does not involve any active components, while the advanced BLE beacon for the DoD estimation presented in [160] did. It made use of an active microwave single pole 4 through (SP4T) switch with an associated micro-controller unit (MCU) and beam-forming network (BFN) to create its four directive switched beams. The frequency-scanning solution described in this chapter takes advantage of the channel hopping scheduling inherent in the BLE advertising protocol to avoid any active beam scanning method. Thus, it reduces the smart antenna cost and control complexity. These wide 120° FoV experiments using the FS-LWA and BLE beacons showed a RMSE of 3.7° and a MAE (mean accumulative error) of 2.7°.

If we compare this performance with the one obtained for the Wi-Fi version of this FS-LWA direction-finding system, it is found that the BLE version shows better angular estimation accuracy. As the table in Figure 4.31b indicates, the DoA estimation error in a FoV of 120° using $N = 11$ Wi-Fi channels presented a RMSE of 12.7°, while for the BLE version it is only 3.7°. This improvement can be attributed to the much more stable power transmission of the BLE beacons, which do not exhibit the RSSI variations observed in the received Wi-Fi power levels when sweeping between Wi-Fi channels (see Figure 4.29b). The BLE beacons are IoT devices actually dedicated to serve as stable signal transmitters. They automatically implement the channel-hopping scheme between its three advertising channels to optimize the ranging and DoD estimation. A more detailed description of the stability of the BLE RSSI datagrams and the robustness of this BLE LWA DoD system can be found in [161]. All of these issues make the channel-scanning RSSI-based DoA technique be more robust with BLE beacons than with Wi-Fi frames.

4.4.3 Two-Dimensional Localization

In order to scan the beam in two perpendicular directions, the crossed-LWA array topology developed in [162] can be used. As shown in Figure 4.35b, two FS-LWAs are arranged in perpendicular directions. Recall that a single FS-LWA can only scan in its longitudinal direction, e.g. the x-direction in Figure 4.35a, to estimate the DoD in only one dimension (θ_X). The two perpendicular FS-LWAs shown in Figure 4.35b produce frequency-beam scanning in the two perpendicular, x and y, directions and provide for 2D DoD estimation (θ_X and θ_Y). This topology has been recently developed for 2D DoD estimation in BLE [163]. Using the $N = 3$ BLE advertising channels and the four ports of the crossed-array, a total of $4{*}N = 12$ fan beams can scan the angular space in the

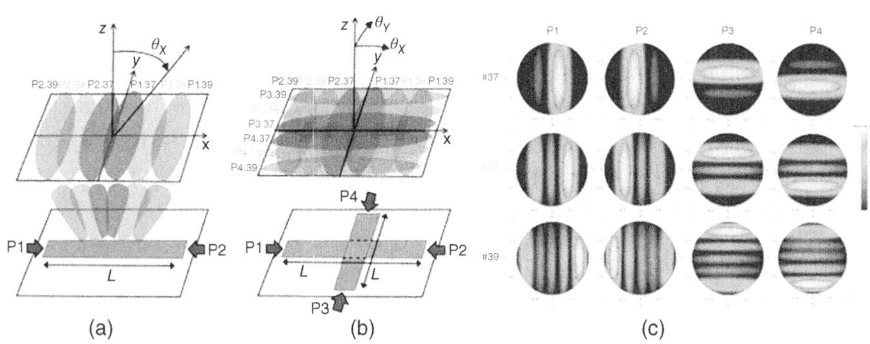

Figure 4.35 Scanning FS-LWA systems. (a) 1D scanning with a single FS-LWA. (b) 2D scanning with a crossed-array of two perpendicular FS-LWAs. (c) Radiation patterns (u–v coordinates) generated by the crossed-array.

two perpendicular directions. Figure 4.35c shows the radiation patterns in u–v coordinates for this crossed-array of FS-LWAs as functions of the BLE frequency channels: #37, #38, or #39, and the array ports where each BLE beacon is located, i.e. 1, 2, 3, or 4 according to Figure 4.35b. Six fan beams are frequency-scanned along the x-direction using ports 1 and 2, and six perpendicular are steered along the y-direction using ports 3 and 4.

Using this 2D-frequency-scanning configuration, the DoD can be estimated with respect to the two principal perpendicular planes, i.e. both θ_X and θ_Y can be estimated. To achieve this, a 2D extension of the frequency-scanning MUSIC algorithm can be applied as explained in [163]. Consequently, the developed smart FS-LWAs can also be used for two-dimensional angular estimation. This result might find useful applications if the smart antenna is located in the ceiling of a room (as it was represented in Figures 4.14 and 4.15). It could then provide two-dimensional localization of IoT mobile devices at the floor level.

4.4.4 Near-Field Effects

Another important aspect to consider when dealing with FS-LWAs for angular estimation is related to the near-field zone. As Eq. (4.3) indicates, the LWA length L_A determines the HPBW. Undoubtedly, an FS-LWA must be sufficiently long in order to synthesize narrow directive scanned beams and, hence, provide angular accuracy. On the other hand, it is well-known that the zone surrounding any antenna can be divided in three distinct regions called the reactive near-field, mid-range (sometimes referred to as the Fresnel zone) and the Fraunhofer or far-field [164] regions. The Fresnel region actually encompasses both the mid-range and far-field regions. These three zones are defined by the outer boundary distance of the near field region and the inner distance (Rayleigh range) of the far-field boundary as:

$$d_{\text{near fields}} = 0.62 \cdot \sqrt{\frac{S^3}{\lambda}} \tag{4.9}$$

$$d_{\text{far fields}} = 2 \cdot \frac{S^2}{\lambda} \tag{4.10}$$

where S stands for the maximum size of the antenna array and λ is the free-space wavelength. In our case with $S = L_A = 45$ cm and $\lambda = 12.24$ cm for the 2.45 GHz ISM band, we obtain a far-field distance of 3.3 m and a near-field distance of 0.53 m = 53 cm.

For the IoT, Smart Areas, and indoor localization situations that employ wireless networks [84, 85], it is very likely that the mobile devices will actually lie in the Fresnel zone between 50 cm and 3 m from the smart antenna. On the other hand, the smart antenna will eventually operate in the Fresnel zone for proximity sensing applications [165–167]. The directive angular radiation patterns in the Fresnel zone are not stable with distance as their shape and particularly their beam width vary with the radial distance. As was shown for the Wi-Fi and BLE DoA estimation examples using FS-LWAs (see Figs 4.25b and 4.32d, respectively), the digital radiation patterns must be carefully calibrated to construct the MUSIC steering vectors and apply the corresponding DoA signal processing. Therefore, it is of key importance to characterize these frequency-scanned patterns as a function of the distance for accurate DoA estimation. This was done in [148] for a fixed-frequency RSSI-based monopulse array used for angle-sensitive proximity sensing using Wi-Fi in the 2.45 GHz band. Similarly, near-field characterization is requested for optimum operation of UHF RFID readers [168, 169].

A preliminary study was performed in [170] for an FS-LWA used for DoD estimation in proximity BLE sensing. These results are summarized here to understand the near-field effects for RSSI-based

DoA estimation using LWAs and channel-hopping schemes. This analysis can be extended to any FS-LWA operating in an iridescent wireless network in the 2.45 GHz ISM band (i.e. Wi-Fi, Bluetooth, or Zigbee), as well as for WSNs in the UHF ISM band such as LoRa or RFID. For the BLE case studied in [170], the FS-LWA had a length $L_A = 76$ cm ($\approx 6\lambda$ at 2.45 GHz) in order to obtain six scanned beams with a HPBW $\Delta\theta = 8.5°$, and a 50° FoV (from $\phi = -25°$ to $\phi = +25°$), using the three advertising channels of BLE. These six directive scanned beams are represented with the far-field patterns plotted in Figure 4.36a. They confirm the narrow HPBW and the desired frequency-scanning range when the BLE channels #37, #38, and #39 and ports P1 and P2 of the FS-LWA are used. However, this HBPW expression is only valid in the far-field zone of the antenna.

Figure 4.36b represents the near-field distribution in the region surrounding the FS-LWA (represented by the light gray rectangle) for the three BLE frequencies and using port P1. It is clear that the directive scanned beams are only defined at a distance sufficiently far from the aperture. To understand the distortion of the near-field radiation patterns, an angular cut at a distance of $d = 50$ cm from the LWA is illustrated in Figure 4.36c.

In contrast to the results shown in Figure 4.36a, the narrow scanned beams are no longer well defined and they become wider closer to the antenna's aperture, losing their directive nature in Figure 4.36c. This behavior makes it impossible to create steering vectors that would capture the DoA of radio signals since the beams are no longer able to filter information from different angles in the near-field zone. Therefore, the LWA is only focused in the far-field zone (far-field focused leaky-wave antenna, FFF LWA) as shown in Figure 4.36a, and the near-zone fields are defocused as shown in Figure 4.36c.

To correct this situation, the LWAs can be focused in the Fresnel zone as first demonstrated with the near-field focused (NFF) LWA designs in the modulated waveguide technology [171]. These NFF LWA designs can also be obtained using modulated (tapered) MLWAs [172] and holographic SIW LWAs [173], thus making the NFF LWA affordable in low-cost PCB (printed circuit board) technologies. As an example, Figure 4.37 shows the radiation patterns obtained for a NFF LWA optimized to focus its radiated near fields at a distance of $d = 50$ cm. The same BLE channels were used to scan a similar FoV from $\phi = -25°$ to $\phi = +25°$ at this short focal distance. This aspect of the results can be seen from the near-field distribution presented in Figure 4.37b. These near fields show a frequency-scanned focused spot at the desired focal distance $d = 50$ cm (represented by the light gray semicircle), which is clearly scanned as the frequency is increased. The near-field angular patterns obtained at this focal distance are plotted in Figure 4.37c, confirming that six narrow frequency-scanned beams were generated that covered the desired FoV from $\phi = -25°$ to $\phi = +25°$ as the BLE channel was swept.

The drawback, as noted previously, is that the NFF radiated fields diverge as they travel out from this short distance, creating defocused patterns for longer distances due to the converging-diverging effect [174]. This behavior can be seen in the near-field distribution in Figure 4.37b, and more clearly in the far-field radiation patterns plotted in Figure 4.37a. The far-field results show how the scanned beams lose directivity in the far-field zone and present higher side lobe levels. Consequently, it is necessary to characterize the radiated fields as a function of distance for accurate location of mobile devices operating in both the Fresnel and far-field zones.

Moreover, NFF modulated LWA designs can be tuned to generate directive frequency-scanned beams at the desired focal distance even in the Fresnel zone, which is quite important for proximity sensing applications. Taking all these effects into account, it has been demonstrated that the RSSI-based DoD-estimation technique using frequency-scanned smart LWAs and channel hopping can be applied to BLE wireless networks, as it was previously shown for Wi-Fi WLANs. The near-field effects must be considered when operating in the Fresnel zone of the smart antennas.

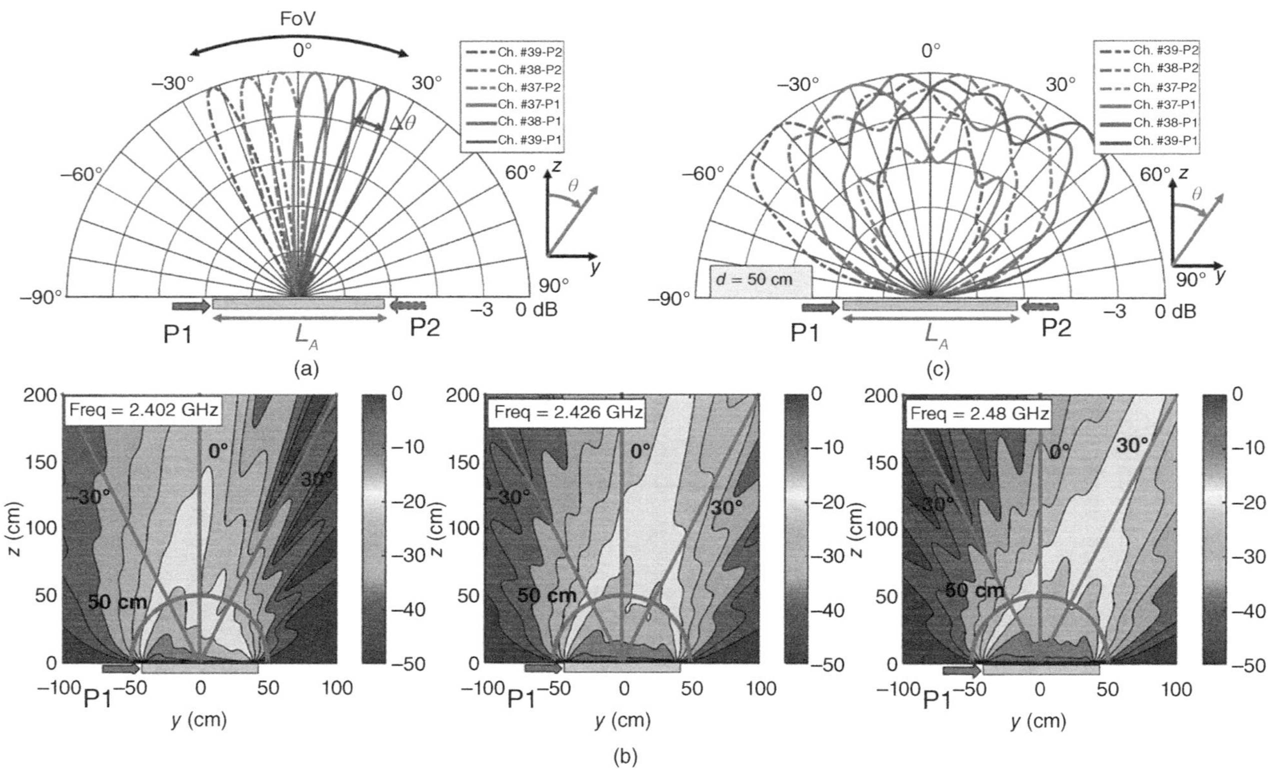

Figure 4.36 Radiated fields radiated by a far-field focused BLE FS-LWA with $L_A = 76$ cm. (a) Far-field radiation patterns. (b) Near-zone normalized field distributions in a region near the LWA. (c) Near-field patterns at $d = 50$ cm.

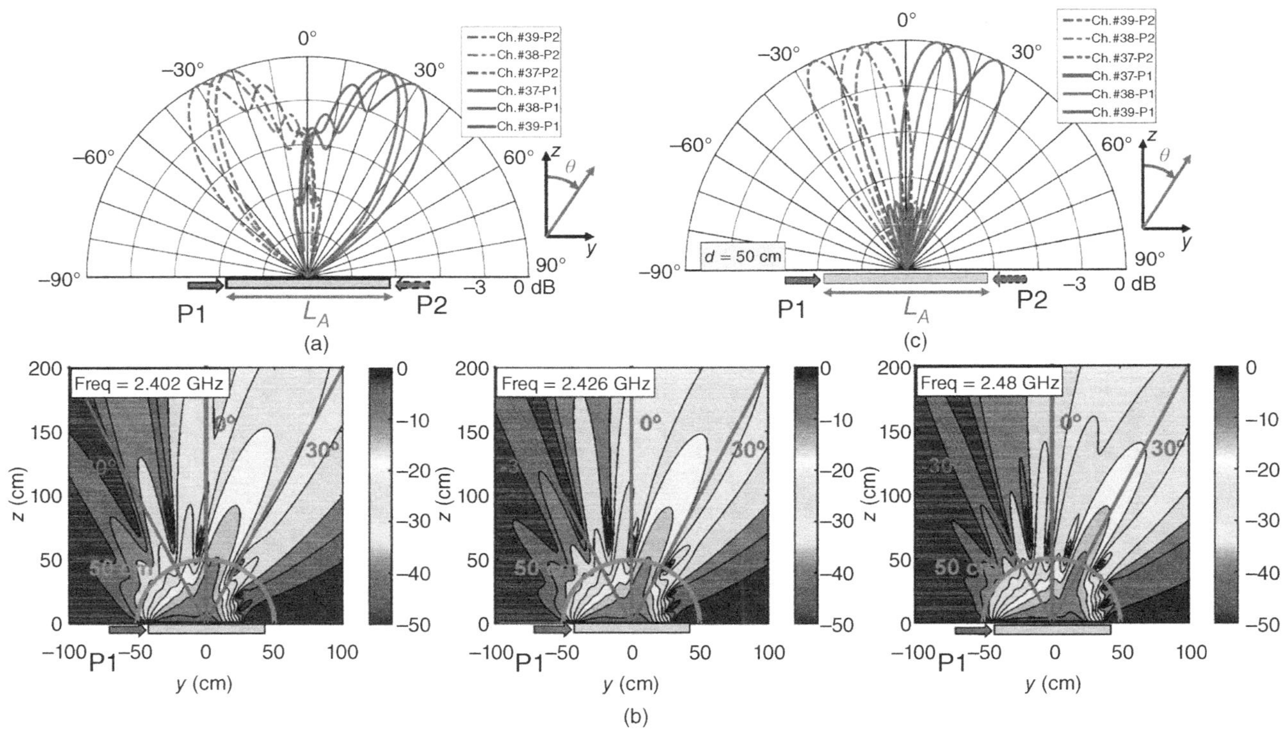

Figure 4.37 Radiated fields radiated by a near-field focused BLE FS-LWA with $L_A = 76$ cm. (a) Far-field radiation patterns. (b) Near-zone normalized field distributions in a region near the LWA. (c) Near-field patterns at $d = 50$ cm.

Any extension to further near-zone distances depends on the frequency of operation and the maximum size of the radiating aperture.

4.4.5 Use of FS-LWA in WSNs

Similarly, smart antennas offer advanced functionalities for Zigbee WSNs operating in the 2.45 GHz ISM band [55, 56]. Smart antennas with directive scanning radiation patterns and reconfigurable capabilities are required to improve the efficient use of energy, increase the limited ranges associated to Zigbee WSNs, and improve their link quality [59–76]. Also, directive antennas are used in these WSNs to improve the security of the radio link [175–178]. Smart antennas are also used for localization in Zigbee WSNs [179–188]. Regarding localization systems in these WSNs [88], the fingerprinting technique is widely employed [96]. Again, angulation techniques avoid the extensive and time consuming calibration of the fingerprinting radio maps and, therefore, are much more efficient. Most of the smart antennas used in Zigbee WSNs for angulation and localization rely of active phased-arrays [179, 180], ESPAR antennas [121, 122], or SBAs [111, 116, 117, 181–183]. Moreover, mechanically steerable antennas have been applied for localization in active Zigbee WSNs [116].

The passive FS-LWAs in conjunction with iridescent channel-hopping schemes can again be used to demonstrate the use of a Zigbee smart WSN for more efficient DoA estimation without the need of active, complex, and expensive smart antennas solutions based on electronic/mechanic scanning. In this sense, Zigbee directly implements time-slotted channel hopping and frequency-diversity [77–83] using the 15 available channels from 2.405 to 2.48 GHz with their 2 MHz bandwidth and 5 MHz spacing (see Figure 4.5). This makes this WSN protocol very suitable for using FS-LWAs to conceive iridescent WSNs with advanced functionalities.

Similarly, iridescent WSNs in the UHF ISM band based on LoRa and passive RFID tags can also be conceived. For the case of passive UHF RFIDs, fingerprinting is not widely employed because of the variations of the radio maps when using passive RFID tags [184]. Most localization schemes are based on phased-array techniques [89, 184–192] or phased-comparing monopulse systems [193–196]. In all of these cases, the phase information between the elements of the antenna array forming the smart reader is required. Furthermore, RSSI-based amplitude monopulse RFID readers have been developed for passive tag localization in the 868 MHz band [197]. Finally, mechanically steered antennas can be also used [113, 114, 198].

To avoid power-consuming motors and phased-scanned arrays, the passive FS LWAs can also be used for DoA estimation in battery-less tags RFID systems. In this case, as described in Table 4.1, the ISM UHF band from 902 to 928 MHz with a total of 25 channels [47] can be used for channel hopping to conceive smart iridescent RFID readers for WSNs using passive tags. Similarly, the available 8 downlink channels (channels 0–7 from 923.3 up to 927.5 MHz) and the 64 uplink channels (channels 0–63 from 902.3 to 914.9 MHz) can be used to realize active WSNs using LoRa WAN [46]. They would have a similar purpose, i.e. to provide iridescent or goniometric angulation in LoRA WSNs using FS-LWAs tuned to this UHF band. In this way, an iridescent UHF LoRA gateway could be used for smart localization, improving both the link efficiency and security, as was done for Zigbee WSNs in the 2.45 GHz band.

For the case of FS-LWAs applied to UHF localization in RFID or LoRa WSNs, it is of great importance to consider the near-field effects described earlier for the Wi-Fi [148] and BLE systems [170] operating in the 2.45 GHz band. The Fresnel zone extends to larger distances using the higher wavelengths in the UHF band [169, 170]. This necessarily forces one to consider near-field effects for the correct estimation of the DoA as has been reported in RFID phased-array readers

[191, 199, 200]. As one can now understand, the concept of iridescent WSNs can be applied for direction finding in any WSN that allows the agile use of channel hopping schemes, as recently demonstrated with FS-LWA UHF designs for passive RFID [221, 222], and active LoRa [223] sensor networks. In Section 4.5, the iridescent wireless topology will be applied to efficient wireless power transfer (WPT).

4.5 LWAs for Efficient Wireless Power Transfer

Far-field, radiative wireless power transfer (RWPT) is a promising technique to remotely power mobile devices over relatively long distances [201–204]. Due to safety power restrictions and low conversion efficiencies, RWPT is mainly considered nowadays for low-power devices and WSNs [205]. As for the case of smart WSNs for efficient radio links, simultaneous wireless information and power transfer (SWIPT) needs smart adaptive antennas able to scan their directive beams toward the location of the device to wirelessly power them using radio signals [206]. Normally, adaptive phased-array antennas are used for RWPT [206, 207], although other solutions such as retro-directive arrays [208], time-modulated arrays (TMA) [209], metasurface antennas [210], frequency-diverse arrays [211], or Rotman-lens SBAs [212] have been developed. All of these smart antennas systems for RWPT rely on active arrays in which their beams are reconfigured using electronic signals to control their tunable RF components. As a result, the use of dedicated expensive smart antennas for RWPT it not usually considered. Consequently, it would be desirable to use the same low-cost smart antenna considered in this chapter for SWIPT in future 5G wireless networks [213], including commercial WLANs such as Wi-Fi [214, 215].

As for the case of active smart antennas applied for efficient wireless connections, we briefly show that FS-LWAs can be applied as a low-cost passive antenna solution for RWPT in iridescent WSNs [216–218]. This is illustrated in Figures 4.38 and 4.39, which represent the power density created by a typical directive panel antenna (Figure 4.38a) and an FS-LWA (Figures 4.38b–d) in a WSN scenario. As illustrated in Figure 4.39a, the WSN is composed of 25 sensors located at floor level (x–y plane), distributed in a 5×5 grid, and with a separation of 30 cm between adjacent sensors. Thus, the total WSN area is $120 \times 120\,\text{cm}^2$. These sensors are illuminated with a directive antenna with gain G_{TX} located at a height $z = H = 1.0\,\text{m}$. The antenna radiates microwave energy with a transmission power level $P_{\text{TX}} = 1.0\,\text{W}$, and generates a power density distribution at the floor level $S\,(\text{W/m}^2)$ that is given by the relation:

$$S(r, \theta, \varphi) = \frac{P_{\text{TX}}}{4\pi r^2} G_{\text{TX}}(\theta, \varphi) \tag{4.11}$$

where r is the distance and θ, φ are the subtending angles between the transmitting antenna and any receiving sensor node position. The DC power transferred to any node of the WSN can be estimated with the link-budget equation used in RWPT [205]:

$$P_{\text{DC}} = \eta_{\text{RF-DC}} \cdot \eta_{\text{rad}}^{\text{RX}} \cdot A_{\text{eff}}^{\text{RX}} \cdot S(r, \theta, \varphi) \tag{4.12}$$

where $\eta_{\text{RF}-\text{DC}}$ is the RF-to-DC conversion efficiency and $A_{\text{eff}}^{\text{RX}}$ and $\eta_{\text{rad}}^{\text{RX}}$ are, respectively, the effective area and the radiation efficiency of the receiving antenna. Unlike long-range point-to-point WPT systems where large collecting antennas are used to maximize the capability to gather the incident power at the receiver [219], $A_{\text{eff}}^{\text{RX}}$ and $\eta_{\text{rad}}^{\text{RX}}$ cannot be maximized in a practical WSN because the nodes must be low-cost and small in size. Similarly, the conversion efficiency $\eta_{\text{RF}-\text{DC}}$ is fixed once the harvester circuitry at the sensor node is designed. Transmitting antennas with high gain G_{TX} are used in WSN RWPT systems [201, 205] to intensify the RF power density at the node (Eq. (4.11)) and therefore enhance the converted DC power level (Eq. (4.12)). However, a high-gain transmitting antenna

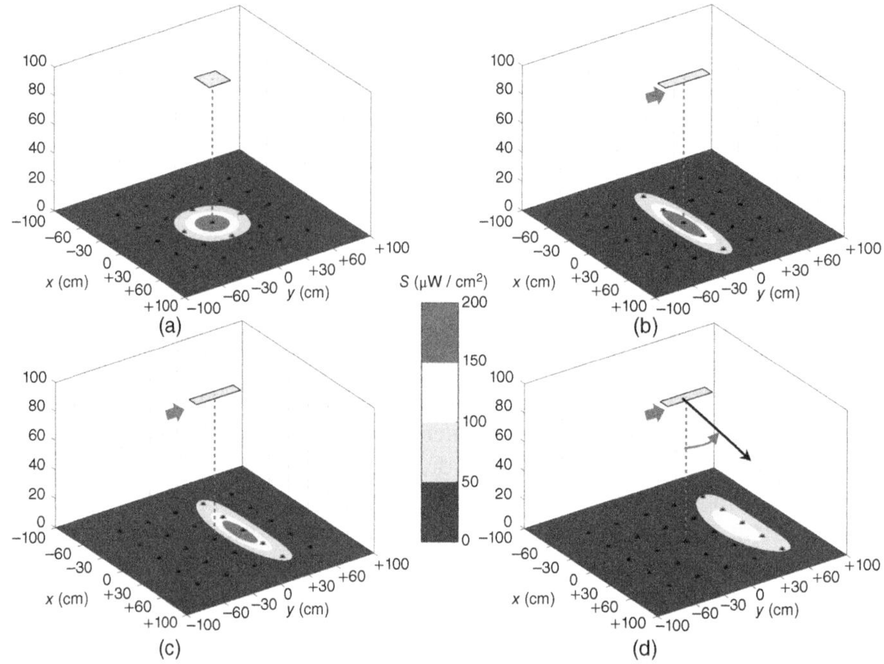

Figure 4.38 WPT in a WSN scenario. (a) Panel antenna. (b–d) FS-LWAs.

implies the use of a directive, narrow radiated beam. It can only efficiently illuminate a small region of the WSN zone. This drawback is illustrated in Figure 4.38a, which show the power density generated by a panel antenna with G_{TX} = 14 dBi. As shown with the dark gray in Figure 4.38a, power densities above 200 μW/cm^2 are obtained just below the transmitting antenna. However, this power radially decays away from it. If we assume a minimum value of 50 μW/cm^2 as the threshold to power our sensor nodes [201], we find that only the central five nodes within a radius of 30 cm are empowered (this limit is shown with the light grey contour curve in all the plots in Figure 4.38). In order to power the 20 remaining nodes of the WSN, the directive beam must be redirected toward other directions. As previously noted, beam steering has been accomplished in RWPT using active 2D beamforming antennas such as adaptive phased-arrays [206, 207], retro-directive arrays [208], TMA [209], metasurface antennas [210], frequency-diverse arrays [211], and SBAs [212].

Using an FS-LWA together with a channel hopping scheme, directive RWPT beaming with iridescent frequency-to-space mapping can be obtained. This outcome is illustrated in Figures 4.38b–d. A fan beam is scanned toward the y-direction of the WNS grid as frequency is increased from 2.4 to 2.5 GHz. It must be emphasized that the FS-LWA provides a G_{TX} = 14 dBi transmitting gain similar to the panel antenna in Figure 4.38a, but with a modified beam shape. The panel antenna effectively creates a circular illuminated region since the radiated pencil beam it radiates has equal HPBWs in both principal planes, i.e. $\Delta\theta_X = \Delta\theta_Y = 30°$. In contrast, the fan beam radiated by the LWA covers a similar stero-angular area, but is redistributed with a narrow HPBW in the scanning y-direction ($\Delta\theta_Y = 15°$), and a much wider beam in the transverse x-direction ($\Delta\theta_Y = 60°$). As a result, the covered stereo-angle ($\Delta\theta_X \Delta\theta_Y$) is similar in both cases, but with different shapes. This way, while the panel antennas in Figure 4.38a illuminates the five sensor nodes in a central circle with $S >$ 50 μW/cm^2, the LWA in Figure 4.38b also illuminates five sensors but in the same row, i.e. five sensors lying along the x-direction with the same y position. Thus, all of the rows at different

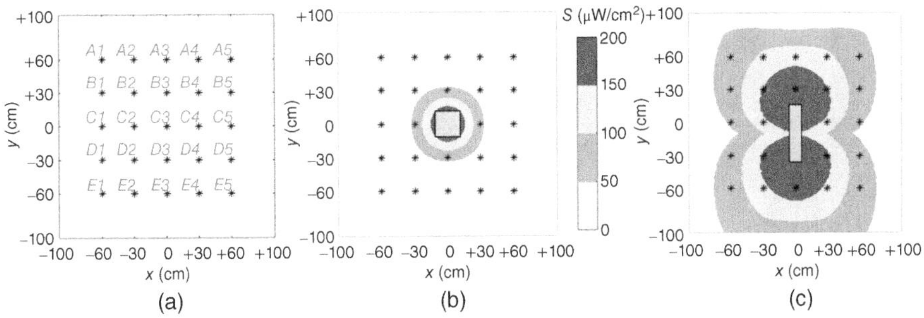

Figure 4.39 Wirelessly powered WSN. (a) Distribution of sensors. (b) Power density generated by a panel antenna. (c) Total accumulated power density generated by an FS-WLA.

y positions can be efficiently illuminated earlier the threshold $S > 50\ \mu W/cm^2$ by simply frequency scanning the fan beam along the y-direction. As the beam is scanned toward higher angles and, hence, distances, the power density decreases as shown in Figure 4.38d. Nevertheless, the row with the five nodes located further from the central point is also powered over the minimum threshold. The total coverage is plotted in Figure 4.39c. As can be seen, all of the 25 nodes are powered using an FS-LWA without the need of an active scanning antenna.

Figure 4.40a shows the radiation patterns in the YZ plane for three frequencies: 2.4, 2.45, and 2.5 GHz. Figure 4.40b displays the corresponding power density profiles along the scanning y-direction for the central column ($x = 0$). The HWM FS-LWA array is designed to generate the required frequency-beam scanning response. A picture of the fabricated antenna is shown in Figure 4.40c. Low-cost FR4 substrate ($\varepsilon_r = 4.48$, $H = 1.0$ mm) was used, and the radiating strip width and length were optimized to $W = 14.75$ mm and $L = 52$ cm, respectively. The frequency

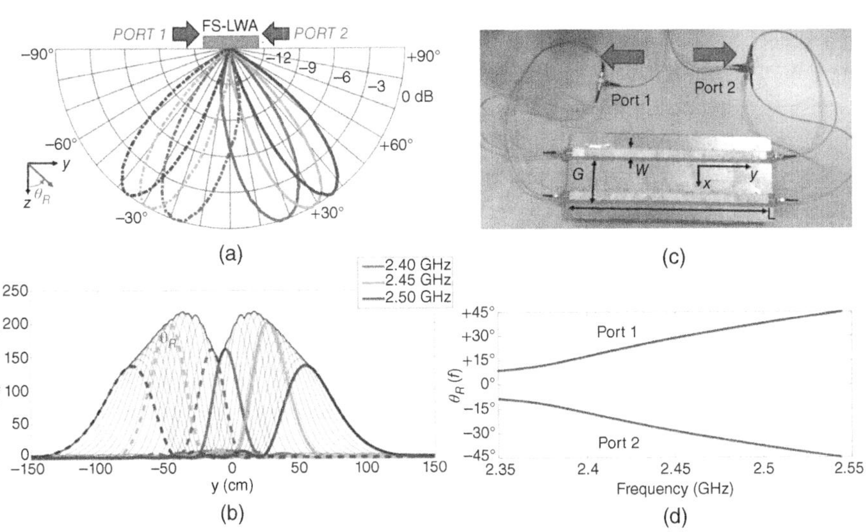

Figure 4.40 Array of FS-LWAs for RWPT. (a) Far-field radiation patterns for three frequencies. (b) Power density generated on the array by the panel antenna along the y-direction for the central column of sensors ($x = 0$) in the WSN for the frequencies in the Zigbee band. Three results are coordinated with the same three frequencies in (a). (c) Photo of the fabricated array of FS-LWAs for RWPT. (d) Frequency-scanning response in the Zigbee band.

Figure 4.41 Experimental set-up of the RWPT using the HWM FS-LWA and a IEEE 802.15.4 Zigbee WSN.

scanning response plotted in Figure 4.40d was obtained. As shown, the FS-LWA scans from $\theta_R = 15°$ at 2.4 GHz up to $\theta_R = 35°$ at 2.5 GHz. As was demonstrated earlier in this chapter, a substrate with a higher dielectric permittivity would have provided a wider angular range for a similar scanning bandwidth. More details about the design of this HWM FS-LWA for RWPT can be found in [216]. The power density distributions in Figure 4.40b demonstrate that the entire length of the WSN zone, from $y = -1$ m to $y = +1$ m, can be scanned using this bandwidth with sufficient radiated microwave power to be above the threshold, i.e. $S > 50\,\mu\text{W/cm}^2$. The beam results for other intermediate frequencies are also plotted. The total power coverage taking into account all of the bandwidth is apparent.

An experimental demonstration of the RWPT using an IEEE 802.15.4 Zigbee WSN and the designed HWM FS-LWA array was performed. As illustrated in Figure 4.4, Zigbee provides 15 frequency channels from channel #11 centered at 2.405 GHz to channel #26 at 2.48 GHz with a 2 MHz channel bandwidth and a 5 MHz inter-channel spacing. The experimental setup in the anechoic chamber is shown in Figure 4.41. A 1.0 W output power amplifier is connected to a microwave signal generator which sweeps the frequencies in the 2.45 GHz band. The amplifier output is switched between port 1 and port 2 of the FS-LWA using an RF SPDT (single pole double throw) switch. A WPT harvester from Powercast [220] is used to convert the incoming RF power to DC at different locations inside the anechoic chamber. Those locations are the node locations sketched in the configuration shown in Figure 4.39a. The harvester device transforms the received RF power to a pulsed signal whose duty cycle is proportional to the converted DC power. The latter is used to charge the sensor's batteries.

To demonstrate the concept of optimum RWPT by frequency sweeping, Figure 4.42a shows the measured DC duty cycle for the five positions in the central column in Figure 4.39a, i.e. A3, B3, C3, D3, and E3. The transmitting RF frequency is swept from 2.38 to 2.55 GHz, and the two feed ports of the FS-LWA are switched to obtain scanning in both the positive and negative angular quadrants

As shown in Figure 4.42a, the nodes at the positive *y*-positions: A3, B3, and C3, are powered when port 1 is used, while the negative *y*-positions: D3 and E3, must be powered using port 2. Also, it is demonstrated in Figure 4.42a that there is an optimum frequency for each node location that maximizes the collected DC power. This pulsed waveform was captured with an oscilloscope; it is shown in Figure 4.42b. To illustrate the results of the experiments, consider the node located at D3 in Figure 4.42a. It shows a maximum duty cycle of 1% when the FS-LWA is fed from port 2 and radiates power at 2.42 GHz. This corresponds to a scan angle $\theta_R = -15°$ as shown in Figure 4.40d. At lower or higher frequencies, the FS-LWA scans at lower or higher angles and, thus, the radiated beam is not optimally directed toward the exact spot of the sensor. Consequently, the collected RF power is reduced and, hence, the output DC power is. For this fixed location D3, Figure 4.42b shows

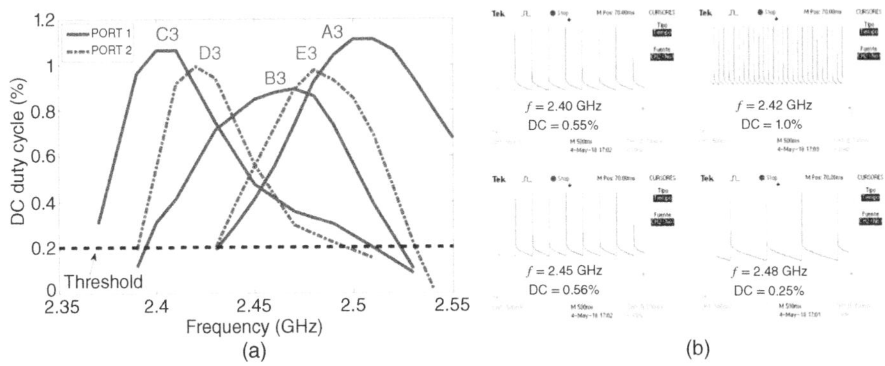

Figure 4.42 RWPT results using the HWM FS-LWA and a IEEE 802.15.4 Zigbee WSN. (a) Measured DC duty cycle at the central column as a function of frequency and switched port. (b) Pulsed waveforms at the RF-DC converter position D3 for several frequencies.

the waveforms measured at the RF-DC converter in four-second time intervals and for several different transmission frequencies. As noted earlier, the waveform shows a higher number of pulses (a lower period between each pulse and thus a higher duty cycle) as the collected RF-to-DC power increases. As shown in Figure 4.42b, the signal presents the highest duty cycle at 2.42 GHz, i.e. 24 pulses in four seconds. The beaming efficiency between the LWA and the node at D3 is maximized at this frequency. The rectified signal shows lower duty cycles for other frequencies because the associated beams are not optimally focused on this location.

The experimental results in Figure 4.42a demonstrated that the five sensor locations in the central column, i.e. $x = 0$, positions A3, B3, C3, D3, E3 in Figure 4.39a, show a peak frequency inside the 2.4–2.48 GHz band at which the duty cycle above the minimum threshold (0.2%) is maximized. This peak RF-to-DC duty cycle is above 0.8% for the five positions in the central column that covers a total length in the y-direction of 1.2 m (see the grid in Figure 4.39a). For other uncentered columns, i.e. other x positions from $x = -60$ cm to $x = +60$ cm, the power decreases with respect to the central column. Nevertheless, the peak power density is above the threshold as theoretically anticipated in Figure 4.39c. Thus, the RF-DC duty cycle is above 0.2% for all 25 node locations and, hence, covers the whole WSN area of $1.2 \times 1.2\ \text{m}^2$. This coverage is illustrated in Figure 4.43b. These experimental results demonstrate that the 25 nodes can be efficiently powered using this frequency-scanning

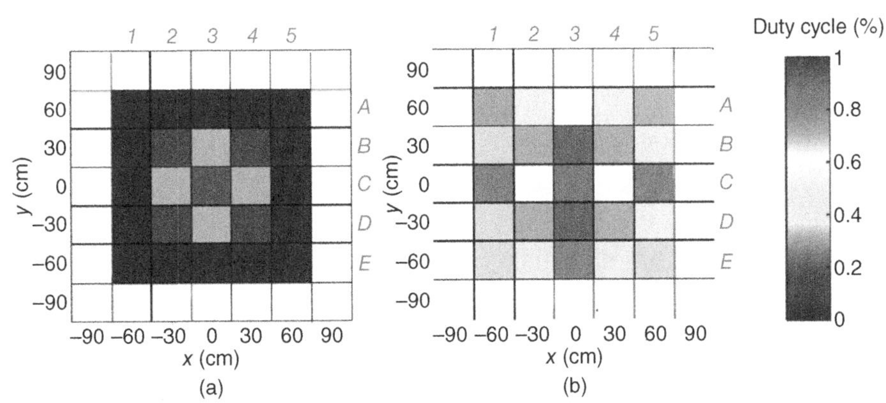

Figure 4.43 Measured RF-DC duty cycle for each node of the WSN of the power received from the smart antenna. (a) Panel antenna. (b) FS-LWA.

RWPT method. The measured duty cycle when using a static panel antenna with similar gain is shown in Figure 4.43a. It was found that only the five central node locations are powered as was anticipated in Figure 4.39b. The theoretical power density distributions in Figure 4.39 and the measured RF-to-DC conversion duty cycle in Figure 4.43 are clearly in very good agreement.

4.6 Conclusion

In this chapter, passive *FS-LWAs* were presented as a low-cost alternative to active-beaming scanning smart antennas. The main advantage of an FS-LWA is that a directive narrow beam can be generated from a *single-feed* radiating aperture and it can be scanned in one dimension by simply *frequency sweeping*. It was emphasized that this approach avoids expensive and power consuming active scanning technologies such as phased-array antennas or SBAs. The HWM LWA was introduced as one of the simplest FS-LWA topologies since it can be fabricated using printed-circuit board techniques. Moreover, it was shown how this HWM LWA can be tuned to provide the desired frequency-beam-scanning responses. This performance characteristic was demonstrated to be of key importance in order to cover a desired angular region using the frequency channels available in practical wireless networks. The concept of an *iridescent wireless network* was introduced as a system in which the smart FS-LWA creates a colored assignation between the available channels and the angular space. This spatial sectorization and agile channel assignation was optimized considering a *cross-layer* procedure. The efficacy of the HWM LWA based iridescent wireless network was exemplified for several practical cases including Wi-Fi WLANs, Bluetooth WPANs, Zigbee WSNs in the ISM 2.45 GHz band, and LoRa and passive RFID WSNs in the ISM 900 MHz band.

Three demanding performance applications of the FS-LWAs were presented:

1) Effective boost of the wireless radio link quality
2) Direction finding using only RSS (Received Signal Strength) data
3) Efficient RWPT

Millions of WLANs and interconnected mobile devices and sensors in current and future wireless ecosystems require efficient systems in terms of cost, complexity, and power consumption. The optimum design of low-cost passive frequency-scanning antennas for channel-agile iridescent wireless networks may offer an interesting technology for them in the context of more sustainable IoT environments.

References

1 T. K. Sarkar, M. C. Wicks, M. Salazar-Palma, and R. J. Bonneau, *Smart Antennas,* 1st ed. Wiley-IEEE Press, 2003.
2 L. C. Godara, *Smart Antennas*. CRC Press LLC, 2004.
3 C. A. Balanis and P. I. Ioannides, *Introduction to Smart Antennas*. Morgan & Claypool Publishers, 2007.
4 A. A. Oliner and D. R. Jackson, "Leaky-wave antennas," in *Antenna Engineering Handbook*, 4th ed., J. L. Volakis, Ed. New York, NY, USA: Mc-Grow-Hill, 2007.
5 A. A. Oliner, "Leakage from higher modes on microstrip line with application to antennas," *Radio Sci.*, vol. 22, pp. 907–912, Nov 1987.

6 G. M. Zelinski, G. A. Theile, M. L. Hastriter, M. J. Havrilla, and A. J. Terzuoli, "Half width leaky wave antennas," *IET Microwaves Antennas Propag.*, vol. 1, no. 2, pp. 341–348, 2007.

7 A. J. Martinez-Ros, J. L. Gomez-Tornero, and G. Goussetis, "Planar leaky-wave antenna with flexible control of the complex propagation constant," *IEEE Trans. Antennas Propag.*, vol. 60, no. 3, pp. 1625–1630, Mar 2012.

8 A. J. Martínez-Ros, J. L. Gómez-Tornero, and F. Quesada-Pereira, "Efficient analysis and design of novel SIW leaky-wave antenna," *IEEE Antennas Wireless Propag. Lett.*, vol. 12, pp. 496–499, Mar 2013

9 M. García-Vigueras, J. L. Gómez-Tornero, G. Goussetis, A. R. Wiley, and Y. J. Guo, "Enhancing frequency-scanning response of leaky-wave antennas using high impedance surfaces," *IEEE Antennas Wireless Propagat. Lett.*, vol. 10, pp. 7–10, Mar 2011.

10 C. Caloz, "Metamaterial dispersion engineering concepts and applications," *Proc. IEEE*, vol. 99, no. 10, pp. 1711–1719, Oct 2011.

11 A. Gil-Martinez, M. Poveda-Garcia, and J. L. Gómez-tornero, "Direct synthesis of frequency-scanned monopulse half-width microstrip leaky-wave antennas," in *Proceedings of European Conference on Antennas and Propagation*, pp. 1–4, 2020.

12 A. J. Martinez-Ros, J. L. Gómez-Tornero, and G. Goussetisy, "Frequency scanning leaky wave antenna for positioning and identification of RFID tags," in *Proceedings of 2011 IEEE International Conference on RFID-Technologies and Applications*, Sitges, Spain, 2011, pp. 451–456.

13 V. R. Komanduri, D. R. Jackson, and S. A. Long, "Radiation characteristics of finite-length 1D-uniform leaky wave antennas radiating at broadside," in *Proceedings of 2010 IEEE Antennas and Propagation Society International Symposium*, Toronto, ON, Canada, 2010, pp. 1–4.

14 M. Poveda-García, J. L. Gómez-Tornero, and D. Cañete-Rebenaque, "Study of the efficiency of half-width substrate integrated waveguide leaky-wave antennas in FR4," in *Proceedings of 2018 2nd URSI Atlantic Radio Science Meeting* (AT-RASC), Gran Canaria, Spain, 2018, pp. 1–4.

15 D. Xie and L. Zhu, "Effective approach to reduce variation of quasi-e-plane beamwidth of eh1-mode microstrip leaky-wave antennas," *IEEE Antennas Wireless Propag. Lett.*, vol. 17, no. 9, pp. 1732–1735, Sep 2018.

16 A. Gil-Martínez, Y. El Gholb, M. Poveda-García, J. L. Gómez-Tornero, and N. El Amrani El Idrissi, "An array of leaky wave antennas for indoor smart wireless access point applications," in *Proceedings of the 7th International Conference on Wireless Networks and Mobile Communications (WINCOM'19)*, Fez, Morocco, Nov 2019.

17 Y. F. Wu, Y. J. Cheng, and Z. X. Huang, "Ka-band near-field-focused 2-D steering antenna array with a focused rotman lens," *IEEE Trans. Antennas Propag.*, vol. 66, no. 10, pp. 5204–5213, Oct 2018.

18 "Beamforming in 802.11ac," Chapter 4 de 802.11ac: A Survival Guide. by Matthew S. Gast, Ed. O'Reilly Media, Inc, Aug 2013, ISBN: 9781449357702, available online at: https://www.oreilly.com/library/view/80211ac-a-survival/9781449357702/ch04.html.

19 Wi-Fi CERTIFIED 6: Next generation Wi-Fi® connectivity, available online at: https://www.wi-fi.org/discover-wi-fi/wi-fi-certified-6.

20 "802.11ac MU-MIMO: Bridging the MIMO Gap in Wi-Fi," available online at: https://www.qualcomm.com/documents/80211ac-mu-mimo-bridging-mimo-gap-wi-fi.

21 Edgewater Wireless Wi-Fi SPECTRUM SLICING, available online at: https://www.edgewaterwireless.com/mcsr-explained/.

22 QSR10GU-AX Plus Chipset Solution, available online at: http://www.quantenna.com/products/qsr10g-ax/.

23 Edgewater Wireless Technical Overview, available online at: https://www.edgewaterwireless.com/wp-content/uploads/2017/12/Edgewater-Wireless-Technical-Overview.pdf.

24 Z. Hao, H. Wang, and W. Hong, "A novel planar reconfigurable monopulse antenna for indoor smart wireless access points' application," *IEEE Trans. Antennas Propag.*, vol. 64, no. 4, pp. 1250–1261, April 2016.

25 J. Hu, G. Q. Luo, and Z.-C. Hao, "A wideband quad-polarization reconfigurable metasurface antenna," *IEEE Access*, vol. 6, pp. 6130–6137, 2018.

26 Y. Yang and X. Zhu, "A wideband reconfigurable antenna with 360°beam steering for 802.11ac WLAN applications," *IEEE Trans. Antennas Propag.*, vol. 66, no. 2, pp. 600–608, Feb 2018.

27 J. Keh, M. Chou, Z. Zhang, Q. An, and W. Liao, "A beamforming-network-based four-squint-beam array antenna for ceiling-mount access point," *IEEE Antennas Wireless Propagat. Lett.*, vol. 18, no. 4, pp. 707–711, April 2019.

28 M. Chou, J.-F. Ke, Z.-C. Zhang, and W.-J. Liao, "A 2 × 2 monopole antenna array for ceiling-mount wireless access point," in *Proceedings of Radio Science and Wireless Technology Conference (CSQRWC) 2019 Cross Strait Quad-Regional*, pp. 1–3, 2019.

29 J. Frigon, C. Caloz, and Y. Zhao, "Dynamic radiation pattern diversity (DRPD) MIMO using CRLH leaky-wave antennas," in *Proceedings of 2008 IEEE Radio and Wireless Symposium*, Orlando, FL, USA, 2008, pp. 635–638.

30 D. Piazza, M. D'Amico, and K. R. Dandekar, "Performance improvement of a wideband MIMO system by using two-port RLWA," *IEEE Trans. Antennas Propag.*, vol. 8, pp. 830-834, 2009.

31 X. Li and J. Frigon, "Capacity analysis of MIMO systems with dynamic radiation pattern diversity," in *Proceedings of IEEE 69th Vehicular Technology Conferences (VTC Spring 2009)*, Barcelona, Spain, 2009, pp. 1–5.

32 D. Patron, "Compact reconfigurable antennas for wireless systems and wearable applications," PhD thesis. Drexler University, May 2015.

33 D. Zelenchuk *et al.*, "W-band planar wide-angle scanning antenna architecture," *J. Infrared Millimeter Terahertz Waves*, vol. 34, no. 2, pp. 127–139, Feb 2013.

34 M. Steeg and A. Stöhr, "High data rate 6 Gbit/s steerable multibeam 60 GHz antennas for 5G hot-spot use cases," in *Proceedings of 2017 IEEE Photonics Society Summer Topical Meeting Series (SUM)*, San Juan, 2017, pp. 141–142.

35 Y. El Gholb, M. Poveda-García, J. L. Gómez-Tornero, J. M. Molina-Garcia-Pardo, and N. El Amrani El Idrissi, "Terminal leaky-wave antenna for efficient 5G mobile communication," in *Proceedings of the 7th Mediterranean Congress of Telecommunications* (CMT 2019), Fez, Morocco, Nov 2019.

36 M. Steeg, M. Szczęsny, and A. Stöhr, "Multiuser 5G hot-spots based on 60 GHz leaky-wave antennas and WDM radio-over-fiber," *Electronic Navigation Research Institute (eds) Air Traffic Management and Systems III. Lecture Notes in Electrical Engineering*, vol. 555. Singapore: Springer, 2019.

37 N. AL-fadhali *et al.*, "Review on frequency reconfigurable antenna using substrate-integrated waveguide for cognitive radio application," *Journal Electromagn. Waves Appl.* Jan 2021, vol. 35, no. 7, pp. 958–990. https://doi.org/10.1080/09205071.2020.1865207.

38 Zigbee Alliance, available online at: https://zigbeealliance.org/.

39 IEEE 802.15.4a-2007 IEEE Standard for PART 15.4: Wireless MAC and PHY Specifications for Low-Rate Wireless Personal Area Networks (LR-WPANs): Amendment 1: Add Alternate PHY, available online at: http://standards.ieee.org/getieee802/download/802.15.4a-2007.pdf.

40 Bluetooth specifications, available online at: https://www.bluetooth.com/specifications/.

41 Wi-Fi IEEE 802.11n 5 GHz band physical specifications. IEEE 802.11n-2009 - IEEE Standard for Information technology-- Local and metropolitan area networks-- Specific requirements-- Part 11: Wireless LAN Medium Access Control (MAC) and Physical Layer (PHY) Specifications Amendment 5: Enhancements for Higher Throughput, https://standards.ieee.org/standard/802_11n-2009.html.

42 Y. Ghasempour, C. R. C. M. da Silva, C. Cordeiro, and E. W. Knightly, "IEEE 802.11ay: next-generation 60 GHz communication for 100 Gb/s Wi-Fi," *IEEE Commun. Mag.*, vol. 55, no. 12, pp. 186–192, Dec 2017.

43 T. Adame, A. Bel, B. Bellalta, J. Barcelo, and M. Oliver, "IEEE 802.11AH: The WiFi approach for M2M communications," *IEEE Wireless Commun.*, vol. 21, no. 6, pp. 144–152, Dec 2014.

44 Wi-Fi IEEE 802.11ah 900 MHz band physical specifications. 802.11ah-2016 - IEEE Standard for Information technology--Telecommunications and information exchange between systems - Local and metropolitan area networks--Specific requirements - Part 11: Wireless LAN Medium Access Control (MAC) and Physical Layer (PHY) Specifications Amendment 2: Sub 1 GHz License Exempt Operation.

45 LoRaWAN UHF WSN specifications, available online at: https://lora-alliance.org/about-lorawan/.

46 LoRaWAN USA frequencies, channels and sub-bands for IoT devices, available online at: https://www.baranidesign.com/faq-articles/2019/4/23/lorawan-usa-frequencies-channels-and-sub-bands-for-iot-devices.

47 Passive UHF RFID specifications, available online at: https://rfid4u.com/rfid-basics-resources/basics-rfid-regulations/.

48 P. Papadimitratos, A. Mishra, and D. Rosenburgh, "A cross-layer design approach to enhance 802.15.4," *IEEE MILCON 2005*, vol. 3, pp. 1719–1726, June 2005.

49 F. Foukalas, V. Gazis, and N. Alonistioti, "Cross-layer design proposals for wireless mobile networks: A survey and taxonomy," *IEEE Commun. Surv. Tutorials*, vol. 10, no. 1, pp. 70–85, First Quarter 2008.

50 M. A. Burlington, *Cognitive Radio Communications and Networks: Principles and Practice.*: Academic Press. 2010.

51 A. Karmokar, "Energy-efficient green radio communications for delay tolerant applications," *Handbook of Green Information and Communication Systems*, 2012. https://www.elsevier.com/books/handbook-of-greeninformation-andcommunicationsystems/obaidat/978-0-12-415844-3.

52 S. Shabdanov, P. Mitran, and C. Rosenberg, "Cross-layer optimization using advanced physical layer techniques in wireless mesh networks," *IEEE Trans. Wireless Commun.*, vol. 11, no. 4, pp. 1622–1631, April 2012.

53 G. Miao and G. Song, *Energy and Spectrum Efficient Wireless Network Design*. Cambridge University Press, 2014.

54 A. Khattab, D. Perkins, and M. Bayoumi, *Cognitive Radio Networks. Analog Circuits and Signal Processing*. New York: Springer, 2013.

55 T. Y. Kheng, Ed., *Smart Wireless Sensor Networks*. IntechOpen, Dec 2010.

56 *Smart Sensors Networks. Communication Technologies and Intelligent Applications*, Series Editor Fatos Xhafa, Academic Press Elsevier Ed. 2017. https://www.amazon.es/Smart- Sensors-Networks- Communication-Technologies/dp/0128098597.

57 E. Felemban *et al.*, "SAMAC: A cross-layer communication protocol for sensor networks with sectored antennas," *IEEE Trans. Mobile Comput.*, vol. 9, no. 8, pp. 1072–1088, Aug 2010.

58 C. Lin and T. Kokkinos, "Cross-layer solutions for extended-range wireless sensor networks," *IEEE Sens. J.*, vol. 13, no. 3, pp. 1044–1054, Mar 2013.

59 A. Erdogan, A. Coskun, and A. Kavak, "The sectoral sweeper scheme for wireless sensor networks: Adaptive antenna array based sensor node management and location estimation," *Wireless Pers. Commun.*, vol. 39, p. 415, 2006.

60 G. Boudour, A. Lecointre, P. Berthou, D. Dragomirecu, and T. Gayraud, "On designing sensor networks with smart antennas," *Proc. IFAC Volumes*, vol. 40, no. 22, pp. 349–356, 2007.

61 G. Giorgetti, A. Cidronali, S. K. S. Gupta, and G. Manes, "Exploiting low-cost directional antennas in 2.4 GHz IEEE 802.15.4 wireless sensor networks," in *Proceedings of 2007 European Conference on Wireless Technologies*, Munich, 2007, pp. 217–220.

62 H.-N. Dai, K.-W. Ng, M. Li, and M.-Y. Wu, "An overview of using directional antennas in wireless networks," *Int. J. Commun. Syst.*, vol. 26, no. 4, pp. 413–448, 2011.

63 K. Kucuk and A Kavak, "Connectivity analysis for wireless sensor networks with antenna array integrated central node," *Wireless Pers. Commun.*, vol. 72, p. 1361, 2013.

64 L. Catarinucci, S. Guglielmi, L. Patrono, and L. Tarricone, "Switched-beam antenna for wireless sensor network nodes," *Progr. Electromagn. Res. C*, vol. 39, pp. 193–207, 2013.

65 L. Catarinucci, S. Guglielmi, R. Colella, and L. Tarricone, "Compact switched-beam antennas enabling novel power-efficient wireless sensor networks," *IEEE Sens. J.*, vol. 14, no. 9, pp. 3252–3259, Sep 2014.

66 B. Liang, B. Sanz-Izquierdo, J. C. Batchelor, and A. Bogliolo, "Active FSS enclosed beam-switching node for wireless sensor networks," in *Proceedings of the 8th European Conference on Antennas and Propagation*, The Hague, 2014.

67 R. Wang, B.-Z. Wang, W.-Y. Huang, and X. Ding, "Compact reconfigurable antenna with an omnidirectional pattern and four directional patterns for wireless sensor systems," *Sensors*, vol. 16, no. 4, p. 552, 2016.

68 T. N. Le, A. Pegatoquet, T. Le Huy, L. Lizzi, and F. Ferrero, "Improving energy efficiency of mobile WSN using reconfigurable directional antennas," *IEEE Comms Lett.*, vol. 20, no. 6, pp. 1243–1246, Jun 2016.

69 Q. Wang *et al.*, "On connectivity of wireless sensor networks with directional antennas," *Sensors (Basel, Switzerland)*, vol. 17, no. 1, p. 134, 2017.

70 A. R. Chandran *et al.*, "Microstrip patch based switched beam antenna at 2.45 GHz for wireless sensor network applications," *J. Electromagn. Waves Appl.*, 2017. https://www.tandfonline.com/doi/abs/10.1080/09205071.2017.1348260?journalCode=tewa20.

71 L. Selavo, D. Vyas, M. Yahyazadeh, and O. Chipara, "PHASER – A phase-shifting antenna for low-power directional communication," in *Proceedings of 2017 13th International Conference on Distributed Computing in Sensor Systems (DCOSS)*, Ottawa, ON, 2017, pp. 65–72.

72 L. Santamaria, F. Ferrero, and L. Lizzi, "Radiation pattern agile antenna for smart IoT gateways," in *Proceedings of 2018* IEEE International Symposium on Antennas and Propagation & USNC/URSI National Radio Science Meeting, Boston, MA, 2018, pp. 977–978.

73 A. A. N. Dihissou, A. Diallo, P. Le Thuc, and R. Staraj, "Directive and reconfigurable loaded antenna array for wireless sensor networks," *Progr. Electromagn. Res., PIER C, IEEE*, vol. 84, pp. 103–117, 2018.

74 A. A. Niyi Dihissou, A. Diallo, P. L. Thuc, and R. Staraj, "Directive and reconfigurable antenna for wireless sensor network to improve link quality between nodes," in *Proceedings of 2019 IEEE Sensors Applications Symposium* (SAS), Sophia Antipolis, France, 2019, pp. 1–5.

75 S. A. P. Mathur, A. R. Chandran, N. Timmons, J. Morrison, and S. Raman, "2.45 GHz pattern reconfigurable antenna for wireless sensor network applications," in *Proceedings of 2019 URSI Asia-Pacific Radio Science Conference (AP-RASC)*, New Delhi, India, 2019, pp. 1–4.

76 S. Ciccia, G. Giordanengo, and G. Vecchi, "Energy efficiency in IoT networks: Integration of reconfigurable antennas in ultra low-power radio platforms based on system-on-chip," *IEEE Internet Things J.*, vol. 6, no. 4, pp. 6800–6810, Aug 2019.

77 B. A. Nahas, S. Duquennoy, V. Iyer, and T. Voigt, "Low-power listening goes multi-channel," in *Proceedings of* 2014 IEEE International Conference on Distributed Computing in Sensor Systems, Marina Del Rey, CA, 2014, pp. 2–9.

78 H. Luo, Z. Huang, and T. Zhu, "A survey on spectrum utilization in wireless sensor networks," *J. Sensors*, vol. 2015, Article ID 624610, 13 pages, 2015.

79 L. Varga, M. Heusse, R. Guizzetti, and A. Duda, "Why is frequency channel diversity so beneficial in wireless sensor networks?," in *Proceedings of 2016 Wireless Days (WD)*, Toulouse, 2016, pp. 1–3.

80 J.-H. Lee, S.-J. Kho, U.-J. Jang, and Y.-S. Hong, "An improved method of avoiding rf congestion in indoor environments," *Procedia Eng.*, vol. 154, pp. 223–228, 2016,

81 S. Maddio, *et al.*, "Multipath robust azimuthal direction of arrival estimation in dual-band 2.45–5.2 GHz networks," *IEEE Trans. Microwave Theory Tech.*, vol. 65, no. 11, pp. 4438–4449, Nov 2017.

82 J.-H. Lee, S.-J. Kho, U.-J. Jang, and Y.-S. Hong, "A method of improving PRR for WiFi interference avoidance in ZigBee networks in indoor environments," *Int. J. Adv. Intell. Paradigm.*, vol. 11, p. 176, 2018.

83 E. Fernández de Gorostiza, J. Berzosa, J. Mabe, and R. Cortiñas, "A method for dynamically selecting the best frequency hopping technique in industrial wireless sensor network applications," *Sensors*, vol. 18, p. 657, 2018.

84 P. Spachos, I. Papapanagiotou, and K. N. Plataniotis, "Microlocation for smart buildings in the era of the Internet of Things: A survey of technologies, techniques, and approaches," *IEEE Signal Process Mag.*, vol. 35, no. 5, pp. 140–152, Sep 2018.

85 F. Khelifi, A. Bradai, A. Benslimane, P. Rawat, and M. Atri, "A survey of localization systems in Internet of Things," *Mob. Networks Appl.*, vol. 24, no. 3, pp. 761–785, Jun 2019.

86 F. Liu *et al.*, "Survey on WiFi-based indoor positioning techniques," *IET Commun.*, vol. 14, no. 9, pp. 1372–1383, Jun 2020.

87 K. E. Jeon, J. She, P. Soonsawad, and P. C. Ng, "BLE beacons for Internet of Things applications: Survey, challenges, and opportunities," *IEEE Internet Things J.*, vol. 5, no. 2, pp. 811–828, Apr 2018.

88 J. Kuriakose , S. Joshi , R. Vikram Raju, and A. Kilaru, "A review on localization in wireless sensor networks," in *Advances in Signal Processing and Intelligent Recognition Systems. Advances in Intelligent Systems and Computing,* S. Thampi, A. Gelbukh, and J. Mukhopadhyay, Eds., vol. 264. Springer, 2014,

89 C. Li, L. Mo, and D. Zhang, "Review on UHF RFID localization methods," *IEEE J. Radio Freq. Ident.*, vol. 3, no. 4, pp. 205–215, Dec 2019.

90 A. Bensky, *Wireless Positioning Technologies and Applications.* Norwood, MA, USA: Artech House, 2016.

91 Y. Gu and F. Ren, "Energy-efficient indoor localization of smart handheld devices using Bluetooth," *IEEE Access*, vol. 3, pp. 1450–1461, 2015.

92 S. Sadowski and P. Spachos, "RSSI-based indoor localization with the Internet of Things," *IEEE Access*, vol. 6, pp. 30149–30161, 2018.

93 L. Chen, L. Pei, H. Kuusniemi,Y. Chen, T. Kröger, and R. Chen, "Bayesian fusion for indoor positioning using Bluetooth Fingerprints," *Wireless Pers. Commun.*, vol. 70, no. 4, pp. 1735–1745, Jun 2013.

94 R. Faragher and R. Harle, "Location fingerprinting with Bluetooth low energy beacons," *IEEE J. Sel. Areas Commun.*, vol. 33, no. 11, pp. 2418–2428, Nov 2015.

95 S. He and S.-H. G. Chan, "Wi-Fi fingerprint-based indoor positioning: Recent advances and comparisons," *IEEE Commun. Surv. Tutorials*, vol. 18, no. 1, pp. 466–490, 2016.

96 G. Qing, H. Wang, and T. Zhang, "Radio frequency fingerprinting identification for Zigbee via lightweight CNN," *Phys. Commun.*, vol. 44, 2021.

97 D. Han, S. Jung, M. Lee, and G. Yoon, "Building a practical Wi-Fi-based indoor navigation system," *IEEE Pervasive Comput.*, vol. 13, no. 2, pp. 72–79, 2014.

98 J. A. Lopez-Pastor, A. J. Ruiz-Ruiz, A. S. Martinez-Sala, and J. Luis Gomez-Tornero, "Evaluation of an indoor positioning system for added-value services in a mall," in *Proceedings of 2019 International Conference on Indoor Positioning and Indoor Navigation* (IPIN), 2019, pp. 1–8.

99 J. Xiong and K. Jamieson, "ToneTrack: Leveraging frequency-agile radios for time-based indoor wireless localization categories and subject descriptors," in *Proceedings of the Annual International Conference on Mobile Computing and Networking,* MOBICOM, pp. 537–549, 2015.

100 M. Kotaru, K. Joshi, D. Bharadia, and S. Katti, "SpotFi: Decimeter level localization using WiFi," in *Proceedings of the 2015 ACM Conference on Special Interest Group on Data Communication*. SIGCOMM '15. New York, NY, USA: ACM, pp. 269–282.

101 W. Gong and J. Liu, "SiFi: Pushing the Limit of time-based WiFi localization using a single commodity access point," *Proc. ACM Interact. Mobile Wearab. Ubiquitous Technol.*, vol. 2, no. 1, 2018, pp. 1–21.

102 "IEEE 802.11-2016 - IEEE Standard for Information technology--Telecommunications and information exchange between systems Local and metropolitan area networks--Specific requirements - Part 11: Wireless LAN Medium Access Control (MAC) and Physical Layer (PH)," Available Online at: https://standards.ieee.org/standard/802_11-2016.html.

103 M. Ibrahim *et al.*, "Verification: Accuracy evaluation of WiFi fine time measurements on an open platform," in *Proceedings of the 24th Annual International Conference on Mobile Computing and Networking* - MobiCom '18, 2018, pp. 417–427.

104 S. Xu, R. Chen, Y. Yu, G. Guo, and L. Huang, "Locating smartphones indoors using built-in sensors and Wi-Fi ranging with an enhanced particle filter," *IEEE Access*, vol. 7, pp. 95140–95153, 2019.

105 Y. Yu *et al.*, "Precise 3D indoor localization based on Wi-Fi FTM and built-in sensors," *IEEE Internet Things J.*, vol. 4662, no. C, pp. 1–1, 2020.

106 J. A. López-Pastor, P. Arqués-Lara, J. J Franco-Peñaranda, A. J. García-Sánchez, and J. L. Gómez-Tornero, "Wi-Fi RTT-based active monopulse RADAR for single access point localization," *IEEE Access*, vol. 9, pp. 34755–34766, 2021.

107 L. C. Godara, "Application of antenna arrays to mobile communications. II. Beam-forming and direction-of-arrival considerations," *Proc. IEEE*, vol. 85, no. 8, pp. 1195–1245, Aug 1997.

108 H. Liu, H. Darabi, P. Banerjee, and J. Liu, "Survey of wireless indoor positioning techniques and systems," *IEEE Trans. Syst. Man Cybern. Part C Appl. Rev.*, vol. 37, no. 6, pp. 1067–1080, Nov 2007

109 R. O. Schmidt, "Multiple emitter location and signal parameter estimation," *Adapt. Antennas Wireless Commun.*, no. 3, pp. 190–194, 2009.

110 R. Roy, A. Paulraj, and T. Kailath, "ESPRIT—A subspace rotation approach to estimation of parameters of cisoids in noise," *IEEE Trans. Acoust.*, vol. 34, no. 5, pp. 1340–1342, 1986.

111 L. Brás, N. B. Carvalho, P. Pinho, L. Kulas, and K. Nyka, "A review of antennas for indoor positioning systems," *Int. J. Antennas Propag.*, vol. 2012, pp. 1-14, Oct 2012.

112 N. Dragos and B. Nath, "VOR base stations for indoor 802.11 positioning," in *Proceedings* of the *10th Annual International Conference on Mobile Computing and Networking, MobiCom '04*, pp. 58–69, Sep 2004.

113 J. Vongkulbhisal and Y. Zhao, "An RFID-based indoor localization system using antenna beam scanning," in *Proceedings of the 9th International Conference on Electrical Engineering/Electronics, Computer, Telecommunications and Information Technology*, Phetchaburi, Thailand, 2012, pp. 1–4.

114 L. Shen, Q. Zhang, J. Pang, H. Xu, P. Li, and D. Xue, "ANTspin: Efficient absolute localization method of RFID tags via spinning antenna" *Sensors* vol. 19, no. 9, p. 2194, 2019.

115 G. Zhang, D. Zheng, J. Guo, Y. Hou, and D. Zeng, "Dynamic modeling and mobility analysis of the 3-R(RRR)R R antenna mechanism," *Robotica,* vol. 1–19, 2021.

116 G. Giorgetti, S. Maddio, A. Cidronali, S. K. S. Gupta, and G. Manes, "Switched beam antenna design principles for angle of arrival estimation," in *Proceedings European Wireless Technology Conference*, Sep 2009, pp. 5–8.

117 S. Maddio, A. Cidronali, and G. Manes, "RSSI/DoA based positioning systems for wireless sensor network," in *New Approach Indoor Outdoor Localization System.* Rijeka, Croatia: InTech, 2012, Chapter 7, pp. 139–162.

118 M. Passafiume, S. Maddio, G. Collodi, and A. Cidronali, "An enhanced algorithm for 2D indoor localization on single anchor RSSI-based positioning systems," in *Proceedings of 2017 European Radar Conference (EURAD)*, 2017, vol. 2018, Jan, pp. 287–290.

119 Y. Li, Z. He, Y. Li, Z. Gao, R. Chen, and N. El-sheimy, "Cost-effective localization using RSS from single wireless access point," *IEEE Trans. Instrum. Meas.*, vol. 69, no. 5, pp. 1860–1870, May 2020. https://doi.org/10.1109/TIM.2019.2922752.

120 E. Taillefer, W. Nomura, J. Cheng, M. Taromaru, Y. Watanabe, and T. Ohira, "Enhanced reactance-domain ESPRIT algorithm employing multiple beams and translational-invariance soft selection for direction-of-arrival estimation in the full azimuth," *IEEE Trans. Antennas Propag.*, vol. 56, no. 8, pp. 2514–2526, Aug 2008.

121 M. Plotka, M. Tarkowski, K. Nyka, and L. Kulas, "A novel calibration method for RSS-based DoA estimation using ESPAR antennas," in *Proceedings of 22nd International Microwave and Radar Conference* (MIKON), Poznan, Poland, 2018, pp. 65–68.

122 L. Kulas, "RSS-based DoA estimation using ESPAR antennas and interpolated radiation patterns," *IEEE Antennas Wireless Propag. Lett.*, vol. 17, no. 1, pp. 25–28, Jan 2018.

123 S. M. Sherman and D. K. Barton, *Monopulse Principles and Techniques*. Boston, MA, USA: Artech House, 2011.

124 J.-C. Wu, C.-C. Chang, T.-Y. Chin, S.-F. Chang, M.-C. Chiu, C.-Y. Hsu, and R.-H. Lee, "Wireless indoor localization using dynamic monopulse receiver," in *Proceedings of the 7th European Radar Conference*, Sep/Oct 2010, pp. 69–72.

125 M. Del Prete, D. Masotti, N. Arbizzani, and A. Costanzo, "Remotely identify and detect by a compact reader with mono-pulse scanning capabilities," *IEEE Trans. Microwave Theory Tech.*, vol. 61, no. 1, pp. 641–650, Jan 2013.

126 S. Abielmona, H. V. Nguyen, and C. Caloz, "Analog direction of arrival estimation using an electronically-scanned CRLH leaky-wave antenna," *IEEE Trans. Antennas Propag.*, vol. 59, no. 4, pp. 1408–1412, Apr 2011.

127 D. Patron, H. Paaso, A. Mämmelä, D. Piazza, and K. R. Dandekar, "Improved design of a CRLH leaky-wave antenna and its application for DoA estimation," *in Proceedings of IEEE-APS Topical Conference on Antennas and Propagation in Wireless Communications* (APWC), Sep 2013, pp. 1343–1346.

128 H. Paaso, A. Mämmelä, D. Patron, and K. R. Dandekar, "DoA estimation through modified unitary MUSIC algorithm for CRLH leaky-wave antennas," in *Proceedings of the IEEE 24th Annual International Symposium on Personal Indoor and Mobile Radio Communication (PIMRC)*, Sep 2013, pp. 311–315.

129 J. Werner *et al.*, "Sectorized antenna-based DoA estimation and localization: Advanced algorithms and measurements," *IEEE J. Sel. Areas Commun.*, vol. 33, no. 11, pp. 2272–2286, Nov 2015.

130 H. Paaso *et al.*, "Experimental results of novel DoA estimation algorithms for compact reconfigurable antennas," *Int. J. Antennas Propag.*, vol. 2017, Art. no. 1613638, Jul 2017.

131 H. Paaso *et al.*, "DoA estimation using compact CRLH leaky-wave antennas: Novel algorithms and measured performance," *IEEE Trans. Antennas Propag.*, vol. 65, no. 9, pp. 4836–4849, Sep 2017.

132 A. J. Martinez-Ros, J. L. Gómez-Tornero, and G. Goussetis, "Frequency scanning leaky wave antenna for positioning and identification of RFID tags," *in Proceedings of IEEE International Conference RFID-Technology and Applications* (RFID-TA), Sep 2011, pp. 451–456.

133 X. Yu and H. Xin, "Direction of arrival estimation utilizing incident angle dependent spectra," in *Proceedings of IEEE MTT-S International Microwave Symposium Digest*, Jun 2012, pp. 1–3.

134 Q. Liu, V. Fusco, D. Linton, and D. Zelenchuk, "Frequency scanning antenna for target location applications," in *Proceedings of the 8th European Conference on Antennas and Propagaion (EuCAP)*, Apr 2014, pp. 2059–2061.

135 M. A. Tehrani, J. J. Laurin, and Y. Savaria, "Multiple targets direction-of-arrival estimation in frequency scanning array antennas," *IET Radar Sonar Navig.*, vol. 10, no. 3, pp. 624–631, 2016.

136 B. Husain, M. Steeg, and A. Stöhr, "Estimating direction-of-arrival in a 5G HotSpot scenario using a 60 GHz leaky-wave antenna," in *Proceedings of IEEE International Conference on Microwaves, Antennas, Communications and Electronic Systems* (COMCAS), 2017, pp. 1–4.

137 M. Poveda-García, D. Cañete-Rebenaque, and J. L. Gómez-Tornero, "Frequency-scanned monopulse pattern synthesis using leaky-wave antennas for enhanced power-based direction-of-arrival estimation," *IEEE Trans. Antennas Propag.*, vol. 67, no. 11, pp. 7071–7086, Nov 2019.

138 Y. Ning, S. Ma, F. Meng, and Q. Wu, "DOA estimation based on ESPRIT algorithm method for frequency scanning LWA," *IEEE Commun. Lett.*, vol. 24, no. 7, pp. 1441–1445, Jul 2020.

139 H. Jiang, C. Cai, X. Ma, Y. Yang, and J. Liu, "Smart home based on WiFi sensing: A survey," *IEEE Access*, vol. 6, pp. 13317–13325, 2018.

140 J. Xiong and K. Jamieson, "ArrayTrack: A fine-grained indoor location system," in *Proceedings of USENIX Symposium Networked Systems Design and Implementation*, no. 279976, 2013, pp. 71–84.

141 S. Sen, J. Lee, K.-H. Kim, and P. Congdon, "Avoiding multipath to revive inbuilding WiFi localization," in *Proceedings of the 11th Annual Internaional Conference on Mobile Systems, Applications, and Services* - MobiSys '13, 2013, p. 249.

142 J. Gjengset, J. Xiong, G. McPhillips, and K. Jamieson, "Phaser: Enabling phased array signal processing on commodity WiFi access points," in *Proceedings of the 20th Annual International Conference on Mobile Computing and Networking, MobiCom* '14, Sep 2014, pp. 153–163.

143 M. Kotaru, K. Joshi, D. Bharadia, and S. Katti, "Spotfi: Decimeter level localization using WiFi," in *Proceedings of ACM SIGCOMM*, 2015.

144 A. Tzur, O. Amrani, and A. Wool, "Direction finding of rogue Wi-Fi access points using an off-the-shelf MIMO–OFDM receiver," *Phys. Commun.*, vol. 17, pp. 149–164, 2015.

145 Y. Zheng, M. Sheng, J. Liu, and J. Li, "OpArray: Exploiting array orientation for accurate indoor localization," *IEEE Trans. Commun.*, vol. 67, no. 1, pp. 847–858, Jan 2019.

146 Y. Zhang and K. Psounis, "Efficient indoor localization via switched-beam antennas," *IEEE Trans. Mob. Comput.*, vol. 19, no. 9, pp. 2101–2115, 2020.

147 J. L. Gomez-Tornero, D. Canete-Rebenaque, J. A. Lopez-Pastor, and A. S. Martinez-Sala, "Hybrid analog-digital processing system for amplitude-monopulse RSSI-based MIMO WiFi direction-of-arrival estimation," *IEEE J. Sel. Top. Sign. Process.*, vol. 12, no. 3, 2018.

148 J. A. López-Pastor, A. Gómez-Alcaraz, D. Cañete-Rebenaque, A. S. Martínez-Sala, and J. L. Gómez-Tornero, "Near-field monopulse doa estimation for angle-sensitive proximity WiFi readers," *IEEE Access J.* Special Section on Emerging Trends, Issues and Challenges for Array Signal Processing and Its Applications in Smart City, vol. 7, no. 1, pp. 88450–88460, June 2019.

149 A. Gil-Martínez, M. Poveda-García, J. A. López-Pastor, J. C. Sánchez-Aarnoutse, and J. L. Gómez-Tornero, "Wi-Fi Direction Finding With Frequency-Scanned Antenna and Channel-Hopping Scheme," *IEEE Sens. J.*, vol. 22, no. 6, pp. 5210–5222, Mar 2022.

150 D. Vasisht, S. Kumar, and D. Katabi, "Decimeter-level localization with a single WiFi access point," in *Proceedings of the 13th USENIX Symposium on Networked Systems Design and Implementation, NSDI 2016*, 2016, pp. 165–178.

151 S. Han, Y. Li, W. Meng, C. Li, T. Liu, and Y. Zhang, "Indoor localization with a single Wi-Fi access point based on OFDM-MIMO," *IEEE Syst. J.*, vol. 13, no. 1, pp. 964–972, 2019.

152 S. Alletto, R. Cucchiara, G. Del Fiore, L. Mainetti, V. Mighali, L. Patrono, and G. Serra, "An indoor location-aware system for an IoT-based smart museum," *IEEE Internet Things J.*, vol. 3, no. 2, pp. 244–253, 2016.

153 M. S. Gast, *Building Applications with iBeacon: Proximity and Location Services with Bluetooth Low Energy*. Sebastopol, CA, USA: O'Reilly Media, 2014.

154 *Specification of the Bluetooth System. Core Version 4.0.* Bluetooth. Kirklnad, WA, USA, Jun 2010.

155 *Specification of the Bluetooth System. Core Version 5.1.* Bluetooth. Kirklnad, WA, USA, Jan 2019.

156 K. Kalliola, "High accuracy indoor positioning based on BLE," in *Nokia Research Center Presentation*, 2011.

157 N. Honma, R. Tazawa, K. Kikuchi, A. Miura, Y. Sugawara, and H. Minamizawa, "Indoor-positioning using RSSI : DOD-based technique versus RSSI-ranging technique," in *Proceedings of the 8th International Conference on Indoor Positioning and Indoor Navigation, IPIN2017*, 2017, p. 171.

158 S. Monfared, T. H. Nguyen, L. Petrillo, P. De Doncker, and F. Horlin, "Experimental demonstration of BLE transmitter positioning based on AOA estimation," in *Proceedings of IEEE International Symposium on Personal, Indoor and Mobile Radio Communications PIMRC*, vol. 2018-Sep, 2018, pp. 856–859.

159 N. Honma, R. Tazawa, A. Miura, Y. Sugawara, and H. Minamizawa, "RSS-based DOA / DOD estimation using bluetooth signal and its application for indoor tracking," in *Proceedings of 2018 International Conference on Indoor Positioning and Indoor Navigation (IPIN)*, 2018, pp. 1–7.

160 R. Tazawa, N. Honma, A. Miura, and H. Minamizawa, "RSSI-based localization using wireless beacon with three-element array," *IEICE Trans. Commun.*, vol. E101.B, no. 2, pp. 400–408, 2018.

161 M. Poveda-García, A. Gómez-Alcaraz, D. Cañete-Rebenaque A. S. Martínez-Sala, J. L. Gómez-Tornero, "RSSI-based direction-of-departure estimation in Bluetooth Low Wnergy using an array of frequency-steered leaky-wave antennas," *IEEE Access J.* Special Section on Emerging Trends, Issues and Challenges for Array Signal Processing and Its Applications in Smart City, vol. 8, pp. 9380–9394, Jan 2020.

162 Y. B. Li, L. L. Li, B. G. Cai, Q. Cheng, and T. J. Cui, "Holographic leaky-wave metasurfaces for dual-sensor imaging," *Sci. Rep.*, vol. 5, pp. 1–7, 2015.

163 M. Poveda-García, E. Andreu-García, J. García-Fernández, D. Cañete-Rebenaque, and J. L. Gómez-Tornero, "Frequency-scanned leaky-wave antenna topologies for two-dimensional direction of arrival estimation in IoT wireless networks," in *Proceedings of the 15th European Conference on Antennas Propagation* (EuCAP 2021), Mar 2021, pp. 1–4.

164 C. A. Balanis, *Antenna Theory*. New York: Wiley, 1982.

165 S. Saloni and A. Hegde, "WiFi-aware as a connectivity solution for IoT pairing IoT with WiFi aware technology: Enabling new proximity based services," in *Proceedings of 2016 International Conference on Internet of Things and Applications (IOTA)*, Pune, 2016, pp. 137–142.

166 M. Hadian, T. Altuwaiyan, X. Liang, B. Sheng, and K. Zhang, "SmartEye: Mobile device proximity monitoring via wireless signal analysis," in *Proceedings of 2019 International Conference on Computing, Networking and Communications* (ICNC), Honolulu, HI, USA, 2019, pp. 365–369.

167 T. J. Pierson, T. Peters, R. Peterson, and D. Kotz, "Proximity detection with single-antenna IoT devices," in *Proceedings of the ACM International Conference on Mobile Computing and Networking (MobiCom)*, ACM Press, Oct 2019.

168 A. Michel and P. Nepa, "UHF-RFID desktop reader antennas: Performance analysis in the near-field region," *IEEE Antennas Wireless Propag. Lett.*, vol. 15, pp. 1430–1433, 2016.

169 A. Michel, P. Nepa, X. Qing, and Z. N. Chen, "Considering high-performance near-field reader antennas: Comparisons of proposed antenna layouts for ultrahigh-frequency near-field radio-frequency identification," *IEEE Antennas Propag. Mag.*, vol. 60, no. 1, pp. 14–26, Feb 2018

170 M. Poveda-Garcia, A. G. Martinez, and J. L. Gomez-Tornero, "Frequency-scanned focused leaky-wave antennas for direction-of-arrival detection in proximity BLE sensing applications," in *Proceedings of the 14th European Conference on Antennas Propagation* (EuCAP 2014), Mar 2020, pp. 1–4.

171 J. L. Gomez-Tornero, F. Quesada-Pereira, A. Alvarez-Melcon, G. Goussetis, A. R. Weily, and Y. J. Guo, "Frequency steerable two dimensional focusing using rectilinear leaky-wave lenses," *IEEE Trans. Antennas Propag.*, vol. 59, no. 2, pp. 407–415, 2011.

172 A. J. Martínez-Ros, J. L. Gómez-Tornero, F. J. Clemente-Fernandez, and J. Monzó-Cabrera, "Microwave near-field focusing properties of width-tapered microstrip leaky-wave antenna," *IEEE Trans. Antennas Propag.*, vol. 61, no. 6, pp. 2981–2990, 2013.

173 A. J. Martínez-Ros, J. L. Gómez-Tornero, and G. Goussetis, "Holographic pattern synthesis with modulated substrate integrated waveguide line-source leaky-wave antennas," *IEEE Trans. Antennas Propag.*, vol. 61, no. 7, pp. 3466–3474, Jul 2013.

174 J. L. Gomez-Tornero, A. R. Weily, and Y. J. Guo, "Rectilinear leaky-wave antennas with broad beam patterns using hybrid printed circuit waveguides," *IEEE Trans. Antennas Propag.*, vol. 59, no. 11, pp. 3999–4007, 2011.

175 M. L. Mendoza and V. H. Z. Silva, "Self-protection strategies in WSN using geometrical arrays of directional antennas and incremental RSSI measuments," in *Proceedings of 2009 First International Conference on Computational Intelligence, Communication Systems and Networks*, Indore, 2009, pp. 131–136.

176 S. S. Rana and N. H. Vaidya, "A new 'direction' for source location privacy in wireless sensor networks'," in *Proceedings of 2012 IEEE Global Communications Conference* (GLOBECOM), Anaheim, CA, 2012, pp. 342–347.

177 S. Shirazipourazad, A. Sen, and S. Bandyopadhyay "Fault-tolerant design of wireless sensor networks with directional antennas," in *Distributed Computing and Networking. ICDCN 2013*, D. Frey, M. Raynal, S. Sarkar, R. K. Shyamasundar, and P. Sinha, Eds., *Lecture Notes in Computer Science*, vol. 7730. Berlin: Springer-Verlag, 2013.

178 D.-I. Curiac, "Wireless sensor network security enhancement using directional antennas: State of the art and research challenges," *Sensors (Basel, Switzerland)*, vol. 16, no. 4, p. 488, 2016.

179 N. Malhotra, M. Krasniewski, C. Yang, S. Bagchi, and W. Chappell, "Location estimation in Ad Hoc networks with directional antennas," in *Proceedings of the 25th IEEE International Conference on Distributed Computing Systems* (ICDCS'05), Columbus, OH, 2005, pp. 633–642.

180 K. Kucuk, A. Kavak, H. Yigit, and C. Ozdemir, "A novel localization technique for wireless sensor networks using adaptive antenna arrays," in *Proceedings of 2008 IEEE Radio and Wireless Symposium*, Orlando, FL, 2008, pp. 483–486.

181 A. Cidronali, S. Maddio, G. Giorgetti, and G. Manes, "Analysis and performance of a smart antenna for 2.45-GHz single-anchor indoor positioning," *IEEE Trans. Microwave Theory Tech.*, vol. 58, no. 1, pp. 21–31, Jan 2010.

182 A. Kundu, S. Roy, and S. K. Parui, "Performance of smart antennas in wireless sensor networks," in *Proceedings of International Conference on Communication, Circuits and Systems* 2012 iC3S(6), pp. 1–4, July 2013.

183 S. Nagaraju, L. J. Gudino, B. V. Kadam, R. Ookalkar, and S. Udeshi, "RSSI based indoor localization with interference avoidance for Wireless Sensor Networks using anchor node with sector antennas," in *Proceedings of 2016 International Conference on Wireless Communications, Signal Processing and Networking* (WiSPNET), Chennai, 2016, pp. 2233–2237.

184 S. Nosović, A. Ascher, J. Lechner, and B. Bruegge, "2-D localization of passive UHF RFID tags using location fingerprinting," in *Proceedings of* 2016 *8th International Congress on Ultra Modern Telecommunications and Control Systems and Workshops* (ICUMT), 2016, pp. 207–212.

185 Y. Zhang, M. G. Amin, and S. Kaushik, "Localization and tracking of passive RFID tags based on direction estimation," *Int. J. Antennas Propagat.*, vol. 2007, Article ID 17426, pp. 1–9, 2007.

186 C. Angerer, R. Langwieser, and M. Rupp, "Direction of arrival estimation by phased arrays in RFID," in *Proceedings of EURASIP Workshop on RFID Technology*, pp. 1–5, Sep 2010.

187 S. Azzouzi, M. Cremer, U. Dettmar, R. Kronberger, and T. Knie, "New measurement results for the localization of UHF RFID transponders using an Angle of Arrival (AoA) approach," in *Proceedings of 2011 IEEE International Conference on RFID*, 2011, pp. 91–97.

188 A. Ascher, J. Lechner, S. Nosovic, P. Eschlwech, and E. Biebl, "Localization of UHF RFID transponders regarding Industry 4.0 scenarios using DoA estimation techniques," in *Proceedings of Smart SysTech 2016; European Conference on Smart Objects, Systems and Technologies*, 2016, pp. 1–6.

189 L. Qiu, X. Liang, and Z. Huang, "PATL: A RFID tag localization based on phased array antenna," *Sci. Rep.*, vol. 7, no. 1, pp. 1–12, 2017.

190 A. Ascher, J. Lechner, S. Nosovic, P. Eschlwech, and E. Biebl, "3D localization of passive UHF RFID transponders using direction of arrival and distance estimation techniques," in *Proceedings of 2017 IEEE 2nd Advanced Information Technology, Electronic and Automation Control* Conference (IAEAC), 2017, pp. 1373–1379.

191 Y. Mohamedatnia, B. Fergania, J.-M. Laheurteb, and B. Poussotb "New methodology for the short range localization of UHF RFID tags using a linear uniform array," *Int. J. RF Technol.*, vol. 10, no. 1–2, pp. 9–26, Jan 2019.

192 P. Eschlwech and E. Bieb, "Target simulation for UHF RFID DoA estimation systems," *Adv. Radio Sci.*, vol. 17, pp. 109–118, Sep 2019.

193 C. Loibl and E. Biebl, "Localization of passive UHF RFID tagged goods with the Monopulse principle for a logistic application," in *Proceedings of Smart SysTech 2012; European Conference on Smart Objects, Systems and Technologies*, 2012, pp. 1–5.

194 R. Parada, A. Carreras, J. Meila-Segui, and R. Pous, "Study of a monopulse system with RFID antennas for applications oriented to retail industry," in *Proceedings of UbiComp*'13, Sep 8–12, 2013.

195 C. Alcaraz, J. V. Balbastre, and F. Vega, "Monopulse RFID reader for enhanced intelligent transportation systems applications," in *Proceedings of* 2015 *IEEE Brasil RFID*, 2015, pp. 1–4.

196 K. Aoki, T. Sakogawa, and F. Kuroki, "An antenna having wide radiation pattern but narrow beam in operation based on monopulse system for security gate applications at UHF band," in *Proceedings of 2017 IEEE Radio and Wireless Symposium* (RWS), 2017, pp. 132–134.

197 Y. Álvarez-López, M. E. Cos-Gómez, and F. Las-Heras-Andrés, "A received signal strength RFID-based indoor location system," *Sens. Actuators, A*, vol. 255, pp. 118–133, 2017.

198 D. Laker, G. Hassan, R. Dobson, and R. Wu "RFID localisation using special antenna technique," *Int. J. Radio Freq. Identif. Technol. Appl.*, vol. 5, no. 1, pp. 77–92, 2018.

199 J. Huiting, A. B. J. Kokkeler, and G. J. M. Smit, "Near field phased array DOA and range estimation of UHF RFID tags," in *Proceedings of 2015 International EURASIP Workshop on RFID Technology* (EURFID), 2015, pp. 103–107.

200 Y. Mohamedatni, M. Benbaghdad, B. Fergani, J. Laheurte, and B. Poussot, "AOA/range estimation method for UHF RFID tags located in near-field region of reader antennas," in *Proceedings of the 12th European Conference on Antennas and Propagation* (EuCAP 2018), 2018, pp. 1–4.

201 Z. Popovic, E. A. Falkenstein, D. Costinett, and R. Zane, "Low-power far-field wireless powering for wireless sensors," *Proc. IEEE*, vol. 101, no. 6, pp. 1397–1409, 2013.

202 K. Huang and V. K. N. Lau, "Enabling wireless power transfer in cellular networks: Architecture, modeling and deployment," *IEEE Trans. Wireless Commun.*, vol. 13, no. 2, pp. 902–912, 2014.

203 K. Huang and X. Zhou, "Cutting the last wires for mobile communications by microwave power transfer," *IEEE Commun. Mag.*, vol. 53, no. 6, pp. 86–93, 2015.

204 N. Shinohara, "History of research and development of beam wireless power transfer," in *Proceedings of 2018 IEEE Wireless Power Transfer Conference (WPTC)*, 2018, pp. 1–4.

205 L. Xie, Y. T. Hou, and W. Lou, "Wireless power transfer and applications to sensor networks," *IEEE Wireless Commun.*, vol. 20, no. 4, pp. 140–145, Aug 2013.

206 Z. Ding, C. Zhong, D. Wing, K. Ng, M. Peng, H. A. Suraweera, R. Schober, H. V. Poor, and I. T. Dec, "Application of smart antenna technologies in simultaneous wireless information and power transfer," *IEEE Commun. Mag.*, vol. 53, no. 4, pp. 86–93, 2015.

207 K. W. Choi *et al.*, "Toward realization of long-range wireless-powered sensor networks," *IEEE Wireless Commun.*, vol. 26 pp. 184–192, 2019.

208 X. Wang, S. Sha, J. He, L. Guo, M. Lu, and S. Member, "Wireless power delivery to low-power mobile devices based on retro-reflective beamforming," *IEEE Antennas Wireless Propagat. Lett.*, vol. 13, pp. 919–922, 2014.

209 D. Masotti, A. Costanzo, M. Del Prete, and V. Rizzoli, “Time-modulation of linear arrays for real-time reconfigurable wireless power transmission,” *IEEE Trans. Microwave Theory Tech.*, vol. 64, no. 2, pp. 331–342, Feb 2016.

210 D. Smith *et al.*, “An analysis of beamed wireless power transfer in the Fresnel zone using a dynamic, metasurface aperture,” *J. Appl. Phys.*, vol. 121, no. 1, pp. 1–14, 2017.

211 E. Fazzini, M. Shanawani, A. Costanzo, and D. Masotti, “A logarithmic frequency-diverse array system for precise wireless power transfer,” in *Proceedings of the 50th European Microwave Conference* (EuMC), Utrecht, Netherlands, 2021, pp. 646–649.

212 A. Eid, J. G. D. Hester, and M. M. Tentzeris, “5G as a wireless power grid,” *Sci. Rep.* vol. 11, p. 636, 2021.

213 A. Costanzo and D. Masotti, “Energizing 5G: Near- and far-field wireless energy and data trantransfer as an enabling technology for the 5G IoT,” *IEEE Microwave Mag.*, vol. 18, no. 3, pp. 125-136, May 2017.

214 O. Georgiou, K. Mimis, D. Halls, W. H. Thompson, and D. Gibbins, “How many Wi-Fi APs does it take to light a lightbulb?,” *IEEE Access*, vol. 4, pp. 3732–3746, 2016.

215 V. Talla, et al., “Powering the next billion devices with Wi-Fi,” *Commun. ACM*, vol. 60, no. 3, pp. 83–91, 2017.

216 M. Poveda-García, J. Oliva-Sánchez, R. Sanchez-Iborra, and D. Cañete-Rebenaque, J. L. Gómez-Tornero, “Dynamic wireless power transfer for cost-effective wireless sensor networks using frequency-scanned beaming,” *IEEE Access J. Spec. Sect. Wirelessly Power. Netw.*, vol. 7, pp. 8081-8094, Jan 2019.

217 J. L. Gómez-Tornero, M. Poveda-García, R. Guzmán-Quirós, and J. C. Sánchez-Arnause, “Design of Ku-band wireless power transfer system to empower light drones,” in *Proceedings of* 2016 *IEEE Wireless Power Transfer Conference* (WPTC2016), Aveiro, Portugal, 5–6 May, 2016.

218 M. Poveda-García and J. L. Gómez-Tornero, “Beaming efficiency of 1-D frequency-scanned based radiative WPT system for wireless sensor networks,” in *Proceedings of* 2019 *IEEE Wireless Power Transfer Conference* (WPTC 2019), 17–21 Jun 2019.

219 A. Massa, G. Oliveri, F. Viani, and P. Rocca, “Array designs for long-distance wireless power transmission: State-of-the-art and innovative solutions,” *Proc. IEEE*, vol. 101, no. 6, pp. 1464–1481, 2013.

220 Powercast’s P21XX Powerharvester Evaluation Kit, Available online: http://www.powercastco.com/products/development-kits/.

221 A. Gil-Martínez, M. Poveda-García, D. Cañete-Rebenaque, and J. L. Gómez-Tornero, “Frequency-scanned monopulse antenna for RSSI-based direction finding of UHF RFID tags,” *IEEE Antennas Wirel. Propagat. Lett.*, vol. 21, no. 1, pp. 158–162, Jan 2022. https://ieeexplore.ieee.org/document/9585386.

222 A. Gil-Martínez, M. Poveda-García, J. García-Fernández, M. Campo-Valera, D. Cañete-Rebenaque, and J. L. Gómez-Tornero, “Direction finding of RFID tags in UHF band using a passive beam-scanning leaky-wave antenna,” *J. Radio Freq. Identif.*, in press, 2022.

223 J. L. Gómez-Tornero, A. Gil-Martínez, M. Poveda-García, and D. Cañete-Rebenaque, “ARIEL: passive beam-scanning Antenna terminal for iridescent and efficient LEO satellite connectivity,” *IEEE Antennas Wirel. Propagat. Lett.*, in press 2022.

5

Antenna-in-Package Design for Wireless System on a Chip

Y. P. Zhang

School of Electrical and Electronic Engineering, Nanyang Technological University, Singapore 639798, Singapore

5.1 Introduction

Since the mid-1990s, complementary metal oxide semiconductor (CMOS) based devices have been the technology driver for the wireless revolution [1]. CMOS processes have enabled full system-on-chip (SoC) integration of nomadic wireless devices such as Bluetooth and Wi-Fi radios. Figure 5.1 shows the die micrographs of a Bluetooth SoC in 0.25-μm CMOS with the die area of 40 mm^2 and a Wi-Fi radio SoC in 0.13-μm CMOS with the die area of 36 mm^2 [2, 3]. Note that the sophisticated Wi-Fi radio comprises two identical dual-band (2.4 and 5 GHz) transceivers. Hence, it can operate as a 2×2 multi-input multi-output (MIMO) system to provide increased data rate, robustness, and range.

Full SoC integration is not suitable in all cases. In fact, the radio chip is separate in many cases, especially in cellular mobile networks. Figure 5.2 shows the die micrograph of a phased array transceiver in 28-nm CMOS with the die area of 2.8×2.6 mm^2 for 5G NR (New Radio) at 28 GHz [4]. Note that the die integrates 8 transmit and receive paths and 86 pads for power, ground, and signals.

The arrival of wireless SoCs or single-chip radios has called for compatible antenna solutions. Naturally, one would consider the solution of integrating an antenna (or antennas) with other radio circuits on the same die, but then would quickly realize that this solution is not viable for wireless SoCs operating in several or tens of GHz, due to cost and performance reasons [5]. It is found that the antenna-on-chip (AoC) solution is more suitable for wireless SoCs operating at terahertz (THz) frequencies [6–10].

A die, i.e. a small block of semiconductor material on which a functional circuit is fabricated, must be packaged for practical use. Figure 5.3 shows the photograph of a ball grid array (BGA) package carrying a die (the square black body). Note that the package is larger than the die and has an array of solder balls to facilitate it being surface mounted on a printed circuit board (PCB). The larger package provides more space to implement an antenna (or antennas) for the wireless SoC it carries. This idea triggered the early developments of antenna-in-package (AiP) technology in the late of 1990s [11]. AiP technology has subsequently enriched the system-in-package (SiP) concept with the addition of the unique function of radiation to the package [12].

AiP technology balances performance, size, and cost well. Hence, it has been widely adopted by chip makers for highly integrated radios and radars, for example, for 5G NR operating in the 28

Antenna and Array Technologies for Future Wireless Ecosystems, First Edition.
Edited by Y. Jay Guo and Richard W. Ziolkowski.

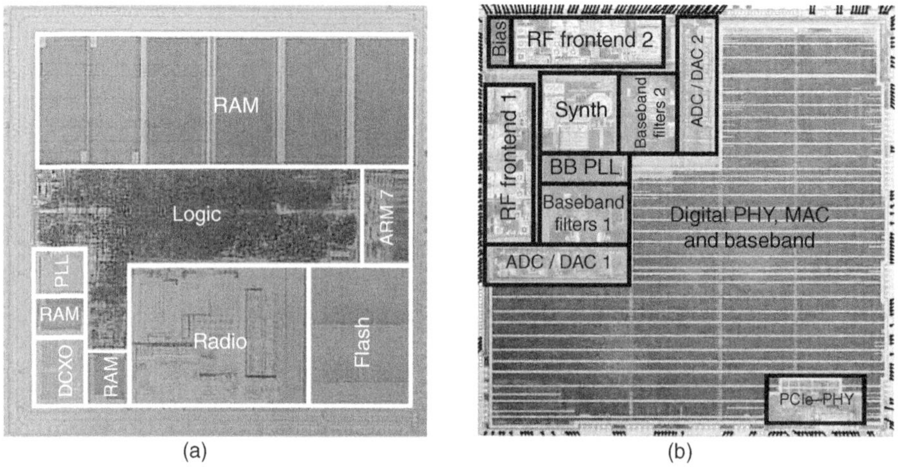

Figure 5.1 Die micrographs: (a) Bluetooth radio. (b) Wi-Fi radio.

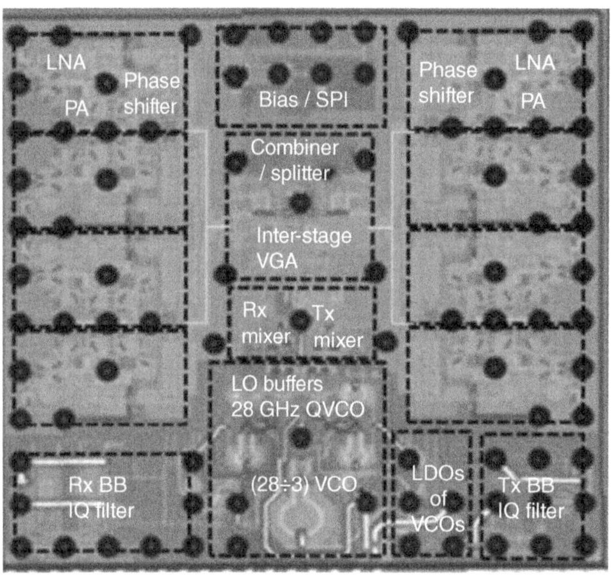

Figure 5.2 Die micrograph of a phased array transceiver for 5G NR. 5G NR, The fifth-generation new radio.

band and automotive radars in the 79-GHz band [13, 14]. This chapter presents the AiP technology for wireless SoCs or highly integrated wireless devices. Section 5.2 describes the high-volume manufacturing (HVM) associated with it. Section 5.3 highlights design considerations for an AiP radiating element, feeding network, ground plane, metal fill, shielding structure, and cooling method. Section 5.4 introduces over-the-air (OTA) testing of an AiP system. Section 5.5 gives three AiP examples. Finally, concluding remarks are made in Section 5.6.

Figure 5.3 Photo of a BGA package. Source: SkyHigh Memory.

5.2 High-Volume Manufacturing of an AiP Module

The success of AiP technology has justified the fabrication of new materials and the development of new processes for HVM of AiP designs. Flip-chip antenna-in-package (FC-AiP) and fan-out antenna-in-package (FO-AiP) are approaches competing for the mass production of AiP systems. An FC-AiP needs a substrate to flip the die. The substrate can be ceramic or organic. An FC-AiP is manufactured with low temperature co-fired ceramic (LTCC) or high-density interconnect (HDI) processes. An FO-AiP relies on an epoxy molding compound (EMC) to embed the die and the redistribution layers (RDLs) for interconnect routing. An FO-AiP is manufactured by fan-out wafer level packaging (FOWLP). Table 5.1 lists the typical material properties and process capabilities in terms of the LTCC, HDI, and FOWLP processes. The choice of either an FC-AiP or FO-AiP is a trade-off among constraints: electrical performance, thermo-mechanical reliability, compactness, manufacturability, and cost. For example, if the AiP design needs higher interconnect density, an FO-AiP realized with an FOWLP process is the best choice as seen from the line width and spacing row in Table 5.1.

5.2.1 LTCC

LTCC uses ceramic tapes and metal pastes. It is easily accessible for prototypes, and foundry services are available for mass production. Table 5.1 shows the thickness, dielectric constant, and loss tangent of fired ceramic materials that can vary from 12.5 to 250 μm, from 4 to 75, and from 0.0004 to 0.006, respectively [15]. The metallic pastes can be chosen from copper, silver, or gold. The metallization layers are screen printed one by one onto the unfired "green" ceramic tapes. Once the printed tapes have been stacked and pressed, they are fired in an oven with a sintering temperature below 900 °C. The standard process can yield a conductor line width and spacing of 100 μm; a via diameter and via-line pitch of 100 μm; and a via pitch of 300 μm. Hence, LTCC has been a popular

Table 5.1 Material properties and process capabilities.

Parameter	LTCC	HDI	FOWLP
Layer	5–70	3–14	2–6
Layer thickness (μm)	100/200	20/520	5/400*
Dielectric constant	5.5–9.5	2.1–4.4	3–3.5
Loss tangent	0.003–0.004	0.003–0.01	0.004–0.01
Metal	Au, Ag, Cu	Cu	Cu
CPE (ppm/°C)	6	0–40	
Metal thick/roughness (μm)	10/1.4	18/3	7.5/0.3
Pad pitch (μm)	250	80	80
Line W/S (μm)	100/100	20/20	5/5
Via diameter (μm)	100	60/100	25
Capture pad (μm)	200	130/200	50
Via pad clearance (μm)	100	50	10
Work size	Small	Large	Fair
Manufacturability	Good	Excellent	Good
Package size	Large	Large	Small

choice for fabricating FC-AiP systems [16–19]. However, LTCC has a small work size and is relatively expensive. In fact, the higher cost is prohibitive for many consumer electronics applications.

5.2.2 HDI

HDI has a variety of dielectric materials that can be used. The choices include, for instance, epoxy/glass flame retardant 4 (FR4), bismaleimide-triazine (BT) resin, polytetrafluoroethylene (PTFE), liquid crystal polymer (LCP), and polyimide (PI) [15]. Among them, FR4 and PI have larger dielectric constants and loss tangents. Both are found to be more suitable for AiP systems operating at a lower frequency. As the operating frequency increases, LCP becomes more popular because of its superiority as both a low-loss and low-cost material. Its dielectric constant is also low, about 2.9. HDI metal is nearly pure copper with its thickness ranging from 18 to 64 μm. HDI allows one to sequentially add additional pairs of thin prepreg dielectric layers on and beneath the thick core dielectric layer. Vias are drilled, blind, or buried; filled; and then plated. HDI currently provides standard 65-μm line widths and spaces, and a 50-μm capability for 18-μm metal layers in high volume situations. The diameters of the vias and their capture pads are typically 60 and 130 μm, respectively, with a pitch of 400 μm [20]. To support the design of FC-AiP systems, HDI has evolved more rapidly in recent years. Now, 20-μm line widths and spaces can be produced with confidence [21].

5.2.3 FOWLP

Unlike LTCC or HDI, FOWLP eliminates the need for a laminate substrate and replaces it with EMCs and RDLs. Polymers like benzocyclobutene (BCB) or PI are used for the electrical isolation between the RDLs. Polymers are a few microns thick, and EMCs are hundreds of microns thick. Both polymers and EMCs have a similar dielectric constant, i.e. 3.0–3.5; but the polymers are lossier with the loss tangent being 0.01 as compared with 0.004 for the EMCs. The metal routings are deposited by a combination of sputtering and electroplating with a thin film process, which currently delivers both line widths and spaces less than 10 μm. In the near future, the line width and space are expected to be down to 2 μm. The most well-known FOWLP is embedded wafer level ball (eWLB) grid array. It has historically been developed for millimeter-wave (mmWave) automotive radar systems; and therefore, it has naturally been used for mass production of mmWave AiP versions [22]. However, because eWLB is restricted to only one RDL, its application for FO-AiP systems is limited. SJ semiconductor (SJSemi) and Taiwan Semiconductor Manufacturing Company (TSMC) recently announced SmartAiP and InFO-AiP, which have multiple metallic and EMC layers to enable the design of more sophisticated FO-AiP systems [23, 24].

5.3 Design Consideration

Co-design of the antenna and the package facilitates the maximization of the AiP performance. The locations of the radiating elements dictate the optimal port locations of the power and low noise amplifiers on the die to minimize the routing losses. Hence, it would be much better if the circuit were also included in the design flow. In addition, an AiP design must consider the thermo-mechanical reliability, die bonding, board assembly, testing strategies, etc. This section highlights design considerations and presents some techniques that have been specifically adopted or developed for AiP technology.

5.3.1 Radiating Element

The radiating element for an AiP module is usually chosen from popular types such as dipole, monopole, loop, slot, Yagi-Uda, grid array, and microstrip patch antennas [25]. Among them, the microstrip patch antennas have been widely adopted due to the availability of abundant design techniques and their versatility to meet AiP design requirements. Figure 5.4 shows the exploded view of a stacked microstrip patch antenna. As can be seen from the figure, the driven patch is embedded between a grounded substrate and a covering superstrate of relative permittivity ε_r, loss tangent δ, length a, and width b. There is a parasitic patch printed on the top surface of the superstrate. Both patches are centered relative to the ground plane. The driven patch of width w_1 and length l_1 is fed by a coaxial probe whose center conductor penetrates through the ground plane. The parasitic patch has the width w_2 and length l_2. The thicknesses of the substrate and superstrate are denoted by h_1 and h_2, respectively.

Recently, Shao *et al.* shed new insight into the principle of operation of the stacked microstrip patch antenna in terms of even and odd modes [26]. Figure 5.5 shows the electric field distributions

Figure 5.4 Structure of the stacked microstrip patch antenna and the associated coordinate system.

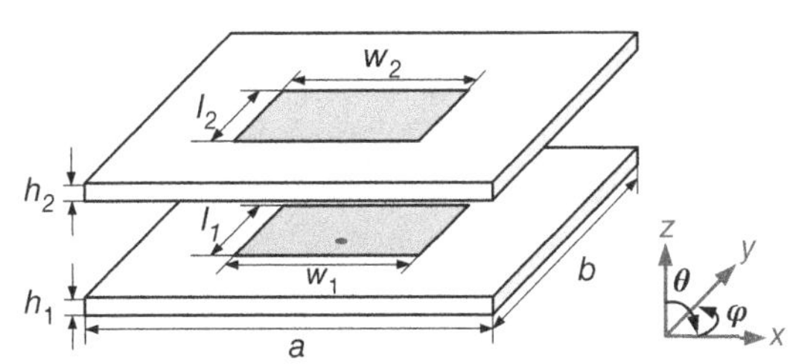

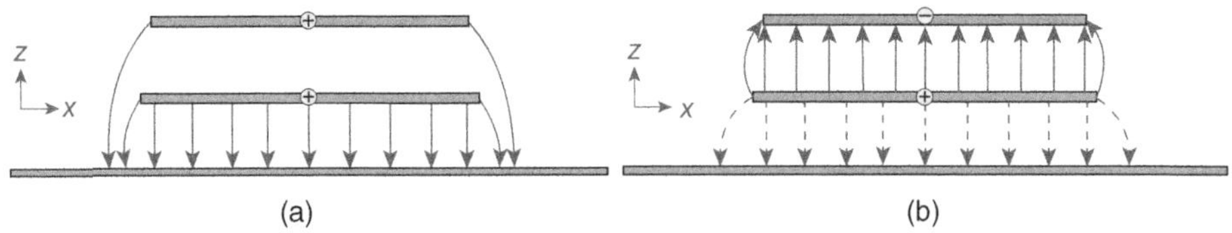

Figure 5.5 Illustration of the quasi electric fields on the cross section of the stacked microstrip patch antenna. (a) Even mode. (b) Odd mode.

of both modes on the cross-section of the stacked microstrip patch antenna. As can be seen, the charge on the two patches is of the same sign for the even mode, resulting in a weak electric field distribution between the two patches. Thus, the resonance of the even mode is dominated by the microstrip antenna, which consists of the top patch, the substrates, and the ground. Therefore, the resonance frequency can be estimated by:

$$f_{r1} = \frac{c_0}{2\sqrt{\varepsilon_{re1}}(l_1 + 2\Delta l_{even})} \tag{5.1}$$

$$\varepsilon_{re1} = \frac{\varepsilon_r + 1}{2} + \frac{\varepsilon_r - 1}{2}\left[1 + \frac{12(h_1 + h_2)}{l_2}\right]^{-\frac{1}{2}} \tag{5.2}$$

where Δl_{even} accounts for the extension of the radiating patch in the even mode and ε_{re1} is the corresponding effective dielectric constant of the substrate. On the contrary, the charge on the two patches is of opposite signs for the odd mode, leading to a strong electric field distribution between the two patches. Thus, the resonance of the odd mode is dominated by the microstrip antenna that consists of the top patch and bottom patch. Therefore, the corresponding resonance frequency can be estimated by:

$$f_{r2} = \frac{c_0}{2\sqrt{\varepsilon_{re2}}(l_1 + 2\Delta l_{odd})} \tag{5.3}$$

$$\varepsilon_{re2} = \frac{\varepsilon_r + 1}{2} + \frac{\varepsilon_r - 1}{2}\left[1 + \frac{12h_2}{l_1}\right]^{-\frac{1}{2}} \tag{5.4}$$

where Δl_{odd} accounts for the extension of the radiating patch in odd mode and ε_{re2} is the corresponding effective dielectric constant of the substrate.

The stacked microstrip patch antenna fed by a microstrip line was first reported by Hall, Wood, and Garrett in 1979 [27]. They showed that a substantial bandwidth improvement could be achieved but at the expense of additional manufacturing complexity and some increase in its overall height. Subsequently, Corq and Pozar demonstrated that a stacked microstrip patch antenna fed though aperture coupling can achieve wider impedance bandwidth while naturally isolating the feed line from the patches [28]. Waterhouse described the stacked microstrip patch antenna fed by a coaxial probe [29]. Liu *et al.* presented a method of designing a stacked microstrip patch antenna with proximity feeding to compensate for the probe inductance [30]. Gao *et al.* discussed how to design a stacked microstrip patch antenna to achieve dual polarization. [31].

The design of stacked microstrip patch antennas for an LTCC-based AiP module was first attempted by Li *et al.* [32]. To simplify the design procedure, the driven and parasitic patches were assumed to be square with the same dimensions $w_1 = l_1 = w_2 = l_2 = l$. Usually, the total thickness of the stacked microstrip patch antenna, $h = h_1 + h_2$, is fixed for a certain bandwidth requirement. Wider bandwidths simply demand thicker substrates. The design procedure is as follows.

(1) Fix the total thickness h of the substrate according to the needed bandwidth.
(2) Let $h_1 = h/4$.
(3) Set $l = 1/2f_c\sqrt{\mu_0\epsilon_0\epsilon_r}$.
(4) Set $2h_1 < h_2 < 2h_1$.
(5) Adjust h_1, h_2, l, and the feed location with a full-wave simulator to meet the design requirements.

This design method is simple but inflexible. Zhou *et al.* reported another method to design a stacked microstrip patch antenna [33]. The design is achieved by means of the synthesis, diagnosis, and tuning of a two-order band pass filter. The input port of the filter is the feed, while the output port of the filter is the far-field radiation of the stacked microstrip patch antenna. The method involves the following steps.

(1) Fix the total thickness h of the substrate according to the needed bandwidth.
(2) Set $h_1 < h_2$.
(3) Set the initial dimensions for the driven and parasitic patches according to Eqs. (5.1)–(5.4).
(4) Construct the **S** matrix from a full-wave simulation and vector fitting.
(5) Transform the **S** matrix to the **Y** matrix and get the coupling matrix **M**.
(6) Calculate the sensitivity of the coupling matrix to the initial dimensions and get the new physical dimensions.
(7) Repeat the aforementioned steps until the design specifications are satisfied.

It is found that this method is efficient. It takes only a few iterations to finish the design of a stacked microstrip patch antennas for an AiP module [34]. For example, it took five iterations to design a stacked microstrip patch antenna with a 45^0 polarization on an LTCC substrate whose dielectric constant $\varepsilon_r = 5.9$ and loss tangent $\delta = 0.002$ in the 28-GHz band. It used six ceramic layers and the thickness of each layer was 96 μm. There were two ceramic layers between the driven patch and the ground plane and four ceramic layers between the driven and parasitic patches. In addition, there were four ceramic layers beneath the ground plane and two metal layers to realize a stripline to feed the antenna from a coplanar waveguide (CPW) port. The outer and inner metal layers were gold and silver, respectively, with their thicknesses being from 8 to 10 μm. The size of the antenna was $a \times b = 6.0 \times 6.0\,\text{mm}^2$ and it achieved an impedance bandwidth from 26.43 to 29.32 GHz or 10.3% at 28 GHz. The gain was 5.0 dBi at 28 GHz and its 3-dB gain bandwidth was from 25.5 to 30.0 GHz or 16.1% at 28 GHz.

5.3.2 Ground Plane

A ground plane is generally a flat conducting surface. The ground plane in an AiP module is a part of both the antenna and the package. For the antenna, the ground plane serves as a reflecting surface for radio waves. It needs not be a continuous surface, but its planar dimensions must be at least a quarter of the wavelength of the radio wave. For the package, the ground plane is connected to the power supply ground terminal and serves as a return path for the currents from the carried die. A continuous surface of large area is preferred to conduct the large return current without significant voltage drops, ensuring that the ground connection of all the circuits on the die are at the same reference potential. Moreover, the ground plane affects the mechanical reliability of the AiP module because thermal effects cause a stress mismatch between the conductor and dielectric layers, which will result in micro factures and warpage of the package. Hence, a full ground plane is not preferred, especially for larger AiP modules. This preference has led to the development of meshed, defected, and patterned ground planes as described in the following text.

5.3.2.1 Meshed Ground Plane

A meshed ground plane is a grid of less than a certain percentage of the conductor coverage. Figure 5.6 shows the geometry of a meshed ground plane [35]. A thermal-mechanical reliability analysis will specify the conductor coverage percentage. It can be implemented with mesh lines of width w and separation s. The separation s is chosen to be larger than the width w so that a clearance c can be easily made between the feed through via and the ground plane conductor to avoid a shorted circuit. Let s be 1/10 of the operating wavelength to minimize the leakage of electromagnetic waves. The highest operating frequency of the meshed ground plane can then be determined. Taking an AiP module in LTCC as an example, we choose $w = 250\ \mu m$ and $s = 500\ \mu m$ for a conductor coverage of less than 50%. The clearance $c = 200\ \mu m$ can be made for a via of 100 μm in diameter. The highest operating frequency is 60 GHz.

5.3.2.2 Defected Ground Plane

A defected ground plane is comprised of a limited number of defects with periodic or aperiodic arrangements. Figure 5.7 shows the geometry of a defected ground plane etched with dumbbell-shaped defects [36]. The defected ground plane makes a longer path for a return

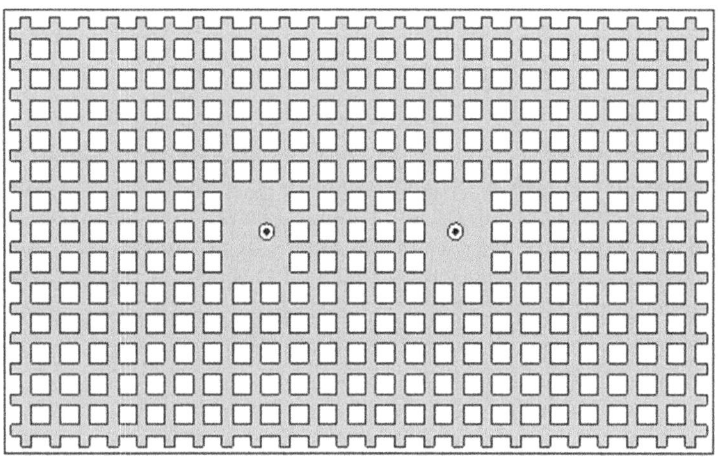

Figure 5.6 Geometry of a meshed ground plane.

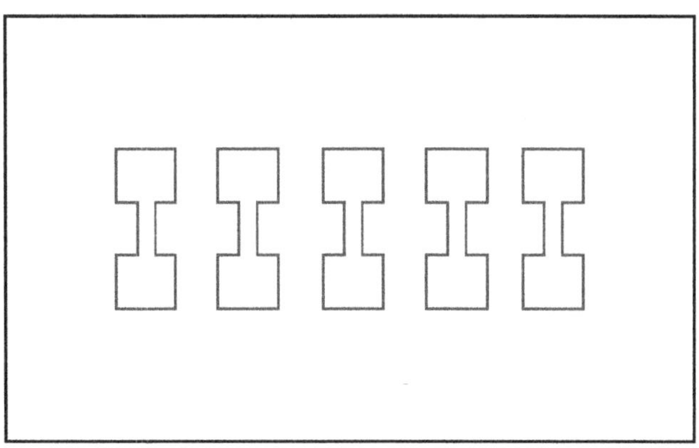

Figure 5.7 Geometry of a defected ground plane.

current, which is equivalent to increasing the ground equivalent inductance. Also, the defected ground plane contains narrow gaps, which is equivalent to increasing the ground equivalent capacitance. The increased inductance and capacitance explain why the defected ground plane exhibits the bandgap characteristics, which can be explored to enhance the performance of an antenna or array in the AiP module. For example, the defected ground plane can be designed to have a proper bandgap to suppress the surface wave to reduce the mutual coupling between microstrip patch antennas in an array.

5.3.2.3 Patterned Ground Plane

A patterned ground plane keeps its portion underneath the radiating elements as solid as possible, while the complement can be meshed or defected. Figure 5.8 shows the geometry of a patterned ground plane used for a microstrip grid array in an AiP module [35]. It was found that this patterned ground plane shifts the resonance frequency downward, expands the impedance bandwidth, and reduces the antenna gain penalty. The patterned ground plane perturbs the current return path and leads to the excitation of an electric field across the rectangular openings. The consequent reactive energy stored near the discontinuities is responsible for the downward frequency shift. The large metal patches formed on the patterned ground plane below the radiating elements reduce the radiation through the ground plane openings and, therefore, helps to reduce the gain penalty.

5.3.3 Feed Network

The most critical aspect in the design of an AiP module is the signal reflection and loss that appear at the die–package–antenna interface of its feed network. The mismatch and loss deteriorate the overall AiP performance and, hence, must be minimized. Typically, the feed network involves pads, bonds, traces, and vias. The pads on the die or on the package are usually in a ground-signal-ground (GSG) or ground-signal-ground-signal-ground (GSGSG) configuration for a single-ended or differential antenna, respectively. The bonds that join the pads on the die and on the package together are wires or bumps, which are implemented with the wire-bonding or flip-chip technique. Wires are over-molded, and bumps are under-filled with epoxy encapsulant. The traces that connect the pads on the package typically form a CPW. The CPW may need to be converted into a microstrip line or a stripline to feed the antenna with a quasi-coaxial via or through an opening in the ground plane.

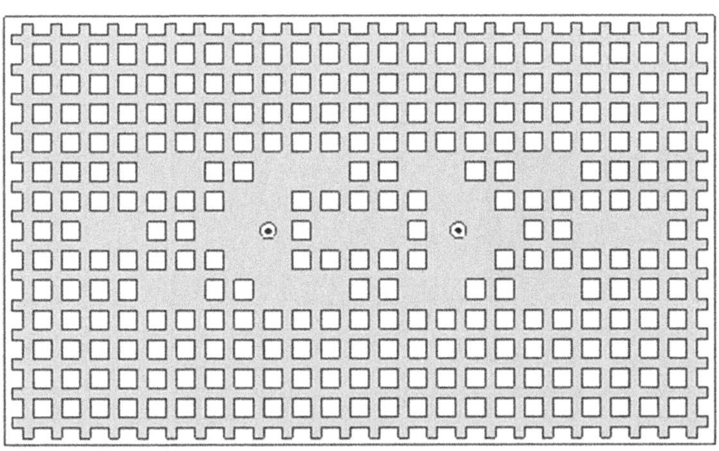

Figure 5.8 Geometry of a patterned ground plane.

A coax via is the most reliable vertical transition between the traces on different layers [37]. The coax via consists of a signal via and several grounded vias in the layered medium. It is a quasi-coaxial line with the signal via as the inner conductor and the ground vias as the outer conductor to support the transmission of the dominant transverse electromagnetic (TEM) mode. The coax via should match not only the impedance but also the mode distribution of the connected lines. It is known that neither CPW nor microstrip lines support a true TEM mode; they sustain a hybrid mode. However, the longitudinal electric and magnetic components are small; and, hence, the hybrid mode is referred to as a quasi-TEM mode. A stripline, like a planar coaxial line, supports a true TEM mode. Hence, the key in the design of traces with coax vias is to maintain the smooth transmission of the dominant true or quasi-TEM mode. There are many designs which employ the same or different types of lines with a coax-via transition [38, 39]. A comprehensive analysis and design of various types of lines with vertical transitions using coax vias for the feed network of the four microstrip grid array antennas that form an AiP module can be found in [40]. As a supplement to the earlier, the remainder of this subsection will discuss the bonds that interconnect the die with the package and that can be used to classify the feed network of an AiP module.

5.3.3.1 Wire

The wire-bonding interconnect, well established in consumer electronics, remains a very attractive solution since it is robust and inexpensive. Moreover, it has the advantage of being tolerant to chip thermal expansion, an important requirement for many applications.

Figure 5.9 is an illustration of a wire-bonding interconnect [41]. Note that three bonding wires connect the GSG pads on the die with the CPW traces on the package. A bonding wire presents high impedance at high frequency because of its large inductance and small capacitance. Thus, keeping the length of the bonding wires to a minimum and using the bonding wires with the largest possible diameter are critical to minimizing their disruptive influence on the electrical signals. The GSG bonding wire arrangement improves the situation by providing a return path in close proximity to the wire and thus slightly reducing the inductance. Typically, a bonding wire is gold and has a diameter of either 25.4 or 32.5 μm. Figure 5.10 shows a photograph of the wire-bonding interconnect of an mmWave AiP module [42]. The diameter and length of the bonding wires were 25.4 and 100 μm, respectively. The insertion loss was found to be less than 1.4 dB from 135 to 155 GHz.

The 100 μm bond wire length is too short for attachment with an automatic wire bonder; the length should be at least 400 μm [43]. Compensation with capacitance may allow using longer bonding wires. Symmetric compensation consumes additional die area and often results in

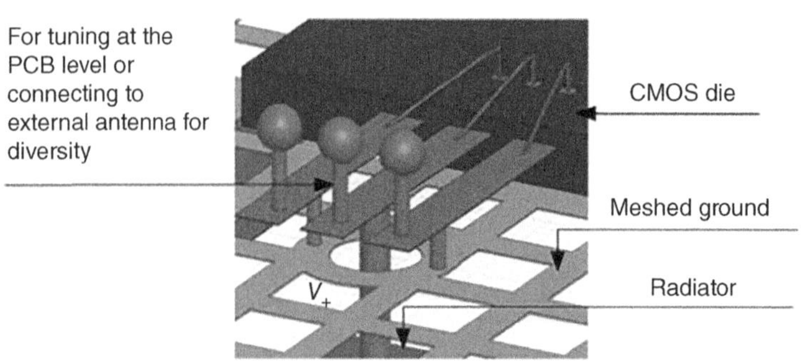

Figure 5.9 Illustration of a wire-bonding interconnect. Source: From [41] / with permission from IEEE.

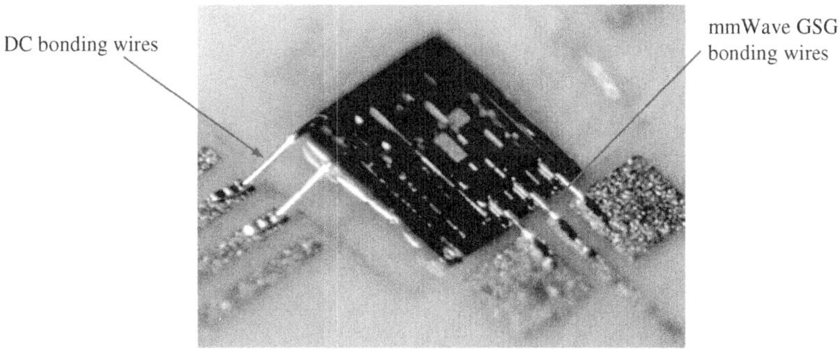

Figure 5.10 Photograph of the bonding-wire interconnect of an AiP module.

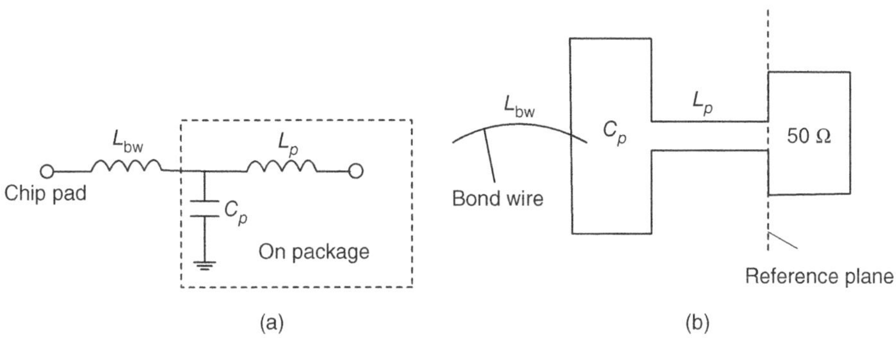

Figure 5.11 Low pass compensation. (a) Circuit model. (b) Layout.

output impedance different from the usual reference. Therefore, asymmetric compensation of only one terminal of a signal wire in a package is preferable for AiP realizations and many other cases [16].

Figure 5.11 shows a low pass compensation for the series inductance introduced by the bonding wire. The inductor L_{bw} models the bonding wire inductance. Inductor L_p and capacitor C_p are realized by the package. This compensation has two drawbacks. One is the large on-package area required to implement, and the other is that it is not suitable for on-package CPW designs [44]. Furthermore, the DC blocking capacitor needs to be implemented on the die to facilitate connecting the radio die to the antenna.

Figure 5.12 shows a band pass compensation for the series inductance introduced by the bonding wire. As illustrated, a series capacitor is used to tune the inductance of the bonding wire to a resonance condition, thus compensating for the high inductance of the bonding wire at the resonance frequency. In the mmWave frequency range, the form factor of the capacitor for compensation is on the order of tens of femtofarads, making the structure very compact. Moreover, the compensation structure enjoys the properties of manufacturing reliability and cost effectiveness. It could be employed successfully for the commonly used chip-to-package connections at mmWave frequencies. Consequently, it will be very desirable for highly integrated mmWave wireless devices, which call for the properties of miniaturization, manufacturing reliability, and mass-production cost effectiveness.

The band pass compensation design involves the following steps [16]:

(1) Identify the bonding wire to be compensated.
(2) Set an operating frequency and bandwidth.

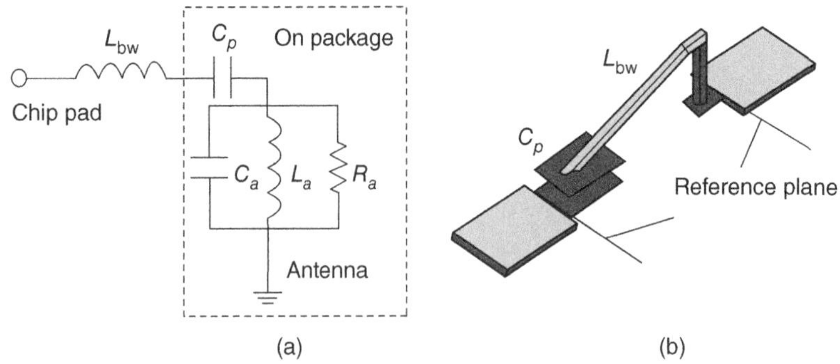

Figure 5.12 Band pass compensation, (a) Circuit model. (b) Layout.

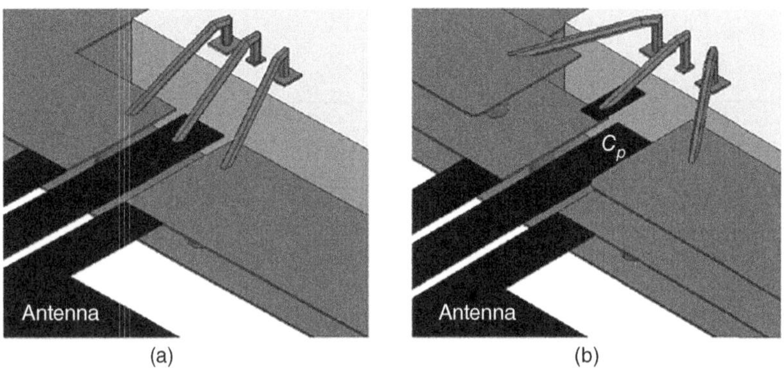

Figure 5.13 Band pass compensation: (a) without and (b) with compensation.

(3) Estimate the bonding wire inductance with the empirical formula in [15].
(4) Calculate the capacitance needed to achieve resonance at the operating frequency with the inductance value obtained from the previous step. Estimate the capacitor dimensions to obtain this capacitance.
(5) Simulate the compensation structure with the AiP module as illustrated in Figure 5.13 in a full-wave electromagnetic solver first and then optimize its frequency response through simulation by only adjusting the compensation structure.

Instead of increasing the length of a bonding wire with the low or band pass compensation, Beer *et al.* treated the GSG bonding wires as an air CPW transmission line [43]. They proposed to keep the length of the transmission line as half a wavelength long at the operating frequency for self-matching. It was noted that this technique is suitable for operating frequencies from 100 to 300 GHz and requires the lengths of bonding wires to be only from 1.5 to 0.5 mm. For such lengths, the diameter and pitch of the bonding wires are not critical, and the assembly process is eased. They demonstrated that the bonding wires attached with an automatic wire bonder had high repeatability and achieved good performance at 130 GHz with a fractional bandwidth up to 13%.

5.3.3.2 Bump

The flip-chip interconnect has better electrical performance than the wire-bonding one because the height of a bump is smaller than the length of a wire and the diameter of a bump is thicker than

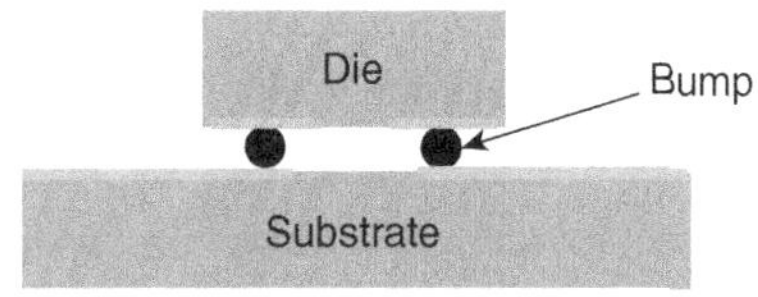

Figure 5.14 Illustration of a feed network with bumps.

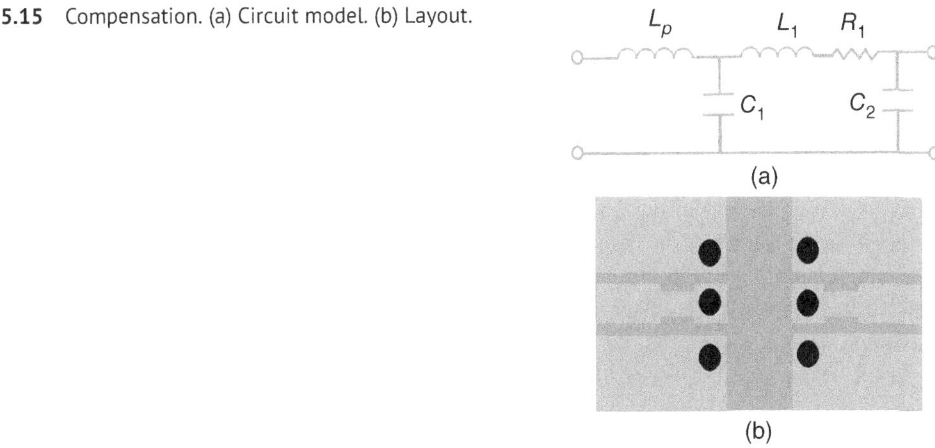

Figure 5.15 Compensation. (a) Circuit model. (b) Layout.

that of a wire. Figure 5.14 is an illustration of a flip-chip interconnect. There are three bumps that connect the GSG pads on the die with the CPW traces on the package. The flip-chip interconnect with the GSG bumps presents low impedance at high frequency because of its large capacitance and small inductance. Thus, keeping a smaller pad, a larger pitch, and a smaller bump diameter is critical to minimizing the flip-chip disruptive influences on the electrical signals. Bumps may be solder balls or copper pillars. Currently, the diameter and pitch of solder balls can be made as small as 75 and 150 μm, respectively. The diameter, height, and pitch of copper pillars can be made as small as 50, 50, and 100 μm, respectively. Kam *et al.* reported the design and optimization of the feed network with solder balls from the die to the antenna for a 60-GHz phased-array AiP module. The diameter of the pads on both the die and the package was set to be 100 μm. The pad pitch was selected to be 250 μm for differential and 225 μm for single-ended antennas. The pads on the package interfaced with a CPW line, which was then converted to a microstrip line. The microstrip line was vertically connected by a coax via to another microstrip line to feed the patch antenna though a resonant opening on the antenna's ground plane. The whole interconnect was well matched to 50 Ω and the insertion loss was better than 1.2 dB in the 60-GHz band [17, 45].

Although the flip-chip interconnect can be optimized by adjusting the size and pitch of the bumps, it is difficult to change those values to a large extent due to the limitations of the process. Hence, it is necessary to compensate the capacitance caused by the overlaps in the transition region. Figure 5.15 shows a circuit model of the flip-chip interconnect. Note that C_1 and C_2 are the pad capacitances on the package and die, respectively; L_1 and R_1 denote the inductance and resistance of the signal solder ball; and L_2 represents the serial inductance for compensation. In particular, note that the parasitic inductance and resistance of the grounded solder balls can also be included to improve the accuracy of the model. Figure 5.15 also shows the compensation structure. Note that the series inductance L_p in the circuit model is realized in the CPW line on the package by a narrow and short section to tune the capacitance of the pads to a resonance condition, thus compensating the capacitance of the pads at the resonance frequency. The compensation design involves the following steps:

(1) Identify the bumps to be compensated.
(2) Set an operating frequency and bandwidth.
(3) Estimate the pad compacitances C_1 and C_2 with the empirical formula in [15].
(4) Calculate the inductance to resonate at the operating frequency with the capacitance values obtained from the previous step. Estimate the dimensions of the inductor L_p to obtain this inductance.
(5) Simulate the compensation structure in a full-wave electromagnetic solver first and then optimize its frequency response through simulations by only adjusting the compensation structure.

5.3.3.3 Via

Vias can be seen in advanced packaging as a replacement for conventional wires or bumps to interconnect the die with the package. Figure 5.16 shows an scanning electron microscope (SEM) of the cross-sectional view of an FOWLP [46]. Note that its vias are formed as a part of three RDLs realized with thin-film technology. They have different diameters, but a similar height of a few microns. Tsai *et al.* simulated the loss from the die pad to the antenna port through such vias and traces on copper RDL [47]. The total loss was found to be 0.7 dB at 60 GHz over a 50-Ω CPW line whose length was 1.6 mm. To understand the parasitic effect of the vias, a circuit model was proposed. It consisted of a series resistance and inductance as well as a shunt capacitance. The series resistance was negligible, the series inductance was 25 pH, and the shunt capacitance was 5.0 fF. Compared with C4 solder balls, it had a 70% reduction in the inductance and a 73% reduction in the capacitance.

In addition, the through-silicon-via (TSV) interconnect creates the shortest possible path from one side of the die to the other. While it has also been proposed for AiP technology, it really has not been demonstrated yet even though there have been several attempts to implement TSVs in a silicon interposer or wafer as a part of the feed network to an antenna (or antennas) in a wafer-level package [48–50]. Nevertheless, the TSV interconnect holds promise to push the AiP technology to applications in the mmWave to THz bands [51].

5.3.4 Metal Fill

Metamaterials and metasurfaces have shown the possibility of manipulating electromagnetic waves and enhancing the performance of antennas. A metasurface, such as soft surface or high impedance surface, can be used to suppress surface waves [52]. The use of a high impedance

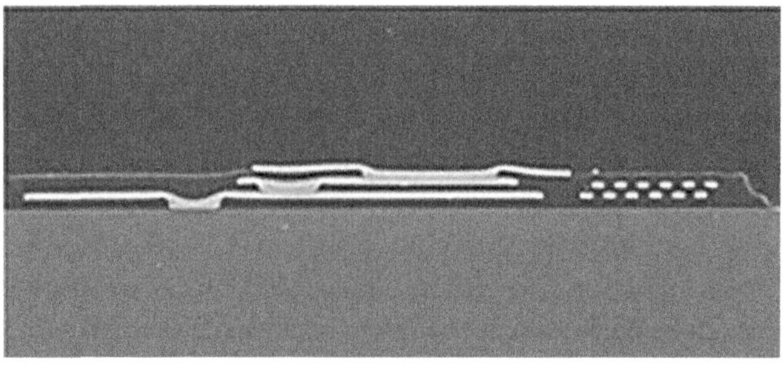

Figure 5.16 SEM of the cross-sectional view of an FOWLP.

surface as an artificial magnetic conductor (AMC) to improve the performance of microstrip dipole antennas and arrays for 60-GHz AiP technology was proposed by Kyriazidou, Contopanagos, and Alexopoulos [53, 54]. The AMC was based on spiral unit cells on a single metallization layer with no vias, which allowed for simultaneously projecting the AMC behavior directly on the antenna metallization layer eliminating the surface waves and obeying the package design rules. Although enhancements in antenna performance have obtained, neither metamaterials nor metasurfaces have been adopted well for practical AiP modules, probably due to the associated increased cost in fabrication.

A certain metal density is required in a package to improve its planarity for manufacturing. The integration of an antenna or array in the package makes it more difficult to pass the design rule checks (DRC) in a manufacturing process. It is therefore suggested to explore dummy metal fill according to the metamaterial concept since a meta-metal fill could enhance the electrical performance of an AiP module without extra costs being incurred. The fences realized with vias and patterned ground planes are examples of meta-metal fill [35].

5.3.5 Shielding Structure

A major concern with an AiP module is the risk of the unwanted coupling between the antenna and the die. On-die passive components such as inductors and capacitors occupy substantial die area and thus are more susceptible to the coupling from the antenna than tiny active devices such as transistors are. An understanding of the impact of the process technology, grounding, guarding, shielding, and decoupling is necessary to reduce this coupling with on the die components as was suggested in [38]. On the other hand, better radiation nulling toward the die or innovative AiP shielding structures need to be devised to minimize the coupling with the in-package antenna. The shielding structures for an AiP antenna include guard rings, fences of vias, conformal shields (CFS), and compartment shields (CPS). The design guidelines for the guard ring and fence of vias approaches can be found in [55]. In what follows, the coupling in an AiP module and the versions of the CFS and CPS structures used to mitigate it are presented.

5.3.5.1 Coupling Mechanism

Taking an on-die inductor as an example, Zhang *et al.* simulated the coupling level between the in-package antenna and the on-die inductor [16]. The on-die spiral inductor was implemented using two metal layers ($\sigma = 2.8 \times 10^7$ S/m) with 10 μm wide lines and an outer diameter of 311 μm in a 9-μm-thick silicon dioxide layer ($\varepsilon_r = 4$) on a 451-μm-thick silicon ($\varepsilon_r = 4$, $\sigma = 10$ S/m) substrate. Figure 5.17a and b shows the coupling simulation models without and with the compensation capacitor, respectively. The on-die inductor has one terminal connected to the package ground by a bond wire and the other terminal together with another grounded bond wire are used as the receiving port, port 2. The feed port of the antenna is located at the signal pad on the die and excites the antenna as the transmitting port, port 1, through the GSG bond wires.

The coupling level was simulated for the worst case, i.e. with no guard ring, with a pattern grounded shield being used for the inductor, and with the distance between the inductor and the GSG pads being the shortest allowed by the on-die layout rule, i.e. with $c = 0.1$ mm. The coupling level was quantified by the magnitude of S_{21}, which was found to be lower than −27.5 dB for the 60-GHz band. The coupling was attributed to inductive coupling from the GSG bond wires to the inductor, radiative coupling from the antenna to the inductor, and resistive coupling between the antenna feed port and the inductor receive port over the low resistivity die substrate.

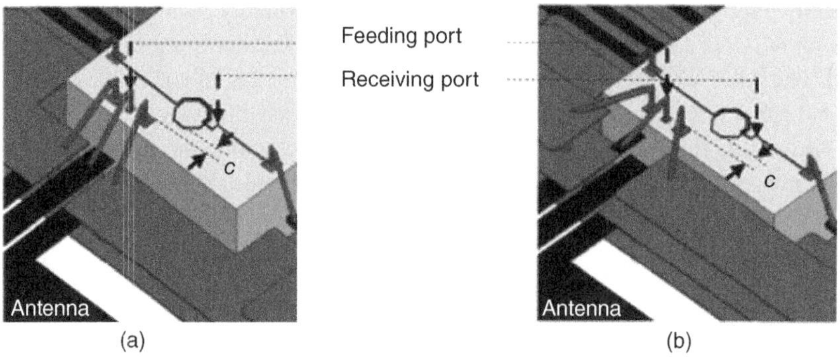

Figure 5.17 Coupling simulation models (a) without, and (b) with the compensation capacitor.

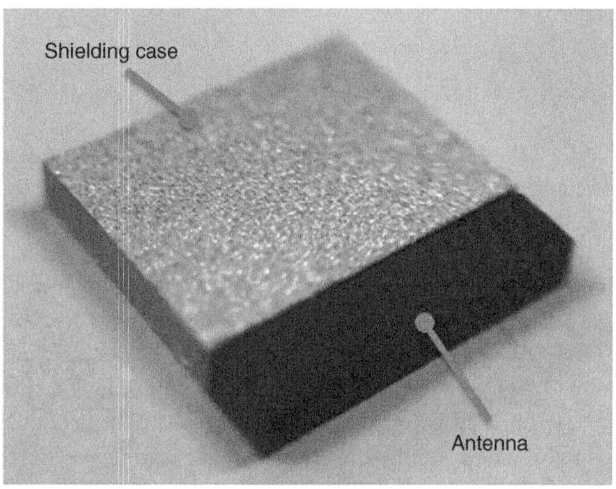

Figure 5.18 Photo of an AiP with a CPS.

5.3.5.2 CFS and CPS

A CFS is an outer conductive coating layer on a selected surface of an AiP module. A CPS is a conductive partition inside the CFS of an AiP. Figure 5.18 shows a photo of an AiP with both the CFS and CPS in place with in a size of $8 \times 7 \times 1\ mm^3$ [56]. It was constructed with a laser trench, metal frame, and surface coating. First, the antenna was integrated into the system, and the die was bonded on the package's substrate. Then, a metal frame was soldered on the package substrate between the antenna and the die. Next, the antenna, metal frame, and die were over-molded together. The molded package was then cut with a laser to open a trench deep enough that the embedded metal frame was exposed. Finally, the sputtering method was used to coat a conductive layer on the selected surface of the package, including the exposed area of the metal frame, to shield the die from the fields radiated by the antenna. The coating layer, around 4 μm thick, was formed with copper and a stainless alloy.

A test spiral inductor was printed on the package substrate to emulate the coupling level between the antenna and the die. It was found that the coupling level was lower than −57 dB at the design frequency, 2.4 GHz. The coupling level would increase to −27 dB at 2.4 GHz without the CFS and CPS being present. Hence, the CFS and CPS achieved a shielding effectiveness (SE) of 30 dB.

The construction of the CPS using a laser trench and a paste filling process can avoid mounting a metal frame around the die. An AiP module that had its CPS implemented in this manner was introduced in [57]. The AiP module had a size of 650 mm^2. The partition wall was made with silver paste. It was 120-μm thick and 1.4-mm high; it connected the coating layer to the package ground. The quality of the coating and filling, as well as the adhesion to the EMC surfaces, was good and no delamination occurred. The measured SE of this CPS was over 35 dB in the frequency band from 25 to 40 GHz.

5.3.6 Cooling Method

Thermal issues arise with high-level integration in CMOS. This is particularly true for wireless SoC devices because of their low power conversion efficiency from DC to radio frequency (RF) which leads to most of the supplied DC power being converted into heat. A direct consequence is a rise of the die junction temperature T_j. Hence, how to manage thermal issues becomes critical for wireless SoC implementations. One must keep T_j within specific limits.

The die junction temperature T_j can be expressed as:

$$T_j = T_a + R_{ja} P_d \tag{5.5}$$

where T_a is the ambient temperature of the die (°C), R_{ja} is the junction to ambient thermal resistance (°C/W), and P_d is the power dissipation in the die (W). Generally, thermal management aims to reduce the thermal resistance R_{ja} and to transfer heat from the die being cooled to the exterior ambient environment while minimizing the power consumed by the cooling mechanisms. It is well known that R_{ja} can be reduced, and the heat removal can be enhanced with the use of thermal interface materials, heat spreaders, and sinks.

5.3.6.1 Passive Cooling

The passive cooling method transfers heat to the ambient air through natural conduction or convection phenomena. It is suitable for a die having a low power density, say, 1.0 W/cm^2. Two passive cooling configurations have been specifically developed for AiP modules. One relies on the natural conduction of heat from the carried die by means of metal grounded traces and metal stamped antennas [58]. The other explores the codesign of the antenna array and 3D-printed heat sink fins. Figure 5.19 shows the photos of an AiP fabricated in LTCC and the AiP module with its 3D-printed heat sink. The sink fins not only remove heat, but they also enhance the antenna

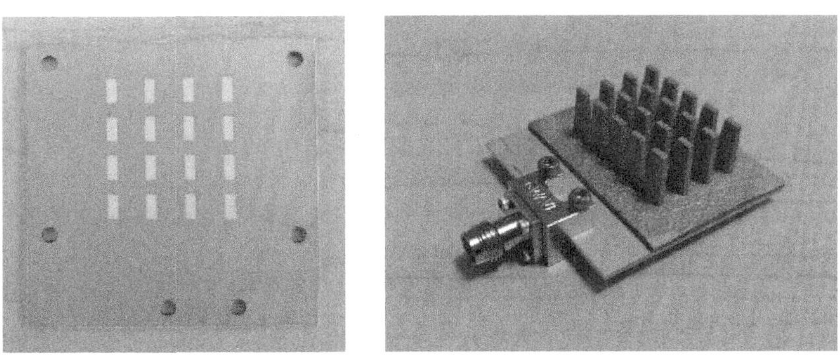

Figure 5.19 Photos of an AiP fabricated in LTCC (left) and the AiP module with its 3D-printed heat sink (right).

radiation performance [59]. A temperature reduction of 47.0 °C can be achieved, indicating an excellent cooling effect by the 3D-printed heat sink.

5.3.6.2 Active Cooling

The active cooling method uses various mechanical devices that consume energy to dissipate heat. It is required for a die having a high-power density, say, 10 W/cm^2. Traditionally, active cooling uses a fan to force air flowing across the heat sink at a desired speed. For the dissipation of 1.0 W of heat from the die, the power consumption by the fan is typically less than 0.2 W. A fully integrated 384-element phased array AiP system was cooled by this method [60]. It was found that the maximum die temperature was 49.8 °C when it had to manage a high-power density of 22 W/cm^2.

To improve the cooling performance and to reduce the cooling power consumption, the replacement of air cooling with liquid cooling has been suggested to remove heat from an AiP module. A disruptive cooling method is to integrate microfluidic channels inside the die to allow cooling liquid to flow inside of it [61]. The microfluidic channels can be placed very close to the hot spots in the die so that the required amount of heat can be extracted in exactly the right place and can prevent it from spreading throughout the die. This revolutionary cooling approach has the potential to keep a lower T_j as well as to reduce the AiP size, weight, and power consumption.

5.4 Testing AiP

Testing of an AiP module occurs as it runs through the design, production, and application phases. In the design phase, testing provides deeper understanding of the characteristics of the AiP module [62]. In the production phase, testing shifts the emphasis from a comprehensive characterization to faster functional verification of the AiP characteristics [63–65]. In the application phase, testing checks compliance with known standards. For example, the AiP modules in 5G smart phones must pass standard compliance testing set by the 3rd generation partnership project (3GPP) [66]. OTA testing is essential in all phases. However, only OTA testing of the AiP system with automated test equipment (ATE) in its production phase is new to the antenna community. Thus, it is introduced here. Interested readers are referred to [62] for descriptions of OTA testing of AiP systems in anechoic chambers in their design and application phases.

5.4.1 Test System

A test system in the production phase consists of an ATE, a test head, and a handler. The ATE is simply a tester that can conduct the testing of DC, digital, analogue, mixed, and RF signals. The test head is actually a socket mounted on a load board, which in turn interfaces with the ATE. The handler is a robot, which first provides three predefined temperatures to heat or chill a device under test (DUT). It then picks up the DUT and places it in the socket while the ATE performs the necessary tests. Finally, it sorts the DUT into groups depending on those test results.

The OTA testing of an mmWave AiP module drives the test equipment beyond its norms and blurs the boundaries between the tester, test head, and handler. The tester can add on mmWave modules that allow a more traditional ATE test plan to occur. The test head has now become a critical component and must be designed holistically with the tester and handler.

Standard spring pins for DC power supply, ground, control, low and intermediate frequency (IF) connections to the package solder balls can be used for an AiP in the socket. It will radiate mmWave signals into the surrounding air region which will be captured by a test antenna. This facilitates measuring the EIRP (effective isotropic radiated power), gain, compression point (P1dB), error vector magnitude (EVM), the third-order intercept point (IP3), adjacent power leakage ratio (APLR), spurious free dynamic range (SFDR), power added efficiency (PAE), etc. Depending on the testing strategies, the test antenna can be positioned either in the far or near-field region of the AiP in the socket. The broadside or end-fire radiation direction of the AiP indicates where the test antenna should be located.

Mroczkowski and Campion reported configurations that integrate the test antenna in the socket both above and beside the AiP being measured [62]. Figure 5.20 illustrates the integration of the test antenna in the socket. Note that the test antenna can be a microstrip patch or a dipole antenna integrated in the socket lid for the AiP's broadside radiation and a horn or waveguide antenna for its end-fire radiation. The test head can use coaxial cables or waveguides for interconnection with the tester.

Shiota *et al.* proposed three options to integrate the test antenna in the handler [65]. Figure 5.21 illustrates one option for which the test antenna can be directly integrated with a standard handler. However, it requires the addition of a highly reliable blind mating mechanism in the socket to provide the mmWave interconnect for the test antenna.

5.4.2 Testing Strategy

Strategies for testing the AiP module in its production phase have been discussed by Moreira [64]. He has described three possible OTA options: far-field testing, radiating near-field testing, and reactive near-field testing.

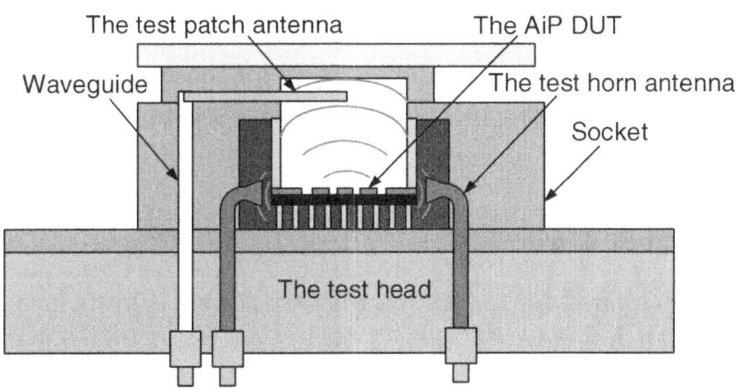

Figure 5.20 Illustration of the integration of the test antennas with the socket.

Figure 5.21 Illustration of the integration of the test antenna with the handler.

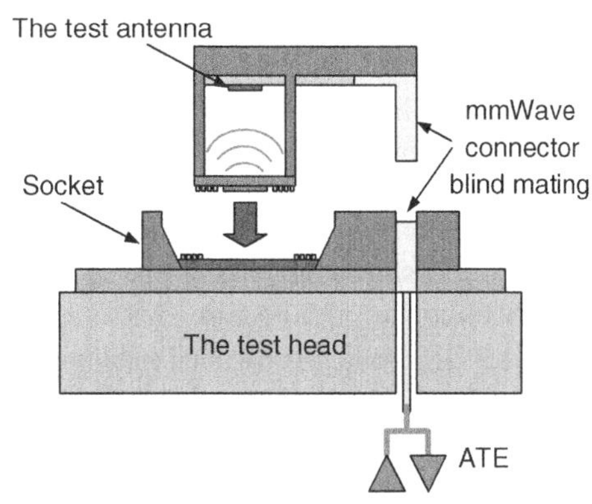

OTA far-field testing usually means that although the test antenna is in the far-field region, neither the AiP DUT nor the test antenna move. They stay static at a certain predefined distance, which means that no traditional antenna radiation measurement is possible. Due to large separation distance required between the test antenna and the AiP DUT, the integration of the test antenna into the test system is difficult. The associated multisite implementation is complex and expensive.

OTA radiating near-field testing moves the test antenna into the radiating near-field region. The integration of the test antenna into the test system is easy. Multisite implementation is simple and cheap. However, it suffers from two important drawbacks. The first is that the test antenna is now so close to the AiP DUT that it will detune the AiP and can even result in a standing-wave effect. The second is that there is no way to test an individual antenna element of an array in the AiP. A golden (standard reference) AiP is needed for comparison. The performance differences between the golden AiP and the AiP DUT will determine whether the AiP DUT passes or fails.

The test antenna must be replaced with a test probe in OTA reactive near-field testing because its very close proximity to the AiP DUT would have detrimental effects its performance characteristics. The test probe needs to be very small for minimal impact on the AiP DUT. It can then test an individual antenna element of an array in the AiP DUT. The integration of the test probe in the test system is easy. Moreover, multisite implementation is possible. However, it has higher non-recurring engineering costs. The coupling between the test probe and an individual antenna element of the array is weak, which requires a more sensitive tester for accurate results.

5.5 Three AiP Examples

Proprietary mmWave AiPs have been demonstrated at the industrial level for consumer wireless applications. One can classify them as wire-bond, flip-chip, and FO-AiPs. This section gives three AiP examples, each representing one of these types.

5.5.1 A Wire-Bond AiP for a Receiver at 60 GHz

The link budget for 60-GHz radios is rather limited due to their low transmit power (10 mW), large propagation losses (68 dB at 1 m), and high data rates (≥2 Gbit/s). Hence, the use of directional antennas is necessary, especially for line-of-sight (LoS) applications. Assuming that the gain values of both the transmit and receive antennas are G_t and G_r, respectively, one can express their combination from the link budget equation for the LoS path in dBi's as [8]:

$$G_t + G_r = 2 + C/N + 20\log_{10}d \quad (5.6)$$

where C/N represents the carrier-to-noise ratio required for demodulation and d is the distance between the transmit and receive antennas. Taking the case of QPSK (quadrature phase shift keying) with the data rate of 2 Gbit/s at $d = 5$ m as an example, one can find from Eq. (5.6) that the combined antenna gain required for the LOS path is 27 dBi.

Figure 5.22 shows the micrograph of a 60-GHz receiver die. The receiver uses a dual-conversion superheterodyne architecture with a 57–64 GHz RF range, 8.1–9.1 GHz IF range, and 16.3–18.3 GHz voltage controlled oscillator (VCO) range. The bandwidth of the in-phase/quadrature (IQ) baseband signal is currently limited by an IF bandwidth of ±1.0 GHz. The critical input/output (I/O) pads are in GSG coplanar configurations. The input pads require a package interconnection with a 50-Ω characteristic impedance from 57 to 64 GHz. While the output pads also require a 50-Ω

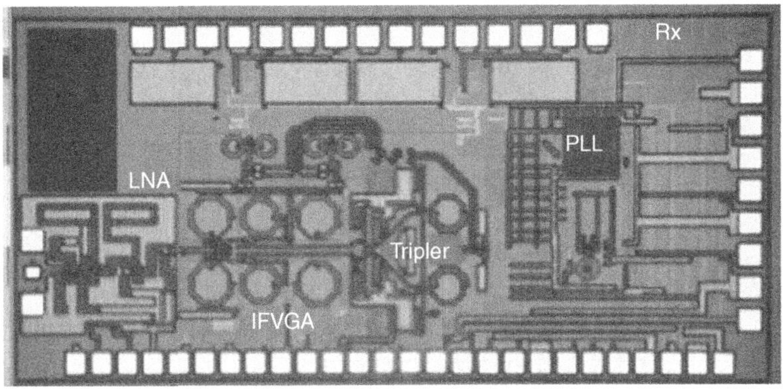

Figure 5.22 Micrograph of a 60-GHz receiver die. LNA, low noise amplifier; PLL, phase-locked loop; RX, receiver; IFVGA, intermediate frequency variable gain amplifier.

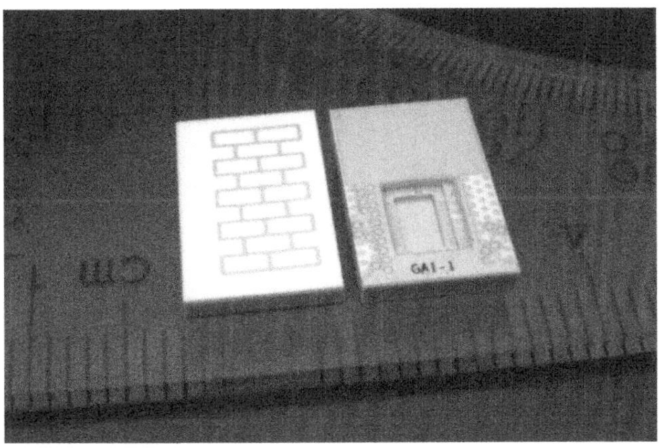

Figure 5.23 Photo of a wire-bond AiP for the 60-GHz receiver.

characteristic impedance, they do so at much lower frequencies, up to 2 GHz, as determined by the maximum bandwidth of the I/Q baseband signals. The receiver die size is $3.4 \times 1.7 \times 0.46\ \text{mm}^3$ [67].

Figure 5.23 shows a photo of the wire-bond AiP for the 60-GHz receiver in Ferro A6 LTCC. Note that the antenna is a microstrip grid array formed with 14 rectangular loops [68]. For the sake of low cost and easy fabrication, the width and thickness of the microstrip lines were kept uniform as 0.15 mm and 0.01 mm, respectively. The length of each loop was 2.5 mm, which is about one guided wavelength. The width of each loop was 1.365 mm, which is about half of the guided wavelength at 60 GHz. The AiP has four co-fired laminated ceramic layers. The first ceramic layer is 0.385 mm thick; the second ceramic layer is 0.285 mm thick with an opening of $3.8 \times 2\ \text{mm}^2$; the third layer is 0.21 mm thick with an opening of $5 \times 3.2\ \text{mm}^2$; and the fourth layer is 0.385 mm thick with an opening of $5 \times 3.8\ \text{mm}^2$. These openings form a three-tier cavity that can house the 60-GHz receiver die. There are also five metallic layers for the AiP. The top layer provides the metallization for the microstrip grid array antenna. The first buried layer provides the metallization for the solid ground plane. The second buried layer provides the metallization for the antenna feed and signal traces. The third buried layer provides the metallization for the signal traces. Finally, the bottom exposed layer provides the metallization for the package ground plane and solder ball pads. The package has 58 I/Os with a Joint Electron Device Engineering Council (JEDEC) standard solder ball pitch

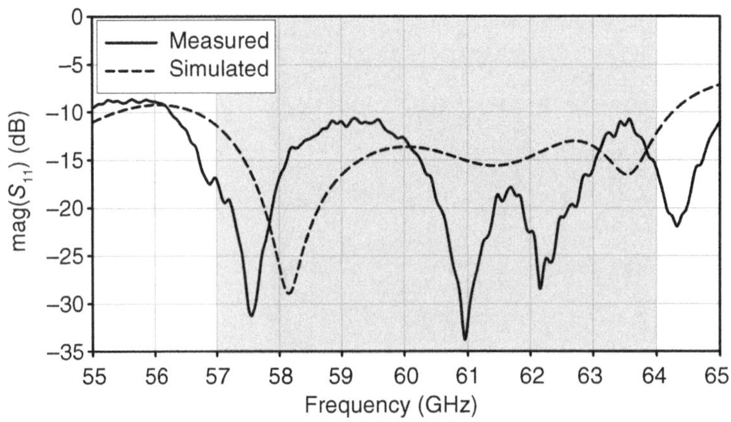

Figure 5.24 Simulated and measured $|S_{11}|$ values of the wire-bond AiP as functions of the source frequency.

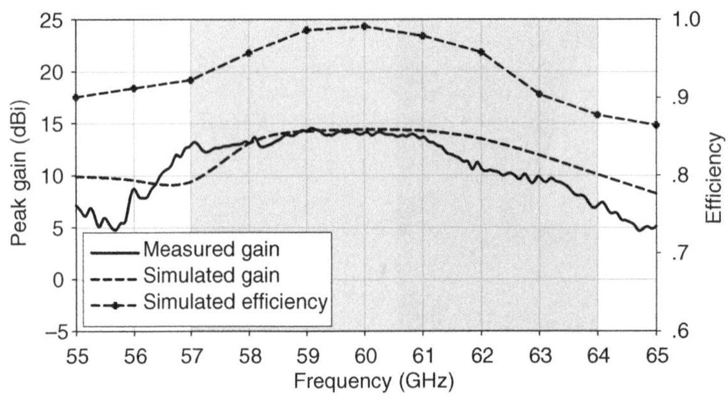

Figure 5.25 Simulated and measured peak realized gain and simulated efficiency of the wire-bond AiP as functions of the source frequency.

of 0.65 mm. Two dummy solder balls are attached to two corners of the package for an enhanced attachment to the system's PCB. The size of the whole package is $13.5 \times 8 \times 1.265\ \text{mm}^3$.

The design specifications for this AiP were an impedance bandwidth of 7 GHz, a radiation efficiency of more than 80%, and a maximum gain of 15 dBi. Figure 5.24 shows that an excellent match to a 50-Ω source was achieved over an impedance bandwidth of 8.7 GHz, from 56.3 to 65.0 GHz. Figure 5.25 shows that the measured and simulated peak gain values are both 14.5 dBi with an estimated efficiency better than 95% at 60 GHz. No de-embedding was made between the post-layout simulated and measured results.

Figure 5.26 compares the simulated with the measured radiation patterns of the AiP at 60 GHz. They reveal that both the *E*- and *H*-plane patterns have their main beam pointed in the boresight direction and have small cross polarization components.

5.5.2 A Flip-Chip AiP for a 5G NR at 28 GHz

The advent of 5G NR mmWave has created unprecedented opportunities for AiP systems. At least two FC-AiP modules have been used in each 5G NR mmWave smartphone and base station.

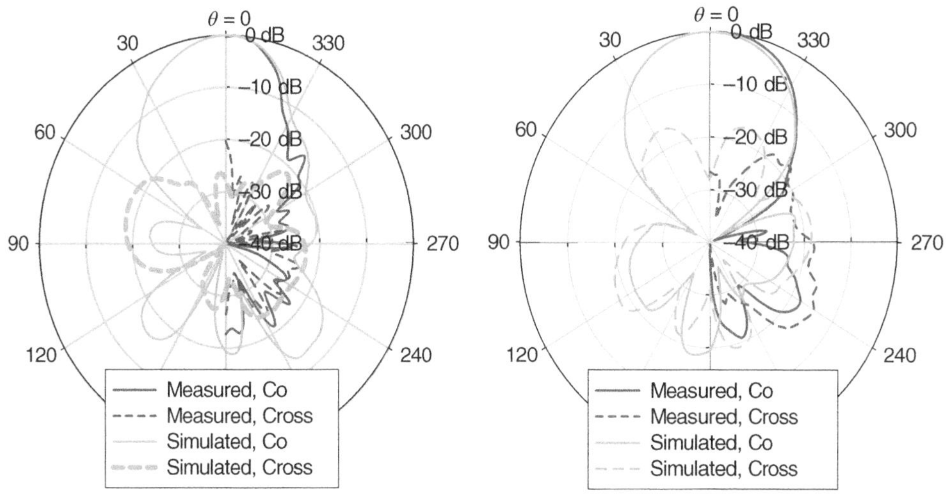

Figure 5.26 Simulated and measured E- and H-plane radiation patterns of the wire-bond AiP at 60 GHz.

Figure 5.27 Beamforming design diagram.

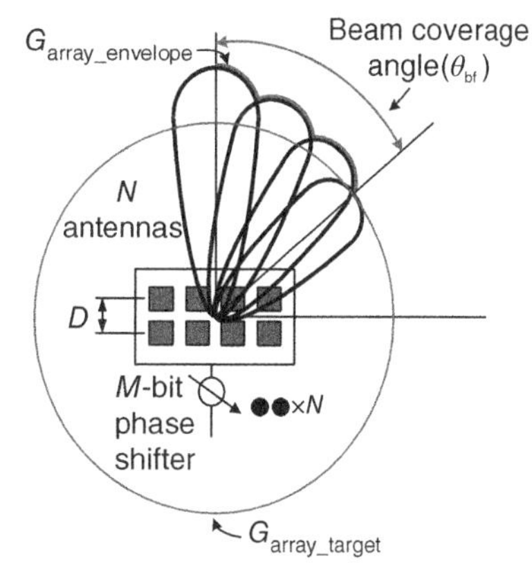

Kim *et al.* reported the design of a phased array transceiver for 5G NR operating at 28 GHz [69]. Following the phased array beamforming design diagram shown in Figure 5.27, they first set the user equipment (UE) maximum EIRP within the target beam coverage angle as

$$20\ \log N + P_{\text{Element}} \geq 23\ \text{dBm} \tag{5.7}$$

where 23 dBm is the maximum output power allowed for the UE and N is the number of antenna elements in the array. Assuming that the output power of a power amplifier is 3 dBm and the gain of an antenna element is 4 dBi, one can find from Eq. (5.7) that N should be more than 7. In order that the phased array envelope gains $G_{\text{array-envelop}}$ are larger than that of the target gain $G_{\text{array-target}}$ within the target beam coverage angle θ_{bf}, the next step is to determine the number of antenna elements N, the number of phase shifter bits M, and the distance between adjacent antenna elements D. The pointing error ($G_{\max} - G_{\min}$) decreases as the number of phase shifter bits increases, but the noise figure (NF) of the receiver (Rx) degrades due to the phase shifter losses. Therefore, there is a specific value of M for an optimal Rx signal-to-noise ratio (SNR). Considering this point and Eq. (5.7), they chose $N = 8$, $M = 3$, and $D =$ half-wavelength.

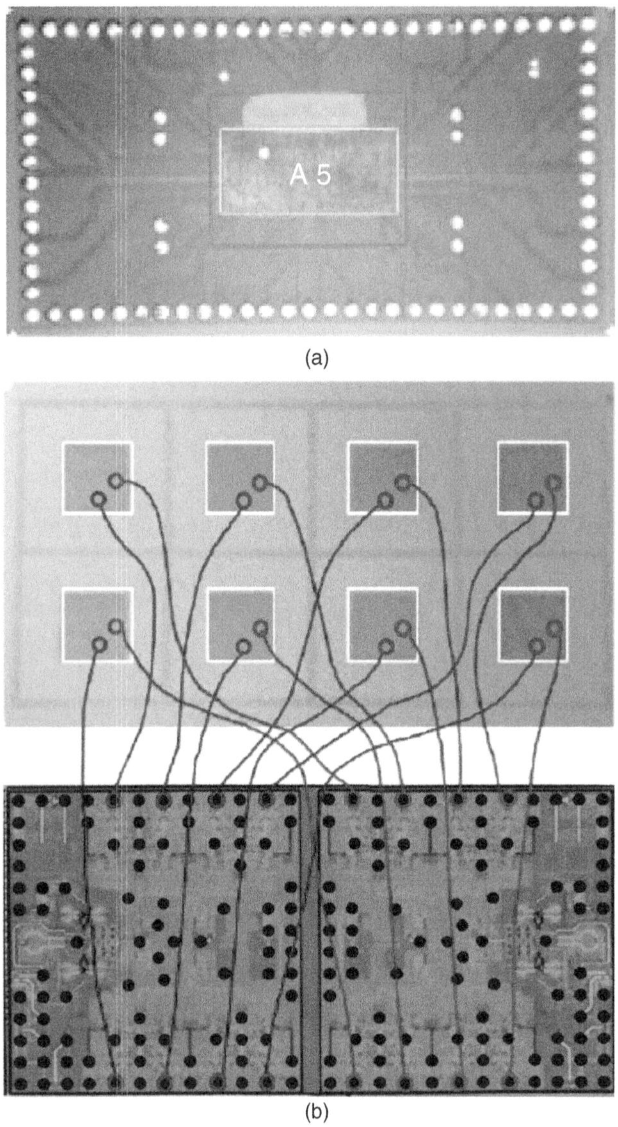

Figure 5.28 The bottom view of the AiP (a) and an illustration of the connections from the I/O pads of the die to the feeds of the patch elements (b).

The micrograph of the die of a single phased array transceiver was shown in Figure 5.2. An FC-AiP was designed to carry a die of two such phased array transceivers. One transceiver fed an antenna array consisting of 2 × 4 patch elements for the horizontal polarization, and the other fed the same antenna array for the vertical polarization. Figure 5.28 shows the bottom view of the AiP and an illustration of the connections from the I/O pads of the die to the feeds of the patch elements. The locations of the patch elements dictate the optimal port locations of the power and low noise amplifiers on the die for minimum loss routing. All the feed lines of the patch elements should have an equal length, which is important to ensure equal loss and phase responses across the operational frequency band. Moreover, it also simplifies or even eliminates the need for calibration.

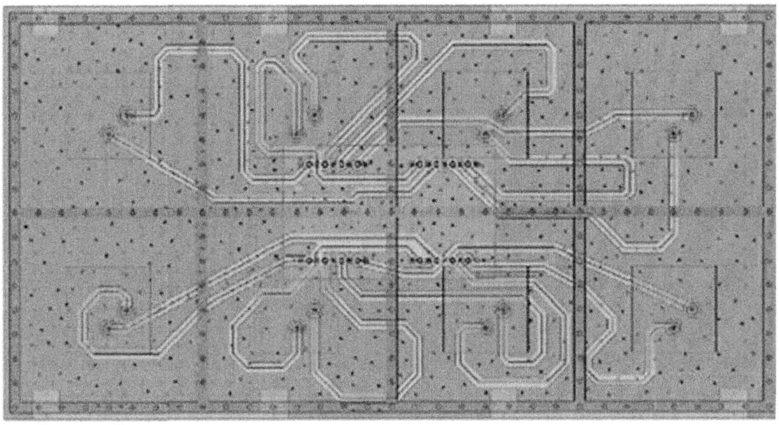

Figure 5.29 The layout of the FC-AiP.

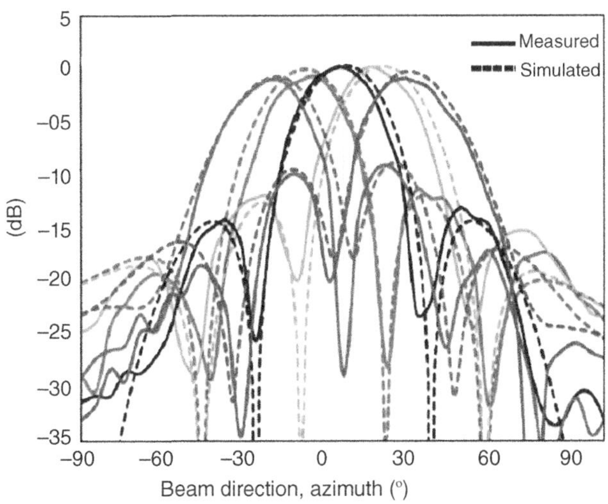

Figure 5.30 The simulated and measured array beam patterns.

Figure 5.29 shows the layout of the FC-AiP. It was fabricated in an HDI process with eight copper layers, one core ($\varepsilon_r = 4.4$, $\tan\delta = 0.006$), six prepreg ($\varepsilon_r = 3.7$, $\tan\delta = 0.007$), and two solder resist ($\varepsilon_r = 3.5$, $\tan\delta = 0.015$) dielectric layers. The total substrate thickness was 0.756 mm. Each square patch element yielded dual polarizations and had an area of $2.35 \times 2.35\ \text{mm}^2$. These patches were separated from the ground plane by 0.548 mm and were surrounded by grounded vias to effectively reduce the coupling between the patch elements. The ports of the dual feeds were equally 0.5 mm away from the center of the patch. The distance between adjacent patch elements was 5.25 mm, a half wavelength in air. The total area of the AiP was $11.6 \times 22\ \text{mm}^2$. The length of the strip line connecting the feed of a patch element and an I/O pad of the die was about 10 mm. Its loss was about 1.0 dB at 28 GHz.

Figure 5.30 shows the simulated and measured array beam patterns. They are in good agreement, validating the design. The maximum array gain including a trace line loss of 1.0 dB was 13.0 dBi. It can scan from −45° to +45° in the azimuth direction. It can scan from −30° to +30° in the elevation direction.

5.5.3 A Fan-Out AiP for a Drone Radar at 122 GHz

Drones are finding more and more applications. The successful operation of a drone strongly depends on its positioning, communicating, and sensing capabilities. GPS and Wi-Fi devices are commonly used in drones for their location and communication functions. Radars for drones to avoid collisions are under development. Figure 5.31 shows a 122-GHz radar die embedded in an FO-AiP [70]. It is seen that the AiP has integrated two differential rhombic arrays with one for transmission and the other for reception. Each rhombic array has 43 elements. The side length of an element is approximately a quarter guided wavelength. The side width of the element is optimized to improve the impedance bandwidth. The element is spaced approximately a half-guided wavelength from the other element for broadside radiation. The maximum number of elements and the overall shape of the array is limited by the maximum dimensions of the eWLB package that can be fabricated. The metallic patterns on the package top surface are present to enhance the array's performance.

Figure 5.32 shows the simulated and measured radiation patterns of the AiP at 122.5 GHz in the E- and H-planes; they have a very good correlation. The radiation pattern is much wider in the

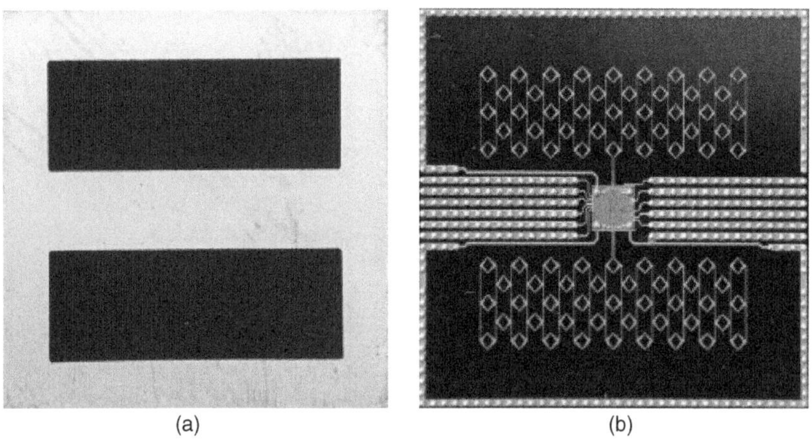

Figure 5.31 Photographs of a 122 GHz radar sensor. (a) Top view. (b) Bottom view.

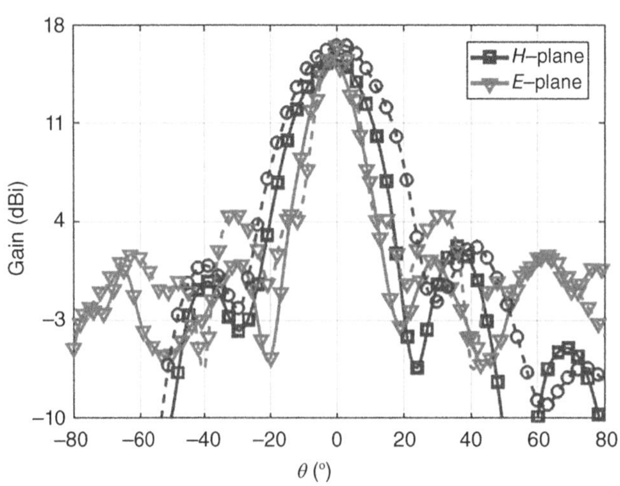

Figure 5.32 Measured (solid lines) and simulated (dashed lines) gain values of the antenna array at 122.5 GHz.

H-plane than it is in the *E*-plane because of the embedded die and the non-symmetrical array configuration in the horizontal and vertical axes. The measured peak side lobe level, a very important parameter for radars to avoid false target detection, was around 14.0 dB and 13.5 dB lower than the main lobe level in the *E*- and *H*-planes, respectively. The measured realized gain of the array was about 15.5 dBi.

5.6 Concluding Remarks

This chapter presented the AiP technologies that enable wireless SoCs and other highly integrated wireless devices. First, the HVM of an AiP module in LTCC, HDI, and FOWLP were described. It was noted that the HDI has been more popular than the other two for AiP for mmWave 5G NR because of its lower costs. Design considerations for the radiating elements, feed network, ground plane, metal fill, shielding structures, and cooling methods employed in AiP systems were highlighted. It was noted that stacked patch antenna, flip-chip feeding, patterned ground plane, CFS, and passive cooling have been widely adopted for them. Next, special emphasis was given to OTA testing of an AiP module in its production phase. It was stated that OTA radiating near-field testing is promising for the functional verification of the AiP DUT by comparing it with a golden AiP sample. Finally, the classification of AiP realizations into their wire-bond, flip-chip, and fan-out types was illustrated with practical examples. In the future, AiP technology will continue to provide direct antenna solutions to highly integrated wireless systems that will operate at even higher mmWave frequencies. AiP modules will also act as parasitic radiators to enhance the performance of THz antennas used in AoC technology. The combination of both AiP and AoC technologies will pave the way for future very large-scale antenna integration.

References

1 A. Abidi *et al.*, "The future of CMOS wireless transceivers," in *ISSCC Dig. Tech. Papers*, Feb 1997.

2 F. Eynde *et al.*, "A fully integrated single-chip SoC for Bluetooth," in *ISSCC Dig. Tech. Papers*, Feb 2001.

3 L. Nathawad *et al.*, "A dual-band CMOS MIMO radio SoC for IEEE 802.11n wireless LAN," in *ISSCC Dig. Tech. Papers*, Feb 2008.

4 H. T. Kim *et al.*, "A 28GHz CMOS direct conversion transceiver with packaged antenna arrays for 5G cellular system," in *RFIC Dig. Tech. Papers*, Jun 2017.

5 O. Kenneth *et al.*, "On-chip antennas in silicon ICs and their application," *IEEE Trans. Electron Devices*, vol. 52, no. 7, pp. 1312–1323, Jul 2005.

6 Y. P. Zhang, M. Sun, and L. H. Guo, "On-chip antennas for 60 GHz radios in silicon technology," *IEEE Trans. Electron Devices*, vol. 52, no. 7, pp. 1664–1668, July 2005.

7 M. Sun and Y. P. Zhang, "100-GHz quasi-Yagi antenna in silicon technology," *IEEE Electron Device Lett.*, vol. 28, no. 5, pp. 455–457, May 2007.

8 Y. P. Zhang and D. Liu, "Antenna-on-chip and antenna-in-package solutions to highly-integrated millimeter-wave devices for wireless communications," *IEEE Trans. Antennas Propag.*, vol. 57, no. 10, pp. 2830–2841, Oct 2009.

9 E. Seok, "A 410GHz CMOS push-push oscillator with an on-chip patch antenna," in *ISSCC Dig. Tech. Papers*, Feb 2008.

10 R. Han *et al.*, "Active terahertz imaging using Schottky diodes in CMOS: Array and 860-GHz pixel," *IEEE J. Solid-State Circuits*, vol. 48, no. 10, pp. 2296–2308, Oct 2013.

11 Y. P. Zhang, "Antenna-in-package technology: Its early development," *IEEE Antennas Propag. Mag.*, vol. 61, no. 3, pp. 111–118, Jun 2019.

12 Y. P. Zhang and J. F. Mao, "An overview of the development of antenna-in-package technology for highly-integrated wireless devices," *Proc. IEEE*, vol. 107, no. 11, pp. 2265–2280, Nov 2019.

13 J. D. Dunworth *et al.*, "A 28GHz Bulk-CMOS dual-polarization phased-array transceiver with 24 channels for 5G user and base station equipment," in *ISSCC Dig. Tech. Papers*, Feb 2018.

14 K. Dandu *et al.*, "High-performance and small form-factor mm-Wave CMOS radars for automotive and industrial sensing in 76-to-81GHz and 57-to-64GHz bands," in *ISSCC Dig. Tech. Papers*, Feb 2021.

15 M. P. Gaynor, *System-in-Package RF Design and Applications,* London, UK: Artech House, 2006.

16 Y. P. Zhang, M. Sun, K. M. Chua, L. L. Wai, and D. Liu, "Antenna-in-package design for wire-bond interconnection to highly-integrated 60-GHz radios," *IEEE Trans. Antennas Propag.*, vol. 57, no. 10, pp. 2842–2852, Oct 2009.

17 D. G. Kam, D. Liu, A. Natarajan, S. K. Reynolds, H. C. Chen, and R. A. Floyd, "LTCC packages with embedded phased-array antennas for 60-GHz communications," *IEEE Microwave Wireless Compon. Lett.*, vol. 21, no. 3, pp. 142–144, Mar 2011.

18 W. B. Hong, A. Goudelev, K. H. Baek, V. Arkhipenkov, and J. Y. Lee, "24-element antenna-in-package for stationary 60-GHz communication scenarios," *IEEE Antennas Wireless Propag. Lett.*, vol. 10, pp. 738–741, 2011.

19 H. J. Kim, J. H. Lee, I. S. Song, and C. S. Park, "Compact LTCC Yagi-Uda type end-fire antenna-in-package for 60 GHz wireless communications," in *IEEE MTT-S Int. Microwave Symp. Dig.*, Orlando, FL, USA, pp. 1–6 Jun 2014.

20 W. B. Hong, K. H. Baek, and A. Goudelev, "Grid assembly-free 60-GHz antenna module embedded in FR-4 transceiver carrier board," *IEEE Trans. Antennas Propag.*, vol. 60, no. 12, pp. 1573–1580, Apr 2013.

21 T. Kamgaing, A. A. Elsherbini, S. N. Oster, and E. Cohen, "Low-profile fully integrated 60 GHz 18 element phased array on multilayer liquid crystal polymer flip chip package," in *Proceedings of IEEE Electronic Components and Technology Conference*, San Diego, CA, USA, 26-29 May 2015.

22 M. Wojnowski *et al.*, "Embedded wafer level ball grid array (eWLB) technology for millimeter-wave applications," in *Proceedings of IEEE Electronic Packaging Technology Conference*, Singapore, 7–9 Dec 2011.

23 B. Yu, Z. Y. Qian, C. C. Lin, J. Lin, Y. P. Zhang, G. L. Yang, and Y. Luo, "A wideband mmWave antenna in fan-out wafer level packaging with tall vertical interconnects for 5G wireless communication," to appear in *IEEE Trans. Antennas Propag.*, vol. 69, no. 10, pp. 6906-6911, October 2021.

24 C. H. Tsai *et al.*, "Fabrication and characterization of millimeter wave 3D InFO dipole antenna array integrated with CMOS front-end circuits," in *Proceedings of IEEE IEDM*, San Francisco, CA, USA, 7–11 Dec 2019.

25 D. Liu and Y. P. Zhang, *Antenna-in-Package Technology and Applications*, IEEE-Wiley, 2020.

26 Z. J. Shao, L. F. Qiu, and Y. P. Zhang, "Design of wideband differentially fed multilayer stacked patch antennas based on bat algorithm," *IEEE Antennas Wireless Propag. Lett.*, vol. 19, no. 7, pp. 1172–1176, Jul 2020.

27 P. S. Hall and C. Wood, "Wide bandwidth microstrip antennas for circuit integration," *Electron. Lett.*, vol. 15, no. 15, pp. 458–460, Jul 1979.

28 F. Croq and D. M. Pozar, "Millimeter wave design of wide-band aperture coupled stacked microstrip antennas," *IEEE Trans. Antennas Propag.*, vol. 39, no. 12, pp. 1770–1776, Dec 1991.

29 R. B. Waterhouse, "Stacked patches using high and low dielectric constant material combinations," *IEEE Trans. Antennas Propag.*, vol. 47, no. 12, pp. 1767–1771, Dec 1999.

30 Z. F. Liu, P. S. Kooi, L. W. Li, and M. S. Leong, "A method for designing broad-band microstrip antennas in multilayered planar structures," *IEEE Trans. Antennas Propag.*, vol. 47, no. 9, pp. 1416–1420, Sep 1999.

31 S. Gao, L. W. Li, M. S. Leung, and T. S. Yeo, "A broad-band dual-polarized microstrip patch antenna with aperture coupling," *IEEE Trans. Antennas Propag.*, vol. 51, no. 4, pp. 898–900, Apr 2003.

32 R. L. Li *et al.*, "Design of compact stacked-patch antennas in LTCC multilayer packaging modules for wireless applications," *IEEE Trans. Adv. Packag.*, vol. 27, no. 4, pp. 581–589, Nov 2004.

33 H. H. Zhou, L. S. Wu, L. F. Qiu, J. F. Mao, and M. Tang, "Diagnosis and tuning of filtering antenna based on extracted coupling matrix," in *Proceedings of IEEE Electrical Design of Advanced Packaging and Systems*, Honolulu, HI, USA, 14–16 Dec 2016.

34 G. Guo, L. S. Wu, Y. P. Zhang, and J. F. Mao, "Stacked patch array in LTCC for 28 GHz antenna-in-package applications,"in *Proceedings of IEEE Electrical Design Advanced Packaging and Systems*, Haining, Zhejiang, China, 14–16 Dec 2017.

35 Y. P. Zhang, M. Sun, D. X. Liu, and Y. L. Lu, "Dual grid array antennas in a thin-profile package for flip-chip interconnection to highly-integrated 60-GHz radios," *IEEE Trans. Antennas Propag.*, vol. 59, no. 4, pp. 1191–1199, Apr 2011.

36 C. S. Kim, J. S. Park, D. Ahn, and J. B. Lim, "A novel 1-D periodic defected ground structure for planar circuits," *IEEE Microwave Wireless Compon. Lett.*, vol. 10, no. 4, pp. 131–133, Apr 2000.

37 E. R. Pillai, "Coax via—a technique to reduce crosstalk and enhance impedance match at vias in high-Frequency multilayer packages verified by FDTD and MoM modeling," *IEEE Trans. Microwave Theory Tech.*, vol. 45, no. 10, pp. 1981-1985, Oct 1997.

38 C. C. Tsai, Y. S. Cheng, T. Y. Huang, Y. P. Hsu, and R. B. Wu, "Design of microstrip-to-microstrip via transition in multilayered LTCC for frequencies up to 67 GHz," *IEEE Trans. Compon. Packag. Manuf. Technol.*, vol. 1, no. 4, pp. 595–601, Apr 2011.

39 D. M. Nair, W. E. McKinzie, B. A. Thrasher, M. A. Smith, E. D. Hughes, and J. M. Parisi, "10 MHz to 100 GHz LTCC CPW-to-stripline vertical transition," in *Proceedings of IEEE MTT-S Int. Microwave Symposium Digest*, Seattle, WA, USA, 2–7 Jun 2013.

40 B. Zhang, D. Titz, F. Ferrero, C. Luxey, and Y. P. Zhang, "Integration of quadruple linearly-polarized microstrip grid array antennas for 60-GHz antenna-in-package application," *IEEE Trans. Compon. Packag. Manuf. Technol.*, vol. 3, no. 8, pp. 1293–1300, Aug 2013.

41 Y. P. Zhang, M. Sun, and W. Lin, "Novel antenna-in-package design in LTCC for single-chip RF transceivers," *IEEE Trans. Antennas Propag.*, vol. 56, no. 7, pp. 2079–2088, Jul 2008.

42 B. Zhang *et al.*, "Integration of a 140 GHz packaged LTCC grid array antenna with an InP detector," *IEEE Trans. Compon. Packag. Manuf. Technol.*, vol. 5, no. 8, pp. 1060–1068, Aug 2015.

43 S. Beer *et al.*, "A self-compensating 130-GHz wire bond interconnect with 13% bandwidth," in *Proceedings of IEEE Antenna Propagation Symposium*, Orlando, FL, USA, 7–13 Jul 2013.

44 Y. Sun, S. Glisic, F. Herzel, K. Schmalz, E. Grass, W. Winkler, and R. Kraemer, "An integrated 60 GHz transceiver front end for OFDM in SiGe: BiCMOS," presented at the Wireless World Research Forum 16, Shanghai, China, 26–28 Apr 2006.

45 D. G. Kam, D. Liu, A. Natarajan, S. K. Reynolds, and R. A. Floyd, "Organic packages with embedded phased-array antennas for 60-GHz wireless chipsets," *IEEE Trans. Compon. Packag. Manuf. Technol.*, vol. 1, no. 11, pp. 1806–1814, Nov 2011.

46 http://www.chipex.cn/english/jsyy_RDLzxbxjsbd3_03.html.

47 C. H. Tsai *et al.*, "Array antenna integrated fan-out wafer level packaging (InFO-WLP) for millimeter wave system applications," in *Proceedings of IEEE IEDM*, Washington, DC, USA, 9–11 Dec 2013.

48 C. Jin, V. N. Sekhar, X. Bao, B. Chen, B. Zheng, and R. Li, "Antenna-in-package design based on wafer-level packaging with through silicon via technology," *IEEE Trans. Compon. Packag. Manuf. Technol.*, vol. 3, no. 9, pp. 1498–1505, Sep 2013.

49 Y. Lamy *et al.*, "A compact 3D silicon interposer package with integrated antenna for 60 GHz wireless applications," in *Proceedings of IEEE 3D System Integration Conference*, San Francisco, CA, USA, 2–4 Oct 2013.

50 W. A. Vitale *et al.*, "Fine pitch 3D-TSV based high frequency components for RF MEMS applications," in *Proceedings of IEEE Electronic Components and Technology Conference*, San Diego, CA, USA, 26–29 May 2015.

51 H. J. Song, "Packages for terahertz electronics," *Proc. IEEE*, vol. 105, no. 5, pp. 1–18, May 2017.

52 D. Sievenpiper, L. Zhang, R. F. J. Broas, N. G. Alexopolous, and E. Yablonovitch, "High-impedance electromagnetic surfaces with a forbidden frequency band," *IEEE Trans. Microwave Theory Tech.*, vol. 47, no. 11, pp. 2059–2074, Nov 1999.

53 C. Kyriazidou, H. Contopanagos, and N. G. Alexopoulos, "Space-frequency projection of planar AMCs on integrated antennas for 60 GHz radios," *IEEE Trans. Antennas Propag.*, vol. 60, no. 4, pp. 1899–1909, 2012.

54 H. F. Contopanagos, C. A. Kyriazidou, A. P. Toda, F. D. Flaviis, and N. G. Alexopoulos, "On the projection of curved AMC reflectors from physically planar surfaces," *IEEE Trans. Antennas Propag.*, vol. 63, no. 2, pp. 646–658, 2015.

55 Y. P. Zhang, "Enrichment of package antenna approach with dual feeds, guard ring, and fences of vias," *IEEE Trans. Adv. Packag.*, vol. 32, no. 3, pp. 612–618, Aug 2009.

56 C. C. Hus *et al.*, "3-D Antenna-in-package design with metallic coating technique," in *Proceedings of IEEE Electronic Components and Technology Conference*, San Diego, CA, USA, 29 May -1 Jun 2018.

57 J. Li *et al.*, "EMI shielding technology in 5G RF system in package module," in *Proceedings of IEEE Electronic Components and Technology Conference*, Virtual Event June 2020.

58 J. H. Park, D. S. Choi, and W. B. Hong, "Millimeter-wave phased-array antenna-in-package (AiP) using stamped metal process for enhanced heat dissipation," *IEEE Antennas Wireless Propag. Lett.*, vol. 18, no. 11, pp. 2355–2359, Nov 2019.

59 J. W. Qian, M. Tang, L. Zhou, Y. P. Zhang, and J. F. Mao, "A 28 GHz heatsink antenna array with linearly tapered fins," Submitted to *IEEE Trans. Antennas Propag.*, 2021.

60 S. Shahramian, M. Holyoak, A. Singh, B. Jalal, and Y. Baeyens, "A fully integrated 384-element scalable W-band phased-prray module with integrated antennas, self-alignment and self-test," *ISSCC Dig. Tech. Papers*, vol. 61, pp. 72–74, Feb 2018.

61 R. van Erp *et al.*, "Co-designing electronics with microfluidics for more sustainable cooling," *Nature*, vol. 585, pp. 211–216, 2020.

62 Z. Y. Zheng, Y. P. Zhang, L. Shi, L. S. Wu, and J. F. Mao, "An overview of probe-based millimeter-wave/terahertz far-field antenna measurement setups," *IEEE Antennas Propag. Mag.*, vol. 63, no. 2, pp. 63–118, Apr 2021

63 J. Mroczkowski and D. Campion, "Production test interface solutions for mmWave and antenna in package (AiP)," *Chip Scale Rev.*, vol. 23, no. 1. pp. 10–13, Jan 2019.

64 N. Shiota, A. Kikuchi, H. Mineo, J. Moreira, and H. Takasu, "Socket design and handler integration challenges in over the air testing for 5G applications," Presented at *TestConX Workshop*, Virtual Event, 11–13 May 2020.

65 J. Moreira, “Testing AiP modules in high-volume production for 5G applications,” *Chip Scale Rev.*, vol. 24, no. 3. pp. 20–24, May 2020.

66 https://www.ni.com/documentation/en/mmwave-ota-validation-test-reference-architecture/latest/manual/manual-overview/.

67 U. Pfeiffer *et al.*, “A chip-scale packaging technology for 60-GHz wireless chipsets,” *IEEE Trans. Microwave Theory Tech.,* vol. 54, no. 8, pp. 3387–3397, Aug 2006.

68 M. Sun, Y. P. Zhang, D. Liu, K. M. Chua, and L. L. Wi, “A ball grid array package with a microstrip grid array antenna for a single-chip 60-GHz receiver,” *IEEE Trans. Antennas Propag.*, vol. 59, no. 6, pp. 2134–2140, Jun 2011.

69 H. T. Kim *et al.*, “A 28-GHz CMOS Direct conversion transceiver with packaged 2 times 4 antenna array for 5G cellular system,” *IEEE J. Solid-State Circuits*, vol. 53, no. 5, pp. 1245–1259, May 2018

70 M. Furqan, F. Ahmed, and A. Stelzer, “A SiGe 122-GHz FMCW radar sensor with 21.5 dBm EIRP based on a 43-element antenna array in an eWLB package,” in *Proceedings of IEEE MTT-S International Conference on Microwaves for Intelligent Mobility (ICMIM)*, Munich, Germany, 15–17 Apr 2018.

6

Terahertz Lens Antennas

Geng-Bo Wu[1], Kwun Wing Cheung[1], Ka Fai Chan[1], and Chi Hou Chan[1,2]

[1] *State Key Laboratory of Terahertz and Millimeter Waves (City University of Hong Kong), Lau Ming Wai Academic Building, 83 Tat Chee Avenue, Kowloon, Hong Kong SAR 999077, China*
[2] *City University of Hong Kong, Department of Electrical Engineering, Yeung Kin Man Academic Building, 83 Tat Chee Avenue, Kowloon, Hong Kong, SAR 999077, China*

6.1 Introduction

While global commercialization of 5G is gradually taking off, many large-scale initiatives on defining the performance requirements and technological challenges for future 6G wireless communications are already in place. One example is the eight-year, €250 million 6G Flagship program established in 2018 by the Academy of Finland. The program is hosted by the University of Oulu and it published the first white paper on 6G in September 2019. It envisages that by 2030, our society will be "data-driven, enabled by near instant unlimited wireless connectivity" via 6G [1, 2]. "Beam me up, Scotty" may become a reality through telepresence and mixed reality enabled by ultrahigh-speed communications, and high-resolution 3D imaging and sensing. The terabit-per-second rates and microsecond latencies in 6G are achieved through the advancements in data collection, processing, transmission, and networking. The supported link distance for 0.3–3 THz is less than 10 m and the system bandwidth is up to 300 GHz [2].

In the seminal review article published in 2007 [3], M. Tonouchi predicted 100 Gbit/s for both indoor and outdoor communications for 2015. German scientists established that earlier in 2013 with a 20 m single-input-single-output (SISO) wireless communications link with a speed of 100 Gbit/s using three channels operating near 237.5 GHz [4]. Doubling the range to 40 m, the date rate was reduced to 75 Gbit/s with a single channel and quadrature amplitude modulation (QAM)-8 signaling. In 2015, French researchers in collaboration with Tektronix pushed the frequency of the wireless link up to 0.4 THz, achieving 32 Gbit/s for a distance of 0.4 m using quadrature phase shift keying (QPSK) [5]. On the other hand, multiplexing eight orbital-angular-momentum (OAM) channels, 32 Gbit/s data stream was transmitted for a short range of 2.5 m at 28 GHz by a team from the University of Southern California, University of Glasgow, and Tel Aviv University in 2014 [6]. This speed would go up if implemented in THz as evidenced by the 1.37 Tbps transmission rate at optical frequencies demonstrated by the same group in 2012 [7]. Innovations on high-gain THz antenna design are required to tackle new challenges in meeting the link budget as well as the different multiplexing schemes.

Polarimetric imaging has an advantage over conventional intensity imaging in terms of contrast enhancement. It has a plethora of applications in remote sensing [8, 9], target detection and recognition [10, 11], and biomedical imaging [12]. As one of the polarimetric imaging modalities,

Antenna and Array Technologies for Future Wireless Ecosystems, First Edition.
Edited by Y. Jay Guo and Richard W. Ziolkowski.

Mueller matrix imaging works well not only in terms of contrast enhancement but also for providing an exclusive insight into the detailed scattering properties of the imaged sample. Enhanced contrast has been demonstrated with Mueller polarimetry where there is little contrast in the raw intensity imagery. Linear and circular polarization measurements of maple leaves are showed, and the corresponding Mueller matrices and their decomposition are interpreted in [13]. Different combinations of the polarizations (LP and CP) of incident and reflected/transmitted fields are measured to construct the Mueller matrix. A smaller beam spot size leads to a finer resolution of the image constructed. The THz band is attractive for bioimaging due to its sensitivity to the water absorption of some biomolecules which arises from their nonionizing property and their vibrational mode resonances. We need antennas that generate LP and CP focusing beams to realize all of the advantages that THz Mueller matrix imaging offers.

An antenna is an indispensable component of wireless communications and imaging systems. Expanding the operating frequency range toward the THz band imposes many challenges in antenna design and implementation. First, due to the high atmospheric absorption at THz frequencies, high-gain antennas are in urgent demand, especially considering the low output power of current THz sources [14]. Second, the intrinsic metal and dielectric material losses at THz frequencies are considerably high, reducing the radiation efficiency of the antenna. Moreover, the wavelength of the electromagnetic (EM) waves at THz frequencies is extremely small. Conventional fabrication techniques such as computer numerical control (CNC) and printed circuit board (PCB) technologies cannot effectively achieve the fabrication accuracy required for the THz range. Although microfabrication processes, typically used for integrated circuit (IC) fabrications, including lithography, wet etching, dry etching, deposition, plating, and sputtering, can provide micro- or even nano-scale manufacturing precision, they are extremely complicated and costly [15]. All these constraints impose many limitations on the performance and functionality of THz antennas.

A lens antenna is a space-fed aperture antenna that utilizes a low-gain primary feed source to illuminate the lens to enhance the antenna gain. Because it does not require a feeding network as does an array antenna, a THz lens antenna is an effective means to achieve high-gain. For example, a silicon lens fed by a leaky-wave enhanced waveguide operating at 545 GHz can achieved a directivity of 28.7 dBi [16]. Nevertheless, fabrication of this silicon lens by CNC machining is expensive, especially if the lens does not possess rotational symmetry.

3D printing, which is alternatively known as additive manufacturing (AM), creates 3D objects layer by layer using relevant materials such as polymers, metals, ceramics, and even cells. This disruptive technology is set to transform industries in construction, tissue engineering, aerospace, and many others [17]. It is no doubt shaping our future. Antenna design has also profited from the evolution of 3D printing. The *Proceedings of the IEEE*, the premium journal of IEEE, published a special issue on AM of radio-frequency components in April 2017 in which several groups summarized their respective long-term research on 3D printed antennas in their review papers [18–26]. A special cluster "3D Printed Antennas and Electromagnetics Structures," was published by the *IEEE Antennas and Wireless Propagation Letters* in November 2018. On both occasions, the papers addressed 3D printed antennas up to the W-band (75–110 GHz) [24–26] with a few exceptions in the H-band (220–325 GHz) [24, 27, 28] due to the limited resolution of the printers and the adopted printing technologies. Some of these antennas, especially the few exceptions of H-band conical horns, were simply fabricated with alternative cost-effective processes rather than exploiting the additional degrees of design freedom offered by 3D printing.

In the rest of this chapter, we will present our recent works on 3D printed lens antennas operating at 300 GHz using inexpensive 3D printers, namely, Form2 and Form3 printers from Formlabs. The lens antennas include LP and CP far-field and near-field focusing antennas, Bessel beam launders, and higher order Bessel beam launchers carrying OAM.

6.2 Printing with 3D Printers from Formlabs

Prior to the lens design, we need to understand the printing process of a 3D printer as well as the electrical properties of the printing materials since there is no one-size-fit-all approach in 3D printing. The millimeter-wave and THz lenses in [29] were printed with a Stratasys Objet30 Scholar system that prints 3D objects from their bottom up with water soluble support materials. In contrast, the Formlabs Form2 printer prints objects upside down without a filling material. Therefore, it is important to investigate its printing mechanisms. Its printing of High Temp Resin is available for 25, 50, and 100 μm layer thicknesses. The resin is chosen since the printed object has a high heat deflection temperature of 238 °C and high resistance to scratching and deformation. For the THz lens, we will use the 25 μm layer thickness for the best performance.

Formlabs recommends using a raft and supports to avoid the adhesion problem between the printed object and the build platform as shown in Figure 6.1a. The software accompanying their printers, PreForm, allows adding the raft and supporting struts automatically. As the unsupported overhang length is limited to 1.0 mm [30], we needed to put in the many supporting struts depicted in Figure 6.1a. We can print the object vertically as shown in Figure 6.1b provided that the overhang structure circled is no more than 1.0 mm. Restrictions on the minimum radius and maximum length of the struts also need to be observed [30]. Alternatively, we can print the object in a slanted fashion as shown in Figure 6.1c. In principle, the structure can be printed with one supporting strut to the far left. However, other struts are needed to support the weight of the structure. In contrast, without the struct to the far right, the circled unsupported part could not be printed. It should be pointed out that the structure is dipped into a pan of resin to be cured at intended positions to build up the object. There are occasions in which resin is trapped and forms residues, as indicated by the dots in Figure 6.1c, due to the finite precision of the laser spot having a diameter of 140 μm. The spot size is reduced in the Form3 printer to 85 μm for better printing resolution. Furthermore, if the lenses are printed in a slanted fashion, uniform fineness of the printing across the lens surfaces could not be guaranteed. To address the adhesion issue, we proposed to modify the base of the structure as shown in Figure 6.2a such that the printed object can be detached from the build platform easily. Similarly, we can modify the object shape for improving the printing support for right-angled edges with overhang length larger than 1.0 mm as shown in Figure 6.2b.

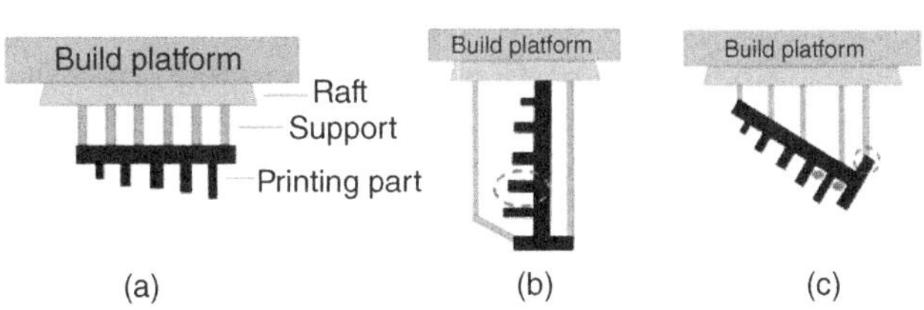

Figure 6.1 Printing mechanism of the Form2 printer. (a) Horizontal printing. (b) Vertical printing. (c) Slanted printing.

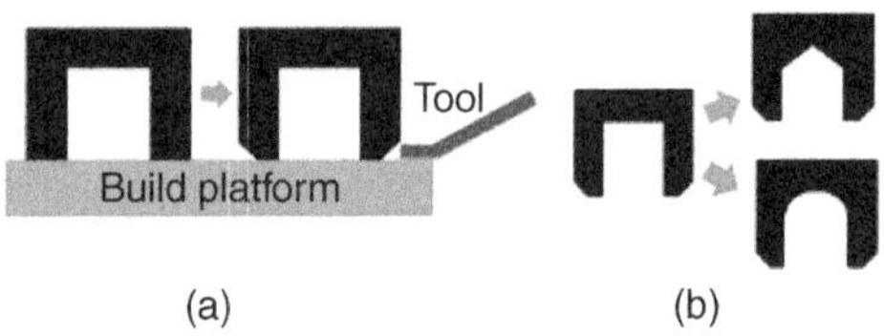

Figure 6.2 Modifications to the sample geometry. (a) Ease of detachment from the build platform. (b) Observing the overhang rule.

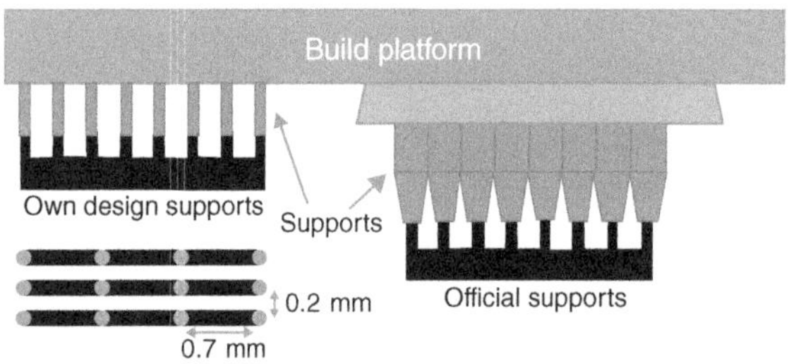

Figure 6.3 Printing approach for fine structures.

When printing the CP lenses, e.g. the CP Fresnel lens in [31], a dielectric grating polarizer is integrated into the lens. For 300 GHz, the air spacing between gratings is 0.25 mm and thus, there is no overhang issue. We can print the lens horizontally as shown in Figure 6.3. The strut provided by PreForm has a minimum diameter of 0.3 mm. On the right side of Figure 6.3, we can observe that the struts attached to the raft are thicker and then gradually reduce to their minimum size. The gratings have a width of 0.2–0.25 mm, which makes the detachment from the struts difficult. We can also choose to add the support manually as shown on the left side of Figure 6.3. The diameter of the struts is 0.2–0.25 mm, depending on the width of the gratings. The struts are 2.0 mm long. If 1.0 mm is used, the resin may be trapped in between the two grating walls and residues fill up the space between them. As these struts are very thin, the printed structure can be easily detached from the build platform. These thin struts can be removed from the printed main structure with a brush. It should be pointed out that the black grating walls are uniform into the page whereas the grey supports are periodic along each black grating wall with a spacing of 0.7 mm, satisfying the 1.0 mm maximum overhang rule. The understanding of the printing mechanisms of the chosen printer allows us to formulate the best printing strategies to realize the proposed THz lenses and to deal with their frequency limits when the 3D printer is not built for our intended purpose. To determine the dielectric constant and loss tangent of a printed sample, readers are referred to the measurement and retrieval approaches described in [32, 33]. The dielectric constant and loss tangent of High Temp Resin are retrieved as 2.66 and 0.03 at 300 GHz, respectively.

Figure 6.4 shows the fabricated CP Fresnel lenses with a magnified view of the lens surface under a microscope. It can be seen in Figure 6.4a that the fabricated lens suffers from limited printing accuracy and uniformity using the printing method recommended by Formlabs. This is caused by the trapped resin and residues that are formed because of the relatively large spot size of the laser beam. As the lens is printed in a slanted fashion, uniform fineness of the printing across the lens surfaces could not be guaranteed. Using the in-house developed approach, the printed lens in Figure 6.4b achieved higher printing accuracy and uniformity.

6.3 Measurement Platforms for Lens Antennas

To characterize the designed lens antennas, we built two THz antenna measurement platforms as shown in Figure 6.5. Figure 6.5a shows the near-field system. The vector network analyzer (Agilent

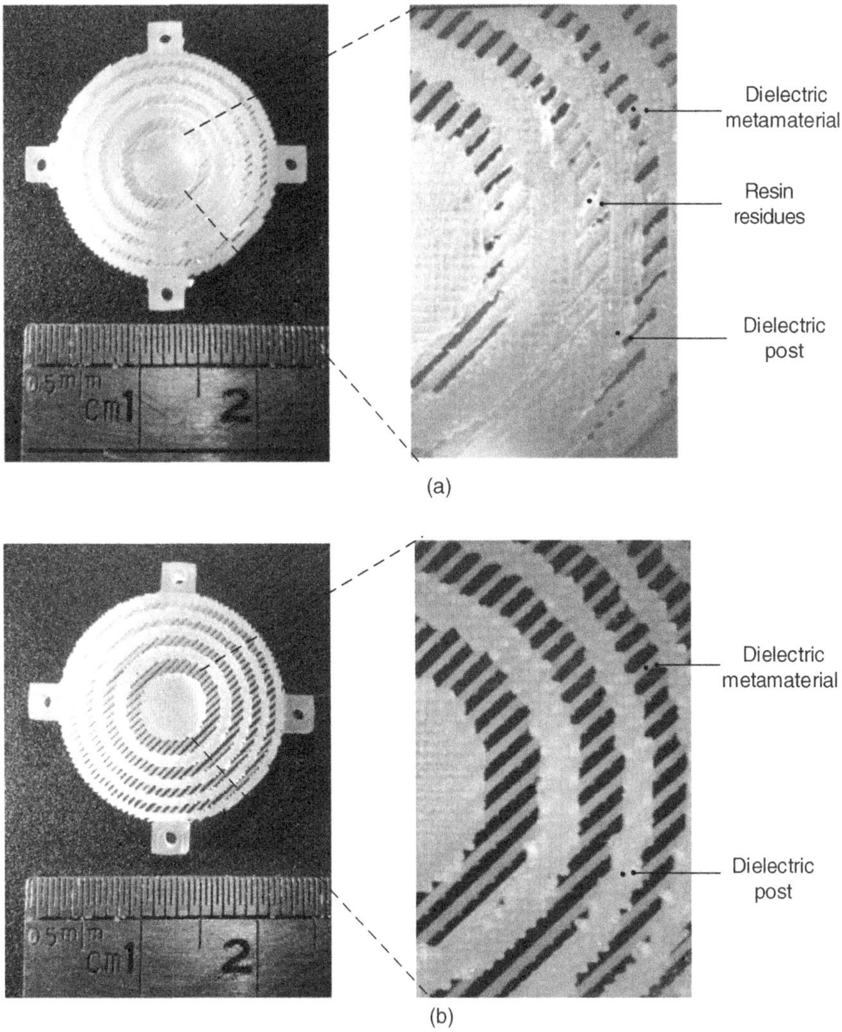

Figure 6.4 Photograph of the fabricated modified Fresnel lens using the (a) Formlabs-recommended and (b) in-house-developed approaches. Source: From [31] / with permission from IEEE.

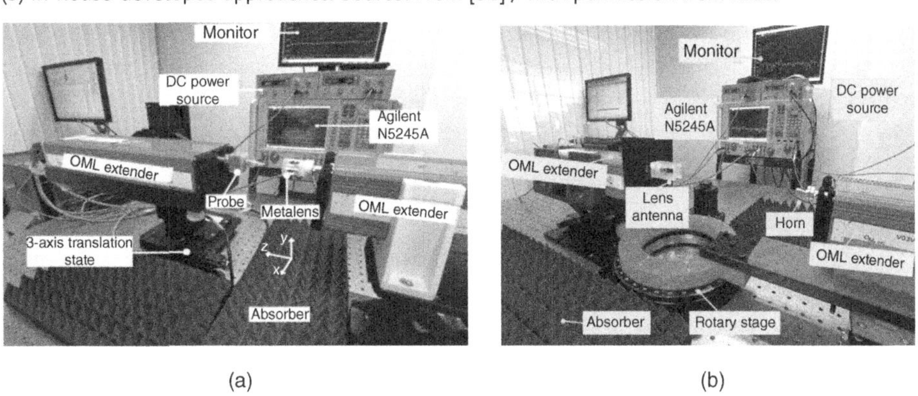

Figure 6.5 THz lens antenna measurement platforms. (a) Near-field. (b) Far-field.

N5245A) with a pair of frequency extenders (OML V03VNA2-T/R) operating from 220 to 325 GHz is adopted. A pair of dc power sources is employed as the power supply for the two OML extenders. The left extender on a three-axis translational stage is used to measure the field from the receiving probe, while the right one is connected to the designed 3D printed lens antenna fed by a horn source. A 3D printed fixture is also fabricated to assemble the lens antenna. The lens antenna acts as the transmitting antenna and the THz probe on the left of Figure 6.5a acts as the receiver. A piece of Eccosorb LS-30 absorber was placed on the front metallic surface of each of the two extenders to create a stable measurement environment. The extenders and translational stage are placed on an optical table with vibration control. Additional absorbers are used to prevent wave reflection from the optical table. The platform allows a 2D scan from which the far-field radiation patterns of the antenna can be computed. It can also scan the near-field radiation in a 3D space to locate the focal spot of a focusing lens. The far-field measurement platform in Figure 6.5b is essentially the same as that in Figure 6.5a except that the receiving horn mounted on the right OML extender is on a rotary stage. The center of the antenna aperture must be on the rotational axis of the rotary stage. Alignment of the transmitting antenna and receiving probe can be performed by locating the maximum receiving power with minimum phase.

6.4 Pixel Design of the Discrete Dielectric Lenses

We presented 3D printed millimeter-wave and THz lenses at 60 and 250 GHz based on the Discrete dielectric lens (DDL) approach in [29]. The resolution of the Stratasys Objet30 Scholar 3D printer used, however, does not permit us to design lenses working at 300 GHz. The dielectric lens in [29] was discretized into a uniform grid of square dielectric posts with different heights for aperture phase modulation. In addition, a square void was created at each end of the dielectric post to serve as an antireflection surface. In other words, the air void was designed as a quarter-wave transformer for impedance matching between air and the dielectric post. An antireflection structure was added only on the top air-dielectric interface in the two pixel designs described in this chapter due to the fabrication limit in the THz band, as well as the top-down printing approach of the Form2 printer. Nevertheless, the ripples appearing in the transmission coefficient without any voids that were shown in [29] are removed here. The pixel element, surrounded by PBCs, was simulated in the full-wave EM simulator ANSYS HFSS (high-frequency structure simulator) to evaluate its transmission response.

The pixel design for the THz DDL is shown in Figure 6.6a. The size of the pixel element is $P = 0.5$ mm, corresponding to $0.5\lambda_0$, where λ_0 is the free-space wavelength at 300 GHz. The employed 3D printing dielectric material is High Temp Resin with a measured relative dielectric constant (ε_r) of 2.66 and loss tangent ($\tan\delta$) of 0.03 at 300 GHz. A rectangular air void ($0.38 \times 0.38 \times 0.225$ mm^3) on the top of the dielectric post performed as an impedance transformer to reduce the multi-reflection at the air-dielectric interface. Here $h_1 = \frac{\lambda_0}{4\sqrt{\varepsilon_r}}$ and $W_1 \approx P\sqrt{\frac{\varepsilon_r - \sqrt{\varepsilon_r}}{\varepsilon_r - 1}}$.

The transmission coefficient as a function of the dielectric post h is shown in Figure 6.6b. A full transmission phase range spanning over 360° was achieved by tuning h from 0.2 to 1.8 mm. The desired compensation phase of each pixel element for the DDL was obtained by sampling the aperture phase distribution for a desirable radiation characteristic. Subsequently, the geometric parameters of each pixel element were determined by using the phase-dimension curve in Figure 6.6b. Since the employed dielectric post element of the DDL is isotropic, the polarization

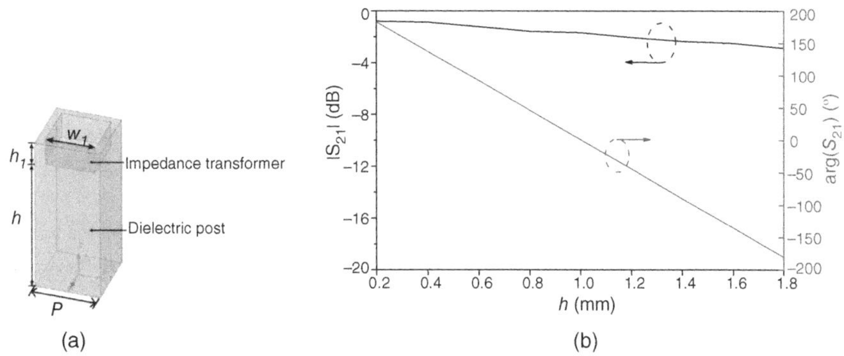

Figure 6.6 The square dielectric pixel element. (a) 3D model. (b) Simulated transmission coefficient as a function of the height of the dielectric post h.

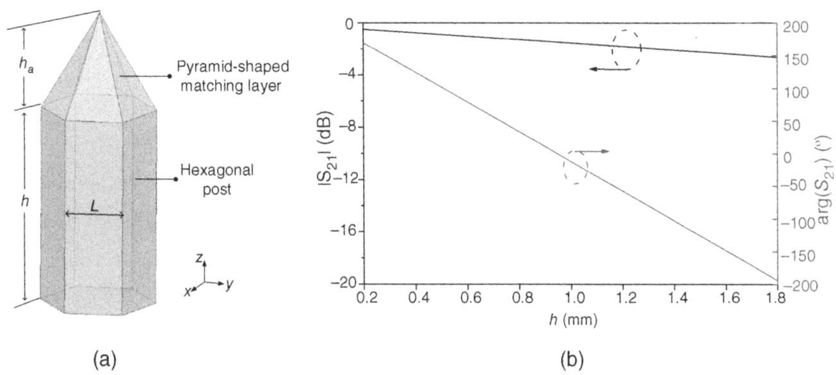

Figure 6.7 The hexagonal dielectric pixel element. (a) 3D model. (b) Simulated transmission coefficient of as a function of the height of the dielectric post h.

of the DDL antenna depends on the polarization of the feed source; and, hence, the beams can be synthesized using both polarizations of the feeder. Due to the dielectric loss of the employed 3D printing material, the simulated $|S_{21}|$ values gradually decrease as h increases.

An alternative for the pixel design is shown in Figure 6.7a when the pixel elements of the DDL are arranged in a skewed grid. The element consists of two components, i.e. the top hexagonal pyramid shaped antireflection structure and the bottom phase compensator. The top hexagonal pyramid with width $L = 0.3$ mm (corresponding to $0.3\lambda_0$) and height $h_a = 1.0$ mm works as a gradient impedance transformer [34] to reduce the multi-reflection at the air-dielectric interface. The dimensions of the antireflection structure are fixed for all elements, while the height of each element h is different from pixel to pixel to compensate for the required transmission phase.

The simulated transmission phase as a function of the height h is shown in Figure 6.7b. It is observed that a full transmission phase spanning over 360° can be achieved as h varies from 0.2 to 1.8 mm with transmission amplitudes all better than −2.6 dB. Since the adopted element structure is isotropic, the polarization of the generated beam depends on the polarization of the incident fields, i.e. the polarization of the feed source. There are no significant differences between the two pixel designs, but the hexagonal pixel has a slightly smaller insertion loss.

6.5 Integration of a Dielectric Polarizer with a Lens for CP Radiation

To generate CP radiation from an LP lens, we can conveniently integrate a dielectric polarizer with the lens via 3D printing. The configuration of the dielectric polarizer adopted here is shown in Figure 6.8. It is composed of periodic sheets of dielectric and air.

The dielectric CP polarizer can be traced back to the artificial dielectric quarter-wave plates used in a waveguide in 1957 [35] and recently it was developed to design a CP antenna at millimeter-wave frequencies [36, 37]. Metamaterial terminology is used for this artificial dielectric structure shown in Figure 6.8 since it is a periodic structure along the y-axis and features an anisotropic characteristic. For the x-polarization (x-pol) incident field, the sheets can be represented by two capacitors connected in parallel. While for the y-polarization (y-pol) incident field, the layers can be viewed as two series capacitors. Therefore, the equivalent permittivities along the x- and y-axes (ε_x and ε_y) can be calculated with [38]:

$$\varepsilon_x = \varepsilon_0 q_0 + \varepsilon_d q_d \tag{6.1a}$$

$$\varepsilon_y = \left(\frac{q_0}{\varepsilon_0} + \frac{q_d}{\varepsilon_d}\right)^{-1} \tag{6.1b}$$

where ε_0 and ε_d are the permittivities of the air and the 3D printed dielectric sheets, respectively. The terms q_0 and q_d are the fractional volume of the air and dielectric sheets, respectively, and whose relationship satisfies:

$$q_o + q_d = 1 \tag{6.2}$$

The equivalent permittivities ε_x and ε_y only depend on the dielectric fractional volume q_d. Taking into account the multiple reflections and transmissions from both ends of the anisotropic material along the z-direction, the transmission coefficient of the dielectric anisotropic metamaterial can be calculated as [31]:

$$T_x(q_d, h) = \frac{4\sqrt{1 + (\varepsilon_r - 1)q_d}\, e^{-jk_{xd}h}}{(1 + \sqrt{1 + (\varepsilon_r - 1)q_d})^2 - (1 - \sqrt{1 + (\varepsilon_r - 1)q_d})^2 e^{-j2k_{xd}h}} \tag{6.3a}$$

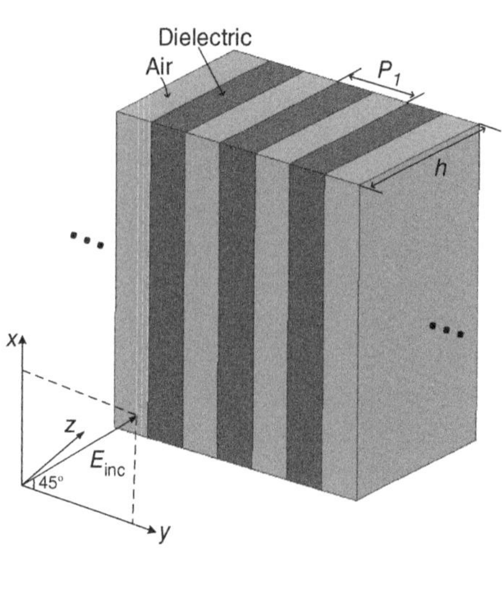

Figure 6.8 Configuration of the dielectric polarizer regarded as an anisotropic metamaterial.
Source: From [31] / with permission from IEEE.

$$T_y(q_d, h) = \frac{4\sqrt{\frac{\varepsilon_r}{(1-q_d)\varepsilon_r+q_d}}e^{-jk_{yd}h}}{\left(1+\sqrt{\frac{\varepsilon_r}{(1-q_d)\varepsilon_r+q_d}}\right)^2 - \left(1-\sqrt{\frac{\varepsilon_r}{(1-q_d)\varepsilon_r+q_d}}\right)^2 e^{-j2k_{yd}h}} \tag{6.3b}$$

where h is the thickness of the dielectric metamaterial. The terms k_{xd} and k_{yd} are the equivalent wavenumbers in the dielectric metamaterial along the x- and y-axes, respectively. They are given as:

$$k_{xd} = k_0\sqrt{1+(\varepsilon_r-1)q_d} \tag{6.4a}$$

$$k_{yd} = k_0\sqrt{\frac{\varepsilon_r}{(1-q_d)\varepsilon_r+q_d}} \tag{6.4b}$$

From (6.3) to (6.4), it is seen that the transmission coefficients are different for the x- and y-polarized incident waves.

When the polarization of the incident LP waves is along the diagonal direction of the metamaterial as shown in Figure 6.8, those fields can be decomposed into x- and y-components with equal amplitude and phase. After passing through the metamaterial polarizer, the x-pol and y-pol transmitted waves will have a 90° phase difference with equal amplitude; and, hence, a CP transmitted wave can be generated. The dielectric fractional volume q_d and thickness of the metamaterial polarizer h should be optimized to satisfy the phase condition of CP waves:

$$|\arg[T_x(q_d, h)] - \arg[T_y(q_d, h)]| = \frac{\pi}{2} \tag{6.5}$$

Combining (6.3)–(6.5), we obtain the relationship between q_d and h for the dielectric CP polarizer depicted in Figure 6.9. To minimize the thickness of the metamaterial polarizer, the dielectric fraction q_d is chosen to be 0.55 and the corresponding thickness h is 1.73 mm. The periodicity along the y-axis, P_1, is 0.5 mm to satisfy the fabrication precision. The transmission phase difference between the x- and y-pol incident waves at different frequencies can be calculated using (6.3) and (6.4). The result is plotted in Figure 6.10. It can be seen that the phase difference is 90° at the center frequency of 300 GHz. The phase difference variation is smaller than 10° from 260 to 320 GHz, indicating a good linear-to-circular (LP-to-CP) transformation performance. Consequently, the dielectric anisotropic metamaterial can be added to both the far-field and focusing LP lenses to realize CP radiation.

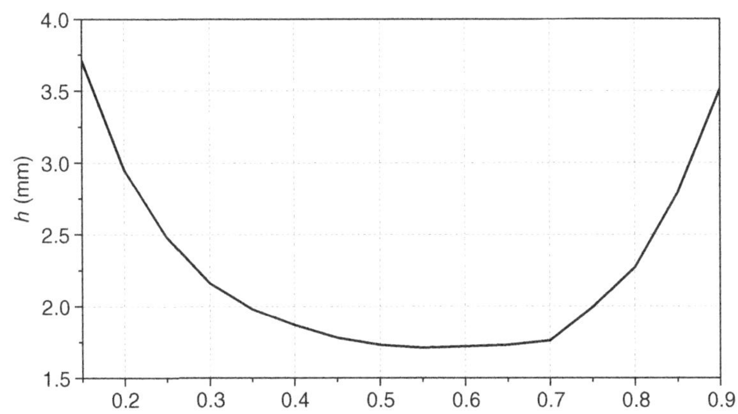

Figure 6.9 Calculated relationship between the dielectric fractional volume q_d and the thickness of the dielectric anisotropic metamaterial h. Source: From [31] / with permission from IEEE.

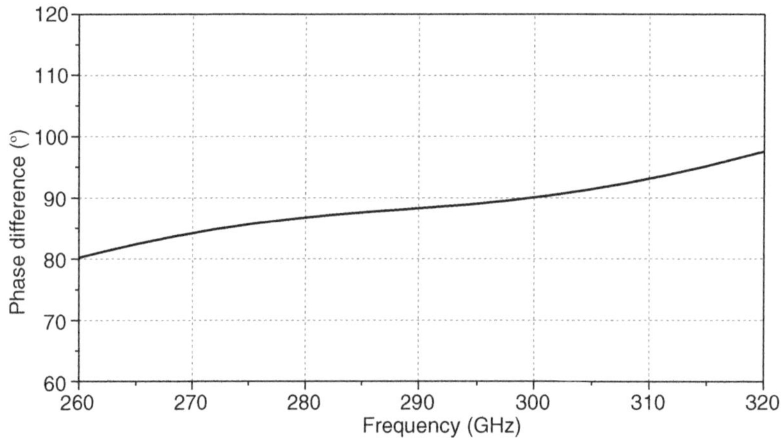

Figure 6.10 Calculated transmission phase difference of the dielectric anisotropic metamaterial between the *x*- and *y*-pol incident waves at different frequencies. Source: From [31] / with permission from IEEE.

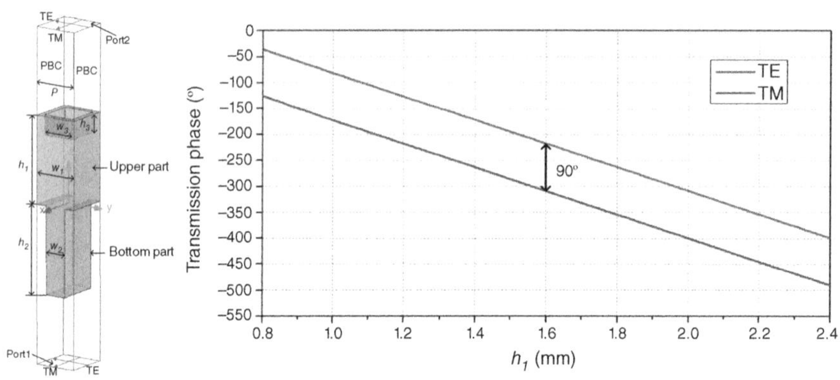

Figure 6.11 Simulated transmission phase for TE and TM incident waves versus h_1. Source: From [39] / with permission from IEEE.

By combining the pixel element in Figure 6.6 for phase modulation and the anisotropic pixel element in Figure 6.8, we have the pixel element design for CP. Figure 6.11 shows the composite CP pixel element and the transmission phase curve for TM (E_x^{inc}) and TE (E_y^{inc}) incident waves. The height of the polarizer h_2 is fixed and the height of the phase compensation part h_1 varies from 0.8 to 2.4 mm. The phase difference between the two orthogonal polarizations at 300 GHz remains at 90° for different values of h_1.

6.6 Design, Fabrication, and Testing of THz Lens Antennas

The goal of using 3D printing for antenna fabrication is not only for performance enhancements, but also for affecting the paradigm shift of antenna designs through exploitation of the superior geometrical flexibility it provides. In each of the following subsections, we will present various 3D printed THz lens antennas with increasing design complexity and design details starting from simple far-field and focusing lenses to more complicated Bessel beam launchers with OAM. All of these designs make use of the pixel element in Figure 6.6 for LP and that in Figure 6.11 for CP radiation. Unless stated otherwise, all of the lenses are fed by an H-band (220–325 GHz) pyramidal horn (Millitech standard gain horn [SGH]-03) with 23 dBi gain. The horn is assumed to be a point source for simplicity.

6.6.1 Far-Field and Near-Field Focusing LP Lenses

To exploit the higher printing resolution of the Form2 printer over that of the Stratasys Objet30 Scholar, we repeated the design of a simple far-field DDL operating at 300 GHz. Referring to Figure 6.12a, since the feed radiates a spherical wave, the pixel elements are designed such that their transmission phases compensate for the different phases corresponding to the different path lengths from the feed to the pixel elements. The phase for broadside radiation will be uniform across the aperture of the lens. There is a progressive phase difference of θ between two adjacent pixels in the x-direction for a beam tilted at an angle θ away from the broadside direction. The focal length (F) to diameter (D) ratio is chosen as 2 to make a tradeoff between the illumination efficiency and the spillover efficiency [40]. The phase center of the horn is not at the open aperture of the horn. The required phase compensation at the mnth pixel is obtained as:

$$\varphi_{mn} = \frac{2\pi f}{c}\left\{\sqrt{F^2 + (x_{mn} - x_o)^2 + (y_{mn} - y_o)^2} - F - x_{mn}\sin\theta\right\} + \Phi_o \tag{6.6}$$

where $(x_o, y_o, 0)$ and (x_{mn}, y_{mn}, F) are the locations of the phase center of the feed and the mnth pixel element, respectively. The angle θ is the direction of the radiation beam, f is the operating frequency, and c is the speed of the light. The term Φ_o is a reference phase and it should be chosen such that the variation of the pixel height should be kept to a minimum.

Figure 6.12b shows the compensated phase distribution among the pixel elements for $\theta = 0$. Figure 6.12c shows the fabricated THz lens with 40×40 pixels and $\Phi_o = 0$ for convenience. We also designed the 3D printed fixture shown in Figure 6.13 for testing of the lens antenna.

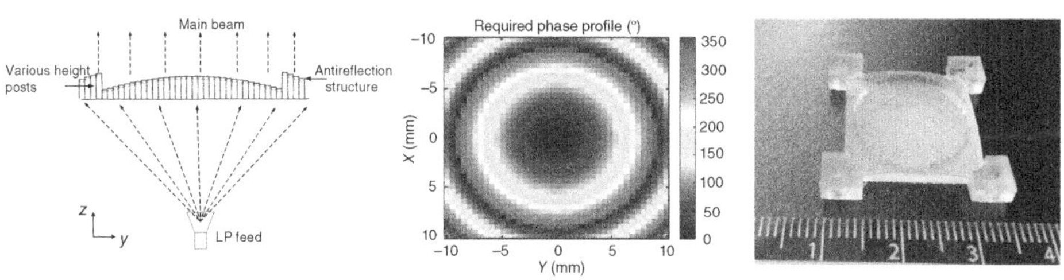

Figure 6.12 Far-field lens design. (a) Phase modulation mechanism. (b) Required phase profile of the lens at broadside radiation. (c) Fabricated lens using Form2 printer.

Figure 6.13 3D printed fixture to hold the lens and the feed horn for testing the antenna. (a) Simulation model. (b) Assembled antenna.

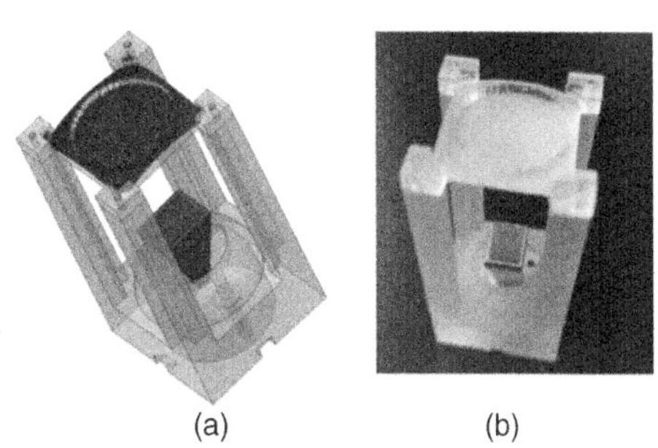

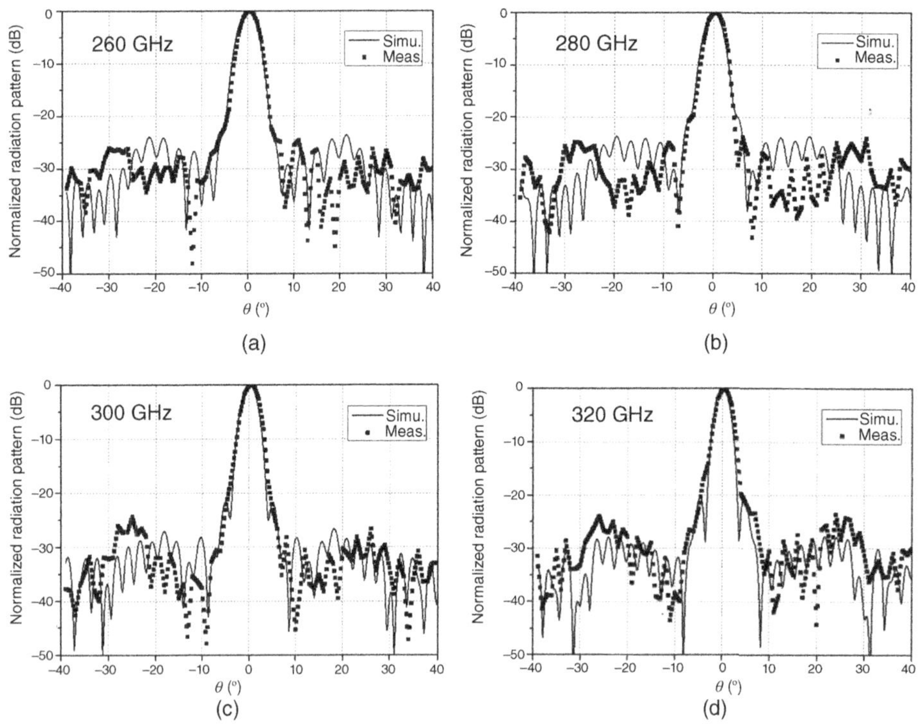

Figure 6.14 H-plane radiation patterns for the 3D printed far-field lens antenna at 260, 280, 300, and 320 GHz.

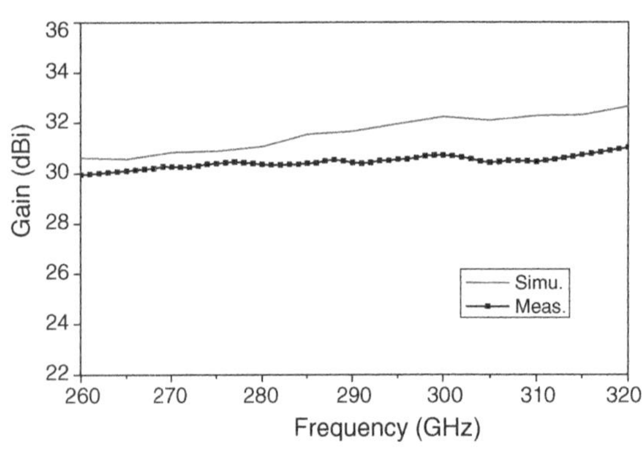

Figure 6.15 Simulated and measured broadside gains versus the source frequency.

Using the far-field measurement platform shown in Figure 6.5b, the far-field radiation pattern of the LP lens antenna can be measured. Figure 6.14 shows the normalized H-plane radiation patterns at 260 to 320 GHz with a 20 GHz increment. It can be seen that excellent agreement is obtained between the simulated and measured results. Figure 6.15 shows the comparison of the simulated and measured broadside gains versus frequency. It can be seen that the discrepancy is less than 2 dB. It should be pointed out that the dielectric constant 2.66 and loss tangent 0.03 at 300 GHz are used in the simulation.

As stated in the Introduction, we need focusing lenses with different polarized radiation for polarimetric imaging. To illustrate the design flexibility offered by 3D printing, we also present a Teflon near-field focusing lens with linear polarization which can be fabricated by CNC machining.

However, it would be difficult to further incorporate the other phase modulations required for the lens antennas to be presented later. Teflon is chosen because of its low loss tangent. In designing the Teflon lens and performing the simulations, a dielectric constant of 2.1 and loss tangent of 0.001 are utilized. The convex Teflon near-field focusing lens consists of two spherical surfaces with the top surface having a radius of R_1 (25.15 mm) and that of the lower surface being R_2 (18.86 mm). The distance between the phase center of the horn and lower surface of the lens is 51.7 mm and that between the focal point and the top surface of the lens is 30.8 mm. Employing the paraxial approximation [41, 42], the transmission coefficient of the thin spherical lens is given by

$$\varphi(x,y) = \exp\left[-jk(n-1)\frac{x^2+y^2}{2}\left(\frac{1}{R_1}+\frac{1}{R_2}\right)\right] \tag{6.7}$$

where n is the refractive index of the lens material, which is 1.45 for our design. Figure 6.16a shows the fabricated Teflon lens. The lens is tested using the near-field measurement platform shown in Figure 6.5a. As the receiving probe and extender module are mounted on a three-axis translational stage, we can search for the focal plane of the lens easily. It is found that the highest measured field along the longitudinal axis (z-axis) is located at 37 mm from where the lens has the maximum diameter of 28 mm. Figures 6.16b,c show the simulated and measured field distributions at 300 GHz. Very favorable agreement is achieved. The simulated and measured field distributions along x-axis and y-axis are compared in Figure 6.17a,b, respectively. The measured spot size (−3 dB) is 1.86 mm which is $1.86\lambda_0$ at 300 GHz. The measured near-field sidelobe levels (SLLs) are −16.5 dB and −14.2 dB in the H- and E-plane, respectively.

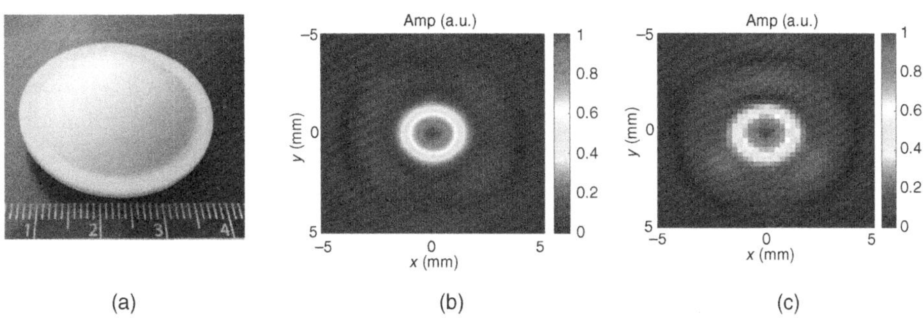

Figure 6.16 Teflon near-field focusing lens. (a) Fabricated sample. (b) Simulated field distribution at the focal plane. (c) Measured field distribution at the focal plane.

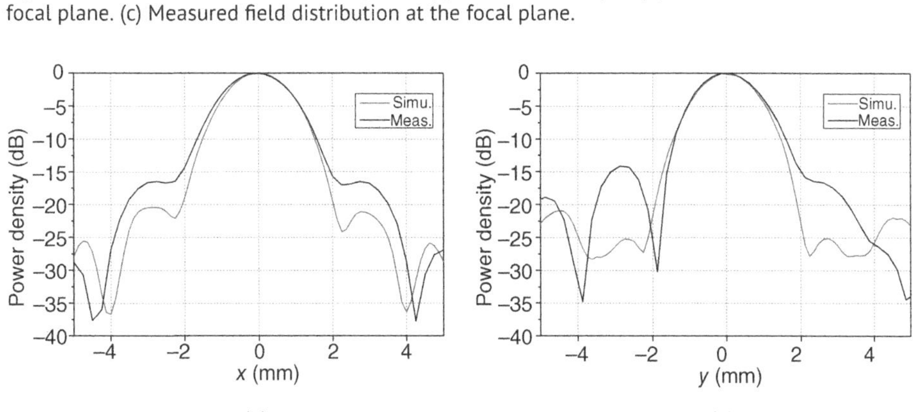

Figure 6.17 Comparison of the simulated and measured results on the measurement plane. (a) Along the x-axis (H-plane). (b) Along the y-axis (E-plane).

Through the results shown earlier, one realizes that CNC machining is applicable for focusing lens operating at 300 GHz. However, when the applications call for CP lenses, CNC machining may be intractable or too expensive when the designed lenses do not possess rotational symmetry. In Section 6.6.2, we present both far-field and near-field focusing CP lenses making use of the pixel element and design phase curve in Figure 6.6 or Figure 6.11.

6.6.2 Far-Field and Near-Field Focusing CP Lenses

The phase modulation mechanism for the CP lens is illustrated in Figure 6.18a. The lens works as a linear-to-circular polarization converter and provides the required transmission phase compensation with spatial phase delays from the phase center of the feed horn to different pixels. The desired transmission phase shift of the *mn*th pixel can be calculated with (6.6) and the various pixel heights are determined from the phase curve in Figure 6.11, where the dimensions of the CP polarizer have been determined in Section 6.4. Figure 6.18b shows the fabricated far-field CP lens antenna with a magnified view of the lens surface. It should be pointed out again that the orientation of the horn is 45° from that of the grating.

The far-field CP antenna can be tested using the near-field measurement platform in Figure 6.5a. The probe is set 7 mm away from the lens aperture. The measured near-field magnitude and phase of the E_x and E_y components over the scanning plane at 300 GHz are plotted in Figure 6.19. Approximately a 90° phase difference between the measured phase of the E_x and E_y components in the center region indicates a good CP radiation of the 3D printed lens.

The far-field radiation patterns can be obtained from the measured near-field data using a fast Fourier transformation. The measured and simulated normalized radiation patterns in the vertical and horizontal planes at 280, 300, and 320 GHz are plotted in Figure 6.20. A good agreement between the measured and simulated patterns is observed. Some discrepancies can be attributed to fabrication tolerance and measurement errors. The measured and simulated gains and axial ratios (ARs) are plotted in Figure 6.21. The measured 1.0 dB gain bandwidth is about 13% and the 3 dB AR bandwidth is better than 18.8% for this right-handed CP lens. The measured gain of the CP lens at 300 GHz is 30.8 dBic, corresponding to an aperture efficiency of 25.63%. The relatively low aperture efficiency is primarily caused by the high dielectric loss of the 3D printing material at THz frequencies. The full-wave simulated radiation efficiency of the designed lens antenna is 50.9%, corresponding to a dielectric loss of 2.94 dB. Simulated results show that the radiation efficiency of the lens can be slightly improved to 55.6% with a gain improvement of 0.3 dB using the reference phase optimization process developed in [44].

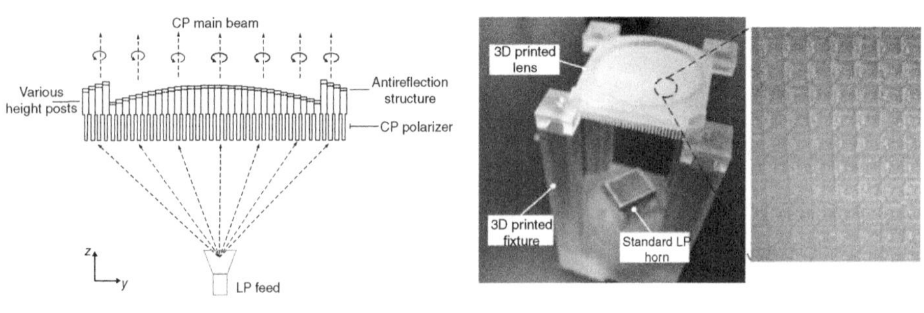

Figure 6.18 Design of the far-field CP lens antenna. (a) Schematic diagram and cross-sectional view. (b) 3D fabricated prototype. Source: From [43] / with permission from IEEE.

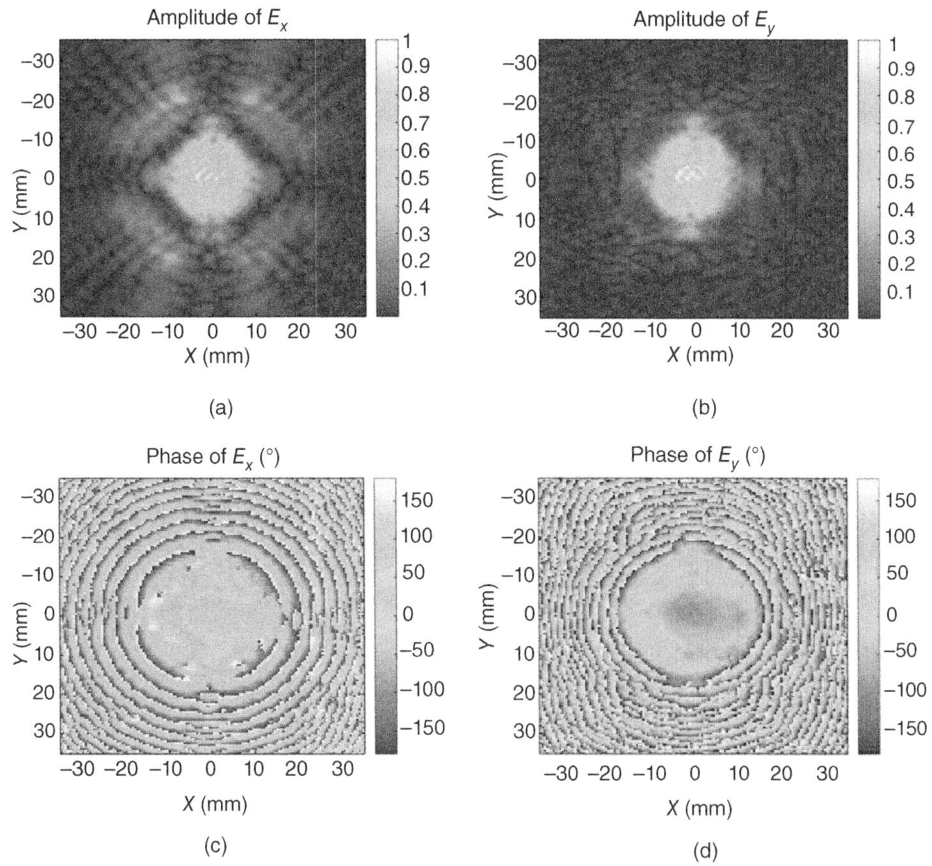

Figure 6.19 Measured near-field distributions over the scanning plane at 300 GHz. (a) Magnitude of E_x. (b) Magnitude of E_y. (c) Phase of E_x. (d) Phase of E_y. Source: From [43] / with permission from IEEE.

Table 6.1 shows a performance comparison of our 3D printed CP THz lens antennas with counterparts fabricated by other means. It is observed that the 3D printed lenses have much higher peak gain values and much wider AR bandwidths. The impedance bandwidth of the lens antenna is basically determined by the impedance bandwidth of the horn since the presence of the lens does not alter the reflection coefficient of the feed significantly.

To generate a CP focusing beam, we need to alter the phase modulation on the lens aperture of the far-field lens. Referring to Figure 6.22a, the height of the dielectric pixel needs to compensate for the path length difference not only from the phase center of the horn to the pixel element, but also from the pixel element to the focal point. The desired transmission phase on the aperture of the lens can be calculated by:

$$\varphi_{mn} = \frac{2\pi f}{c}\{R_{1mn} + R_{2mn}\} + \Phi_o \tag{6.8}$$

where R_{1mn} is the spatial distance between the phase center of the feed antenna and the mnth element, and R_{2mn} is the spatial distance between the focal point and the mnth element. The term Φ_o is again a constant reference phase. The height profile of the lens is again designed using the phase curve in Figure 6.11. The dimensions of the grating remain the same as given in Section 6.4. Figure 6.22b shows the desired transmission phase distributions on the lens aperture for the TE

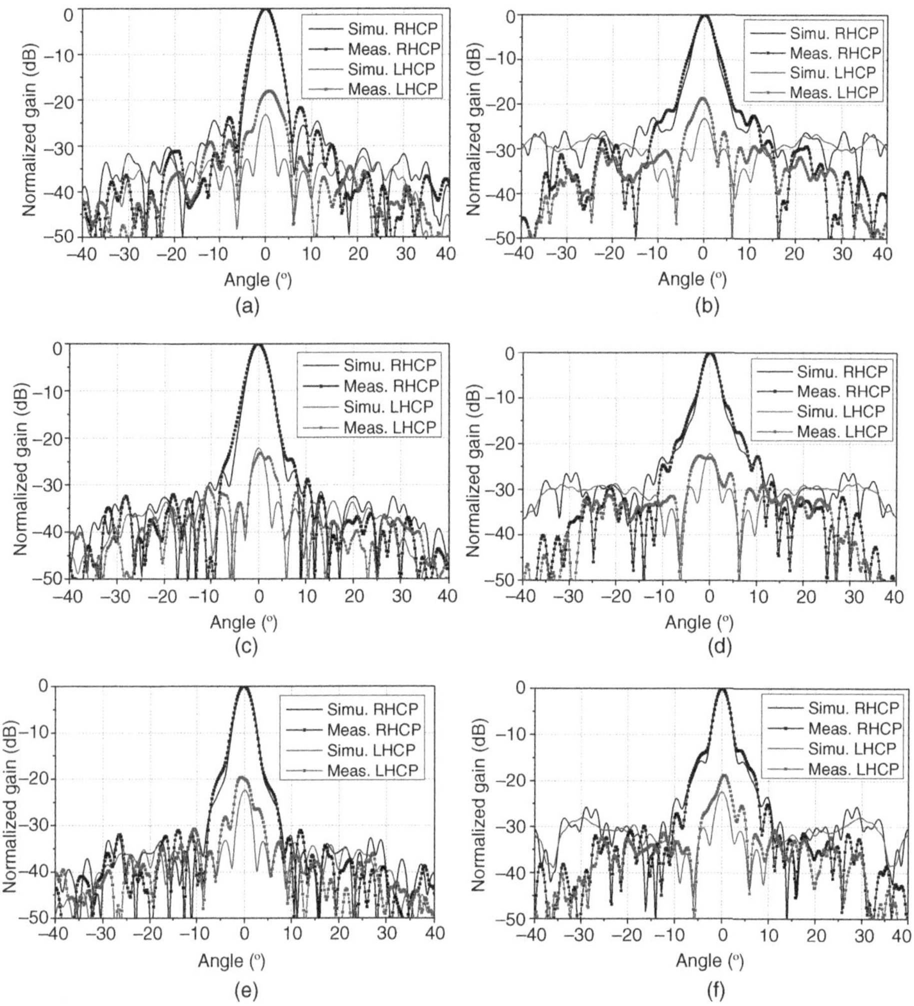

Figure 6.20 Measured and simulated radiation patterns in the *xz*-plane at (a) 280 GHz, (b) 300 GHz, and (c) 320 GHz, and in the *yz*-plane at (d) 280 GHz, (e) 300 GHz, and (f) 320 GHz. Source: From [43] / with permission from IEEE.

and TM mode components. The lens is composed of 40 × 40 elements. The distance between the phase center of the feed horn and the lens is 40 mm and the designed focal point is 60 mm away from the lens. As stated in Section 6.4, TE refers to the electric field component perpendicular to the orientation of the grating of the polarizer (E_y) and TM refers to that parallel to the orientation of the grating (E_x). The fabricated near-field focusing CP lens is shown in Figure 6.22c. The near-field platform in Figure 6.5a was utilized for the measurements.

The measured magnitude and phase of the E_x and E_y components on the focal plane ($z = 60$ mm) for 300 GHz are plotted in Figures 6.23 and 6.24, respectively. It can be seen that the transmitted EM power was concentrated into a small spot on the focal plane. The measured −3 dB spot diameter is around 3.0 mm. Meanwhile, the phase difference between the E_x and E_y components are plotted in Figure 6.25. One finds it is approximately equal to 90° in the central focusing region, indicating that a good CP focusing was achieved.

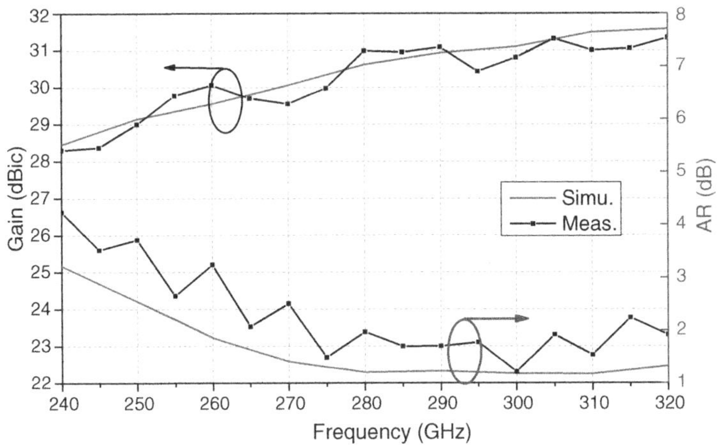

Figure 6.21 Simulated and measured gains and axial ratios at different source frequencies. Source: From [43] / with permission from IEEE.

Table 6.1 Comparison with Other CP THz Antennas.

References	Frequency (GHz)	Antenna type	Fabrication technique	Peak gain (dBic)	Imp. BW ($\|S_{11}\| < -10$ dB) (%)	AR BW (AR < 3 dB) (%)
[45]	500	Antipodal curvedly tapered slot antenna	Micromachining	12.25	>4	>4
[46]	500	Double-fan-shaped slot antenna	Micromachining	12.5	>4	>4
[47]	300	Modulated metasurface antenna	Micromachining	20	~11	NA
[48]	300	Horn	Wire electrical discharge machining	18.4	>20	2.2
Our work [31]	300	Modified Fresnel lens	3D printing	27.4	>18.8	>18.8
Our work [43]	300	Discrete dielectric lens	3D printing	30.8	>18.8	>18.8

"NA" indicates that the parameter is not available.

The right-hand circular polarized (RHCP) and left-hand circular polarized (LHCP) near-fields can be calculated with [49]:

$$E_{\text{RHCP}} = (E_x - jE_y)/\sqrt{2} \tag{6.9a}$$

$$E_{\text{LHCP}} = (E_x + jE_y)/\sqrt{2} \tag{6.9b}$$

The calculated RHCP and LHCP fields on the focal plane along the lines $x = 0$ and $y = 0$ are plotted in Figure 6.26. The cross polarization (LHCP) are all below −17 dB, which validates the CP

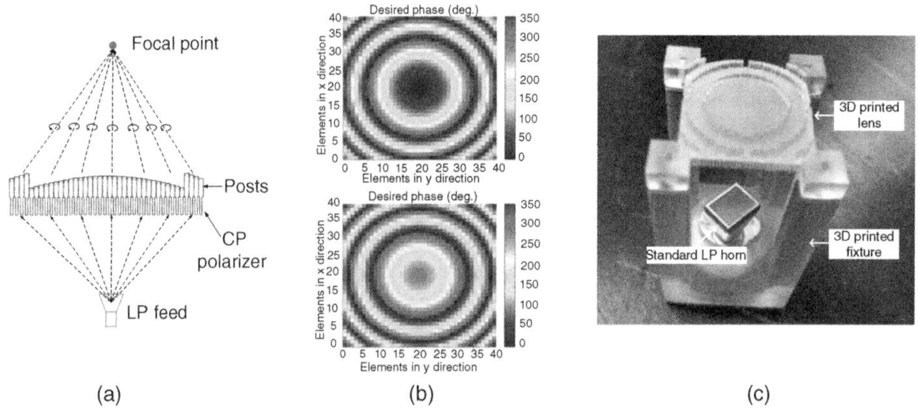

Figure 6.22 Design of the far-field CP lens antenna. (a) Schematic diagram and cross-sectional view. (b) Desired transmission phase distribution on the aperture: TE (top) and TM (bottom) mode components. (c) 3D fabricated prototype. Source: From [39] / with permission from IEEE.

focusing capability of the 3D printed lens. The spot size of this DDL (3 mm) is bigger than that of the Teflon lens (1.86 mm) as the diameter of the Teflon lens (28 mm) is larger than that of the DDL (20 mm). The DDL is CP, and the Teflon lens is LP.

A nondivergent beam over a range of distance is preferred for some applications. We can achieve such beams through the design of Bessel beam launchers by incorporating a more complicated phase modulation over the lens aperture.

6.6.3 Bessel Beam LP Lenses

A Bessel beam is a kind of nondiffractive beam with a Bessel-type intensity distribution along the radial direction, i.e. the direction along the plane transverse to the beam axis. It is a natural solution to the Helmholtz equation in free space. A practical, realizable truncated Bessel beam, i.e. a pseudo-Bessel beam, still features a nondiffracting property, maintaining a high intensity focus over a considerable distance [50]. First proposed for optical applications, the Bessel beam has been extended to lower microwave bands with applications of wireless power transmission and secure short-range point-to-point communications [51, 52].

Various types of Bessel beam launchers have been reported in published literature in the past few decades. Refractive conical axicon lenses are the most commonly used method in the optical region [53]. Other Bessel beam generation techniques include diffracting rings placed in the focal plane of a convergent lens [54] and computer-generated holograms [55]. For microwave frequencies, a Bessel beam can be generated by radial line slot arrays [56], sub-sampled distributed arrays [51], and phase shifting surface antennas [52]. Most research on Bessel beam launchers are focused on the optical and microwave bands. Few works have been conducted in the THz band.

Figure 6.27a shows a schematic illustration of the DDL introduced for Bessel beam generation. The DDL is designed to transform the spherical wave from the feed horn to a transmitted Bessel beam whose wave vectors lie on a cone. Using geometric optics, the required transmission phase profile of the lens can be calculated as:

$$\varphi_{mn} = \frac{2\pi f}{c}\left\{\sqrt{(x_{mn}-x_f)^2+(y_{mn}-y_f)^2+z_f^2}+\sqrt{x_{mn}^2+y_{mn}^2}\sin\delta\right\}+\Phi_o \tag{6.10}$$

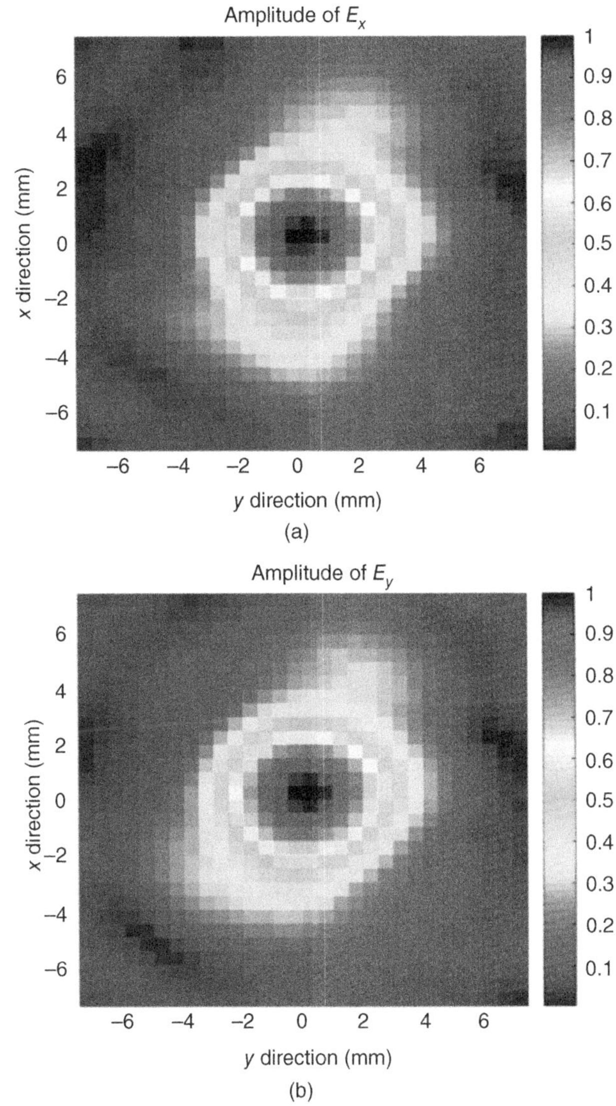

Figure 6.23 Measured amplitude of the (a) E_x and (b) E_y components. Source: From [39] / with permission from IEEE.

where $(x_{mn}, y_{mn}, 0)$ and (x_f, y_f, z_f) are the coordinates of the mnth lens pixel element and the phase center of the feed horn, respectively. The term δ is the axicon angle, which is set at 20° in this design. The term Φ_0 is a reference phase constant added to all the pixels on the aperture. The lens is 15 mm in diameter and the distance between the phase center of the horn and the lens is set at 24 mm to provide the proper illumination of the aperture. The phase curve in Figure 6.6 was utilized to determine the height profile of the lens based on the desired phase distribution shown in Figure 6.27b. The fabricated Bessel beam launcher is shown in Figure 6.27c.

The Bessel beam launcher was measured using the near-field measurement platform shown in Figure 6.6a. Figure 6.28 depicts the full-wave simulation and measured intensity profiles at different transversal planes at 300 GHz. Good agreement between the simulated and measured results is

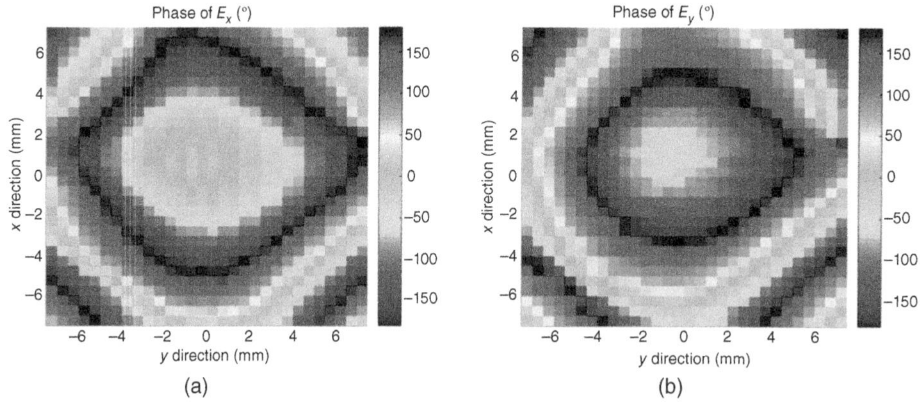

Figure 6.24 Measured phase of the (a) E_x and (b) E_y components. Source: From [39] / with permission from IEEE.

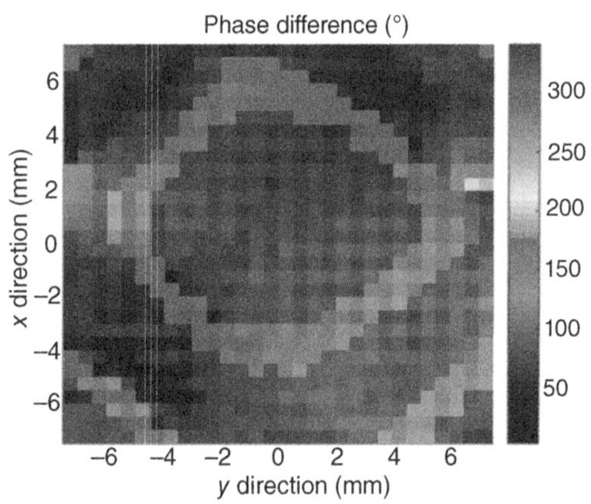

Figure 6.25 Phase difference between the measured phases of the E_x and E_y components. Source: From [39] / with permission from IEEE.

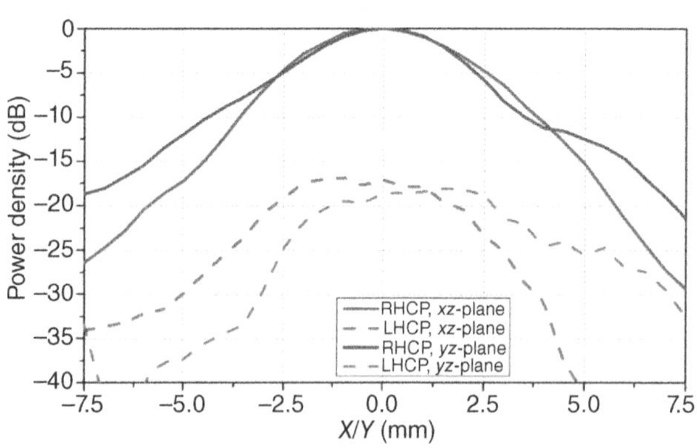

Figure 6.26 Measured RHCP and LHCP near-field on the focal plane. Source: From [39] / with permission from IEEE.

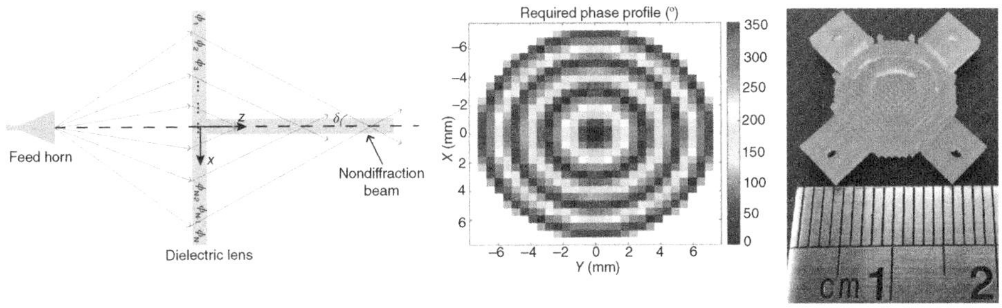

(a) (b) (c)

Figure 6.27 3D printed Bessel beam launcher. (a) Schematic diagram. (b) Desired phase modulation profile. (c) Realized 3D printed lens. Source: From [57] / with permission from IEEE.

observed. Some deviations are mainly caused by fabrication errors and non-ideal measurement environment. A well-defined aperture Bessel distribution is observed in Figure 6.28. The simulated and measured power densities along the z-axis at 300 GHz are plotted in Figure 6.29. It can be seen that a large axial intensity distance (within a −3 dB ripple) over 11.2 mm ($11.2\lambda_0$) can be achieved. The measured spot size is about 0.96 mm ($0.96\lambda_0$) which is smaller than that of the near-field Teflon focusing lens. All these results demonstrate that the design of the 3D printed DDL for THz Bessel beam generation was successful.

Adding a progressive phase shift θ along the x-direction, the Bessel beam can be tilted to an angle θ away from the broadside direction. The phase profile in (6.10) can be modified as

$$\varphi_{mn} = \frac{2\pi f}{c}\left\{\sqrt{(x_{mn}-x_f)^2+(y_{mn}-y_f)^2+z_f^2}+\sqrt{x_{mn}^2+y_{mn}^2}\sin\delta - x\sin\theta\right\}+\Phi_o \qquad (6.11)$$

to achieve this beam tilt. The desired phase distribution obtained with (6.11) is plotted in Figure 6.30a for the same axicon angle $\delta = 20°$ and the tilt angle θ also at 20°. The realized DDL lens using the phase curve in Figure 6.6 is shown in Figure 6.30b. The distance between the phase center of the horn and lens is again set at 24 mm and there were 30×30 pixels with each pixel being $0.5 \times 0.5\,\text{mm}^2$ in size. The simulation model used in the HFSS simulations is shown in Figure 6.30c.

The simulated results on the xoz plane are shown in Figure 6.31a and those of the measured results using the near-field measurement platform of Figure 6.5a are shown in Figure 6.31b. Both the simulated and measured results demonstrate successful beam tilting of the Bessel beam with it pointing at 20° away from the broadside direction. It should be noted that the plotting regions of the simulated and measured results are slightly different.

We further modulate the phase distribution on the lens aperture to incorporate the desired OAM feature in the launched radiation in Section 6.6.4.

6.6.4 Bessel Beam LP Lenses Carrying OAM

OAM multiplexing has been demonstrated to increase channel capacity and spectral efficiency [6, 7]. However, vortex waves carrying OAM suffer from diffraction-based divergence. Furthermore, the power captured by a receiver reduces significantly as a function of the distance between it and the transmitter because of its finite size. As shown in Figures 6.29 and 6.31, a Bessel beam can maintain a high intensity focus over a considerable range. By virtue of its non-diffracting nature, a

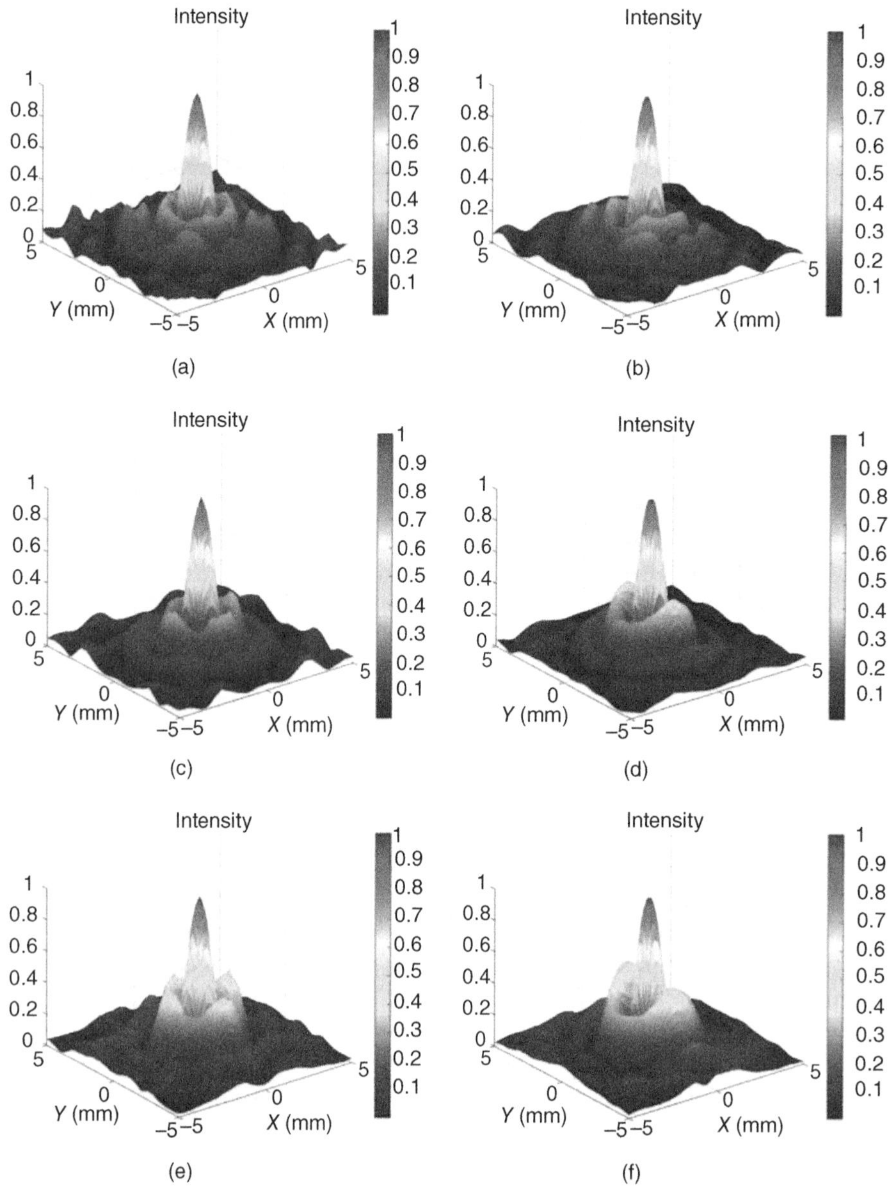

Figure 6.28 Simulated and measured power densities on different transversal planes at 300 GHz. Simulated results: z = (a) 4, (c) 7, and (e) 10 mm. Measured results: z = (b) 4, (d) 7, and (f) 10 mm. Source: From [57] / with permission from IEEE.

Bessel beam carrying OAM, also known as a higher-order Bessel beam, is a potential approach to overcome the diffraction divergence problem associated with vortex beams.

Based on the higher-order Bessel beam concept, we present an extended version that can cover a custom-defined longitudinal range and can be matched to a practical application scenario. An OAM signal that covers a specific longitudinal service range reduces the interference with other channels and increases the signal-to-noise ratio for the receiver. In contrast, conventional higher-order Bessel beams radiate OAM waves simply starting from the antenna aperture and begin to diffract

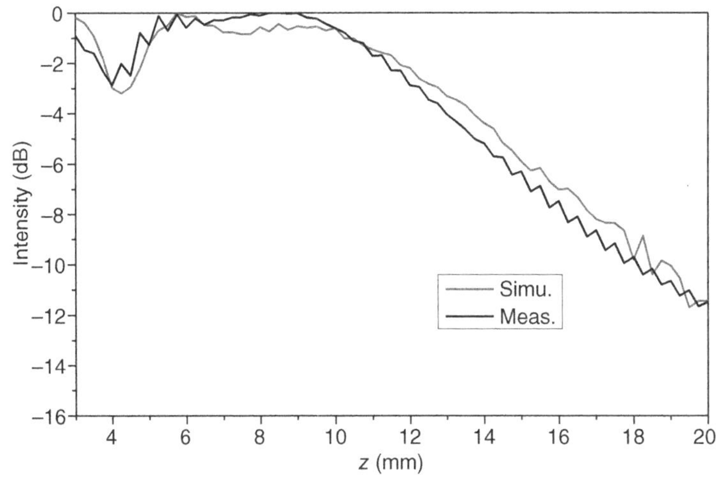

Figure 6.29 Simulated and measured power densities along the *z*-axis for the Bessel launcher at 300 GHz. Source: From [57] / with permission from IEEE.

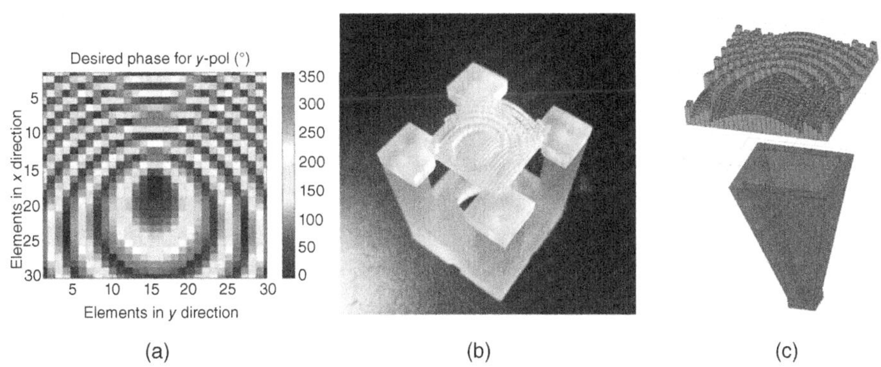

Figure 6.30 Bessel beam launcher with a tilted beam 20° away from the broadside. (a) Desired phase modulation. (b) Realized DDL. (c) Simulation model with the designed lens and feed horn.

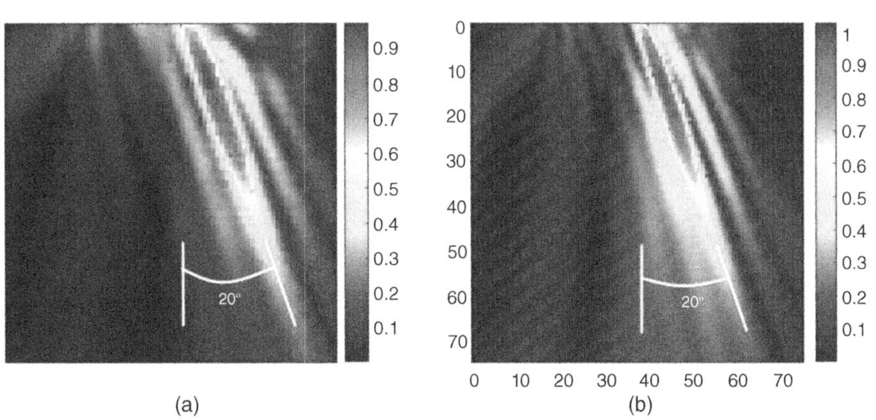

Figure 6.31 Simulated and measured results of the tilted Bessel beam launcher. (a) Simulation. (b) Measurement.

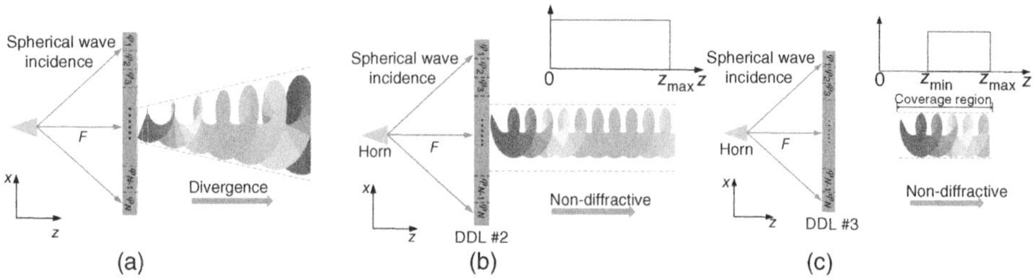

Figure 6.32 DDL antennas. (a) DDL 1 generates a diffractive OAM beam. (b) DDL 2 generates a higher-order Bessel beam. (c) DDL 3 generates an extended higher-order Bessel beam. Source: From [63] / with permission from IEEE.

as the distance reaches the Raleigh length [58–60]. The DDL for the extended high-order Bessel beam is synthesized using the alternating projection method (APM) [61, 62]. A more desirable field distribution with better uniformity in the longitudinal coverage region is attained.

Figure 6.32 shows the 3 DDLs to be presented here. The DDL 1 and DDL 2 designs were developed for launching an OAM beam and a non-diffracting higher-order Bessel beam, respectively. The extended higher-order Bessel beam synthesized by the APM is labelled as DDL 3.

The field distribution in cylindrical coordinates of a beam propagating along the longitudinal direction (z-axis) can be expressed at the observation point (ρ, φ, z) as:

$$E(\rho, \phi, z, t) = E_o J_\iota(k_\rho \rho) e^{(-jk_z z - j\iota\phi + j\omega t)} \tag{6.12}$$

where E_0 is the amplitude and $J_\iota(k_\rho \rho)$ is the ιth order Bessel function of the first kind. The wavenumber components k_ρ and k_z are the transversal and longitudinal propagation constants; they satisfy $\sqrt{k_\rho^2 + k_z^2} = k_o = 2\pi f/c$. The extended Bessel beam (6.12) has a vortex phase term of $e^{(-j\iota\phi)}$. When $\iota = 0$, the Bessel does not carry any OAM. When $\iota \neq 0$, it is a higher-order Bessel beam carrying OAM with a mode number of ι.

Figure 6.33a shows the configuration of a conventional OAM-carrying Bessel beam launcher. It consists of a bottom dielectric hyperbolic lens, a middle dielectric conical axicon, and a top spiral surface plane. Their corresponding functions are transforming the spherical phase front from the feed source into a plane wave, generating a non-diffractive zero-order Bessel beam and adding a helical phase distribution to the Bessel beam. Together, they form the OAM-carrying Bessel beam launcher. However, this launcher suffers from a bulky size, weight, and high dielectric loss, especially at THz frequencies. To address these issues, a single DDL shown in Figure 6.32b was developed to mimic the functionalities of these three bulky refractive dielectric lenses through phase modulation. The required transmission phase compensation of the mnth element of the DDL is given by:

$$\varphi_{mn}(\rho_{mn}, \phi_{mn}) = k_o \sqrt{F^2 + \rho_{mn}^2} + k_o \sin\delta - \iota\phi_{mn} \tag{6.13}$$

where (ρ_{mn}, ϕ_{mn}) are the cylindrical coordinates of the corresponding mnth element. The term F is the focal length, which is the spatial separation from the phase center of the feed horn to the DDL aperture. The term δ is the axicon angle of the Bessel beam and $\tan\delta = k_\rho/k_z$. The first term in (6.13) on the right-hand side mimics the function of the hyperbolic lens in Figure 6.32a. The second and third terms emulate the functions of the axicon and the spiral surface plane, respectively. Each of the three DDLs in Figure 6.32 has a diameter of 30 mm and focal length $F = 75$ mm. They are designed to have an OAM mode index $\iota = +2$.

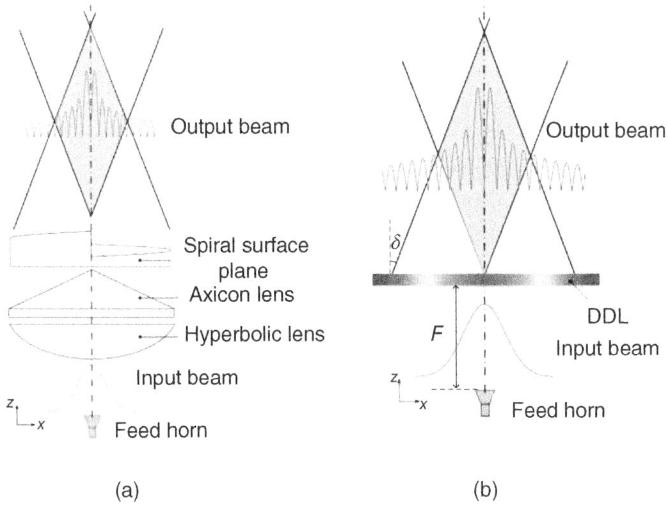

Figure 6.33 Schematic of the higher-order Bessel beam generation using (a) three dielectric structures and (b) a low profile DDL. Source: From [63] / with permission from IEEE.

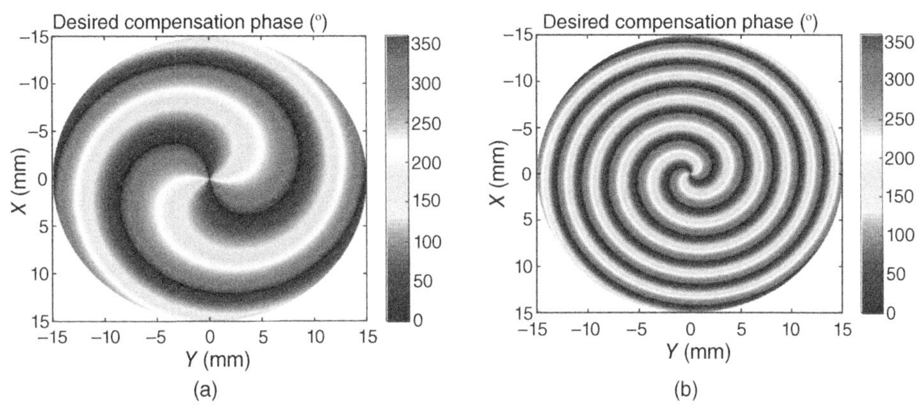

Figure 6.34 Desired compensation phase distributions for OAM carrying beams with mode index $\iota = +2$. (a) DDL 1 for a conventional diffracting OAM beam. (b) DDL 2 for a non-diffracting higher-order Bessel beam carrying OAM. Source: From [63] / with permission from IEEE.

When the axicon angle $\delta = 0°$ is set for DDL 1, the second term on the right-hand side of (6.13) becomes zero, corresponding to removing the middle axicon lens in Figure 6.33a and generating a conventional vortex wave. For DDL 2, $\delta = 15°$, which corresponds to a Bessel beam carrying OAM with mode index $\iota = +2$. The desired compensation phase distributions for DDL 1 and DDL 2 are shown in Figure 6.34a,b, respectively.

Utilizing the phase curve in Figure 6.6, DDL 1 and DDL 2 were designed and fabricated. Since the dielectric post element of the DDLs is isotropic, the polarization of the DDL antenna depends on the polarization of the antenna feed. The beams can be synthesized using both polarizations of the feed. As the DDL antennas are electrically large structures, full-wave simulation of them is challenging in terms of high computational costs. A simple analysis method based on the field equivalence principle has been developed to quickly evaluate the near-field radiation characteristics of the DDL antennas. Readers are referred to [63] for the details of this analysis. Due to the dielectric loss of the 3D printing material, the simulated radiation efficiency of the DDL 2 antenna is 71.1%. Alternately, the simulated efficiency of the conventional conical axicon using the same dielectric material with an identical non-diffractive range is 53.3%. The DDL antenna shows a great advantage in transmission efficiency at THz frequencies even though it has a thinner profile.

A diagonal horn (VDI WR-3.4) with a gain of 26 dBi at 300 GHz was adopted as the feed source for the DDL antenna. For accurate modeling, the incident fields illuminating on the DDL aperture are obtained by full-wave simulation of the diagonal horn. The co-pol of the diagonal horn is along the x-axis. The calculated magnitude of E_x in the longitudinal plane (xz-plane), and the magnitude and phase of E_x in the transversal plane (xy-plane) at $z = 30$ mm are plotted for both DDL 1 and DDL 2 in Figure 6.35. It is not difficult to see that both DDL 1 and DDL 2 can generate OAM radiation with mode index $\imath = +2$. However, the field radiated by DDL 1 spreads out along the propagation direction while that of DDL 2 is concentrated around the longitudinal axis over a long propagation distance, approximately 40 mm ($40\lambda_0$). The divergence effect of the vortex wave can be overcome by DDL 2, which maintains the diameter of its transversal energy ring nearly constant at 3.5 mm ($3.5\lambda_0$) over the whole propagation distance.

From Figure 6.35, it is not difficult to see that the radiated power confinement of the higher-order Bessel beam begins at the aperture of DDL 2. We can further optimize the phase modulation at the lens aperture to realize an extended non-diffracting high-order Bessel beam covering an extended longitudinal region using the APM.

The APM, also known as the intersection approach [61], has been used for array antenna far-field beam-forming applications [62]. We adopt it here to synthesize a desired near-field longitudinal field distribution. The schematic diagram of the APM is shown in Figure 6.36. The sets M and R are the longitudinal field distributions that fulfill the mask requirement and that can be realized by the DDL, respectively. Set R depends on the aperture field distribution of the DDL. The APM is an iterative process that searches for the intersection of the two sets, $M \cap R$, as depicted in Figure 6.36. Set M is defined as:

$$M \equiv \{E(\rho = 0, z) : M_L(z) \leq |E(\rho = 0, z)| \leq M_U(z)\}, \quad z > 0 \tag{6.14}$$

where $E(\rho = 0, z)$ is the field distribution along the longitudinal direction generated by the DDL. The terms M_L and M_U are the lower and upper bounds, respectively, between which the longitudinal field distribution must be located.

The aperture field distribution over the DDL, $E_{n+1}(\rho = 0, z)$ in the $(n + 1)$th iteration is updated as:

$$E_{n+1}(\rho = 0, z) = \mathcal{P}_R \mathcal{P}_M E_n(\rho = 0, z) \tag{6.15}$$

The operator $\mathcal{P}_M$ is a forward projector casting the longitudinal field distribution radiated by the DDL onto the set of longitudinal fields that comply with the set M. On the other hand, the operator $\mathcal{P}_R$ is the backward projector, which projects the longitudinal field distribution that satisfies the mask specification onto the set of longitudinal fields that can be realized by the DDL. Specifically, $\mathcal{P}_M$ utilizes the upper and lower bounds of the mask to correct the longitudinal field distribution realized by the DDL as:

$$\mathcal{P}_M E(\rho = 0, z) = \begin{cases} M_U(z)\dfrac{E(\rho = 0, z)}{|E(\rho = 0, z)|}, |E(\rho = 0, z)| > M_U(z) \\ E(\rho = 0, z), M_L(z) \leq |E(\rho = 0, z)| \leq M_U(z) \\ M_L(z)\dfrac{E(\rho = 0, z)}{|E(\rho = 0, z)|}, |E(\rho = 0, z)| < M_L(z) \end{cases} \tag{6.16}$$

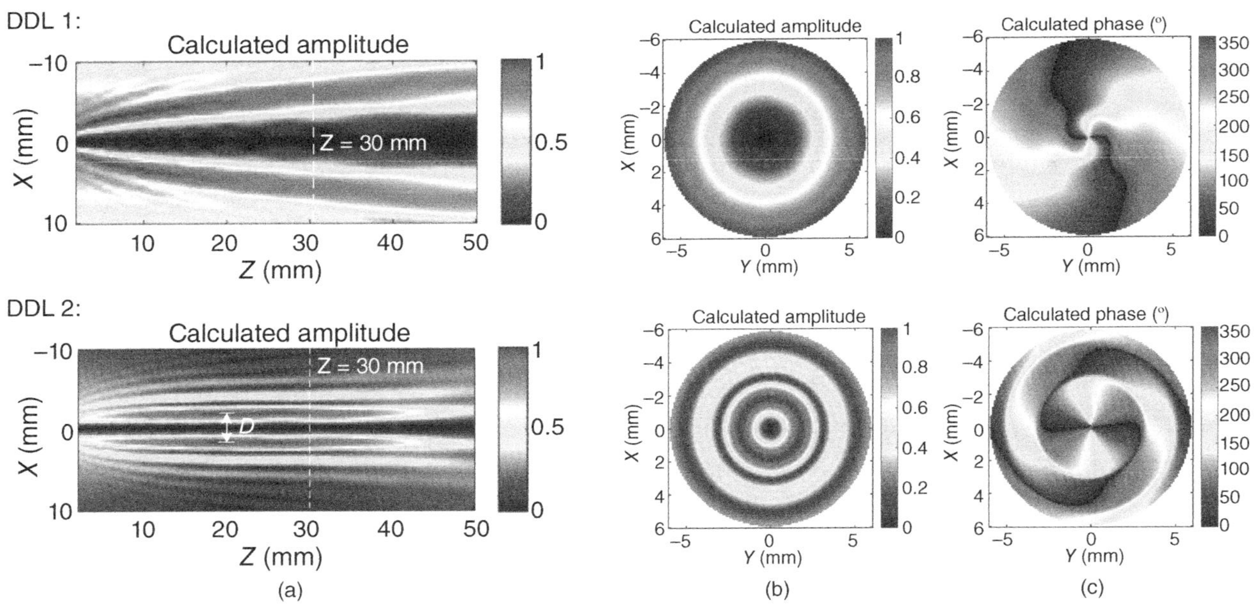

Figure 6.35 Calculated (a) amplitude distribution in the longitudinal plane (*xz*-plane), (b) amplitude and (c) phase distributions in the transversal plane ($z = 30$ mm). Source: From [63] / with permission from IEEE.

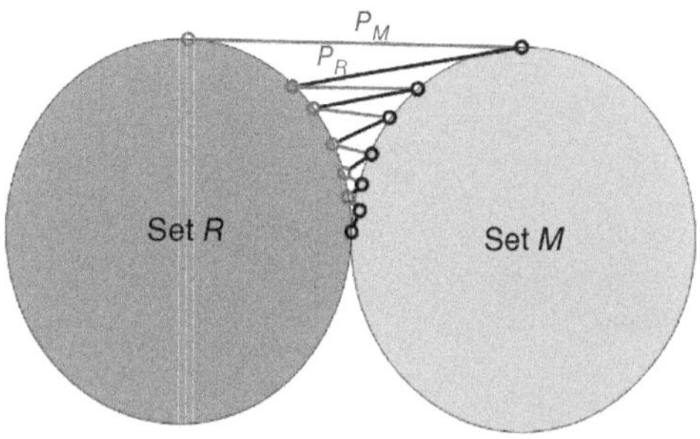

Figure 6.36 Schematic diagram of the APM. Source: From [63] / with permission from IEEE.

The backward projector $\mathcal{P}_R$ in the conventional far-field shaped-beam array antennas is defined as [61]:

$$\mathcal{P}_R = \mathcal{F}\mathcal{J}_C\mathcal{J}_I\mathcal{F}^{-1} \tag{6.17}$$

where $\mathcal{F}$ and $\mathcal{F}^{-1}$ are the Fourier and inverse Fourier transform operators, respectively, and $\mathcal{J}_C$ and $\mathcal{J}_I$ will be defined shortly. In far-field applications, the Fourier transform corresponds to the decomposition of a transverse aperture field $E(x, y, z = 0)$ into its spatial frequency spectrum, (k_x, k_y, k_z) in Cartesian coordinates, with each spatial frequency representing a plane wave. The spiral phase information from (6.13) can be incorporated into the phase distribution for the generation of the higher-order Bessel beam. For the azimuthally invariant high-order Bessel beam, the transverse aperture field $E(\rho, z = 0)$is transformed into its spatial frequency spectrum, (k_ρ, k_z) in cylindrical coordinates. The Fourier transform in (6.17) can be written as:

$$S(k_\rho, z = 0) = \int_0^\infty E(\rho, z = 0) J_0(k_\rho \rho)\rho d\rho \tag{6.18}$$

where S is the spatial frequency spectrum. From the spatial frequency spectrum of the aperture field, the radiated longitudinal field distribution of the DDL can be calculated as [64]:

$$E(\rho = 0, z) = \int_0^{k_0} S(k_\rho, z = 0) e^{jk_z z} k_z dk_z \tag{6.19}$$

Conversely, if the longitudinal field distribution $E(\rho = 0, z)$ is known, the field on the DDL aperture $E(\rho, z = 0)$ can be recovered. Therefore, the inverse Fourier transform in (6.17) can be replaced by:

$$S(k_\rho, z = 0) = \frac{1}{2\pi k_z} \int_{-\infty}^{\infty} E(\rho = 0, z) e^{-jk_z z} dz \tag{6.20}$$

$$E(\rho, z = 0) = \int_0^\infty S(k_\rho, z = 0) J_0(k_\rho \rho) k_\rho dk_\rho \tag{6.21}$$

It is observed from (6.18) to (6.21) that the transverse aperture field distribution $E(\rho, z = 0)$ and the longitudinal field distribution $E(\rho = 0, z)$ are linked by the spatial frequency spectrum. The operators $\mathcal{J}_I$ and $\mathcal{J}_C$ in (6.17) are identical to those of the APM employed in conventional far-field shaped-beam antenna analyses [62, 64, 65]. Specifically, $\mathcal{J}_I$ selects all the transversal field distributions within the DDL aperture, while $\mathcal{J}_C$ introduces constraints in the aperture amplitude distribution imposed by the feed horn as follows:

$$\arg\left[E'_n(\rho, z = 0)\right] = \arg[E_n(\rho, z = 0)] \tag{6.22}$$

$$\left|E'_n(\rho, z = 0)\right| = A_1(\rho) \tag{6.23}$$

where A_I is the aperture amplitude distribution created by the fields radiated from the feed. Once the iterative process is converged, the required phase modulation of the DDL for generating an extended higher-order Bessel beam carrying OAM is given by:

$$\varphi(\rho, \phi) = \arg[E_n(\rho, z = 0)] + k_0\sqrt{F^2 + \rho^2} - \iota\phi \tag{6.24}$$

The procedure described earlier is also employed to synthesize DDL 3 with the same size and focal length as DDL 1 and DDL 2. The desired longitudinal coverage is designated as ($z \in [20$ mm, 35 mm]). A proper starting aperture phase distribution belonging to M_R can help to avoid converging to non-desired local minima and, hence, reduce the number of iterations [61, 62, 65, 66]. The axicon phase distribution in DDL 2 is adopted as the initial aperture phase distribution for DDL 3. The iterative process converges in 25 iterations. Figure 6.37a shows the obtained compensation phase distribution of DDL 3 and its height profile can be determined by the height-phase curve in Figure 6.6. The calculated field distribution $|E_x|$ in the longitudinal plane is given in Figure 6.37b. The calculated non-diffracting range, defined as the longitudinal range at which the power density is 3 dB below its maximum value is [21 mm, 36 mm] against the target range of [20 mm, 35 mm]. The amplitude and phase of E_x in the transversal plane ($z = 30$ mm) are shown in Figure 6.37c. One finds from Figure 6.37b,c that the non-diffractive radiated field is concentrated around the longitudinal axis and the diameter of its transversal energy ring is nearly kept constant at 3.2 mm ($3.2\lambda_0$). An OAM mode $\iota = +2$ is generated.

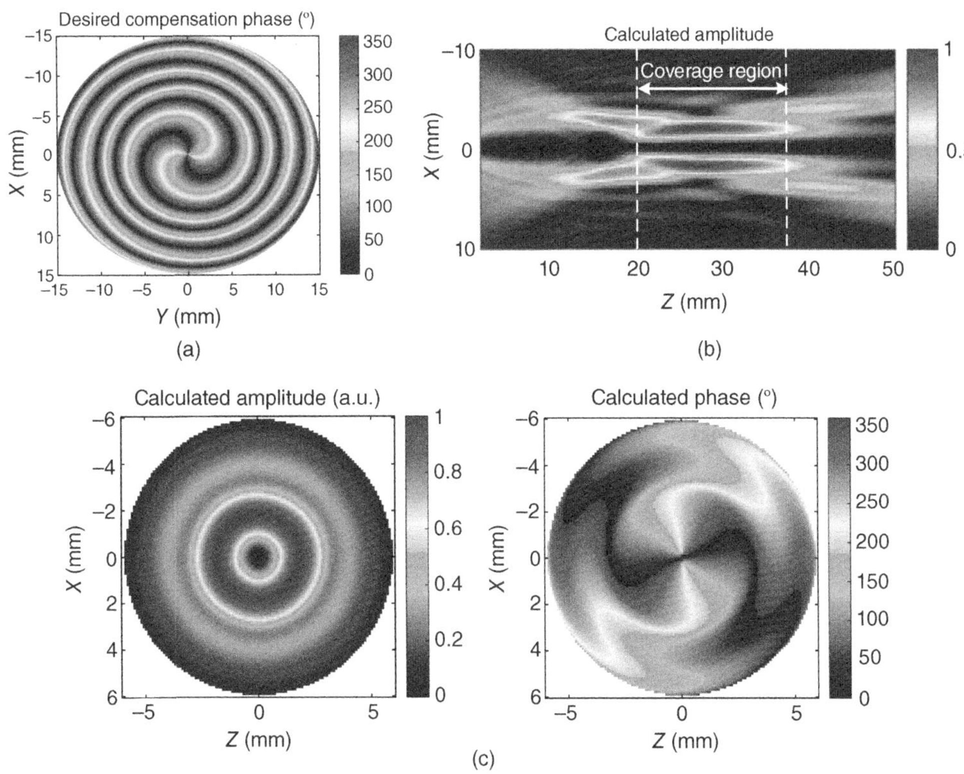

Figure 6.37 Phase modulation and calculated results of DDL 3. (a) Desired phase compensation distribution. (b) Magnitude of E_x in the longitudinal plane (xz-plane). (c) Amplitude and phase of E_x in the transversal plane ($z = 30$ mm). Source: From [63] / with permission from IEEE.

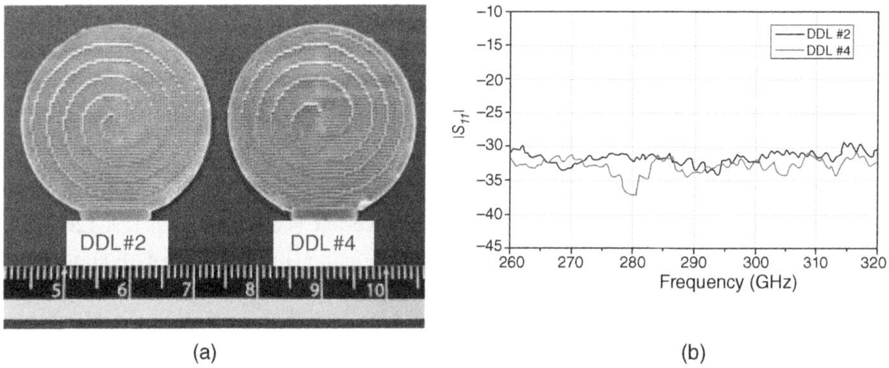

Figure 6.38 Launchers of higher-order Bessel beam carrying OAM. (a) 3D printed prototypes. (b) Measured reflection coefficients versus frequency. Source: From [63] / with permission from IEEE.

To validate our designs, DDL 2 and DDL 3 were 3D printed and tested using the near-field measurement platform shown in Figure 6.5a. Figure 6.38a shows the fabricated DDL 2 and DDL 3. Each of the two DDLs is fed by a diagonal horn to form a Bessel beam launcher carrying OAM. Figure 6.38b shows the reflection coefficients of these DLL launchers. It is observed that they are smaller than −28 dB within the whole frequency range from 260 to 320 GHz, which is attributed to the space feeding nature of these DDL launchers.

Figure 6.39 shows the measured $|E_x|$ distributions in the longitudinal plane (xz-plane) at 280, 300, and 320 GHz for both DDL 2 and DDL 3. A central amplitude singularity along the propagation axis is formed for both DDLs. Well defined non-diffracting beams are observed from 280 to 320 GHz, showing the wideband nature of the two launchers for the generation of high-order and extended higher-order Bessel beams. The measured radiated power of DDL 3 is mainly located in the longitudinal coverage from 21 to 36 mm. In contrast, the circular Airy OAM beams reported in [67] have a maximum intensity at the autofocusing point. The extended higher-order Bessel beam maintains a high intensity within the intended longitudinal coverage range, which offers more flexibility in short range communications and detections.

For comparison purposes, the 1D radial cut of the calculated and measured $|E_x|$ distributions at $z = 30$ mm and $y = 0$ mm for DDL 2 and DDL 3 are shown in Figure 6.40a,b, respectively. Good agreement between the calculated and measured results is observed. Some discrepancies are present and are primarily attributed to the simplified model in the analysis method in which the edge-diffracted field of the lens and shadowing effects of the neighboring elements are not considered [63]. Moreover, fabrication errors, possible misalignments, and a non-ideal measurement environment may also account for these slight discrepancies. Figure 6.41a,b depict the measured phase of E_x in the transversal plane at $z = 30$ mm for DDL 2 and DDL 3 at 300 GHz, respectively. A consecutive 4π angular phase variation along full concentric circles with increasing phase in the clockwise direction is observed. These measured results demonstrate the successful design of non-diffractive higher-order and extended higher-order Bessel beams carrying OAM.

6.7 Summary

In this chapter, we have demonstrated the designs of a variety of THz lens antennas, which employ phase modulation of the lens aperture. We have described our investigation of 3D printing

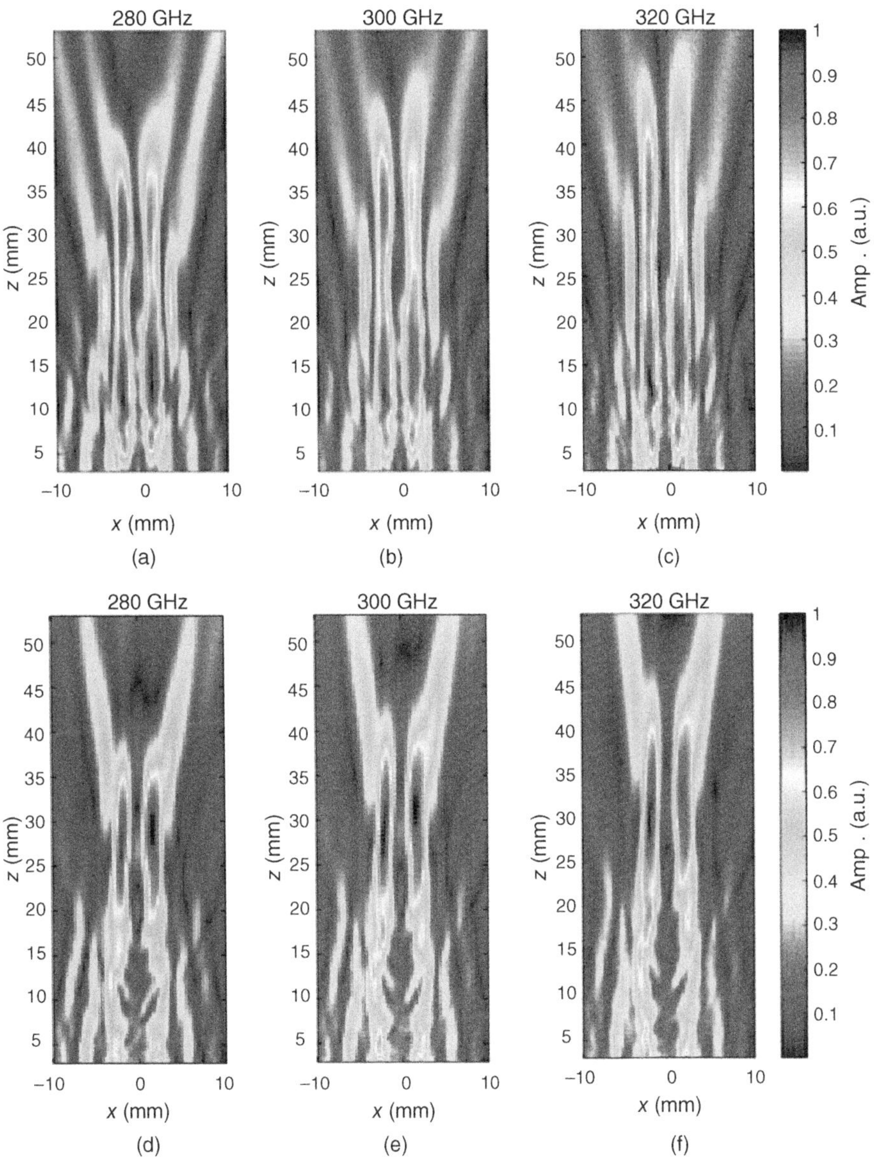

Figure 6.39 Measured $|E_x|$ distributions in the longitudinal plane (xz-plane). (a)–(c) DDL 2 for high-order Bessel beam launcher. (d)–(f) DDL 3 for extended high-order Bessel beam launcher. Source: From [63] / with permission from IEEE.

approaches to push the performance envelopes of inexpensive 3D printers toward the realization of the designed THz antennas. We also took full advantage of the additional design degrees of freedom provided by 3D printing to realize various THz lens antennas, including LP and CP high-gain antennas and focusing antennas, and Bessel beam and higher-order Bessel beam launchers. Using two cascaded DDLs with in plane rotation, a 2D beam scanning Bessel beam launcher and an OAM mode reconfigurable DDL antenna operating at 300 GHz have been realized in [68] and [69], respectively.

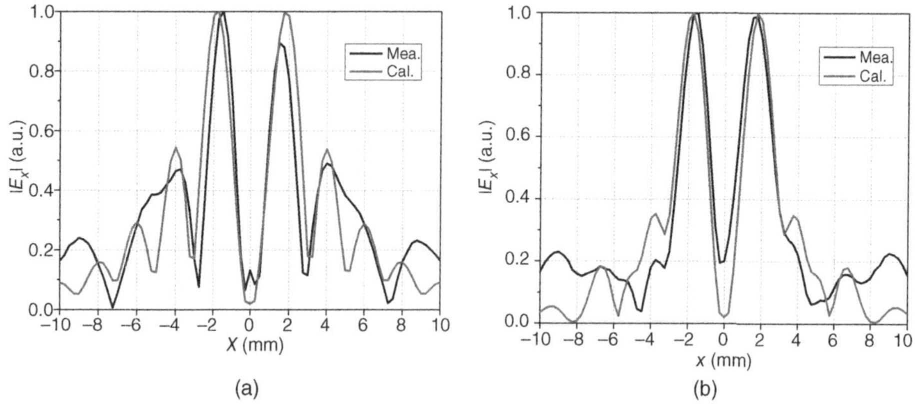

Figure 6.40 Measured and calculated distributions located at $z = 30$ mm and $y = 0$ mm of the (a) higher-order Bessel beam generated by DDL 2, and (b) extended higher-order Bessel beam generated by DDL 3, respectively, at 300 GHz.

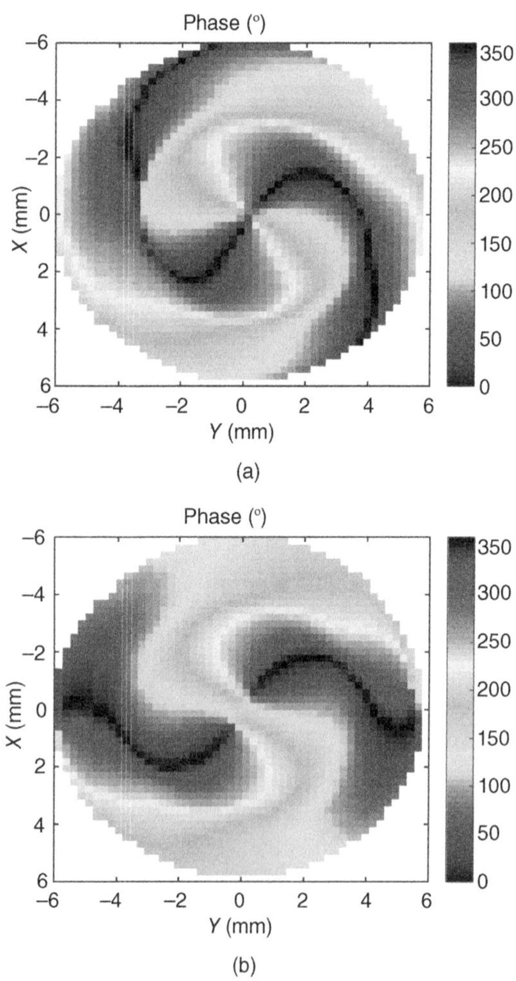

Figure 6.41 Measured phase of distributions in the transversal plane at $z = 30$ mm for (a) DDL 2 and (b) DDL 3, respectively, at 300 GHz. Source: From [63] / with permission from IEEE.

We should point out that the upper frequency we can reach is not only limited by the resolution of the 3D printers, but also the loss tangent of the printing materials. At this point, our operating frequency is capped at 300 GHz. On the other hand, if the lens has rotational symmetry like an elliptical dielectric lens, it can be fabricated by CNC machining as we have demonstrated in the focusing LP lens. As Teflon is a low-loss material even at THz frequency, a Teflon lens can be an alternative to the more expensive silicon lens. Incorporating an elliptical Teflon lens with a 4×4 array radiator IC, we were able to boost the EIRP (effective isotropic radiated power) up to 29.1 dBm with more than a 20 dB directivity boost at 458.3 GHz. Similarly, 27.8 dBm in EIRP with a directivity around 30 dBi was achieved for another radiator IC operating at 699 GHz.

Finally, we firmly believe that there are many further opportunities for the design, simulation, fabrication, and measurement of THz DDL antennas. As 3D printing materials and machines are refined and other cost-effective fabrication techniques are developed, THz lens antennas will become readily available to meet the needs of 6G and beyond wireless and imaging systems.

Acknowledgment

This work was supported in part by the Hong Kong Research Grants Council under the Theme-Based Research Scheme (T42-103/16-N) and General Research Fund (CityU 11250216 and CityU 11205819).

References

1 6G Flagship. Retrieved from https://www.oulu.fi/6gflagship/inbrief on May 20, 2021.

2 M. Latva-aho and K. Leppänen, Eds., *Key Drivers and Research Challenges for 6G Ubiquitous Wireless Intelligence*. Finland: 6G Flagship, University of Oulu, Sep 2019.

3 M. Tonouchi, "Cutting-edge terahertz technology," *Nat. Photonics*, vol. 1, pp. 97–105, Feb 2007.

4 S. Koening, D. Lopez-Diaz, J. Antes, F. Boes, R. Henneberger, A. Leuther, A. Tessmann, R. Schmogrow, D. Hillerkuss, R. Palmer, T. Zwik, C. Koos, W. Freude, O. Ambacher, J. Leuthold, and I. Kallfass, "Wireless sub-THz communication system with high data rate," *Nat. Photonics*, vol. 7, pp. 977–981, Dec 2013.

5 G. Ducournau, K. Engenhardt, P. Szriftgiser, D. Bacquet, M. Zaknoune, R. Kassi, E. Lecomte, and J.-F. Lampin, "32 Gbit/s QPSK transmission at 385 GHz using fibre-optic technologies and THz double heterodyne detection," *Electron. Lett*, vol. 51, no. 12, pp. 915–917, Jun 2015.

6 Y. Yan, G. Xie, M. P. J. Lavery, H. Huang, N. Ahmed, C. Bao, Y. Ren, Y. Cao, L. Li, Z. Zhao, A. F. Molisch, M. Tur, M. J. Padgett, and A. E. Willner, "High-capacity millimeter-wave communications with orbital angular momentum multiplexing," *Nat. Commun.*, 5, Art. no. 4816, Sep 2014.

7 J. Wang, J.-Y. Yang, I. M. Fazal, N. Ahmed, Y. Yan, H. Huang, Y. Ren, Y. Yue, S. Dolinar, M. Tur, and A. E. Willner, "Terabit free-space data transmission employing orbital angular momentum multiplexing," *Nat. Photonics*, vol. 6, pp. 488–496, Jul 2012.

8 J. S. Tyo, D. L. Goldstein, D. B. Chenault, and J. A. Shaw, "Review of passive imaging polarimetry for remote sensing applications," *Appl. Opt.*, vol. 45, no. 22, pp. 5453–5469, 2006.

9 P. Terrier, V. Devlaminck, and J. M. Charbois, "Segmentation of rough surfaces using a polarization imaging system," *J. Opt. Soc. Am. A*, vol. 25, no. 2, pp. 423–430, 2008.

10 C. S. Chun and F. A. Sadjadi, "Polarimetric imaging system for automatic target detection and recognition," Phys. Innov., Inc., Inver Grove Heights, MN, USA, Tech. Rep. ADA392865, Mar 2000.

11 Y. Gu, C. Carrizo, A. A. Gilerson, P. C. Brady, M. E. Cummings, M. S. Twardowski, J. M. Sullivan, A. I. Ibrahim, and G. W. Kattawar, "Polarimetric imaging and retrieval of target polarization characteristics in underwater environment," *Appl. Opt.*, vol. 55, no. 3, pp. 626–637, 2016.

12 A. Van Eeckhout, A. Lizana, E. Garcia-Caurel, J. J. Gil, A. Sansa, C. Rodríguez, I. Estévez, E. González, J. C. Escalera, and I. Moreno, "Polarimetric imaging of biological tissues based on the indices of polarimetric purity," *J. Biophotonics*, vol. 11, no. 4, Art. no. e201700189, 2018.

13 C. H. L. Patty, D. A. Luo, F. Snik, F. Ariese, W. J. Buma, I. L. Ten Kate, R. J. M. van Spanning, W. B. Sparks, T. A. Germer, and G. Garab, "Imaging linear and circular polarization features in leaves with complete Mueller matrix polarimetry," *Biochim. Biophys. Acta*, vol. 1862, no. 6, pp. 1350–1363, 2018.

14 H.-J. Song, and T. Nagatsuma, "Present and future of terahertz communications," *IEEE Trans. THz Sci. Technol.*, vol. 1, no. 1, pp. 256–263, Sep 2011.

15 K. M. Luk, S. F. Zhou, Y. Li, F. Wu, K. B. Ng, C. H. Chan, and S. Pang, "A microfabricated low-profile wideband antenna array for terahertz communications," *Sci. Rep.*, 7, p. 1268, Dec 2017.

16 N. Llombart, G. Chattopadhyay, A. Skalare, and I. Mehdi, "Novel terahertz antenna based on a silicon lens fed by a leaky wave enhanced waveguide," *IEEE Trans. Antennas Propag.*, vol. 59, no. 6, pp. 2160–2168, Jun 2011.

17 A. Bolwell, 3D printing is transforming these industries. (July 30, 2018) Retrieved from https://hpmegatrends.com/3d-printing-is-transforming-these-industries-f94661c6b1b3 on May 20, 2021.

18 T.-H. Chio, G.-L. Huang, and S.-G. Zhou, "Application of direct metal laser sintering to waveguide-based passive microwave components, antennas, and antenna arrays," *Proc. IEEE*, vol. 105, no. 4, pp. 632–644, Apr 2017.

19 J. Maas, B. Liu, S. Hajela, Y. Huang, X. Gong, and W. J. Chappell, "Laser-based layer-bylayer polymer stereolithography for high-frequency applications," *Proc. IEEE*, vol. 105, no. 4, pp. 645–654, Apr 2017.

20 J. S. Silva, M. Garcia-Vigueras, T. Debogovic, J. R. Costa, C. A. Fernandes, and J. R. Mosig, "Stereolithography-based antennas for satellite communications in Ka-band," *Proc. IEEE*, vol. 105, no. 4, pp. 655–667, Apr 2017.

21 A. I. Dimitriadis, T. Debogovic, M. Favre, M. Billod, L. Barloggio, J.-P. Ansermet, and E. de Rijk, "Polymer-based additive manufacturing of high-performance waveguide and antenna component," *Proc. IEEE*, vol. 105, no. 4, pp. 668–676, Apr 2017.

22 K. H. Church, B. B. Crane, P. L. Deffenbaugh, T. P. Ketterl, C. G. Neff, P. B. Nesbitt, J. T. Nussbaum, C. Perkowski, H. Tsang, J. Castro, J. Wang, and T. M. Weller, "Multimaterial and multilayer direct digital manufacturing of 3-D structural microwave electronics," *Proc. IEEE*, vol. 105, no. 4, pp. 688–701, Apr 2017.

23 S. A. Nauroze, J. G. Hester, B. K. Tehrani, W. Su, J. Bito, R. Bahr, J. Kimionis, and M. M. Tentzeris, "Additively manufactured RF components and modules: toward empowering the birth of cost-efficient dense and ubiquitous IoT implementations," *Proc. IEEE*, vol. 105, no. 4, pp. 702–722, Apr 2017.

24 B. Zhang, Y.-X. Guo, H. Zirath, and Y. P. Zhang, "Investigation on 3-D-printing technologies for millimeter-wave and terahertz applications," *Proc. IEEE*, vol. 105, no. 4, pp. 723–736, Apr 2017.

25 H. Xin and M. Liang, "3-D-printed microwave and THz devices using polymer jetting techniques," *Proc. IEEE*, vol. 105, no. 4, pp. 737–755, Apr 2017.

26 W. J. Otter and S. Lucyszyn, "Hybrid 3-D-printing technology for tunable THz applications," *Proc. IEEE*, vol. 105, no. 4, pp. 756–767, Apr 2017.

27 J. Romeu, S. Blanch, N. Vidal, J. M. Lopez-Villegas, and A. Augasca,"Assessment of 3-D printing technologies for millimeter-wave reflectors," *IEEE Antennas Wireless Propag. Lett.*, vol. 17, no. 11, pp. 2061–2064, Nov 2018.

28 M. D. Wu, B. Li, Y. Zhou, D. L. Guo, Y. Liu, F. Wei, and X. Lv, "Design and measurement of a 220 GHz wideband 3-D printed dielectric reflectarray," *IEEE Antennas Wireless Propag. Lett.*, vol. 17, no. 11, pp. 2094–2098, Nov 2018.

29 H. Yi, S.-W. Qu, K. B. Ng, C. H. Chan, and X. Bai, "3D printed terahertz and millimeter-wave lenses with fixed and scanning beam," *IEEE Trans. Antennas Propag.*, vol. 64, no. 2, pp. 442–449, Feb 2016.

30 Formlabs Design Guide. Retrieved from https://hpmegatrends.com/3d-printing-is-transforming-these-industries-f94661c6b1b3 on May 20, 2021.

31 G.-B. Wu, Y.-S. Zeng, K. F. Chan, S.-W. Qu, and C. H. Chan, "3D printed circularly polarized modified Fresnel lens operating at terahertz frequencies," *IEEE Trans. Antennas Propag.*, vol. 67, no. 7, pp. 4429–4437, Jul 2019.

32 K. B. Ng and C. H. Chan, "On the dielectric properties of substrates with different surface conditions for submillimeter-wave and terahertz applications," *THz Sci. Technol.*, vol. 9, no. 2, pp. 45–59, Jun 2016.

33 X. Ruan and C. H. Chan, "Terahertz free-space dielectric property measurements using time- and frequency-domain setups," *Int. J. RF Microwave Comput. Aided Eng.*, vol. 29, p. e21839, May 2019.

34 J. Bomba, J. Suszek, M. Makowski, A. Sobczyk, and M. Sypek, "3-D printed anti-reflection structures for the terahertz region," *J. Infrared Millimeter Terahertz Waves*, vol. 39, no. 1, pp. 24–35, Sep 2017.

35 H. S. Kirschbaum, and S. Chen, "A method of producing broad-band circular polarization employing an anisotropic dielectric," *IRE Trans. Microwave Theory Tech.*, vol. 5, no. 3, pp. 199–203, Mar 1957.

36 K. X. Wang, and H. Wong, "A wideband millimeter-wave circularly polarized antenna with 3-D printed polarizer," *IEEE Trans. Antennas Propag.*, vol. 65, no. 3, pp. 1038–1046, Mar 2017.

37 K. X. Wang, and H. Wong, "Design of a wideband circularly polarized millimeter-wave antenna with an extended hemispherical lens," *IEEE Trans. Antennas Propag.*, vol. 66, no. 8, pp. 4303–4308, Aug 2018.

38 Y. V. Serdyuk, A. D. Podoltsev, and S. M. Gubanski, "Numerical simulations of dielectric properties of composite material with periodic structure," *J. Electrostat.*, vol. 63, no. 11, pp. 1073–1091, 2005.

39 G.-B. Wu, Y.-S. Zeng, K. F. Chan, S.-W. Qu, and C. H. Chan, "3-D printed terahertz lens with circularly polarized focused near field," in *Proceedings of the 13th European Conference on Antennas and Propagation*, EuCAP2019, Krakow, Poland, March 31-April 5, 2019. Article number 8739330.

40 A. Yu, F. Yang, A. Z. Elsherbeni, J. Huang, and Y. Rahmat-Samii, "Aperture efficiency analysis of reflectarray antennas," *Microwave Opt. Technol. Lett.*, vol. 52, no. 2, pp. 364–372, Feb 2010.

41 J. W. Goodman, *Introduction to Fourier Optics*. Roberts and Company Publishers, 2005.

42 B. E. A. Saleh, and M. C. Teich, *Fundamentals of Photonics*. John Wiley & sons, 2019.

43 G. B. Wu, Y.-S. Zeng, K. F. Chan, S.-W. Qu, and C. H. Chan, "High-gain circularly polarized lens antenna for terahertz applications," *IEEE Antennas Wireless Propag. Lett.*, vol. 18, no. 5, pp. 921–925, May 2019.

44 P. Nayeri, M. Liang, R. A. Sabory-Garcι, M. Tuo, F. Yang, M. Gehm, H. Xin, and A. Z. Elsherbeni, "3D printed dielectric reflectarrays: low-cost high-gain antennas at sub-millimeter waves," *IEEE Trans. Antennas Propag.*, vol. 62, no. 4, pp. 2000–2008, Apr 2014.

45 Y. Liu, H. Lu, Y. Wu, M. Cui, B. Li, P. Zhao, and X. Lv, "Millimeterwave and terahertz waveguide-fed circularly polarized antipodal curvedly tapered slot antennas," *IEEE Trans. Antennas Propag.*, vol. 64, no. 5, pp. 1607–1614, May 2016.

46 P. Zhao, Y. Liu, H. Lu, Y. Wu, and X. Lv, "Experimental realization of terahertz waveguide-fed circularly polarized double-fan-shaped slot antenna," *IEEE Antennas Wireless Propag. Lett.*, vol. 16, pp. 2066–2069, Nov 2017.

47 D. González-Ovejero, C. Jung-Kubiak, M. Alonso-delPino, T. Reck, and G. Chattopadhyay, "Design, fabrication and testing of a modulated metasurface antenna at 300 GHz," in *Proceedings of the 11th European Conference on Antennas and Propagation* (EUCAP), 2017, pp. 3416–3418.

48 B. Aqlan, M. Himdi, L. Le Coq, and H. Vettikalladi, "Sub-THz circularly polarized horn antenna using wire electrical discharge machining for 6G wireless communications," *IEEE Access*, vol. 8, pp. 117245–117252, Jun 2020.

49 J. L. Gómez-Tornero, D. Blanco, E. Rajo-Iglesias, and N. Llombart, "Holographic surface leaky-wave lenses with circularly-polarized focused near-fields - Part I: concept, design and analysis theory," *IEEE Trans. Antennas Propag.*, vol. 61, no. 7, pp. 3475–3485, Jul 2013.

50 J. Durnin, "Exact solutions for nondiffracting beams, I. The scalar theory," *J. Opt. Soc. Am. A*, vol. 4, pp. 651–654, 1987.

51 P. Lemaitre-Auger, S. Abielmona, and C. Caloz, "Generation of Bessel beams by two-dimensional antenna arrays using sub-sampled distributions," *IEEE Trans. Antennas Propag.*, vol. 61, no. 4, pp.1838–1849, Apr 2013.

52 Y. C. Zhong and Y. J. Cheng, "Ka-band wideband large depth-of-field beam generation through a phase shifting surface antenna," *IEEE Trans. Antennas Propag.*, vol. 64, no. 12, pp. 5038–5045, Dec 2016.

53 R. M. Herman and T. A.Wiggins, "Production and uses of diffractionless beams," *J. Opt. Soc. Am. A*, vol. 8, pp. 932–942, 1991.

54 J. Durnin, J. J. Miceli, Jr., and J. H. Eberly, "Diffraction-free beams," *Phys. Rev. Lett.*, vol. 58, pp. 1499–1501, 1987.

55 J. Turunen, A. Vasara, and A. T. Friberg, "Holographic generation of diffraction-free beams," *Appl. Opt.*, vol. 27, pp. 3959–3962, 1988.

56 A. Mazzinghi, M. Balma, D. Devona, G. Guarnieri, G. Mauriello, M. Albani, and A. Freni, "Large depth of field pseudo-Bessel beam generation with a RLSA antenna," *IEEE Trans. Antennas Propag.*, vol. 62, no. 8, pp. 3911–3919, Aug 2014.

57 G. Wu, K. F. Chan, W. C. Mok, and C. H. Chan, "3-D printed terahertz lens for Bessel beam generation," in *Proceedings of 2019 IEEE Asia-Pacific Microwave Conference* (APMC), Singapore, Singapore, 2019, pp. 637–639.

58 X. Wei, C. Liu, L. Niu, Z. Zhang, K. Wang, Z. Yang, and J. Liu, "Generation of arbitrary order Bessel beams via 3D printed axicons at the terahertz frequency range," *Appl. Opt.*, vol. 54, no. 36, pp. 10,641–10,649, Dec 2015.

59 N. Kou, S. Yu, and L. Li, "Generation of high-order Bessel vortex beam carrying orbital angular momentum using multilayer amplitude-phase-modulated surfaces in radio frequency domain," *Appl. Phys. Exp.*, vol. 10, no. 1, 2017.

60 D. Comite, G. Valerio, M. Albani, A. Galli, M. Casaletti, and M. Ettorre, "Exciting vorticity through higher order Bessel beams with a radial-line slot-array antenna," *IEEE Trans. Antennas Propag.*, vol. 65, no. 4, pp. 2123–2128, Apr 2017.

61 O. M. Bucci, G. Franceschetti, G. Mazzarella, and G. Panariello, "Intersection approach to array pattern synthesis," *IEE Proc.-H Microwaves Antennas Propag.*, vol. 137, no. 6, pp. 349–357, Dec 1990.

62 O. M. Bucci, G. Mazzarella, and G. Panariello, "Reconfigurable arrays by phase-only control," *IEEE Trans. Antennas Propag.*, vol. 39, no. 7, pp. 919–925, Jul 1991.

63 G.-B. Wu, K. F. Chan, and C. H. Chan, "3-D printed terahertz lens to generate higher-order Bessel beams carrying OAM," *IEEE Trans. Antennas Propag.*, vol. 69, no. 6, pp. 3399-3408. 2021.

64 T. Čižmár, and K. Dholakia, "Tunable Bessel light modes: engineering the axial propagation," *Opt. Express,* vol. 17, no. 18, pp. 15558–15570, 2009.

65 J. A. Encinar, M. Arrebola, F. Luis, and G. Toso, "A transmit-receive reflectarray antenna for direct broadcast satellite applications," *IEEE Trans. Antennas Propag.*, vol. 59, no. 9, pp. 3255–3264, Sep 2011.

66 P. Nayeri, F. Yang, and A. Z. Elsherbeni, "Design and experiment of a single-feed quad-beam reflectarray antenna," *IEEE Trans. Antennas Propag.*, vol. 60, no. 2, pp. 1166–1171, Feb 2011.

67 N. K. Efremidis, and D. N. Christodoulides, "Abruptly autofocusing waves," *Opt. Lett.*, vol. 35, no. 23, pp. 4045–4047, 2010.

68 G.-B. Wu, K. F. Chan, S.-W. Qu, and C. H. Chan, "A 2-D beam-scanning Bessel launcher for terahertz applications," *IEEE Trans. Antennas Propag.*, vol. 68, no. 8, pp. 5893–5903, Aug 2020.

69 G.-B. Wu, K. F. Chan, S.-W. Qu, K. F. Tong, and C. H. Chan, "Orbital angular momentum (OAM) mode-reconfigurable discrete dielectric lens operating at 300 GHz," *IEEE Trans. THz Sci. Technol.*, vol. 10, no. 5, pp. 480–489, Sep 2020.

7

Photonics-Based Millimeter-Wave Band Remote Beamforming of Antenna Arrays Integrated with Photodiodes

Shigeyuki Akiba and Jiro Hirokawa

Tokyo Institute of Technology, Department of Electrical and Electronics Engineering, 2-12-1 Ookayama, Meguro-ku, Tokyo 152-8550, Japan

7.1 Introduction

In recent wireless communication systems represented by long term evolution (LTE)-Advanced and 5G, various use cases such as inter-vehicle communication and the Internet of things (IoT) are expected to expand. It is estimated that the amount of mobile data traffic in the future will become much larger compared with today's data traffic. In order to accommodate such explosive growth of data traffic, it is required to use higher frequencies and to densely install mobile base stations (BSs) covering small areas, which are so-called "small cells" [1, 2]. In such a situation, however, the radio-interference among plural BSs degrades signal quality and the new installation of many BSs results in the increase of the total cost and power consumption of radio access networks.

Beamforming by an antenna array is an effective method to ensure high signal quality against the radio-interference and to realize lower power consumption. There have been many studies of beamforming so far [3–5]. In many cases, however, beamforming techniques need radio frequency (RF) amplifiers to compensate for the RF propagation loss and RF phase shifters to control the relative phases between RF signals to be supplied to each antenna element. In order to implement the beamforming function cost-effectively and power-effectively, a radio-over-fiber (RoF) transmission technique, where RF signals can be provided to antennas through low-loss optical fibers without changing signal format, can be an attractive solution [6–11]. It is possible to design an RoF transmission system so that it has remote antenna beamforming capability as well as remote power feeding capability at a control site (CS). Such a feature may enable us to eliminate RF amplifiers and phase shifters for beamforming at the antenna site (AS) and the equivalent functionalities can be installed at the CS. Thus, it contributes to significant simplification of the AS. Furthermore, when we consider the recent advances of photonic devices, such as semiconductor-based tunable lasers [12] and high-speed/high-output power photodiodes (PDs) [13], such photonics-based radio technologies have a high potential to realize intelligent and sophisticated high-speed radio access systems for mobile communication networks in the future.

We describe in the following text a concept of photonics-based remote beamforming of antenna arrays. This approach yields suitable configurations of antenna arrays in the millimeter-wave band in which each antenna element is integrated with a photodiode; delay control is attained by utilizing the chromatic dispersion (CD) characteristics of optical fibers; and direct delay control is achieved with variable optical delay devices.

Antenna and Array Technologies for Future Wireless Ecosystems, First Edition.
Edited by Y. Jay Guo and Richard W. Ziolkowski.

The rest of this chapter is organized as follows. In Section 7.2, the basic concept and operation principle of photonics-based remote antenna beamforming with the RoF technique are introduced. In Section 7.3, the detailed design and fabrication of a 4 × 2 antenna array integrated with photodiodes and its fundamental beamforming characteristics, including digital signal transmission over the 60 GHz band, are described. Section 7.4 presents a way toward more flexible array patterns and proposes a compact antenna module integrated with a photodiode. Its design and fabrication, as well as its beamforming characteristics at 60 GHz, are shown. In both Sections 7.3 and 7.4, the CD property of an optical fiber is utilized to change the relative RF phases of the RoF signals supplied to the antenna elements. Section 7.5 shows the use of an optical variable delay line (VDL) to control the RF phases of those RoF signals. Utilization of a tunable laser monolithically integrated with an electro-absorption modulator is also demonstrated in Section 7.6. Sections 7.3–7.6 are concerned with techniques handling the main traffic carrying data of users and relevant perspectives are described in Section 7.7. In Section7.8, we describe an estimation of the direction of user equipment (UE), which is necessary for effective beamforming. Section 7.9 summarizes the chapter with some prospective remarks.

7.2 Configuration of Photonics-Based Antenna Beamforming Utilizing the RoF Technique

7.2.1 Basic Concept and Configuration

Figure 7.1a illustrates the basic concept of antenna arrays integrated with photodiodes for RF beamforming. RoF signals are transmitted through optical fibers and are converted to RF signals by photodiodes and they are directly fed to the antenna array elements. A photodiode and an antenna array element are supposed to be integrated as schematically shown in Figure 7.1b,c. A 4 × 4 microstrip antenna array as an example is depicted in Figure 7.1b, while Figure 7.1c is an illustration of the cross section of one of those antenna elements. A microstrip antenna and a PD chip are mounted on the same substrate. The RoF signal carried through an optical fiber will be fed to the PD chip directly. Actual implementation will be presented in the later sections. Carrying RF power through an optical fiber in the form of RoF is very meaningful in the millimeter-wave band because the loss of a coaxial cable is very large, while the loss of an optical fiber is negligibly small. In addition, because of the short wavelengths in the millimeter waveband, the overall size of the integrated device could be small even with many array elements on the substrate. Each RF signal generated

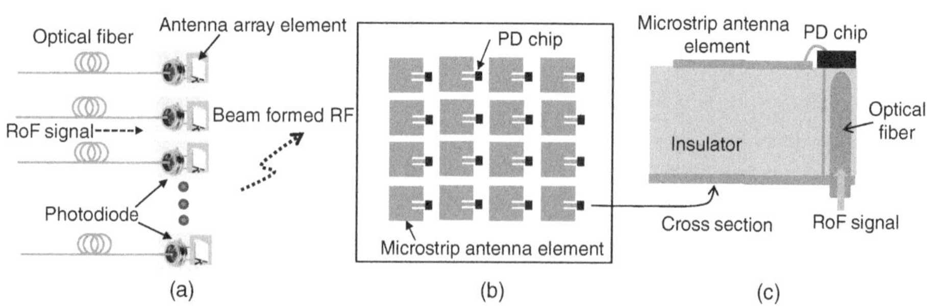

Figure 7.1 Antenna arrays integrated with photodiodes for RF beamforming. (a) Basic concept. (b) Schematic illustration of a 4 × 4 microstrip antenna array integrated with PD chips. (c) Cross section of an antenna element.

by a photodiode is required to have a different relative phase for beamforming. How one realizes such a different RF phase is presented and explained in the following text in detail. Beamformed RF is radiated from the antenna array to a specific direction that is determined by the relative RF phases of the antenna array elements as inferred in Figure 7.1a.

7.2.2 Generation of RoF Signals with Different RF Phases to be Fed to Antenna Element

It is straight forward to use a VDL to change the RF phase of the RoF signal as shown in Figure 7.2. An intensity modulator (IM) copies an RF downlink signal onto the optical power emitted from the light source (LS) and generates an RoF signal. The use of VDL, which could be constructed by a movable corner cube as shown in the inset, is very attractive in the millimeter wavelength range RF application. For example, the wavelength of a 60 GHz RF signal is 5.0 mm and therefore a change in the optical path of 2.5 mm corresponds to a phase change of 180° in the RoF signal. It means that we can build a very compact VDL. On the other hand, the physical delay change is usually accomplished by mechanical equipment such as a motor. We may then have some restriction on how quickly the beamforming can be carried out. If a VDL is realized that does not rely on mechanical components, for example, by making use of piezoelectricity, and it works in practice, such restrictions will be eliminated.

Another useful method to achieve a phase change is to utilize the CD characteristics of an optical fiber, which induces a wavelength dependence of the propagation delay of an RoF signal. The different propagation group delays caused by the CD are schematically shown in Figure 7.3. A standard single mode fiber has finite CD values in the 1.5 μm range as depicted in the inset. The light output from a tunable light source (TLS) is modulated by the downlink RF signal through an IM and is transmitted through an optical fiber. When the wavelength λ_1 is changed to λ_2, the phase of the RoF signal arriving at the photodiode is shifted due to the CD characteristics. The wavelength change $\Delta\lambda$ required for the RF phase shift of π is given by the relation:

$$\Delta\lambda = 1/(2LDf_{RF}) \tag{7.1}$$

where L is the length of the optical fiber; D is the CD parameter and is typically about 17 ps/nm/km for a standard single mode fiber (SML); and f_{RF} is the RF frequency. The wavelength change in the TLS is made electrically and we are then able to have much faster beamforming compared with the previous optical VDL method.

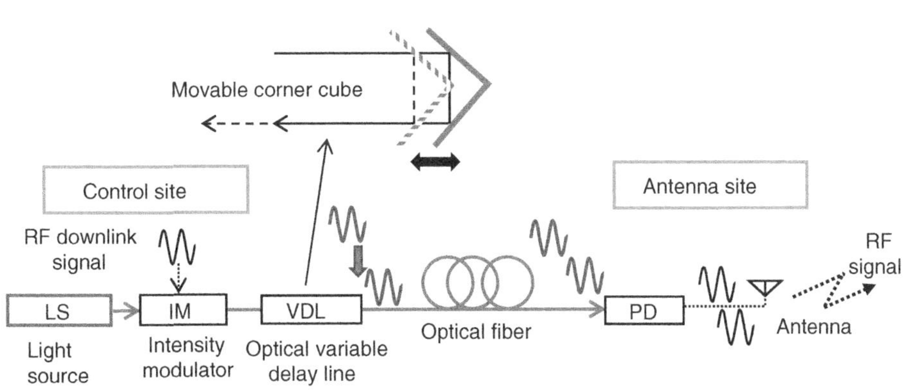

Figure 7.2 Optical variable delay line (VDL) to change the RF phase of the RoF signal.

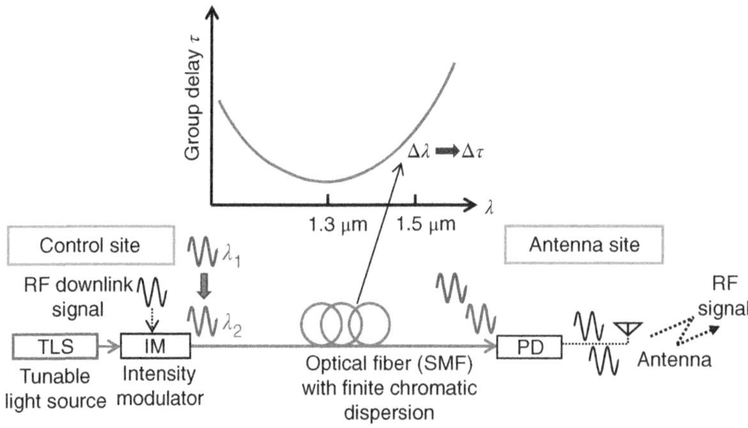

Figure 7.3 Chromatic dispersion of an optical fiber induces a phase shift to the RoF signal when the wavelength is changed.

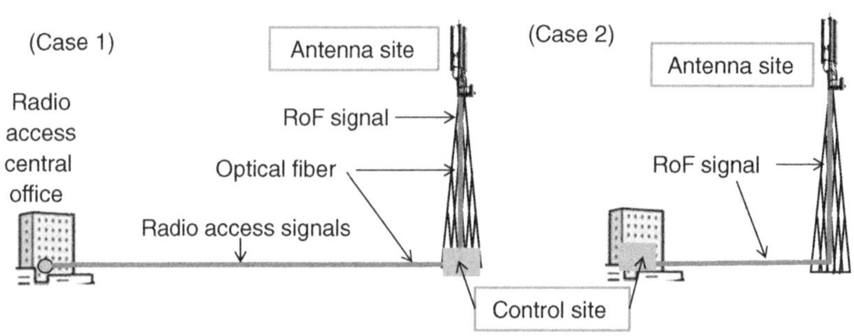

Figure 7.4 Assumed location of the "control site" as well as the "antenna site."

7.2.3 Transmission of Multiple RoF Signals with Different RF Phases

We consider transmission of RoF signals from a "control site (CS)" to an "antenna site (AS)" through optical fibers. The AS is trivial and is the place where antennas for RF radiation, specifically integrated with photodiodes in the present case, are located. On the other hand, the CS in this case means a place where multiple RoF signals with controlled RF phases are generated and launched into the optical transmission fibers. It can be located at a place close to the AS, for example, in a cubicle set up underneath an antenna tower or in a basement of a building where the AS is located. It can also be located a little bit further away from the AS, possibly in a remote radio access central office. Figure 7.4 illustrates the two cases. (Case 1) corresponds to the former, and (Case 2) corresponds to the latter. We think that (Case 1) is a normal case in which various radio access signals including baseband (BB) information are transmitted from the radio access central office to the CS.

In both cases multiple RoF signals have to be transmitted through optical fibers. The following three kinds of fiber technologies are candidates to accomplish this task:

(1) *Bundled fiber:* An easy and simple option would be to use a bundled fiber in which many fibers are compactly bundled in the form of a cable. As the number of fibers is increased, the overall volume, weight, and cost are also increased, although recent massive deployments of optical fibers all over the world are improving the situation and alleviating such drawbacks. It is

considered that such a bundled fiber can be used mainly in (Case 1) where the length between the CS and the AS would be within the range of a few tens of meters or a hundred meters.

(2) *Multi-core fiber* [14]: A multi-core fiber (MCF) is another candidate. As an evolving technology, some MCFs with four or seven cores having the same diameter as a conventional optical fiber are already available for practical applications with proven reliability. The main purpose to use an MCF is to increase the transmission capacity while keeping the installation space the same as the case using a single core fiber. Although the diameter of a fiber is desired to be unchanged, it is considered to be acceptable to have a much larger number of cores in a fiber with a larger diameter compared with the conventional optical fiber for many applications. This perspective is especially valid when we consider that the length between the CS and the AS in (Case 1) may be at most a hundred meters and in (Case 2) a few kilometers.
(3) *Wavelength division multiplexing (WDM) transmission:* A more photonic method would be to utilize WDM, in which different RoF signals are transmitted through different wavelengths in an optical fiber. The WDM scheme is rather well known as frequency division multiplexing (FDM) in the field of a radio technology, and it is already an established technology in the large capacity long distance optical fiber transmission. It is likely that some advanced WDM devices will be required and may bring about some cost increase.

We have studied all three methods and have demonstrated that the beamforming functionality was realized as theoretically predicted. They are explained in the following sections and can also be found in the published papers [15–19], including some relevant technical items.

7.3 Two-Dimensional (4 × 2) Antenna Array Integrated with Photodiodes for 60 GHz Band Applications

7.3.1 Design and Fabrication

At first, we describe a 4 × 2 antenna array for 60 GHz band application. Figure 7.5a,b are illustrations of the overall structure of such an integrated photonic array-antenna (IPA). The pictures are shown in Figure 7.6a,b. The IPA consists of a PD-integrated 60 GHz-band microstrip antenna substrate (Figure 7.5a) and an optical signal feeding jig (Figure 7.5b). In Figure 7.1, the microstrip antenna and the PD are supposed to be fabricated on the same surface of the substrate. However, in this case the PD is mounted on the surface opposite to the microstrip antenna. The details are explained in the following text.

The detailed structure of the PD-integrated antenna substrate is depicted in Figure 7.7. It consists of eight 60 GHz-band microstrip antennas, an antenna substrate, chip carriers, co-planar lines, PD bias lines, and PD chips. On the antenna side, there are 4 × 2 microstrip antennas. The separation between each adjacent antenna element is 2.5 mm (0.5 λ_{RF} at 60 GHz) in the direction of the four microstrip antennas and 3.0 mm (0.6 λ_{RF} at 60 GHz) in the direction of the two microstrip antennas. The size of each antenna element is 0.6 mm × 2.1 mm, and the feed point is shifted by 0.27 mm from its center toward its long edge. Wideband applications such as multi-gigabit transmissions are expected at 60 GHz and, therefore, it is desirable for an antenna to have a wide bandwidth. Compared with a conventional square microstrip antenna, the bandwidth is increased as the difference between the length and the width of the antenna increases, but with a loss in the sharpness of the resonance. In the present case, we adopted the rectangular shape aiming for a larger bandwidth.

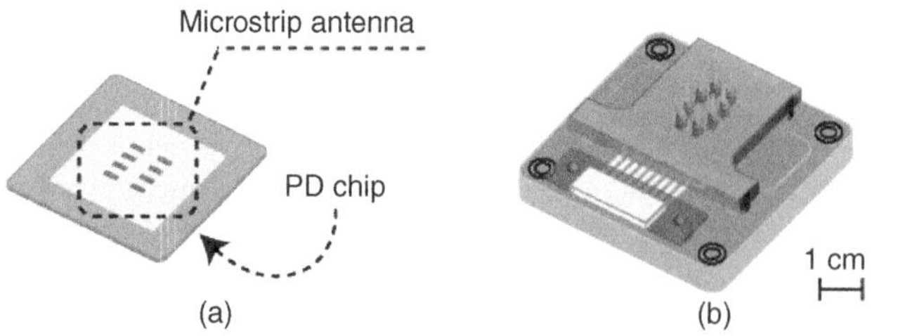

Figure 7.5 Structure of the integrated photonic array-antenna (IPA). (a) PD-integrated antenna substrate. (b) Jig for the optical signal feed.

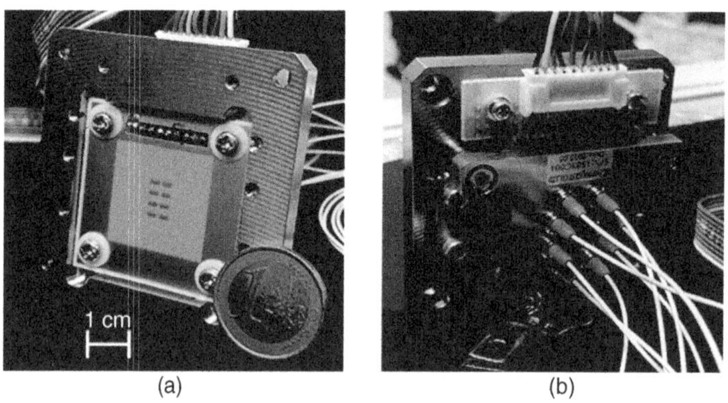

Figure 7.6 Photographs of the fabricated IPA. (a) RF output side. (b) Optical input side.

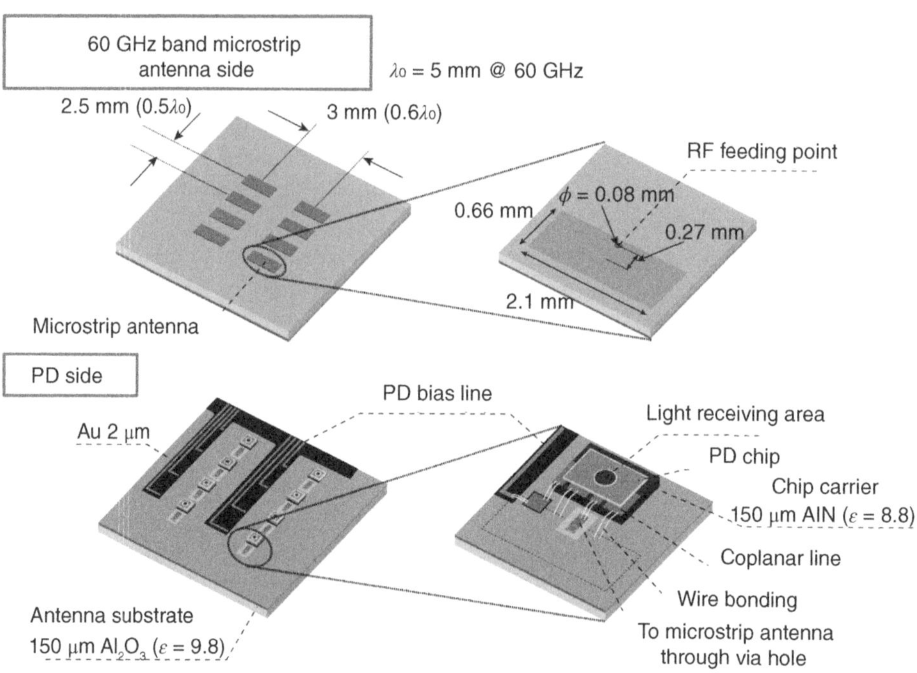

Figure 7.7 Detailed structure of the PD-integrated antenna substrate.

On the PD side, the chip carrier is made of aluminum nitride (AlN), which has a dielectric constant ε of 8.8. It is mounted on the microstrip antenna's ground (GND) side, which is gold (Au). A co-planar line adjusted to 50 Ω at 60 GHz is formed on the chip carrier. The PD chip is bonded on each chip carrier via a co-planar line by a flip chip bonding method. The co-planar line and antenna feeding point are connected electrically by wire-bonding. The feed point on the PD side is connected electrically to the microstrip antenna on the antenna side through a via-hole.

Figure 7.8 shows the characteristics of the array-antenna designed with the ANSYS high frequency structure simulator (HFSS), which is a three-dimensional electromagnetic field simulator for antenna engineering. In the figure we define the A-plane, which contains the direction perpendicular to the antenna surface and the direction of the shorter size of the antenna. The B-plane contains the direction perpendicular to the antenna surface and the direction of the longer size of the antenna. Figure 7.8a,b show the realized gain when all 4 × 2 RF signals are driven in an in-phase condition. Figure 7.8a shows the realized gain at 60 GHz as a function of the radiation angle, i.e. the realized gain pattern of the antenna. Figure 7.8b shows the RF frequency response in the front direction of the antenna substrate. As shown in Figure 7.8b, the phased antenna array is designed properly to operate in the 60 GHz band. We note that the direct measurement of the frequency response of an IPA, e.g. Figure 7.8b, is not easy to obtain because its RF feed includes the electrical to optical (E/O) conversion by an IM and the optical to electrical (O/E) conversion by a PD.

Figure 7.9 shows the structure of the jig that facilitates the optical signal feed. It consists of optical fibers, prisms, and lenses. The optical signal through a fiber is incident on each prism and lens combination and is focused onto the light receiving area of the PD to produce an RF signal for each antenna element. The coupling efficiency between a PD and an optical fiber output in an actual device was 70% or more.

Figure 7.10 shows the dependence of the bias voltage V_{bias} of the photodiode used in the IPA shown earlier, when one of its eight elements was used for the measurement. Figure 7.10a shows the measurement setup. The output from an LS at the wavelength of 1.55 μm was modulated by a lithium niobate (LN) based IM at the RF frequency of 60 GHz and was then boosted by an erbium-doped fiber amplifier (EDFA). The PC is a polarization controller. The input power to the photodiode P_{in} was changed by the variable optical attenuator (VOA) whose adjustable parameter was the bias voltage applied to the photodiode. The RF power P_{RF} was received by a horn antenna and its output was measured by a spectrum analyzer (SA). Because a uni-traveling-carrier photodiode (UTC-PD) [13] has the unique property that only electrons are involved in the high

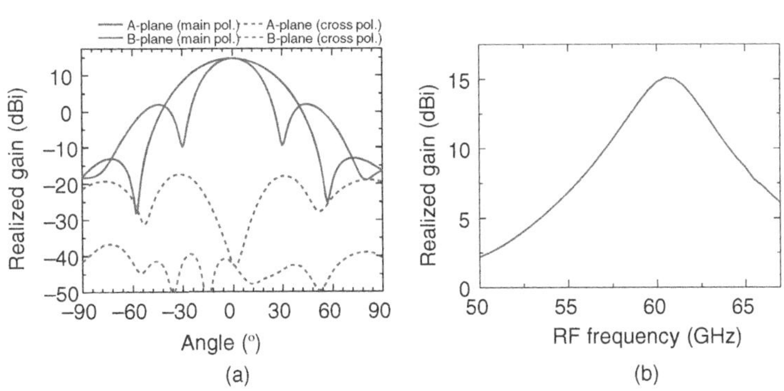

Figure 7.8 Realized gain of the 4 × 2-element array designed with HFSS. (a) Radiation pattern. (b) RF frequency response. The maximum gain is 15.1 dBi at 60.7 GHz and the 3 dB bandwidth of the realized gain is 7.4% (58.4–62.9 GHz).

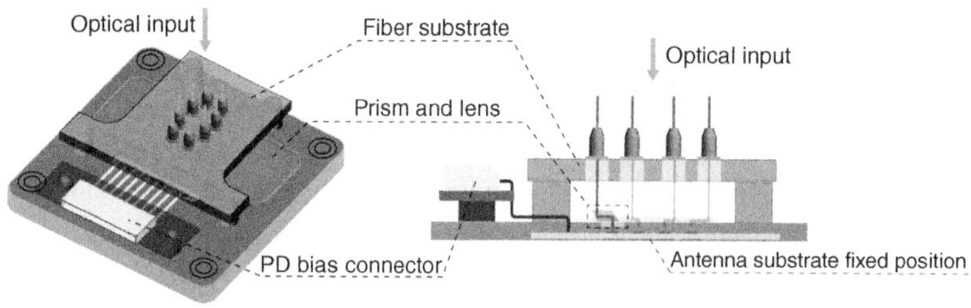

Figure 7.9 Structure of the jig facilitating the optical signal feed.

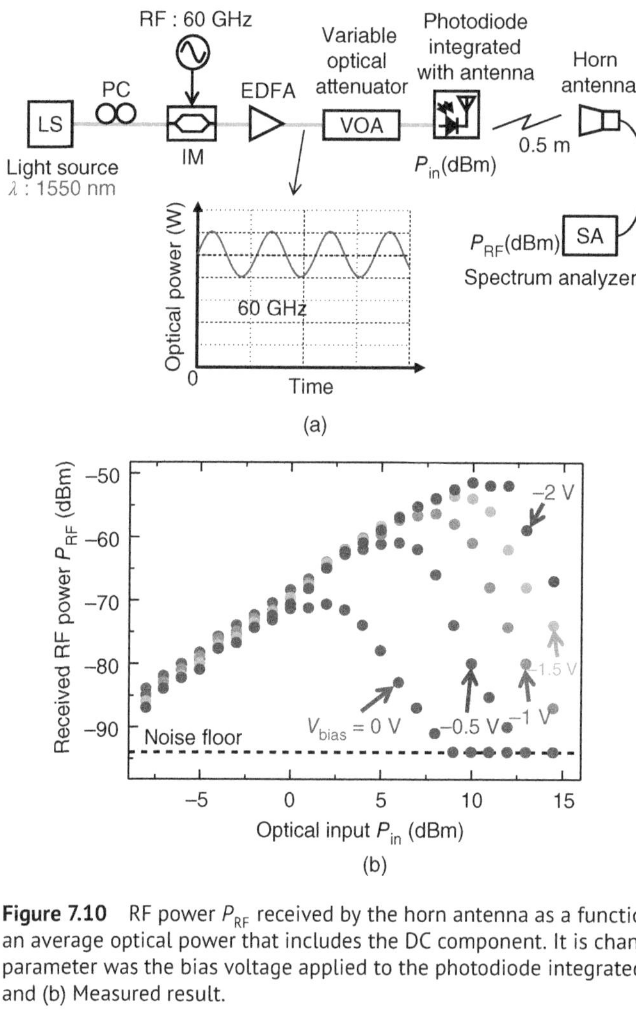

Figure 7.10 RF power P_{RF} received by the horn antenna as a function of the optical input power P_{in}. P_{in} is an average optical power that includes the DC component. It is changed by the VOA whose adjustable parameter was the bias voltage applied to the photodiode integrated with antenna (a) Measurement setup and (b) Measured result.

frequency photo-response, it has a high responsivity in the millimeter waveband. It is noted that the RF signal can be generated from the photodiode even without the bias voltage, i.e. $V_{bias} = 0$. This feature is due to the built-in potential formed in the photodiode. As the bias voltage is increased to the photodiode, the RF power is increased. It is considered that a bias voltage of −2.0 V is almost sufficient to have a high O/E conversion efficiency with a high speed response.

This property may be understood from the following explanation. When the input power to the PD is 10 dBm, i.e. 10 mW, the photo-current is 10 mA assuming a typical responsivity of 1.0 A/W for a PD in the 1.5 μm range. The resultant voltage induced across the 50 Ω load is then 0.5 V. The actual voltage may be smaller than that because the RoF signal contains a DC component as depicted in Figure 7.10a and the 60 GHz component is smaller than 10 mW.

7.3.2 Beamforming Utilizing WDM RoF Transmission

Figure 7.11 shows the experimental setup of the demonstration of 60 GHz antenna beamforming in an RoF system using the IPA presented earlier. Two optical signals with the wavelengths 1550 and 1545 nm were generated by the TLSs and were simultaneously amplitude-modulated by an IM at 60 GHz. The input RF power was set to be 10.0 dBm. After being boosted by an EDFA, the RoF signals were propagated over a 2 km-long SMF that had a typical CD parameter of 16–18 ps/nm/km in the wavelength range of 1.5 μm. The two RoF signals were demultiplexed and divided into four signals in each wavelength, and then fed to the IPA. In order to compensate for an initial RF phase difference between the RF signals radiated from each antenna element, VDLs were inserted just before the integrated photonic antenna. The initial RF phase offset could also have been compensated by tuning the wavelength of the LS. Hence, the VDLs will not be necessary when the system is configured so that the RF signal to each antenna element is fed through a distinctive LS.

The eight RoF signals fed to the integrated photonic antenna were in-phase at the initial condition ($\lambda_1 = 1550$ nm, $\lambda_2 = 1545$ nm). We measured the RF received power at the observation angle $\theta = 0°$, while λ_1 was varied from 1549.5 to 1550.5 nm. In order to obtain an RF phase shift of π, the wavelength variation was theoretically estimated to be 0.25 nm from Eq. (7.1).

Figure 7.12 schematically shows the arrangement of the IPA as a transmitter and the horn antenna as a receiver. The beam steering was performed by changing the wavelength on the A plane. As shown in Figure 7.12, the RoF signal λ_1 was assigned to the left four microstrip antenna elements and λ_2 was assigned to the right four microstrip antenna elements. We note that the UTC-PDs were operated with zero bias voltage for simplicity. Therefore, no power supply to the AS was required. Although beam steering in this experiment was operated in one dimension as an initial study, we are also expecting two-dimensional beam steering by assigning different RoF signals to each antenna element.

Figure 7.13 shows the result of the beamforming operation. The gray dots show the measurement results as a function of $\Delta\lambda$. The solid line is the theoretical curve of the array factor calculated from the following formula:

$$\sum_{k=1}^{N} \exp\left\{-j\left(\frac{2\pi}{\lambda_{\mathrm{RF}}} d_k \sin\theta + \beta_k\right)\right\} \tag{7.2}$$

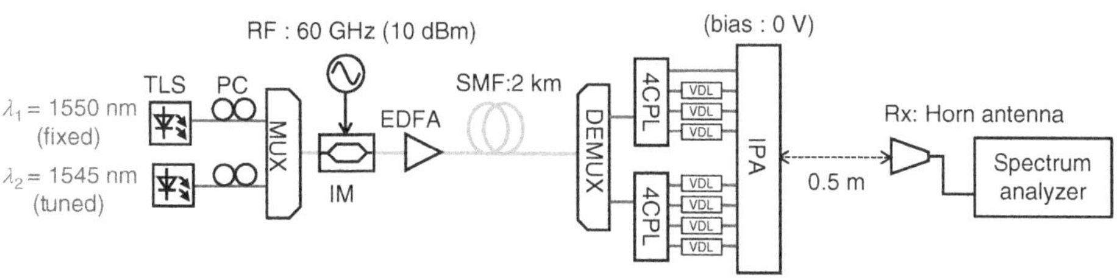

Figure 7.11 Remote antenna beamforming system utilizing WDM transmission of two RoF signals.

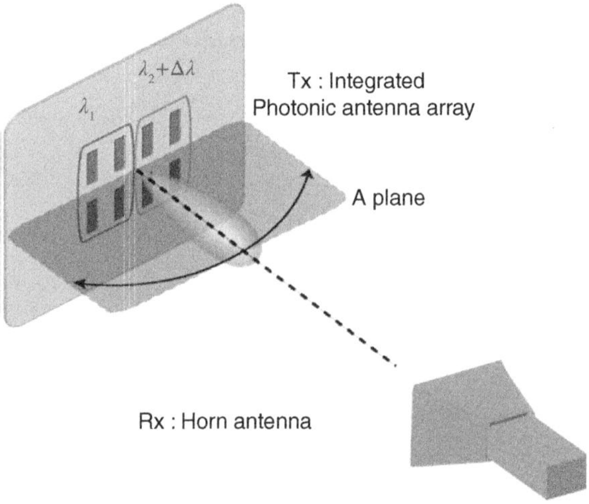

Figure 7.12 Schematic illustration of the measurement of the RF received power.

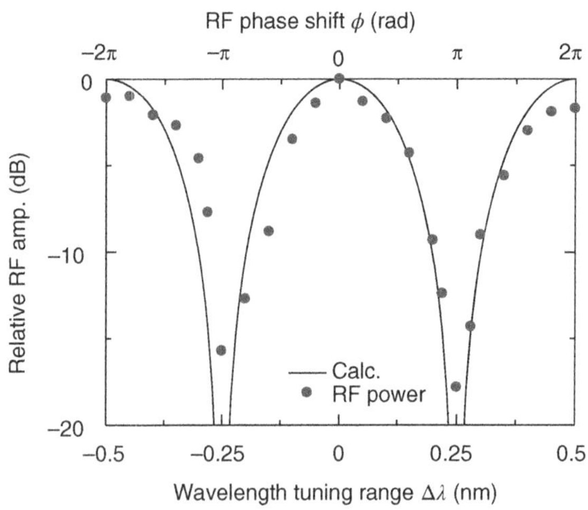

Figure 7.13 Received RF power in the observation angle $\theta = 0$ with changing the wavelength.

where λ_{RF} is the wavelength of the RF carrier, d_k is the separation between the neighboring antenna elements, θ is the observation angle, and β_k is the relative RF phase shift between the RF signals into each antenna element. In the present case β_k is derived as $\pi\ \Delta\lambda/0.25$, i.e. $\beta_1 = \beta_2 = 0$ for $\Delta\lambda = 0$, and $\beta_1 = 0$ and $\beta_2 = \pi$ for $\Delta\lambda = 0.25$ nm. We notice that the measured values almost agree with the calculated array factor. Although we expected that the relative RF amplitude would recover to the same value as its initial condition at the wavelength tuning range of +0.5 and −0.5 nm, it was slightly decreased. Because of the measurement circumstances, we had to use optical components with different lengths of the pigtail fibers. We have deduced that the reason why the relative RF amplitude does not recover its initial condition is mainly due to the differences in those pigtail fiber lengths.

Figure 7.14 shows the measured RF radiation patterns at $\Delta\lambda_{opt} = 0$ and 0.25 nm. In the case of $\Delta\lambda_{opt} = 0$ nm, the relative RF power was maximum at the observation angle 0°. In the case of $\Delta\lambda = 0.25$ nm, the maximum was at the angle 30°.

The measurement results almost coincide with the theoretical lines in Figures 7.13 and 7.14. Thus, the beamforming operation utilizing the proposed antenna array was confirmed.

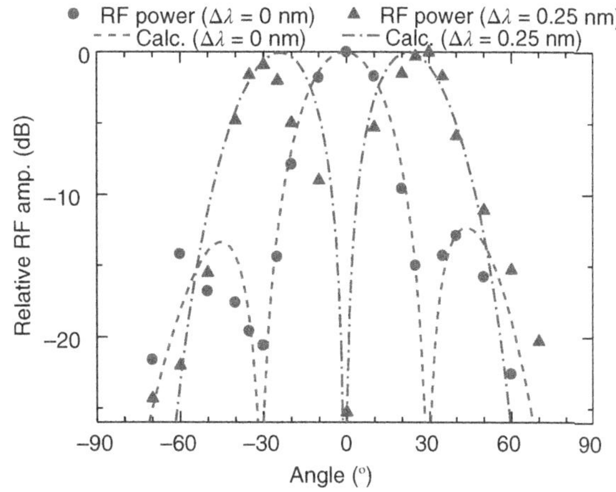

Figure 7.14 RF radiation pattern. The 3 dB radiation bandwidth is 26° (−14° to 12°) when $\Delta\lambda = 0$ nm.

7.3.3 Fading of an Amplitude Modulated RoF Signal Due to Chromatic Dispersion

When an RoF signal is generated by means of amplitude modulation, as described earlier, and has double sidebands, the CD induces a relative phase shift between the two sidebands of the RoF signal. This causes a fading effect, i.e. a reduction of the RoF signal. The RF output power P_{RF} taking into account the relative RF phase shift between the double sidebands is given by the following equation [20, 21]:

$$P_{RF} \propto \cos^2\left(\frac{\pi L D \lambda_o^{\,2} f_{RF}^{\,2}}{c}\right) \tag{7.3}$$

where λ_o is the wavelength of the optical carrier, c is the light velocity in the SMF, L is the fiber length, and D is the CD parameter. The RF modulation components are canceled out when the phase shift reaches π. Therefore, it is important to consider the issue mentioned earlier when we design a beam steering system.

Figure 7.15 shows the CD effect, i.e. the fading effect on an RoF signal amplitude in the 1.5 μm wavelength range. The experimental setup is shown in Figure 7.15a for transmission through SMFs with lengths of 1, 2, and 3 km when the EDFA was used to boost the optical signal. Optical spectra without and with the optical filter are also presented. It is clear that the optical filter eliminated the longer wavelength side band and yielded a single side band. The relative RF power as a function of the fiber length is shown in Figure 7.15b. The solid and broken lines are the results calculated from Eq. (7.3) for the RF frequency of 60 and 30 GHz, respectively.

The experimental results at 60 GHz that were measured for the fiber lengths of 0, 1, 2, and 3 km are also plotted and almost agree with the calculated curve. Considering a standard single mode fiber for the CD parameter, 17 ps/nm/km, it was adopted for D. Since the fiber length was 2 km and the RF frequency was 60 GHz in the experiment shown in Figure 7.11, the fading effect of the relative RF power is almost negligible and the relative amplitude is almost the same as the case for the fiber length of 0 km.

Figure 7.15 indicates that when we consider the (Case 2) in Figure 7.4 as the configuration of the RoF signal transmission, careful attention has to be paid to the choice of the fiber length and RF frequency. If it was the (Case 1), i.e. if the separation between the CS and the AS was small, for example, less than 100 m, then the reduction of the RF power can be ignored. If we use the wavelength range of 1.3 μm, D is very small as depicted in Figure 7.3 and the fading effect is negligible.

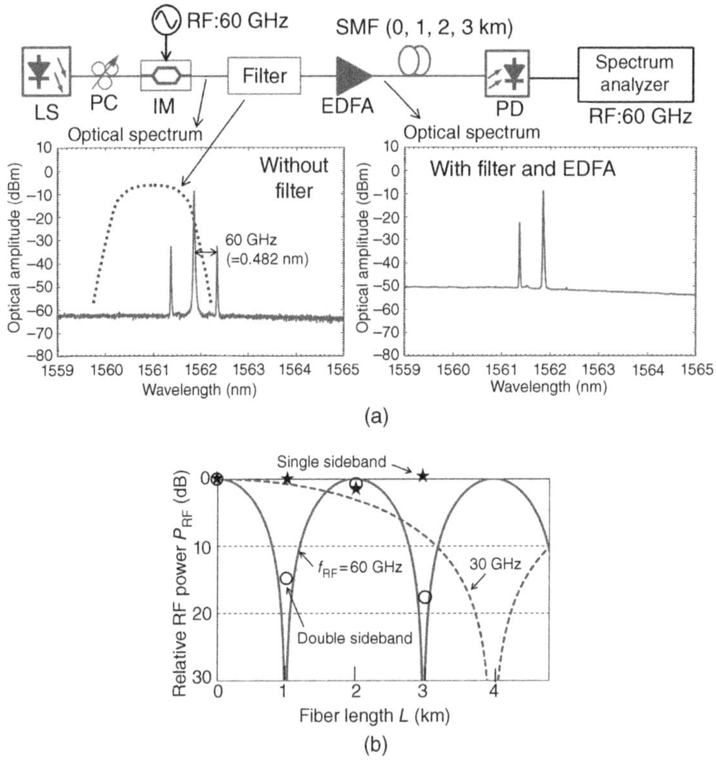

Figure 7.15 Chromatic dispersion effect on the RoF signal amplitude in the 1.5 μm range. (a) Experimental setup for the measurement of the effect and the spectra for double sideband (without filter) and single sideband (with filter and EDFA) transmission through a SMF whose lengths were 1, 2, and 3 km. (b) Relative RF power calculated from Eq. (7.3) as a function of the fiber length for the RF frequency of 60 GHz (solid line) and 30 GHz (broken line) as well as the plots for 60 GHz double side band and single side band transmission in fiber lengths of 0, 1, 2, 3 km.

If we can generate an RoF signal with single sideband instead of double sideband as demonstrated earlier, the fading effect would also be negligible.

7.3.4 Digital Signal Transmission at 3.5-Gbit/s

We describe the digital signal transmission experiment that was carried out in order to confirm the feasibility of IPA-based beamforming.

Figure 7.16 shows the experimental setup to measure the 3.5 Gbit/s quaternary phase shift keying (QPSK) signal transmission of the fabricated IPA with antenna beamforming. The optical path is basically the same as in Figure 7.11. The RF carrier frequency emitted from the IPA was set to be 60.48 GHz, which corresponded to the center frequency of channel 2 in the standardized 60 GHz band [22]. In the experiment, two RoF signals with different optical wavelengths, λ_1 and λ_2, were used as the LSs. A 1.76 Gbaud QPSK BB signal with the intermediate frequency (IF) of 4 GHz was generated by an arbitrary waveform generator (AWG). It was then converted up to 60.48 GHz by being mixed with a 56.48-GHz RF tone signal from a local oscillator (LO). After passing through a band-pass filter (BPF), which was inserted to reject the 56.48 GHz RF tone signal,

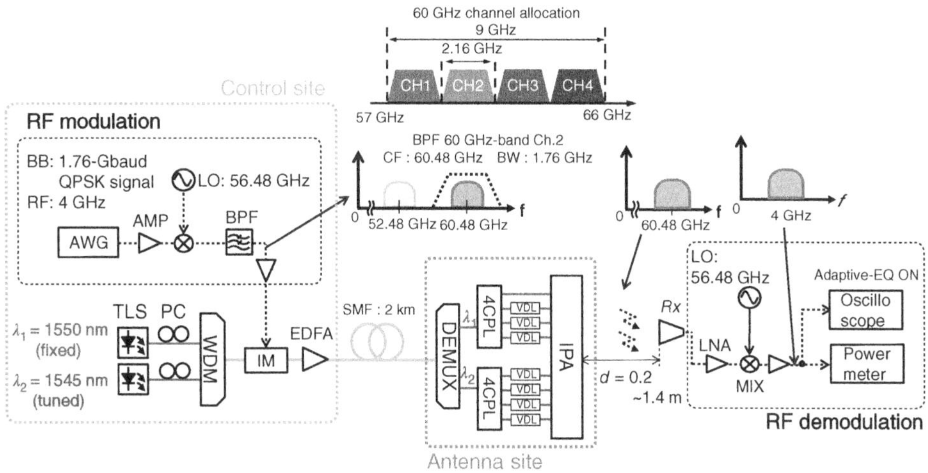

Figure 7.16 Experimental setup to measure the 3.5-Gbit/s QPSK signal transmission.

the 60.48 GHz QPSK signal was routed to an IM. Two optical signal outputs from the TLSs, whose wavelengths were $\lambda_1 = 1550$ nm and $\lambda_2 = 1545$ nm, were multiplexed by a WDM coupler and amplitude-modulated by the IM at the 60.48 GHz RF signal. After being boosted by an EDFA, the two RoF signals were propagated over a 2 km long SMF and de-multiplexed to each wavelength by a WDM coupler in the AS. The two RoF signals were then divided into four signals by optical couplers (CPLs) and fed to each antenna element of the IPA.

In order to compensate for an RF phase shift due to the difference of the two initial optical wavelengths, seven optical VDLs were inserted just before the IPA. Recall that such an initial RF phase offset can also be compensated by tuning the wavelength of the LS, thus making the VDLs unnecessary when the system is configured so that the RF signal to each antenna element is fed through individualized LSs. The RoF signals were converted to the 60.48 GHz QPSK RF signals by the PDs of the IPA, and the RF signal was radiated from the antenna array. It should be noted that the experiment was conducted without supplying any bias voltages to any of the PDs, i.e. it was conducted with a fully passive condition for the AS. After 0.2–1.4 m long radio propagation, the 60.48 GHz RF signal was received by a horn antenna, and then converted down to 4 GHz. The 4 GHz RF signal was injected into the oscilloscope and the RF power meter to measure the signal-to-noise ratio (SNR) of the QPSK signal and the received RF power, respectively. The measured SNR was calculated from the error vector magnitude (EVM) value by the software installed in the oscilloscope.

Figure 7.17a,b shows the RF radiation patterns at $\Delta\lambda = 0$ and 0.25 nm, respectively, where the change of the relative RF power is plotted against the observation angle. The constellations for the corresponding SNR and the EVM values are also shown in Figure 7.17. In the case of $\Delta\lambda = 0$ nm, the RF power was maximum at the observation angle of 0°. It changed to a null value in the case of $\Delta\lambda = 0.25$ nm. The transmission length L and the actual CD parameter D of the SMF between the CS and the AS were 2.0 km and 16.7 ps/nm/km, respectively. From these parameters, the $\Delta\lambda$ required to obtain a π RF phase shift was theoretically estimated to be 0.25 nm from Eq. (7.1). The broken lines in Figure 7.17a,b were calculated by HFSS. The measured RF powers and SNRs almost coincide with the dotted lines except for the measurement points where the received RF signals were too small to be clearly discriminated from noise. The constellations were also changed according to the observation angle. A SNR of more than 9.8 dB, which corresponds

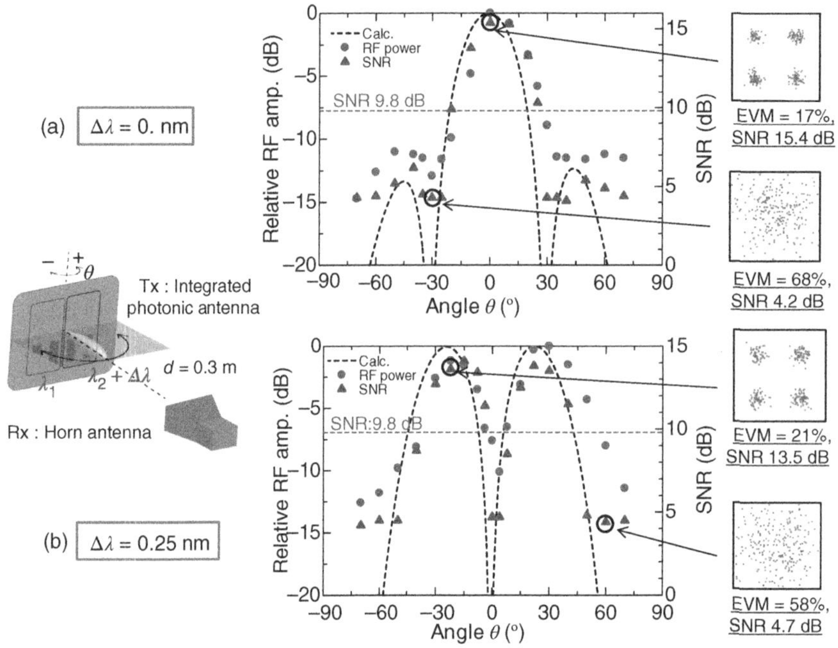

Figure 7.17 RF radiation patterns with corresponding SNRs when the IPA beam is controlled (a) $\Delta\lambda = 0$ nm and (b) $\Delta\lambda = 0.25$ nm.

to the bit error rate of 1.0×10^{-3} that is required for error-free transmission with a low-density parity-check (LDPC) code, was successfully obtained. Thus, the signal quality of the 3.5 Gbit/s QPSK signal in the IPA-based beamforming operation was experimentally confirmed.

7.3.5 Wireless Communication Range

As shown in Figure 7.16, the distance between the IPA as a transmitter and the horn antenna as a receiver ranges from 0. 2 to 1.4 m. However, from a practical point-of-view it is obvious that the distance should be greater, as long as a 100 m or more even for small cell systems.

Figure 7.18 shows the change of the received RF powers and associated SNRs for the 3.5-Gbit/s QPSK signal as the RF propagation distance between the antenna array and the receiver horn antenna is changed from 0.2 to 1.4 m. The achievable RF propagation distance in this experimental setup was about 1.0 m considering the required SNR of 9.8 dB. In order to extend the propagation distance to 10 or 100 m, an RF propagation loss of another 20 and 40 dB must be compensated.

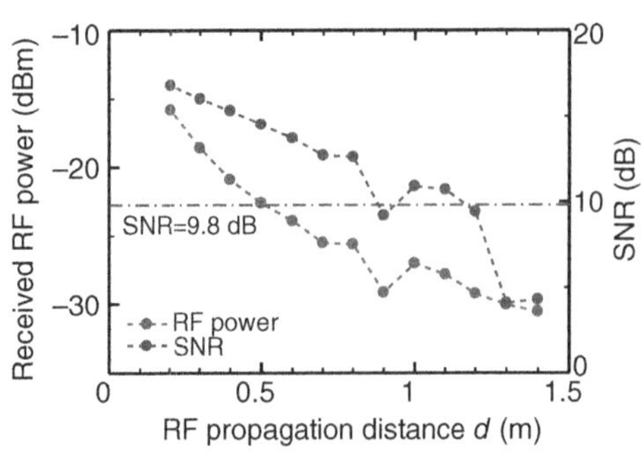

Figure 7.18 Change of the received RF power and SNR as functions of the RF propagation distance.

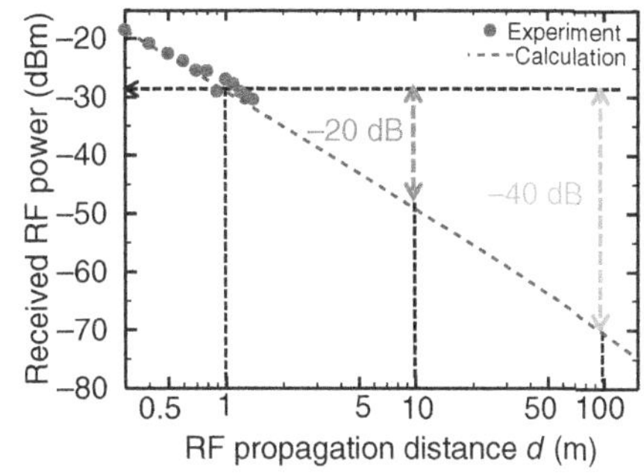

Figure 7.19 Calculated result of the received RF power as a function of the RF propagation distance.

We can reasonably argue that the transmission range can indeed be extended to 100 m or more. In the experiments described earlier, the received RF power was restricted by the performance of the IMs and PDs. The 60 GHz frequency response of the IM used in those experiments was actually worse by 11 dB than that obtained in the lower frequency band. In the future, we may be able to use much better modulators whose 3-dB bandwidths will be as large as 60 GHz [23]. In that case, we can improve the frequency response by 8 dB. In addition, the O/E conversion efficiency of the UTC-PD used in this experiment was as low as 0.3 A/W at 60 GHz. If the PD is optimally designed and engineered, the efficiency of the O/E conversion can be improved to be twice as large as the present value [24]. As a consequence, we may be able to achieve a 5 dB improvement for the RF signal component. Moreover, the RF output power from the UTC-PDs can be higher when we utilize the latest devices, where the RF output power can be as large as −6 dBm without any bias voltages and +12 dBm with the bias voltage [13]. As the output power from the UTC-PD in the experiment was about −30 dBm at zero bias voltage and −10 dBm with the bias voltage, we will be able to improve the undesired loss by 37 dB (= 8 + 5 + 24) at zero bias voltage and 55 dB (= 8 + 5 + 42) with the bias voltages. Therefore, as shown in Figure 7.19, the RF propagation distance could be extended to 20 m at zero bias voltage, and 160 m with the bias voltages. Much longer distances can be attained when the number of antenna elements in the IPA is increased.

7.4 Compact Antenna Module Integrated with Photodiodes to Achieve a Flexible Array Pattern

The IPA described in Section 7.3 is advantageous in terms of its size and space because it is integrated in a single small board. However, it may be difficult to arrange a number of these arrays arbitrarily and create the beam pattern flexibly because the number of antenna and PD arrays is fixed uniquely. This is especially true because the structure of the jig for the optical signal feed shown in Figure 7.9 is restricted to the $(m \times n)$ arrangement. Our preliminary study of the structural design has revealed that the arrangement $(m \times 3)$, where m is 3 or larger, is difficult to fabricate, while $(m \times 2)$ is feasible as shown in Figure 7.9. We do not exclude the possibility that in the future we can build an arbitrary arrangement by realizing a much more compact optical signal feed jig with improved designs by incorporating a lens and a prism. We propose in this section a novel 60-GHz compact antenna module in order to alleviate the aforementioned difficulties.

7.4.1 Design and Fabrication of an Integrated Photonic Antenna Module

Figure 7.20 shows a schematic overview of the 60 GHz compact antenna module as well as a photograph of the whole module with the optical fiber being present. The module consists of the antenna

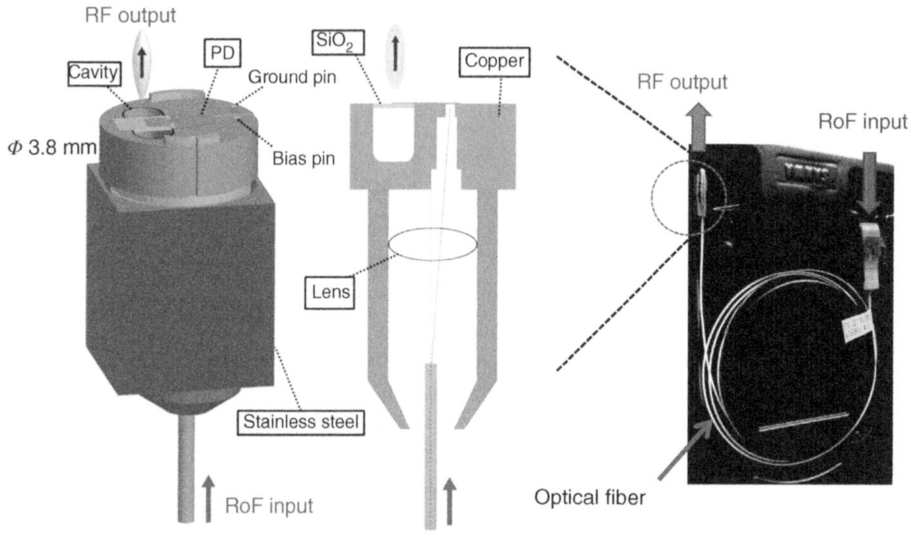

Figure 7.20 Schematic overview of the 60-GHz compact antenna module and a photograph of the whole module including the optical fiber.

element, which is basically a cavity antenna, integrated with a UTC-PD and an optical fiber. The RoF signal through the fiber is focused onto the light receiving area of the PD chip by a lens. The output RF signal from the PD chip is fed to an RF strip line formed on the top surface of the aperture of the cavity antenna.

Figure 7.21 shows a photograph and an illustration of the top surface of the antenna module. The diameter of the module is 3.8 mm. The RF strip line to connect the PD to the aperture is a coplanar line formed on a SiO_2-based dielectric substrate. The dielectric substrate is buried by 0.1 mm in the top surface and stretched out to the other side of the aperture so that the coplanar line cannot slip off. The aperture is a cavity of 1.0 mm in the *x*-axis direction, 2.5 mm in the *y*-axis direction, and 1.2 mm in the *z*-axis direction, respectively. It is designed with the resonance in the 60 GHz band and with the RF signal being radiated toward the normal of the aperture (front direction).

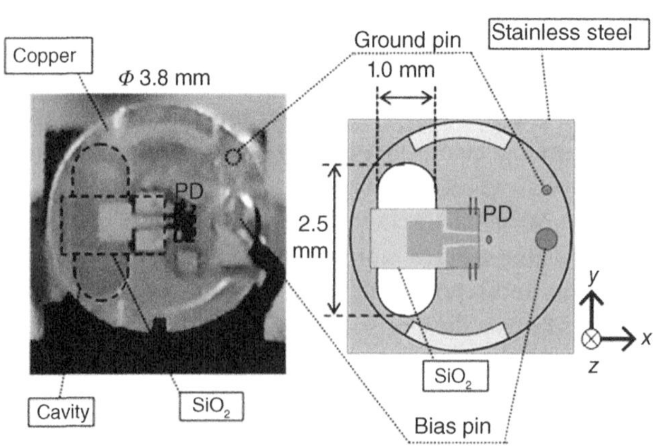

Figure 7.21 Photograph and figure of the top surface of the 60 GHz compact antenna module.

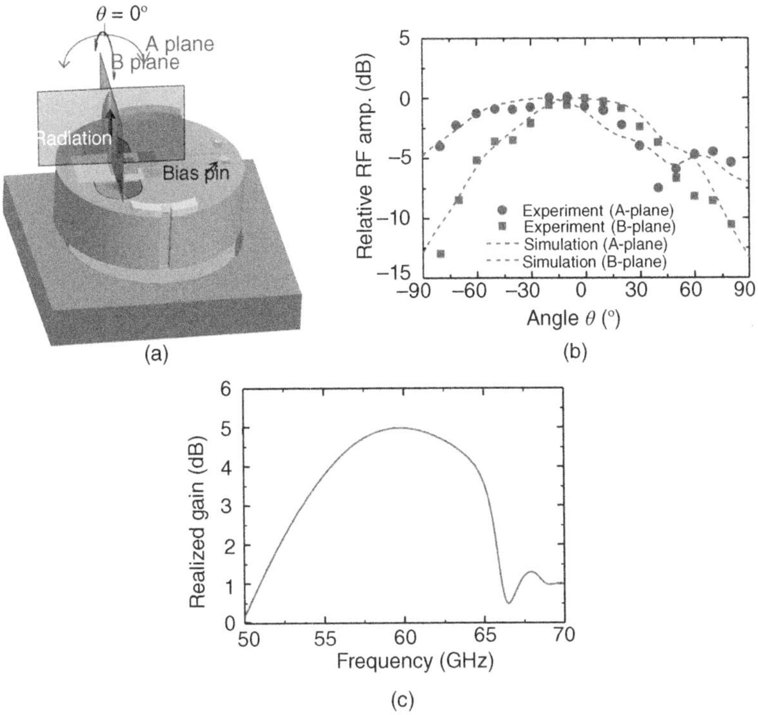

Figure 7.22 60 GHz compact antenna module. (a) HFSS simulation model. (b) Radiation patterns measured and calculated with HFSS at 60 GHz. (c) Frequency dependence of the realized gain calculated with HFSS.

Various antenna arrays can be formed by arranging the square stainless steel outer cover, which can be seen in Figure 7.20 illustration, side-by-side.

The radiation pattern and the frequency response of the module are shown in Figure 7.22. Figure 7.22a is a schematic illustration of the radiation part of the module; Figure 7.22b shows the radiation patterns of a fabricated module at 60 GHz. The circular and square plots are the measured results and the broken lines are the calculated results obtained by using HFSS. We notice that both results almost agree, verifying that the device was fabricated as designed. The frequency response, i.e. the realized gain calculated by HFSS is depicted in Figure 7.22c. It has been deduced that the dip seen around 66 GHz is attributed to the presence of the bias pins. It is noted again that the direct measurement of the frequency response for the antenna itself is not easy to obtain in an IPA experiment because the measurement results include the characteristics of the E/O conversion by the IM and the O/E conversion by the PD. The actual measured frequency response was somewhat varied from module to module and we noticed such a dip appeared for of them.

7.4.2 Beamforming by an Array of the Antenna Array Modules

Figure 7.23 shows the experimental setup for the RoF-based beamforming with an array of the 60-GHz array modules. Four optical signals output from the TLSs, whose wavelengths were $\lambda_1 = 1560$, $\lambda_2 = 1553$, $\lambda_3 = 1550$, and $\lambda_4 = 1545$ nm, were selected to be the central wavelengths of the WDM filters used in the experiment. They were simultaneously amplitude-modulated by a 60 GHz RF signal from the LO at the IM. After being boosted by the EDFA, the RoF signals propagated through a 2 km long SMF. The four RoF signals were de-multiplexed to each wavelength by

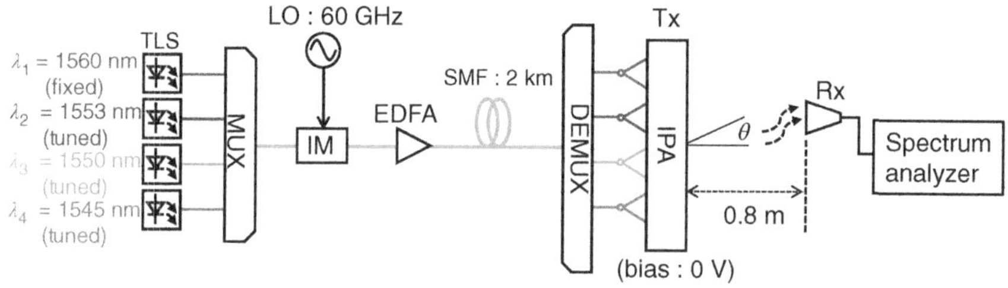

Figure 7.23 Experimental setup for the RoF-based beamforming of the arrayed modules.

a WDM coupler at the AS. The four RoF signals were then divided into two signals by CPLs and fed to each antenna module. The experimental measurement was conducted without supplying any bias voltages for the PDs. After a 0.8 m long radio propagation distance, the 60 GHz RF signal was received by a horn antenna receiver (Rx), and the signal was then observed with an RF SA. In this experiment, we arranged eight antenna modules into a 4 × 2 array. The element separation was 3.9 mm, which corresponded to 0.78λ at 60 GHz.

Figure 7.24 shows the arrangement of the 4 × 2 compact array of antenna modules for the radiation pattern measurement. The wavelengths λ_2, λ_3, and λ_4 were changed to attain five patterns. The corresponding RF phase shifts were changed so that the main beam angle was (i) 0°, (ii) 10°, (iii) 20°, (iv) 30°, and (v) 40°, as shown in Table 7.1.

Figure 7.25 shows the RF radiation patterns in each case of the five wavelength tuning ranges. The solid lines were the results calculated with HFSS. The measured RF powers almost coincided with the simulated results. The slight deviations between the simulated and measured values may be caused by some output power differences in the individual modules. If the number of lasers and modules are increased, we expect to be able to create a large variety of beam patterns.

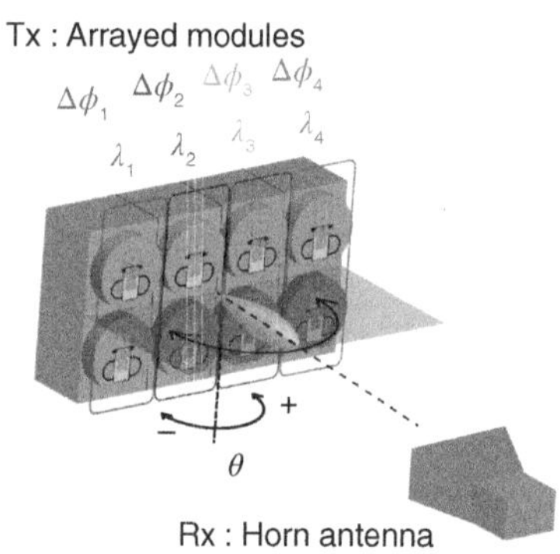

Figure 7.24 Setup to measure the radiation pattern radiated by the arrayed modules.

Table 7.1 Relationship between the wavelength shift and the RF phase shift.

Wavelength shift $\Delta\lambda$ (nm) ⇒ ($\Delta\lambda_1, \Delta\lambda_2, \Delta\lambda_3, \Delta\lambda_4$)	**RF phase shift ⇒ ($\Delta\phi_1, \Delta\phi_2, \Delta\phi_3, \Delta\phi_4$)**
① (0, 0, 0, 0)	① (0°, 0°, 0°, 0°)
② (0, −0.07, −0.14, −0.20)	② (0°, −48.8°, −97.5°, −146°)
③ (0, −0.13, −0.23, 0.10)	③ (0°, −96°, −192°, −288°)
④ (0, 0.25, 0, 0.25)	④ (0°, 180°, −360°, −540°)
⑥ (0, −0.19, 0.11, −0.08)	⑤ (0°, −140°, −280°, −420°)

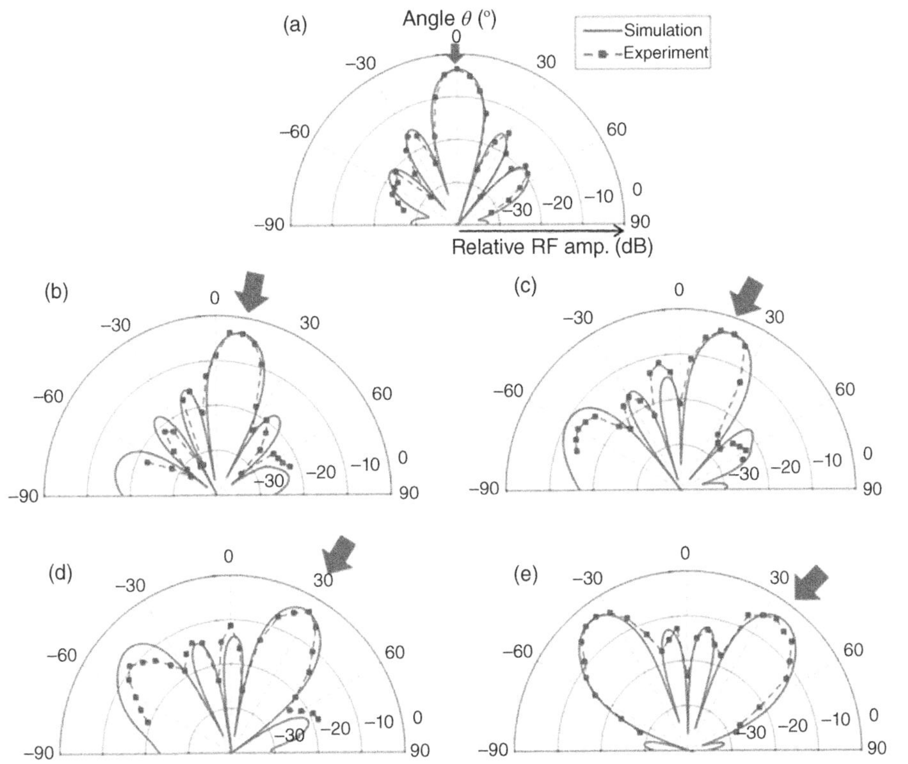

Figure 7.25 Measured RF radiation patterns radiated by the 2 × 4 compact module antenna array.

7.5 Direct Delay Control for Beamforming by Variable Optical Delay Devices with 10 Gbit/s Class Data Transmission

7.5.1 Direct Delay Control by Optical Variable Line

In this section, instead of the previous method utilizing the CD characteristics of single mode fibers to controlling the phase of the RoF signals, we adopt the use of a VDL for the RF phase shift with the aim to simplify the RoF signal generation for beamforming at a CS. When using a VDL to control the RF phases, the delay length variation required for realizing a π phase shift is a half wavelength. In the case of micro-wave bands such as 2 GHz, the delay length is 7.5 cm and it is considered to be too large for practical use. On the other hand, in the case of a millimeter-wave band such as 60 GHz, the delay length variation is 2.5 mm and we are able to realize very compact and cost-effective VDL devices such as the movable corner cube shown earlier in Figure 7.2.

First of all, we would like to show by a simple experiment that the VDL works as expected. Despite the restriction in terms of the number of cores and available lengths, an MCF was used as a trial for the transmission of an RoF signal. Figure 7.26 shows the experimental setup for the 60-GHz beamforming antenna employing an RoF system with VDLs.

The optical signal emitted from an LS at the wavelength of 1.55 μm whose polarization was properly adjusted by a PC was modulated by an LN IM with a 60-GHz continuous wave RF signal. The intensity-modulated optical signal, i.e. the RoF signal, was boosted by an EDFA inserted after the IM and was divided into eight signals by an optical coupler. One of those signals was

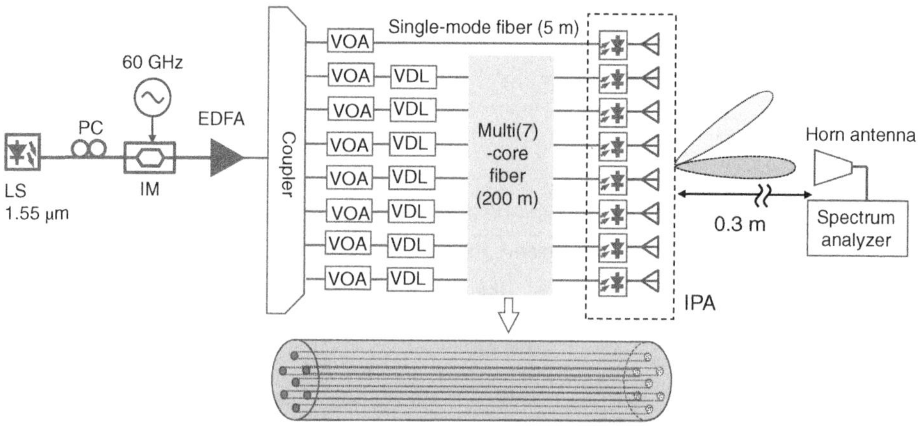

Figure 7.26 Experimental setup to show the fundamental operation of a VDL for beamforming. A multi-core fiber (MCF) was used for the transmission of the RoF signals.

propagated over a 5.0 m long SMF and fed to an element of the 60 GHz band 4 × 2 IPA described in Section 7.3. The other signals were propagated over a 200 m long MCF (seven cores in this case) and fed to each antenna element. We would like to mention that an optical MCF with more than eight cores was not available. In order to adjust the optical power and the RF phases, VOAs and VDLs were inserted after the optical coupler. The difference in the lengths of the SMF and the MCF did not matter in this case because the modulation signal was a continuous RF wave. After a 0.3 m long radio propagation length, the 60 GHz RF signal was received by a standard horn antenna and then observed with an RF SA. We note that the experimental setup was shielded with RF absorption materials.

Figure 7.27 shows the change of the RF received power at the observation angle of 0° versus the optical delay. The delay variation needed to obtain the π RF phase shift was theoretically estimated to be 8.3 ps for 60 GHz RF signal. Consequently, the delay was varied from −16.6 to 16.6 ps with the VDL. The solid line in Figure 7.27 is the theoretical result calculated from the array factor expressed by Eq. (7.2). The closed circles and triangles are the measured results of the A plane and

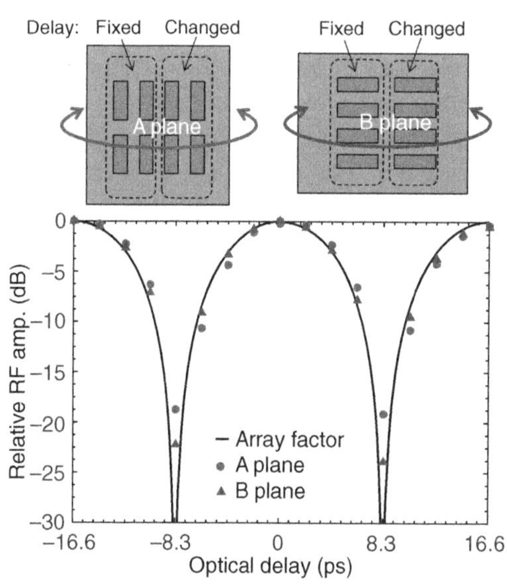

Figure 7.27 Relative RF power as a function of the optical delay change.

B plane, respectively, as indicated in the figure. The measurement was conducted by tuning the optical delay of four elements on the right side for both the beamforming in the A and B planes. It was confirmed that the measured RF powers coincided with the calculated results and that the measured RF power was periodically changed when the optical delay was changed. The measured radiation patterns were also almost coincident with the theoretical curves calculated with HFSS.

7.5.2 Demonstration of Beamforming of a Digital Signal

We arranged eight of the 60 GHz band compact antenna modules described in Section 7.4 into a 1×8 array as shown in Figure 7.28. The experiment was conducted with a bias voltage of −2.0 V applied to the UTC-PDs.

The experimental setup for 7 Gbit/s 16QAM data transmission is shown in Figure 7.29. The process for generating and recovering the digital signals is similar to that described in Section 7.4. On the transmitter side, the 1.76 Gbaud 16QAM signal (corresponding bit rate: 7.04 Gbit/s) with the IF of 4.0 GHz was generated from an AWG. The generated 4.0 GHz signal was then frequency-up-converted to 64.8 GHz by being mixed with the 60.8 GHz RF continuous wave

Figure 7.28 Schematic illustration of the arrangement for the 1×8 antenna module array.

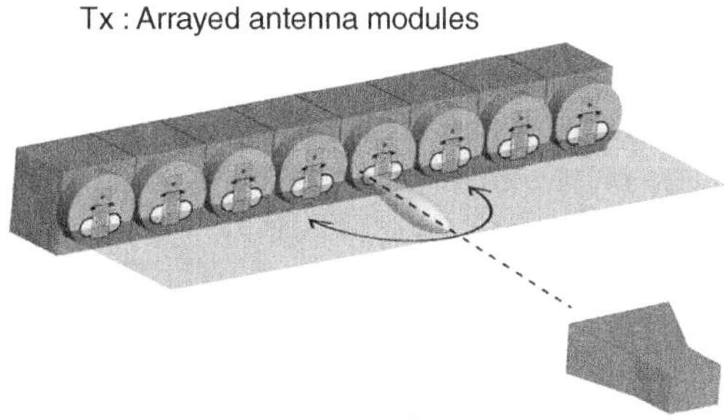

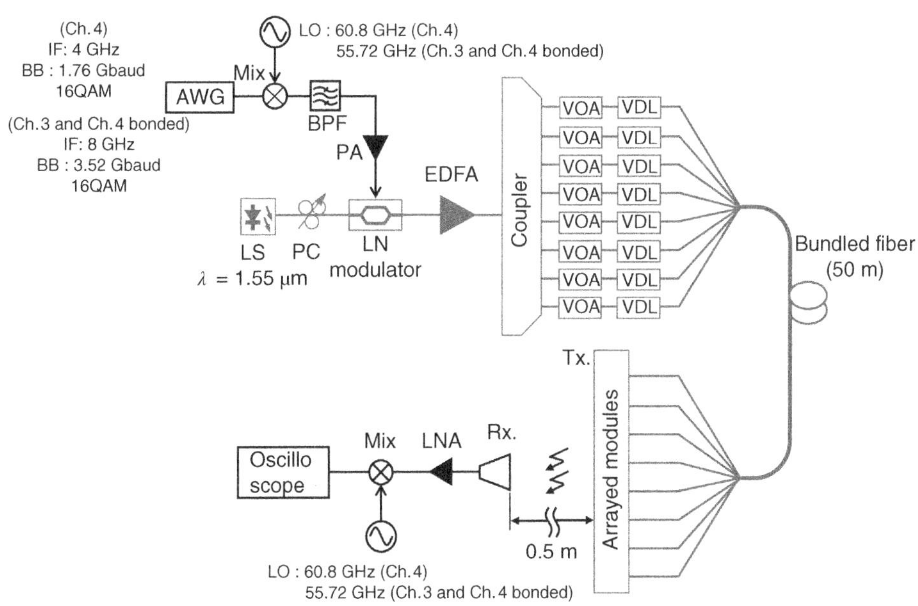

Figure 7.29 Experimental setup of the 10-Gbit/s class data transmission experiments with the beamforming array of 60-GHz band compact antenna modules.

produced by an LO. The frequency 64.8 GHz is the IEEE-standardized central frequency of the 60-GHz band channel 4 as depicted in Figure 7.16. The mixed signal was passed through the BPF to suppress the undesired frequency components and noise. The center frequency and the pass-bandwidth were 64.8 and 1.76 GHz, respectively. The 64.8 GHz 16QAM RF signal was then amplified by the RF power amplifier (PA) and applied to the LN optical IM. After being boosted by the EDFA, the generated 64.8 GHz 16QAM RoF signal was divided into eight signals. The RoF signals passing through the VOA and VDL propagated over a 50 m long bundled fiber and were fed to the 60 GHz band compact antenna modules.

We would like to note that the bundled fiber was used for these experiments. Since the data symbols in the digital signal transmission have to be synchronized among the eight RoF signals, the combination of a MCF with seven cores and a SMF, as shown in Figure 7.26, is therefore not appropriate. As a plausible application scenario, we may consider placing a photodiode-integrated antenna array on the rooftop of a building or at the top of a radio station tower (AS). One would then install the LSs with the VDLs and VOAs, as well as the electrical devices for the signal processing and BB-to-RF conversion, in the basements of the buildings or at cubicles (CSs) in the vicinity of the radio station towers. It corresponds to the (Case 1) in Figure 7.4. If an MCF with more than eight cores were available, it would be a much better choice and would correspond to the (Case 2) in Figure 7.4.

The radiated 64.8 GHz 16QAM RF signal was received by a standard horn antenna, where the propagation distance of the RF signal was set to 0.5 m. After being received by the horn antenna, the received signal was amplified by an LNA and then frequency-down-converted to 4.0 GHz by being mixed with the 60.8 GHz RF continuous wave from the LO. The 4.0 GHz signal was then input into a real-time oscilloscope. The lengths of all of the eight optical signal paths were adjusted to be equal in advance so as to synchronize the divided eight data symbols. The radiation pattern was controlled by changing the optical path length using the VDL. The received RF power and the SNR of the 16QAM signal were measured by the oscilloscope at each observation angle.

Figure 7.30a–c shows the relative RF amplitude and SNR values at each observation angle in the cases where the peak angle of the RF amplitude was 0°, 10°, and 30°, respectively. The solid lines are the simulated results calculated with HFSS. The closed circles and triangles are the measured relative RF amplitude and SNR values, respectively. The obtained constellations are also shown in the insets. In all cases, the measured RF powers and SNRs almost coincided with the simulated results for the main lobes where the received RF power was relatively large. On the other hand, the measured data did not agree well with the simulated results for the side lobes where the received RF power level was relatively small. This was especially true where the signal levels were smaller than −10 dB compared with the peak values of the main lobes and close to the noise level. It can be emphasized, however, that we could obtain a SNR level of more than 16.6 dB in the vicinity of the main lobe. It corresponds to the bit error rate (BER) of 1.0×10^{-3}, which is required for error-free data transmission with a LDPC code. Thus, the 7 Gbit/s digital data transmission was successfully realized with an array of 60 GHz band compact antenna modules with antenna beamforming capability. However, the propagation distance was only 0.5 m, which, of course, is too short for practical applications. Nevertheless, as mentioned before in Section 7.4, the distance could be extended to longer than 100 m by improving the characteristics of the LN modulators and the UTC PDs. We would like to note that when the signal format changes, the required SNR and, therefore, the maximum transmission distance for error free transmission would change. For example, the required SNR for QPSK is 6.7 dB smaller than that for 16QAM. This difference corresponds to almost doubling the transmission distance.

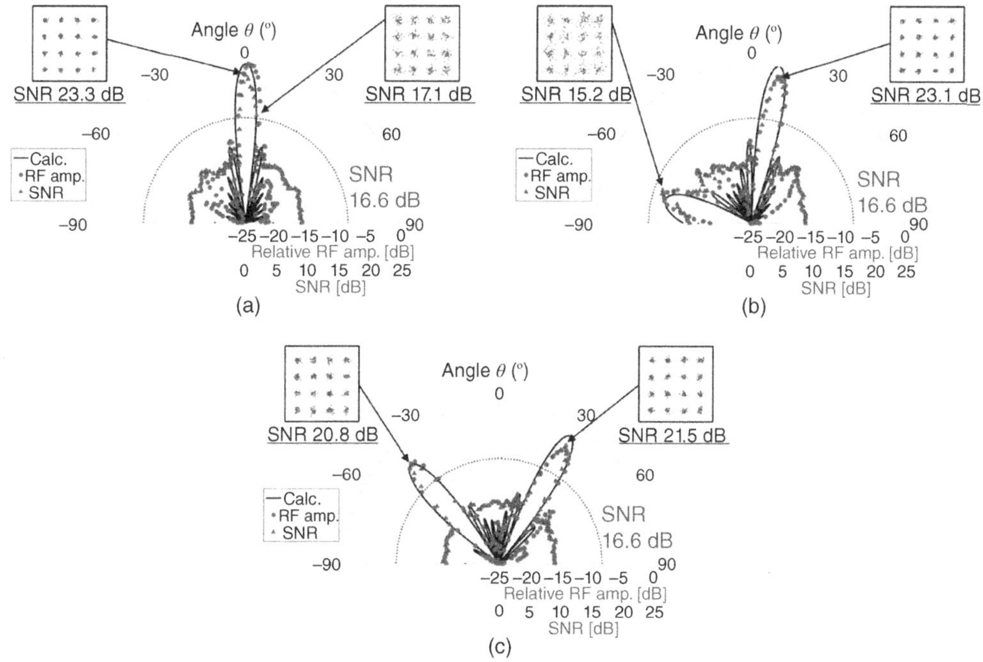

Figure 7.30 Relative RF amplitude, SNR and constellation at each observation angle when 7 Gbit/s (1.76 Gbaud, 16QAM) data was transmitted when the peak angle of the RF amplitude was specified to be (a) 0°, (b) 10°, and (c) 30°.

To demonstrate further higher speed transmissions, we attempted 14 Gbit/s 16QAM data transmission using channel-bonding of the 60 GHz band channel 3 and channel 4. Those channels can be seen in the channel allocation diagram in Figure 7.16. The experimental setup is almost the same as the aforementioned experiment. The values of the IF and the BB that were used are shown in Figure 7.29; they are marked as (Ch. 3 and Ch. 4 bonded). A 3.52 Gbaud 16QAM BB signal (corresponding bit rate:14.08 Gbit/s) with an IF of 8.0 GHz was generated from the AWG. The 8.0 GHz IF signal was then frequency-up-converted to 63.72 GHz, which is the IEEE-standardized central frequency of the 6 GHz band of the bonded channel 3 and channel 4. It was achieved by being mixed with a 55.72 GHz RF continuous wave from the LO. The signal was then passed through the WR-10 waveguide as a BPF to eliminate unwanted frequency components. The 63.72 GHz 16QAM RF signal was amplified by the PA and applied to the LN modulator. The rest of the procedure for the measurements was the same as mentioned previously.

Figure 7.31a–c shows the relative RF amplitude and SNR at each observation angle in the case when the angle of the peak RF amplitude was specified to be 0°, 10°, and 30°, respectively. The solid lines are the simulated results calculated with HFSS and the closed circles and triangles are the measured relative RF amplitude and SNR values, respectively. The obtained constellations are also shown in the insets. In all cases, we obtained SNR values that were more than 16.6 dB in the vicinity of the main beam. Thus, the 14 Gbit/s digital data transmission was successfully realized with an array of 60 GHz band compact antenna modules with antenna beamforming capability. The measured data at the side lobes seemed to be more disturbed due to an increase of the noise floor caused by the wider frequency band.

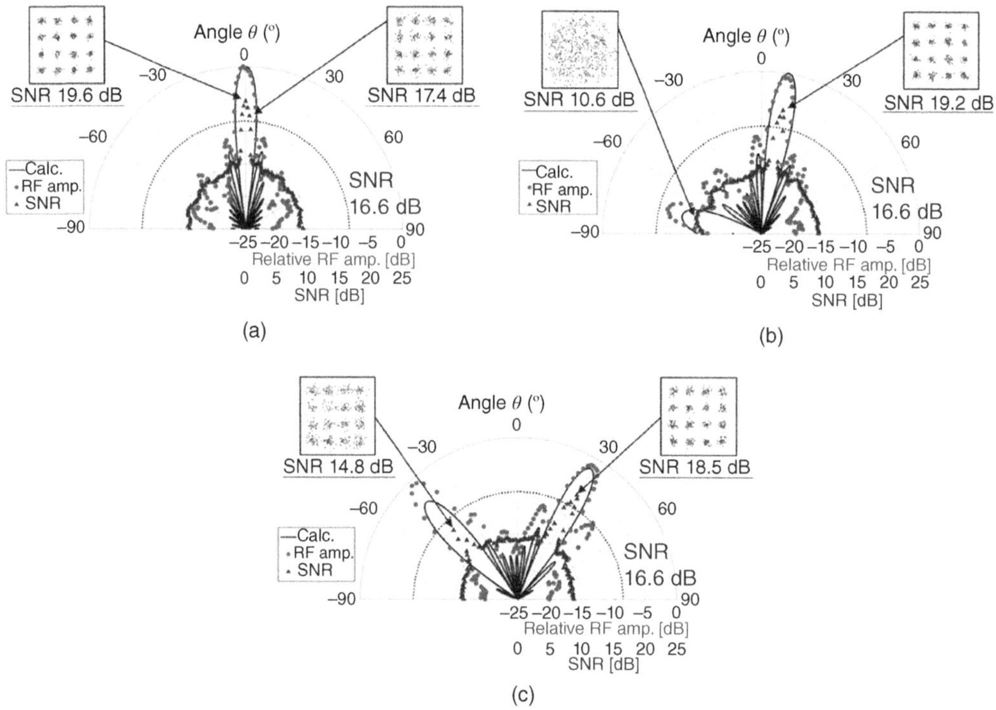

Figure 7.31 Relative RF amplitude, SNR and constellation at each observation angle when 14 Gbit/s 16QAM data was transmitted when the peak angle of the RF amplitude was specified to be (a) 0°, (b) 10°, and (c) 30°.

7.6 Antenna Array Beamforming Using a 1.3-μm Band EML as the Light Source

Techniques that could simplify the RoF signal generation for the array of IPA modules have been explored. We have generated a 40 GHz RoF signal using a 1.3 μm band electro-absorption modulator-integrated DFB-LD (EML), which is widely used in Ethernet applications, and an optical tunable filter. This experiment is described in this section.

Figure 7.32 shows the experimental setup. The 40 GHz module has a structure similar to the 60 GHz version. The module has the same diameter, but its cavity part is Al_2O_3 instead of air as shown in the figure. The array configuration is 1 × 8 as in the previous section. The EML used in this experiment is designed for operation at 25 Gbit/s and, therefore, the modulation efficiency was quite low at 40 GHz. In order to overcome such low modulation efficiency, we tried to increase the modulation index by utilizing the edge of the transmission characteristics of the optical filter passband to reduce the carrier component. We note that the whole experimental setup was covered by RF absorbing materials in order to avoid any leakage of the 40 GHz radiation into the region exterior to the setup.

Figure 7.33a shows the optical spectrum of the 40 GHz RoF signal generated from the EML; the transmission characteristics of the optical filter are indicated by the dashed line. Figure 7.33b shows the optical spectrum of the RoF signal after the filter and a praseodymium-doped fiber amplifier (PDFA). By using this technique, the sideband to carrier ratio of the RoF signal was improved from −25 to −11 dB. This improvement is associated with the 9 dB increase of the 40 GHz RF signal.

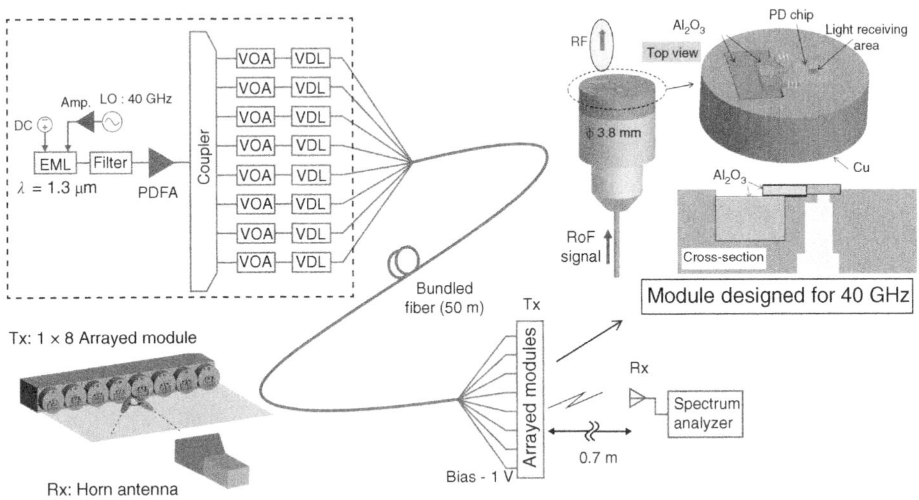

Figure 7.32 Experimental setup used to demonstrate the generation of various beam patterns at 40 GHz by using an EML as the light source for a 40 GHz version of the array of compact antenna modules.

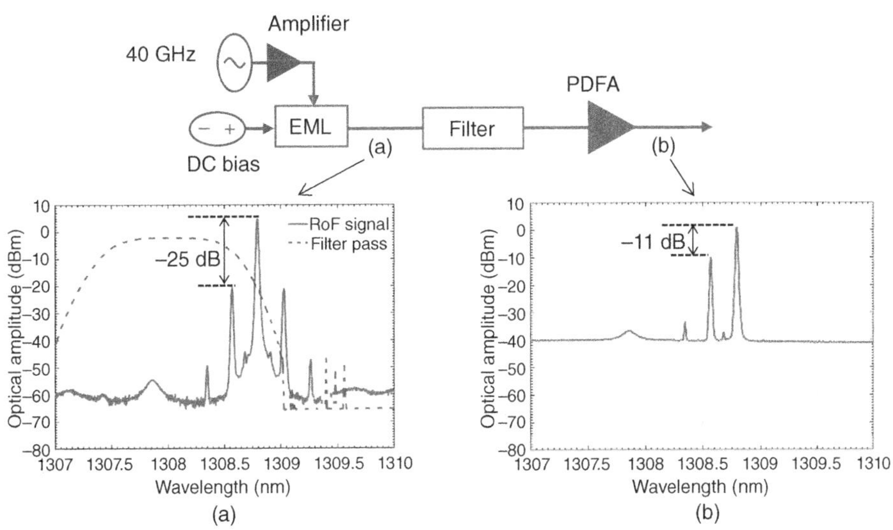

Figure 7.33 Optical spectrum associated with the EML. (a) EML-generated 40 GHz RoF signal and the transmission characteristics of the optical filter. (b) RoF signal after it has passed through the optical filter and has been boosted by the PDFA.

Figure 7.34a–f shows the measured radiation patterns when the main beam direction is 0°, 20°, wide-angle single, ±10°, ±60°, and wide-angle dual, respectively. The wide-angle single beam is aimed at covering a wide set of directions, for example, from −50° to +50° as shown in Figure 7.34c. On the other hand, the wide-angle dual beam is aimed at covering two wide sets of directions, for example, from −50° to −20° as well as from 20° to 50° as shown in Figure 7.34f. The solid lines are the simulated results calculated with HFSS and the closed circles are the measured relative RF amplitudes. Good agreement was observed between main lobes of the measured and simulated patterns. The deviations observed in their side lobes may be attributed to some output differences among the individual modules.

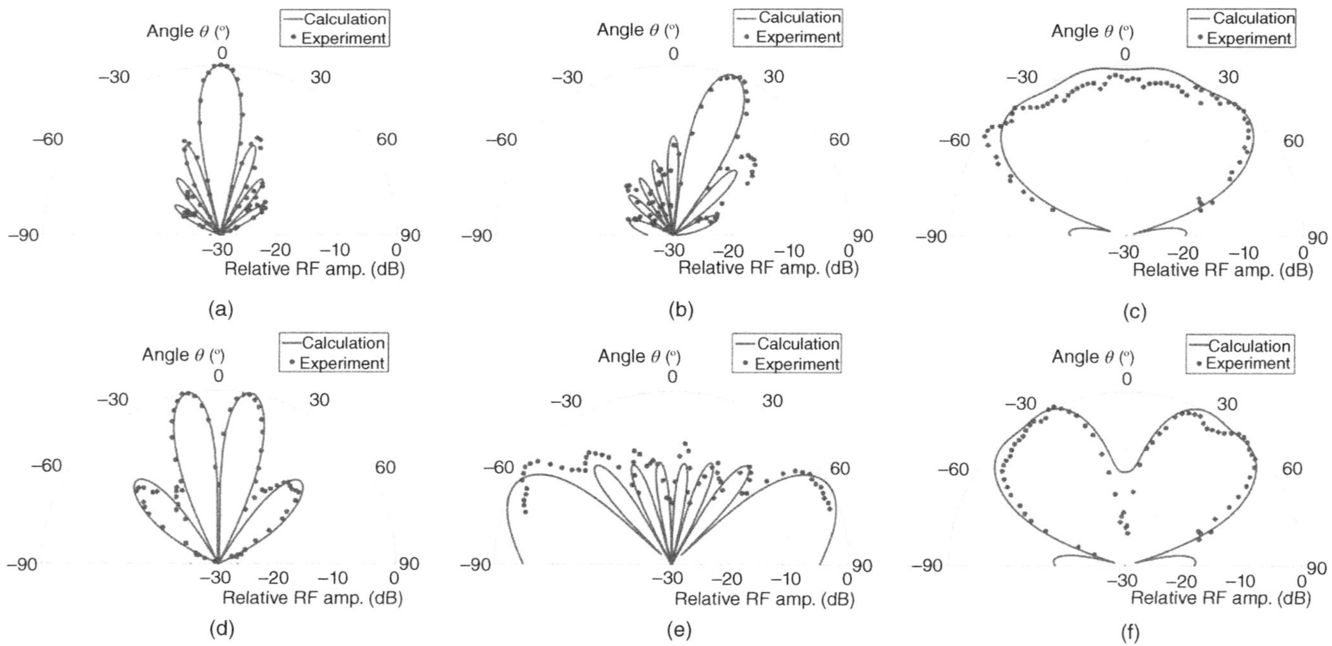

Figure 7.34 Measured radiation patterns when the main beam direction is (a) 0°, (b) 20°, (c) wide-angle single, (d) ±10°, (e) ±60°, and (f) wide-angle dual.

7.7 Perspectives

It was demonstrated in Sections 7.5 and 7.6 that antenna beamforming using a VDL as the phase shifter leads to the simplification of the transmission CS. Digital data transmission experiments at 7 and 14 Gbit/s with a beamforming array of 60 GHz compact antenna modules were described. Additionally, it was shown that by introducing a 1.3 μm band EML, which is widely used in Ethernet applications, and an optical tunable filter, a 40 GHz RoF signal can be generated and various beam patterns were successfully demonstrated utilizing VDLs and VOAs. The VDL used in these studies is based on mechanical control and, therefore, the delay variation speed is on the order of a millisecond, which is considered to be short enough for normal beamforming circumstances, but may be too long for some specific applications.

We would also like to note that the presented beamforming operations should work similarly in a different architecture. For example, we may install the LS and the electrical devices at a location far apart from the AS, for example, a few kilometers away. This configuration corresponds to the (Case 2) in Figure 7.4. In that case, the RoF signal and the control signals for the VDLs and VOAs will be transmitted via a fiber to the basements of the buildings or the cubicles where the CS is. The RoF signal will then be divided there to be fed to the VDLs and VOAs. We do not have to take care about any fading effects of the RoF signals as long as we employ a 1.3 μm wavelength and a standard SMF. A variety of indoor applications can be another scenario, especially when the 60 GHz band is utilized.

We think that an array arrangement with more elements, for example, a 4 × 8 array, would be possible for practical use. Although the actual demonstrations were restricted to a 1 × 8 array due to the limited number of available devices. We would rather emphasize and pursue the advantages of a simple and compact architecture using VDLs in the millimeter-wave band. We would also like to expect that MCFs with a larger number of cores will become available as the transmission fiber. We could then arrange various arrays with many antenna elements and directed beams.

We also need to make a comment on the power consumption. The array-antenna can concentrate power toward a desired direction. For example, the 1 × 8 array antenna described earlier has a gain of about 14 dBi. Although there would be extra power consumption in order to control the beam direction, the total energy consumed would be reduced in many cases. From the viewpoint of a photonics-based architecture, the loss of an optical fiber is only 0.2 dB/km at 1.55 μm, which is much smaller than that of millimeter wave coaxial cables which is several dB/m. Moreover, our system can realize low-loss transmission. Furthermore, the insertion loss of the VDLs and VOAs used to facilitate the beamforming can be as small as 1.0 dB or much less by using state-of-the-art optical assembling technologies.

7.8 Estimation of the Direction of a User Terminal

While the beamforming operation was described in the previous sections, a technique for accurately estimating the direction of the UE is indispensable for properly pointing the antenna beam. In recent mobile systems the UE location can be estimated from the information collected from the mobile BSs in the vicinity of the UE. In this section, however, we try to estimate a user direction by conducting a bidirectional RoF transmission.

Figure 7.35 shows the experimental setup used to estimate the direction of a user terminal. The 60 GHz band IPA presented in Section 7.3 was adopted for RF downlink transmission at the AS.

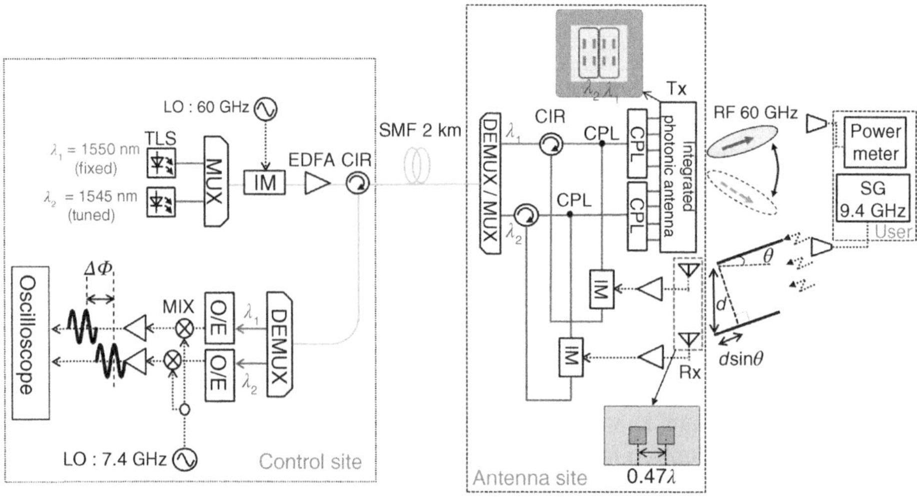

Figure 7.35 Experimental setup for estimating the direction of user equipment (UE).

Two 9.4 GHz microstrip antenna elements, which are separated by 0.47λ, were used as the RF uplink receiver.

Using each 9.4 GHz antenna as a receiver, it was easy to use them to form a pair of two element antenna arrays. The consequent arrangement shown in Figure 7.35 resulted in the simplification of the experimental setup. As the required data rate is not very high in actual use cases, it was sufficient to use the 9.4 GHz antennas for the RF and optical uplink transmissions. Two optical signals (λ_1 = 1550 nm, λ_2 = 1545 nm) were multiplexed by a WDM coupler and modulated by a 60 GHz RF signal in an IM. As a result, two RoF downlink signals at λ_1 and λ_2 were created at the CS. After a 2 km long SMF transmission, the two RoF signals were de-multiplexed with a WDM coupler and passed through an optical circulator (CIR) at the AS. These RoF downlink signals were divided into two signal parts by CPLs. One part was injected into the IPA (Tx) and the 60 GHz RF downlink signal was radiated. While the λ_1 was fixed at 1550 nm, the λ_2 was changed from 1545 to 1545.25 nm. The other part of the RoF downlink signal tapped by the CPL was used for the optical uplink transmission. These tapped RoF signals were injected into the IMs and were re-modulated by the 9.4 GHz RF uplink signal that was sent from a horn antenna transmitter. This transmitter acted as the UE. The signal it radiated was received by the 9.4 GHz antennas.

The re-modulated 9.4 GHz uplink RoF signal was merged by the CIR and, thus, the uplink RoF signal used to search for the UE direction was generated. A certain amount of 60 GHz component was included in the uplink RoF signals, but it did not affect the following processing. After a 2.0 km SMF propagation from the AS to the CS, the two uplink RoF signals were down converted to 2.0 GHz RF signals by being mixed with a 7.4 GHz RF tone after their O/E conversion.

The phase difference $\Delta\Phi$ between the two RF uplink signals was measured by changing the angle θ of the 9.4 GHz antenna-arrays from −70° to 70°. The phase difference $\Delta\Phi$ measured at the CS corresponds to the summation of the phase difference $\Delta\phi_1$, which is caused by the RF path difference of the received signals at the two microstrip antenna elements, and the phase difference $\Delta\phi_2$, which was caused by the CD effect of the two RoF uplink signals during the 2.0 km SMF propagation. Therefore, the $\Delta\Phi$ can be expressed as

$$\Delta\Phi = \Delta\phi_1 + \Delta\phi_2 = 2\pi\, d\, \sin\theta/\lambda_{RF} + 2\pi\, \Delta\lambda\, LD f_{RF} \tag{7.4}$$

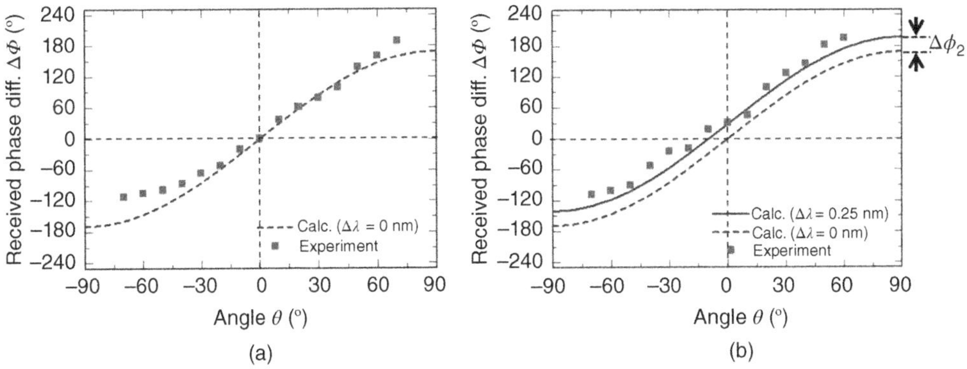

Figure 7.36 Measured phase difference of RF uplink signal $\Delta\Phi$ as a function of the incident angle θ. (a) $\Delta\lambda = 0$ nm. (b) $\Delta\lambda = 0.25$ nm.

where d is the separation between the elements in the two element array, θ is the incident angle of the uplink signal, and $\Delta\lambda$ is the wavelength shift of λ_2. Figure 7.36a,b show the measured $\Delta\Phi$ as a function of the incident angle θ of the RF uplink signal when $\Delta\lambda = 0$ and 0.25 nm, respectively, where the relative RF phases between the two RoF downlink signals were 0 and π, respectively.

The dashed and solid lines in Figure 7.36 were calculated with Eq. (7.4) in the cases with $\Delta\lambda = 0$ and 0.25 nm, respectively. The measured $\Delta\Phi$ almost coincided with the calculation except for the measurement points at the large incident angles. In Figure 7.36b, the phase difference $\Delta\phi_2$ due to the CD effect can be observed. The results indicate that the incident angle of the RF uplink signal (i.e. the direction of the user terminal) can be estimated. By increasing the number of antenna elements that receive the RF uplink signal, we may be able to estimate the user direction more accurately.

7.9 Summary

Beamforming techniques and small cell architectures are considered to be essential aspects of future wireless access mobile systems including those of 5G. Some 5G systems have already been deployed and service providers have begun starting wireless services. Nevertheless, they employ relatively lower frequencies, and highly advanced state-of-the-art semiconductor technologies are utilized to facilitate multiple antennas and massive data processing. However, much higher frequency bands, such as 60 GHz, will be used in many future applications whereby the envisioned features of 5G are fully realized. When we consider that such systems should meet the requirements of cost and power consumption effectiveness, it remains a significant issue as to how such systems will be realized in practice. We may face difficulty in continuing to rely mainly on the advances in semiconductor technologies.

As emphasized in this chapter, we believe that the presented photonics-based millimeter waveband beamforming, where RoF signals are transmitted through low-loss optical fibers to a remote antenna array integrated with high-speed photodiodes, will be a practical candidate as a new approach to future antenna array architectures. The detailed designs and fabrication methods described have been studied in order to realize such new beamforming schemes. In the various beamforming demonstrations we have presented, the RF phases in the RoF signals have been controlled by utilizing the CD characteristics of optical fibers or variable optical delay devices.

These demonstrations and experimental trials are mainly aimed at wireless traffic from the BS to the UE, i.e. the so-called downlink traffic. A question may be raised whether the presented techniques are applicable to the uplink traffic as well. The configuration shown in Figure 7.35 may be a possible solution. However, we have to study further the overall technical details by taking into account the uplink RF frequencies and the methods used to create the uplink RoF signals. Note that since the processing of uplink signals at the BS usually does not require high-power devices, power consumption may not be a big issue and the choice of electronics-based signal processing components may also be a solution.

We hope that the presented studies and experimental results will be beneficial and helpful in exploring new architectures for future mobile wireless systems.

Abbreviations

AS	antenna site
AWG	arbitrary waveform generator
BB	baseband
BPF	band pass filter
BS	base station
CD	chromatic dispersion
CPL	coupler
CIR	circulator
CS	control site
EML	electro-absorption modulator integrated with laser diode
EVM	error vector magnitude
HFSS	high frequency structural simulator
IF	intermediate frequency
IM	intensity modulator
IPA	integrated photonic array-antenna
LD	laser diode
LDPC	low-density parity-check
LN	Lithium Niobate
LNA	low noise amplifier
LO	local oscillator
LS	light source
MCF	muli-core fiber
PA	power amplifier
PC	polarization controller
PD	photo diode
PDFA	praseodymium doped fiber amplifier
QAM	quadrature amplitude modulation
RF	radio frequency
RoF	radio on fiber
SA	spectrum analyzer

SMF	single mode fiber
SNR	signal-to-noise ratio
TLS	tunable light source
UE	user equipment
UTC-PD	uni-traveling carrier photo diode
VDL	variable delay line
VOA	variable optical attenuator
WDM	wavelength division multiplexing

References

1 R. Razavi and H. Claussen, "Urban small cell deployments: impact on the network energy consumption," in *Proceedings of IEEE Wireless Communications and Networking Conference Workshops (WCNCW)*, Apr 2012, pp. 47–52.

2 S. F. Chou, T. C. Chiu, Y. J. Yu, and A. C. Pang, "Mobile small cell deployment for next generation cellular networks," in *Proceedings of IEEE Global Communications Conference (GLOBECOM)*, Dec 2014, pp. 4852–4857.

3 I. Chiba, R. Yonezawa, and K. Kihira, "Adaptive array antenna for mobile communication," in *Proceedings of 2000 IEEE International Conference on Phased Array Systems and Technology*, May 2000, pp. 109–112.

4 P. G. Richardson, F. J. Adams, K. J. Lewis, and J. C. Medley, "PACER (Phased Array Concepts Evaluation Rig): design, development and adaptive beamforming experiments," in *Proceedings of 2000 IEEE International Conference on Phased Array Systems and Technology*, May 2000, pp. 421–424.

5 K. Wu, Y. J. Cheng, T. Djerafi, X. P. Chen, N. Fonseca, and W. Hong, "Millimeter-wave integrated waveguide antenna arrays and beamforming networks for low-cost satellite and mobile systems," in *Proceedings of the 4th European Conference on Antennas and Propagation (EuCAP)*, C33P2-5, April 2010.

6 M. Y. Frankel and R. D. Esman, "True time-delay fiber-optic control of an ultrawideband array transmitter/receiver with multibeam capability," *IEEE Trans. Microwave Theory Tech.*, vol. 43, no. 9, pp. 2387–2394, Sep 1995.

7 M. Tadokoro, T. Taniguchi, and N. Sakurai, "Optically-controlled beamforming for 60 GHz-ROF system using dispersion of optical fiber and DFWM," in *Proceedings of 2007 IEEE/OSA Optical Fiber Communication Conference (OFC)*, paper OWN2, Anaheim, CA, Mar 2007.

8 J. Liu, W. Noonpakdee, H. Takano, and S. Shimamoto, "A novel RF signal beamforming scheme over optical wireless communications," in *Proceedings of* 2011 International Conference on Space Optical Systems and Applications (ICSOS), May 2011, pp. 346–350.

9 S. Akiba, M. Oishi, and H. Matsuno, "Effects of chromatic dispersion on RF power feeding to array antenna through fiber," in *Proceedings of the 17th Opto-Electronics and Communications Conference (OECC)*, Busan, Korea, July 2012, pp. 323–324.

10 M. Oishi, H. Matsuno, K. Nishimura, and S. Akiba, "Experimental study of chromatic dispersion effects on antenna beamforming by RF over fiber," in *Proceedings of 2012 IEEE Int'l. Topical Meeting on Microwave Photonics (MWP)*, Noordwijk, The Netherlands, Sep 2012, pp. 140–143.

11 T. P. McKenna, J. A. Nanzer, and T. R. Clark, "Photonic beam steering of a millimeter-wave array with 10-gb/s data transmission," *IEEE Photonics Technol. Lett.*, vol. 26, no. 14, pp. 1407–1410, July 2014.

12 L. A. Colden, S. C. Nicholes, L. Johansson, S. Ristic, R. S. Guzzon, E. J. Norberg, and U. Krishnamachari, "High performance InP-based photonic ICs," *IEEE/OSA J. Lightwave Technol.*, vol. 29, no. 4, pp. 554–570, Feb 2011.

13 H. Ito, S. Kodama, Y. Muramoto, T. Furuta, T. Nagatsuma, and T. Ishibashi, "High-speed and high-output InP-InGaAs unitraveling-carrier photodiodes," *IEEE J. Sel. Top. Quantum Electron.*, vol.10, no.4, pp. 709–727, July-Aug 2004.

14 T. Mizuno, H. Takara, A. Sano, and Y. Miyamoto, "Dense space-division multiplexed transmission systems using multi-core and multi-mode fiber," *IEEE J. Lightwave Technology*, vol. 34, no. 2, pp. 582–592, Jan 2016.

15 K. Minoguchi, Y. Nishikawa, M. Oishi, S. Akiba, J. Hirokawa, and M. Ando, "Beam steering scheme of photonic array-antennas for 60 GHz RF signals generated by optical two-tone technique," in *Proceedings of the 19th Opto-Electronics and Communications Conference (OECC)*, paper ThP10A-3, July 2014.

16 M. Oishi, T. Hirasawa, K. Furuya, S. Akiba, J. Hirokawa, and M. Ando, "3.5-Gbit/s QPSK signal radio-over-fiber transmission with 60-GHz integrated photonic array-antenna beamforming," *IEEE J. Lightwave Technol.*, vol. 34, no. 20, pp. 4758–4764, Oct 2016.

17 K. Furuya, T. Hirasawa, M. Oishi, S. Akiba, J. Hirokawa, and M. Ando, "60 GHz-band photonic-integrated array-antenna and module for radio-over-fiber-based beamforming," *IEICE Trans. Commun.*, vol. E100-B, no. 10, pp.1717–1725, Oct 2017.

18 T. Nagayama, S. Akiba, T. Tomura, and J. Hirokawa, "Photonics-based millimeter-wave band remote beamforming of array-antenna integrated with photodiode using variable optical delay line and attenuator," *IEEE J. Lightwave Technol.*, vol. 36, no. 19, pp. 4416–4422, Mar 2018.

19 T. Hirasawa, S. Akiba, J. Hirokawa, and M. Ando, "Fabrication and evaluation of integrated photonic array-antenna system for RoF based remote antenna beamforming," *IEICE Trans. Electron.*, vol. E102-C, no. 3, pp. 235–242, Mar 2019.

20 G. J. Meslener, "Chromatic dispersion induced distortion of modulated monochromatic light employing direct detection," *IEEE J. Quantum Electron.*, vol. 20, no. 10, pp. 1208–1216, Oct 1984.

21 U. Gliese, S. Norskov, T. N. Nielsen, "Chromatic dispersion in fiber-optic microwave and millimeter-wave links," *IEEE Trans. Microwave Theory Tech.*, vol. 44, no. 10, Oct 1996, 1716–1724.

22 T. Baykas, C.-S. Sum, Z. Lan, J. Wang, M. A. Rahman, H. Harada, and S. Kato, "IEEE 802.15.3c: The first IEEE wireless standard for data rates over 1 Gb/s," *IEEE Commun. Mag.*, vol. 49, no. 7, pp. 114–121, Jul 2011.

23 H. N. Klein, H. Chen, D. Hofmann, S. Staroske, A. G. Steffan, and K.-O. Velthaus, "1.55 μm Mach-Zehnder modulators on InP for optical 40/80 Gbit/s transmission networks," in *Proceedings of Int'l. Conference on Indium Phosphide and Related Materials*, paper TuA2.4, Princeton, NJ, May 2006.

24 A. Umbach, "RF photodiodes and modulators," in *Proceedings of 2015 IEEE Int'l. Topical Meeting on Microwave Photonics (MWP)*, paper TuD-1, Paphos, Cyprus, Oct 2015.

8

Contemporary Array Analysis Using Embedded Element Patterns

David B. Davidson[1] *and Karl F. Warnick*[2]

[1] *Curtin University, International Centre for Radio Astronomy Research, Perth, WA, Australia*
[2] *Brigham Young University, Department of Electrical and Computer Engineering, Provo, UT, USA*

8.1 Introduction

Antenna arrays have a long history, dating back over a century. They are becoming ever more important in communications and other applications. The German physicist Karl Ferdinand Baum constructed a three element, switchable array in 1909. The demands of especially radar during WWII saw the deployment of a large number of RF and microwave systems using antenna arrays. Terrestrial TV transmissions (both the older analogue and the newer digital standards) routinely use the Yagi-Uda or log-periodic arrays, or variants thereof. Antenna arrays are core components of mobile communication base stations. Fifth generation (5G) systems and beyond will increasingly use beamforming for diversity gain, multiple input multiple output (MIMO) communications, null steering, and interference reduction.

Much of the design work in these areas has been done using the classical array theory found in most antenna textbooks, for example [1–4]. Fundamental to this approach is that the array pattern is the product of an array factor and the radiation pattern of an individual element. The array factor and isolated element pattern decomposition has allowed the development of some powerful and elegant analytical array synthesis methods.

A key assumption underlying the array factor method is that mutual coupling is negligible. In many cases this is a reasonable assumption. Mutual coupling may impact the behavior of the array for general communications and beamforming applications, but not in ways that are especially detrimental to the overall system performance. Papers continue to be published on classical array design methods for applications with moderate design requirements, but the underlying ideas are mature and interest in the research community has largely shifted to other topics.

In application areas with stringent performance requirements, on the other hand, mutual coupling must be included in the array design process. Areas of particular interest include high-sensitivity receiving arrays for medical imaging, satellite communications, remote sensing, and radio astronomy [5]. Tightly packed arrays for space-constrained MIMO systems also fall into this category. For these applications, classical array theory and the array factor technique are useful in developing notional designs, but are not adequate in fully characterizing and optimizing the overall system performance.

Mutual coupling impacts the performance of an array in several ways. It changes the active impedance of the array elements, which influences the excitations that must be applied to the elements to achieve a specified beam pattern. These may be excitation phases in traditional phased

Antenna and Array Technologies for Future Wireless Ecosystems, First Edition.
Edited by Y. Jay Guo and Richard W. Ziolkowski.

array analogue systems, or beamformer weights in receive-only digital systems. As the beam scan angle changes, mutual coupling causes the active impedances presented at array element ports to vary, which complicates impedance matching for antenna arrays. Mutual coupling distorts individual element patterns. As a consequence, the overall array pattern is no longer simply the product of the array factor and an element pattern. While the effects of mutual coupling are traditionally seen as deleterious, mutual coupling impacts positively for some antenna elements (such as Vivaldi arrays) by reducing the size of the elements for a given operating bandwidth requirement.

For high-sensitivity array applications, the effects of mutual coupling, be they positive or negative, can have a substantial effect on the system performance. Mutual coupling can be incorporated approximately using analytical formulations or rigorously using numerical modeling. This is particularly important for arrays with moderate numbers of elements (tens to hundreds), where infinite array approximations using unit cells are not sufficiently accurate.

Analytical methods for analyzing mutual coupling were dominant before the 2000s. With the great strides made in computational electromagnetics (CEM) over the last few decades, mutual coupling can be accurately predicted using simulation tools. Many papers have been published on numerical results for coupling in MIMO arrays, astronomical arrays, and other types of multi-antenna systems.

Despite the large body of work on the topic, it is not easy to tease out the effects of coupling on overall performance. Simply being able to compute mutual coupling does not translate into incorporating its effects into a design procedure. This chapter outlines a suitable framework for modeling mutually coupled arrays and combining the array with connected electronics, such as low noise amplifiers (LNAs) and beamforming networks. Key to this are the *embedded element patterns* (EEPs), also known as active, or scan, element patterns. The theory presented in this chapter summarizes a more extensive treatment available in [5]. We introduce methods from beamforming theory, in particular the maximum directivity beamformer, which require the array overlap and signal response matrices. These matrices are derived from the EEPs. We also briefly address receive-only systems, where again, the EEPs and derived quantities play an important role.

8.2 Design Methods: Classical Array Factor

8.2.1 Array Theory: The Uniform Linear Array

Classical array theory is best illustrated using an introductory example encountered in most antenna textbooks, the uniform linear array (ULA). We assume that the array is comprised of N identical elements spaced d apart with the $n = 1$ element located at the origin and with a progressive phase shift δ applied in some fashion between the driving signals at the element feeds. The angle off broadside is ϕ; see Figure 8.1.

The array pattern $\vec{E}(\vec{r})$ is found by summing the fields radiated by each element, using the far-field approximation which retains only the phase difference term between each element:

$$\begin{aligned}
\vec{E}(\vec{r}) &= \sum_{n=1}^{N} \frac{i_n}{I_0} e^{j\vec{k}\cdot\vec{r}} \vec{E}_{\text{elem}}(\vec{r}) \\
&= \vec{E}_{\text{elem}}(\vec{r}) \sum_{n=1}^{N} \frac{i_n}{I_0} e^{j\vec{k}\cdot\vec{r}} \\
&= \vec{E}_{\text{elem}}(\vec{r}) AF(\theta, \phi)
\end{aligned} \tag{8.1}$$

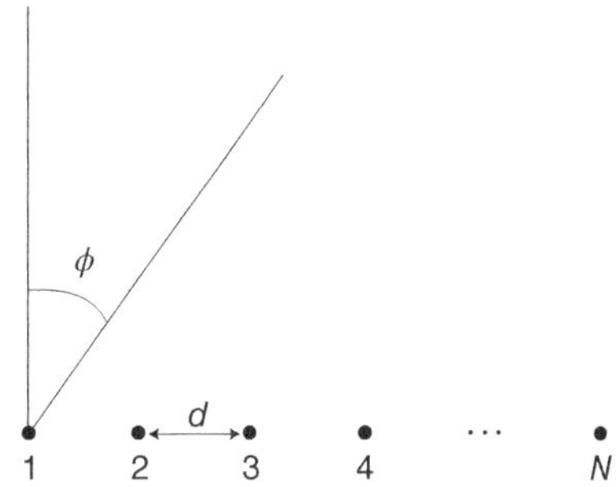

Figure 8.1 The uniform linear array, after [6]. Source: From [6] / with permission from McGraw-Hill.

where $AF(\theta, \phi)$ is the *array factor*. This analysis is often done assuming isotropic radiators – or equivalently, point sources – but for identical elements, it can be generalized as earlier. Since both the overall array pattern and the element pattern have a $1/r$ decay, the array factor is independent of r. Moreover, in the plane as shown in Figure 8.1, it is a function of the angle ϕ only:

$$\vec{E}(\phi) = \vec{E}_{\text{elem}}(\phi) AF(\phi) \tag{8.2}$$

For the particular case of the ULA, we obtain using the formula for a geometric series:

$$\begin{aligned} E(\phi) &= E_{\text{elem}}(\phi) \frac{1 - e^{jN\psi}}{1 - e^{j\psi}} \\ &= E_{\text{elem}}(\phi) \frac{\sin(N\psi/2)}{\sin(\psi/2)} \angle (N-1)\frac{\psi}{2} \end{aligned} \tag{8.3}$$

with $\psi = \beta d \sin\phi + \delta$ and $\beta = 2\pi/\lambda$. We have further simplified the notation by retaining only the relevant vector component. By placing the phase reference at the center of the array, the phase term is eliminated, and the array factor is simply the term

$$AF(\phi) = \frac{\sin(N\psi/2)}{\sin(\psi/2)} \tag{8.4}$$

The major limitation in the aforementioned analysis is the exclusion of *mutual coupling*; the element patterns are assumed to be identical. These would usually be isolated element patterns. Mutual coupling can be brought into the analysis in the case of an infinite array, by taking the element pattern $\vec{E}_{\text{elem}}$ to be the pattern of an element embedded in it. For a finite array, the element patterns are not identical; and the approximation in Eq. (8.2) breaks down. For a finite, mutually coupled array, the analysis requires computing the patterns for each individual array element, the topic of Section 8.3.

A typical further assumption is that *active impedances*[1] are the *isolated element impedances.*[2] Mutual coupling in general causes the elements to present different impedances at the ports. Note that infinite array approximations are often used instead, but they may not be sufficiently accurate for finite arrays.

While design methods based on the array factor lead to values for the driving point currents on the array elements, most transmission-line feed networks are *voltage* sources. The port currents are given in terms of the port voltages as

$$\mathbf{i}_{\text{port}} = \mathbf{Y}_A \mathbf{v}_{\text{port}} \tag{8.5}$$

where $\mathbf{Y}_A$ is the array mutual admittance matrix. As a result, the port currents differ from the applied port voltages. Since $\mathbf{Y}_A$ is nondiagonal for a mutually coupled array, the driving currents

1 The impedance in the array environment.
2 The impedance of the array element in the absence of neighboring elements.

will be incorrect if the voltages are taken to be the excitations for a given array factor without accounting for this relationship in the beamforming network. This point is illustrated later in this chapter.

8.3 Embedded Element Patterns

Since EEPs form a core part of this chapter, it is appropriate to discuss these in some detail. EEPs were first described in the literature in the early 1990s [7, 8]. In the early literature the terms "active" and "scan" patterns are also used. An EEP represents the radiation pattern of an array with one element driven (the active element) and all other elements being passive with their ports terminated. The key point is that the EEP is the radiation pattern taking the full array into account, including scattering by elements near the driven element.

We will start with an intuitive example before moving on to more general definitions. Figure 8.2 shows the EEPs for a two-element thin dipole array whose elements are parallel to each other. The dipoles are both $\ell = 0.5\lambda$ long, spaced $d = 0.5\lambda$ apart, and have the radius 0.0025λ. These patterns depend on the port terminations on the undriven ports. We show the EEPs for open, matched (50 Ω) and short circuited (SC) terminations. The EEPs for the open circuit load mean that one element port is driven and the other element port is open circuited (OC). The EEPs for the matched and short circuit terminations are computed similarly, with one port driven and the other matched or short circuited, respectively.

The EEPs for the OC terminations are almost unchanged from an isolated element pattern. The field radiated by the driven element toward the OC element are scattered by two short $d = \lambda/4$ wires, which are nonresonant and interact weakly with the radiated field from the driven dipole. This idea will be exploited later as a minimum scattering antenna (MSA). Note that there are some receive-only systems in use that have high port impedances, so this can have utility.

The opposite extreme is a SC EEP. As the short circuit termination is easy to implement in an integral equation solver, this is a typical EEP computed by a method of moments (MoM) simulation program. In this case, the driven element sees one resonant $d = \lambda/2$ wire, which interacts strongly with the radiated field from the driven dipole. The intermediate case, i.e. EEPs with a matched port termination, is also shown.

All of these sets of EEPs have their uses, as will be discussed. The point to note is how the mutual coupling interacts with the port terminations to produce quite different patterns. Fortunately, the differences between the EEPs for different loading conditions on the undriven element can be easily taken into account using network theory.

The next goal is to represent the EEPs mathematically. It is convenient to work with one polarization at a time. We choose a polarization defined by the unit vector $\hat{p}$ (usually $\hat{\theta}$ or $\hat{\phi}$) and denote the corresponding component of the EEP as

$$E_{p,n}(\bar{r}) = \hat{p} \cdot \overline{E}_n(\bar{r}) \tag{8.6}$$

This is the $\hat{p}$ component of the electric field radiated[3] by the array at the far field point $\bar{r}$ with the nth element of the array excited and the other elements loaded with some arbitrary impedance Z_g or admittance Y_g.

For some applications, the EEPs need to be computed over the full radian sphere, i.e. for $0 \leq \theta \leq 180°$ and $0 \leq \phi \leq 360°$. For an array located above an infinite ground, it will be a hemisphere. A full

3 We address the receiving case later in the chapter.

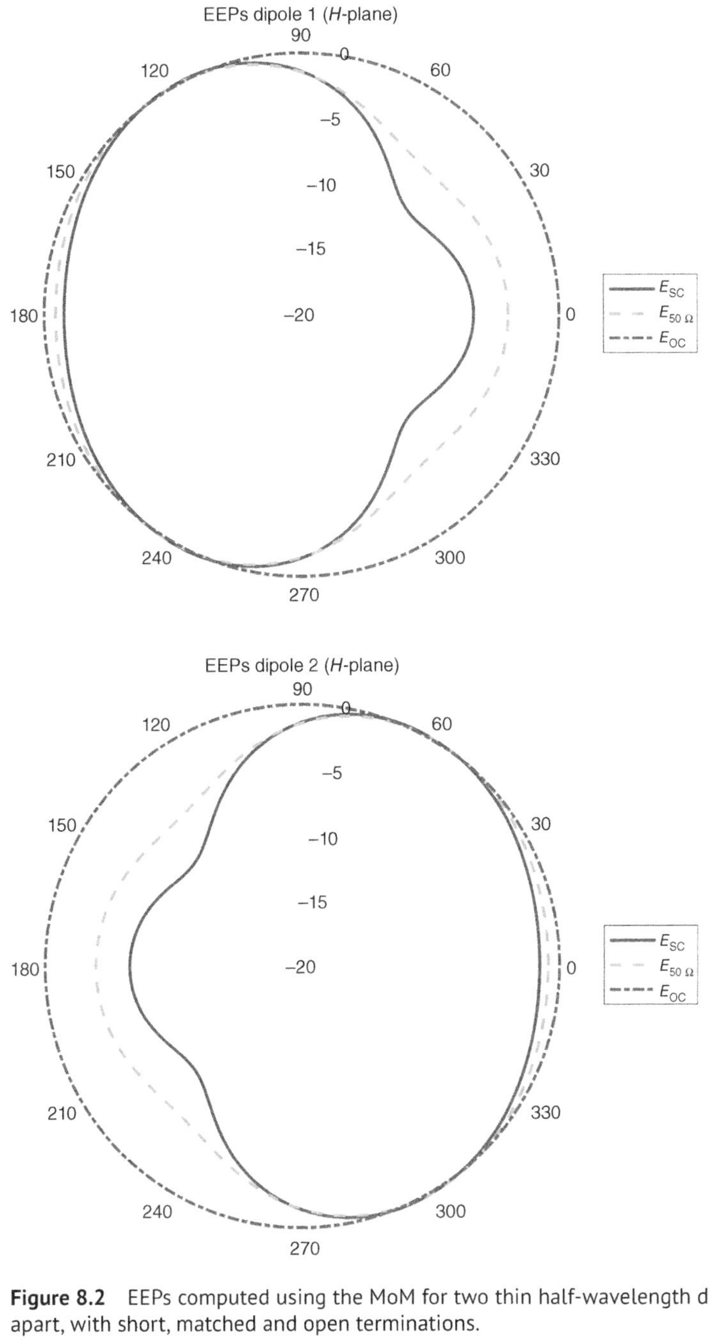

Figure 8.2 EEPs computed using the MoM for two thin half-wavelength dipoles, one half wavelength apart, with short, matched and open terminations.

set of EEPs at a specific frequency will be stored as a three-dimensional (3D) array, with dimensions $N_\theta \times N_\phi \times N_{\text{elem}}$, where N_θ is the number of θ angles, N_ϕ is the number of ϕ angles, and N_{elem} is the number of elements in the array. It should be appreciated that this can become a significant amount of data for a large array with finely sampled EEPs. For a rapidly varying pattern, it may be necessary to sample at approximate degree intervals.

For large arrays, the EEP can be computed relative to a local origin near the nth element. This choice reduces the variation with angle and the number of required samples. The phase offset relative to the global origin must be included when the EEP is used in a computation of a derived quantity, such as the overlap matrix given in the following text. If this is done, the EEPs are smoother with respect to angle. Then the required number of samples increases with the electrical size of the element, not the size of the full array.

The EEP samples are assembled into the N-element column vector

$$\mathbf{E}_p = \begin{bmatrix} \hat{p} \cdot \overline{E}_1 \\ \vdots \\ \hat{p} \cdot \overline{E}_N \end{bmatrix} \tag{8.7}$$

where we have suppressed the dependence of the EEP on the position vector $\overline{r}$. We use an overline to denote 3D fields as in $\overline{E}_n$ and bold type to denote vectors with one field sample, voltage, or current phasor per array element, as in $\mathbf{E}_p$. The bold vectors have N elements, where N is the number of array elements, and the vectors with overline are three-dimensional.

One of the properties derived from the EEPs is the *element pattern overlap matrix* $\mathbf{A}$, of dimension $N_{\text{elem}} \times N_{\text{elem}}$. This is computed from the element patterns using[4]

$$A_{mn} = \frac{1}{2\eta} \int \overline{E}_m(\overline{r}) \cdot \overline{E}_n^*(\overline{r})\, r^2 d\Omega \tag{8.8}$$

The asterisk indicates the complex conjugate. The angular integral requires the EEPs to be computed over the full angular range, sampled appropriately as discussed earlier. For accurate integration, low order quadrature schemes usually suffice.

When first introduced, EEPs were most commonly defined for matched loads terminating all of the passive elements, and this usage persists. However, EEPs can be computed using *any* convenient terminating impedance. These can be readily converted from one to another reference mathematically [9], and an example will be shown later in this chapter.

8.4 Approximate Analysis Methods: The Lossless, Resonant, Minimum Scattering Approximation

Before addressing the full-wave, rigorous solutions using CEM techniques, it is worth asking whether mutual coupling can be approximately included in some way to provide better solutions than the array factor method, but does not require a computationally expensive CEM solution. It turns out that the answer is yes, via a method called the *lossless, resonant, minimum scattering approximation* (LRMSA). LRMSA includes mutual coupling effects in the array model in an approximate way. The accuracy of this method depends on the element type as well as the array type, and it can vary significantly.

The LRMSA is based on approximating the element in an array as an MSA. An MSA is an antenna that does not perturb the incident field and produces no scattered field. As we saw in Section 8.3, an OC half-wave dipole is approximately an MSA. The EEPs for an MSA with an open circuit element loading condition are equal to the isolated element pattern with phase shifts representing the offset of each element from a global origin.

---

4 EEPs are defined with respect to the global origin. As noted earlier, it may be convenient to compute these with respect to a local origin, in which case the phase offset relative to the global origin must be included.

Central to the LRMSA is the following interesting property of the element pattern overlap matrix **A** defined in Section 8.3. By conservation of energy, the overlap matrix and the mutual impedance matrix are related by

$$\mathbf{Z}_A = \frac{2}{|I_0|^2}\mathbf{A} + \mathbf{R}_{\text{loss}} + j\mathbf{X}_A \tag{8.9}$$

where $\mathbf{R}_{\text{loss}}$ is the part of the mutual impedance matrix corresponding to losses in the antenna elements and $\mathbf{X}_A$ is the array mutual reactance matrix. If the antenna losses and mutual reactances are small, then the array mutual impedance matrix can be approximated by

$$\mathbf{Z}_A \simeq \frac{2}{|I_0|^2}\mathbf{A} \tag{8.10}$$

This important relationship allows mutual coupling effects to be brought into the array model by calculating overlap integrals of the EEPs. If the array is lossless ($\mathbf{R}_{\text{loss}} = 0$) and resonant ($\mathbf{X}_A = 0$), the approximation (8.10) is exact.

Combining both aspects of the LRMSA, the MSA assumption allows the open circuit loaded EEPs to be found from the isolated element pattern. The latter is often available in analytical form; and the lossless, resonant assumption then allows the mutual coupling matrix to be found from the overlap integrals of the approximate EEPs. Thus the mutual coupling can be brought into the array model without a time-consuming numerical simulation or full-wave CEM model. The approximate mutual impedance matrix Eq. (8.10) is then used to transform the open circuit loaded EEPs (isolated element pattern with phase shifts) to EEPs with matched loads or with the loading that the array experiences in its operational state using the formulas in [9]. When gain and other performance quantities are computed from the resulting array model, the results satisfy the conversation of energy principle. Consequently, each element in a receiving array modeled with the LRMSA experiences a reduced element gain and the overall array has an effective area that is commensurate with the array aperture size.

8.4.1 LRMSA Examples

We first compare a rigorous MoM computation for the two-element dipole array having the element lengths and spacing both 0.5λ and radius 0.0025λ with the pattern overlap integral method using isolated elements. The MoM codes are the authors' own [10, 11], verified against the literature and commercial software for this specific problem. The LRMSA approximation uses the analytical pattern for an isolated $\lambda/2$ dipole. Table 8.1 compares the array mutual impedance matrix $\mathbf{Z}_A$ computed using the MoM and LRMSA. The matrix is symmetrical, so only one row need be reported. The LRMSA is surprisingly accurate in approximating the real (resistive) part of the array mutual impedance matrix, but it cannot compute the reactive part. The reason is that the pattern overlap integrals are based on the far fields.

Table 8.1 A comparison of the MoM and LRMSA.

MoM		LRMSA	
$\mathbf{Z}_A(1,1)$	$\mathbf{Z}_A(1,2)$	$\mathbf{Z}_A(1,1)$	$\mathbf{Z}_A(1,2)$
$79.96 + j38.38$	$-16.72 - j30.92$	$73.13 + j0$	$-12.53 + j0$

a) Source: The authors.

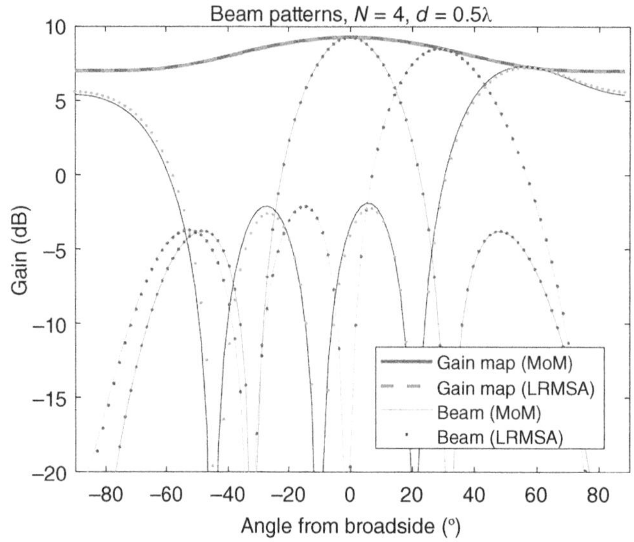

Figure 8.3 Comparison of the LRMSA with MoM methods [12]. The beam patterns are shown steered to 0°, 30°, and 60°. The gain map – the gain in the steering direction – is also shown.

Our next example is a four element dipole array with the same elements and half wavelength spacing. The beam patterns computed using full MoM solutions and the LRMSA are shown in Figure 8.3. They show excellent agreement.

Even though the LRMSA uses *isolated* element patterns, the approximately computed array mutual coupling matrix allows more accurately computed driving point currents. For system parameters that are sensitive to the full array mutual coupling matrix, the LRMSA may have more limited utility. This is another a topic that we return to later in this chapter.

8.4.2 Isotropic Radiators

An interesting result follows for an array of isotropic radiators. The pattern overlap integrals can be evaluated analytically, so that

$$A_{mn} = P_{\text{rad}} \frac{\sin(kr_{mn})}{kr_{mn}} \tag{8.11}$$

Here, r_{mn} is the distance between elements m and n. The diagonal element represents the power radiated by each element. Mutual coupling between elements oscillates, decaying inversely with the element spacing. For $\lambda/2$ spacing, the off-diagonal elements are zero and there is no mutual resistance. Real array elements often demonstrate qualitatively similar behavior in their mutual resistances.

8.5 Exact Design Methods: Full-Wave Computational Electromagnetics Modeling

CEM codes have become standard design tools in the RF and microwave industry. The most widely used methods are the MoM, finite-difference time-domain (FDTD),[5] and finite element method

5 The closely-related finite volume method is included here.

(FEM). As implemented in widely-used commercial codes, the MoM and FEM are frequency domain formulations. The FDTD operates in the time domain.

When commercial CEM packages first became widely available during the 1990s, each major software package was closely associated with a specific numerical method. Most commercial CEM packages now offer a variety of solvers within the same user interface, allowing users to select the most appropriate solver for the problem at hand. For thin dipoles, the MoM is normally the method of choice since the problem becomes one-dimensional and no absorbing boundary condition or other form of mesh closure is required. The FEM and FDTD methods mesh 3D space with volumetric elements (typically tetrahedral or cuboidal elements, respectively) and require the imposition of some form of absorbing boundary condition or other type of mesh closure such as the hybrid FEM/MoM approach.

8.5.1 MoM Modeling of Thin Wires

With the development of low frequency arrays for radio astronomy, such as LOFAR in Europe, HERA, and SKA-Low prototypes [13], there has been a modest resurgence of interest in thin wire modeling codes. The MoM solution of thin wires was one of the first problems successfully addressed with computational techniques during their introduction in the 1960s. Thin wire codes were adapted to modeling 3D structures decades ago by the commonly used Numerical Electromagnetics Code (NEC) and similar software tools. The thin wire approach to 3D structures has long since been superseded by the 3D MoM, FEM, and FDTD methods. Here, we are returning to the original purpose of the thin wire method and using it to model dipole antenna arrays.

In the present context of array analysis, the full array mutual impedance matrix, $\mathbf{Z}_A$, and the full set of EEPs are required. For an N element array, $\mathbf{Z}_A$ will be of dimension $N \times N$; and there will be N EEPs, sampled in angle. If the EEPs are used to compute overlap integrals, they are sampled over a full sphere. However, when they are used to generate beam patterns, they often need only to be sampled in one plane. For a dual-polarized array, as commonly used in radio astronomy, there is a port for each polarization. Therefore, N in the aforementioned text is replaced by $2N$. If cross-polarization effects are important, the EEPs must include both field components. It may be convenient from a coding perspective to use a global coordinate system and compute three vector components for the EEPs.

With a typical thin-wire code, where the ports are delta-gap voltage sources, not exciting a port with a voltage or exciting it with zero voltage corresponds to short-circuiting the port. As such, one can proceed with a sequence of linear system solves, one per array element, with all ports zeroed except the port on the element n under consideration. For each analysis, the nth column of the array mutual admittance matrix $\mathbf{Y}_A$ is obtained. At the same time, the SC loaded EEP $\overline{E}_{\text{SC}}$ can be found by calculating the far field radiation from the current induced on the array elements. When the solves are completed for all the ports, the full array mutual impedance matrix $\mathbf{Z}_A = \mathbf{Y}_A^{-1}$.

The OC patterns can be obtained from [9]

$$\overline{E}^{\text{OC}} = \mathbf{Z}_A \overline{E}^{\text{SC}} \frac{I_0}{v_{\text{OC}}} \tag{8.12}$$

where I_0 and v_{OC} are normalizing terms. These are the port currents and voltages used in computed the OC and SC patterns, respectively, and are often unity. Unless it is diagonal, i.e. there is no mutual coupling, the array mutual impedance matrix spreads the SC EEPs across the array.

The $\overline{E}^{OC}$ is an array containing all N patterns, per polarization. Because the radiated fields from the array for a given set of element port input current excitations can be computed from the open circuit loaded EEPs, it can be unpacked into N EEPs, $\overline{E}_n^{OC}$, indexed by element number n. The total radiated electric field, with beamformer weight vector $\mathbf{w}$, is then given by

$$\overline{E}(\overline{r}) = \frac{1}{I_0} \sum_{n=1}^{N} w_n^* \overline{E}_n^{OC}(\overline{r}) \tag{8.13}$$

In Section 8.5.2, we discuss how beams may be formed and steered using both the classical array factor method and a method fully leveraging EEPs.

8.5.2 Beamforming and Beam Steering

8.5.2.1 Limitations of the Array Factor Method

The beam of a phased array can be steered by using a progressive phase delay along the array's spatial extent chosen to reflect the geometric phase delay between the element phase centers of a plane wave in the desired steering direction. For an array with uniform amplitude excitation in the xy plane, this is given as

$$AF(\theta, \phi) = I_0 \sum_{m=1}^{N_x} e^{j(m-1)(kd_x \sin\theta \cos\phi + \alpha_x)} \sum_{n=1}^{N_y} e^{j(n-1)(kd_y \sin\theta \sin\phi + \alpha_y)} \tag{8.14}$$

where

$$\alpha_x = -kd_x \sin\theta_0 \cos\phi_0$$
$$\alpha_y = -kd_y \sin\theta_0 \sin\phi_0$$

Here, (θ_0, ϕ_0) is the main beam pointing direction, d_x and d_y are the spacings along x and y, respectively, and α_x and α_y are the progressive phase shifts required along x and y to steer the beam to the desired pointing direction.

These phase shifts do not take mutual coupling into account. Mutual coupling in general presents different impedances at the ports. Design methods based on the array factor require the correct driving point currents on dipole-type antennas because the radiated fields are proportional to current. However, most transmission-line feed networks are effectively *voltage* sources. Furthermore, the ports defined in most MoM codes are also voltage, not current, sources. As noted earlier, the port currents, in terms of the port voltages, are given by

$$\mathbf{i}_{\text{port}} = \mathbf{Y}_A \mathbf{v}_{\text{port}} \tag{8.15}$$

where $\mathbf{Y}_A$ is the array mutual admittance matrix. As a result, the port currents differ from the applied port voltages, resulting in the incorrect excitation being applied. Knowledge of the array mutual admittance matrix permits this to be corrected. The required port voltages can be computed from the desired currents by the inverse of the aforementioned equation. The same process that produces that matrix can also generate all the EEPs, permitting more sophisticated schemes to be employed. They include, for instance, the beamforming method to be described next. Even if the correct driving point currents are computed, the AF method still does not take into account variations in individual element patterns. The method to be described next also circumvents this limitation.

8.5.2.2 The Maximum Directivity Beamformer

Given knowledge of the EEPs, more sophisticated algorithms may be used to form steered beams. By combining the total radiated power and the power density as outlined in [[5], Eq. (4.63)], the directivity can be expressed in matrix form as

$$D(\Omega) = \frac{4\pi r^2 \mathbf{w}^H \mathbf{B}(\vec{r})\ \mathbf{w}}{\mathbf{w}^H \mathbf{A}\ \mathbf{w}} \tag{8.16}$$

The H superscript is the Hermitian (complex conjugate) transpose. $\mathbf{B}$ is the array signal response matrix, with elements

$$B_{mn} = \frac{1}{2\eta} \vec{E}_m \cdot \vec{E}_n^* \tag{8.17}$$

It is computed from the EEPs in the direction of the desired scan angle, and as such, is angle-dependent. $\mathbf{A}$ is the pattern overlap matrix defined in Eq. (8.8). The dimension of each of these matrices is $N_{\text{elem}} \times N_{\text{elem}}$, where N_{elem} remains the number of array elements.

It can be shown [[5], Section 4.9] that a set of beamformer weights that optimizes the directivity in a particular direction and, hence, the effective receiving area, can be obtained from the generalized eigenvalue problem

$$\lambda \mathbf{A}\mathbf{w} = \mathbf{B}\mathbf{w} \tag{8.18}$$

The directivity is optimized by choosing the maximum eigenvalue $\lambda_{\max}$. The corresponding eigenvector is the desired beamformer weight vector:

$$\lambda_{\max} \mathbf{A}\mathbf{w} = \mathbf{B}\mathbf{w} \tag{8.19}$$

In this equation, both $\mathbf{A}$ and $\mathbf{B}$ are frequency dependent via the EEPs. Additionally, $\mathbf{B}$ is scan angle dependent. For both of these matrices, accurately computed or measured EEPs are required over the frequency band of interest.

Before moving off this topic, it should be noted that there exists a rich theory around beamforming. One example is the linearly constrained minimum variance beamformer, which permits one to incorporate additional constraints such as placing nulls in the directions of fixed interferers. An extensive discussion of this topic may be found in [[5], Chapter 10].

8.5.2.3 An Example: A Base Station Array

Many wireless systems use dipole arrays, often above a ground plane, for base station applications. In this section, a four-by-two array of horizontal dipoles above a ground plane is analyzed. It is illustrated in Figure 8.4. Each dipole is 0.47λ in length, has a 0.0025λ radius, and is spaced 0.5λ in both the parallel and collinear directions. The array is positioned 0.25λ above a PEC ground plane located at $x = 0$. The dipoles are appropriately spaced to avoid grating lobes, which results in a tightly coupled array. Each dipole and its negative image in the ground plane result in an element pattern that is null in the direction of the ground plane.

Results for directivities computed using MoM thin-wire codes developed by the authors [10, 11] are shown in Figure 8.5. In the figures, the beam is scanned in the H-plane (xy-plane) from broadside (along the x-axis) to 60° off broadside. Each figure shows four results. The first (Max Dir BF) uses weights computed with the maximum directivity beamformer. The second (MoM AF) uses the MoM code, with progressive voltage phasing derived from the array factor. As noted previously, this approach does not account for mutual coupling. Verification results computed with the

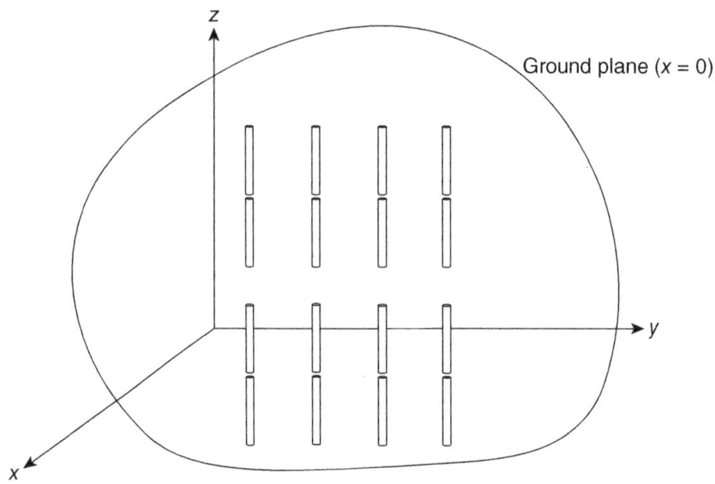

Figure 8.4 The four-by-two array of approximately half-wavelength dipoles above a ground plane analyzed in the text.

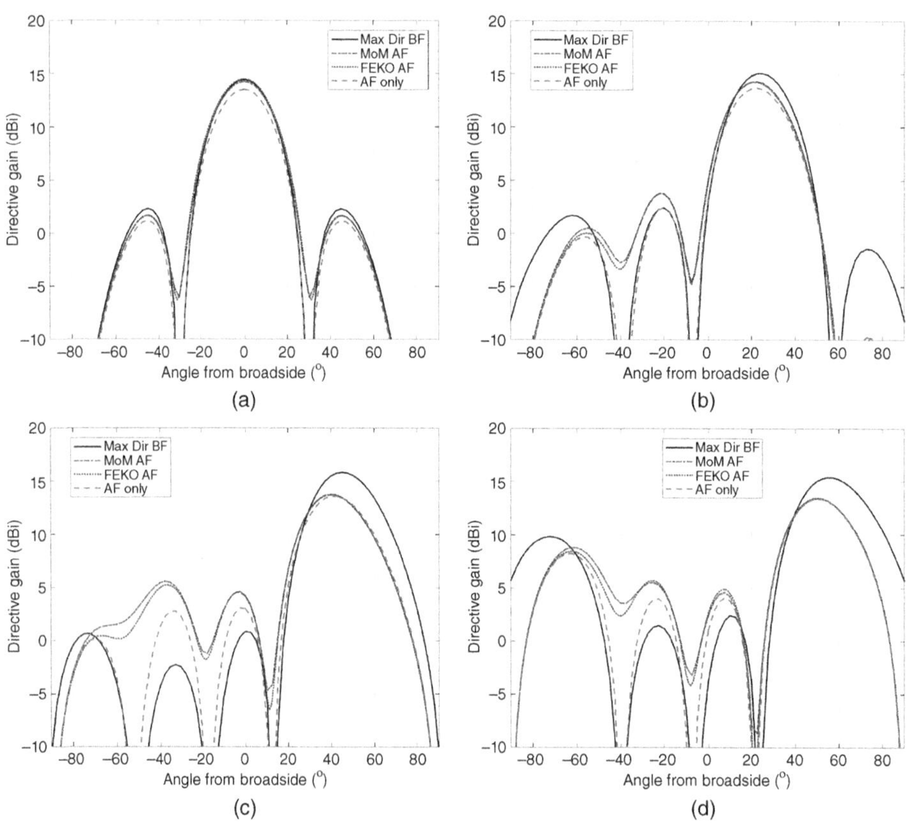

Figure 8.5 Directivity as a function of the scan angle from broadside for the 4 × 2 dipole array in the text, comparing the maximum directivity beamformer with array factor based methods. (a) Broadside. (b) Scan angle = 22.5°. (c) Scan angle = 45°. (d) Scan angle = 60°.

commercial software FEKO, particularly FEKO's thin-wire Galerkin MoM (FEKO AF) implementation are also shown. Finally, the array factor-only (AF-only) results are plotted. For broadside (no scanning), all four results are in excellent agreement. Note that the slightly lower directivity of the AF-only method is due to the element pattern not being included. As the scan angle is progressively increased, the results diverge. As shown for the 45° scan angle, a beam squint has developed. The peak of all the beams using the AF in one form or another is at about 40°. The peak directivity has reduced by around 2.4 dB. Nevertheless, the maximum directivity beamformer is still working well at 45°. This effect is even more pronounced at 60°. Larger scan angles are limited by the null imposed by the ground plane at 90° with respect to the broadside direction.

8.5.3 Summarizing Array Analysis Options

In terms of simplicity, computational cost, and accuracy, the following considerations apply to the three approaches to array modeling discussed earlier. The classical array factor method is the simplest and cheapest. Nevertheless, the patterns realized may be significantly in error for arrays with strong mutual coupling. This is especially true as an array is scanned. By incorporating mutual resistance, the LRMSA can provide a significant improvement in accuracy at minimal computational overhead. However, it is limited to array elements which approximate an MSA to some degree. Finally, full-wave EM modeling provides a numerically rigorous solution, but potentially at high computational cost for large array with complex elements.

Another option not considered in this chapter is the use of infinite array solutions for large arrays. The array performance can then be calculated with formulas, but for practical reasons usually still involves CEM tools. The accuracy of the analysis depends on how well the infinite array approximates the actual finite arrays. For most communication applications, arrays are quite modest in size; and hence, infinite array simulations may be of limited utility.

8.6 Receive-Only Systems

Most telecommunication systems are both transmit and receive. Nonetheless, there is an extensive class of high-sensitivity, receive-only systems which have unique design considerations. These include radio telescope, passive remote sensing, and satellite receiving systems. In spite of the interesting research issues surrounding receiving arrays, these systems have received fairly limited coverage in the literature. Central to receive-only systems is the reciprocity principle for obtaining antenna receiving characteristics from radiated EEPs and an appropriate network model for the array. Since the traditional focus in antenna theory is on transmitters, a number of long-standing figures of merit for antennas need to be generalized in the context of receive-only array systems.

8.6.1 Array Receivers: Network Model

The network model shown in Figure 8.6 includes the array elements; the amplifiers, which are usually LNAs; other receiving circuitry, e.g. mixers and analogue to digital converters; and finally a beamformer. The last feature is typically digital.

The amplifier input stage may be matched to the system impedance Z_0, but this is not essential as the power levels after the LNA are somewhat arbitrary when considering their effect on the overall signal-to-noise ratio (SNR). What is important is that the LNA is well noise-matched to the port impedances so that the SNR at the beam output is as high as possible. For a receiving array, the

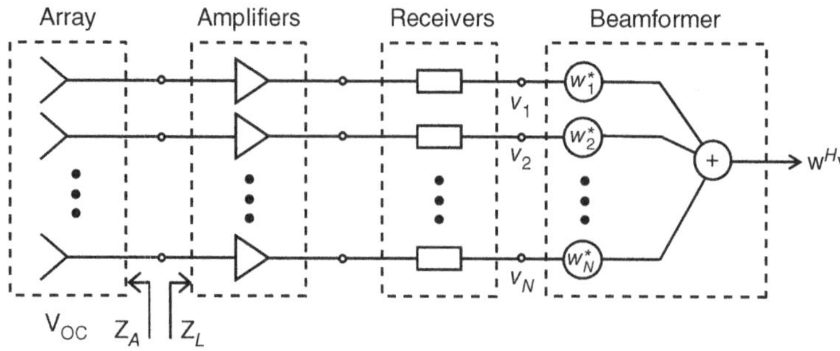

Figure 8.6 Array antenna receiver and beamformer system. Baseband voltage signals at the receiver outputs are arranged into the column vector **v** and combined – after multiplying by a vector of complex beamformer weights **w** – to yield a scalar output voltage $v = \mathbf{w}^H\mathbf{v}$. Source: From [5] / with permission from Cambridge University Press.

impedances presented at the element ports to the LNAs are active and depend on the beamformer coefficients. Active impedances have been understood for many years in the context of transmitting arrays. More recently, the community has developed a deeper understanding of active impedance matching for receivers [14, 15].

8.6.2 Array Receivers: Reciprocity and EEPs

An array receiver system with active electronics and digital signal processing is clearly not reciprocal. In the receiving chain, reciprocity does apply to the antenna elements provided that they are made of reciprocal materials and also that the external medium is reciprocal. These conditions are almost always the case in practice. This aspect of the system permits a connection to be made between the EEPS, which are computed on transmit and the voltages induced by an incident wave at array element ports.

The open circuit voltages at the terminals of the nth array element when the incident field intensity is E^{inc} are [16–19]

$$v_{\text{oc},n} = \frac{4\pi j r e^{jkr}}{\omega\mu I_0} E^{\text{inc}} \left[\hat{p} \cdot \overline{E}_n(\overline{r}) \right] \tag{8.20}$$

Here, $\overline{E}_n(\overline{r})$ is the open circuit loaded embedded element radiated field pattern with the nth array element excited, $\hat{p}$ is a unit vector representing the polarization of the incident wave, and E^{inc} is the phasor intensity. The point $\overline{r} = r\hat{r}$ is such that $\hat{r}$ is in the direction from which the incident wave arrives at the receiving array. For convenience in matrix operations on the array outputs, we assemble the voltages $v_{\text{oc},n}$ into an N-element column vector $\mathbf{v}_{\text{oc}}$.

For element ports terminated with loads $\mathbf{Z}_L$ and with the array mutual impedance matrix $\mathbf{Z}_A$, the loaded voltages at element ports (inputs to LNAs) are

$$\mathbf{v}_A = \mathbf{Z}_L(\mathbf{Z}_L + \mathbf{Z}_A)^{-1}\mathbf{v}_{\text{oc}} \tag{8.21}$$

The receive voltages here include both the load impedance matrix *and* the mutual impedance matrix. If the EEPs are computed using the actual terminating loads rather than the OC patterns, then this equation simplifies to samples of the components of the EEPs aligned with the incident wave polarization and evaluated at a point in the direction of the source of the incident field. The loaded voltages $\mathbf{v}_A$ at the element ports are related by a further linear transformation to the voltages at the outputs of the signal paths before beamforming, labeled as **v** in Figure 8.6.

Combining the transformation from the loaded port voltages to the array output voltages into one linear transformation leads to

$$\mathbf{v} = \mathbf{Q}\mathbf{v}_{\mathrm{oc}} \tag{8.22}$$

where **Q** is a transformation matrix, which incorporates the effects of the amplifier and receiver network. If the receiver chains in the system have identical complex voltage gains g, and the receiver chains are uncoupled after the first stage LNAs. This matrix is then given by

$$\mathbf{Q} = g\mathbf{Z}_L(\mathbf{Z}_L + \mathbf{Z}_A)^{-1} \tag{8.23}$$

8.6.3 Array Receivers: Figures of Merit

For a receiver, the key measure of performance is the SNR value. The SNR with the signal intensity normalized out can be converted into the receiver sensitivity as

$$\frac{A_e}{T_{\mathrm{sys}}} \tag{8.24}$$

or, equivalently, to the ratio G/T_{sys}, where T_{sys} is the equivalent system noise temperature. This quantity is so fundamental that it is referred to in the IEEE standard for antenna terms rather generically as the “figure of merit” [20].

When analyzing or designing a complex system, we desire to express the SNR and sensitivity in terms of efficiency parameters related to the antenna radiation pattern, antenna losses, and noise sources from the external environment and receiver electronics. Because the system cannot readily be operated or modeled as a transmitter, traditional antenna parameters such as gain and antenna efficiency are challenging to apply to an array receiver.

As a step toward expressing the receiver sensitivity in terms of the efficiencies of each component in the system, we will express the signal and noise powers using the array output voltage correlation matrix. The correlation matrix represents the expectation values as the outer product of the complex voltage time series at the array outputs, i.e.

$$\mathbf{R}_{\mathbf{v}} = E[\mathbf{v}\mathbf{v}^H] \tag{8.25}$$

$\mathbf{R}_{\mathbf{v}}$ is an N by N matrix that captures the signal and noise power at each element port and the spatial structure of the signal and noise environment. The array output voltage correlation matrix includes contributions from the signal of interest, $\mathbf{R}_{\mathrm{sig}}$, and the noise from the environment, antenna losses, and receiver electronics, $\mathbf{R}_n$. The sensitivity of the array in terms of the signal and noise correlation matrices is

$$\frac{A_e}{T_{\mathrm{sys}}} = \frac{k_B\, B}{S_{\mathrm{sig}}} \frac{\mathbf{w}^H\, \mathbf{R}_{\mathrm{sig}}\, \mathbf{w}}{\mathbf{w}^H\, \mathbf{R}_{\mathrm{n}}\, \mathbf{w}} \tag{8.26}$$

in units of m^2/K, where k_{B} is the Boltzman constant. The numerator, the signal correlation matrix, is given by

$$\mathbf{R}_{\mathrm{sig}} = 2\eta |c_1|^2 S_{\mathrm{sig}} \mathbf{Q}\, \mathbf{B}_p\, \mathbf{Q}^H \tag{8.27}$$

The single-polarization array signal response matrix is

$$\mathbf{B}_p(\bar{r}) = \frac{1}{2\eta} \mathbf{E}_p(\bar{r})\, \mathbf{E}_p^H(\bar{r}) \tag{8.28}$$

where $\mathbf{E}_p$ is the vector of EEPs evaluated in the direction of arrival of the signal of interest. S_{sig} is the power density in the incident signal of interest and c_1 is the leading scalar factor in (8.20).

The denominator is the noise correlation matrix. It is in general not straightforward to obtain. Subject to some simplifying assumptions including identical lossless antennas; impedance and noise matched ports; real mutual impedances $Z_A = R_A$ and all elements identical, one finds that for receiver noise dominated systems it is given by

$$\mathbf{R}_n \approx \mathbf{R}_{\text{LNA}} = k_B T_{\min} B\, R_A(\mathbf{I} + \tilde{\mathbf{A}}\, \tilde{\mathbf{A}}^H) \tag{8.29}$$

Here $\tilde{\mathbf{A}}$ is the normalized pattern **overlap matrix**, normalized such that the diagonal elements are equal to unity. This approximation shows how coupling in an array, as captured by the pattern overlap matrix, relates to the degree of cross-correlation of the front end amplifier noise. A full noise model, including external thermal noise, antenna loss noise, and receiver noise can be found in [5].

8.6.4 An Example of Receiver System Temperature Calculation

The computation of the system temperature of an array scanned significantly off broadside using the theory earlier is considered in this subsection. We take the opportunity to revisit the LRMSA. It is clear from Eq. (8.29) that the overlap matrix plays an important role in determining the system noise temperature. We will see from the example in the following text that some applications require more accuracy than the LRMSA can provide and a full-wave model of the array is required to accurately model the receiver sensitivity.

The example system is a linear array comprised of four parallel half-wave thin dipoles, spaced a half-wavelength apart. The following LNA parameters are assumed: $T_{\min} = 50$ K, $R_n = 3\ \Omega$ and $Z_{\text{opt}} = 50\ \Omega$. The results in Figure 8.7 show T_{sys} computed using a full MoM code, an EMF approximation to the mutual impedance matrix, and an LRMSA approximation to the overlap matrix

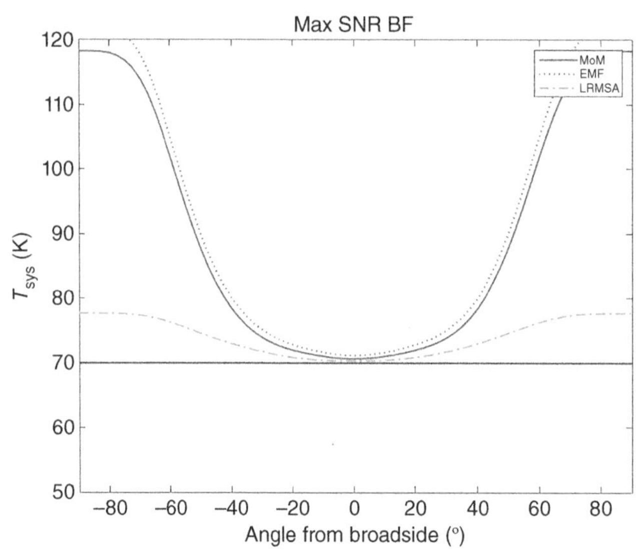

Figure 8.7 Comparison of the LRMSA with MoM methods for the system temperature T_{sys} as a function of the beam scan angle [12]. Source: From [12] / with permission from IEEE.

A and $\text{Re}[Z_A]$. (Details of this LRMSA computation are given in [[5], Section 6.16]). The weights are computed using a maximum SNR beamformer, a generalization of the maximum directivity beamformer discussed earlier.

As the beam steers toward endfire, the LRMSA curve reflects the qualitative variation of the system noise with the beam scan angle caused by the mutual coupling. For steered beams, the active impedance mismatch at the element port and LNA junctions worsens and the system noise increases. Compared to the MoM results, the LRMSA under-predicts the active impedance mismatch. It does not include the reactive part of the array mutual impedance matrix which contributes to the active impedance mismatch in the MoM calculation. This example shows that the LRMSA may be useful when designing the array to reduce mutual coupling effects. While it is much faster than the MoM solution, quantitative prediction of the system noise requires the full-wave model.

8.7 An Open Question Around EEPs

An interesting and open question is to what accuracy do the set of EEPs need to be computed? Using a numerical model, we compute EEPs for an example array as the number of degrees of freedom in the model increases. Figure 8.8 shows the convergence of a set of EEPs as the number of MoM segments is increased. Figure 8.9 shows the convergence of the full array pattern. This is a six element ULA, comprising thin (0.001λ) half-wavelength dipoles. The dipoles are parallel to each other, with inter-element spacing reduced to 0.4λ to deliberately accentuate mutual coupling effects. The results in Figure 8.9 are almost indistinguishable; the array factor has dominated the effects of slightly differing EEPs and effectively "washed" them out. However, this has not been systematically studied to date and a general statement in this regard cannot yet be made.

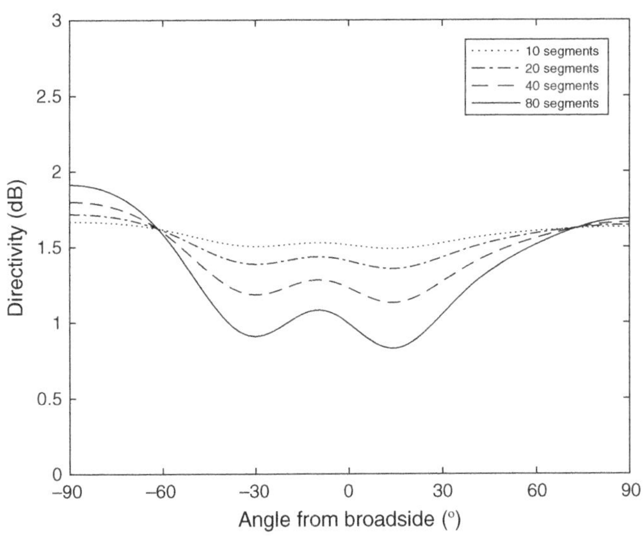

Figure 8.8 Convergence of the EEPs for the six element ULA described in the text. The third array element EEP is shown. The slight asymmetry is due to there being only two elements on one side, but three on the other.

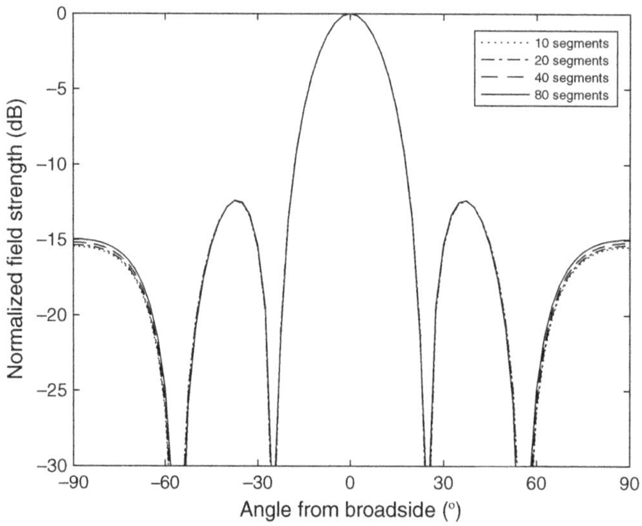

Figure 8.9 Convergence of the full array patterns as a function of the MoM discretization. In this case, the results for the main beam and first side-lobe lie on top of each other. The differences only become noticeable in the second side-lobe.

8.8 Conclusions

This chapter has provided an overview of the use of EEPs and the array impedance (or admittance) matrix in array system modeling. The theoretical framework for array modeling in terms of the overlap matrix and its relationship to signal and noise correlation matrices for a receiving array is known, but has been scattered across the literature. The overlap matrix and network model approach does not appear in standard antennas textbooks, which by and large focus on the traditional array factor approach to array design. Outside a small community well versed in these powerful methods, they have not been widely adopted by antenna engineers. A more detailed exposition of this theory may be found in [5], which to best of the authors' knowledge has been the first to provide a coherent synthesis of this material in a book.

Progress in CEM has made the analysis and design of arrays tractable using full-wave models that rigorously incorporate mutual coupling. Large, finite arrays still challenge CEM packages, and work continues on finding faster and more efficient methods for handling such arrays computationally. For future communication applications, it can be expected that performance requirements will continue to become more demanding. Most dense arrays exhibit significant mutual coupling and design methods rigorously incorporating mutual coupling effects will remain important.

References

1 C. A. Balanis, *Antenna Theory: Analysis and Design*. Hoboken, NJ: John Wiley and Sons, 3rd ed., 2005.

2 J. D. Kraus and R. J. Marhefka, *Antennas for All Applications*. New York, NY: McGraw-Hill, 3rd ed., 2003.

3 R. J. Mailloux, *Phased Array Antenna Handbook*. Boston, MA, USA: Artech House, 3rd ed., 2018.

4 W. L. Stutzman and G. A. Thiele, *Antenna Theory and Design*. New York, USA: John Wiley and Sons, 2nd ed., 2012.

5 K. F. Warnick, R. Maaskant, M. V. Ivashina, D. B. Davidson, and B. D. Jeffs, *Phased Arrays for Radio Astronomy, Remote Sensing, and Satellite Communications*. Cambridge, UK: Cambridge University Press, 2018.

6 J. D. Kraus, *Radio Astronomy*. New York, USA: McGraw-Hill, 1966.

7 D. F. Kelley and W. L. Stutzman, "Array antenna pattern modeling methods that include mutual coupling effects," *IEEE Trans. Antennas Propag.*, vol. 41, no. 12, pp. 1625–1632, Dec 1993.

8 D. M. Pozar, "The active element pattern," *IEEE Trans. Antennas Propag.*, vol. 42, no. 8, pp.1176–1178, 1994. 10.1109/8.310010.

9 K. F. Warnick, D. B. Davidson, and D. Buck, "Embedded element pattern loading condition transformations for phased array modeling," *IEEE Trans. Antennas Propag.*, vol. 69, no. 3, pp.1769–1774, 2021. 10.1109/TAP.2020.3027593.

10 D. B. Davidson, *Computational Electromagnetics for RF and Microwave Engineering*. Cambridge, UK: Cambridge University Press, 2nd ed., 2011.

11 K. F. Warnick, *Numerical Methods for Engineering: An Introduction Using MATLAB and Computational Electromagnetics Examples*. Rayleigh, NC: SciTech, 2011.

12 K. F. Warnick and D. B. Davidson, Fast modeling, loading transformations, and impedances from embedded element patterns for mutually coupled antenna arrays. In *North American Radio Science Meeting*, p. 1, July 2020, Montreal, Canada.

13 D. B. Davidson, P. Bolli, M. Bercigli, P. Di Ninni, J. Monari, F. Perini, M. Sokolowski, S. Tingay, D. Ung, M. Waterston, R. Wayth, and F. M. Zerbi, Electromagnetic modelling of the SKA-LOW AAVS2 prototype. In *2020 XXXIIIrd General Assembly and Scientific Symposium of the International Union of Radio Science*, pp. 1–4, Aug 2020. 10.23919/URSI-GASS49373.2020.9232307.

14 R. Maaskant and E. E. M. Woestenburg, Applying the active antenna impedance to achieve noise match in receiving array antennas. In *2007 IEEE Antennas and Propagation Society International Symposium*, pp. 5889–5892. IEEE, 2007.

15 K. F. Warnick, B. Woestenburg, L. Belostotski, and P. Russer, "Minimizing the noise penalty due to mutual coupling for a receiving array," *IEEE Trans. Antennas Propag.*, vol. 57, no. 6, pp.1634–1644, 2009.

16 A. T. de Hoop, "The *n*-port receiving antenna and its equivalent electrical network," *Philips Res. Rep.*, vol. 30, pp. 302–315, 1975.

17 A. T. de Hoop and G. de Jong, "Power reciprocity in antenna theory," *Proc. Inst. Electr. Eng.*, vol. 121, no. 10, pp.1051–1056, 1974.

18 W. A. Davis and S. Licul, "Antennas," in *An Introduction to Ultra Wideband Communication Systems*, J. Reed, Ed. Englewood Cliffs, NJ, USA: Prentice Hall, pp.159–162 2005.

19 P.-S. Kildal, *Foundations of Antenna Engineering: A Unified Approach for Line-of-Sight and Multipath*. Gothenburg, Sweden: Kildal Antenn AB, 2015.

20 IEEE Std 145-2013, "IEEE Standard Definitions of Terms for Antennas," 2014.

9

Angle-of-Arrival Estimation in Large-Scale Hybrid Antenna Arrays

Kai Wu and Y. Jay Guo

Global Big Data Technologies Centre, University of Technology Sydney, Sydney, Australia

9.1 Introduction

Large-scale antenna arrays, or massive antenna arrays, are widely regarded as a critical technology for the emerging fifth-generation (5G) [1–3] and beyond 5G communications (B5G) systems [4]. These large-scale antenna arrays can produce very high antenna gain and needle-like antenna beams, thus enabling long communications range and spectrum reuse in the spatial domain. Another driver for employing large-scale antenna arrays is the frequency increase planned for 5G and B5G systems. The scarcity of spectrum resources has already pushed 5G toward the millimeter-wave (mmWave) frequencies [5] and 6G is expected to occupy part of the terahertz (THz) range. While using higher frequencies leads to greater propagation losses, the related shorter wavelengths fortunately allow for the integration of larger numbers, i.e. up to hundreds and even thousands, of miniaturized antennas in a limited space. The consequent array gain can partially compensate for the significant radio propagation losses associated with mmWave and THz signals [4, 6].

It should be noted that the physical sizes of radio frequency (RF) chains, which consist of analog-to-digital converters (ADC) and digital-to-analog converters (DAC), power amplifiers, and filters, etc., do not shrink as much as the antenna array when the frequency become higher [7, 8]. Furthermore, the digital signal processing of these large-scale arrays would consume a large amount of energy, thus rendering it unacceptable for many applications. Consequently, hybrid antenna arrays have emerged as an important candidate for the implementation of large-scale antenna arrays, particularly for 5G massive MIMO systems [2, 9–19].

A hybrid antenna array consists of an adequate number of analog subarrays with phase controllable antenna elements. These analog subarrays are then combined digitally. They are typically arranged in a localized configuration to facilitate wiring and schematic design [9, 12]. Their antennas are co-located as illustrated in Figure 9.1a. The number of required RF chains in a hybrid antenna array equals the number of its analog subarrays. Moreover, each subarray can be effectuated with a different type of antenna system, e.g. a phase-controlled antenna array or a passive beamforming network (BFN).

As illustrated in the following text, large-scale hybrid antenna arrays can be employed for many important and interesting 5G and B5G applications. Some of the most popular hybrid array structures are presented. As a key to effectively employing large-scale antenna arrays, the challenges of and solutions to angle-of-arrival (AoA) estimation for hybrid arrays are discussed.

Antenna and Array Technologies for Future Wireless Ecosystems, First Edition.
Edited by Y. Jay Guo and Richard W. Ziolkowski.

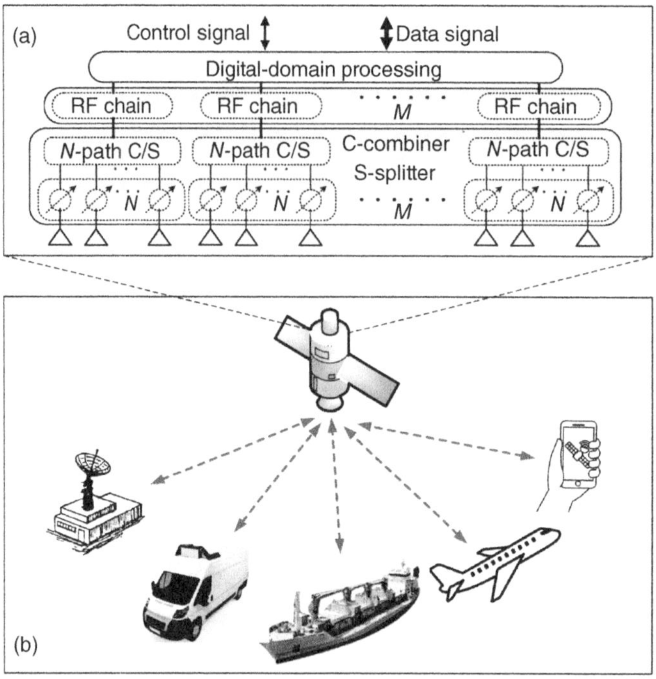

Figure 9.1 Potential applications of hybrid arrays in satellite communications. (a) A one-dimensional linear localized hybrid antenna array. (b) Popular satellite communication scenarios. Source: nerthuz / Adobe Stock.

9.1.1 Satellite Communications

Large-scale antenna arrays are a promising technique for future wideband multi-beam terabit satellite communications [20, 21]. Possible applications include network connectivity to remote or isolated regional areas, as well as to aircraft and vessels typically beyond the coverage of terrestrial wireless networks [22]. This is illustrated in Figure 9.1b.

Due to the fact that the available communications spectrum is becoming increasingly scarce [23], these future terabit satellite communications systems are very likely to operate in the higher Q/V bands, between 33 and 75 GHz [24]. Large-scale antenna arrays are uniquely suited for them. Their large number of antennas can increase substantially the antenna gain-to-noise temperature ratio (G/T). It can also enable spatial multiplexing, thus improving the spectral efficiency of satellite communications [21]. Furthermore, the short wavelengths of the Q/V frequency bands allow for the integration of very large numbers of antennas a limited satellite platform. The associated high beamforming gains of these large-scale antenna arrays can help compensate for the signal attenuation faced by long-distance radio propagation at these higher frequencies [6, 25, 26].

9.1.2 Simultaneous Wireless Information and Power Transfer (SWIPT)

Large-scale antenna arrays have been extensively adopted in the studies of wireless information and power transfer (WIPT). This interest is due to their enormous potential for providing very large spatial degrees-of-freedom (DoF). While RF radiation has been extensively used for wireless communications, it has recently been increasingly considered for providing energy to electronic devices. Wireless power transfer (WPT) is critical to self-sustaining electronic devices and networks, especially in human-unfriendly environments where battery replacement and recharging are difficult or even impossible [27]. In fact, WIPT has a great potential to facilitate environment-friendly and self-sustainable communication networks [28].

In general, there are two types of WIPT designs, namely, (i) simultaneous WIPT in which energy signals are modulated to serve the purpose of wireless information transfer (WIT) [26, 29, 30]; and (ii) concurrent WPT and WIT, where the transmissions of information and energy signals take place at different transmitters [31–33]. Because of the strong attenuation of wirelessly transferred power as a function of distance, power transfer efficiency is the key measure of WIPT. Large-scale antenna arrays are an effective means to enhance it.

9.1.3 High-Speed Railway (HSR) Communications

A critical application of large-scale mmWave antenna arrays is high-speed railway (HSR) communications. Being an efficient and sustainable means of land transport, future HSRs are envisaged to be safer, greener, and more convenient [34, 35]. Given the predicted increased passenger capacity of high-speed trains (HSTs), there is expected to be a substantially increased number of wireless connections between passengers and the Internet. A practical HSR scenario is illustrated in Figure 9.2. A train-top antenna array acts as the proxy to relay in-cabin (WiFi, cellular, or Ethernet) traffic to track-side base stations (BSs) or communications satellites.

Given the scarcity of frequencies and the demand for high data rates of tens of gigabits per second (Gbps) [34, 36], high carrier frequencies with broad bandwidths, such as mmWave and THz, are likely to be used. The mmWave frequency range: 24.25–52.6 GHz has already been specified for 5G new radio and considered for HST scenarios [37]. Feasibility studies, including extensive measurement campaigns, have been conducted to verify the efficacy of this approach. For example, a 30 GHz large-scale hybrid array has been tested over an HST link between an onboard relay and track-side infrastructure [37]. A significant improvement of the spectral efficiency compared with legacy LTE-configured HST has been demonstrated at a train speed of 500 km/h [37, Fig. 5]. While there are still issues in achieving THz antenna arrays with low loss and high gains [34], they have nevertheless started to show promise. For example, a recent work [38] has reported a 400-GHz antenna array with a measured gain of 33.66 dBi.

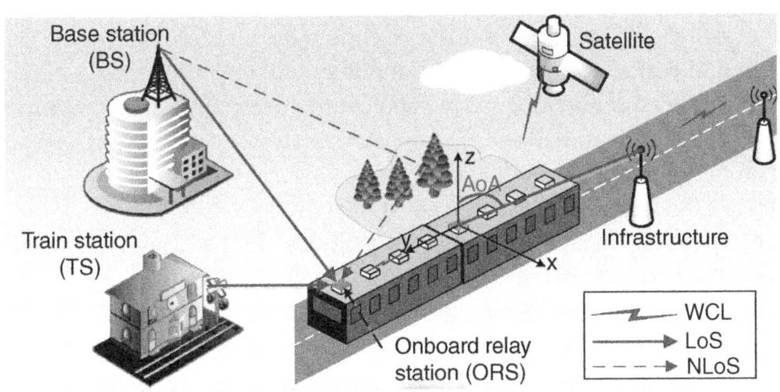

Figure 9.2 Schematic diagram of the "smart" railway communication system with the interconnections between the HST and BS, TS, infrastructures, as well as satellite. WCL refers to wireless communication link.

9.1.4 Angle-of-Arrival (AoA) Estimation

AoA estimation is a crucial technique for the operation of large-scale antennas arrays in many mmWave and THz applications. It enables in situ configuration of the arrays to quickly capture the impinging signals and then form narrow and high gain beams, thus achieving high signal-to-noise ratio (SNR) and throughput. Therefore, it is indispensable for efficient operation of large-scale hybrid arrays. A fast and accurate AoA estimation can also help avoid the long round-trip delay of the channel estimation and feedback associated with conventional systems, thus speeding up beamforming and guaranteeing seamless, reliable connectivity. However, the conventional AoA estimation techniques become inadequate for hybrid antenna arrays. This drawback occurs because these techniques are typically designed for digital arrays in which the numbers of antennas and RF chains are assumed to be the same. For hybrid arrays, however, the number of RF chains is much smaller than that of the antennas elements. Important channel information that is accessible to digital arrays, such as the phase difference between antennas, is not readily accessible with mmWave hybrid antenna arrays. To address these challenges, significant progress has been made in recent years to attain fast and accurate AoA estimation for large-scale hybrid antenna arrays operating at mmWave frequencies. This chapter aims to provide an overview of this increasingly important research topic.

In particular, we focus on the state-of-the-art AoA estimation methods in large-scale hybrid antenna arrays. Given their high popularity, two types of hybrid structures are mainly discussed: discrete subarrays and lens/Butler matrix based arrays. For each type of hybrid structure, low-complexity AoA estimation methods are reviewed for narrowband and wideband scenarios, respectively. Given their abilities to tackle challenging issues, such as estimation ambiguities from which many other methods suffer, some distinct AoA estimation breakthroughs in hybrid arrays of phased subarrays are presented. Moreover, some intrinsic features/properties that enable fast and accurate AoA estimations in discrete Fourier transform (DFT) beam antenna-based hybrid arrays are described.

9.2 Popular Hybrid Structures of Large-Scale Antenna Arrays

A hybrid structure becomes a preferable and cost-effective design when the scale of an antenna array is large. A hybrid antenna array consists of a large number of antennas grouped into a much smaller number of analog subarrays. Each antenna has an individual reconfigurable phase shifter; each subarray is connected to a single RF chain. As mentioned earlier, the physical sizes of the RF chains, which consist of ADCs and DACs, power amplifiers, and filters, do not shrink much in the mmWave frequency band, in contrast to lower frequency band counterparts. In such a structure, hardware impairments to optimal performance, like different delays in the RF components, can be calibrated and suppressed before baseband the digital signal processing stage.

Some one-dimensional linear large-scale hybrid arrays are shown in Figure 9.3. The previously noted localized arrangement facilitate the wiring and schematic designs [39], as depicted in Figure 9.3a, in which subarrays consist of discrete antenna elements and, hence, are referred to discrete antenna arrays (DAAs). They are generally a uniform linear or rectangular array. In the 3GPP HST evaluation [37], a 30-GHz large-scale hybrid array with 256 antennas was installed as an onboard relay. The array was rectangular with four 8×8 analog subarrays and four RF chains. The antenna spacing was half of a wavelength.

The hybrid array structure shown in Figure 9.3a is often referred to as a partially-connected hybrid array in the communications literature. This arrangement is in contrast to the

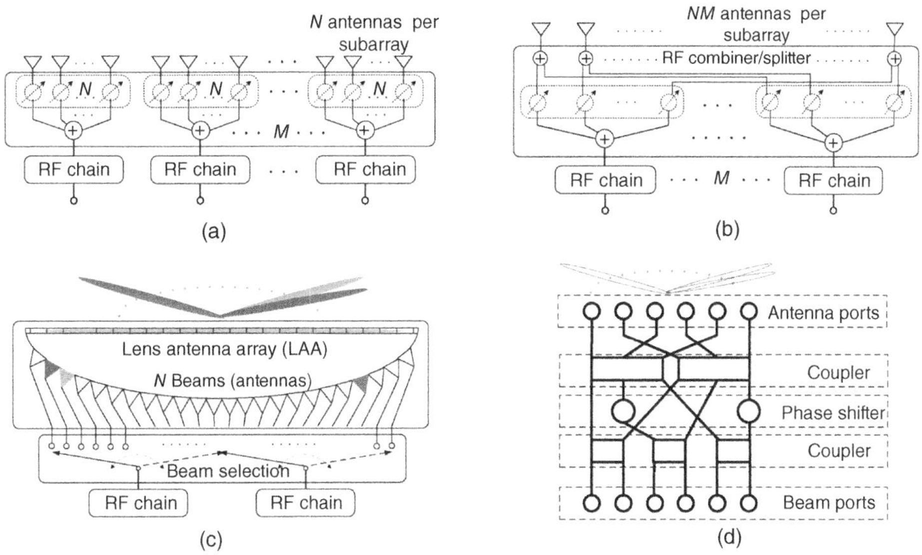

Figure 9.3 Popular hybrid array structures. (a) Partially-connected hybrid arrays. (b) Fully-connected hybrid arrays. (c) Multi-beam lens antenna array. (d) Schematic of a Butler matrix-based antenna array.

fully-connected counterpart shown in Figure 9.3b. The fully-connected one has a larger beamforming gain per subarray in comparison with the partially-connected array because many more antennas are accommodated in each subarray. However, it is easy to see that the fully-connected hybrid array requires $(M-1)N$ more phase shifters than the partially-connected one. Moreover, the power combiner/splitter in the fully-connected hybrid array needs MN-paths, while the partially-connected one requires only N-paths. Thus, the insertion loss and power consumption of a partially-connected hybrid array are much lower than in the fully-connected one.

In order to increase the energy efficiency and the level of integration, as well as to reduce the cost, a subarray can be replaced with a lens antenna or a Butler matrix [40, 41]. These structures are shown in Figures 9.3c,d, respectively. In a lens antenna array (LAA), the lens focuses the electromagnetic energy onto its parabolic focal surface. By placing its antennas on the focal surface of the lens, a complete set of DFT beams can be produced with sinc-shape beam patterns that point in directions that are evenly spaced within $[0,2\pi)$. Each antenna, referred to as a *beam port*, can be excited to activate a DFT beam. A switch can be used to select a beam port and connect it to the RF chain that is hardwired to the switch [40].

A Butler matrix is a passive analog beamforming circuit. From a signal processing perspective, a Butler matrix-based antenna array is similar to an LAA in the sense that they both generate DFT beams. Each beam port can activate an individual DFT beam, like the lens antenna. Furthermore, like the LAA, a switch can be used to select and connect a beam port to an RF chain. In addition to Butler matrix, there are also other passive BFNs, such as the Blass matrix and the Nolen matrix, that can be used as subarrays in a large-scale hybrid array. Both of these BFNs also generate DFT beams. We refer interested readers to [42] for more details on these BFNs.

The lens antenna and Butler matrix-based arrays can be much more energy-efficient than the phase shifter based DAAs due to their high integration level. Because the DFT beams can be decoupled losslessly between the azimuth and elevation [11], the one-dimensional Butler matrix-based antenna arrays and lens antennas can both be readily extended to two-dimensional uniform planar arrays. The DAAs can also be readily extended to two-dimensional uniform planar arrays if DFT beams are adopted. It should be pointed out that fully-connected hybrid arrays are less tractable because they involve too many crossover points and, hence, are less useful in practice.

9.3 AoA Estimation – State of the Art

An important issue to be addressed in localized hybrid arrays is the accurate estimation of the AoA, which is critical to beamforming, interference suppression or localization [15, 17]. The hybrid nature and the sheer scale of massive MIMO hinder the use of conventional estimation techniques [9–12]. Earlier research was focused on extending the results of analog or digital arrays to hybrid arrays.

9.3.1 Narrowband Hybrid Array of Phased Subarrays

Let N be the number of N-dimensional orthonormal bases used to be the phase shifts of a subarray over N symbols as was considered in [16]. The steering matrix of the localized hybrid array becomes an $NM \times NM$ matrix, where every M columns account for a symbol and each column only contains non-zero entries (acting as beamforming weights) corresponding to a single sub-array. The received signal vector can be constructed accordingly over N symbols.

The covariance of the steering matrix is an identity matrix and, hence, does not distort the signal subspace of the covariance of the received signals. This feature allows the classical MUSIC and ESPRIT algorithms [43] to be applied. However, they require the computationally demanding singular value decomposition (SVD) calculations. The complexity would grow cubically with the total number of antennas NM. This becomes computationally prohibitive since the number of antennas is typically great in a large-scale antenna array.

The received signal of the hybrid array considered in [17] was restructured to be a linear function of a sparse vector collecting path gains in quantized directions. The AoA was estimated through the non-zero element in the vector by using compressive sensing techniques. The coefficient of the vector, depending on a multi-resolution beamforming codebook, was adjusted to scan in different directions with different beamwidths. This algorithm was later extended to use a non-orthogonal codebook in which the beams were generated by different codewords and overlapped [19]. However, these algorithms require the receiver to repeatedly select the codewords for the transmitter to narrow down the beam. Therefore, the estimation delay would be lengthy, and the estimation accuracy would be subject to the resolutions of the codebooks.

More recent approaches take cross-correlations among subarrays for AoA estimation [11–13]. The same phase shifts were deployed in [11] across subarrays at every symbol to scan in one direction. By accumulating the cross-correlations between the received signals of adjacent subarrays, the propagation phase offset between adjacent subarrays, Nu, was estimated to be the phase of the accumulated cross-correlation, denoted by $\widehat{Nu}$. Since all the subarrays have the same complex gain in the direction of the AoA, the accumulation is constructive. The AoA u was estimated to be $\hat{u}_n = \frac{\widehat{Nu}+2k\pi}{N}$ $(k = -\lfloor \frac{N}{2} \rfloor, -\lfloor \frac{N}{2} \rfloor + 1, \ldots, \lfloor \frac{N}{2} \rfloor)$ for an odd number of N or $(N + 1)$ possible estimates due to flooring $\lfloor \cdot \rfloor$. By adjusting the phase shifts, the N or $(N + 1)$ estimates were sequentially scanned in the following N or $(N + 1)$ symbols. The estimate with the highest receive power was recursively scanned. It was updated through the accumulated cross-correlations between the received signals of adjacent subarrays to recalibrate the scanning direction until convergence was achieved. Unfortunately, a long estimation delay occurs, and the approach also suffers from error propagations resulting from an incorrectly identified initial direction. The algorithm was later applied to wideband systems, but the issue of error propagation remained [12]. The authors in [13] endeavored to improve the estimation accuracy in the early symbols of [11] to suppress the error propagation. However, the results, though improved, were still susceptible to error propagation.

Following the cross-correlation methods [11–13], the work in [14] designed subarray-specific time-invariant phase shifts. As proved in [14] for special localized hybrid arrays with N being a multiple of the number of subarrays M, the complex gains of the subarrays in the direction of ϕ formed a Fourier series with a consistent phase difference u between the coefficients, provided that the subarrays adopted evenly spaced phase shifts between $[0,2\pi)$ per symbol. However, the gain of the mth subarray, denoted by $P_m(u)$, was coupled multiplicatively with e^{jmNu} in the received signal of the Mth subarray, i.e. $P_m(u)e^{jmNu}$. Note that this is only an illustrative signal model with the scaling coefficients and noise suppressed. Thus, one needs to estimate Nu and suppress e^{jmNu} in the received signal before taking the inverse discrete Fourier transform (IDFT) of $P_m(u)$ and correlating the Fourier coefficients to unambiguously estimate u. Nevertheless, the estimation of Nu is non-trivial and can incur ambiguities due to the use of subarray-specific phase shifts [14].

The complex gains of subarrays can have different phases in the direction of ϕ. The estimate of Nu, obtained from the cross-correlation of received signals, can be corrupted by the different phases of the subarray gains. This compromises the estimation accuracy of Nu and u. There are also undetectable angles in the case that the AoA happens to coincide with the nulls of all subarrays. Moreover, the phase shifts are limited to a special case with N being a multiple of M. The design is thus inapplicable to general localized hybrid arrays. Therefore, in Section 9.4.1, a robust and unambiguous cross-correlation-based AoA estimation method, addressing the aforementioned issues of the work, is presented in the following text [44].

9.3.2 Wideband Hybrid Array of Phased Subarrays

Accurate and fast estimation of the AoA is the key enabling technology underlying wideband multi-beam Terabit satellite-borne communication systems [45–47]. In the case of widely deployed geosynchronous Earth orbit (GEO) satellites, the a-priori knowledge on the locations of fixed ground stations helps the satellites steer their beams. However, it is still possible that some misalignment exists due to the inclination of the satellites and the orbit perturbations [48, 49]. This situation can become worse for aircrafts and vessels for which the a-priori location knowledge may not be unavailable or is inaccurate at the satellites [50]. On the other hand, low Earth orbit (LEO) satellites have been increasingly deployed for terabit satellite communications. Unlike GEO satellites, the coverage areas of LEO satellites dynamically change [48]. Moreover, the instantaneous information on the trajectory and orbit of an LEO satellite can be inaccurate or even unknown [51]. In both the GEO and LEO satellite cases, the accurate and fast estimation of the AoA is vital for improving the beamforming gain and the radio link quality, as well as increasing the data rate, especially in low SNR situations.

Onboard accurate and efficient wideband AoA estimation in satellite-borne localized hybrid arrays is very challenging. With the hybrid nature and the sheer scale of the arrays, conventional wideband AoA estimation techniques – classical signal subspace based algorithms, such as incoherent signal sub-space method (ISSM) [52], coherent signal-subspace method (CSM) [53] and the subsequent improvement algorithm [54, 55], are inapplicable or unsuitable. This drawback arises because they require a reliable initial AoA estimation and iterative refinements to achieve high accuracy [56]. Since the estimation delay could be lengthy and the initial accuracy cannot be guaranteed, the actual estimation accuracy achieved could be poor.

Most cross-correlation-based AoA estimation methods reviewed in Section 9.3.1, e.g. [9–11, 14], are developed for narrowband systems. The only exception is the algorithm presented in [11] which was later applied to wideband systems in [12] in which narrowband operations, i.e. cross-correlation between subarrays, were independently conducted at each sub-carrier followed by cross-correlations between the sub-carriers. Unfortunately, the issue of error propagation remains [12]. Moreover, the cross-correlation between sub-carriers only enables the estimation of aNu with a denoting a coefficient. Here, Nu is the phase difference between adjacent subarrays in a linear localized hybrid array as shown in Figure 9.3. Since a can be much smaller than one, the error in the estimate aNu will be amplified when dividing it by a to attain the final estimate of Nu. In Section 9.4.2, we will describe a cross-correlation-based wideband AoA estimation method [57] that reduces the loss of the estimated SNR caused by the cross-correlation over sub-carriers to the minimum.

9.3.3 Lens Antenna Arrays

Accurate knowledge of the AoA is also critical to the effective operation of LAAs. The AoA allows for the correct selection of the DFT beam pointing at the dominant path [41, 58, 59]. Moreover, channel estimation can be substantially simplified by first estimating the AoA and then the path responses in the consecutive transmission blocks [60, 61]. This simplification can occur because the AoAs of the mmWave/THz channels have a much larger coherent time, i.e. they stay unchanged for a longer time than the path responses [34, 62].

Due to the distinct array physics and structure of the LAA, many existing AoA estimation techniques, which are generally developed for DAAs [18, 44, 57, 63–68], become inapplicable. Although the algorithms presented in [69, 70] are developed for LAAs, they suffer from inefficiency and/or inaccuracy. In [69, 70], the beamspace (spanned by DFT beams) channel estimation techniques were developed for LAAs by enumerating all DFT beams and selecting those with the outputs earlier a predefined threshold. The least square (LS) approximation was performed on the selected beams for channel estimations in [69]. In [71], compressive sensing was employed for LAAs and the beamspace channel structure was used to improve the estimation accuracy of each channel. The estimation errors of these approaches can be as large as half the width of a DFT beam of an LAA [69–71].

An advanced AoA estimation algorithm, referred to as the auxiliary beam pair (ABP) was designed in [72] for DAAs in which DFT beamforming was employed. Specifically, two DFT beams with the two strongest outputs were identified; and the AoA was estimated from the ratio of the power difference of the DFT beams to the total power of the beams. The ABP was able to achieve better estimation accuracy than the other existing methods in [18, 67–71], through the exploitation of the deterministic amplitudes of DFT beams. However, for the AoA estimation of a path, the ABP enumerated the DFT beams to search for the two adjacent beams with the two strongest outputs. Thus, this process incurred long estimation delays.

It is non-trivial to estimate the AoA in wideband LAAs. As discussed in [73–75], a chief reason is that the spatial-wideband effect, also known as the beam squint effect [75], is non-negligible in mmWave/THz wideband massive MIMO. This leads to frequency-dependent spatial responses, as depicted in Figure 9.4.[1] Consider a typical linear mmWave massive MIMO with 128 antennas

1 The spatial-wideband effect refers to the phenomenon that occurs when the beam pointing direction and the beamwidth of a wideband array vary with the frequency. Weighting a wideband large-scale array by the conventional steering vector at θ, i.e. $\left[1, e^{\frac{2\pi d \sin\theta}{\lambda}}, \dots, e^{\frac{2\pi(N-1)d\sin\theta}{\lambda}}\right]^T$, T being the transpose operation, the beam points at

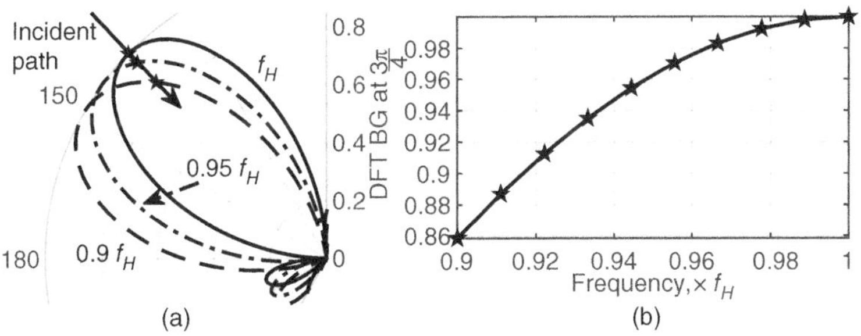

Figure 9.4 Illustration of the spatial-wideband effect for an eight-element linear array. (a) The spatial responses of the fourth DFT beam at different frequencies. (b) The beamforming gain (BG) of the second DFT beam at $\theta = \frac{3\pi}{4}$ over the frequencies plotted (a).

and a bandwidth of 1.0 GHz at the 60 GHz carrier frequency. The maximum delay caused by a cross-aperture propagation can be as large as 1.058 times of the sampling interval [60, 73]. The spatial-wideband effect hinders the effective AoA estimation and beamforming, particularly for passive and analog arrays (such as the LAA) for which their beams cannot be adjusted for different frequencies.

Another reason behind the difficulty in the AoA estimation of LAAs is that the key information for the estimation, the phase difference between adjacent antenna elements, is obscure. This obscuration is because the antenna-received signals are combined through DFT beamforming in LAAs, making the aforementioned key information not directly available. Moreover, since the number of RF chains is typically much smaller than that of the DFT beams [70, 71], only a subset of the combined DFT beams can be selected to feed into the RF chains at any instant.

Only a few studies [41, 69–71, 76] have been carried out on the channel/AoA estimation of LAAs. Most of the studies [41, 69–71] were focused on the narrowband channel/AoA estimation. These methods typically require a large number of training symbols for searching the dominant path [69–71]. Compressive sensing was applied in [76] for the channel estimation in LAAs where the spatial-wideband effect was accounted for by identifying the different paths across sub-carrier frequencies. The accuracy achieved in [76] was shown to be much better than that of the previous methods (developed for DAAs) [77, 78] in which the spatial-wideband effect was overlooked. However, the AoA estimation was not pursued in [76–78]. An accurate estimation of the AoA for wideband signals was achieved in [57], which, despite being developed based on DAAs, performed DFT beamforming. However, the N DFT beams were enumerated at the cost of a long training delay, and the spatial-wideband effect was overlooked. It is worth noting that the few existing methods [60, 76] which did take into account the spatial-wideband effect only took countermeasures to compensate for the effect. This choice prolongs the estimation delay or incurs a high computational complexity. The spatial-wideband effect was exploited in [79], in contrast to counteracting in previous works, to improve the AoA estimation efficiency in LAAs. By selecting DFT beams and sub-carriers according to [79, Theorems 1&2], the spatial-wideband effect actually makes the gains of selected beams present fixed patterns over selected sub-carriers. The patterns can then be used for fast identifying the strongest DFT beams, substantially improving the efficiency of AoA estimation in wideband large-scale hybrid arrays.

θ if $f = f_H$, where N denotes the total number of antenna elements. As f decreases, the beam pointing direction becomes larger than θ and the beamwidth expands as depicted in Figure 9.4a. The term d is the antenna spacing and λ is the wavelength of the largest frequency f_H. Due to this frequency dependent effect, a DFT beam can have different spatial responses across frequencies as depicted in Figure 9.4b.

We remark that all the previous discussions on AoA estimation in LAAs are, in theory, applicable to Butler matrix based antenna arrays and other passive BFN based antennas because they all effectively employ DFT beams. In Section 9.5, we will reveal some new features of these antennas which facilitate efficient and accurate AoA estimations.

9.4 Fast and Accurate AoA Estimation Techniques for Large-Scale Arrays of Phased Subarrays

In this section, we present two fast and accurate AoA estimation methods [44, 57] that have been recently developed for large-scale hybrid mmWave arrays of localized phased subarrays. An example is illustrated in Figure 9.5. We shall deal with the narrowband case first and then move on to the wideband case. These methods are interesting because they disclose some intrinsic features/properties of these arrays.

9.4.1 Narrowband AoA Estimation

Let us start by looking at the basic system and signal model. Figure 9.5 illustrates a rectangular localized hybrid array with $M_xN_x \times M_yN_y$ evenly spaced antenna elements in two dimensions. The array is divided into $M_x \times M_y$ localized rectangular analog subarrays, each being comprised of $N_x \times N_y$ antenna elements. These antennas are connected to adjustable phase shifters, one for each, and aggregated at an ADC. The RF generation and down-converters are omitted in the figure for clarity.

The mmWave/THz frequencies are dominated by directive line-of-sight (LoS) radio links with insignificant multi-path interferences [11, 12, 80, 81]. Consider a single-path narrowband received signal $\tilde{s}(t)$ with wavelength λ_c, elevation angle θ and azimuth angle ϕ, the received signal at the (m_x, m_y)th analog subarray ($m_x = 0,1, \dots, M_x - 1; m_y = 0,1, \dots, M_y - 1$) can be written as [9–11, 14]

$$s_{m_x,m_y}(t) = \tilde{s}(t)P^t_{m_x,m_y}(u_x, u_y)e^{j(m_xN_xu_x+m_yN_yu_y)} \tag{9.1}$$

where the noise term is dropped for brevity, $P^t_{m_x,m_y}(u_x, u_y)$ is the radiation pattern of the (m_x, m_y)th analog subarray at the tth symbol, and

$$P^t_{m_x,m_y}(u_x, u_y) = \sum_{n_x=0}^{N_x-1}\sum_{n_y=0}^{N_y-1} P^{n_x,n_y}_{m_x,m_y}(u_x, u_y) \times e^{j(n_xu_x+n_yu_y+\alpha^t_{m_x,m_y}(n_x,n_y))} \tag{9.2}$$

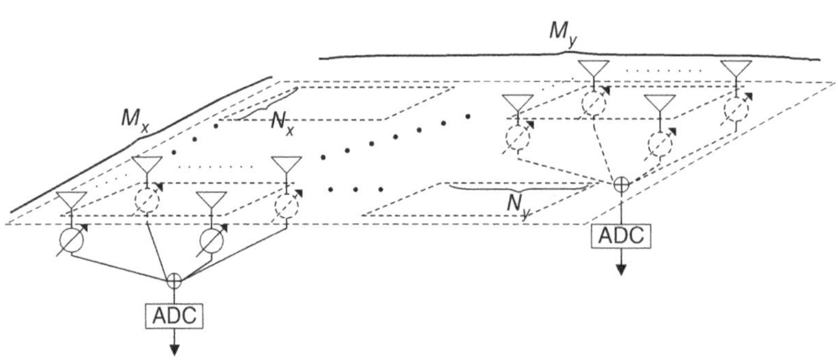

Figure 9.5 A two-dimensional planar localized hybrid antenna array composed of $M_x \times M_y$ subarrays. Each subarray has $N_x \times N_y$ antenna elements, each connected to an analog phase shifter. The outputs at analog subarrays are further processed to generate digital samples. The RF generation and down-converters are suppressed for clarity.

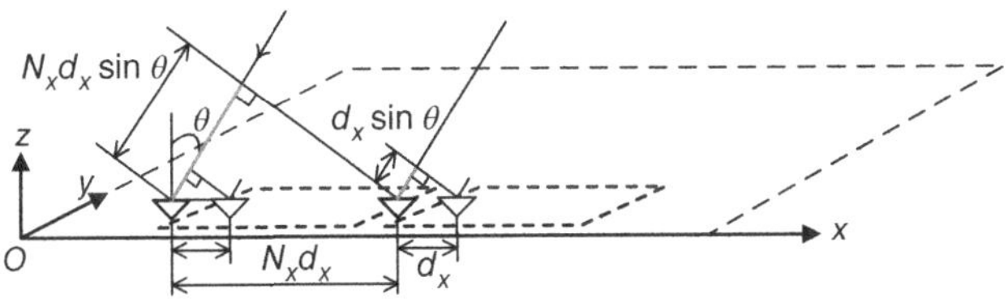

Figure 9.6 Illustrating the relations between the coordinate angles and the beamspace-domain AoA terms.

where $P^{n_x,n_y}_{m_x,m_y}(u_x, u_y)$ is the individual radiation pattern of the (n_x, n_y)th antenna element ($n_x = 0,1,\dots, N_x - 1$, $n_y = 0,1,\dots, N_y - 1$) at the (m_x, m_y)th analog subarray and $\alpha^t_{m_x,m_y}(n_x, n_y)$ is the phase shift at the antenna element and tth symbol. The terms n_x and n_y are the indexes for the antennas along the x- and y-axes of the subarray, respectively. In the AoA estimation literature, $P^{n_x,n_y}_{m_x,m_y}(u_x, u_y) = 1$ is generally assumed [11–14, 16, 82].

Note that u_x and u_y are typically referred to the beamspace-domain AoAs; they are different from the spatial-domain AoAs. In particular, they are defined as

$$u_x = 2\pi d_x \sin\theta \cos\phi / \lambda_c, \quad u_y = 2\pi d_y \sin\theta \sin\phi / \lambda_c \tag{9.3}$$

where d_x and d_y are the distances between two adjacent antenna elements along the x-axis and y-axis, respectively. The relation between the coordinate angles and beamspace-domain terms are indicated in Figure 9.6.

The estimation of the AoA, or more specifically θ and ϕ, is essentially accomplished by estimating u_x and u_y, since

$$\theta = \text{sign}\{u_x\}\sin^{-1}\left(\frac{\lambda_c\sqrt{u_x^2 + u_y^2}}{2\pi\sqrt{d_x + d_y}}\right), \quad \phi = \tan^{-1}\left(\frac{u_y d_x}{u_x d_y}\right) \tag{9.4}$$

Here, sign$\{u_x\}$ takes the sign of u_x. Note from (9.1) and (9.2) that the estimation of u_x can be decoupled from that of u_y. For illustration convenience, we simplify the signal to one dimension by suppressing the subscripts "$_x$" and "$_y$" and consider a linear localized hybrid array with M analog subarrays and N antenna elements per subarray. Thus, the beamspace-domain AoA to be estimated is now given by

$$u = 2\pi d \sin\theta / \lambda_c \tag{9.5}$$

9.4.1.1 Phase Shift Calculation

It was proposed in [44] to perform beamforming by setting the following phase shifts:

$$\alpha^t_m(n) = n\alpha^t_m = 2\pi n(\text{mod}\{m, K\}P + t)/L \tag{9.6}$$

where m, t, and n are the indexes for analog subarrays, training symbols and antennas, respectively ($0 \le m \le M - 1$; $0 \le t \le P - 1$; $0 \le n \le N - 1$). The terms K is the number of different phase shifts adopted per symbol; P is the number of training symbols; L is the total number of different phase shifts in use, and $L = PK$. The term α^t_m is the slope of the phase shifts with respect to the antenna index in subarray m; and hence, it provides the beamforming direction of the subarray. Since N may not be a multiple of M, K is designed to be a factor of N satisfying:

$$2 < K \le M, \ N = QK, \ K, Q \in \mathbb{Z}^+ \tag{9.7}$$

where $\mathbb{Z}^+$ stands for the set of positive integers.

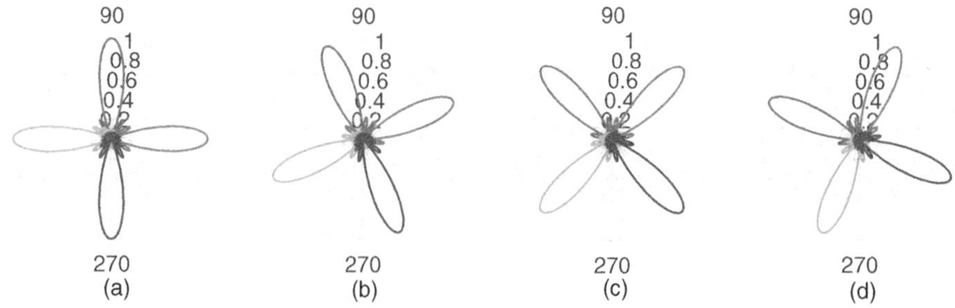

Figure 9.7 Illustrating the beam patterns under the phase shift design given by (9.6). (a) $t = 0$. (b) $t = 1$. (c) $t = 2$. (d) $t = 3$.

This beamforming scheme is illustrated in Figure 9.7, where $L = N = 16, K = P = 4$. There are 16 numbers of DFT beams in total in this configuration. We see from Figure 9.7a that the first symbol steers four DFT beams, one through each subarray. The pointing directions of the four beams are evenly spaced in $[0,2\pi]$ (radian). Then as time increases, the four beams rotate together as one DFT beam. After four symbols, all of the DFT beams have been enumerated. If more than four symbols are used for the AoA estimation, the four sets of beams are cyclically used, i.e. at symbol $t = 4$, the first four beams at $t = 0$ will be used.

9.4.1.2 Estimation of *Nu*

As per the tth symbol, the phase shift at the nth antenna of the mth subarray is $n\alpha_m^t$, according to (9.6). By substituting (9.6) into (9.2), the complex gain of the mth subarray $P_m^t(u)$ is given by

$$\begin{aligned} P_m^t(u) &= \sum_{n=0}^{N-1} e^{jnu} e^{-jn\alpha_m^t} = \sum_{n=0}^{N-1} e^{jnu} e^{-j2\pi \frac{\mathrm{mod}\{m,K\}P+t}{L} n} \\ &= \frac{1-e^{j2N\left(\frac{u}{2}-\pi\frac{\mathrm{mod}\{m,K\}P+t}{L}\right)}}{1-e^{j2\left(\frac{u}{2}-\pi\frac{\mathrm{mod}\{m,K\}P+t}{L}\right)}} = e^{j(N-1)\omega_m^t} \frac{\sin(N\omega_m^t)}{\sin(\omega_m^t)} \end{aligned} \tag{9.8}$$

where $\omega_m^t = \frac{u}{2} - \frac{\alpha_m^t}{2} = \frac{u}{2} - \pi\left(\frac{t}{L}+\frac{m}{K}\right)$ for $m = 0,1,\ldots,K-1$. Because $\alpha_m^t = \frac{2\pi(mP+t)}{L}$ for $m = 0,1,\ldots,K-1$, it can be rewritten as $\alpha_m^t = \frac{2\pi m}{K} + \frac{2\pi t}{L}$ based on $L = PK$. We see from the aforementioned equation that $P_m^t(u)$ is a periodic function of m with the cycle of K. By substituting (9.8) into (9.1), the cross-correlation of the received signals between the mth and the $(m+1)$th analog subarrays at symbol t is given by

$$\rho_m(t) = s_m^*(t)s_{m+1}(t) = |\tilde{s}(t)|^2 G_m^t(u) e^{jNu} \tag{9.9}$$

where $m = 0,1,\ldots,K-1$ for $K < M$. In the case that $K = M$, $s_K(t)$ is replaced by $s_0(t)$, given the periodicity of the phase shifts, and $G_m^t(u) = P_m^{t*}(u)P_{m+1}^t(u)$ can then be written as

$$G_m^t(u) = e^{\frac{-j(N-1)\pi}{K}} \frac{\sin(N\omega_m^t)}{\sin(\omega_m^t)} \frac{\sin(N\omega_{m+1}^t)}{\sin(\omega_{m+1}^t)} \tag{9.10}$$

We see from (9.9) that Nu can be estimated by taking the phase of $\rho_m(t)$. However, there are two problems. First, the amplitude of $G_m^t(u)$ can be small and then the SNR in $\rho_m(t)$ is also small, which affects the accuracy of the Nu estimation. Second, the sign of $G_m^t(u)$ is unknown, meaning that directly taking the angle of $\rho_m(t)$ has an estimation ambiguity. Fortunately, we have proved in [44,

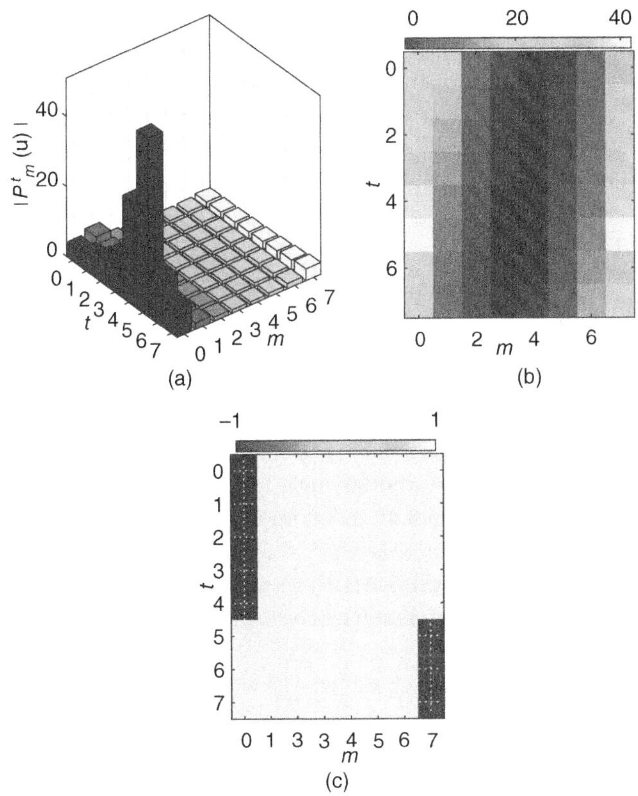

Figure 9.8 Illustrating the sign rule in $G_m^t(u)$, where (a) gives the beamforming gain over t and m; (b) gives the amplitude of $G_m^t(u)$; and (c) gives the signs of $G_m^t(u)$.

Theorem 1] that there is a deterministic sign rule for $G_m^t(u)\ \forall m, t$. In particular, the rule can be stated as: *At a symbol t, only $G_{m'}^t(u) \neq 0$, with the largest amplitude, has the opposite sign to all the rest of $G_m^t(u)$, where $m \in [0, K-1]$, $m' \in [0, K-1]$ and $m \neq m'$.*

The sign rule is illustrated in Figure 9.8, where $L = N = 64$, $K = M = 8$, $P = 8$ and $u = 0.445$ (rad). Jointly observing the same row in Figure 9.8b,c, we can see the sign rule depicted earlier. For example, in the first row, i.e. $t = 0$, the first cross-correlation at $m = 0$ achieves the maximum amplitude, as seen from Figure 9.8b. Then correspondingly, we see from Figure 9.8c, that the first element in the first row has the sign opposite to all the other elements. For another example, let us look at the row $t = 6$. From Figure 9.8b, we see that the cross-correlation result at $m = 7$ has the largest amplitude, and then from Figure 9.8c, we see that the result has the opposite sign to the rest.

With the previous rule, we can align the signs of $\rho_m(t)$ over m and t, and coherently accumulate $\rho_m(t)$ for a high-accuracy estimation of Nu. The sign alignment can be performed as

$$\tilde{\rho}_m(t) = \begin{cases} (-1)^Q \rho_m(t) & \text{if } m \neq m' \\ (-1)^{Q+1} \rho_{m'}(t) & \text{if } m = m' \end{cases}, \quad (m = 0,1,\ldots,K-1,\ t = 0,1,\ldots,P-1) \tag{9.11}$$

where m' is the index of the cross-correlation with the largest amplitude in each symbol. Note that the coefficient $(-1)^Q$ occurs because the numerator of $G_m^t(u)$, as given in (9.10), can be written as $(-1)^Q \sin^2(N\omega_m^t)$ due to the way the beamforming is performed. Interested readers may refer to [44] for all of the details.

9.4.1.3 Estimation of u

Let $n = n' + qK$ $(0 \le n' \le K-1,\ 0 \le q \le Q-1)$. Equation (9.8) becomes

$$P^t_m(u) = \sum_{n'=0}^{K-1}\sum_{q=0}^{Q-1} e^{j(n'+qK)u} e^{-j2\pi \frac{mP+t}{L}(n'+qK)} = \sum_{n'=0}^{K-1} g^t_{n'}(u) e^{-j\frac{2\pi n' m}{K}}, \quad m = 0,1,\dots,K-1 \tag{9.12}$$

where $g^t_{n'}(u)$, as given in the following text, is the Fourier coefficients of $P^t_m(u)$ [14]

$$g^t_{n'}(u) = e^{jn'(u-\frac{2\pi t}{L})} e^{j(N-K)(\frac{u}{2}-\frac{\pi t}{L})} \frac{\sin\left(\frac{Nu}{2} - \frac{N\pi t}{L}\right)}{\sin\left(\frac{Ku}{2} - \frac{K\pi t}{L}\right)}, \quad n' = 0,1,\dots,K-1 \tag{9.13}$$

From (9.13), we notice that the cross-correlation of $g^t_{n'}(u)$ and $g^t_{n'+1}(u)$ can be used for a unique estimation of u. This is on the condition that the beamforming gains $P^t_m(u),\ m = 0,1,\dots,M-1,\ \forall t$ are available.

Revisiting (9.1), the subarray beamforming gains can be obtained by compensating the phases of subarray outputs. This compensation can be performed by multiplying $s_m(t)$ given in (9.1) by $e^{-jm\widehat{Nu}}$. Here, $\widehat{Nu}$ denotes the Nu estimation attained in Section 9.4.1.2. Taking the IDFT of $s_m(t)e^{-jm\widehat{Nu}}$, $m = 0,1,\dots,K-1$, we obtain $a_{n'}(t) = \tilde{s}(t)g^t_{n'}(u)$, $n' = 0,1,\dots,K-1$.

A cross-correlation can be taken between any two consecutive IDFT results, i.e. $a_{n'}(t)$ and $a_{n'+1}(t)$, to achieve an unambiguous estimate of u. This cross-correlation can be expressed as

$$r_{n'}(t) = a^*_{n'}(t)a_{n'+1}(t) = e^{j(u-\frac{2\pi t}{L})} \left| \frac{\sin\left(\frac{Nu}{2} - \frac{N\pi t}{L}\right)}{\sin\left(\frac{Ku}{2} - \frac{K\pi t}{L}\right)} \right|^2 |\tilde{s}(t)|^2, \quad n' = 0,1,\dots,K-2 \tag{9.14}$$

We see that a coherent accumulation of $r_{n'}(t)$ can be readily performed over n' and t before taking the angle for the estimation of u.

9.4.2 Wideband AoA Estimation

The method reviewed in Section 9.4.1 can be extrapolated to wideband AoA estimation in mmWave large-scale hybrid arrays of localized subarrays. However, there is an interesting trade-off between the accumulated gain and error amplification when combining multiple sub-carriers. To understand this, it is necessary to establish the signal model for the wideband AoA estimation.

Let $\tilde{s}(t)$ denote the complex incoming time-domain signal in the baseband. Assume that the frequency range of $\tilde{s}(t)$ is $[f_l, f_h]$ and $B = f_h - f_l$ is the frequency bandwidth. The spacing between adjacent antenna elements, denoted by d, is half the wavelength of the highest frequency, i.e. $d = \frac{\lambda_l}{2}$, where $\lambda_l = \frac{c}{f_h}$ and c is the light speed. When a wideband symbol is digitalized, I samples are generated. At symbol t, the ith sample $\tilde{s}(t)$ is taken at $t_i = \frac{tI+i}{f_s}$, where f_s is the sampling frequency, t is the index for symbols $(t \ge 0)$ and i is the index for samples within a symbol and, hence, the index for the sub-carriers $(0 \le i \le I-1)$.

Suppose that the incoming direction of $\tilde{s}(t_i)$ is θ. The time-domain output of the mth subarray can be written as [12]

$$s_m(t_i) = \sum_{n=0}^{N-1} P^n_m(\theta)\tilde{s}(t_i - \tau^m_n(\theta))e^{jn\alpha^t_m} \tag{9.15}$$

where $P^n_m(\theta)$ is the radiation pattern of the nth antenna element at the mth subarray $(0 \le n \le N-1, 0 \le m \le M-1)$; $\tau^n_m(\theta) = \frac{(nd+mNd)\sin\theta}{c}$ is the propagation delay of the nth antenna element at the mth subarray with reference to the left-most antenna element; and α^t_m is the phase shift value of the mth subarray at symbol t. The phase shifts are assumed to be frequency flat, i.e. the same phase shifts for all I samples of symbol t. Again, we drop the noise term for brevity.

For the tth wideband symbol, the I-point DFT of $s_m(t_i)$, denoted by $S_m(f_i)$, $0 \le i \le I-1$, is given by

$$S_m(f_i) = \sum_{n=0}^{N-1} P_m^n(\theta)\tilde{S}(f_i)e^{j2\pi f_i \tau_m^n(\theta)}e^{jn\alpha_m^t} = \tilde{S}(f_i)P_m(\rho_i, u)e^{jm\rho_i Nu},$$

$$\text{s.t. } P_m(\rho_i, u) = \sum_{n=0}^{N-1} P_m^n(\theta)e^{jn\rho_i u}e^{jn\alpha_m^t} \tag{9.16}$$

where $\tilde{S}(f_i) = \text{DFT}\{\tilde{s}(t_i)\}$ and $u = \frac{2\pi d \sin\theta}{\lambda_l} = \pi \sin\theta$ since $d = \frac{\lambda_l}{2}$. The term $P_m(\rho_i, u)$ is the beamforming gain of the mth subarray under α_m^t. The term ρ_i is the normalized frequency:

$$\rho_i = \frac{f_i + f_l}{f_h} \in \left[\frac{f_l}{f_h}, 1\right] \tag{9.17}$$

where $f_i = \frac{iB}{I}$, $0 \le i \le I-1$, is the centroid frequency of the ith sub-carrier in the baseband. Different from the narrowband scenarios, $P_m(\rho_i, u)$ depends on both u and f_i through ρ_i. The estimation of the AoA, namely, θ, is essentially the estimate of u, since u and θ are related by a one-to-one mapping, i.e.

$$\theta = \sin^{-1}\left(\frac{u\lambda_l}{2\pi d}\right) \tag{9.18}$$

Substituting the phase shifts designed in (9.6) into (9.16), the beam pattern of the mth subarray at any symbol t and sub-carrier i ($0 \le m \le K-1$, $0 \le t \le P-1$, and $0 \le i \le I-1$), i.e. $P_m(\rho_i, u)$, can be given by

$$P_m(\rho_i, u) = \sum_{n=0}^{N-1} e^{jn\rho_i u}e^{-jn\alpha_m^t} = \sum_{n=0}^{N-1} e^{jn\rho_i u}e^{-j2\pi\frac{\text{mod}(m,K)P+\text{mod}(t,P)}{L}n} = e^{j(N-1)\omega_m^i}\frac{\sin(N\omega_m^i)}{\sin(\omega_m^i)} \tag{9.19}$$

where $\omega_m^i = \frac{\rho_i u}{2} - \pi\left(\frac{t}{L} + \frac{m}{K}\right)$. As typically assumed in the AoA estimation literature [11–14], we set $P_m^n(\theta) = 1$; i.e. the antenna elements are omni-directional with unitary antenna gains. We can see from (9.19) that $P_m(\rho_i, u)$ is periodic over m and t, with cycles of K and P, respectively. For illustration convenience, we take $m = 0,1,\dots,K-1$ and $t = 0,1,\dots,P-1$ in the following.

Similar to the narrowband case illustrated in Section 9.4.1, we can take the cross-correlation between adjacent subarray outputs to estimate Nu. A major difference here is that the cross-correlation is taken per sub-carrier (versus per symbol earlier). In particular, we have

$$R_m(f_i) = S_m^*(f_i)S_{m+1}(f_i) = |\tilde{S}(f_i)|^2 G_m(\rho_i, u)e^{j\rho_i Nu}, \text{ s.t. } G_m(\rho_i, u) = P_m^*(\rho_i, u)P_{m+1}(\rho_i, u) \tag{9.20}$$

Based on (9.19), we further have

$$G_m(\rho_i, u) = e^{\frac{-j(N-1)\pi}{K}}\frac{\sin(N\omega_m^i)}{\sin(\omega_m^i)}\frac{\sin(N\omega_{m+1}^i)}{\sin(\omega_{m+1}^i)} \tag{9.21}$$

Despite that sub-carrier ρ_i is involved earlier, the expression of $G_m(\rho_i, u)$ is rather similar to $G_m^t(u)$ in the narrowband case, see (9.10). In fact, as proved in [57, Theorem 1], the sign rule stated for $G_m^t(u)$ is similarly applied to $G_m(\rho_i, u)$. Specifically, we have: *At any symbol t and sub-carrier i ($i = 0,1,\dots,I-1$), only $G_{m'}(\rho_i, u)$ with the largest non-zero amplitude has the opposite sign to all $G_m(\rho_i, u)$, $m \ne m'$, where m, $m' \in [0, K-1]$.* Enabled by the rule, we can adjust the signs of the $R_m(f_i)$ in (9.20) at every sub-carrier i, as given by

$$R_m(f_i) = \begin{cases} (-1)^Q R_m(f_i) & \text{if } m \ne m' \\ (-1)^{Q+1} R_m(f_i) & \text{if } m = m' \end{cases} \tag{9.22}$$

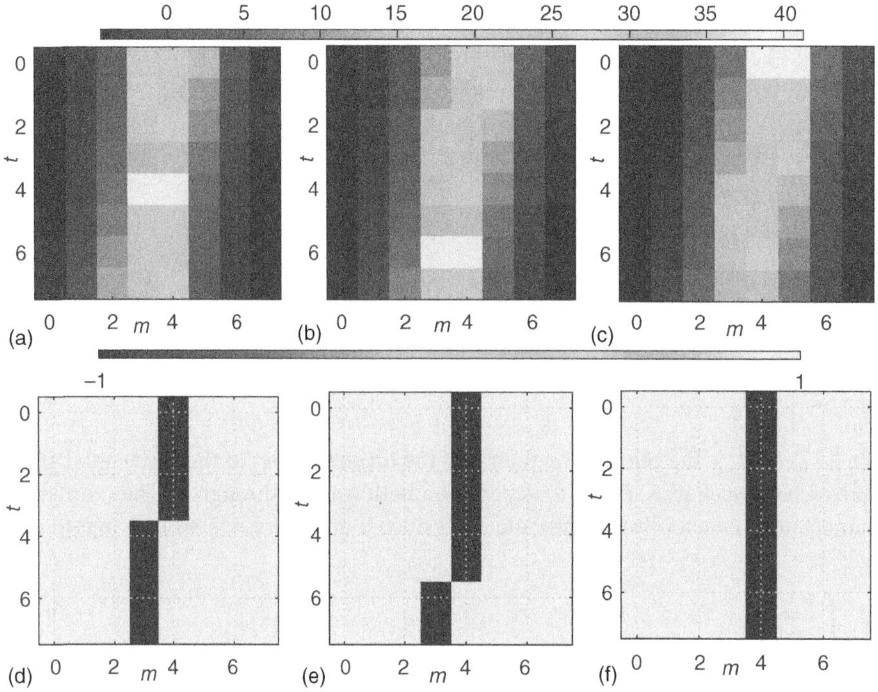

Figure 9.9 Illustrating the sign rule in $G_m(\rho_i, u)$. The top row shows the amplitudes. The bottom row shows the signs. The left, middle, and right columns are for $i = 0, I/2$, and $I - 1$, respectively.

where $m = 0,1, \ldots, K-1$, $i = 0,1, \ldots, I-1$, and m' is the index to the cross-correlation with the largest amplitude per sub-carrier.

Figure 9.9 illustrates the sign rule, where $L = N = 64$, $K = M = 8$, $P = 8$, $I = 256$, $B = 0.1f_H$ and $u = 3.9$ (rad). Let us jointly look at the two sub-figures in the first column. We see that at any row, e.g. $t = 0$, when the amplitude achieves the maximum, the sign of the element becomes opposite to the others. Another observation is that as i varies, the indexes of the sign-changing elements can change as well. This shows the non-trivial impact of the bandwidth on AoA estimation.

With the sign of $R_m(f_i)$ adjusted, a second cross-correlation of $R_m(f_i)$ over f_i can be taken to further exploit the frequency-domain DoF. This leads to

$$\begin{aligned}\tilde{R}_m(f_i) &= R_m^*(f_i)R_m(f_{i+\Delta_i})\\ &= |\tilde{S}(f_i)|^2|\tilde{S}(f_{i+\Delta_i})|^2 G_m^*(\rho_i, u)G_m(\rho_{i+\Delta_i}, u)e^{j\delta_{\rho_i}Nu}, \quad 0 \le i \le I - \Delta_i - 1\end{aligned} \tag{9.23}$$

where Δ_i is the difference of indexes for sub-carriers and $\delta_{\rho_i} = \rho_{i+\Delta_i} - \rho_i = \frac{\Delta_i B}{If_h}$. Since the signs of $G_m(\rho_i, u)$, $\forall m, t, i$ have been aligned, $|\tilde{S}(f_i)|^2|\tilde{S}(f_{i+\Delta_i})|^2 G_m^*(\rho_i, u)G_m(\rho_{i+\Delta_i}, u)$ in (9.23) becomes consistent in sign. Coherent accumulation can be performed over m, f_i, and t to improve the estimation SNR and accuracy; it is given by

$$\tilde{R} = \sum_{m=0}^{M-2}\sum_{t=0}^{P-1}\sum_{i=0}^{I-\Delta_i-1} \tilde{R}_m(f_i) \tag{9.24}$$

As shown in (9.23), Nu can be estimated by evaluating the phase of $\tilde{R}$, as given by

$$\widehat{Nu} = \arg\{\tilde{R}\}/\delta_{\rho_i} \tag{9.25}$$

We see from (9.25) that the estimation error of $\widehat{Nu}$ can be amplified by $\delta_{\rho_i}(= \rho_{i+\Delta_i} - \rho_i = \frac{\Delta_i B}{If_h}) \ll 1$. To this end, $\delta_{\rho_i}N < 1$ is expected to be as close to 1 as possible. This requires Δ_i to be large. However,

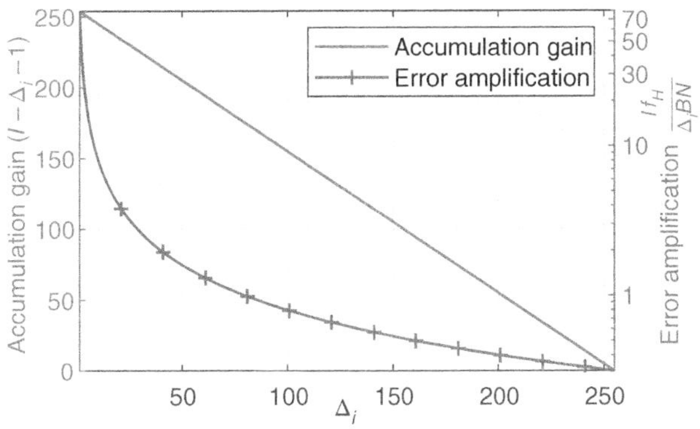

Figure 9.10 Illustrating the trade-off between the accumulation gain and the error amplification factor, caused by Δ_i, when $N = 32$, $I = 256$, and $f_H = 0.1B$.

a large value of Δ_i can compromise the gain of the coherent accumulation of (9.24), degrading the estimation of $\arg\{\tilde{R}\}$ and $\widehat{Nu}$. This trade-off is illustrated in Figure 9.10.

An intriguing question is: "*Now that the accumulation gain and error amplification factor reduce with substantially different speeds in terms of the sub-carrier interval, is there an optimal value of Δ_i that can minimize the overall estimation error of $\widehat{Nu}$?*" This question has been explored and answered in [57]. In particular, the mean squared error lower bound (MSELB) of $\widehat{Nu}$ was first derived therein,[2] which turns out to be a convex function of Δ_i. Then the optimal value of Δ_i that minimizes the MMSE is solved, as given by $\Delta_i^* = 2I/3$.

With Nu estimated, we can suppress the impact of Nu in (9.16) and estimate u at each sub-carrier based on a unique relation between the subarray beamforming gains and the AoA. Such a relation can be established as shown in (9.12). Interested readers can either try to establish the relation based on (9.19) and (9.12), or refer to [57] for the results.

Parameters	Values	Parameters	Values	Parameters	Values
M, N	4,8	K	4	f_s	B
I	256	P	3	AoA	$U_{[-\pi,\pi]}$
f_c	70 GHz	B	$0.1f_c$	Δ_i	85

Next, we provide a set of simulations to demonstrate the estimation performance of the reviewed methods and designs. The simulation parameters are given in the following table, where $U_{[-\pi,\pi]}$ denotes the uniform distribution in $[-\pi, \pi]$.

Figure 9.11a plots the mean squared error (MSE) of $\widehat{Nu}$ with the growth of γ_e (antenna-level SNR), where $\Delta_i = 1$, $\frac{2I}{3}$ and $I - 1$ are taken to illustrate the effect of Δ_i on the Nu estimation. We can confirm from the figure that $\Delta_i = \frac{2I}{3}$ gives the best Nu estimation. We also see that the selection $\Delta_i = \frac{2I}{3}$ allows for the estimate of Nu asymptotically approaching the MSELB. The bound is plotted based on [57, Eq.(25)].

Figure 9.11b plots the MSE of $\hat{u}$, which is obtained under the three different estimations of Nu shown in Figure 9.11a. Note that different numbers of sub-carriers are combined for $\hat{u}^1$, $\hat{u}^2$, and $\hat{u}^3$

2 Note that the MSELB is a modified Cramŕ–Rao lower bound (CRLB) which, like the standard CRLB, provides the theoretical limit, i.e. a lower bound, to the error variance of any parameter estimator [82, 83]. The modified CRLB is particularly useful when, in addition to the parameter to be estimated, the observed data also depend on other unwanted/unknown parameters [83], such as channel gains.

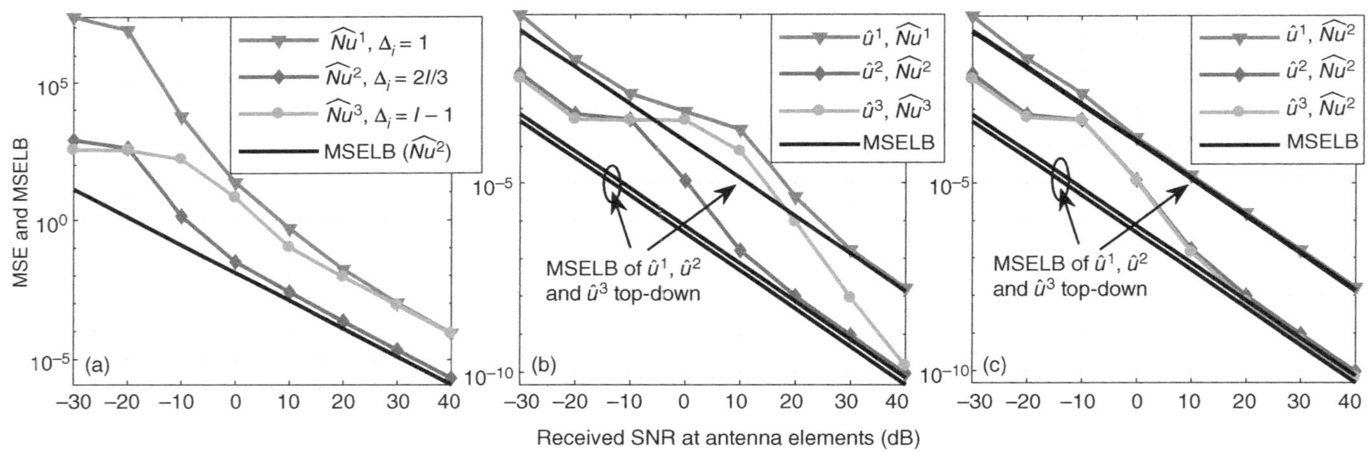

Figure 9.11 Estimation of the AoA. (a) MSE of $\widehat{Nu}$ versus the receive SNR at antenna element, γ_e. (b) MSE of $\hat{u}$ versus γ_e, where $\hat{u}^1$, $\hat{u}^2$, and $\hat{u}^3$ stand for $\hat{u}$ obtained with $i_2 - i_1 = 1$, $\frac{2I}{3}$ and $I - 1$, respectively; (c) the best Nu estimation $\widehat{Nu}^2$ is used for generating the three u estimations.

as indicated in the figure caption. Also note that the MSELB of $\hat{u}$ is plotted based on [57, Eq. (41)]. We see that the estimation accuracy of Nu has an obvious impact on the u estimation.

Figure 9.11c correspondingly plots the MSE of $\hat{u}$ upon the best Nu estimation, i.e. $\widehat{Nu}^2$. Moreover, the results are given for three different numbers of accumulated sub-carriers. As expected, we can see that $\hat{u}$ can be increasingly more accurate as the number of accumulated sub-carriers increases.

9.5 Fast and Accurate AoA Estimation

We proceed to introduce some interesting features of DFT beam antennas that are extracted from some recent AoA estimation work [41, 79, 84] on mmWave large-scale hybrid antenna arrays employing DFT beams. We refer here to all lens antennas, Butler matrix based antennas and other multi-beam antennas based on circuit type BFNs that generate DFT beams as *DFT beam antennas*. From a signal processing perspective, the features to be reviewed are applicable to all DFT beam antennas. However, we point out that it may not be easy for some DFT beam antenna to support (yet) the realizations of these features in practice. This, on the other hand, may inspire new designs of DFT beam antennas.

9.5.1 Wide-Beam Synthesis

Being able to synthesize wide-beam flexibly is desirable for the AoA estimation in mmWave large-scale DFT beam antenna arrays. Let us elaborate on this comment by considering the example given in Figure 9.12. When using DFT beam antennas, we often need to search for the beam with the maximum beamforming gain. The number of beam ports, denoted by N, in mmWave large-scale arrays can be tens to hundreds, while that of the RF chain, denoted by K, can be much smaller. If each beam is enumerated by identifying the strongest beam, at least $\lceil N/K \rceil$ numbers of symbols would be required. This can be too time-consuming. Nevertheless, if wide beams can be synthesized and used for multi-sectioning search, the number of symbols required

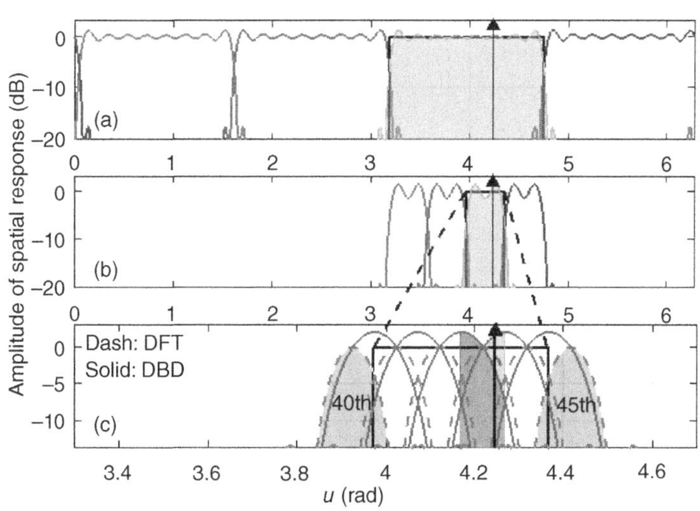

Figure 9.12 Illustration of the beam search using a DFT beam antenna array with $N = 64$, $K = 4$, and $u = 4.25$ (rad).

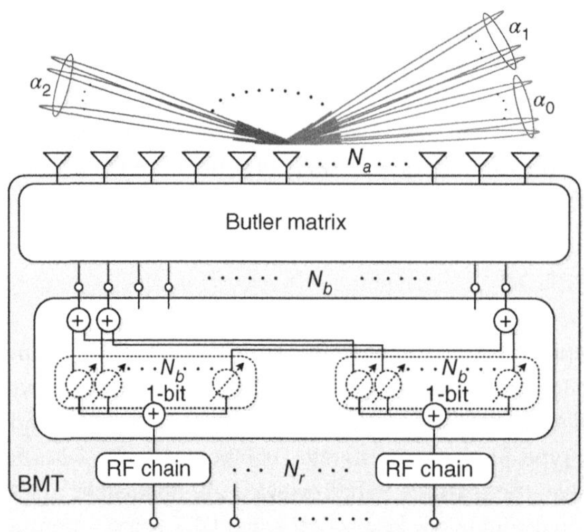

Figure 9.13 Schematic diagram of a single Butler matrix-based transceiver (BMT), where N_a, N_b, N_r are the numbers of antennas, beam ports and RF chains per Butler matrix, respectively. In mmWave large-scale arrays, N_b can be smaller than N_a. For ease of illustration, we consider here $N_a = N_b$.

for identifying the strongest beam will be reduced to $\lceil \log_K N \rceil$ as depicted. The number of symbols in this example thus reduces from $16(= \lceil N/K \rceil)$ to only $3(= \lceil \log_K N \rceil)$.

With the acknowledgment of the benefit of wide beams, the question then becomes how to synthesize such beams in mmWave large-scale DFT beam antenna-based hybrid arrays. To answer this question, one needs to be clear about the DoFs available for the beam synthesis. In DFT beam antenna-based hybrid arrays, e.g. the one shown in Figure 9.13, the number of RF chains, denoted by N_r, can be much smaller than N_b (the number of DFT beams), especially in the case of massive MIMO [71, 76]. Therefore, a beam selection network is required to connect the N_b DFT beams and N_r RF chains.

A beam selection can be built on switches; namely, each RF chain would be equipped with a switch that can be connected to any beam port. A more flexible beam selection network is based on 1-bit phase shifters [71], as shown in Figure 9.13. It was further proposed in [41] that each 1-bit phase shifter can be individually enabled or disabled.

Denote the beam selection network by $\mathbf{S} \in \mathbb{R}^{N_b \times N_r}$. The (n_b, n_r)th element of $\mathbf{S}$ can take one of the following values according to the beam selection network designed in [41],

$$\mathbf{S}(n_b, n_r) = \begin{cases} e^{j\frac{2\pi\times 0}{2}} = 1, & \text{bit 0} \\ e^{j\frac{2\pi\times 1}{2}} = -1, & \text{bit 1} \\ 0, & \text{disabled} \end{cases} \tag{9.26}$$

where the 1-bit phase shifter between the n_b-th beam port and the n_rth RF chain is disabled when choosing $\mathbf{S}(n_b, n_r) = 0$. As illustrated, we see that only the beam selection network can be designed to synthesize a wide beam from the DFT beam antenna array. Each column of $\mathbf{S}$ in this particular example can be designed to steer a wide beam for a single RF chain.

Despite the very limited values that an entry of $\mathbf{S}$ can take, an easy wide-beam synthesis method was developed in [41, Theorem 1]. In particular, *a wide beam can be synthesized by selecting consecutive DFT beams covering the angular region, reversing the signs of every second beam, and simply adding the beams together*. This can be translated into the following mathematical language: *Given*

N_a DFT beams, denoted by $g(u, n_a)$ ($n_a = 0, 1, \dots, N_a - 1$), and the $N_a \times 1$ beam selection vector connected to the n_rth RF chain:

$$\mathbf{S}(:, n_r) = \left[\overbrace{\underbrace{0, \dots, 0}_{m},\ \underbrace{(-1)^m, \dots, (-1)^{m+K-1}}_{K},\ 0, \dots, 0}^{N_a} \right]^T \tag{9.27}$$

the beam synthesized as $\sum_{n_a=0}^{N_a-1} g(u, n_a)\mathbf{S}(n_a, n_r)$ approximately has the following beam pattern,

$$P(u) = \begin{cases} N_a e^{j\frac{N_a}{2}u}, & \frac{2\pi m}{N_a} \leq u \leq \frac{2\pi(m+K-1)}{N_a} \\ 0, & \text{otherwise} \end{cases} \tag{9.28}$$

where m is an integer number satisfying $1 \leq m \leq N_a - K$, and $\mathbf{S}(n_a, n_r)$ is the beam selection state connecting the n_ath DFT beam port and the n_rth RF chain (e.g. see (9.26)).

We remark that the aforementioned beam selection allows a wide-beam with any beamwidth and pointing direction to be synthesized in real time. In addition, multiple wide beams can be synthesized, covering non-contiguous angular regions. Figure 9.14 illustrates the wide beam synthesized using (9.27) and the WDFT[3] [68]. We see that the ripples in the mainlobes synthesized with (9.27) can be slightly more severe than those with the WDFT because the WDFT is designed to minimize the variances of the mainlobe responses. Nevertheless, the mainlobe-to-sidelobe ratio of (9.27) is higher than that of the WDFT. By aligning the first sidelobes of the synthesized beams, we see that the beamforming gain obtained with (9.27) is much larger than that with the WDFT. Moreover, the sidelobe levels under (9.27) decrease much faster than those obtained with the WDFT, and its beams achieve gains of up to 16 dB and 22 dB for $N_r = 8$ and 16, respectively. It is noteworthy that (9.27) linearly combines the DFT beams with the coefficients having only three values, i.e. ± 1 and 0. In contrast, the WDFT optimizes the phase rotational speed through an exhaustive search [68]; hence, it requires more complexity than using (9.27).

Figure 9.15 further demonstrates the flexibility of the wide-beam synthesis method using (9.27). We see that it can synthesize not only a beam with a wide mainlobe, but also a beam with two

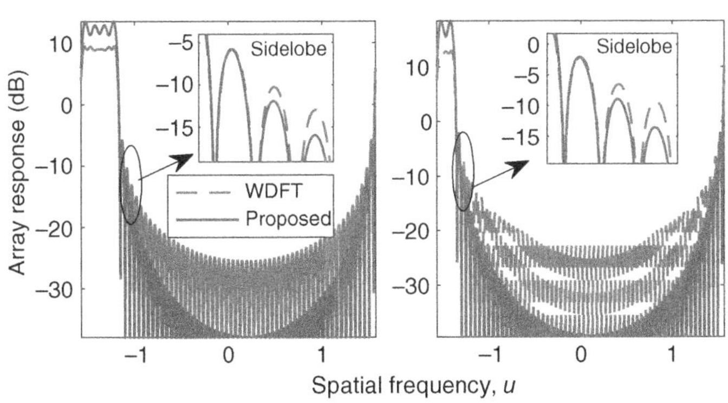

Figure 9.14 Comparisons of the WDFT [68] and the proposed method, where $N_a = N_b = 64$, $N_r = 8$. On the left, the wide beam covering the first eight DFT beams is synthesized and $N_r = 16$. On the right, the wide beam covering the first four DFT beams is synthesized.

3 This method synthesizes a wide beam based on the DFT beams by optimizing the phase-only coefficients of the underlying beams to minimize the mainlobe ripples.

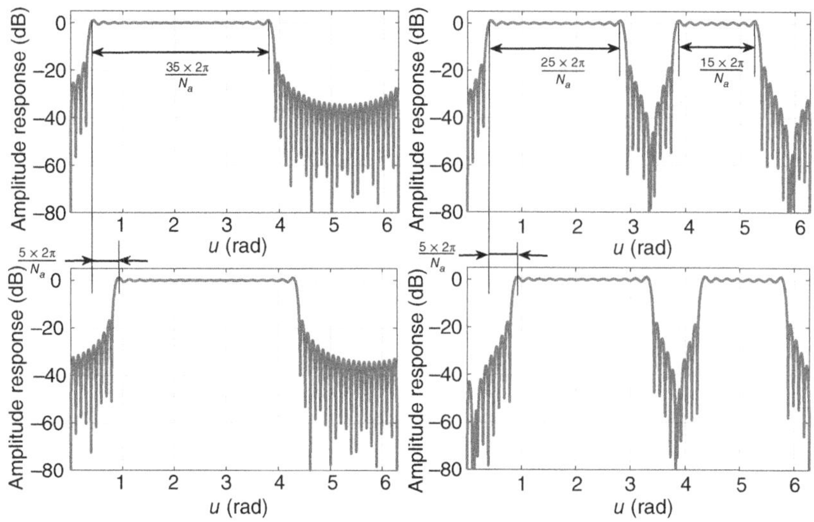

Figure 9.15 Illustrating the flexibility of the wide-beam synthesis.

separate mainlobes shifted away from each other. Note that the multi-mainlobe wide beams are also beneficial in mmWave massive MIMO communications. They can be used to speed up the beam search for channel and AoA estimations [19].

9.5.2 DFT Beam Difference

Next, we introduce the DFT beam difference (DBD) that is another beneficial property of a DFT beam antenna. It can be used for improving the AoA estimation accuracy. The DBD can be obtained by taking the difference of two adjacent DFT beams. It is illustrated in Figure 9.16, where we see that the maximum amplitude gain of the DBD is larger than that of the two underlay DFT beams. The intuitive reason is that the phase of the two DFT beams at their intersection has the same magnitude but opposite signs. This feature is also illustrated in Figure 9.16. Assuming that the index of the DBD shown in the figure is $n(=0,1,\dots,N-1)$, where N is the number of underlying DFT beams, we see from the figure that the nth DBD has *a unique angular region*, i.e. $(2\pi n/N, 2\pi(n+1)/N)$, where the amplitude of the DBD is larger than any other DBD. Moreover, the AoA can be unambiguously estimated from the received signals of the two DBDs, denoted as x_n and x_{n+1}, provided that the AoA u is covered by the unique angular regions of two adjacent, say the nth and $(n+1)$th, DBDs.

Let $\hat{u}$ denote the AoA estimate. We have [41]

$$\hat{u} = \frac{2\pi(n+1)}{N} - \arcsin\left(\frac{\rho \sin\frac{2\pi}{N}\left(\text{sign}\{\rho\}\sqrt{1-\rho^2}\cos\frac{2\pi}{N} - 1\right)}{\sin^2\frac{2\pi}{N} + \rho^2\cos^2\frac{2\pi}{N}}\right), \text{ s.t. } \rho = \frac{|x_n|^2 - |x_{n+1}|^2}{|x_n|^2 + |x_{n+1}|^2} \tag{9.29}$$

Overall, as shown in Figure 9.13, we can to estimate the AoA of the strongest incident path in the DFT beam antenna-based hybrid array by:

Step 1: Searching for the strongest two DBDs using the multi-sectioning approach illustrated in Figure 9.12;

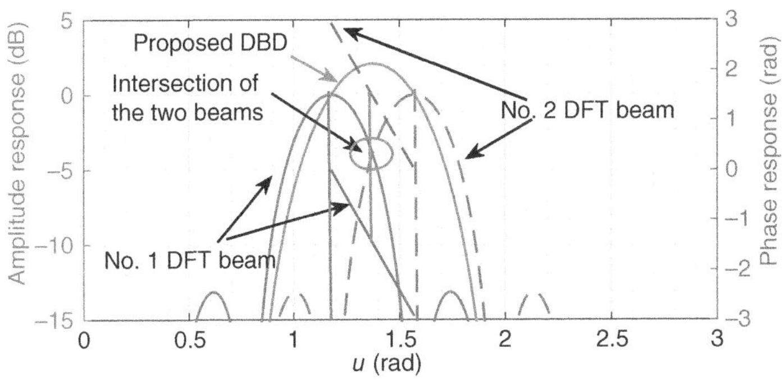

Figure 9.16 Illustrating the formation of a DBD based on two adjacent DFT beams.

Step 2: Collecting the outputs of the two DBDs and constructing the ratio ρ in (9.29); and
Step 3: Solving the estimate of u using the relation between u and ρ, as established in (9.29).

Prior to the DBD, DFT beams were used for the AoA estimation in a way similar to what was described earlier. In particular, one searches for the two strongest DFT beams and their outputs are then used to construct a ratio, as was done in (9.29). Note that we add a tilde symbol earlier a variable to denote its differentiation in the DFT beam-based AoA estimation. Assuming that the nth and $(n+1)$th DFT beams have the strongest outputs, u can be estimated with the following expression [72]:

$$\hat{u} = \frac{2\pi n}{N} - \arcsin\left(\frac{\tilde{\rho}\sin\frac{2\pi}{N}}{\sin^2\frac{2\pi}{N} + \tilde{\rho}^2\cos^2\frac{2\pi}{N}} - \frac{\tilde{\rho}\sqrt{1-\tilde{\rho}^2}\sin\frac{2\pi}{N}\cos\frac{2\pi}{N}}{\sin^2\frac{2\pi}{N} + \tilde{\rho}^2\cos^2\frac{2\pi}{N}} \right), \text{ s.t. } \tilde{\rho} = \frac{|\tilde{x}_n|^2 - |\tilde{x}_{n+1}|^2}{|\tilde{x}_n|^2 + |\tilde{x}_{n+1}|^2} \tag{9.30}$$

Using the DBD for AoA estimation with a DFT beam antenna-based hybrid array has a salient advantage over using only DFT beams. In particular, the SNRs of the two strongest DBD outputs are larger than those in the two strongest DFT beam outputs. As a result, the probability of correctly identifying the two strongest DBDs is higher than that for DFT beams. This is demonstrated in Figure 9.17. The improvement of using DBD over DFT is obvious.

The AoA estimation performance is shown in Figure 9.18, where the AoA is performed based on the beams identified for plotting Figure 9.17. Observing the two figures simultaneously, we notice that correctly detecting the beams for the AoA estimation is critical to its final performance. This caveat is particularly true in the low SNR regions, e.g. $\gamma \leq 16$ dB in the figures. We also see from the two figures that using the DBDs still achieves a better AoA estimation performance than directly using DFT beams in high SNR regions. As explained earlier, this outcome arises because the SNRs in the DBDs are higher than those in the DFT beams.

9.6 Conclusions

In this chapter, we have reviewed the state-of-the-art AoA estimation methods for mmWave/THz large-scale hybrid arrays. The discussion started with an overview of the importance of large-scale arrays in many 5G and B5G applications. Then, a number of popular hybrid array structures were illustrated; the importance and challenges with them were outlined, and the existing AoA estimation methods for hybrid arrays were surveyed. Recent breakthroughs on fast and accurate AoA estimation in large-scale array of phased subarrays followed by DFT beam antennas were presented. In

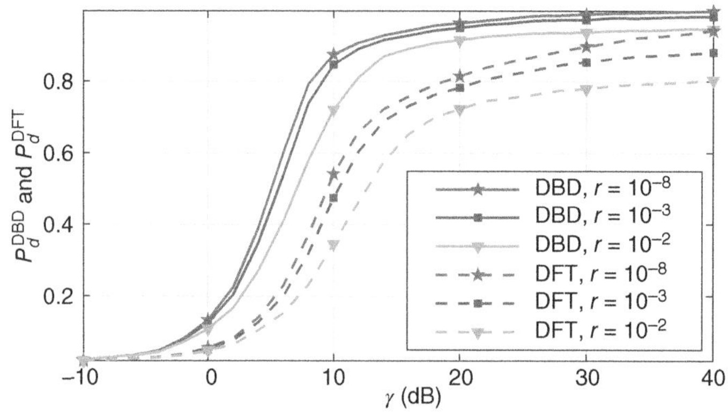

Figure 9.17 Comparison of the probability of detecting the two strongest DBDs, denoted by P_d^{DBD}, and that of detecting the two strongest DFT beams, denoted by P_d^{DFT}. The ratio between the power of the analog noise and that of the digital noise is denoted by r. The analog noise is referred to as the noise generated by analog processing, like the beam selection shown in Figure 9.13. On the other hand, the digital noise is referred to the noise caused by digitization. Here, $N = 64$. The AoA conforms to a uniform distribution in $[-\pi, \pi]$ and is randomly drawn for 2×10^4 independent trials. The term γ is the noise power in the digitized signals, i.e. x_n and x_{n+1}.

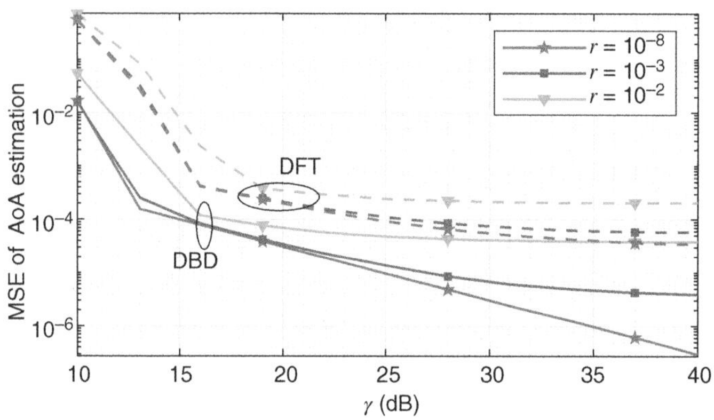

Figure 9.18 Comparison of the AoA estimation performance using the DBDs and the DFT beams.

particular, some intrinsic rules and features of DFT beams were unveiled. Our presentation aimed to inspire new designs and improvements of mmWave/THz large-scale antenna arrays.

References

1 C.-X. Wang, F. Haider, X. Gao, X.-H. You, Y. Yang, D. Yuan, H. Aggoune, H. Haas, S. Fletcher, and E. Hepsaydir, "Cellular architecture and key technologies for 5G wireless communication networks," *IEEE Commun. Mag.*, vol. 52, no. 2, pp. 122–130, Feb 2014. ISSN 0163-6804. 10.1109/MCOM.2014.6736752.

2 W. Roh et al., "Millimeter-wave beamforming as an enabling technology for 5G cellular communications: Theoretical feasibility and prototype results," *IEEE Commun. Mag.*, vol. 52, no. 2, pp. 106–113, Feb 2014. ISSN 0163-6804. 10.1109/MCOM.2014.6736750.

3 Y. Niu, Y. Li, D. Jin, L. Su, and A. V. Vasilakos, "A survey of millimeter wave communications (mmWave) for 5G: Opportunities and challenges," *Wirel. Netw.*, vol. 21, no. 8, pp. 2657–2676, 2015.

4 X. You et al., "Towards 6G wireless communication networks: Vision, enabling technologies, and new paradigm shifts," *Sci. China Inf. Sci.*, vol. 64, no. 1, pp. 1–74, 2021.

5 T. S. Rappaport et al., "Millimeter wave mobile communications for 5G cellular: It will work! *IEEE Access*, vol. 1: 335–349, 2013. ISSN 2169-3536. 10.1109/ACCESS.2013.2260813.

6 T. E. Bogale, L. B. Le, A. Haghighat, and L. Vandendorpe, "On the number of RF chains and phase shifters, and scheduling design with hybrid analog–digital beamforming," *IEEE Trans. Wirel. Commun.*, vol. 15, no. 5, pp. 3311–3326, May 2016. ISSN 1536-1276. 10.1109/TWC.2016.2519883.

7 D. Dawn, P. Sen, S. Sarkar, B. Perumana, S. Pinel, and J. Laskar, "60-GHz integrated transmitter development in 90-nm CMOS," *IEEE Trans. Microw. Theory Tech.*, vol. 57, no. 10, pp. 2354–2367, Oct 2009. ISSN 0018-9480. 10.1109/TMTT.2009.2029028.

8 Y. J. Guo and R. W. Ziolkowski, *Advanced Antenna Array Engineering for 6G and Beyond Wireless Communications*. John Wiley and Sons, 2021.

9 J. A. Zhang, X. Huang, V. Dyadyuk, and Y. J. Guo, "Massive hybrid antenna array for millimeter-wave cellular communications," *IEEE Trans. Wirel. Commun.*, vol. 22, no. 1, pp. 79–87, 2015.

10 Y. J. Guo, X. Huang, and V. Dyadyuk, "A hybrid adaptive antenna array for long-range mm-wave communications [antenna applications corner]," *IEEE Antennas Propag. Mag.*, vol. 54, no. 2, pp. 271–282, 2012.

11 X. Huang, Y. J. Guo, and J. D. Bunton, "A hybrid adaptive antenna array," *IEEE Trans. Wirel. Commun.*, vol. 9, no. 5, pp. 1770–1779, 2010.

12 X. Huang and Y. J. Guo, "Frequency-domain AoA estimation and beamforming with wideband hybrid arrays," *IEEE Trans. Wirel. Commun.*, vol. 10, no. 8, pp. 2543–2553, Aug 2011. ISSN 1536-1276. 10.1109/TWC.2011.062211.100439.

13 J. A. Zhang, X. Huang, and Y. J. Guo, "Adaptive searching and tracking algorithm for AoA estimation in localized hybrid array," in *International Conference on Communications Workshop (ICCW)*, pages 1095–1100. IEEE, June 2015. 10.1109/ICCW.2015.7247323.

14 J. A. Zhang, W. Ni, P. Cheng, and Y. Lu, "Angle-of-arrival estimation using different phase shifts across subarrays in localized hybrid arrays," *IEEE Commun. Lett.*, vol. 20, no. 11, pp. 2205–2208, Nov 2016. ISSN 1089-7798. 10.1109/LCOMM.2016.2602822.

15 S. Kutty and D. Sen, "Beamforming for millimeter wave communications: An inclusive survey," *IEEE Commun. Surv. Tuts.*, vol. 18, no. 2, pp. 949–973, 2016.

16 S.-F. Chuang, W.-R. Wu, and Y.-T. Liu, "High-resolution AoA estimation for hybrid antenna arrays," *IEEE Trans. Antennas Propag.*, vol. 63, no. 7, pp. 2955–2968, 2015.

17 O. El Ayach, S. Rajagopal, S. Abu-Surra, Z. Pi, and R. W. Heath, "Spatially sparse precoding in millimeter wave MIMO systems," *IEEE Trans. Wirel. Commun.*, vol. 13, no. 3, pp. 1499–1513, 2014.

18 A. Alkhateeb, O. El Ayach, G. Leus, and R. W. Heath, "Channel estimation and hybrid precoding for millimeter wave cellular systems," *IEEE J. Sel. Top. Signal Process.*, vol. 8, no. 5, pp. 831–846, Oct 2014. ISSN 1932-4553. 10.1109/JSTSP.2014.2334278.

19 M. Kokshoorn, H. Chen, P. Wang, Y. Li, and B. Vucetic, "Millimeter wave MIMO channel estimation using overlapped beam patterns and rate adaptation," *IEEE Trans. Signal Process.*, vol. 65, no. 3, pp. 601–616, Feb 2016. ISSN 1053-587X. 10.1109/TSP.2016.2614488.

20 R. C. de Lamare, “Massive MIMO systems: Signal processing challenges and future trends,” *URSI Radio Sci. Bull.*, vol. 86, no. 4, pp. 8–20, Dec 2013. ISSN 1024-4530. 10.23919/URSIRSB.2013.7909827.

21 P. D. Arapoglou, K. Liolis, M. Bertinelli, A. Panagopoulos, P. Cottis, and R. De Gaudenzi, “MIMO over satellite: A review,” *IEEE Commun. Surv. Tuts.*, vol. 13, no. 1, pp. 27–51, First 2011. ISSN 1553-877X. 10.1109/SURV.2011.033110.00072.

22 B. G. Evans, “The role of satellites in 5G,” in *2014 7th Advanced Satellite Multimedia Systems Conference and the 13th Signal Processing for Space Communications Workshop (ASMS/SPSC)*, pp. 197–202, Sept 2014. 10.1109/ASMS-SPSC.2014.6934544.

23 W. Ni and I. B. Collings, “A new adaptive small-cell architecture,” *IEEE J. Sel. Areas Commun.*, vol. 31, no. 5, pp. 829–839, May 2013. ISSN 0733-8716. 10.1109/JSAC.2013.130502.

24 E. Cianca, T. Rossi, A. Yahalom, Y. Pinhasi, J. Farserotu, and C. Sacchi, “EHF for satellite communications: The new broadband frontier,” *Proc. IEEE*, vol. 99, no. 11, pp. 1858–1881, Nov 2011. ISSN 0018-9219. 10.1109/JPROC.2011.2158765.

25 Z. Fang, W. Ni, F. Liang, P. Shao, and Y. Wu, “Massive MIMO for full-duplex cellular two-way relay network: A spectral efficiency study,” *IEEE Access*, vol. 5: 23288–23298, 2017. 10.1109/ACCESS.2017.2766079.

26 C. Qin et al., “Joint beamforming and user selection in multiuser collaborative MIMO SWIPT systems with non-negligible circuit energy consumption,” *IEEE Trans. Veh. Technol.*, vol. 99, pp. 3909–3923, June 2017. ISSN 0018-9545. 10.1109/TVT.2017.2719051.

27 Y. Huang and B. Clerckx, “Waveform design for wireless power transfer with limited feedback,” *IEEE Trans. Wirel. Commun.*, vol. 17, no. 1, pp. 415–429, Jan 2018. ISSN 1536-1276. 10.1109/TWC.2017.2767578.

28 Y. Zeng, B. Clerckx, and R. Zhang, “Communications and signals design for wireless power transmission,” *IEEE Trans. Commun.*, vol. 65, no. 5, pp. 2264–2290, May 2017. ISSN 0090-6778. 10.1109/TCOMM.2017.2676103.

29 R. Zhang and C. K. Ho, “MIMO broadcasting for simultaneous wireless information and power transfer,” *IEEE Trans. Wirel. Commun.*, vol. 12, no. 5, pp. 1989–2001, May 2013. ISSN 1536-1276. 10.1109/TWC.2013.031813.120224.

30 T. D. P. Perera, D. N. K. Jayakody, S. K. Sharma, S. Chatzinotas, and J. Li, “Simultaneous wireless information and power transfer (SWIPT): Recent advances and future challenges,” *IEEE Commun. Surv. Tuts.*, vol. 99, pp. 264–302, 2017. 10.1109/COMST.2017.2783901.

31 K. Huang and X. Zhou, “Cutting the last wires for mobile communications by microwave power transfer,” *IEEE Commun. Mag.*, vol. 53, no. 6, pp. 86–93, June 2015. ISSN 0163-6804. 10.1109/MCOM.2015.7120022.

32 L. Yang, Y. Zeng, and R. Zhang, “In-band wireless information and power transfer with lens antenna array,” *IEEE Commun. Lett.*, vol. 21, no. 1, pp. 100–103, Jan 2017. ISSN 1089-7798. 10.1109/LCOMM.2016.2612193.

33 G. Zhu and K. Huang, “Analog spatial cancellation for tackling the near-far problem in wirelessly powered communications,” *IEEE J. Sel. Areas Commun.*, vol. 34, no. 12, pp. 3566–3576, Dec 2016. ISSN 0733-8716. 10.1109/JSAC.2016.2611961.

34 K. Guan, G. Li, T. Kürner, A. F. Molisch, B. Peng, R. He, B. Hui, J. Kim, and Z. Zhong, “On millimeter wave and THz mobile radio channel for smart rail mobility,” *IEEE Trans. Veh. Technol.*, vol. 66, no. 7, pp. 5658–5674, July 2017. ISSN 0018-9545. 10.1109/TVT.2016.2624504.

35 B. Ai et al., “Challenges toward wireless communications for high-speed railway,” *IEEE Trans. Intell. Transp. Syst.*, vol. 15, no. 5, pp. 2143–2158, Oct 2014. ISSN 1524-9050. 10.1109/TITS.2014.2310771.

36 K. Guan et al., "Towards realistic high-speed train channels at 5G millimeter-wave band-part II: Case study for paradigm implementation," *IEEE Trans. Veh. Technol.*, vol. 67, no. 10, pp. 9129–9144, Oct 2018. ISSN 0018-9545. 10.1109/TVT.2018.2865530.

37 F. Hasegawa et al., "High-speed train communications standardization in 3GPP 5G NR," *IEEE Commun. Stand. Mag.*, vol. 2, no. 1, pp. 44–52, March 2018. ISSN 2471-2825. 10.1109/MCOM-STD.2018.1700064.

38 Z. Miao et al., "A 400-GHz high-gain quartz-based single layered folded reflectarray antenna for Terahertz applications," *IEEE Trans. Terahertz Sci. Technol.*, vol. 9, no. 1, pp. 78–88, Jan 2019. ISSN 2156-342X. 10.1109/TTHZ.2018.2883215.

39 C. Lin and G. Y. L. Li, "Terahertz communications: An array-of-subarrays solution," *IEEE Commun. Mag.*, vol. 54, no. 12, pp. 124–131, Dec 2016. ISSN 0163-6804. 10.1109/MCOM.2016.1600306CM.

40 X. Gao et al., "Low RF-complexity technologies to enable millimeter-wave MIMO with large antenna array for 5G wireless communications," *IEEE Commun. Mag.*, vol. 56, no. 4, pp. 211–217, April 2018. ISSN 0163-6804. 10.1109/MCOM.2018.1600727.

41 K. Wu, W. Ni, T. Su, R. P. Liu, and Y. J. Guo, "Efficient Angle-of-Arrival estimation of lens antenna arrays for wireless information and power transfer," *IEEE J. Sel. Areas Commun.*, vol. 37, no. 1, pp. 116–130, Jan 2019. ISSN 0733-8716. 10.1109/JSAC.2018.2872363.

42 Y. J. Guo, M. Ansari, and N. J. G. Fonseca, "Circuit type multiple beamforming networks for antenna arrays in 5G and 6G terrestrial and non-terrestrial networks (Invited Paper)," *IEEE J. Microw.*, vol. 1, no. 2, pp. 704–722, April 2021.

43 L. C. Godara, "Application of antenna arrays to mobile communications. II. Beam-forming and direction-of-arrival considerations," *Proc. IEEE*, vol. 85, no. 8, pp. 1195–1245, 1997.

44 K. Wu, W. Ni, T. Su, R. P. Liu, and Y. J. Guo, "Robust unambiguous estimation of angle-of-arrival in hybrid array with localized analog subarrays," *IEEE Trans. Wirel. Commun.*, vol. 17, no. 5, pp. 2987–3002, May 2018. ISSN 1536-1276. 10.1109/TWC.2018.2805713. accepted.

45 V. Joroughi, M. Á. Vázquez, A. I. Pérez-Neira, and B. Devillers, "Onboard beam generation for multibeam satellite systems," *IEEE Trans. Wirel. Commun.*, vol. 16, no. 6, pp. 3714–3726, June 2017. ISSN 1536-1276. 10.1109/TWC.2017.2687924.

46 V. Joroughi, M. Á. Vázquez, and A. I. Pérez-Neira, "Generalized multicast multibeam precoding for satellite communications," *IEEE Trans. Wirel. Commun.*, vol. 16, no. 2, pp. 952–966, Feb 2017. ISSN 1536-1276. 10.1109/TWC.2016.2635139.

47 M. Hasan and C. Bianchi, "Ka band enabling technologies for high throughput satellite (HTS) communications," *Int. J. Sat. Commun. Netw.*, vol. 34, no. 4, pp. 483–501, Nov 2016. ISSN 1542-0981. 10.1002/sat.1161. http://dx.doi.org/10.1002/sat.1161. SAT-15-0016.R2.

48 Y. Vasavada, R. Gopal, C. Ravishankar, G. Zakaria, and N. BenAmmar, "Architectures for next generation high throughput satellite systems," *Int. J. Sat. Commun. Netw.*, vol. 34, no. 4, pp. 523–546, Jan 2016. ISSN 1542-0981. 10.1002/sat.1175. http://dx.doi.org/10.1002/sat.1175. sat.1175.

49 O. Montenbruck and E. Gill, *Satellite Tracking and Observation Models*. Berlin, Heidelberg: Springer, pp. 193–232, 2000. ISBN 978-3-642-58351-3. 10.1007/978-3-642-58351-3_6. https://doi.org/10.1007/978-3-642-58351-3_6.

50 S. D. Ilcev, *Global Mobile Satellite Communications for Maritime, Land, and Aeronautical Applications*. Dordrecht: Springer, 2005. ISBN 1402027842 (electronic bk.). http://rave.ohiolink.edu/ebooks/ebc/1402027842.

51 M. A. Cervera, A. Ginesi, and K. Eckstein, "Satellite-based vessel automatic identification system: A feasibility and performance analysis," *Int. J. Sat. Commun. Netw.*, vol. 29, no. 2, pp. 117–142, 2011. ISSN 1542-0981. 10.1002/sat.957. http://dx.doi.org/10.1002/sat.957.

52 G. Su and M. Morf, "The signal subspace approach for multiple wide-band emitter location," *IEEE Trans. Acoust. Speech Signal Process.*, vol. 31, no. 6, pp. 1502–1522, Dec 1983. ISSN 0096-3518. 10.1109/TASSP.1983.1164233.

53 H. Wang and M. Kaveh, "Coherent signal-subspace processing for the detection and estimation of angles of arrival of multiple wide-band sources," *IEEE Trans. Acoust. Speech Signal Process.*, vol. 33, no. 4, pp. 823–831, Aug 1985. ISSN 0096-3518. 10.1109/TASSP.1985.1164667.

54 H. Hung and M. Kaveh, "Focussing matrices for coherent signal-subspace processing," *IEEE Trans. Acoust. Speech Signal Process.*, vol. 36, no. 8, pp. 1272–1281, Aug 1988. ISSN 0096-3518. 10.1109/29.1655.

55 M. A. Doron and A. J. Weiss, "On focusing matrices for wide-band array processing," *IEEE Trans. Signal Process.*, vol. 40, no. 6, pp. 1295–1302, Jun 1992. ISSN 1053-587X. 10.1109/78.139236.

56 F. Sellone, "Robust auto-focusing wideband DOA estimation," *Signal Process.*, vol. 86, no. 1, pp. 17–37, 2006.

57 K. Wu, W. Ni, T. Su, R. P. Liu, and Y. J. Guo, "Fast and accurate estimation of angle-of-arrival for satellite-borne wideband communication system," *IEEE J. Sel. Areas Commun.*, vol. 36, no. 2, pp. 314–326, Feb 2018. ISSN 0733-8716. 10.1109/JSAC.2018.2804047.

58 A. Shahmansoori et al., "Tracking position and orientation through millimeter wave lens MIMO in 5G systems," *CoRR*, abs/1809.06343, 2018. http://arxiv.org/abs/1809.06343.

59 S. A. Shaikh and A. M. Tonello, "Localization based on angle of arrival in EM lens-focusing massive MIMO," in *2016 IEEE 6th ICCE-Berlin*, pp. 124–128, Sep. 2016. 10.1109/ICCE-Berlin.2016.7684736.

60 B. Wang, F. Gao, S. Jin, H. Lin, and G. Y. Li, "Spatial- and frequency-wideband effects in millimeter-wave massive MIMO systems," *IEEE Trans. Signal Process.*, vol. 66, no. 13, pp. 3393–3406, July 2018. ISSN 1053-587X. 10.1109/TSP.2018.2831628.

61 Z. Guo, X. Wang, and W. Heng, "Millimeter-wave channel estimation based on 2-D beamspace MUSIC method," *IEEE Trans. Wirel. Commun.*, vol. 16, no. 8, pp. 5384–5394, Aug 2017. ISSN 1536-1276. 10.1109/TWC.2017.2710049.

62 M. R. Akdeniz et al., "Millimeter wave channel modeling and cellular capacity evaluation," *IEEE J. Sel. Areas Commun.*, vol. 32, no. 6, pp. 1164–1179, June 2014. ISSN 0733-8716. 10.1109/JSAC.2014.2328154.

63 G. Yang, C. K. Ho, and Y. L. Guan, "Dynamic resource allocation for multiple-antenna wireless power transfer," *IEEE Trans. Signal Process.*, vol. 62, no. 14, pp. 3565–3577, July 2014. ISSN 1053-587X. 10.1109/TSP.2014.2328980.

64 X. Chen, C. Yuen, and Z. Zhang, "Wireless energy and information transfer tradeoff for limited-feedback multiantenna systems with energy beamforming," *IEEE Trans. Veh. Technol.*, vol. 63, no. 1, pp. 407–412, Jan 2014. ISSN 0018-9545. 10.1109/TVT.2013.2274800.

65 K. W. Choi, D. I. Kim, and M. Y. Chung, "Received power-based channel estimation for energy beamforming in multiple-antenna RF energy transfer system," *IEEE Trans. Signal Process.*, vol. 65, no. 6, pp. 1461 1476, March 2017. ISSN 1053-587X. 10.1109/TSP.2016.2637320.

66 S. Abeywickrama, T. Samarasinghe, C. K. Ho, and C. Yuen, "Wireless energy beamforming using received signal strength indicator feedback," *IEEE Trans. Signal Process.*, vol. 66, no. 1, pp. 224–235, Jan 2018. ISSN 1053-587X. 10.1109/TSP.2017.2764854.

67 J. Seo, Y. Sung, G. Lee, and D. Kim, "Training beam sequence design for millimeter-wave MIMO systems: A POMDP framework," *IEEE Trans. Signal Process.*, vol. 64, no. 5, pp. 1228–1242, March 2016. ISSN 1053-587X. 10.1109/TSP.2015.2496241.

68 S. Noh, M. D. Zoltowski, and D. J. Love, "Multi-resolution codebook and adaptive beamforming sequence design for millimeter wave beam alignment," *IEEE Trans. Wireless Commun.*, vol. 16, no. 9, pp. 5689–5701, Sept 2017. ISSN 1536-1276. 10.1109/TWC.2017.2713357.

69 L. Yang, Y. Zeng, and R. Zhang, "Efficient channel estimation for millimeter wave MIMO with limited RF chains," in *2016 IEEE ICC*, pp. 1–6, May 2016. 10.1109/ICC.2016.7510952.

70 J. Hogan and A. Sayeed, "Beam selection for performance-complexity optimization in high-dimensional MIMO systems," in *2016 Annual CISS*, pp. 337–342, March 2016. 10.1109/CISS.2016.7460525.

71 X. Gao, L. Dai, S. Han, I. Chih-Lin, and X. Wang, "Reliable beamspace channel estimation for millimeter-wave massive MIMO systems with lens antenna array," *IEEE Trans. Wirel. Commun.*, vol. 16, no. 9, pp. 6010–6021, Sept 2017. ISSN 1536-1276. 10.1109/TWC.2017.2718502.

72 D. Zhu, J. Choi, and R. W. Heath, "Auxiliary beam pair enabled AoD and AoA estimation in closed-loop large-scale millimeter-wave MIMO systems," *IEEE Trans. Wirel. Commun.*, vol. 16, no. 7, pp. 4770–4785, July 2017. ISSN 1536-1276. 10.1109/TWC.2017.2702617.

73 B. Wang, F. Gao, S. Jin, H. Lin, G. Y. Li, S. Sun, and T. S. Rappaport, "Spatial-wideband effect in massive MIMO with application in mmWave systems," *IEEE Commun. Mag.*, vol. 56, no. 12, pp. 134–141, December 2018. ISSN 0163-6804. 10.1109/MCOM.2018.1701051.

74 M. Cai, K. Gao, D. Nie, B. Hochwald, J. N. Laneman, H. Huang, and K. Liu, "Effect of wideband beam squint on codebook design in phased-array wireless systems," in *2016 IEEE GLOBECOM*, pp. 1–6, Dec 2016. 10.1109/GLOCOM.2016.7841766.

75 J. H. Brady and A. M. Sayeed, "Wideband communication with high-dimensional arrays: New results and transceiver architectures," in *2015 IEEE ICCW*, pp. 1042–1047, June 2015. 10.1109/ICCW.2015.7247314.

76 X. Gao, L. Dai, S. Zhou, A. M. Sayeed, and L. Hanzo, "Beamspace channel estimation for wideband millimeter-wave MIMO with lens antenna array," in *IEEE ICC*, pp. 1–6, May 2018. 10.1109/ICC.2018.8422131.

77 Z. Gao et al., "Channel estimation for millimeter-wave massive MIMO with hybrid precoding over frequency-selective fading channels," *IEEE Commun. Lett.*, vol. 20, no. 6, pp. 1259–1262, June 2016. ISSN 1089-7798. 10.1109/LCOMM.2016.2555299.

78 K. Venugopal et al., "Channel estimation for hybrid architecture-based wideband millimeter wave systems," *IEEE J. Sel. Areas Commun.*, vol. 35, no. 9, pp. 1996–2009, Sep. 2017. ISSN 0733-8716. 10.1109/JSAC.2017.2720856.

79 K. Wu, W. Ni, T. Su, R. P. Liu, and Y. J. Guo, "Exploiting spatial-wideband effect for fast AoA estimation at lens antenna array," *IEEE J. Sel. Top. Signal Process.*, vol. 13, no. 5, pp. 902–917, Sep. 2019. ISSN 1941-0484. 10.1109/JSTSP.2019.2937691.

80 T. S. Rappaport, G. R. MacCartney, M. K. Samimi, and S. Sun, "Wideband millimeter-wave propagation measurements and channel models for future wireless communication system design," *IEEE Trans. Commun.*, vol. 63, no. 9, pp. 3029–3056, Sept 2015. ISSN 0090-6778. 10.1109/TCOMM.2015.2434384.

81 C. Han, A. O. Bicen, and I. F. Akyildiz, "Multi-ray channel modeling and wideband characterization for wireless communications in the Terahertz band," *IEEE Trans. Wirel. Commun.*, vol. 14, no. 5, pp. 2402–2412, May 2015. ISSN 1536-1276. 10.1109/TWC.2014.2386335.

82 X. Huang and Y. J. Guo, "Closed-form MSE performance for phase estimation from Gaussian reference signals," in *Commununications and Information Technologies (ISCIT), 2011 11th International Symposium on*, pp. 154–158. IEEE, 2011.

83 R. Reggiannini, "A fundamental lower bound to the performance of phase estimators over Rician-fading channels," *IEEE Trans. Commun.*, vol. 45, no. 7, pp. 775–778, Jul 1997. ISSN 0090-6778. 10.1109/26.602582.

84 K. Wu, W. Ni, T. Su, R. P. Liu, and Y. J. Guo, "Expeditious estimation of angle-of-arrival for hybrid Butler matrix arrays," *IEEE Trans. Wirel. Commun.*, vol. 18, no. 4, pp. 2170–2185, April 2019. ISSN 1558-2248. 10.1109/TWC.2019.2900613.

10

Electrically Small Antenna Advances for Current 5G and Evolving 6G and Beyond Wireless Systems

Richard W. Ziolkowski

Global Big Data Technologies Centre, School of Electrical and Data Engineering, University of Technology Sydney, Ultimo, New South Wales, Australia

10.1 Introduction

The Fourth Industrial Revolution, Industry 4.0, is upon us. It represents the integration of humans and technological systems, i.e. cyber-physical systems. It promises many advances including, for instance, biometric sensors to help us maintain our health; sensor networks to enable smart infrastructure and buildings; smart agriculture to monitor crops and manage operations; autonomous vehicles to enhance our standards of living; wireless power transfer (WPT) to reduce device weight and size, eliminating short-life batteries and reducing environmental wastes; and marvelous mobile communication platforms and satellite constellations to enhance social interactions, collaborations, and cohesiveness. Electromagnetics, including antennas, propagation and measurement research, has played, is playing and will continue to play a pivotal role in realizing these anytime-wireless-everywhere (AWE) inspired dreams.

Antennas are the central technology that empowers all aspects of the Industry 4.0 wireless ecosystems, particularly those associated with the much anticipated Internet of things (IoT) and its ubiquitous devices. Seemingly contradictory desires for radiating and receiving elements with superior multifunctional and reconfigurable performance characteristics in smaller, easier-to-fabricate and cheaper packages abound. Over 2400 journal, magazine and conference papers have been published on electrically small antennas alone in IEEE XPlore over the last decade. Needless to say, AWE-inspired electrically small antennas are an active research area for academic, commercial, and government investigators who are stakeholders in advancing wireless technologies.

This chapter on electrically small antennas will begin with a brief discussion of the standard figures of merit associated with the performance characteristics of electrically small radiating and scattering systems. It will then focus on what some of the recent innovations in this area are. Systems that have overcome many of the traditional performance trade-offs associated with electrically small elements will be discussed. Novel applications benefiting from these developments and consequent future opportunities will be highlighted.

Given the ginormous number of contributions made in the ESA area by so many researchers, I will apologize up-front to all whose work is not mentioned. I will naturally emphasize the efforts associated with my group and collaborators from around the world. There are numerous books one can consult specifically for many more details on electrically small antennas, e.g. [1–3], as well as for more general concepts in the traditional antenna textbooks used worldwide, e.g. [4–6].

Antenna and Array Technologies for Future Wireless Ecosystems, First Edition.
Edited by Y. Jay Guo and Richard W. Ziolkowski.

All of the simulations to be discussed were performed analytically (and implemented in MatLab), with in-house developed finite difference time domain software, and with commercial software environments. A majority of the antenna designs were modeled with different versions of the ANSOFT high frequency structure simulator (HFSS). More recent designs were obtained with the corresponding HFSS tools now in the ANSYS Electromagnetics Suite. All simulations were performed with realistic materials unless otherwise noted.

10.2 ESA Figures of Merit

H. A. Wheeler basically provided the generally used definition of an electrically small antenna (ESA) to our technical community in 1947 [7] and later carefully clarified it in 1959 [8]. The now well-known radiansphere is the smallest sphere of radius a that completely surrounds the antenna. An electrically small transmitting or receiving antenna is one with $a < \lambda/2\pi$ or $ka < 1$. As Wheeler indicated, the radiansphere surface can be considered to be the boundary between the near and far fields of a small antenna. An associated interpretation leads to the Friis transmission equation [4, 8]. Let P_{trans} be the input power to an antenna whose effective area is A_{trans} and that is driven at the source wavelength λ. Let P_{rec} be the power received by an antenna whose effective area is A_{rec} and that is separated from the transmitting antenna by the distance r. The surface area of the sphere of radius r centered at the transmitting antenna is $S_r = 4\pi r^2$. The cross-section of the radiansphere, C_{RS}, is given by the expression $C_{\text{RS}} = \pi a^2 = \lambda^2/(4\pi)$. Then

$$\frac{P_{\text{rec}}}{P_{\text{trans}}} = \frac{A_{\text{rec}}A_{\text{trans}}}{C_{\text{RS}}S_r} = \frac{A_{\text{rec}}A_{\text{trans}}}{\lambda^2 r^2} \tag{10.1}$$

The Wheeler cap measurement method [9, 10] in fact uses a radiation shield [8], i.e. a conducting radiansphere (hemisphere when ground plane is present) to measure the radiation efficiency (RE) of an ESA.

When most engineers hear the term "electrically small antenna," they automatically think inefficient, narrow bandwidth, minimal directivity, and a need to design an impedance matching network. Moreover, they would assume that maybe one of these performance figures-of-merit could be improved with tradeoffs amongst the others. As will be demonstrated in the following text, these common assumptions have been proven to be no longer valid within the last two decades.

10.2.1 Bandwidth

In many ways the bandwidth story starts with the well-known paper by Chu [11] in which a lower-bound on the fundamental quantity, the quality factor Q, of the antenna was related to the electrical parameter, ka. The original expression was refined by McClean [12] into the generally accepted form for a passive antenna with a simple, single resonance:

$$Q_{\text{Chu}} = \left[\frac{1}{(ka)^3} + \frac{1}{ka}\right] \tag{10.2}$$

The seminal paper by Yaghjian and Best provided a precise discussion on how to calculate Q at a resonance (reactance is zero at f_{res} and the derivative of the reactance at f_{res} is positive) and an anti-resonance (reactance is zero at f_{res} and the derivative of the reactance at f_{res} is negative). They also related it to the derived quantity, the 3-dB fractional bandwidth (FBW), from which the generally quoted –10-dB FBW, i.e. those frequencies surrounding the simple resonance frequency with $|S_{11}| \leq -10\,\text{dB}$, is readily obtained. These quantities are approximately related to Q as:

$$\text{FBW}_{3\text{dB}} \sim 2/Q$$
$$\text{FBW}_{10\text{dB}} = \frac{1}{3}\text{FBW}_{3\text{dB}} \tag{10.3}$$

Thus, the FBW values are essentially maximized when the lower bound on Q is achieved.

There has since been a never ending stream of reports that have attempted to define a more precise lower bound on Q. They have taken into account an antenna's geometric and other design parameters; have offered yet better understanding on how to properly calculate its reactive power; and have even incorporated time variations in its material, structural and circuit element properties, e.g. [13–22].

There are several useful rules of thumb to keep in mind that have arisen from the chatter. As has been discussed, for instance, in [23, 24], the lower bound on Q should take into account the RE η_{rad}:

$$Q_{\text{lb}} = \eta_{\text{rad}} Q_{\text{Chu}} = \eta_{\text{rad}} \left[\frac{1}{(ka)^3} + \frac{1}{ka}\right] \tag{10.4}$$

i.e. the lower bound is smaller than the Chu limit. If the antenna is circularly polarized (CP) rather than linearly polarized (LP), this relation becomes:

$$Q_{\text{lb,CP}} = \frac{1}{2}\eta_{\text{rad}} \left[\frac{1}{(ka)^3} + \frac{2}{ka}\right] \tag{10.5}$$

These lower bound expressions point out why larger intrinsic losses and, hence, lower radiation efficiencies are acceptable in practice because larger bandwidths ensue.

Another useful improvement was derived in [13]. It was shown that the lower bound of an electric ESA with $ka \ll 1$ is more correctly $Q_{\text{lb},e} \sim 1.5\ Q_{\text{lb}}$ and that of a magnetic ESA is $Q_{\text{lb},m} \sim 3\ Q_{\text{lb}}$. Thus, an electric ESA will generally have twice the FBW in comparison with that of a magnetic ESA of similar electrical size.

Yet another important fact is that more bandwidth is accessible by a three-dimensional (3D) antenna that efficiently fills the radiansphere rather than a planar one. In fact, $Q_{\text{lb,planar}} \sim (9\pi/8)\ Q_{\text{lb}}$ [14, 17], i.e. the FBW of a planar antenna is about 3.5 times smaller than a 3D one that fills the radiansphere. There have been many studies that have suggested antenna designs that truly approach the fundamental lower bound, e.g. [23–27]. A stellar example is the self-resonant four-arm spherical folded helix antenna at $ka = 0.265$ with the measured $Q = 1.52\ Q_{\text{lb}}$ [23–25]. A review of hundreds of ESA designs in relation to the Q-based lower bound is given in [28].

Finally, defining the electrical size ka when a ground plane is present unfortunately becomes a bit nebulous. It has been shown that if the currents on the ground plane are localized near the radiator, the ground plane size has a minimal effect on the impedance matching, e.g. [29]. Thus, the often used electrically small criterion for a radiator integrated with a ground plane, $ka \leq 0.5$, is quite reasonable. Nevertheless, it has also been shown that it fails to account for the directivity which changes dramatically depending on the size of the ground plane relative to that of the radiator [29, 30]. On the other hand, the radian sphere radius in the compilation [28] was chosen expeditiously to be the sphere which encloses the entire ground plane if its radius is smaller than $\lambda/4$, and to be the radius of the radiansphere totally enclosing the radiating elements otherwise.

10.2.2 Directivity and Efficiency

Directivity represents how much real power is radiated by an antenna into a particular direction when it is excited by a source relative to how much would be if the same source was driving a hypothetical isotropic electromagnetic radiator. The elemental radiators, i.e. infinitesimal electric Hertzian dipole (EHD) and loop (equivalent to a magnetic dipole) antennas are well-known to have a maximum directivity equal to $D_{max} = 1.5$ (1.76 dB) [4–6].

Electrical small radiating systems with enhanced directivity have been realized recently. These Huygens dipole antennas (HDAs), i.e. balanced pairs of electric and magnetic dipole elements, have attained directivities in access of the infinitesimal system value, 3 (4.77 dB), as will be discussed more in the following text, i.e. note that with $N = 1$ in (10.9), the maximum $D_{max} = 3.0$. The fact that those electrically small systems have their maximum directivity in access of this limiting value is simply due to the fact that they are finite in size and, consequently, have bits of other modes present which make small contributions to the overall radiated power.

More generally, recall that if the total radiated power is P_{rad} and the power supplied to the antenna system is P_{input}, then the total overall efficiency of an antenna would be

$$\text{Total overall efficiency} = 100\% \times \frac{P_{rad}}{P_{input}} \tag{10.6}$$

The radiation efficiency is then

$$\text{Radiation efficiency} = 100\% \times \frac{P_{rad}}{P_{accepted}} \tag{10.7}$$

where the total power accepted by the antenna is $P_{accepted} = (1 - |S_{11}|^2)\, P_{input}$. The better matched an antenna is to its source, the more power it has to potentially radiate.

If the radiating system has an effective transverse area A_{eff} and if it is uniformly driven at the excitation wavelength λ, then its maximum directivity and realized gain are [4]

$$\begin{aligned} D_{max} &= \frac{4\pi}{\lambda^2} A_{eff} \\ RG_{max} &= e_{total} \times D_{max} \end{aligned} \tag{10.8}$$

where the total efficiency of the radiating system, i.e. taking into account the material losses, mismatch losses, etc., is e_{total}. Consequently, a larger effective radiating aperture will provide a higher directivity. Moreover, these relations clearly indicate that an ESA, by nature, is directivity challenged. Note that $e_{total} = (1 - |S_{11}|^2)\, e_{em}$, where the term e_{em} represents the overall efficiency associated with the radiator (all of the efficiencies tied to losses other than the mismatch loss) and, hence, the accepted power. Therefore, the maximum gain is $G_{max} = e_{em} \times D_{max}$. At a given frequency, the gain of a time-harmonic (continuous wave [CW]) antenna on transmit and receive is identical.

Is there a means to achieve directivities well beyond the dipole values with a single radiator? By taking into account both the transverse electric and magnetic modes, Harrington demonstrated that the maximum directivity from a source region as a function of the maximum number of multipole modes, N, that it can radiate is [31, 32]:

$$D_{max} = N^2 + 2N \tag{10.9}$$

Therefore, by exciting many higher order modes, one can in principle achieve very high maximum directivity from a single element. There has been much interest recently in the physics literature to use high permittivity dielectrics to enable the overlap of dipole and higher order modes in nano-systems to attain high directivity. A Huygens nanoparticle laser based on this concept was reported in [33].

Another approach, in principle, is a superdirective system. A useful operational definition of superdirectivity, e.g. as emphasized by Hansen [34, 35], is to achieve a directivity greater than that obtained with the same antenna configuration being uniformly excited (constant amplitude and phase). While this is basically an array definition, a precise one for a single element is a bit stickier because even a tiny ideal electric Hertzian dipole is "superdirective" by it, i.e. because its directivity is 1.5, its effective electromagnetic area from (10.8) is $A_{\text{eff}} = 3\lambda^2/8\pi$, which is much larger than its physical area. The concept of superdirectivity has permeated the physics and applied physics literature repeatedly since Oseen discussed the concept of "needle radiation" nearly a century ago [36] and continues to be an active research field [37]. Nevertheless, it is extremely challenging in practice. There have been and continue to be many attempts to overcome the ill-posedness associated with superdirectivity, especially taking into account system constraints, e.g. [38, 39]. A superdirective electrically small radiating system was recently reported in [40] based on a metamaterial (MTM)-inspired multilayered nanoparticle.

10.3 Overcoming Conventional ESA Stigmas and Trade-Offs

As noted earlier, experimentally verified examples of ESAs that have overcome the conventional wisdom of their drawbacks have been reported. Many were initially associated with the concept of electromagnetic MTMs [41–43], artificial materials whose electromagnetic responses can be designed to have properties not attainable with naturally occurring media. Many more recent examples have arisen from the fundamental electromagnetic physics and engineering principles they imbue.

10.3.1 Metamaterial-Based ESAs

The adaptation of a variety of epsilon-negative (ENG), mu-negative (MNG), and double negative (DNG) MTMs to achieve miniaturized antennas followed immediately after the initial surge of publications investigating their exotic physics properties. There are several review papers available with extensive reference lists that cover many of the designs in the first MTM decade [44–49]. I simply call your attention to several small MTM-based antennas reported by some of the early research teams in this area: [50–68]. Small communications and cross-dipole antennas were reviewed, respectively, in [69] and [70]. A variety of compact composite right/left-handed (CRLH)-based antenna systems have been reviewed very recently [71, 72], as have antennas augmented with meta-structures, e.g. [73–77].

My team's initial MTM-based ESA concept consisted of a center-fed electrically small dipole antenna centered in and surrounded by either a DNG [50] or an ENG [51] shell. For simplicity, I will denote it as a core–shell antenna. The driven dipole radiator is a highly capacitive element. The double positive (DPS) air region itself acts as a capacitive element. The ENG shell, for example, is also electrically small and is excited by the fields radiated by the dipole. Consequently, it too is a capacitive element. However, because it is filled with a negative permittivity medium, it acts like an inductor. The combination of the capacitive and inductive elements forms an electrically small LC resonator. Nearly complete impedance matching to the source without any matching network and radiation efficiencies approaching 100% by properly tuning them together was demonstrated [51]. It was later proved that the driven element does not have to be surrounded by the ENG shell.

The consequent electrically small LC resonator formed by it and the ENG shell is strongly excited by the driven dipole as long as the dipole is in close proximity to it [78]. The basic principles of this core–shell antenna concept were verified experimentally with the dual structure, a coax-fed semi-loop antenna surrounded by an MNG hemisphere on a ground plane [79].

It must also be emphasized that there were many in the IEEE Antennas and Propagation (AP) and Microwave Theory and Techniques (MTT) professional societies who were adamant that many of these MTM-based concepts were not credible, even despite existing experimental results. Monographs were even published by well-known colleagues labeling some of the work as "voodoo science" or the materials with exotic and even simple properties (e.g. empathically denying the existence of media with negative parameters) as "unobtainium" [80, 81]. Some negative feelings still persist because, like many persisting and emerging areas, e.g. magnetic fusion, graphene, quantum computers, ..., there is much hype about the potential outcomes to acquire recognition and funding. Others who have jumped on the bandwagon have misused the terminology and, unfortunately, have nurtured them.

We must all understand and continue to be reminded that it really does take years to properly vet innovative ideas and truly demonstrate their unique advantages over existing systems. For instance, it was suggested that the dipole-ENG shell system could be realized based on a Drude model of a plasma in a glass shell whose density was small enough that the operating frequency of the driven dipole was below its plasma frequency. Analytical and numerical models [82, 83] suggested this approach was realistic. More than a decade later, the concept has been verified with some outstanding experimental efforts [84].

As a historical side note discussed in [85], my team had considered passive and active materials constructed with artificial molecules before the turn of the century. As they would be currently termed, these passive and active "meta-atoms" were developed as complex passive and active loads connected to very electrically small dipole and loop antennas [86–91]. The fields re-radiated, i.e. scattered, by these designed electric (magnetic) inclusions in a substrate superimposed to yield a variety of dispersion engineered permittivity (permeability) responses. Over a decade later, the electric (meanderline antenna loaded with an inductor) and magnetic (loop antenna loaded with a capacitor) unit cells shown in Figure 10.1 were successfully combined into one of the electrically smallest DNG unit cells still today and a block of the resulting low loss DNG MTM was tested. The results experimentally verified their simulated performance characteristics [92, 93], i.e. a refractive index $n = -3.1$ with less than 1.0 dB/cm loss at 400 MHz with unit cells whose overall size was $\sim \lambda/75$.

Finally, I would be remiss not to highlight one of the first MTMs, i.e. artificial magnetic conductors (AMCs), that were developed to enhance the performance of antennas. The high-impedance surface (HIS) mushroom [94] and UC-PBG [95] structured ground planes mitigated surface wave effects and facilitated low-profile antenna systems. Many antennas using these and other structured ground planes are nicely reviewed in [96]. An AMC based solely on capacitively-loaded loop (CLL) unit cells without the presence of a ground plane [97] realized the magnetic wall considerations in [86]. One of the first investigations on how long in time it takes for a MTM to realize its properties employed a related CLL-based AMC [98]. Similar time domain considerations facilitated the realization of a zero-index MTM superstrate [99] for a high-gain antenna in a 60 GHz high-data-rate system. Predicted multi-Giga-bit information transfer based on an on-off-key (OOK) modulation was confirmed experimentally [100].

An electrically small near-field resonant parasitic (NFRP) antenna system that used a structured ground plane to achieve a high directivity was designed in [101]. However, we found it a severe challenge in general to fit the required periodic structure within in a Wheeler sphere smaller than

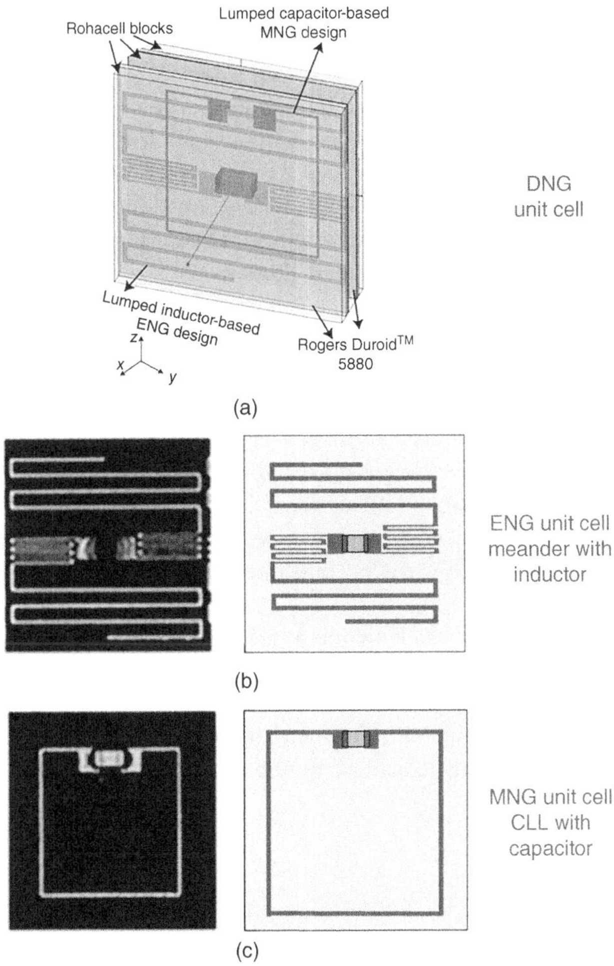

Figure 10.1 DNG Metamaterial with $n_{real} = -3.1$ at 400 MHz [92, 93]. (a) Isometric view of the composite DNG unit cell. (b) Meanderline antenna-based ENG layer of the unit cell. Left: Photo, Right: Simulation model. (c) Capacitively-loaded loop (CLL) antenna-based MNG layer of the unit cell. Left: Photo, Right: Simulation model. Source: From [93] / with permission from AIP Publishing LLC.

$ka = 1.5$. As a consequence, we turned to other approaches to attain higher directivity ESAs as will be illustrated in the following text.

10.3.2 Metamaterial-Inspired ESAs

Even though very small unit cells were realized, a MTM, i.e. a block of artificial material consisting of enough layers of these unit cells in the wave propagation direction to attain homogenized values of the permittivity and permeability (basically 3–5) is still quite thick if one desires an antenna system that has its $ka = 0.5$, i.e. $a \sim \lambda/12.6$. Moreover, the unit cells were planar; contained lumped elements; and redesigning them for a curved geometry, e.g. the core–shell antenna system, was and remains very challenging. The three-dimensional lego-design used in [79] was bulky, and we found that precise registration (alignment) of the elements in the various layers was required for the best performance. Nevertheless, other teams successfully combined a single layer of MTM unit cells both as substrates and superstrates, which would now be denoted as metasurfaces, with radiating elements for enhanced performance characteristics [53, 59, 61].

Luckily, a paradigm shift associated with the MTM-based ESAs and ESAs in general arose from those efforts with the antennas reported in [102]. We found that exciting only a single MTM unit cell that was resonant near the operating frequency of interest with a small driven element was sufficient to achieve a highly efficient ESA without the need for any matching network. Because no MTM was involved, but its success followed from MTM designs, we coined the term "MTM-inspired" to denote those original ESAs. Many of the original MTM-inspired designs were reviewed in [44]. There have been numerous MTM-inspired ESAs recently reported, e.g. [103–110].

As they were investigated further, it became clear that the fundamental radiation physics and engineering of the original designs is best captured with the designation: NFRP antennas. The NFRP antennas consist of a simple driven element and one or more NFRP elements. The concept is illustrated in Figure 10.2a,b with field and circuit concepts, respectively. By adjusting the sizes and shapes of the driven and NFRP elements, as well as the distance and material between them, one can engineer the imaginary part of the antenna's input impedance, i.e. its reactance, to be zero and its real part, i.e. its resistance, to be matched to that of the source, as well as emitting fields that are well matched to the impedance of free space. Nearly complete impedance matching and a high radiation efficiency are thus attained [44].

A simple example is shown in Figure 10.3a. The driven element is a small curved top-hat dipole radiator; the NFRP element is an Egyptian axe dipole (EAD) [101]. The surface currents on the driven and NFRP elements illustrated in Figure 10.3b illustrate one of the advantages of the design, i.e. the main radiating currents occur primarily on the NFRP element, not the driven one. This aspect allows for the presence of more NFRP elements to achieve multiple functions with only a single driven element.

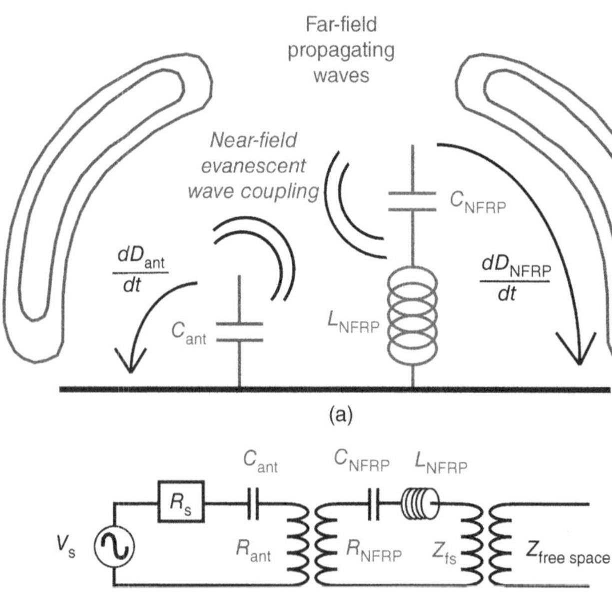

Figure 10.2 Fundamental operating principles of NFRP antennas [44]. (a) Electromagnetic concepts. (b) Equivalent circuit aspects. Source: From [44] / with permission from IEEE.

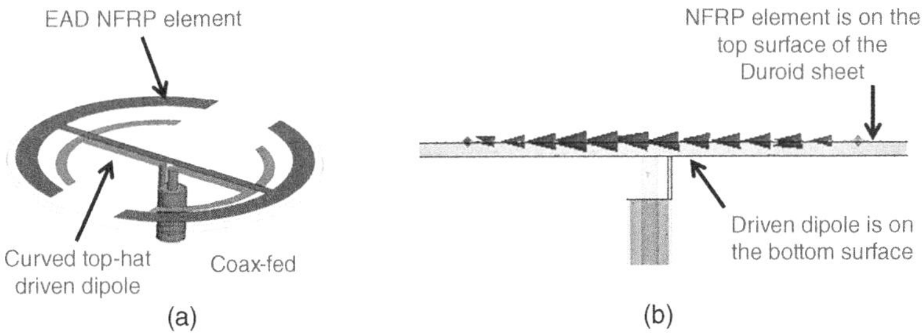

Figure 10.3 Egyptian axe dipole (EAD) NFRP antenna. Source: Adapted from [101]. (a) HFSS model. (b) Simulated surface currents.

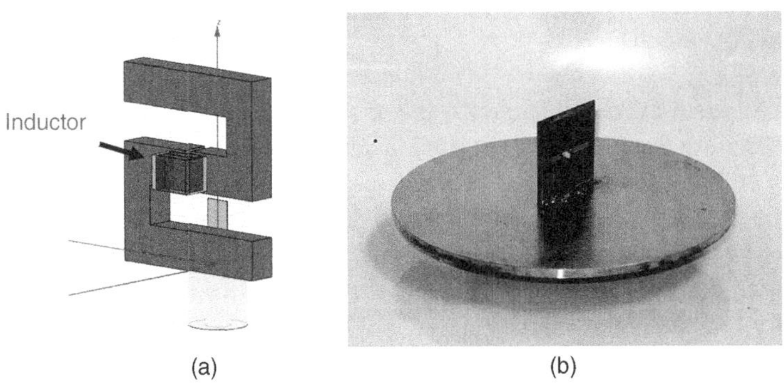

Figure 10.4 Electrically small Z-antenna. (a) HFSS model. (b) 300 MHz prototype. Source: From [111] / with permission from IEEE.

Two of the early NFRP antenna designs and their prototypes serve as a prelude to more recent examples. They were developed during a Defense Advanced Research Projects Agency (DARPA) sponsored Multidisciplinary Research Initiative (MURI) program that was led by the Boeing Aerospace Company [85]. Boeing Phantom Works fabricated our designs and they were tested at the National Institute of Standards and Technology (NIST) in Boulder, CO.

The Z-antenna shown in Figure 10.4 is an electric NFRP design. A coax-fed printed monopole acts as the driven element. The inductor loaded Z structure is the NFRP element. The design served several purposes at the time. It demonstrated the flexibility to realize versions as small as $ka = 0.046$ at very high frequency (VHF) to ultra-high frequency (UHF) frequencies; the ability to tune the resonances of one design over a very large bandwidth, i.e. to demonstrate frequency agility; and to understand how the overall efficiency of this type of antenna varied, for example, when ka was varied by almost an order of magnitude, i.e. from 0.016 (~30%) to 0.14 (>90%) [112]. Prototypes of the UHF designs were fabricated and tested [44, 111]. They were quite valuable since they taught us how to tailor the parameters to account for the lumped element being introduced into the NFRP element rather than into a circuit. The 570 MHz, $ka = 0.40$ design shown in Figure 10.4 had an measured overall efficiency greater than 80%.

The 3D magnetic EZ antenna shown in Figure 10.5 was a magnetic NFRP antenna. The driven element was a coax-fed semi-loop. The NFRP element was an extruded, 3D CLL structure. Prototypes of the original design [102] were fabricated for operation at the UHF frequency of 300 MHz [44, 114] and at the VHF frequency of 100 MHz [44, 113]. The former, which had $ka \sim 0.43$ at 301 MHz, had a measured overall efficiency greater than 94%. The latter is shown in Figure 10.5. It was a low profile antenna (height = $\lambda/25$) whose measured resonance frequency was 105 MHz

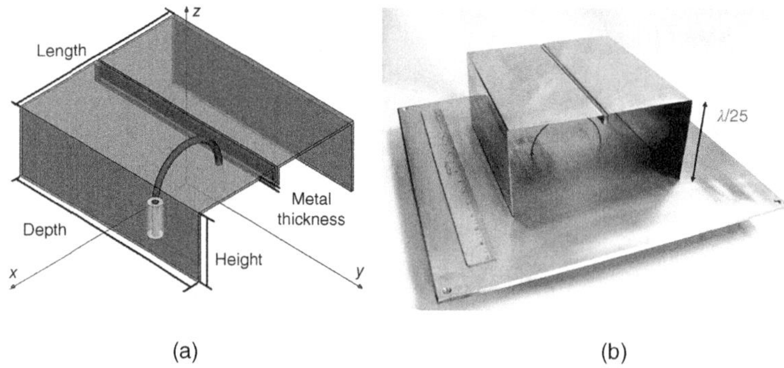

Figure 10.5 Electrically small 3D Magnetic EZ antenna. (a) HFSS model. (b) 100 MHz Prototype. Source: From [113] / with permission from AIP Publishing LLC.

giving $ka = 0.46$ and whose measured overall efficiency was ~95%. The change in the radiation efficiency as ka decreases was also studied [113]. It was found in analogy with the lower bound in Q that the radiation efficiency of the magnetic NFRP designs decreased much more quickly as ka did in comparison with the their electric counterparts. Variations of this 3D magnetic EZ design have proven to be quite resilient as an ESA for high power microwave (HPM) systems [115–117].

The VHF and UHF systems were examined because achieving physically tiny, highly efficient antennas in those bands and even lower ones might have enabled a host of applications. A $\lambda/2$ dipole antenna at 30 MHz is approximately 5.0 m in length. Very interesting robust communications concepts have been reported recently at the lower end of the very low frequency (VLF) band [118] among mobile agents such as robots in complex non-line-of-sight indoor and urban-type scenarios. The need for efficient miniature antennas has driven the realization of several unique designs [119–126].

10.3.3 Non-Foster Circuits

Recalling the lower bounds on Q for passive ESA systems, their bandwidths are naturally narrow. Most of the described early electric- and magnetic-based NFRP ESA systems exhibited FBWs on the order of 3–4% and 1–2%, respectively, in sync with those bounds. Despite the general demand for greater bandwidth, there are many narrow band antenna applications for which ESAs are well-suited. They include, for instance, the low frequency ESAs for communications between mobile agents in complex environments noted earlier [118] and for WPT involving dedicated transmitters, which will discussed in the following text. Furthermore, in conjunction with the rapid growth of the IoT applications, low power wide area networks (LPWAN) and their associated technologies, such as LoRA and Sigfox, have become popular for low-rate (narrow bandwidth) long-range radio communications [127]. LoRA systems are important for rural settings and have promising potential for search and rescue applications [128]. Nevertheless, there is always the demand for greater bandwidths in compact wireless devices.

Because the noted physics-based bounds on Q are directly connected to passive radiating systems, an approach to achieve a large instantaneous bandwidth with an ESA is accomplished by introducing active systems. There are two basic approaches as illustrated in Figure 10.6. The first is to introduce an external active matching network that compensates for both the resistance and reactance of the input impedance of the ESA to match it to the source. The second introduces an active internal reactance element into the antenna itself. Both methods are facilitated by using non-Foster

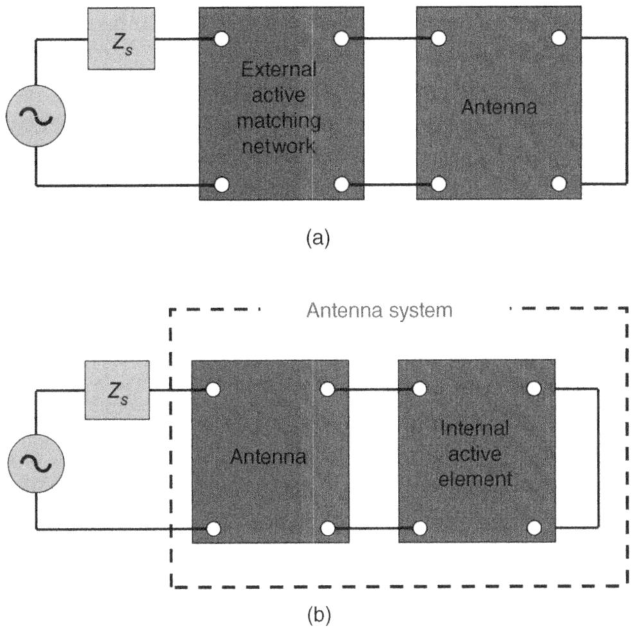

Figure 10.6 Methods to obtain an ESA with a large instantaneous bandwidth. (a) Active matching network. (b) Internal active element.

(NF) elements, i.e. powered transistor-based circuits whose impedance characteristics can decrease with increasing frequency in violation of Foster's reactance theorem [129, 130]. Foster's theorem basically tells us that the stored electric and magnetic field energies of a passive electromagnetic system must remain positive as the frequency changes, i.e. the frequency derivative of its reactance must be positive. NF elements can exist because their power source introduces energy into the system, negating the passive bounds.

There have been a number of successful realizations of wide bandwidth ESAs employing both types of active approaches. External active matching networks have been considered over the last 50 years [131–144]. More recent variants recognize that the Chu bound arises from the fact that passive antennas are LTI (linear time-invariant) systems. A variety of non-LTI (active) systems have been considered to attain wider bandwidths from electrically small systems. As argued in [21], the use of non-linear and time-varying components has a long history back to first wireless telegraphy experiment in 1902 by Marconi which was successfully enabled by a spark-gap ESA. Impedance modulation, usually through the use of time varying inductor and capacitors [145], is a form of parametric amplification (see, e.g. [146], Chapter 11). This time-variation concept is also a fundamental aspect of space-time, time-modulated, and magnetless non-reciprocal metasurfaces to control radiated wavefronts and directional beams [147–153]. Another approach is the use of temporal modulation of the matching network, e.g. through the introduction of switches for direct amplitude modulation (DAM) of an ESA [154–157]. On the other hand, the internal (embedded) active elements have been reported only in the last decade [139, 158–173].

The process that my team has used to achieve a NF-augmented NFRP antenna emphasizes the initial choice of a highly efficient, frequency-agile, passive NFRP antenna that has either an inductor or capacitor incorporated into its NFRP element(s) and has as low a Q value as possible. The presence of such a lumped element facilitates the frequency agility, i.e. it allows the antenna's resonance frequency to be tuned. Since such a low-Q passive radiator will have a wider bandwidth, it relaxes the demands on the active element's required performance.

Sweeping the lumped element's value allows one to obtain a reactance versus resonance frequency (X–f) curve and it will have a NF behavior, i.e. a negative slope with respect to the frequency. A negative impedance convertor (NIC) based on a pair of cross-coupled transistors is then designed to create an active inductor or capacitor that replaces the passive one and matches the X–f curve. The NFRP antenna design with this active element being present is further refined with a co-design process. It is critical that realistic lumped element models and all interconnecting traces in the NIC circuit implementation with the NFRP element be included in the co-design process to account for the parasitic capacitances, inductances, and resistances associated with them. The final optimal design should have the resistance of the NF element as small as possible across the entire extended bandwidth to maintain the high radiation efficiency of the passive design.

Several frequency agile versions of the basic CLL-based protractor NFRP antenna [44] have been obtained and their prototypes tested [174–176]. This magnetic dipole type of NFRP antenna has been an attractive choice in practice because one can incorporate a varactor into its NFRP element, e.g. across its gap, and tune its resonance frequency simply by changing the voltage applied to the varactor. This choice was encouraged because of the successful use of varactors for tunable magnetic MTM unit cells in the physics literature, e.g. [177]. The analysis, simulation, and testing of the NIC-capacitor augmented version was described in the articles [159, 161, 163].

Nevertheless, the first NF-augmented ESA design we developed was actually an electric one, the canopy antenna [178]. Its structure is illustrated in Figure 10.7. The NFRP element is the metallic spherical cap connected to a ground plane by four inductor posts. It is excited by the coax-fed monopole. The electric choice was associated with the noted fact that passive electric dipole-based ESAs naturally have more bandwidth than the magnetic ones. Moreover, of the various electric-based NFRP ESAs considered in [178], the Q value of the canopy design came closest to the Q lower bound. The electrical size was purposely picked to be small, i.e. $ka = 0.0467$ ($a \sim \lambda/137$), for a 297.4 MHz operating frequency. The passive design had $Q = 1.75\ Q_{\text{Chu,lb}} = 1.17\ Q_{\text{Thal,lb}}$ and a $\text{FBW}_{10\text{dB}} = 0.0133\%$. Replacing the passive inductors with NIC-versions, the FBW was greater than 10% [158]. This stellar result actually encouraged the subsequent NIC-inductor augmented EAD-based NFRP ESA prototypes [159, 160, 163], one of which is shown in Figure 10.8a. The measured results confirmed a bandwidth several times the fundamental passive upper bound. Similar outcomes have been reported with embedded NF elements in [166, 167].

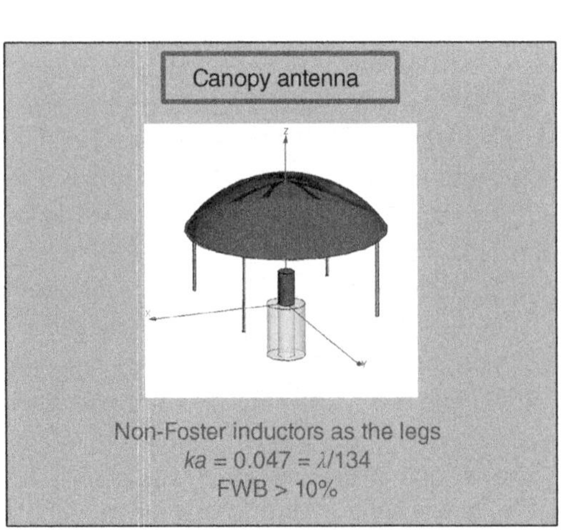

Figure 10.7 The canopy antenna augmented with four non-Foster elements, which are the four NIC-based inductor posts that support the metal canopy. Source: From [158] / with permission from IEEE.

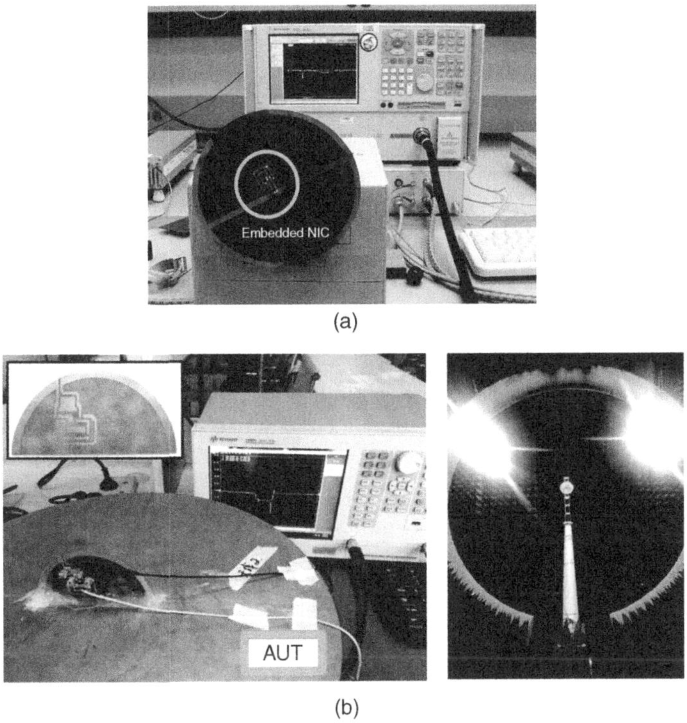

Figure 10.8 Experimentally validated NFRP ESAs whose NFRP elements were augmented with non-Foster elements. (a) EAD example (NIC-inductor). Source: From [160] / with permission from AIP Publishing LLC. (b) CLL example (NIC-capacitor). Source: From [172] / with permission from IEEE.

Performance comparisons of the two-dimensional (2D) NF EAD version in Figure 10.8a and a three-dimensional (3D) variation were considered in [168]. The results confirmed that initial passive antenna designs that have wider instantaneous bandwidths (i.e. smaller Q values), regardless of their 2D or 3D nature, will perform better than narrower bandwidth ones when they are augmented with the NF elements. This outcome is due to their decreased sensitivities to component and fabrication tolerance errors encountered with the NF circuit realizations. The recently tested 3D CLL-based NIC-capacitor augmented magnetic NFRP ESA shown in Figure 10.8b further confirmed these conclusions [172].

Despite the promising results from both the internal and external NF approaches, many challenges remain. Many of them are nicely reviewed in [179]. In particular, it is well known that the cross-coupled transistor realizations, for instance, of NIC elements depend on positive feedback for their operation and, hence, are very prone to instabilities. Stability analyses [136, 180–182] have led to further understanding of the appearance of NIC instabilities and their mitigation. Moreover, the NIC designs are very sensitive to the actual parameters of the physical components used to realize them, as well as to their assembly with manual soldering. A very promising method that avoids component tolerance and assembly issues is to realize the NIC element as a complementary metal-oxide semiconductor (CMOS) processed integrated circuit (IC) [183–185]. Another attractive approach recently reported is to not avoid the instabilities, but to use them to power the radiating elements as self-oscillating antennas [186]. The initial prototype crossed [186] and Huygens [187, 188] dipole self-oscillating antennas have significantly wide measured frequency-agile impedance bandwidths.

Yet another performance issue with active elements, particularly in a receiving antenna, is the signal-to-noise-ratio (SNR). There have only been a few investigations reported. Both [189] and [190] consider the active matching circuit approach. While the former indicates that there may be NF advantages overall, the latter concludes there are none. In contrast, the internal NF element approach SNR study [172] with the 3D NF CLL-based ESA shown in Figure 10.8b demonstrated significant improvements. More studies will need to be performed to attain a conclusive answer in general if there is one. Again, because of the active nature of the systems, the SNR outcome may simply depend on the NF element design and how it is employed.

10.4 More Complex Electrically Small NFRP Antennas

One of the advantages of the NFRP ESA paradigm is that it facilitates incorporating several NFRP elements within an electrically small volume. Because the NFRP elements are not connected directly to the source in any way nor necessarily with each other, the function that each one performs can be developed (in many cases) completely independent from the others. Multifunctional ESAs are an immediate consequence. They are, of course, quite advantageous in many wireless applications desiring more capabilities in ever smaller footprints. Moreover, this separation of intended functions allows one to incorporate additional electronic components that are associated with either the driven or NFRP elements or both. This additional degree of design freedom allows one to achieve a variety of reconfigurable ESA systems.

10.4.1 Multifunctional Designs

A few of the earlier NFRP designs illustrate the multifunctional concept in an electrically small package. By introducing two orthogonal electric NFRP elements of different sizes and a bowtie-shaped driven element that excites both of them, linear polarized (LP) fields at two independent frequencies of operation, GPS-L1 and Global Star (GS), were obtained in [191]. Moreover, CP fields were also obtained at a specified operating frequency f_0 with the same configuration without any phase shifter simply by adjusting the NFRP element shapes and sizes to have one resonance slightly above f_0 and one below it so their reactances naturally provide the requisite 90° phase difference between the LP radiators. A three metal layer, dual-band CP version also made good use of the crossed-dipole resonator configuration [192]. All had high radiation efficiencies.

The two resonator design was extended to four resonators in [193] to obtain CP fields radiated at the GPS-L1 and GPS-L2 frequencies. The tested prototype is shown in Figure 10.9A. Four interleaved magnetic-based protractor NFRP elements were driven by two printed monopoles fed with a single coax. The electric resonator concept was further extended to eight elements in [194]. The optimized design parameters of this compact structure were determined to naturally obtain CP radiation and multi-band operation that covered all five of the GPS bands, L1–L5. The tested prototype is shown in Figure 10.9B.

Other multi-band NFRP ESAs have also been developed. These include monopole [195] and dipole driven [196] elements loaded with CLL NFRP elements as their resonators. Integrating an AMC surface with pairs of crossed-dipole resonators, a triband CP antenna was achieved [197].

Filtennas are another multifunctional antenna concept that leads to more compact front-ends in wireless communication systems. They represent the advantageous integration of the filter(s) with the antenna into a single module. They possess attractive features such as a controllable passband frequency response in both their reflection coefficient and realized gain values. They have become popular for the next generation of wireless devices supporting high-speed data rates and portable

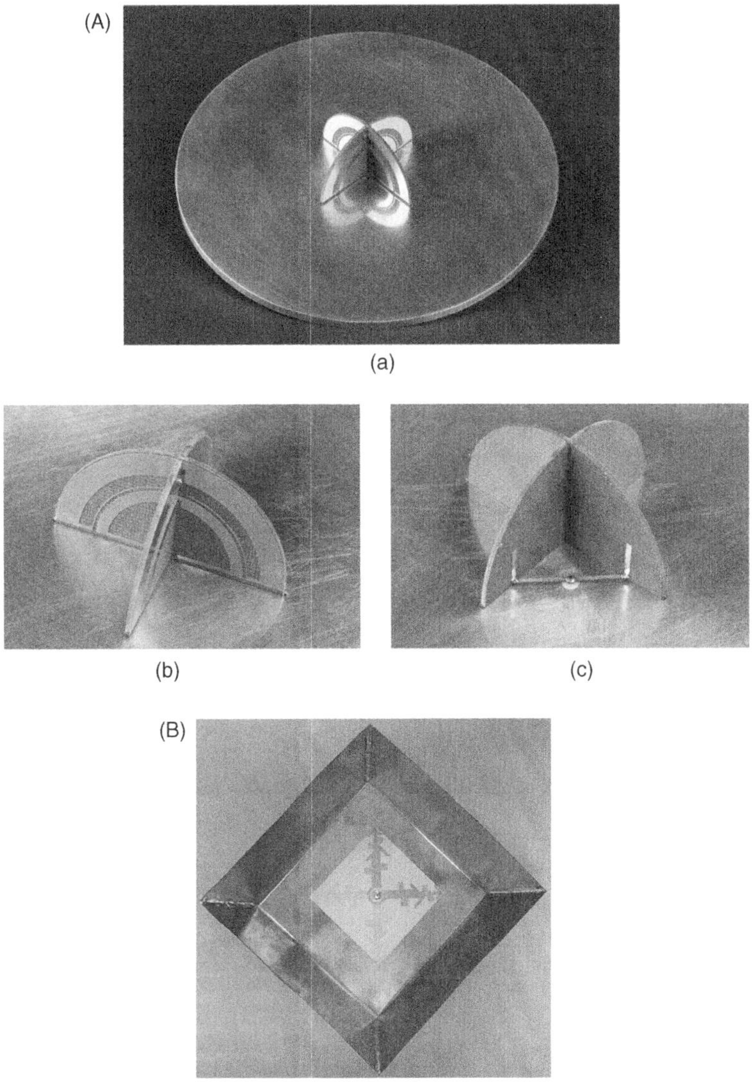

Figure 10.9 Experimentally validated multifunctional NFRP ESAs. (A) Four resonators for dual-band CP operation. (a) Perspective view. (b) Zoom-in. (c) Coax-driven dipoles on the backside of the view in (b). Source: From [193] / with permission from IEEE. (B) Eight resonators to obtain CP operation at all five GPS frequencies, L1–L5. Source: From [194] / with permission from IEEE.

applications [198]. Electrically small filtennas that have relied on CLL-based resonators [199], more complex driven elements to attain LP radiation [200], and more complicated NFRP elements to achieve more bandwidth for both LP and CP performance [201].

10.4.2 Reconfigurable Designs

Reconfigurable antennas are now a standard means to achieve multifunctional performance in wireless mobile terminals [202, 203]. Innovative technologies have advanced their capabilities [204]. Nevertheless, achieving reconfigurability in an electrically small package remains a challenge.

Several reconfigurable NFRP ESAs have been developed. Systems that change both pattern [205–210] and polarization [211] characteristics have been investigated. The NFRP paradigm has been particularly advantageous because the switches can be placed either in the driven elements or in the NFRP elements.

Two examples are illustrated in Figure 10.10. The pattern reconfigurable ESA [207] shown in Figure 10.10A radiates three unidirectional beams along the horizontal plane into separate 120°

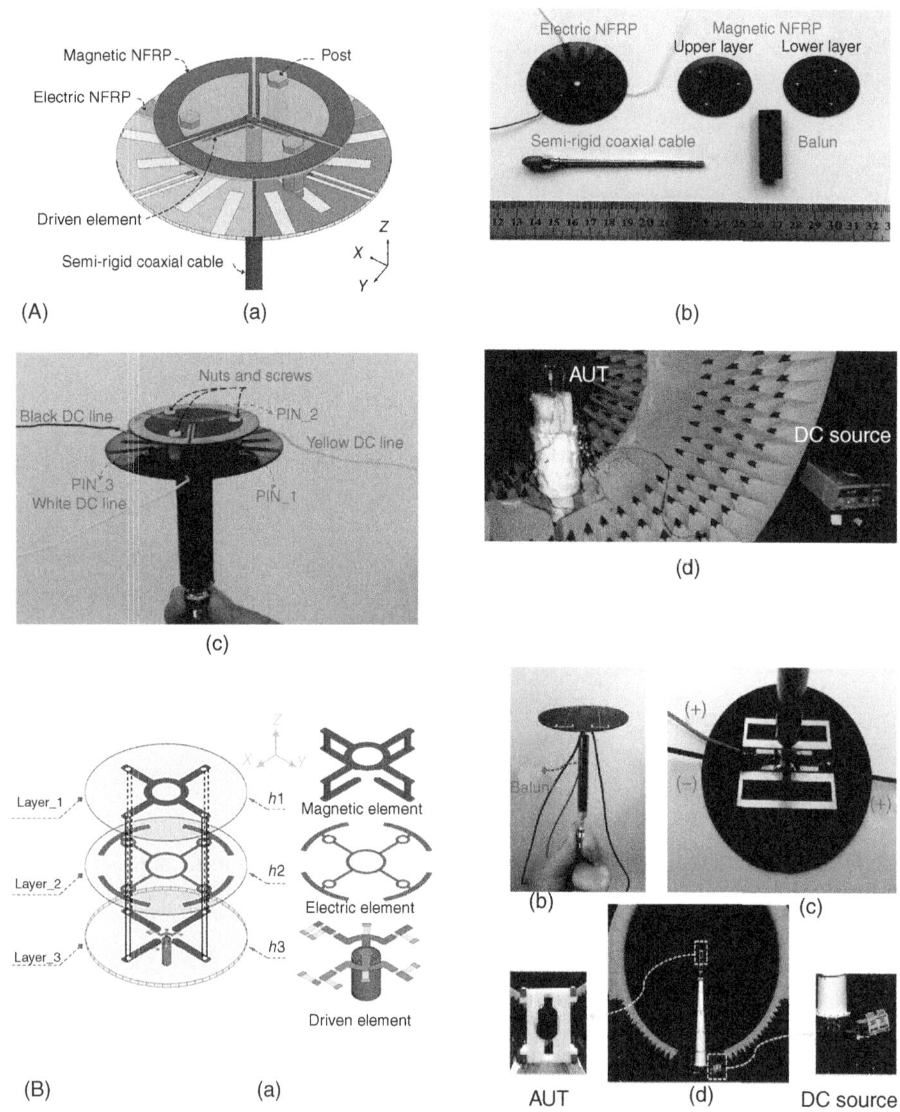

Figure 10.10 Experimentally validated reconfigurable NFRP ESAs. (A) Pattern reconfigurable ESA whose unidirectional 120° sector beams point along the azimuthal plane. (a) HFSS model. (b) Components. (c) Assembled antenna. (d) Antenna under test (AUT). Source: From [207] / with permission of IEEE. (B) Polarization reconfigurable ESA that can radiate four different polarization states: two LP and two CP. (a) HFSS model. (b) Perspective view from assembled antenna. (c) Bottom view of assembled antenna. (d) AUT. Source: From [211] / with permission from IEEE.

sectors with beamwidths completely covering each of them. The driven element is fixed. Diodes are embedded in the electric and magnetic NFRP elements and are switched on and off to direct the beam into the desired sector. Thus, this ESA provides coverage of the entire azimuthal plane.

The polarization reconfigurable ESA [211] shown in Figure 10.10B has its NFRP elements fixed. The diodes are incorporated into its driven element. It radiates unidirectional fields in the broadside direction. Depending upon the ON and OFF states of the diodes, it radiates one of four polarization states: one of two LP states along two orthogonal planes and either LHCP (left-hand circular polarization) or RHCP (right-hand circular polarization) states.

10.5 Directive Electrically Small NFRP Antennas

As noted earlier, electrically small electric and magnetic dipole antennas have a directivity near to 1.5 (1.76 dB). More importantly for the following discussion, they radiate a figure-eight pattern in one principal plane with its null directions being along the dipole's axis and an omnidirectional pattern in the orthogonal one. An ESA that would radiate primarily into one hemisphere rather than both, i.e. that would have a high front-to-back ratio (FTBR), would have many advantages for applications in which maximizing the power radiated (received) toward a receiver (from a transmitter) is desirable. For instance, as a transmitter, it would not waste half of its radiated power in an unwanted direction. Moreover, if this could be achieved without the necessity of a ground plane, an antenna engineer would have more flexibility in integrating the antenna system with a mobile platform. For instance, while a monopole antenna with a finite ground plane has an omnidirectional pattern about it, its peak direction is significantly elevated from it. An NFRP ESA such as the one shown in Figure 10.10A does not have a ground plane to finesse. Its peak is along the horizontal plane and it has a high FTBR with respect to the selected beam direction.

The two main approaches to achieving higher directivity with a NFRP ESA have been the use of a quasi-Yagi configuration in which a NFRP element acts as a director and/or another acts as a reflector. The other is by engineering a Huygens source, i.e. by arranging the driven element in a manner that it effectively excites a balanced pair of in-phase electric and magnetic NFRP elements. Both approaches have achieved unidirectional endfire and broadside radiating sources.

10.5.1 Quasi-Yagi

The quasi-Yagi ESAs come in two main styles, 2D and 3D. For instance, the 2D version [212] is printed on a single copper-cladded substrate. While it had a modest directivity, it was flexible. Its prototype was tested successfully in two bending modes and on cylinders of different radii.

The 3D versions have quite different arrangements. A spiral-shaped driver and director mounted on a ground plane were very closely spaced in [213] and obtained an endfire realized gain of 8.81 dBi at 450 MHz. The LP ESA shown in Figure 10.11 consisted of three EAD elements. Basically an EAD NFRP antenna (driven and NFRP elements) integrated with a EAD parasitic reflector. Two variations of this design were reported [214]. One prototype operated around 0.96 GHz with $ka = 0.76$ and its reflector separated from the tightly coupled pair by approximately $\lambda/10$ at the upper frequency of its 9.4% FBW_{10dB}, 1.02 GHz. Its radiation efficiency was greater than 93% and its peak broadside directivity (along the axis of the system) was 5.05 dB with a 11.4 dB FTBR. The second was $ka = 0.5$ version of the first with a slightly smaller profile operating around 1.1 GHz with a narrower 2.3% FBW_{10dB}. The peak broadside directivity was 4.2 dB with a 5.8 dB FTBR. A related CP ESA [215] consisted of a driven crossed-dipole loaded with NFRP elements and a similarly shaped

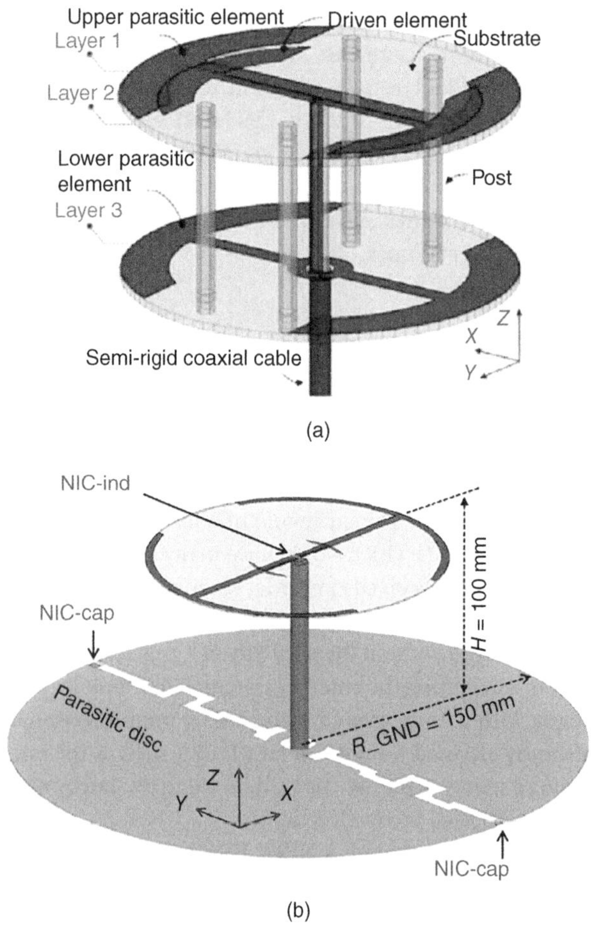

Figure 10.11 Quasi-Yagi NFRP ESAs. (a) Broadside radiating passive LP system. Source: From [214] / with permission from IEEE. (b) Broadside radiating NF-augmented LP system. Source: From [164] / with permission from John Wiley & Sons.

cross-dipole reflector. Its prototype had $ka = 0.71$ and a $\lambda/15$ profile with a 10.56% 3-dB AR bandwidth centered around 1.47 GHz. Its broadside gain was 2.31 dBic with a FTBR of 6.4 dB and a 80% average RE value at 1.39 GHz.

A pattern reconfigurable version of the spiral-shaped system was reported in [216]. A polarization reconfigurable version of the EAD-based system shown in Figure 10.11a is reported in [217]. With diodes integrated into its driven element, it achieves four polarization states, two LP and two CP with peak realized gains around 3 dBic and greater than 10 dB FTBR and 70% RE values.

A 300 MHz NFRP EAD antenna was integrated with a slot-modified parasitic copper disk to obtain an electrically small system with a high directivity and a large FTBR in [218]. The EAD NFRP element acted as the director. The slot-modified disk acted as the reflector. They were separated by $\lambda/10$. The system radiated a broadside unidirectional cardioid pattern with significant directivity and increase of the FTBR value relative to the EAD antenna alone. It had a high radiation efficiency, but very small FBW_{10dB}, <1.0%. In order to achieve more bandwidth, the parasitic disk was augmented with NIC-based capacitors. A wider directivity bandwidth was achieved [165]. Both the EAD element and the parasitic disk were later augmented, respectively, with a NIC-based inductor and NIC-based capacitors [164]. This 300 MHz NF-based ESA is illustrated in Figure 10.11b. It had $ka = 0.5$ and attained peak directivities greater than 6.3 dB with larger FTBR values > 20 dB and RE values > 84% over a much larger realized gain bandwidth, >10%. It was the first example of an ESA that overcame all of the conventional tradeoffs, i.e. it was efficient and had a large bandwidth with high peak directivity and FTBR values over it.

10.5.2 Huygens Dipole Antennas

Much recent work to achieve ESAs with higher directivity has been focused on Huygens sources. A Huygens source is actually well-known originally from reflector antenna developments [4]. A crossed electric and magnetic dipole combination located at the focal point of a parabolic reflector will induce currents on its surface that are parallel everywhere leading to cross-polarization free far-field behavior.

The magneto-electric antennas that originated in the mid-2000s [219] are nicely reviewed in [220]. While the electromagnetics is the same, they have been described in terms of complementary sources. A large number of other authors have reported modifications and improvements of the original design, e.g. [221, 222]. They are distinguished from the electrically small systems to be described next in that they are on the order of $\lambda/2$ in size. Recent works have discussed a miniaturized version by including a CLL-based MTM into the configuration [223] and a reconfigurable system [224].

The NFRP HDAs consist of a driven element and orthogonal electric and magnetic NFRP elements. As shown in Figure 10.12, when the equivalent electric and magnetic currents arising from the driven antenna exciting the NFRP elements are properly balanced, they produce a unidirectional cardioid pattern [225, 226]. The direction of the peak directivity follows from the right-hand rule of the electric current direction crossed with the magnetic one.

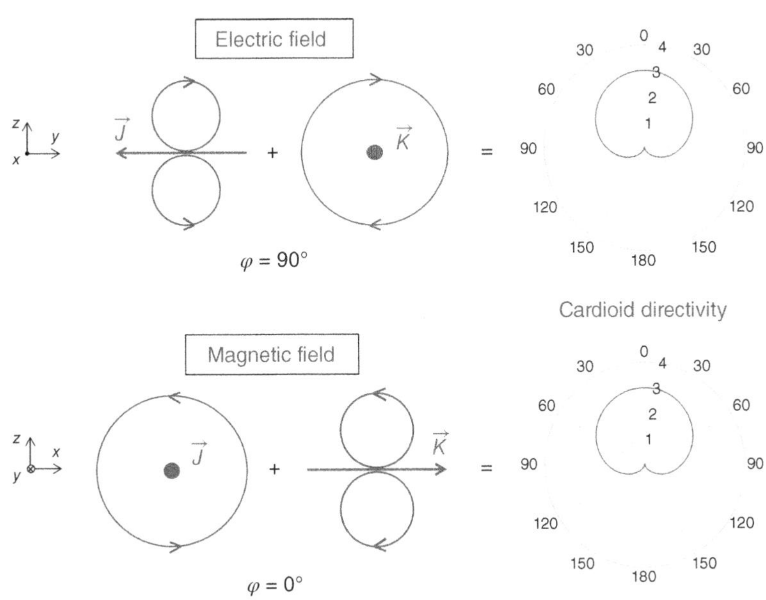

Figure 10.12 The Huygens radiation physics of a pair of balanced, in-phase, orthogonally oriented, infinitesimal electric and magnetic dipoles in free space.

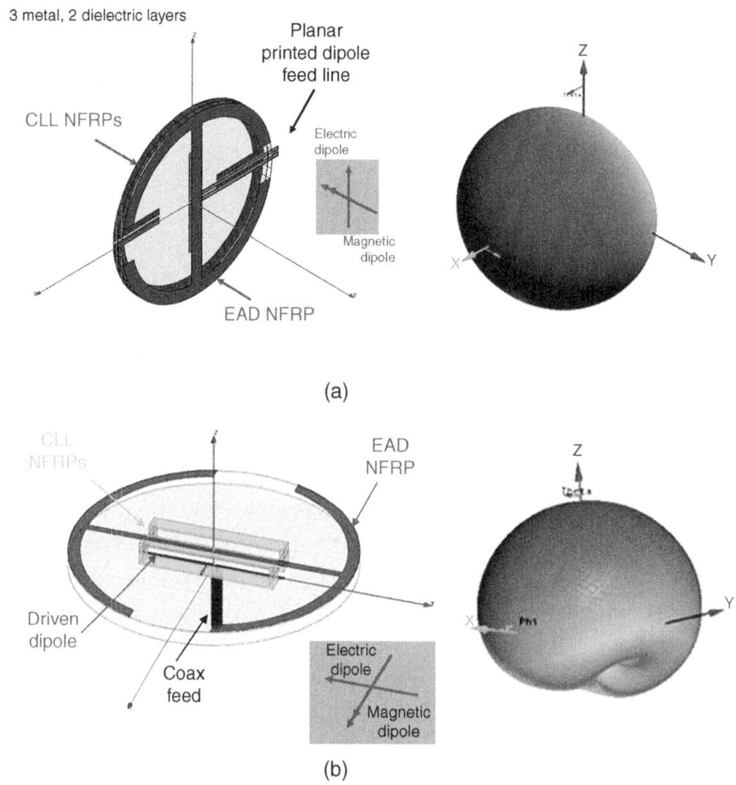

Figure 10.13 Huygens dipole antenna configurations. (a) Endfire radiating. Source: From [227] / with permission from IEEE. (b) Broadside radiating. Source: From [225] / with permission from IEEE.

The initial NFRP ESA was the endfire planar LP design shown in Figure 10.13a [227]. Its electrical size was $ka = 0.46$, and it was resonant at 1.475 GHz. It was designed with two 0.7874 mm thick Rogers Duroid™ 5880 disks and the subsequent three copper layers. A transmission line oriented along the x-axis feeds a dipole antenna oriented along the z-axis on the middle layer. The dipole excites an electric EAD NFRP element printed on one outside surface that is oriented parallel to the z-axis and two magnetic CLL-based NFRP protractor elements printed on the other one whose dipole moments are oriented parallel to the y-axis. As shown, the radiated cardioid pattern had its peak directivity, 4.50 dB, along the $+x$-axis at $f_{res} = 1.54$ GHz and had a 17.1 dB FTBR value. The overall (RE) efficiency was 85.9% (86.6%). The unidirectional emitter shown in Figure 10.10A is of a similar endfire HDA nature.

The first broadside radiating HDA is shown in Figure 10.13b. It also consisted of two of the same copper-clad substrates. A coax-fed dipole is printed on the bottom surface of the lower one; the EAD NFRP element is printed on the upper one. Two 3D copper CLL NFRP elements lie on the top surface of the bottom substrate. The gap at the center of the EAD is traversed with an inductor. The offset gaps of the CLLs are traversed with capacitors. The resonance frequency was 299.17 MHz and its $ka = 0.45$ with a 87.9% RE value. It was very low profile, the overall height being $\sim \lambda_{res}/78$. The maximum directivity of this LP HDA was 4.98 dB with the corresponding FTBR value being 17.64 dB. The first realization of the broadside radiating LP HDA had a different but analogous design [228]. A split EAD element was combined with the CLL element and driven by a coax-fed dipole. The prototype was low-profile with its height being $\sim \lambda_{res}/20$ and with $ka \sim 0.64$

at 1.50 GHz. Its peak directivity was 4.70 dB with the corresponding FTBR being 16.92 dB. The RE value was 71.6%. The cross-polarization value was smaller than −33 dB.

Note that the FBW_{10dB} of Figure 10.13a was 1.57%. Similarly, the FBW_{10dB} of Figure 10.13b was 0.20%. Because of the delicate balance needed between the electric and magnetic resonators, the bandwidths of these HDA systems are narrow. The measured FBW_{10dB} of the prototype was 0.62%, i.e. a 9.3 MHz, −10-dB impedance bandwidth. As a consequence, its design was modified to achieve a better passive performance and the lumped elements were replaced with NIC versions. The resulting NF-augmented HDA had a FBW that was 10 times larger [170]. Most recently the preliminary results of a frequency-agile self-oscillating NF Huygens radiator were presented that has a measured 1 : 2 bandwidth, from 30 to 60 MHz [188].

A different endfire HDA was reported in [229]. It also was a two Rogers Duroid 5880 substrate layer design. A printed spiral acted as the magnetic NFRP element. Meanderline inductor loaded electric dipole driven and NFRP elements completed the design. It had $ka = 0.47$ at the ISM-band frequency, 0.916 MHz, and was very low profile with a height of $\lambda_{res}/103$. The directivity was 4.66 dB and an RE of 66.4%. Its FBW_{10dB} was 1.65%. Another broadside radiating HDA was reported in [230] that evolved from [228]. Two complementary spiral resonators (CSRs) and a CLL structure were used as the electric and magnetic dipoles, respectively.

A CP HDA was developed in [231]. It is illustrated in Figure 10.14. It consists of two orthogonal pairs of EAD-CLL NFRP elements driven by a coax-fed crossed dipole. The electrically small

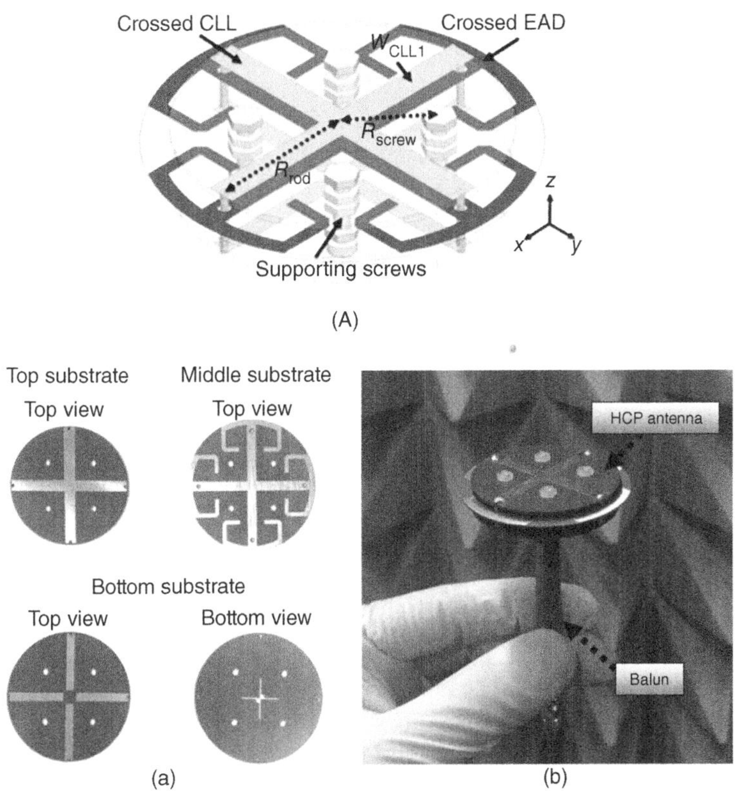

Figure 10.14 CP Huygens dipole antenna. (A) HFSS model. (B) Prototype with a bazooka balun attached for accurate measurements. (a) Fabricated layers. (b) Assembled antenna. Source: From [231] / with permission from IEEE.

($ka = 0.73$) prototype operated at 1584 MHz. It was a low profile design with its height being $\sim \lambda_{res}/25$. The measured peak realized gain was 2.7 dBic with a 17.7 dB FTBR value and a 68% RE value. In a similar manner to the introduction of multiple EAD and CLL elements into an electrically small package, a variety of yet more complex, multifunctional HDA-based systems have been realized. A two LP, one band system was reported in [232]. A dual band LP system was reported in [233]. A two port, dual CP system was reported in [234].

10.5.3 Additional Reported ESAs

As previously noted, ESAs have attracted significant interest in the AP community and beyond. Whether you like or dislike "MTMs," the fact is that they have caused a true disruption in the way that many researchers view artificial materials and think about and understand the physics and engineering associated with electromagnetic-based radiating and scattering systems. Over time, those insights may be the most important contribution that the MTM frenzy has made. As with the already noted examples, they have impacted numerous ESA designs.

Nevertheless, there are many miniaturized antenna systems that have been conceived and explained with conventional approaches. The four-arm spherical helix antenna [25] is an excellent example from the first decade of this century. Another is the omnidirectional monopole-driven slot antenna [235]. This antenna was further showcased in the measurement analysis and campaign in [236]. A very important aspect of this work is that it nicely emphasizes the difficulties associated with measuring ESAs and provides reasons why they occur and the means to mitigate them. The list of potential problems, e.g. currents on measurement cables, misalignment of the surface normal, and polarization mismatch, is rather daunting. You will notice sleeve (bazooka) baluns in the prototype pictures herein which were necessary to help ensure accurate measured results.

As one moves to higher frequencies, these measurement difficulties increase. Discrepancies between simulated and measured results are most often simply blamed on fabrication and assembly errors, especially when the antenna under test (AUT) is simply put into a commercial test range and a button is pushed. While the ESA measurement demands of reviewers continue to be unrelenting, i.e. "no measurements, no publication," this topic has received little attention in the literature despite its importance. It deserves more consideration.

An ESA constructed as a driven monopole coupled to a custom-designed parasitic was developed in [237] for hearing-instruments. A meander dipole-based ESA was tailored for a radio frequency identification (RFID)-based pressure sensor in [238] for glaucoma monitoring. Many other more conventional ESAs have been developed for a variety of medical related applications [239]. They all have important societal benefits.

While there are a vast number of compact/small antenna articles published indicating MTMs as being part of the radiating/receiving structures and none are actually present, there are others that are intimately connected to MTM-inspired concepts if one were to examine them closely. For example, the patch antenna in [240] was made small using a shorting pin and a defected ground structure (DGS) (i.e. DGSs evolved originally from electromagnetic bandgap [EBG] structures, which have often been identified as MTMs [41]). The interesting NFRP ESA developed in [241] is close in design to a 2D magnetic EZ antenna [102]. The RFID antennas considered in [242] involve a driven dipole coupled to a NFRP CLL element. The LC resonators loading a planar inverted-F antenna considered in [243] are well-known MTM unit cells. Directly driving 2D split-ring resonators (SRRs) led to interesting performance characteristics in [244]. Directly driving 3D SRRs led to the excitation of both electric and magnetic responses and an attractive quasi-isotropic ESA in [245]. Does it really matter whose favorite terminology is used to explain an antenna's operation if it is actually correct and helps a reader understand it? Absolutely not! What is really important is if the antenna does what is intended, has good performance characteristics, and has been described adequately for a reader to understand its operation and to be able to reproduce the reported results.

10.6 Forward Looking ESA Applications

Given the anticipated demand for efficient, multifunctional, compact wireless devices for AWE-inspired applications, there are many opportunities for the development of novel practical ESAs for current fifth generation (5G) and evolving 6G and beyond systems. However, to make significant progress, multidisciplinary approaches involving not only further electromagnetic discoveries, but also unfolding material, circuit and measurement advances must be cultivated and nourished. Several topics that my team and collaborators are currently investigating are discussed in the following text. They are but a tiny subset of the multitude of possibilities open to the AP communities.

10.6.1 WPT and Rectennas

It is expected that AWE wireless and electronic devices have the potential to revolutionize many aspects of human life. They are being employed not only in consumer areas like communications, personal entertainment, smart homes, intelligent transportation systems, and smart cities, but also for security and health-care applications. However, the lack of very compact, long-lasting portable power sources has hindered the deployment of many IoT concepts, e.g. current battery and super-capacitor systems are too large, too heavy, or too short-lived. Consequently, WPT technologies (e.g. also termed microwave power transmission [MPT] when restricted to microwave frequencies) have become a major trend to overcome this issue.

WPT systems negate the need for heavy, bulky, short-lived batteries and for their replacement, particularly in devices embedded in the human body or in structural materials such as in building and tunnel walls or bridge road beds. Moreover, they are environmentally friendly in the context that they avoid the need to dispose of those short-lived batteries. Furthermore, many WPT-enabled devices can be powered simultaneously from the same source. The recent pursuits of the IoT paradigm have led to the intense development of a variety of WPT technologies [246, 247].

WPT comes in many forms: near-field, mid-range, and far-field systems. I will emphasize only the far-field systems. There are two main approaches to wirelessly powering devices remotely and robustly over long distances. One exploits already available ambient electromagnetic resources through wireless energy harvesting (WEH), i.e. unintentional WPT or energy recycling, which is discussed nicely, e.g. in [248–250]. Because the magnitudes of the transferred unintentional wireless power are small, WEH takes advantage of many existing frequency bands and necessarily requires broadband or multi-band antennas to capture as much of the available spectrum as is possible.

The other WPT approach is intentional WPT in which sources radiate fields at specified frequencies to wirelessly power a device. Intentional WPT aims to transfer as much power as possible from the wireless source to the targeted devices. Far-field WPT has a long history dating back into the 1960s. There are several very nice reviews of the past and recent developments [251–259].

A major component of any far-field WPT system is a rectenna, i.e. a receiving antenna integrated with a rectifying circuit to convert the incident electromagnetic (AC) power into DC power.

Narrowband operation within specific frequency bands, particularly the industrial, scientific, and medical (ISM) bands, is quite common. However, the realization of high performance, ultra-compact (electrically small) rectennas, i.e. the antenna-rectifying circuit combination, is challenging. While a majority of IEEE publications on WEH and WPT have been associated with the microwave engineering community, a variety of electromagnetic analyses have also been reported. For instance, the theoretical bounds on the power transfer efficiency (PTE) and those associated with practical antennas has been studied, e.g. [260, 261]. Reviews of the key components of rectenna systems and their design requirements can be found, e.g. in [262, 263]. Reviews of antennas for microwave [264] and millimeter-wave (mmWave) [265] WEH and WPT, as well as the applications of metasurfaces to WEH and WPT [266, 267], have very recently appeared.

IoT devices such as sensors, communication devices, and RFID tags require a single rectenna that is electrically small, but has a large energy capture capacity. Consequently, the HDA systems with their cardioid-shaped radiation patterns are notably advantageous. They acquire the maximum amount of power from their driving sources while not wasting power radiated in the opposite direction. A CP system has notable advantages over an LP one in certain scenarios in which the orientation of the rectenna is not guaranteed and the WPT source is either an LP or CP system.

Several of the initial MTM-inspired ESAs were transformed into rectennas by integrating rectifying circuits with them. Both electric [268] and magnetic [269] NFRP antenna prototypes demonstrated their predicted performance characteristics. Moreover, the NFRP HDA systems discussed earlier were recently adapted to realize electrically small ($ka < 0.77$) and low-profile, $0.04\lambda_0$, LP and CP WPT rectennas at 915 MHz in the ISM band [270]. They too were facilitated by the seamless integration of highly efficient rectifiers, i.e. the RF signal to DC power conversion circuits, with the LP and CP HDAs. Their optimized prototypes have cardioid, very wide broadside radiation patterns, and effective capture areas larger than their physical size. Experimental results verified that they achieved a measured 89% and 82% peak AC-to-DC conversion efficiency, respectively, effectively confirming their simulated results.

One important lesson learned with those early electrically small rectenna systems was that the NFRP designs could be modified so that the antenna was matched directly to the rectifier without any intervening matching network, significantly improving their conversion efficiencies. While this aspect of the NFRP design approach was employed in the LP HDA system in [270] to attain its high conversion efficiencies, the original CP HDA rectenna design could not be tuned to be matched directly to the rectifier. The power lost to the necessary matching element was remedied in [271]. By changing the driven dipole to a more inductive form, the antenna was matched directly to the rectifier and the system reached a maximum conversion efficiency of 90.6%. The effectiveness of directly matching the receiving antenna to the rectifier in both WPT and WEH rectenna systems was also emphasized in [272].

The advantage of the NFRP paradigm to facilitate the presence of orthogonal LP HDA systems simultaneously as illustrated in Figure 10.14 was further explored for a simultaneous wireless information and power transfer (SWIPT) in [273]. SWIPT is currently being investigated in a wide range of different systems for both civil and industrial applications [259, 274]. It is one aspect of the green communications concept [275]. The CP HDA antenna and HDA rectenna designs guided the modifications needed to achieve a SWIPT HDA system. The prototype HDA-based SWIPT system had two integrated ports. One LP HDA pair was attached through one port to a rectifier for the WPT performance. The second, orthogonal LP HDA pair and its port were dedicated to communications. Similar performance characteristics to the LP versions of the LP HDA antenna and rectenna systems were obtained with over a 30 dB isolation between the two functions.

10.6.2 Wirelessly Powered Sensors

The next generation of sensors and control systems will be embedded in the AWE-inspired IoT electronic devices and will be consequential if they are powered with WEH and WPT technologies [276–280]. Wirelessly powered sensors and sensor networks for monitoring the health of humans (Wireless Healthcare), crops (Smart Agriculture), buildings (Smart Buildings) and even all systems and activities associated with urban environs (Smart Cities) are but a few of the imagined applications. They all will require large scale, battery-free cooperative wireless sensor networks. The HDA rectennas coupled with sensors are exceptional candidates as nodes for these systems.

Two electrically small rectenna-based wirelessly powered light and temperature sensors were developed in [281] that operate at 915 MHz in the 902-928 MHz ISM bands. One was a NFRP EAD antenna that was seamlessly integrated without any matching network with a highly efficient sensor-augmented rectifier. Its prototype was electrically small with $ka = 0.47$ at its resonance frequency, 906 MHz; was very thin (a single piece of copper-clad substrate); and its peak realized gain was 1.27 dBi. Because the system acts as an electric dipole, its omnidirectional property is ideal for capturing incident AC wireless power from any azimuthal direction and converting it into DC power. Both a photocell as the light sensor and a thermistor as the temperature sensor were demonstrated. The resistive properties of the photocell and thermistor changed the rectifier's output voltage level; an acoustic alarm was activated once a threshold value was attained. A peak 90% AC-to-DC conversion efficiency was obtained and its omnidirectional pattern was confirmed.

An LP version of the three layers electrically small, low-profile NFRP HDA shown in Figure 10.14 was designed and similarly integrated with the same light- and temperature-sensor-augmented rectifiers. Its prototype had $ka = 0.73$ at its resonance frequency, 908 MHz, and was low profile with its height being $\lambda_{res}/25$. Its broadside realized gain cardioid pattern had a 3.8 dBi peak value, more than 3 dB higher than the EAD value. The measured peak AC-to-DC conversion efficiency was 88%.

Measurements of the prototypes of both the light-sensor and temperature-sensor augmented omni- and unidirectional rectenna systems confirmed their predicted performance characteristics. Wireless powering of both the sensor and alarm systems was demonstrated. The omnidirectional nature of the EAD-based sensors is ideal, for example, for ceiling mounted systems in long galleries in a museum. The unidirectional nature of the HDA-based sensors makes it very suitable, for example, for surface-mounted applications, e.g. as temperature and light sensors on a wall in an office building or even on wine barrels resting in a cavern. As demonstrated in [226], the NFRP HDA design can be modified for placement on the surface of such dielectrics and still maintain its unidirectional performance characteristics.

A single Rogers Duroid 5880 substrate version of the LP NFRP HDA shown in Figure 10.15 was developed recently and integrated with the same sensor-augmented rectifiers [283]. Like the original endfire HDA [227], this ultra-thin rectenna system had a large electromagnetic wave capture capability and achieves nearly complete conversion of the incident energy into DC power. The HDA prototype has a realized gain of 4.6 dBi and a half power beamwidth (HPBW) of 134°. The entire rectenna is electrically small with $ka = 0.98$; is low cost and easy to fabricate; and has a measured 88% peak AC-to-DC conversion efficiency. As discussed in [253], large-scale rectenna arrays are usually implemented to capture larger amounts of power. This ultra-thin HDA system is an attractive candidate for a multi-element HDA-based rectenna array. Such a system is currently under development.

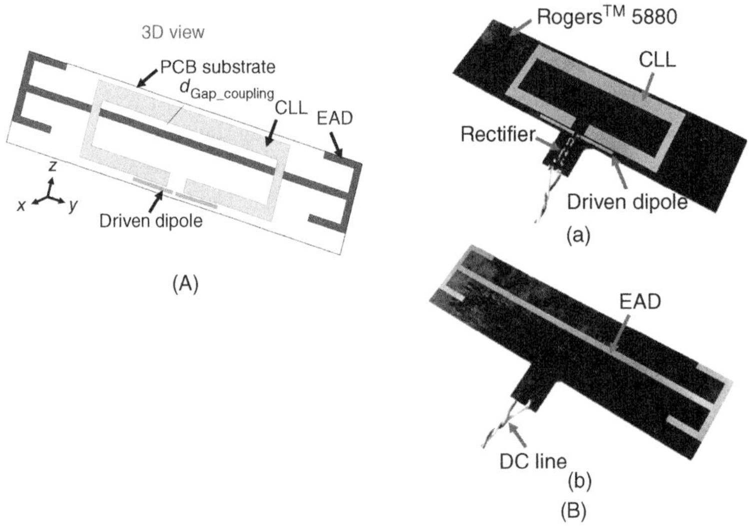

Figure 10.15 Ultra-thin NFRP HDA WPT system. (A) Antenna alone. (B) Antenna augmented with the sensor. (a) Front view. (b) Back view. Source: From [282] / with permission from IEEE.

10.6.3 Pregtronics

The concept of integrating antennas into structural materials has many applications for sensor networks and communication systems. Incorporating such electromagnetic systems into, for example, the actual bodies of land, sea and space vehicles as well as in buildings, bridges, and tunnels, one could monitor many different properties of those structures and the environments in which they operate, as well as perform their electromagnetic functions. It becomes much more than an antenna problem. It necessitates the co-design of the antenna with the materials with further adjustments in the design and implementation as the materials and fabrication processes occur and evolve.

For instance, multifunction, conformal load-bearing antenna structures offer unique structural weight savings and aerodynamic advantages in composite aerospace platforms. Many aircraft communications systems fall within the HF to UHF bands. Efficient antennas in those ranges are physically large and are generally found in the form of blade/surface monopole and/or lone wire antennas. With the advent of new antenna technologies such as the NFRP ESAs, the ability to seamlessly integrate their passive and active versions into aerospace platforms has become increasingly important. They have the key capability to be downsized for integration into small aerospace platforms. This is particularly interesting in the case of smaller unmanned aerial systems (e.g. drones) where there are significant internal volume and weight restrictions [284, 285].

A version of the EAD ESA [285] and versions of the electronics that would be needed for its NF-augmentation [284] have been developed for use in a grade of composites called "pre-impregnated"(pre-preg) materials. These pre-preg materials differ from traditional textiles in that they contain a B-staged epoxy resin that must be baked in an autoclave at temperatures greater than 170 °C, and under pressures upward of 700 kPa to achieve their maximum strength. Before this curing process, they can be conformably molded to any surface. A now everyday example of these composite materials are the carbon fiber reinforced plastics used in tennis rackets and bicycles.

The antenna consisted of a differentially fed dipole element integrated with a NFRP EAD element. To achieve good matching when the antenna was in situ, a meanderline-loaded version

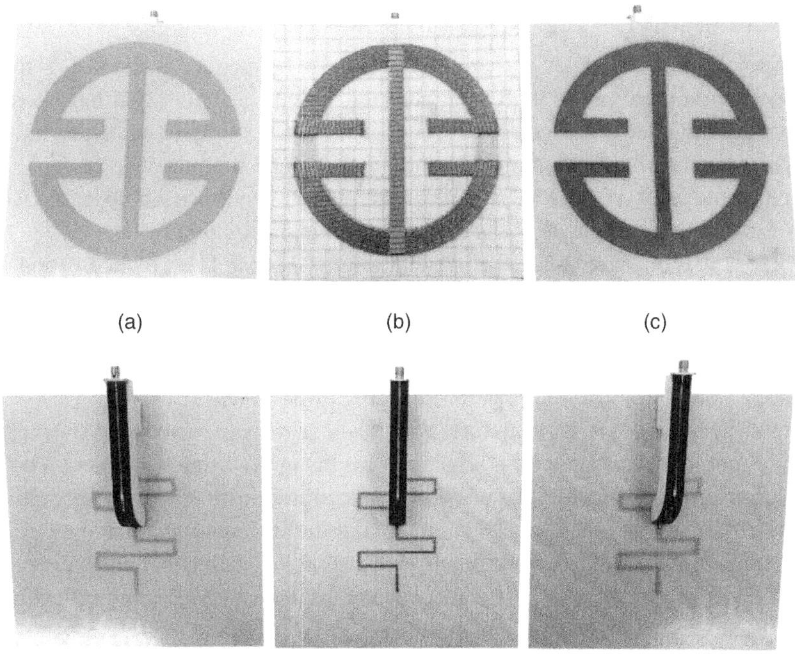

Figure 10.16 NFRP EAD ESA fabricated in a pre-preg structural composite material. (a) Copper version. (b) Embroidered version. (c) Veil version. Source: From [282] / with permission from IEEE.

of the dipole element was found to be necessary. The system was designed and tested at the arbitrary frequency of 300 MHz. Three ESA cases were investigated to ascertain the performance of the manufacturing techniques and material properties used to build them, as well as their performance characteristics. Photos of their prototypes are shown in Figure 10.16. Uniquely, an embroidered conductive thread and a new carbon fiber based, non-woven mat were investigated for use as the conducting elements. Both cases were compared with a copper variant of the EAD antenna. The embroidered version was achieved by learning how to and then sewing the components of the antenna into the pre-preg before curing. It was the original approach and required a programmable, heavy duty sewing machine to sew the desired shapes. The veil version has since become the standard. The required shape can be laser cut out from a large piece of the non-woven mat. All three prototypes were tested. Measurements confirmed that both the non-woven mat and the embroidered versions of the EAD antennas performed similarly to the copper version [285].

Microwave pre-preg electronics, i.e. "pregtronics," was similarly established originally with embroidered transmission line studies and then by incorporating an ultrawideband (UWB) amplifier as the main component of a microwave circuit [284]. Although only a simple UWB microwave amplifier-based circuit was demonstrated, it is clear that this pathfinder prototype circuit with its passive and active components demonstrated the hybridization of composite structures and electronics. Extending the concept to incorporate digital electronics and combining all of the pregtronic concepts with the ESAs have been considered. These co-designed pregtronic-based ESAs allow for the development of smart sensing, conformal structures/skins, which would offer a significant advantage for efficiently utilizing the limited volume available on unmanned aerial vehicles (UAVs) and other mobile platforms.

10.6.4 Millimeter-Wave Designs

One evolutionary technology in 5G cellular systems that is finally happening is the use of the mmWave frequency spectrum to realize the high data rates and system capacities that have been forecast to meet the evergrowing performance demands of consumers. Experimental results have proved that the 28 GHz frequency band is a suitable band for the initial rollouts of some mobile terminals. While much of the chatter about 5G antenna systems has been focused on mmWave arrays, compact antennas continue to be in demand for many IoT applications.

Both LP and CP NFRP HDA ESAs have been developed for the 28 GHz band [286]. The LP model and its prototype are shown in Figure 10.17a. Only simulations of the CP model were provided. The use of both the LP and CP systems for on-body use were evaluated and it was demonstrated that the body had little effect on their electromagnetic performance characteristics. The potential of these HDA systems for on-body SWIPT applications is currently being investigated.

Omnidirectional circularly polarized (OCP) antennas with lower gain are expected to be the most suitable candidates for general device-to-device (D2D) applications in 5G wireless systems. They have much larger radiation coverage and, hence, enable diverse communication links among multiusers. An 28 GHz OCP ESA was realized in [287]. The model and its prototype are shown in Figure 10.17b. Instead of the electric and magnetic dipole elements being orthogonal, the balanced pair is parallel in this OCP design. It is believed to be a very good candidate for D2D communication applications.

As one can see from their photos, these 28 GHz ESAs are really tiny. While it is anticipated that electrically small versions of any antenna system will continue to be desired even as the operating

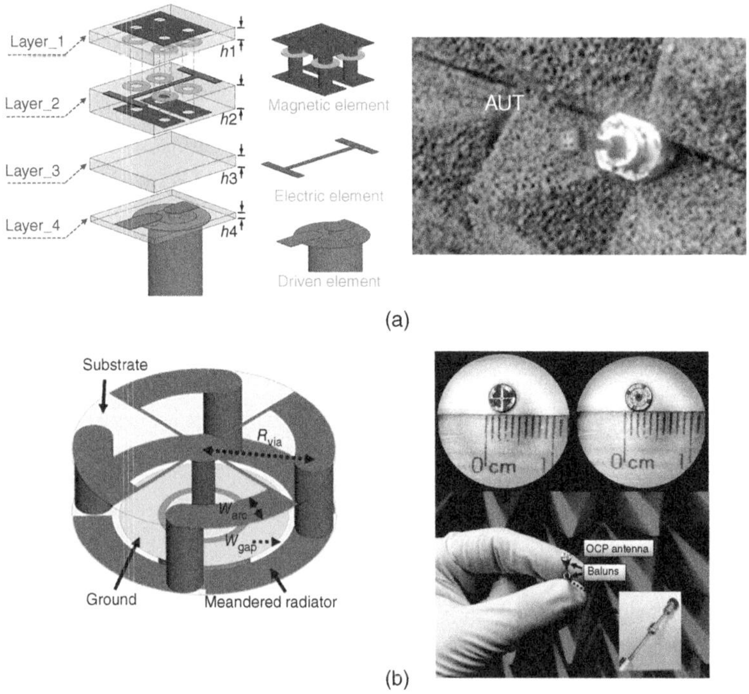

Figure 10.17 Millimeter-wave NFRP ESAs at 28 GHz. (a) LP HDA. Source: From [286] / with permission from IEEE. (b) Omnidirectional circular polarized (OCP) antenna. Source: From [287] / with permission from IEEE.

frequencies increase, their physical sizes may be small enough for many applications. Moreover, the highly directive beams necessary for a variety of 5G and beyond mmWave applications does necessitate arrays. Several examples of end-fire arrays of HDAs in the mmWave bands have been reported at 30 GHz, e.g. [288] and 60 GHz, e.g. [289]. Since it has been shown that the NFRP HDA designs can be tailored for on-chip applications [226], arrays of them for millimeter-wave WPT (mmWPT) are currently under development.

10.6.5 Beam-Steerable Huygens Dipole Antenna Arrays

While electrically small HDAs yield higher directivity and similar wide beamwidths when compared with basic dipole and slot antennas, many current and future wireless applications require long distance operation and multi-user coverage [290–292]. These application-driven features are attainable with arrays. For example, base station antenna arrays that are used for emerging WPT applications should have both high gain and large HPBW performance characteristics in order to broadcast and deliver wireless power to a large number of widely distributed Internet-of-things (IoT) sensors [281]. However, it has been challenging and even believed to be contradictory to achieve both features simultaneously in compact and low-cost designs. One practical approach to this issue is to design a high directivity linear antenna array with a narrow beamwidth in one principal plane and with wide beamwidth in the orthogonal plane [293–298], and then to insist that the array be capable of steering its narrow beam in order to enlarge its coverage.

Linear antenna arrays formed with electric- and magnetic-based elements with beam-steering and multi-beam performance have been extensively studied, e.g. [299–312]. One very popular design methodology is to adopt a Butler matrix (BM) as the beamforming feed network that excites a linear array of antenna elements. Element choices have included horn antennas [299], varieties of slot antennas [300–305], microstrip patch antennas [306–308], modified dipoles [309–312] and magnetoelectric (ME) dipole antennas [313]. A challenge all of these designs share is being able to simultaneously achieve a wide (3-dB) HPBW, e.g. larger than 150° in the principal vertical plane that is orthogonal to the array axis; high radiation efficiency; compact size; and easy, cost-effective implementation on a single piece of printed circuit board (PCB) substrate.

The design and realization of an ultra-thin, beam-steerable, Huygens dipole antenna array (HDAA) that consists of a linear arrangement of the electrically small, linearly-polarized HDAs was reported in [314]. Each element of the array was a version of the ultra-thin Huygens linearly polarized (HLP) antenna shown in Figure 10.15 shifted to operate at 2.45 GHz. Consequently, identical unidirectional cardioid-shaped radiation patterns with wide HPBWs are obtained in both the E- and H-planes from each element. Linear arrays of these HDAs were developed that achieved steerable beams with narrow HPBWs in the E-plane (vertical plane along the array direction) with very small realized gain variation and the desired wide HPBW (larger than 150°) in the H-plane (vertical orthogonal plane). In contrast to the endfire arrays of Huygens sources discussed in [315], these innovative HDAAs radiate relative to the broadside direction.

To appreciate the possibilities of using electrically small antennas to develop compact, high performance arrays, the theoretical aspects of the HDAs and the arrays constructed from them are reviewed briefly. A short description of the prototype HDAA that was fabricated and tested as a proof-of-concept system will be shared. It was a three-element beam-steerable HDAA excited by a 3×3 BM feed network at 2.45 GHz in the 2.4–2.5 GHz ISM and WLAN bands. It verified the attractive high directivity and wide beamwidth performance characteristics one can attain with unidirectional-radiating HDAAs.

10.6.5.1 HDA Theory

As demonstrated in [314], the electric fields generated by a balanced pair of an infinitesimal electric current (J) moment $I_e = I_0\ell\ \hat{y}$ and a magnetic current (K) moment $I_m = -\eta I_0\ell\ \hat{x}$ in the far field are:

$$\begin{aligned}\vec{E}^{ff}_{\omega,J}(\vec{r}) &= j\ \omega\mu\ I_0\ell \frac{e^{-jkr}}{4\pi r}(\hat{r}\times\hat{r}\times\hat{y})\\ \vec{E}^{ff}_{\omega,K}(\vec{r}) &= -j\ \omega\mu\ I_0\ell \frac{e^{-jkr}}{4\pi r}(\hat{r}\times\hat{x})\end{aligned} \tag{10.10}$$

Therefore, the total far field for an infinitesimal electric and magnetic dipole pair is

$$\begin{aligned}\vec{E}^{ff}_{\omega,\text{total}}(\vec{r}) &= j\ \omega\mu\ I_0\ell \frac{e^{-jkr}}{4\pi r}\vec{\mathcal{P}}(\theta,\phi)\\ \vec{H}^{ff}_{\omega,\text{total}}(\vec{r}) &= \frac{1}{\eta}\hat{r}\times\vec{E}^{ff}_{\omega,\text{total}}(\vec{r})\end{aligned} \tag{10.11}$$

where the angular variation of the vector field calculated with the vector product expressions is:

$$\begin{aligned}\vec{\mathcal{P}}(\theta,\phi) = \left[(\hat{r}\times\hat{r}\times\hat{y} - \hat{r}\times\hat{x})\right] &= \sin^2\theta\sin\phi\cos\phi\ \hat{x}\\ &- (\cos^2\theta + \sin^2\theta\cos^2\phi + \cos\theta)\ \hat{y}\\ &+ [\sin\theta\sin\phi + \sin\theta\cos\theta\sin\phi]\ \hat{z}\end{aligned} \tag{10.12}$$

In the yz-plane with $\phi = \pi/2$ one then has

$$\vec{\mathcal{P}}(\theta,\phi=\pi/2) = -(1+\cos\theta)\left[\cos\theta\ \hat{y} - \sin\theta\ \hat{z}\right] \tag{10.13}$$

Similarly, in the zx-plane with $\phi = 0$ one also has

$$\vec{\mathcal{P}}(\theta,\phi=0) = -(1+\cos\theta)\ \hat{y} \tag{10.14}$$

The cardioid factor: $(1 + \cos\theta)$, characteristic of a Huygens source field, is recognized immediately, i.e. the total fields are null along the negative z-axis, where $\theta = \pi$, and are increased by a factor of 2 along the $+z$-axis.

The corresponding time-averaged Poynting's vector and the total power radiated are given by the expressions

$$\begin{aligned}\vec{S}_{\text{total}}(\vec{r}) &= \frac{1}{2}\ \text{Re}\left\{\vec{E}^{ff}_{\omega,\text{total}}(\vec{r})\times\left[\vec{H}^{ff}_{\omega,\text{total}}(\vec{r})\right]^*\right\}\\ &= \frac{1}{2\eta}|\vec{E}^{ff}_{\omega}(\vec{r})|^2\end{aligned} \tag{10.15}$$

$$P^{\text{rad}}_{\text{total}} = \oiint_{S^2_\infty} d\Omega\ \hat{r}\cdot\vec{S}_{\text{total}}(\vec{r}) \tag{10.16}$$

where S^2_∞ is the sphere (S^2) centered on the origin and whose radius is infinitely large. The directivity follows as

$$D(\theta,\phi) = \frac{4\pi r^2\vec{S}_{\text{total}}(\vec{r})\cdot\hat{r}}{P^{\text{rad}}_{\text{total}}} \tag{10.17}$$

Consequently, the balanced electric and magnetic dipole pair gives

$$\vec{S}(\vec{r}) = \frac{1}{2}(k\ell)^2\eta|I_0|^2\frac{1}{(4\pi r)^2}(1+\cos\theta)^2\,\hat{r} \tag{10.18}$$

$$P^{\text{rad}}_{\text{tot}} = \frac{(k\ell)^2}{6\pi}\eta\,|I|^2 = 2\times\left[\frac{\pi}{3}\left(\frac{\ell}{\lambda}\right)^2\eta\,|I|^2\right] \tag{10.19}$$

Thus, the total power radiated is twice the total power radiated by either dipole individually. The directivity follows as

$$D(\theta,\phi) = \frac{4\pi r^2\vec{S}(\vec{r})\cdot\hat{r}}{P^{\text{rad}}_{\text{tot}}} = \frac{3}{4}(1+\cos\theta)^2 \tag{10.20}$$

Therefore, the maximum directivity of the electric-magnetic dipole pair ($\mathcal{N} = 1$), which is along the positive z-axis, is 3, twice the value of either dipole alone confirming the Harrington result [31, 32]: $D_{\text{max}} = \mathcal{N}^2 + 2\times\mathcal{N} = 1^2 + 2\times 1 = 3$.

A MatLab program was created to compute the performance characteristics of an infinitesimal electric dipole oriented along the y-axis and the basic Huygens dipole pair with its electric and magnetic dipoles oriented, respectively, along the $+y$ and $-x$ axes. The integrals of the total radiated power (10.16) were calculated numerically. The directivity of the electric and HDAs obtained in this manner are shown in Figure 10.18. The figure-eight pattern of the dipole element in the principle vertical plane containing the dipole (E-plane) and the omnidirectional pattern in the principal vertical plane orthogonal to it and the cardioid pattern of the Huygens element in both principal planes are clearly identified. The peak directivity of the electric dipole antenna and the HDA are 1.76 and 4.77 dB, respectively. The algorithms to attain these results were integrated into the array calculations considered next as test cases for them. Thus, as will be illustrated in the following text, a linear HDAA has two immediate advantages over a linear array of electric dipole antennas, i.e. it yields unidirectional rather than bidirectional patterns and has a factor of 3 dB (two times) larger peak directivity.

10.6.5.2 Array of Huygens Dipole Sources

The corresponding results for a finite number of electric and magnetic current sources in three dimensions follows by superposition. In particular, consider a set of N electric and magnetic

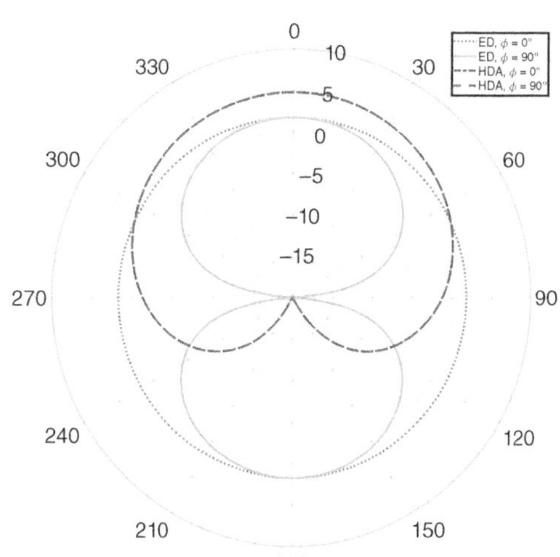

Figure 10.18 The directivity pattern (dBi) in the yz-plane for the electric dipole and the Huygens dipole antennas. $\theta = 0°$ is the broadside direction. Source: From [314] / IEEE / CC BY-4.0.

elemental current sources located at the points $\vec{r}_n = x_n\,\hat{x} + y_n\,\hat{y} + z_n\,\hat{z}$, $n = 1, \ldots, N$, in $\mathbb{R}^3$:

$$\begin{aligned}\vec{J}_n(\vec{r}) &= I_{e,n}\ell_{e,n}\delta(x - x_n)\ \delta(y - y_n)\ \delta(z - z_n)\ \hat{u}_n \\ \vec{K}_n(\vec{r}) &= I_{m,n}\ell_{m,n}\delta(x - x_n)\ \delta(y - y_n)\ \delta(z - z_n)\ \hat{v}_n\end{aligned} \tag{10.21}$$

The electromagnetic fields in the far field of these sources referenced to the coordinate origin are:

$$\begin{aligned}\vec{E}^{ff}_{\omega,J_n}(\vec{r}) &= +j\ \omega\ \mu I_{e,n}\ell_{e,n}\frac{e^{-jkr}}{4\ \pi r}e^{+j\ k\hat{r}\cdot\vec{r}_n}\left(\hat{r}\times\hat{r}\times\hat{u}_n\right) \\ \vec{H}^{ff}_{\omega,K_n}(\vec{r}) &= +j\ \omega\ \varepsilon I_{m,n}\ell_{m,n}\frac{e^{-jkr}}{4\ \pi r}e^{+j\ k\hat{r}\cdot\vec{r}_n}\left(\hat{r}\times\hat{r}\times\hat{v}_n\right)\end{aligned} \tag{10.22}$$

Consequently, with balanced current moments: $I_{m,n}\ell_{m,n} = \eta I_{e,n}\ell_{e,n} = \eta I_n\ell$, the electric fields radiated into the far field by each of the electric and magnetic current sources of an N-element linear Huygens array are:

$$\begin{aligned}\vec{E}^{ff}_{\omega,J_n}(\vec{r}) &= +j\ \omega\ \mu I_n\ell\frac{e^{-jkr}}{4\ \pi r}e^{+j\ k\hat{r}\cdot\vec{r}_n}\left(\hat{r}\times\hat{r}\times\hat{u}_n\right) \\ \vec{E}^{ff}_{\omega,K_n}(\vec{r}) &= +j\ \omega\ \mu I_n\ell\frac{e^{-jkr}}{4\ \pi r}e^{+j\ k\hat{r}\cdot\vec{r}_n}\left(\hat{r}\times\hat{v}_n\right)\end{aligned} \tag{10.23}$$

giving the total electric field

$$\begin{aligned}\vec{E}^{ff}_{\omega,\text{total}}(\vec{r}) = +j\ \omega\ \mu\ell\frac{e^{-jkr}}{4\ \pi r}&\left\{\sum_{1}^{N}\left[I_n e^{+j\ k\hat{r}\cdot\vec{r}_n}\right]\right. \\ &\left.\times\left[\left(\hat{r}\times\hat{r}\times\hat{u}_n\right) + \left(\hat{r}\times\hat{v}_n\right)\right]\right\}\end{aligned} \tag{10.24}$$

Clearly, the standard array factor and element factors can be identified. The total electric field radiated by the N-element Huygens array into the far field becomes:

$$\begin{aligned}\vec{E}^{ff}_{\omega,\text{total}}(\vec{r}) = +j\ \omega\ \mu\ell\frac{e^{-jkr}}{4\ \pi r} \\ \times\sum_{n=1}^{N}\text{AF}_{n,total}(\theta,\phi)\overrightarrow{\text{EVF}}_{n,total}(\theta,\phi)\end{aligned} \tag{10.25}$$

where the array factor and element vector factor are, respectively,

$$\begin{aligned}\text{AF}_{n,total}(\theta,\phi) &= I_n e^{+j\ k\hat{r}\cdot\vec{r}_n} \\ \overrightarrow{\text{EVF}}_{n,total}(\theta,\phi) &= \hat{r}\times\hat{r}\times\hat{u}_n + \hat{r}\times\hat{v}_n\end{aligned} \tag{10.26}$$

Since all of the electric and magnetic dipoles of each Huygens pair are taken to be oriented the same, i.e. the electric dipoles are oriented as $+\hat{y}$ and the magnetic dipoles are oriented as $-\hat{x}$, the element vector factor is the same for all elements and the more standard array far-field expression is attained from (10.25):

$$\begin{aligned}\vec{E}^{ff}_{\omega,\text{total}}(\vec{r}) = +j\ \omega\ \mu\ell\frac{e^{-jkr}}{4\ \pi r} \\ \times\text{AF}_{\text{total}}(\theta,\phi)\times\overrightarrow{\text{EVF}}(\theta,\phi)\end{aligned} \tag{10.27}$$

where the terms:

$$\mathrm{AF}_{\mathrm{total}}(\theta, \phi) = \sum_{n=1}^{N} I_n e^{+j\, k\hat{r}\cdot\vec{r}_n} \tag{10.28}$$

$$\overrightarrow{\mathrm{EVF}}(\theta, \phi) = \hat{r} \times \hat{r} \times \hat{y} - \hat{r} \times \hat{x} = \vec{P}(\theta, \phi)$$

Several idealized HDAA results are now immediately accessible. We assume for simplicity and ease of notation that each array has an odd number of elements. It is assumed that all of the magnitudes of the sources are the same, i.e. $|I_n| = I_0$, and that the Huygens elements are displaced uniformly along the y-axis with the center-to-center distance being d. The element vector factor is known from the single element analysis. We take with no loss of generality the standard element separation distance $d = \lambda/2$.

The patterns in the E-plane (yz-plane with $\phi = \pi/2$) and the H-plane (zx-plane with $\phi = 0$) are presented. The steerable beam direction is specified in the E-plane by the angle $(\theta, \phi) = (\theta_0, \phi_0 = \pi/2)$. Thus, we have $k\hat{r} \cdot \vec{r}_n = nkd \sin\theta \sin\phi - \sin\theta_0$. To achieve a tilted beam in the yz-plane pointing in the θ_0 direction, a constant phase factor is introduced with the source amplitudes. The amplitudes relative to the center of the array take the form: $I_n = I_0 \exp\{j[n - (N-1)/2]\, kd \sin\theta_0 \sin\phi_0\}$ for $n = 1,2, \ldots, N$ with, as noted, $\phi_0 = \pi/2$. Thus, the array factor takes on the conventional form

$$\mathrm{AF}_{\mathrm{total}}(\theta, \phi) = I_0 \sum_{n=1}^{N} e^{+j\, [n-(N-1)/2]kd\, (\sin\theta - \sin\theta_0)} \tag{10.29}$$

The calculated directivity patterns in both principal planes for the single, three, and nine HDA element arrays that generate broadside beams pointing in the direction ($\theta = 0°$, $\phi = 90°$) are shown in Figure 10.19. The maximum directivity for the one, three, and nine element arrays is 4.77 dB (3), 9.17 dB (8.26), and 13.83 dB (24.15). We note that the approximate maximum directivity as obtained with [4] (6–44) for a broadside-radiating array of nine isotropic elements is only $D_{\max} \approx 2N(d/\lambda) = 9 = 9.54$ dB.

Similarly the calculated directivity results for the single, three, and nine HDA element arrays that generate beams pointing in the direction ($\theta = 30°$, $\phi = 90°$) are shown in Figure 10.20. Their peak

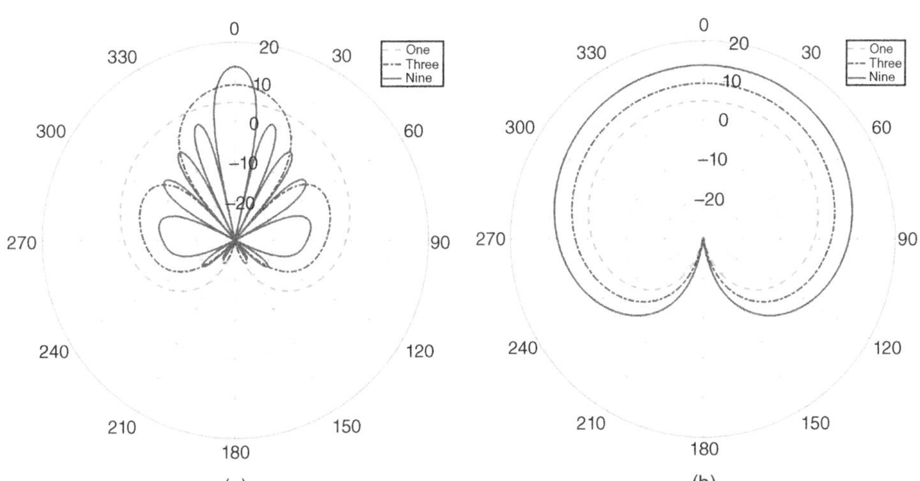

Figure 10.19 Directivity patterns (dBi) in the two principal vertical planes for the one, three, and nine element Huygens dipole antenna array radiating in the broadside direction [314]. (a) $\phi = 90°$. (b) $\phi = 0°$. Source: From [314] / IEEE / CC BY-4.0.

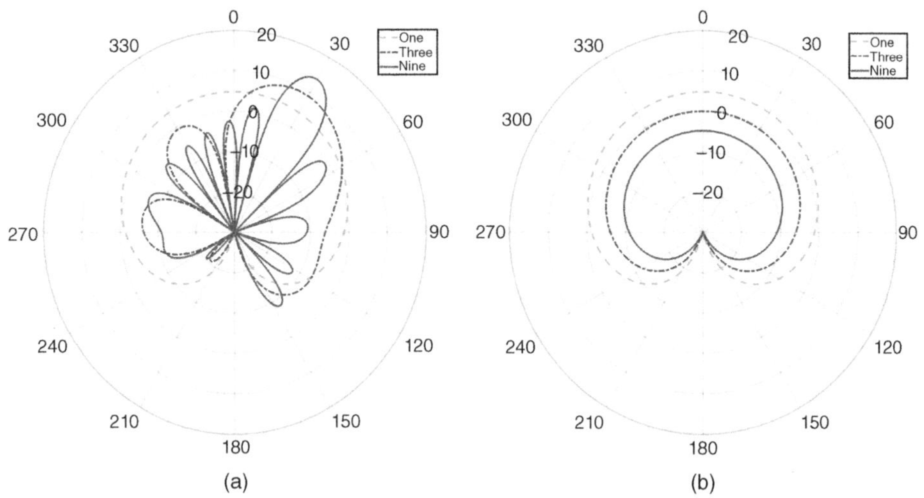

Figure 10.20 Directivity patterns (dBi) in the two principal vertical planes for the one, three, and nine element Huygens dipole antenna array radiating a directive beam pointed at ($\theta = 30^\circ, \phi = 90^\circ$) from in the broadside direction [314]. (a) $\phi = 90^\circ$. (b) $\phi = 0^\circ$. Source: From [314] / IEEE / CC BY-4.0.

directivity values are 4.77 dB (3), 8.94 dB (7.84), and 13.61 dB (22.96). Note the narrowing of the HPBW of the beams in the E-plane and the maintenance of the very wide HPBW in the H-plane in both the broadside and beam-tilted cases as the number of array elements increases.

Comparisons of the nine element HDAA and the corresponding nine element electric dipole array for broadside ($\theta_0 = 0^\circ$) and $\theta_0 = 30^\circ$ directed beams are shown in Figure 10.21a,b, respectively. It is clear then that the Huygens dipole array yields slightly more than a 3 dB increase in the directivity over the array of electric dipole elements. This enhancement occurs because the elements are finite in size; i.e. they are not infinitesimal electric and magnetic dipoles. The peak

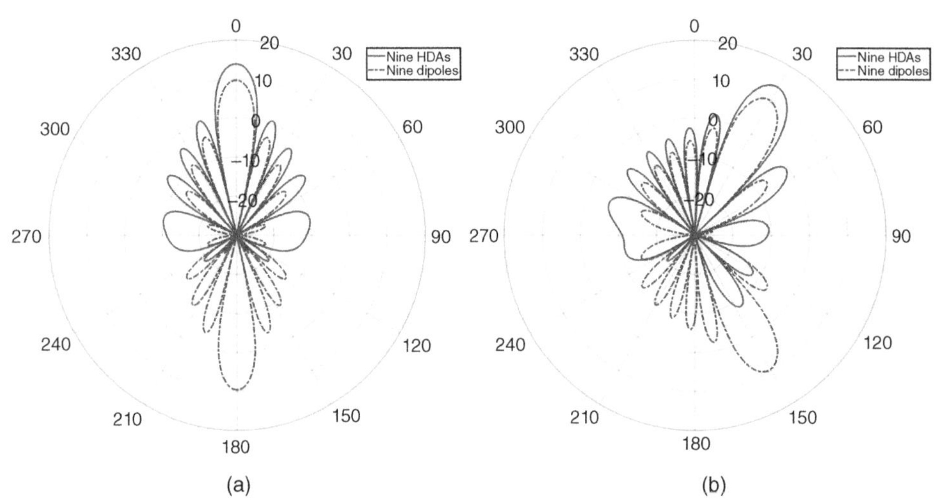

Figure 10.21 Directivity patterns (dBi) in the principal vertical plane $\phi = 90^\circ$ for the nine element HDA and electric dipole arrays. (a) Beam pointed in the broadside direction. (b) Beam pointed away from broadside [314]. (a) $\theta_0 = 0^\circ$. (b) $\theta_0 = 30^\circ$. Source: From [314] / IEEE / CC BY-4.0.

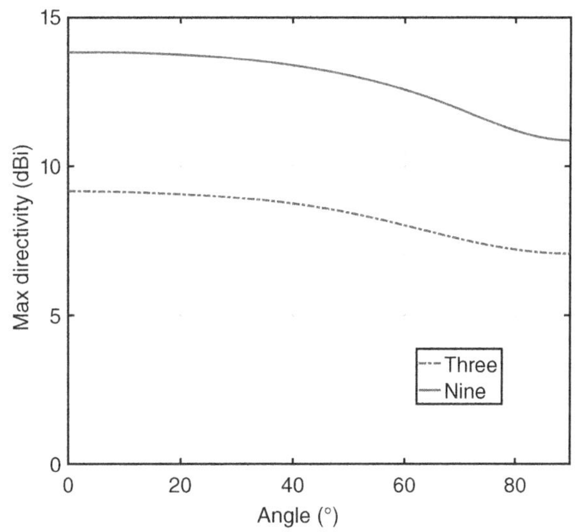

Figure 10.22 The maximum directivity (dBi) in the *yz*-plane of the three and nine element HDAAs as functions of the beam angle θ_0. Source: From [314] / IEEE / CC BY-4.0.

directivity of the HDA and dipole arrays is, respectively, 13.83 and 9.66 dBi in the broadside case and 13.61 and 9.66 dBi in the 30° case. The huge difference between them can be seen in the back direction where the dipole radiates with a FTBR = 0 dB versus FTBR ∼ ∞ for the HDA array.

While its array factor is exactly pointed at $\theta_0 = 30°$, the maximum directivity of the HDAA differs very slightly from that direction simply because of the Huygens element pattern. The calculated maximum directivity of the three and nine element HDA arrays as the beam angle θ_0 is scanned from 0° to 90° is presented in Figure 10.22. The directivity when $\theta_0 = 90°$ is 7.07 and 10.87 dB, for the three and nine element HDAA, respectively. The difference from the peak directivity in each case: 9.17 and 13.83 dB, is 2.10 and 3.56 dB, respectively. Thus, because of the very wide HPBW of an HDA, the 3-dB rolloff point of the HDAAs is pushed far toward the horizon. This is another interesting practical aspect of the HDAAs.

10.6.5.3 Proof-of-Concept Prototype

A proof-of-concept prototype of an ultra-thin three-element HDAA excited by a co-planar twin-line feed structure with differential outputs was realized with a 3 × 3 BM implementation [314]. The entire system was realized on a single sheet of 0.5 oz, 0.508 mm thick, Rogers Duroid 5880 copper-clad substrate. It was fabricated and tested. The measured results were in very good agreement with their simulated values. The entire antenna array configuration is shown in Figure 10.23. The center-to-center distance between each of the elements is $D_{space} = 0.45\lambda_{2.45\ GHz}$. The HDA elements were $ka = 0.97$ in size. Thus, the physical distance of separation between the elements was only $0.14\lambda_{2.45\ GHz}$. The air gaps between each HDA element proved to be a very effective decoupling approach. The 3 × 3 BM consists of three 3 dB couplers realized with microstrip lines. When each port, Port#1 to Port#3, of the BM is excited, the three output ports, Port#4 to Port#6, have the same amplitude, but with a specific phase progression. They were connected to the three HDA elements.

The simulated 3D realized gain patterns of the HDAA when each input port is excited separately are shown in Figure 10.24. When one port is excited, the other two are terminated with a 50-Ω load. Good beam-steering performance is realized in the *E*-plane (left); broad angular coverage is attained. The patterns are very wide in the *H*-plane (right). Thus, the three-state system achieves both high directivity and broad angular coverage.

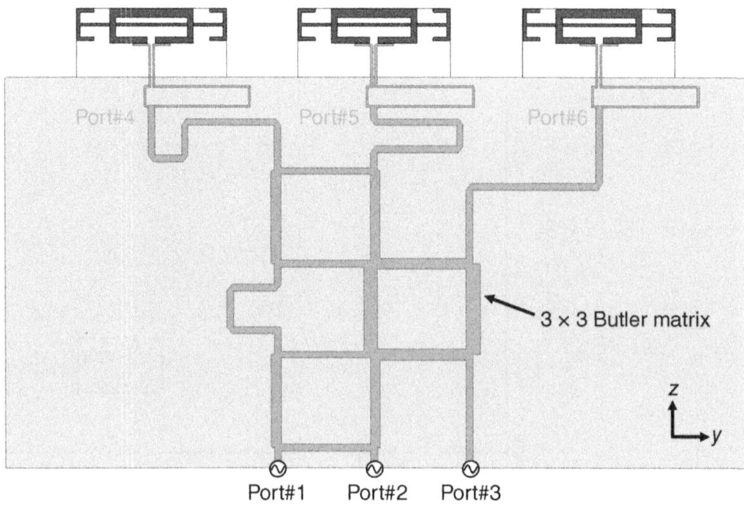

Figure 10.23 Design configuration of the 3 × 3 Butler matrix. Source: From [314] / IEEE / CC BY-4.0.

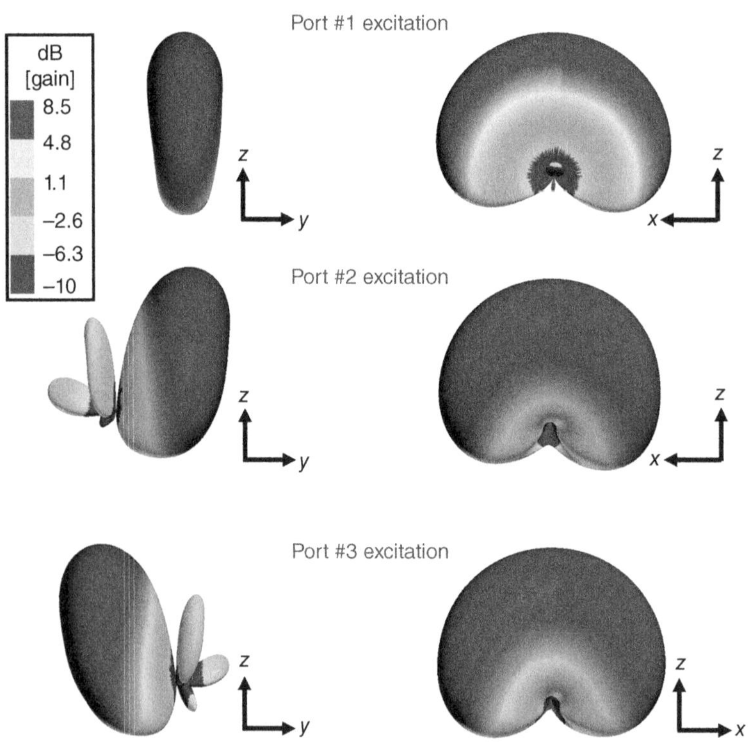

Figure 10.24 Simulated 3D beam-steered radiation patterns of the HDAA when each input port of the BM is excited separately, the other two being terminated in matched loads. Source: From [314] / IEEE / CC BY-4.0.

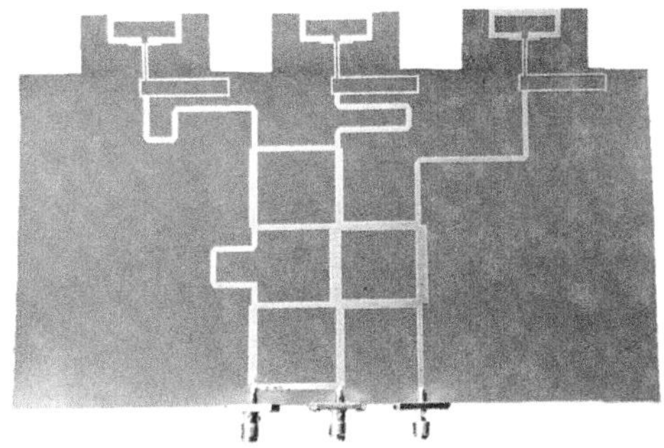

Figure 10.25 Fabricated 3 × 3 Butler matrix-fed three-element linear HDAA. Source: From [314] / with permission from IEEE.

The BM-excited HDAA was fabricated with low-cost PCB manufacturing technology. A photo of it is shown in Figure 10.25. Three female SubMiniature version A (SMA) connectors were adopted as the signal input ports. The $|S_{11}|$ values were measured with a vector network analyzer (VNA) from Keysight Technologies^{TM}. The radiation patterns were measured in a near-field multi-probe anechoic chamber system.

Figure 10.26a–c shows the measured and simulated $|S_{11}|$ and realized gain values as functions of the source frequency when Port#1, Port#2 and Port#3 are excited, respectively. The measured and simulated results are in reasonably good agreement. The overlapped −10-dB impedance

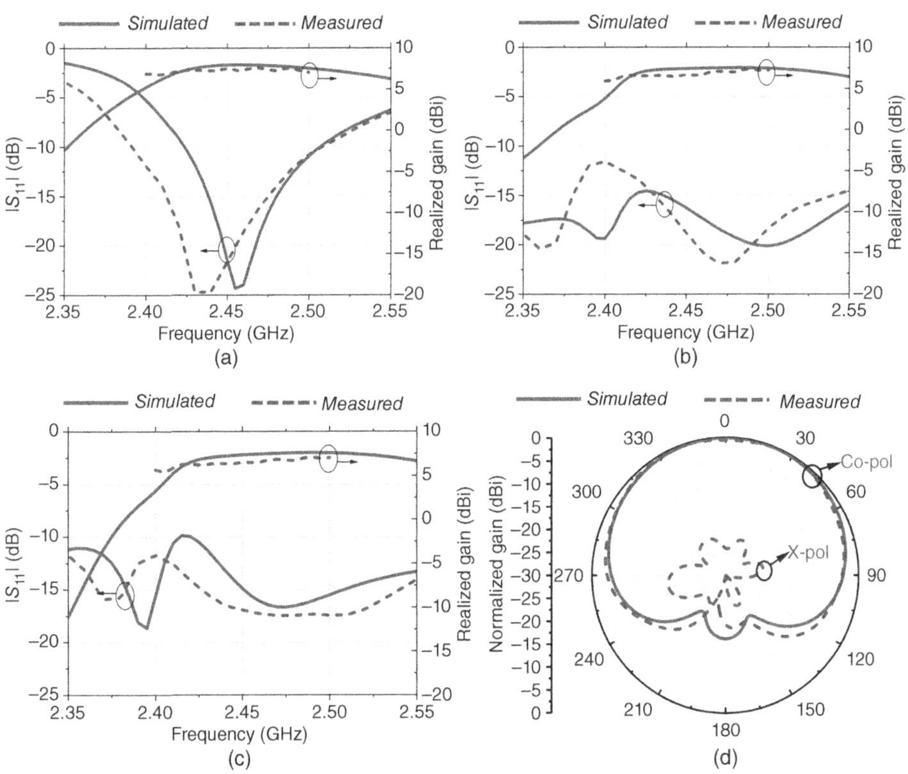

Figure 10.26 Measured and simulated $|S_{11}|$ and realized gain values as a function of the source frequency when (a) Port#1; (b) Port#2; and (c) Port#3 are excited separately. (d) Measured and simulated *H*-plane normalized realized gain patterns when Port#1 (broadside radiated beam) is excited. Source: From [314] / IEEE / CC BY-4.0.

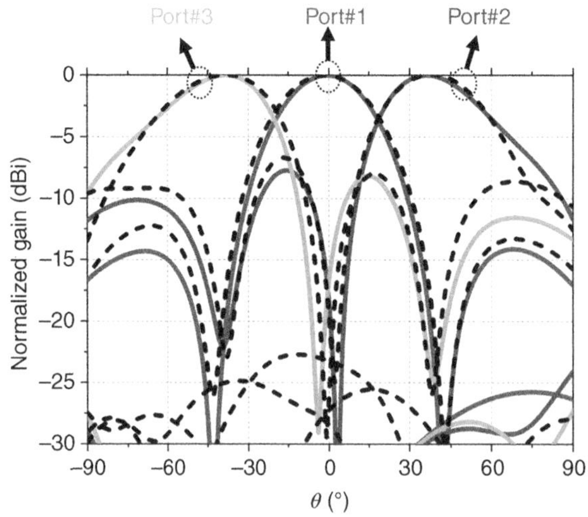

Figure 10.27 Measured (dashed lines) and simulated (solid lines) beam-steered *E*-plane normalized realized gain patterns of the 3×3 BM-fed linear three-element HDAA prototype when its three ports are excited separately. Source: From [314] / IEEE / CC BY-4.0.

bandwidth covers 110 MHz from 2.395 to 2.505 GHz, a 5.5% FBW. The realized gain values are stable over this impedance bandwidth. The measured peak realized gain values are 7.6, 7.3, and 7.1 dBi, when Port#1, Port#2 and Port#3 are excited, respectively. Thus, the peak realized gain variation for all three beams was only 0.5 dBi. Figure 10.26d shows the measured and simulated normalized H-plane realized gain patterns when Port#1 is excited. A very wide HPBW was realized for this broadside radiated beam case that covers 170° from −86° to +84°. A high FTBR, 18 dB, was attained.

Figure 10.27 shows the measured and simulated normalized radiation patterns when the three ports are individually excited. The measured and simulated results were in very good agreement. The combined *E*-plane patterns covered a wide angular range. When the beam angle varies over 121° from −63° to +58°, the measured gain variation was only 3.1 dB. Wide-angle coverage in both the *E*- and *H*-planes was thus demonstrated.

The measured beamwidth in the *H*-plane is larger than 160° for the broadside radiation mode. The corresponding high-directivity beam in the *E*-plane has a relatively narrow 34° HPBW. Good beam-steering performance with small gain variations is also obtained in the *E*-plane and the combined HPBW covers 120°. In addition to its excellent radiation performance, the HDAA prototype was also low-cost, ultra-thin and easily fabricated by standard PCB technology. It is an ideal candidate for various wireless applications that require long distance and multi-user coverage, e.g. as the base station antenna for WPT to a large number of remotely distributed battery-free IoT devices [270].

10.7 Summary

As you will read in several other chapters in this book, many electromagnetic, antenna and MTM researchers are heavily focused on array developments for current 5G and evolving 6G beyond systems. Nevertheless, the usefulness of efficient electrically small multifunctional, reconfigurable antennas as the ideal candidates for many of the emerging AWE-inspired IoT applications cannot be overstated. Given the concerns over how much energy will be needed to enable arrays for the anticipated 5G and beyond wireless ecosystems, any reduction of the power required by individual elements associated with high performance wireless mobile terminals and their enabling antenna systems will be welcomed.

The WPT concepts discussed herein will help make 5G and beyond IoT systems more sustainable. Years of simply designing the electronic circuits and the antennas independently of each other and using intervening matching networks to enable their combinations has to stop. Co-design will only increase performance as well as save time and money. Integration of the electrically small transmitting and receiving systems directly into structural materials will decrease the weight and enhance the aesthetics of IoT devices and enable a host of applications. Co-designing compact antenna systems with novel materials to enable their functions such as reconfigurability at high mmWave frequencies will have to become commonplace. Associated multidisciplinary approaches to realizing these antenna systems will become the norm as will more advanced measurement techniques to verify their performance characteristics. While you may not have been awestruck by the plethora of ESAs developed in the last two decades and described in this chapter, I sincerely hope that the various concepts discussed and illustrated, as well as the awe-inspiring realized examples in it, will stimulate innovative ways of looking at your old and new problems and, hence, prove useful to your current and future research efforts.

Acknowledgments

A very special thank you goes to Prof. Oscar Quevedo-Teruel, KTH Royal Institute of Technology. As the Editor-in-Chief (EIC) of the EurAAP Reviews of Electromagnetics (RoE) journal announced in November 2020, he very kindly invited and encouraged me to put together the first version of this overview for the inaugural RoE issue.

I also wish to extend many special thanks to all of my current and former MS and PhD students; Visiting Students and Scholars; and Post-docs, many of whom still collaborate with me; and my Collaborators for all of their efforts associated with the antenna systems featured in this chapter. There would be little to write about if it were not for them!! In particular, I would like to thank Prof. TANG Ming-Chun, Chongqing University, for his careful proof-reading of an early version of the RoE overview article and for the very satisfying past decade of shared ESA accomplishments.

Finally, I again sincerely apologize to many of you for not highlighting your contributions to the ESA area.

References

1 L. Jofre and M. Martínez-Vázquez, *Handbook on Small Antennas*. Brussels, Belgium: EurAAP, 2012.
2 S. R. Best, *Advances in Electrically Small Antennas*. Hoboken, NJ: John Wiley & Sons, 2012.
3 J. L. Volakis, C.-C. Chen, and K. Fujimoto, *Small Antennas: Miniaturization Techniques & Applications*. New York, NY: McGraw Hill, 2010.
4 C. A. Balanis, *Antenna Theory: Analysis and Design*. Hoboken, NJ: John Wiley & Sons, 4th ed., 2016.
5 J. D. Kraus and R. J. Marhefka, *Antennas for All Applications*. New York, NY: McGraw Hill, 3rd ed., 2002.
6 W. L. Stutzman and G. A. Thiele, *Antenna Theory and Design*. Hoboken, NJ: John Wiley & Sons, 2012.

7 H. A. Wheeler, "Fundamental limitations of small antennas," *Proc. IRE*, vol. 35, no. 12, pp. 1479–1484, Dec 1947.

8 H. A. Wheeler, "The radiansphere around a small antenna," *Proc. IRE*, vol. 47, no. 8, pp. 1325–1331, Aug 1959.

9 R. H. Johnston and J. G. McRory, "An improved small antenna radiation-efficiency measurement method," *IEEE Antennas Propag. Mag.*, vol. 40, no. 5, pp. 40–48, Oct 1998.

10 D. Heberling, M. Geissler, O. Litschke, and M. Martínez-Vázquez, "Improved radiation efficiency measurement of mobile systems and application to a novel multiband antenna," in *Proceedings of 2003 SBMO/IEEE MTT-S International Microwave and Optoelectronics Conference-IMOC 2003*, volume 1, pp. 381–386, Foz do Iguacu, Brazil, 20–23 Sep 2003. IEEE.

11 L. J. Chu, "Physical limitations of omni-directional antennas," *J. Appl. Phys.*, vol. 19, no. 12, pp. 1163–1175, May 1948.

12 J. S. McLean, "A re-examination of the fundamental limits on the radiation Q of electrically small antennas," *IEEE Trans. Antennas Propag.*, vol. 44, no. 5, pp. 672–676, May 1996.

13 H. L. Thal, "New radiation Q limits for spherical wire antennas," *IEEE Trans. Antennas Propag.*, vol. 54, no. 10, pp. 2757–2763, Oct 2006.

14 M. Gustafsson, C. Sohl, and G. Kristensson, "Physical limitations on antennas of arbitrary shape," *Proc. R. Soc. A*, vol. 463, no. 2086, pp. 2589–2607, Jul 2007.

15 A. D. Yaghjian and H. R. Stuart, "Lower bounds on the Q of electrically small dipole antennas," *IEEE Trans. Antennas Propag.*, vol. 58, no. 10, pp. 3114–3121, Oct 2010.

16 G. A. E. Vandenbosch, "Explicit relation between volume and lower bound for Q for small dipole topologies," *IEEE Trans. Antennas Propag.*, vol. 60, no. 2, pp. 1147–1152, Feb 2011.

17 M. Gustafsson, M. Cismasu, and B. L. G. Jonsson, "Physical bounds and optimal currents on antennas," *IEEE Trans. Antennas Propag.*, vol. 60, no. 6, pp. 2672–2681, Jun 2012.

18 M. Capek, L. Jelinek, P. Hazdra, and J. Eichler, "The measurable Q factor and observable energies of radiating structures," *IEEE Trans. Antennas Propag.*, vol. 62, no. 1, pp. 311–318, Jan 2013.

19 M. Capek, M. Gustafsson, and K. Schab, "Minimization of antenna quality factor," *IEEE Trans. Antennas Propag.*, vol. 65, no. 8, pp. 4115–4123, Aug 2017.

20 K. Schab, L. Jelinek, M. Capek, C. Ehrenborg, D. Tayli, G. A. E. Vandenbosch, and M. Gustafsson, "Energy stored by radiating systems," *IEEE Access*, vol. 6, pp. 10553–10568, 2018.

21 M. Manteghi, "Fundamental limits, bandwidth, and information rate of electrically small antennas: Increasing the throughput of an antenna without violating the thermodynamic Q-factor," *IEEE Antennas Propag. Mag.*, vol. 61, no. 3, pp. 14–26, Jun 2019.

22 M. Gustafsson, M. Capek, and K. Schab, "Tradeoff between antenna efficiency and Q-factor," *IEEE Trans. Antennas Propag.*, vol. 67, no. 4, pp. 2482–2493, Apr 2019.

23 S. R. Best, "Low Q electrically small linear and elliptical polarized spherical dipole antennas," *IEEE Trans. Antennas Propag.*, vol. 53, no. 3, pp. 1047–1053, Mar 2005.

24 S. R. Best, "Electrically small resonant planar antennas: Optimizing the quality factor and bandwidth," *IEEE Antennas Propag. Mag.*, vol. 57, no. 3, pp. 38–47, Jun 2015.

25 S. R. Best, "The radiation properties of electrically small folded spherical helix antennas," *IEEE Trans. Antennas Propag.*, vol. 52, no. 4, pp. 953–960, Apr 2004.

26 O. S. Kim, O. Breinbjerg, and A. D. Yaghjian, "Electrically small magnetic dipole antennas with quality factors approaching the Chu lower bound," *IEEE Trans. Antennas Propag.*, vol. 58, no. 6, pp. 1898–1906, Jun 2010.

27 H. R. Stuart and A. D. Yaghjian, "Approaching the lower bounds on Q for electrically small electric-dipole antennas using high permeability shells," *IEEE Trans. Antennas Propag.*, vol. 58, no. 12, pp. 3865–3872, Dec 2010.
28 D. F. Sievenpiper, D. C. Dawson, M. M. Jacob, T. Kanar, S. Kim, J. Long, and R. G. Quarfoth, "Experimental validation of performance limits and design guidelines for small antennas," *IEEE Trans. Antennas Propag.*, vol. 60, no. 1, pp. 8–19, Jan 2011.
29 R. W. Ziolkowski, "The directivity of a compact antenna: An unforgettable figure of merit," *EPJ Appl. Metamater.*, vol. 4, p. 7, 2017.
30 P.-S. Kildal and S. R. Best, "Further investigations of fundamental directivity limitations of small antennas with and without ground planes," in *2008 IEEE Antennas and Propagation Society International Symposium*, pp. 1–4. IEEE, 2008.
31 R. F. Harrington, "On the gain and beamwidth of directional antennas," *IRE Trans. Antennas Propag.*, vol. 6, no. 3, pp. 219–225, Jul 1958.
32 R. F. Harrington, "Effect of antenna size on gain, bandwidth, and efficiency," *J. Res. Nat. Bur. Stand.*, vol. 64, no. 1, pp. 1–12, Jan-Feb 1960.
33 I. Liberal, I. Ederra, R. Gonzalo, and R. W. Ziolkowski, "Induction theorem analysis of resonant nanoparticles: Design of a Huygens source nanoparticle laser," *Phys. Rev. Appl.*, vol. 1, no. 4, 044002, May 2014.
34 R. C. Hansen, "Fundamental limitations in antennas," *Proc. IEEE*, vol. 69, no. 2, pp. 170–182, Feb 1981.
35 R. C. Hansen, "Some new calculations on antenna superdirectivity," *Proc. IEEE*, vol. 69, no. 10, pp. 1365–1366, Oct 1981.
36 C. W. Oseen, "Die einsteinsche nadelstichstrahlung und die maxwellschen gleichungen," *Annalen der Physik*, vol. 374, no. 19, pp. 202–204, 1922.
37 R. W. Ziolkowski, "Using Huygens multipole arrays to realize unidirectional needle-like radiation," *Phys. Rev. X*, vol. 7, no. 3, 031017, Jul 2017.
38 D. Margetis, G. Fikioris, J. M. Myers, and T. T. Wu, "Highly directive current distributions: General theory," *Phys. Rev. E*, vol. 58, no. 2, pp. 2531–2547, 1998.
39 M. Gustafsson and S. Nordebo, "Optimal antenna currents for Q, superdirectivity, and radiation patterns using convex optimization," *IEEE Trans. Antennas Propag.*, vol. 61, no. 3, pp. 1109–1118, Mar 2013.
40 S. Arslanagić and R. W. Ziolkowski, "Highly subwavelength, superdirective cylindrical nanoantenna," *Phys. Rev. Lett.*, vol. 120, no. 23, 237401, Jul 2018.
41 N. Engheta and R. W. Ziolkowski, *Metamaterials: Physics and Engineering Explorations*. Hoboken, NJ: John Wiley & Sons, 2006.
42 G. V. Eleftheriades and K. G. Balmain, *Negative-refraction Metamaterials: Fundamental Principles and Applications*. Hoboken, NJ: John Wiley & Sons, 2005.
43 C. Caloz and T. Itoh, *Metamaterials: Transmission Line Theory and Microwave Applications: The Engineering Approach*. Hoboken, NJ: John Wiley & Sons, 2006.
44 R. W. Ziolkowski, P. Jin, and C.-C. Lin, "Metamaterial-inspired engineering of antennas," *Proc. IEEE*, vol. 99, no. 10, pp. 1720–1731, Oct 2011.
45 C. Caloz, "Metamaterial dispersion engineering concepts and applications," *Proc. IEEE*, vol. 99, no. 10, pp. 1711–1719, Oct 2011.
46 J. L. Volakis and K. Sertel, "Narrowband and wideband metamaterial antennas based on degenerate band edge and magnetic photonic crystals," *Proc. IEEE*, vol. 99, no. 10, pp. 1732–1745, Oct 2011.

47 Y. Dong and T. Itoh, "Metamaterial-based antennas," *Proc. IEEE*, vol. 100, no. 7, pp. 2271–2285, Jul 2012.

48 M. Durán-Sindreu, J. Naqui, F. Paredes, J. Bonache, and F. Martín, "Electrically small resonators for planar metamaterial, microwave circuit and antenna design: A comparative analysis," *Appl. Sci.*, vol. 2, no. 2, pp. 375–395, Apr 2012.

49 E. J. Rothwell and R. O. Ouedraogo, "Antenna miniaturization: Definitions, concepts, and a review with emphasis on metamaterials," *J. Electromagn. Waves Appl.*, vol. 28, no. 17, pp. 2089–2123, 2014.

50 R. W. Ziolkowski and A. D. Kipple, "Application of double negative materials to increase the power radiated by electrically small antennas," *IEEE Trans. Antennas Propag.*, vol. 51, no. 10, pp. 2626–2640, Oct 2003.

51 R. W. Ziolkowski and A. Erentok, "Metamaterial-based efficient electrically small antennas," *IEEE Trans. Antennas Propag.*, vol. 54, no. 7, pp. 2113–2130, Jul 2006.

52 S. A. Arslanagić, R. W. Ziolkowski, and O. Breinbjerg, "Radiation properties of an electric Hertzian dipole located near-by concentric metamaterial spheres," *Radio Sci.*, vol. 42, no. RS6S16, Nov 2007. DOI: 10.1029/2007RS003663.

53 H. Mosallaei and K. Sarabandi, "Antenna miniaturization and bandwidth enhancement using a reactive impedance substrate," *IEEE Trans. Antennas Propag.*, vol. 52, no. 9, pp. 2403–2414, 2004.

54 F. Qureshi, M. A. Antoniades, and G. V. Eleftheriades, "A compact and low-profile metamaterial ring antenna with vertical polarization," *IEEE Antennas Wireless Propag. Lett.*, vol. 4, pp. 333–336, 2005.

55 K. Buell, H. Mosallaei, and K. Sarabandi, "A substrate for small patch antennas providing tunable miniaturization factors," *IEEE Trans. Microw. Theory Techn.*, vol. 54, no. 1, pp. 135–146, Jan 2006.

56 H. R. Stuart and A. Pidwerbetsky, "Electrically small antenna elements using negative permittivity resonators," *IEEE Trans. Antennas Propag.*, vol. 54, no. 6, pp. 1644–1653, Jun 2006.

57 A. Alù, F. Bilotti, N. Engheta, and L. Vegni, "Subwavelength, compact, resonant patch antennas loaded with metamaterials," *IEEE Trans. Antennas Propag.*, vol. 55, no. 1, pp. 13–25, Jan 2007.

58 K. B. Alici and E. Özbay, "Radiation properties of a split ring resonator and monopole composite," *Phys. Status Solidi A*, vol. 244, no. 4, pp. 1192–1196, Apr 2007.

59 E. Sáenz, R. Gonzalo, I. Ederra, J. C. Vardaxoglou, and P. de Maagt, "Resonant meta-surface superstrate for single and multifrequency dipole antenna arrays," *IEEE Trans. Antennas Propag.*, vol. 56, no. 4, pp. 951–960, Apr 2008.

60 M. A. Antoniades and G. V. Eleftheriades, "A folded-monopole model for electrically small NRI-TL metamaterial antennas," *IEEE Antennas Wireless Propag. Lett.*, vol. 7, pp. 425–428, 2008.

61 F. Bilotti, A. Alu, and L. Vegni, "Design of miniaturized metamaterial patch antennas with μ-negative loading," *IEEE Trans. Antennas Propag.*, vol. 56, no. 6, pp. 1640–1647, Jun 2008.

62 C. Caloz, T. Itoh, and A. Rennings, "CRLH traveling-wave and resonant metamaterial antennas," *IEEE Antennas Propag. Mag.*, vol. 50, no. 5, pp. 25–39, Oct 2008.

63 O. S. Kim and O. Breinbjerg, "Miniaturised self-resonant split-ring resonator antenna," *Electron. Lett.*, vol. 45, no. 4, pp. 196–197, Feb 2009.

64 G. Mumcu, K. Sertel, and J. L. Volakis, "Miniature antenna using printed coupled lines emulating degenerate band edge crystals," *IEEE Trans. Antennas Propag.*, vol. 57, no. 6, pp. 1618–1624, Jun 2009.

65 I. K. Kim and V. V. Varadan, "Electrically small, millimeter wave dual band meta-resonator antennas," *IEEE Trans. Antennas Propag.*, vol. 58, no. 11, pp. 3458–3463, Nov 2010.

66 Y. Yu, S. He, and Z. Shen, "Electrically small omni-directional antenna of circular polarization," in *2012 IEEE International Symposium on Antennas and Propagation*, pp. 1–2, Chicago, IL, USA, 2012.

67 H. Nakano, K. Yoshida, and J. Yamauchi, "Radiation characteristics of a metaloop antenna," *IEEE Antennas Wireless Propag. Lett.*, vol. 12, pp. 861–863, 2013.

68 M.-C. Tang and R. W. Ziolkowski, "A study of low-profile, broadside radiation, efficient, electrically small antennas based on complementary split ring resonators," *IEEE Trans. Antennas Propag.*, vol. 61, no. 9, pp. 4419–4430, Sep 2013.

69 H. Wong, K.-M. Luk, C. H. Chan, Q. X., K. K. So, and H. W. Lai, "Small antennas in wireless communications," *Proc. IEEE*, vol. 100, no. 7, pp. 2109–2121, Jul 2012.

70 S. X. Ta, I. Park, and R. W. Ziolkowski, "Crossed dipole antennas: A review," *IEEE Antennas Propag. Mag.*, vol. 57, no. 5, pp. 107–122, Oct 2015.

71 J.-G. Lee and J.-H. Lee, "Compact zeroth-order resonator (ZOR) antennas," *EPJ Appl. Metamater.*, vol. 6, p. 6, 2019.

72 M. Alibakhshikenari, B. S. Virdee, L. Azpilicueta, M. Naser-Moghadasi, M. O. Akinsolu, C. H. See, B. Liu, R. A. Abd-Alhameed, F. Falcone, I. Huynen, T. A. Denidni, and E. Limiti, "A comprehensive survey of metamaterial transmission-line based antennas: Design, challenges, and applications," *IEEE Access*, vol. 8, pp. 144778-144808, 2020.

73 M. A. Antoniades, H. Mirzaei, and G. V. Eleftheriades, "Transmission-line based metamaterials in antenna engineering," in *Handbook of Antenna Technologies*. Singapore: Springer, pp. 377–449, 2016.

74 H. Nakano, T. Abe, and J. Yamauchi, "Quasi-theoretical investigation of four circularly polarized metaloop antennas," *EPJ Appl. Metamater.*, vol. 6, 2, 2019.

75 M. Casaletti, G. Valerio, O. Quevedo-Teruel, and P. Burghignoli, "An overview of metasurfaces for thin antenna applications," *C. R. Phys.*, vol. 21, no. 7–8, pp. 659–676, 2020.

76 J. Wang, Y. Li, Z. H. Jiang, T. Shi, M.-C. Tang, Z. Zhou, Z. N. Chen, and C.-W. Qiu, "Metantenna: When metasurface meets antenna again," *IEEE Trans. Antennas Propag.*, vol. 68, no. 3, pp. 1332–1347, Mar 2020.

77 P. Kumar, T. Ali, and M. M. Manohara Pai, "Electromagnetic metamaterials: A new paradigm of antenna design," *IEEE Access*, vol. 9, pp. 18722–18751, 2021.

78 S. Arslanagić and R. W. Ziolkowski, "Active coated nano-particle excited by an arbitrarily located electric Hertzian dipole-resonance and transparency effects," *J. Opt.*, vol. 12, no. 2, 024014, Feb 2010.

79 R. B. Greegor, C. G. Parazzoli, J. A. Nielsen, M. H. Tanielian, D. C. Vier, S. Schultz, C. L. Holloway, and R. W. Ziolkowski, "Demonstration of impedance matching using a mu-negative (MNG) metamaterial," *IEEE Antennas Wireless Propag. Lett.*, vol. 8, pp. 92–95, 2008.

80 R. C. Hansen, *Electrically Small, Superdirective, and Superconducting Antennas*. Hoboken, NJ: John Wiley & Sons, 2006.

81 R. C. Hansen and R. E. Collin, *Electrically Small, Superdirective, and Superconducting Antennas*. Hoboken, NJ: Wiley-IEEE Press, 2012.

82 R. W. Ziolkowski and A. Erentok, "At and below the Chu limit: Passive and active broad bandwidth metamaterial-based electrically small antennas," *IET Microw. Antennas Propag.*, vol. 1, no. 1, pp. 116–128, Feb 2007.

83 A. Erentok and R. W. Ziolkowski, "A hybrid optimization method to analyze metamaterial-based electrically small antennas," *IEEE Trans. Antennas Propag.*, vol. 55, no. 3, pp. 731–741, Mar 2007.

84 V. Laquerbe, R. Pascaud, A. Laffont, T. Callegari, L. Liard, and O. Pascal, "Towards antenna miniaturization at radio frequencies using plasma discharges," *Phys. Plasmas*, vol. 26, no. 3, 033509, Mar 2019.

85 R. W. Ziolkowski, "Metamaterials: The early years in the USA," *EPJ Appl. Metamater.*, vol. 1, 5, 2014.

86 R. W. Ziolkowski and F. Auzanneau, "Artificial molecule realization of a magnetic wall," *J. Appl. Phys.*, vol. 82, no. 7, pp. 3192–3194, Jun 1997.

87 R. W. Ziolkowski and F. Auzanneau, "Passive artificial molecule realizations of dielectric materials," *J. Appl. Phys.*, vol. 82, no. 7, pp. 3195–3198, Jul 1997.

88 F. Auzanneau and R. W. Ziolkowski, "Theoretical study of synthetic bianisotropic materials," *J. Electromagn. Waves Appl.*, vol. 12, no. 3, pp. 353–370, Apr 1998.

89 F. Auzanneau and R. W. Ziolkowski, "Microwave signal rectification using artificial composite materials composed of diode-loaded electrically small dipole antennas," *IEEE Trans. Microw. Theory Tech.*, vol. 46, no. 11, pp. 1628–1637, Nov 1998.

90 F. Auzanneau and R. W. Ziolkowski, "Explicit matrix formulation for the analysis of synthetic linearly and nonlinearly loaded materials in FDTD," *Prog. Electromagn. Res.*, PIER vol. 24, pp. 139–161, 1999.

91 F. Auzanneau and R. W. Ziolkowski, "Artificial composite materials consisting of nonlinearly loaded electrically small antennas: Operational-amplifier-based circuits with applications to smart skins," *IEEE Trans. Antennas Propag.*, vol. 47, no. 8, pp. 1330–1339, Aug 1999.

92 A. Erentok, R. W. Ziolkowski, J. A. Nielsen, R. B. Greegor, C. G. Parazzoli, M. H. Tanielian, S. A. Cummer, B.-I. Popa, T. Hand, D. C. Vier, and S. Schultz, "Low frequency lumped element-based negative index metamaterial," *Appl. Phys. Lett.*, vol. 91, no. 18, 184104, Nov 2007.

93 A. Erentok, R. W. Ziolkowski, J. A. Nielsen, R. B. Greegor, C. G. Parazzoli, M. H. Tanielian, S. A. Cummer, B.-I. Popa, T. Hand, D. C. Vier, and S. Shultz, "Lumped element-based, highly sub-wavelength, negative index metamaterials at UHF frequencies," *J. Appl. Phys.*, vol. 104, no. 3, 034901, Aug 2008.

94 D. Sievenpiper, L. Zhang, R. F. J. Broas, N. G. Alexopolous, and E. Yablonovitch, "High-impedance electromagnetic surfaces with a forbidden frequency band," *IEEE Trans. Microw. Theory Tech.*, vol. 47, no. 11, pp. 2059–2074, Nov 1999.

95 R. Coccioli, F.-R. Yang, K.-P. Ma, and T. Itoh, "Aperture-coupled patch antenna on UC-PBG substrate," *IEEE Trans. Microw. Theory Tech.*, vol. 47, no. 11, pp. 2123–2130, Nov 1999.

96 F. Yang and Y. Rahmat-Samii, *Electromagnetic Band Gap Structures in Antenna Engineering*. Cambridge, UK: Cambridge University Press, 2009.

97 A. Erentok, P. L. Luljak, and R. W. Ziolkowski, "Characterization of a volumetric metamaterial realization of an artificial magnetic conductor for antenna applications," *IEEE Trans. Antennas Propag.*, vol. 53, no. 1, pp. 160–172, Jan 2005.

98 S. J. Franson and R. W. Ziolkowski, "Numerical studies of the interaction of time-modulated multi-gigabit sequences with metamaterial structures at millimeter-wave frequencies," *Int. J. Numer. Model.*, vol. 19, no. 2, pp. 195–213, 2006.

99 S. J. Franson and R. W. Ziolkowski, "Confirmation of zero-n behavior in a high gain grid structure at millimeter-wave frequencies," *IEEE Antennas Wireless Propag. Lett.*, vol. 8, pp. 387–390, 2008.

100 S. J. Franson and R. W. Ziolkowski, "Gigabit per second data transfer in high-gain metamaterial structures at 60 GHz," *IEEE Trans. Antennas Propag.*, vol. 57, no. 10, pp. 2913–2925, Oct 2009.

101 P. Jin and R. W. Ziolkowski, "High-directivity, electrically small, low-profile near-field resonant parasitic antennas," *IEEE Antennas Wireless Propag. Lett.*, vol. 11, pp. 305–309, 2012.

102 A. Erentok and R. W. Ziolkowski, "Metamaterial-inspired efficient electrically small antennas," *IEEE Trans. Antennas Propag.*, vol. 56, no. 3, pp. 691–707, Mar 2008.

103 S. Yan, P. J. Soh, and G. A. E. Vandenbosch, "Compact all-textile dual-band antenna loaded with metamaterial-inspired structure," *IEEE Antennas Wireless Propag. Lett.*, vol. 14, pp. 1486–1489, 2014.

104 J. C. Myers, P. Chahal, E. Rothwell, and L. Kempel, "A multilayered metamaterial-inspired miniaturized dynamically tunable antenna," *IEEE Trans. Antennas Propag.*, vol. 63, no. 4, pp. 1546–1553, 2015.

105 C. Zhu, T. Li, K. Li, Z.-J. Su, X. Wang, H.-Q. Zhai, L. Li, and C.-H. Liang, "Electrically small metamaterial-inspired tri-band antenna with meta-mode," *IEEE Antennas Wireless Propag. Lett.*, vol. 14, pp. 1738–1741, 2015.

106 J. Zhang, S. Yan, and G. A. E. Vandenbosch, "Realization of dual-band pattern diversity with a CRLH-TL-inspired reconfigurable metamaterial," *IEEE Trans. Antennas Propag.*, vol. 66, no. 10, pp. 5130–5138, 2018.

107 L. Garcia-Gamez, L. Bernard, S. Collardey, H. Covic, R. Sauleau, K. Mahdjoubi, P. Potier, and P. Pouliguen, "Compact GNSS metasurface-inspired cavity antennas," *IEEE Antennas Wireless Propag. Lett.*, vol. 18, no. 12, pp. 2652–2656, Dec 2019.

108 Z. Wang, Y. Dong, and T. Itoh, "Ultraminiature circularly polarized RFID antenna inspired by crossed split-ring resonator," *IEEE Trans. Antennas Propag.*, vol. 68, no. 6, pp. 4196–4207, Jun 2020.

109 Z. Wang and Y. Dong, "Low-profile omnidirectional WIFI antennas with pattern reconfigurability inspired by meta-resonators," *IEEE Trans. Antennas Propag.*, vol. 68, no. 10, pp. 6935–6942, Oct 2020.

110 Y. Shi, Q. Zeng, Y. Shang, S. Wang, S. Zhang, T. Qiu, Y. Wu, and Q. Si, "An overview of wideband metamaterials inspired electrically small antennas," in *Proceedings of 2020 Cross Strait Radio Science & Wireless Technology Conference (CSRSWTC)*, Fuzhou, China, 13–16 December 2020. IEEE.

111 R. W. Ziolkowski, P. Jin, J. A. Nielsen, M. H. Tanielian, and C. L. Holloway, "Experimental verification of Z antennas at UHF frequencies," *IEEE Antennas Wireless Propag. Lett.*, vol. 8, pp. 1329–1333, 2009.

112 R. W. Ziolkowski, "Efficient electrically small antenna facilitated by a near-field resonant parasitic," *IEEE Antennas Wireless Propag. Lett.*, vol. 7, pp. 581–584, 2008.

113 C.-C. Lin, R. W. Ziolkowski, J. A. Nielsen, M. H. Tanielian, and C. L. Holloway, "An efficient, low profile, electrically small, three-dimensional, very high frequency magnetic EZ antenna," *Appl. Phys. Lett.*, vol. 96, no. 10, 104102, 2010.

114 R. W. Ziolkowski, C.-C. Lin, J. A. Nielsen, M. H. Tanielian, and C. L. Holloway, "Design and experimental verification of a 3D magnetic EZ antenna at 300 MHz," *IEEE Antennas Wireless Propag. Lett.*, vol. 8, pp. 989–993, 2009.

115 J. Ng, R. W. Ziolkowski, J. S. Tyo, M. C. Skipper, M. D. Abdalla, and J. Martin, "An efficient, electrically small, three-dimensional magnetic EZ antenna for HPM applications," *IEEE Trans. Plasma Sci.*, vol. 40, no. 11, pp. 3037–3045, Nov 2012.

116 E. S. Ramon, J. S. Tyo, R. W. Ziolkowski, M. C. Skipper, M. D. Abdalla, J. M. Martin, and L. L. Altgilbers, "A compact multi-frequency, high power radiating system combining dual-band, electrically small magnetic EZ antennas and multi-frequency standing wave oscillator sources," *IEEE Trans. Antennas Propag.*, vol. 62, no. 6, pp. 3281–3289, Jun 2014.

117 E. S. Ramon, J. S. Tyo, R. W. Ziolkowski, M. C. Skipper, M. D. Abdalla, and J. Martin, "Integration and operation of an electrically small magnetic EZ antenna with a high-power standing wave oscillator source," *IEEE Antennas Wireless Propag. Lett.*, vol. 15, pp. 642–645, 2015.

118 J. Choi, F. T. Dagefu, B. M. Sadler, and K. Sarabandi, "Low-power low-VHF ad-hoc networking in complex environments," *IEEE Access*, vol. 5, pp. 24120–24127, 2017.

119 S. Lim, R. L. Rogers, and H. Ling, "A tunable electrically small antenna for ground wave transmission," *IEEE Trans. Antennas Propag.*, vol. 54, no. 2, pp. 417–421, Feb 2006.

120 J. Oh, J. Choi, F. T. Dagefu, and K. Sarabandi, "Extremely small two-element monopole antenna for HF band applications," *IEEE Trans. Antennas Propag.*, vol. 61, no. 6, pp. 2991–2999, Jun 2013.

121 J. M. Baker and M. F. Iskander, "A new design approach for electrically small high-frequency antennas," *IEEE Antennas Wireless Propag. Lett.*, vol. 14, pp. 402–405, 2015.

122 J. Choi, F. T. Dagefu, B. M. Sadler, and K. Sarabandi, "Electrically small folded dipole antenna for HF and low-VHF bands," *IEEE Antennas Wireless Propag. Lett.*, vol. 15, pp. 718–721, 2015.

123 K. Ghaemi, R. Ma, and N. Behdad, "A small-aperture, ultrawideband HF/VHF direction-finding system for unmanned aerial vehicles," *IEEE Trans. Antennas Propag.*, vol. 66, no. 10, pp. 5109–5120, Oct 2018.

124 M. R. Nikkhah, F. T. Dagefu, and N. Behdad, "Electrically-small platform-based antennas for an unmanned ground vehicle," *IEEE Trans. Antennas Propag.*, vol. 68, no. 7, pp. 5189–5198, Jul 2020.

125 A. Bouvy and N. Behdad, "A heuristic study of the bandwidth potential of electrically-small, platform-based antennas at the HF band," *IEEE Trans. Antennas Propag.*, vol. 69, no. 2, pp. 623–635, Feb 2021.

126 L. Batel, J.-L. Mattei, V. Laur, A. Chevalier, and C. Delaveaud, "Tunable magneto-dielectric material for electrically small and reconfigurable antenna systems at VHF band," *Ceramics*, vol. 3, no. 3, pp. 276–286, 2020.

127 K. Mekki, E. Bajic, F. Chaxel, and F. Meyer, "A comparative study of LPWAN technologies for large-scale IoT deployment," *ICT Express*, vol. 5, no. 1, pp. 1–7, 2019.

128 G. M. Bianco, R. Giuliano, G. Marrocco, F. Mazzenga, and A. Mejia-Aguilar, "LoRa system for search and rescue: Path loss models and procedures in mountain scenarios," *IEEE Internet Things J.*, vol. 8, no. 3, pp. 1985–1999, Feb 2021.

129 R. M. Foster, "A reactance theorem," *Bell Syst. Tech. J.*, vol. 3, no. 2, pp. 259–267, 1924.

130 W. Geyi, P. Jarmuszewski, and Y. Qi, "The Foster reactance theorem for antennas and radiation Q," *IEEE Trans. Antennas Propag.*, vol. 48, no. 3, pp. 401–408, Mar 2000.

131 A. J. Poggio and P. E. Mayes, "Bandwidth extension for dipole antennas by conjugate reactance loading," *IEEE Trans. Antennas Propag.*, vol. 19, no. 4, pp. 544–547, Jul 1971.

132 J. T. Aberle and R. Loepsinger-Romak, *Antennas with non-Foster Matching Networks*, vol. 2. San Rafael, CA: Morgan & Claypool Publishers, 2007.

133 J. T. Aberle, "Two-port representation of an antenna with application to non-Foster matching networks," *IEEE Trans. Antennas Propag.*, vol. 56, no. 5, pp. 1218–1222, May 2008.

134 S. E. Sussman-Fort and R. M. Rudish, "Non-Foster impedance matching of electrically-small antennas," *IEEE Trans. Antennas Propag.*, vol. 57, no. 8, pp. 2230–2241, Aug 2009.

135 K.-S. Song and R. G. Rojas, "Non-Foster impedance matching of electrically small antennas," in *2010 IEEE Antennas and Propagation Society International Symposium*, pp. 1–4. IEEE, 2010.

136 E. Ugarte-Munoz, S. Hrabar, D. Segovia-Vargas, and A. Kiricenko, "Stability of non-Foster reactive elements for use in active metamaterials and antennas," *IEEE Trans. Antennas Propag.*, vol. 60, no. 7, pp. 3490–3494, Jul 2012.

137 C. R. White, J. S. Colburn, and R. G. Nagele, "A non-Foster VHF monopole antenna," *IEEE Antennas Wireless Propag. Lett.*, vol. 11, pp. 584–587, 2012.

138 Y. Fan, K. Z. Rajab, M. Munoz, and Y. Hao, "Electrically small half-loop antenna design with non-Foster matching networks," in *Proceedings of 2012 6th European Conference on Antennas and Propagation (EUCAP)*, pp. 126–129, Prague, Czech Republic, 26–30 March 2012. IEEE.

139 F. Albarracín-Vargas, V. Gonzalez-Posadas, F. J. Herraíz-Martinez, and D. Segovia-Vargas, "Design method for actively matched antennas with non-Foster elements," *IEEE Trans. Antennas Propag.*, vol. 64, no. 9, pp. 4118–4123, Sep 2016.

140 A. M. Elfrgani and R. G. Rojas, "Biomimetic antenna array using non-Foster network to enhance directional sensitivity over broad frequency band," *IEEE Trans. Antennas Propag.*, vol. 64, no. 10, pp. 4297–4305, Oct 2016.

141 T.-Y. Shih and N. Behdad, "Wideband, non-Foster impedance matching of electrically small transmitting antennas," *IEEE Trans. Antennas Propag.*, vol. 66, no. 11, pp. 5687–5697, Nov 2018.

142 H.-C. Chen, H.-Y. Yang, C.-C. Kao, and T.-G. Ma, "Slot antenna with non-Foster and negative conductance matching in consecutive bands," *IEEE Antennas Wireless Propag. Lett.*, vol. 18, no. 6, pp. 1203–1207, 2019.

143 J. Choi, F. T. Dagefu, B. M. Sadler, and K. Sarabandi, "A miniature actively matched antenna for power-efficient and bandwidth-enhanced operation at low VHF," *IEEE Trans. Antennas Propag.*, vol. 69, no. 1, pp. 556–561, Jan 2021.

144 M. Alibakhshikenari, B. S. Virdee, A. A. Althuwayb, L. Azpilicueta, N. O. Parchin, C. H. See, R. A. Abd-Alhameed, F. Falcone, I. Huynen, T. A. Denidni et al., "Bandwidth and gain enhancement of composite right left handed metamaterial transmission line planar antenna employing a non foster impedance matching circuit board," *Sci. Rep.*, vol. 11, no. 1, pp. 1–11, 2021.

145 H. Li, A. Mekawy, and A. Alù, "Beyond Chu's limit with Floquet impedance matching," *Phys. Rev. Lett.*, vol. 123, no. 16, 164102, 2019.

146 R. E. Collin, *Foundations for Microwave Engineering*. Hoboken, NJ: John Wiley & Sons, 2007.

147 Y. Hadad, D. L. Sounas, and A. Alù, "Space-time gradient metasurfaces," *Phys. Rev. B*, vol. 92, no. 10, 100304, 2015.

148 A. Shaltout, A. Kildishev, and V. Shalaev, "Time-varying metasurfaces and Lorentz non-reciprocity," *Opt. Mater. Express*, vol. 5, no. 11, pp. 2459–2467, 2015.

149 S. A. Tretyakov, "Metasurfaces for general transformations of electromagnetic fields," *Philos. Trans. R. Soc. London, Ser. A*, vol. 373, no. 2049, 20140362, 2015.

150 N. Chamanara, Y. Vahabzadeh, and C. Caloz, "Simultaneous control of the spatial and temporal spectra of light with space-time varying metasurfaces," *IEEE Trans. Antennas Propag.*, vol. 67, no. 4, pp. 2430–2441, 2019.

151 C. Caloz and Z.-L. Deck-Léger, "Spacetime metamaterials-Part I: General concepts," *IEEE Trans. Antennas Propag.*, vol. 68, no. 3, pp. 1569–1582, 2019.

152 C. Caloz and Z.-L. Deck-Léger, "Spacetime metamaterials-Part II: Theory and applications," *IEEE Trans. Antennas Propag.*, vol. 68, no. 3, pp. 1583–1598, Mar 2019.

153 X. Wang, A. Díaz-Rubio, H. Li, S. A. Tretyakov, and A. Alù, "Theory and design of multifunctional space-time metasurfaces," *Phys. Rev. Appl.*, vol. 13, no. 4, 044040, 2020.

154 A. Babakhani, D. B. Rutledge, and A. Hajimiri, "Near-field direct antenna modulation," *IEEE Micro. Mag.*, vol. 10, no. 1, pp. 36–46, Feb 2009.

155 K. Schab, D. Huang, and J. J. Adams, "Pulse characteristics of a direct antenna modulation transmitter," *IEEE Access*, vol. 7, pp. 30213–30219, 2019.

156 K. Schab, D. Huang, and J. J. Adams, "An energy-synchronous direct antenna modulation method for phase shift keying," *IEEE Open J. Antennas Propag.*, vol. 1, pp. 41–46, 2020.

157 K. Schab, "Radiation efficiency limits in direct antenna modulation transmitters," *IEEE Antennas Wireless Propag. Lett.*, vol. 19, no. 11, pp. 1911–1915, Nov 2020.

158 P. Jin and R. W. Ziolkowski, "Broadband, efficient, electrically small metamaterial-inspired antennas facilitated by active near-field resonant parasitic elements," *IEEE Trans. Antennas Propag.*, vol. 58, no. 2, pp. 318–327, Feb 2010.

159 N. Zhu and R. W. Ziolkowski, "Active metamaterial-inspired broad-bandwidth, efficient, electrically small antennas," *IEEE Antennas Wireless Propag. Lett.*, vol. 10, pp. 1582–1585, 2011.

160 N. Zhu and R. W. Ziolkowski, "Broad-bandwidth, electrically small antenna augmented with an internal non-Foster element," *IEEE Antennas Wireless Propag. Lett.*, vol. 11, pp. 1116–1120, 2012.

161 N. Zhu and R. W. Ziolkowski, "Design and measurements of an electrically small, broad bandwidth, non-Foster circuit-augmented protractor antenna," *Appl. Phys. Lett.*, vol. 101, no. 2, 024107, 2012.

162 M. Barbuto, A. Monti, F. Bilotti, and A. Toscano, "Design of a non-Foster actively loaded SRR and application in metamaterial-inspired components," *IEEE Trans. Antennas Propag.*, vol. 61, no. 3, pp. 1219–1227, 2012.

163 N. Zhu and R. W. Ziolkowski, "Broad bandwidth, electrically small, non-Foster element-augmented antenna designs, analyses, and measurements," *IEICE Trans. Commun.*, vol. 96, no. 10, pp. 2399–2409, Oct 2013.

164 R. W. Ziolkowski, M.-C. Tang, and N. Zhu, "An efficient, broad bandwidth, high directivity, electrically small antenna," *Microwave Opt. Tech. Lett.*, vol. 55, no. 6, pp. 1430–1434, Jun 2013.

165 M.-C. Tang, N. Zhu, and R. W. Ziolkowski, "Augmenting a modified Egyptian axe dipole antenna with non-Foster elements to enlarge its directivity bandwidth," *IEEE Antennas Wireless Propag. Lett.*, vol. 12, pp. 421–424, 2013.

166 H. Mirzaei and G. V. Eleftheriades, "A resonant printed monopole antenna with an embedded non-Foster matching network," *IEEE Trans. Antennas Propag.*, vol. 61, no. 11, pp. 5363–5371, Nov 2013.

167 J. Church, J.-C. S. Chieh, L. Xu, J. D. Rockway, and D. Arceo, "UHF electrically small box cage loop antenna with an embedded non-Foster load," *IEEE Antennas Wireless Propag. Lett.*, vol. 13, pp. 1329–1332, 2014.

168 R. T. Cutshall and R. W. Ziolkowski, "Performance characteristics of planar and three-dimensional versions of a frequency-agile electrically small antenna," *IEEE Antennas Propag. Mag.*, vol. 56, no. 6, pp. 53–71, Dec 2014.

169 G. Fu and S. Sonkusale, "Broadband wireless radio frequency power telemetry using a metamaterial resonator embedded with non-Foster impedance circuitry," *Appl. Phys. Lett.*, vol. 106, no. 20, 203504, 2015.

170 M.-C. Tang, T. Shi, and R. W. Ziolkowski, "Electrically small, broadside radiating Huygens source antenna augmented with internal non-Foster elements to increase its bandwidth," *IEEE Antennas Wireless Propag. Lett.*, vol. 16, pp. 712–715, 2017.

171 H. Jaafar, D. Lemur, S. Collardey, and A. Sharaiha, "Parametric optimization of a non-Foster circuit embedded in an electrically small antenna for wideband and efficient performance," *IEEE Trans. Antennas Propag.*, vol. 67, no. 6, pp. 3619–3628, Jun 2019.

172 T. Shi, M.-C. Tang, Z. Wu, H.-X. Xu, and R. W. Ziolkowski, "Improved signal-to-noise ratio, bandwidth-enhanced electrically small antenna augmented with internal non-Foster elements," *IEEE Trans. Antennas Propag.*, vol. 67, no. 4, pp. 2763–2768, Apr 2019.

173 A. A. Althuwayb, M. Alibakhshikenari, B. S. Virdee, F. Falcone, and E. Limiti, "Overcoming inherent narrow bandwidth and low radiation properties of electrically small antennas by using an active interior-matching circuit," *IEEE Access*, vol. 9, pp. 20622–20628, 2021.

174 M.-C. Tang and R. W. Ziolkowski, "A frequency agile, ultralow-profile, complementary split ring resonator-based electrically small antenna," *Microw. Opt. Technol. Lett.*, vol. 55, no. 10, pp. 2425–2428, Oct 2013.

175 M.-C. Tang, R. W. Ziolkowski, S. Xiao, M. Li, and J. Zhang, "Frequency-agile, efficient, near-field resonant parasitic monopole antenna," *IEEE Trans. Antennas Propag.*, vol. 62, no. 3, pp. 1479–1483, Mar 2014.

176 M.-C. Tang and R. W. Ziolkowski, "Frequency-agile, efficient, circularly polarized, near-field resonant antenna: Designs and measurements," *IEEE Trans. Antennas Propag.*, vol. 63, no. 11, pp. 5203–5209, Nov 2015.

177 I. V. Shadrivov, S. K. Morrison, and Y. S. Kivshar, "Tunable split-ring resonators for nonlinear negative-index metamaterials," *Opt. Express*, vol. 14, no. 20, pp. 9344–9349, Oct 2006.

178 P. Jin and R. W. Ziolkowski, "Low-Q, electrically small, efficient near-field resonant parasitic antennas," *IEEE Trans. Antennas Propag.*, vol. 57, no. 9, pp. 2548–2563, Sep 2009.

179 S. Hrabar, "First ten years of active metamaterial structures with "negative" elements," *EPJ Appl. Metamater.*, vol. 5, 9, 2018.

180 S. D. Stearns, "Non-Foster circuits and stability theory," in *Proceedings of 2011 IEEE International Symposium on Antennas and Propagation (APSURSI)*, pp. 1942–1945, Spokane, WA, USA, 2011.

181 S. D. Stearns, "Stable band-pass non-Foster circuits," in *Proceedings of 2015 IEEE International Symposium on Antennas and Propagation USNC/URSI National Radio Science Meeting*, pp. 1386–1387, Vancouver, BC, Canada, 2015.

182 J. Loncar, S. Hrabar, and D. Muha, "Stability of simple lumped-distributed networks with negative capacitors," *IEEE Trans. Antennas Propag.*, vol. 65, no. 1, pp. 390–395, Jan 2017.

183 C. R. White, J. W. May, and J. S. Colburn, "A variable negative-inductance integrated circuit at UHF frequencies," *IEEE Microw. Compon. Lett.*, vol. 22, no. 1, pp. 35–37, Jan 2011.

184 D. J. Gregoire, C. R. White, and J. S. Colburn, "Wideband artificial magnetic conductors loaded with non-Foster negative inductors," *IEEE Antennas Wireless Propag. Lett.*, vol. 10, pp. 1586–1589, 2011.

185 S. Saadat, H. Aghasi, E. Afshari, and H. Mosallaei, "Low-power negative inductance integrated circuits for GHz applications," *IEEE Microw. Compon. Lett.*, vol. 25, no. 2, pp. 118–120, Feb 2015.

186 L. Vincelj, I. Krois, and S. Hrabar, "Toward self-oscillating non-Foster unit cell for future active metasurfaces," *IEEE Trans. Antennas Propag.*, vol. 68, no. 3, pp. 1665–1679, Mar 2020.

187 L. Vincelj, S. Hrabar, and R. W. Ziolkowski, "Non-Foster self-oscillating Huygens radiator," in *Proceedings of 2020 IEEE International Symposium on Antennas and Propagation USNC/URSI National Radio Science Meeting*, pp. 521–522, Montreal, Canada, 5-10 Jul 2020.

188 L. Vincelj, S. Hrabar, and R. W. Ziolkowski, "Experimental demonstration of non-Foster self-oscillating Huygens radiator," in *Proceedings of the 14th Congress on Artificial Materials for*

Novel Wave Phenomena - Metamaterials 2020, pp. X–508–X–510, New York, NY, USA, 28 Sep–3 Oct 2020.

189 Y. Fan, K. Z. Rajab, and Y. Hao, "Noise analysis of broadband active metamaterials with non-Foster loads," *J. Appl. Phys.*, vol. 113, no. 23, 233905, 2013.

190 M. M. Jacob and D. F. Sievenpiper, "Gain and noise analysis of non-Foster matched antennas," *IEEE Trans. Antennas Propag.*, vol. 64, no. 12, pp. 4993–5004, Dec 2016.

191 P. Jin, C.-C. Lin, and R. W. Ziolkowski, "Multifunctional, electrically small, planar near-field resonant parasitic antennas," *IEEE Antennas Wireless Propag. Lett.*, vol. 11, pp. 200–204, 2012.

192 S. X. Ta, K. Lee, I. Park, and R. W. Ziolkowski, "Compact crossed-dipole antennas loaded with near-field resonant parasitic elements," *IEEE Trans. Antennas Propag.*, vol. 65, no. 2, pp. 482–488, Feb 2017.

193 P. Jin and R. W. Ziolkowski, "Multi-frequency, linear and circular polarized, metamaterial-inspired, near-field resonant parasitic antennas," *IEEE Trans. Antennas Propag.*, vol. 59, no. 5, pp. 1446–1459, May 2011.

194 S. X. Ta, H. Choo, I. Park, and R. W. Ziolkowski, "Multi-band, wide-beam, circularly polarized, crossed, asymmetrically barbed dipole antennas for GPS applications," *IEEE Trans. Antennas Propag.*, vol. 61, no. 11, pp. 5771–5775, Nov 2013.

195 O. Turkmen, G. Turhan-Sayan, and R. W. Ziolkowski, "Single-, dual-, and triple-band metamaterial-inspired electrically small planar magnetic dipole antennas," *Microw. Opt. Technol. Lett.*, vol. 56, no. 1, pp. 83–87, Jan 2014.

196 S. Dakhli, H. Rmili, J.-M. Floc'h, M. Sheikh, A. Dobaie, K. Mahdjoubi, F. Choubani, and R. W. Ziolkowski, "Printed multiband metamaterial-inspired antennas," *Microw. Opt. Technol. Lett.*, vol. 58, no. 6, pp. 1281–1289, Jun 2016.

197 S. X. Ta, I. Park, and R. W. Ziolkowski, "Circularly polarized crossed dipole on an HIS for 2.4/5.2/5.8-GHz WLAN applications," *IEEE Antennas Wireless Propag. Lett.*, vol. 12, pp. 1464–1467, 2013.

198 P. P. Shome, T. Khan, S. Koul, and Y. Antar, "Filtenna designs for radio-frequency front-end systems: A structural-oriented review," *IEEE Antennas Propag. Mag.*, early access, 2020, DOI: 10.1109/MAP.2020.2988518.

199 M.-C. Tang, Y. Chen, and R. W. Ziolkowski, "Experimentally validated, planar, wideband, electrically small, monopole filtennas based on capacitively loaded loop resonators," *IEEE Trans. Antennas Propag.*, vol. 64, no. 8, pp. 3353–3360, Aug 2016.

200 X. Chen, M. Tang, D. Yi, and R. W. Ziolkowski, "Wideband, electrically small, near-field resonant parasitic dipole antenna with stable radiation performance," *IEEE Antennas Wireless Propag. Lett.*, vol. 19, no. 5, pp. 826–830, May 2020.

201 M. Tang, D. Li, Y. Wang, K. Hu, and R. W. Ziolkowski, "Compact, low-profile, linearly and circularly polarized filtennas enabled with custom-designed feed-probe structures," *IEEE Trans. Antennas Propag.*, vol. 68, no. 7, pp. 5247–5256, Jul 2020.

202 J. T. Bernhard, *Reconfigurable Antennas*. Williston, VT, USA: Morgan & Claypool, 2007.

203 C. G. Christodoulou, Y. Tawk, S. A. Lane, and S. R. Erwin, "Reconfigurable antennas for wireless and space applications," *Proc. IEEE*, vol. 100, no. 7, pp. 2250–2261, Jul 2012.

204 Y. J. Guo, P. Qin, S. Chen, W. Lin, and R. W. Ziolkowski, "Advances in reconfigurable antenna systems facilitated by innovative technologies," *IEEE Access*, vol. 6, pp. 5780–5794, 2018.

205 M.-C. Tang, T. Shi, and R. Ziolkowski, "Electrically small uniplanar antenna with pattern-agile performance," *Electron. Lett.*, vol. 51, no. 16, pp. 1228–1229, Aug 2015.

206 S. Dakhli, H. Rmili, J.-m. Floc'h, M. Sheikh, K. Mahdjoubi, F. Choubani, and R. W. Ziolkowski, "Capacitively loaded loop-based antennas with reconfigurable radiation patterns," *Int. J. Antennas Propag.*, vol. 2015, 523198, 2015.

207 M.-C. Tang, B. Zhou, and R. W. Ziolkowski, "Low-profile, electrically small, Huygens source antenna with pattern-reconfigurability that covers the entire azimuthal plane," *IEEE Trans. Antennas Propag.*, vol. 65, no. 3, pp. 1063–1072, Mar 2017.

208 M.-C. Tang, Y. Duan, Z. Wu, X. Chen, M. Li, and R. W. Ziolkowski, "Pattern reconfigurable, vertically polarized, low-profile, compact, near-field resonant parasitic antenna," *IEEE Trans. Antennas Propag.*, vol. 67, no. 3, pp. 1467–1475, Mar 2018.

209 M. Tang, Y. Chen, X. Chen, D. Mu, and R. W. Ziolkowski, "Design and testing of a simple, compact antenna with reconfigurable broadside and monopole-like patterns," *IEEE Antennas Wireless Propag. Lett.*, vol. 19, no. 6, pp. 897–901, Jun 2020.

210 Z. Wu, M. Tang, M. Li, and R. W. Ziolkowski, "Ultralow-profile, electrically small, pattern-reconfigurable metamaterial-inspired Huygens dipole antenna," *IEEE Trans. Antennas Propag.*, vol. 68, no. 3, pp. 1238–1248, Mar 2020.

211 M.-C. Tang, Z. Wu, T. Shi, and R. W. Ziolkowski, "Electrically small, low-profile, planar, Huygens dipole antenna with quad-polarization diversity," *IEEE Trans. Antennas Propag.*, vol. 66, no. 12, pp. 6772–6780, Dec 2018.

212 M.-C. Tang, B. Zhou, and R. W. Ziolkowski, "Flexible uniplanar electrically small directive antenna empowered by a modified CPW-feed," *IEEE Antennas Wireless Propag. Lett.*, vol. 15, pp. 914–917, 2016.

213 S. Lim and H. Ling, "Design of electrically small Yagi antenna," *Electron. Lett.*, vol. 43, no. 5, pp. 3–4, 2007.

214 M.-C. Tang, R. W. Ziolkowski, S. Xiao, and M. Li, "A high-directivity, wideband, efficient, electrically small antenna system," *IEEE Trans. Antennas Propag.*, vol. 62, no. 12, pp. 6541–6547, Dec 2014.

215 S. X. Ta, I. Park, and R. W. Ziolkowski, "Broadband electrically small circularly polarized directive antenna," *IEEE Access*, vol. 5, pp. 14657–14663, 2017.

216 S. Lim and H. Ling, "Design of electrically small, pattern reconfigurable Yagi antenna," *Electron. Lett.*, vol. 43, no. 24, pp. 1326–1327, 2007.

217 M. Tang, Q. Lin, M. Li, and R. W. Ziolkowski, "Polarization-reconfigurable Yagi-configured electrically small antenna," *IEEE Trans. Antennas Propag.*, vol. 69, no. 3, pp. 1757–1762, Mar 2021.

218 M.-C. Tang and R. W. Ziolkowski, "Efficient, high directivity, large front-to-back-ratio, electrically small, near-field-resonant-parasitic antenna," *IEEE Access*, vol. 1, pp. 16–28, 2013.

219 K.-M. Luk and H. Wong, "A new wideband unidirectional antenna element," *Int. J. Microw. Opt. Technol.*, vol. 1, no. 1, pp. 35–44, 2006.

220 K.-M. Luk and B. Wu, "The magnetoelectric dipole – a wideband antenna for base stations in mobile communications," *Proc. IEEE*, vol. 100, no. 7, pp. 2297–2307, Jul 2012.

221 H.-J. Seo and A. A. Kishk, "Wideband linear and dual-polarized antenna based on Huygens' source principle," in *Proceedings of 2011 XXXth URSI General Assembly and Scientific Symposium*, pp. 1–4, Istanbul, Turkey, 2011. IEEE.

222 H.-J. Seo and A. A. Kishk, "Wideband magnetic-electric antenna with linear single or dual polarization," *Prog. Electromagn. Res.*, vol. 155, pp. 53–61, 2016.

223 M. Li, K. Luk, L. Ge, and K. Zhang, "Miniaturization of magnetoelectric dipole antenna by using metamaterial loading," *IEEE Trans. Antennas Propag.*, vol. 64, no. 11, pp. 4914–4918, Nov 2016.

224 F. Wu and K. M. Luk, "A compact and reconfigurable circularly polarized complementary antenna," *IEEE Antennas Propag. Mag.*, vol. 16, pp. 1188–1191, 2017.

225 R. W. Ziolkowski, "Low profile, broadside radiating, electrically small Huygens source antennas," *IEEE Access*, vol. 3, pp. 2644–2651, Dec 2015.

226 R. W. Ziolkowski, "Custom-designed electrically small Huygens dipole antennas achieve efficient, directive emissions into air when mounted on a high permittivity block," *IEEE Access*, vol. 7, pp. 163365–163383, Jul 2019.

227 P. Jin and R. W. Ziolkowski, "Metamaterial-inspired, electrically small Huygens sources," *IEEE Antennas Wireless Propag. Lett.*, vol. 9, pp. 501–505, 2010.

228 M.-C. Tang, H. Wang, and R. W. Ziolkowski, "Design and testing of simple, electrically small, low-profile, Huygens source antennas with broadside radiation performance," *IEEE Trans. Antennas Propag.*, vol. 64, no. 11, pp. 4607–4617, Nov 2016.

229 S. Lee, G. Shin, S. M. Radha, J. Choi, and I. Yoon, "Low-profile, electrically small planar Huygens source antenna with an endfire radiation characteristic," *IEEE Antennas Wireless Propag. Lett.*, vol. 18, no. 3, pp. 412–416, Mar 2019.

230 V. Jantarachote, S. Chalermwisutkul, W. Chansumran, and W. Longcharoen, "Electrically small Huygens source antennas based on complementary spiral resonator," in *Proceedings of 2019 16th International Conference on Electrical Engineering/Electronics, Computer, Telecommunications and Information Technology (ECTI-CON)*, pp. 752–755, Pattaya, Thailand, 2019.

231 W. Lin and R. W. Ziolkowski, "Electrically-small, low-profile, Huygens circularly polarized antenna," *IEEE Trans. Antennas Propag.*, vol. 66, no. 2, pp. 636–643, Feb 2018.

232 M. Tang, Z. Wu, T. Shi, H. Zeng, W. Lin, and R. W. Ziolkowski, "Dual-linearly polarized, electrically small, low-profile, broadside radiating, Huygens dipole antenna," *IEEE Trans. Antennas Propag.*, vol. 66, no. 8, pp. 3877–3885, 2018.

233 M. Tang, Z. Wu, T. Shi, and R. W. Ziolkowski, "Dual-band, linearly polarized, electrically small Huygens dipole antennas," *IEEE Trans. Antennas Propag.*, vol. 67, no. 1, pp. 37–47, Jan 2019.

234 Z. Wu, M. Tang, T. Shi, and R. W. Ziolkowski, "Two-port, dual-circularly polarized, low-profile broadside-radiating electrically small Huygens dipole antenna," *IEEE Trans. Antennas Propag.*, vol. 69, no. 1, pp. 514–519, Jan 2021.

235 S. Sufyar and C. Delaveaud, "A miniaturization technique of a compact omnidirectional antenna," *Radioengineering*, vol. 18, no. 4, pp. 373–380, Dec 2009.

236 L. Huitema, C. Delaveaud, and R. D'Errico, "Impedance and radiation measurement methodology for ultra miniature antennas," *IEEE Trans. Antennas Propag.*, vol. 62, no. 7, pp. 3463–3473, Jul 2014.

237 L. Fmrch, S. H. Kvist, J. Thaysen, and K. B. Jakobsen, "Electric-coupled antenna for hearing-instrument applications," in *Proceedings of 2016 Loughborough Antennas Propagation Conference (LAPC)*, pp. 1–4, Loughborough, UK, 14–15 Nov 2016.

238 A. Faul and J. Naber, "A novel 915 MHz, RFID-based pressure sensor for glaucoma using an electrically small antenna," *Analog Integr. Circuits Signal Process.*, vol. 85, no. 1, pp. 167–180, 2015.

239 A. Sani, M. Rajab, R. Foster, and Y. Hao, "Antennas and propagation of implanted RFIDs for pervasive healthcare applications," *Proc. IEEE*, vol. 98, no. 9, pp. 1648–1655, Sep 2010.

240 A. A. Salih and M. S. Sharawi, "A dual-band highly miniaturized patch antenna," *IEEE Antennas Wireless Propag. Lett.*, vol. 15, pp. 1783–1786, 2016.

241 S. Yan and G. A. E. Vandenbosch, "A fully planar near-field resonant parasitic antenna," *Prog. Electromagn. Res.*, vol. 54, pp. 163–169, 2014.

242 P. Turalchuk, I. Munina, M. Derkach, O. Vendik, and I. Vendik, "Electrically small loop antennas for RFID applications," *IEEE Antennas Wireless Propag. Lett.*, vol. 14, pp. 1786–1789, 2015.

243 S. V. Reddy, K. Saurav, D. Sarkar, and K. V. Srivastava, "Design of compact planar inverted-F antennas loaded with LC resonators," in *Proceedings of 2015 IEEE UP Section Conference on Electrical Computer and Electronics (UPCON)*, pp. 1–5, Allahabad, India, 4–6 Dec 2015.

244 S. Zuffanelli, G. Zamora, P. Aguilà, F. Paredes, F. Martín, and J. Bonache, "On the radiation properties of split-ring resonators (SRRs) at the second resonance," *IEEE Trans. Microw. Theory Tech.*, vol. 63, no. 7, pp. 2133–2141, Jul 2015.

245 J. Kim and S. Nam, "A compact quasi-isotropic antenna based on folded split-ring resonators," *IEEE Antennas Wireless Propag. Lett.*, vol. 16, pp. 294–297, 2017.

246 L. Roselli, C. Mariotti, M. Virili, F. Alimenti, G. Orecchini, V. Palazzi, P. Mezzanotte, and N. B. Carvalho, "WPT related applications enabling Internet of Things evolution," in *Proceedings of 2016 10th European Conference on Antennas and Propagation (EuCAP)*, pp. 1–2, Davos, Switzerland, 11–15 April 2016.

247 M. R. Palattella, M. Dohler, A. Grieco, G. Rizzo, J. Torsner, T. Engel, and L. Ladid, "Internet of Things in the 5G era: Enablers, architecture, and business models," *IEEE J. Sel. Areas Commun.*, vol. 34, no. 3, pp. 510–527, Mar 2016.

248 J. A. Hagerty, F. B. Helmbrecht, W. H. McCalpin, R. Zane, and Z. B. Popovic, "Recycling ambient microwave energy with broad-band rectenna arrays," *IEEE Trans. Microw. Theory Tech.*, vol. 52, no. 3, pp. 1014–1024, Mar 2004.

249 Z. Popovic, "Cut the cord: Low-power far-field wireless powering," *IEEE Microw. Mag.*, vol. 14, no. 2, pp. 55–62, Feb 2013.

250 C. R. Valenta and G. D. Durgin, "Harvesting wireless power: Survey of energy-harvester conversion efficiency in far-field, wireless power transfer systems," *IEEE Microw. Mag.*, vol. 15, no. 4, pp. 108–120, Jun 2014.

251 W. C. Brown, "The history of power transmission by radio waves," *IEEE Trans. Microw. Theory Tech.*, vol. 32, no. 9, pp. 1230–1242, Sep 1984.

252 B. Strassner and K. Chang, "Microwave power transmission: Historical milestones and system components," *Proc. IEEE*, vol. 101, no. 6, pp. 1379–1396, Jun 2013.

253 A. Massa, G. Oliveri, F. Viani, and P. Rocca, "Array designs for long-distance wireless power transmission: State-of-the-art and innovative solutions," *Proc. IEEE*, vol. 101, no. 6, pp. 1464–1481, Jun 2013.

254 A. Costanzo, M. Dionigi, D. Masotti, M. Mongiardo, G. Monti, L. Tarricone, and R. Sorrentino, "Electromagnetic energy harvesting and wireless power transmission: A unified approach," *Proc. IEEE*, vol. 102, no. 11, pp. 1692–1711, 2014.

255 N. B. Carvalho, A. Georgiadis, A. Costanzo, H. Rogier, A. Collado, and et al., "Wireless power transmission: R & D activities within Europe," *IEEE Trans. Microw. Theory Tech.*, vol. 62, no. 4, pp. 1031–1045, Apr 2014.

256 A. Costanzo and D. Masotti, "Smart solutions in smart spaces: Getting the most from far-field wireless power transfer," *IEEE Microw. Mag.*, vol. 17, no. 5, pp. 30–45, May 2016.

257 Z. Popovic, "Near- and far-field wireless power transfer," in *Proceedings of 2017 13th International Conference on Advanced Technologies, Systems and Services in Telecommunications (TELSIKS)*, pp. 3–6, Nis, Serbia, 2017.

258 J. G. D. Hester, J. Kimionis, and M. M. Tentzeris, "Printed motes for IoT wireless networks: State of the art, challenges, and outlooks," *IEEE Trans. Microw. Theory Tech.*, vol. 65, no. 5, pp. 1819–1830, May 2017.

259 A. Costanzo and D. Masotti, "Energizing 5G: Near- and far-field wireless energy and data transfer as an enabling technology for the 5G IoT," *IEEE Microw. Mag.*, vol. 18, no. 3, pp. 125–136, 2017.

260 Y. G. Kim and S. Nam, "Determination of the impedance parameters of antennas and the maximum power transfer efficiency of wireless power transfer," *IEEE Trans. Antennas Propag.*, vol. 67, no. 8, pp. 5132–5144, Aug 2019.

261 J. Kim, Y. Lim, and S. Nam, "Efficiency bound of radiative wireless power transmission using practical antennas," *IEEE Trans. Antennas Propag.*, vol. 67, no. 8, pp. 5750–5755, Aug 2019.

262 D. Surender, T. Khan, F. A. Talukdar, A. De, Y. M. M. Antar, and A. P. Freundorfer, "Key components of rectenna system: A comprehensive survey," *IETE J. Res.*, pp. 1–27, May 2020. DOI: 10.1080/03772063.2020.1761268.

263 N. Shinohara, "Trends in wireless power transfer: WPT technology for energy harvesting, millimeter-wave/THz rectennas, MIMO-WPT, and advances in near-field WPT applications," *IEEE Microw. Mag.*, vol. 22, no. 1, pp. 46–59, Jan 2020.

264 M. Wagih, A. S. Weddell, and S. Beeby, "Rectennas for radio-frequency energy harvesting and wireless power transfer: A review of antenna design," *IEEE Antennas Propag. Mag.*, vol. 62, no. 5, pp. 95–107, Oct 2020.

265 M. Wagih, A. S. Weddell, and S. Beeby, "Millimeter-wave power harvesting: A review," *IEEE Open J. Antennas Propag.*, vol. 1, pp. 560–578, 2020.

266 L. Li, X. Zhang, C. Song, and Y. Huang, "Progress, challenges, and perspective on metasurfaces for ambient radio frequency energy harvesting," *Appl. Phys. Lett.*, vol. 116, no. 6, 060501, 2020.

267 A. A. Eteng, H. H. Goh, S. K. A. Rahim, and A. Alomainy, "A review of metasurfaces for microwave energy transmission and harvesting in wireless powered networks," *IEEE Access*, vol. 9, pp. 27518–27539, 2021.

268 N. Zhu, R. W. Ziolkowski, and H. Xin, "A metamaterial-inspired, electrically small rectenna for high-efficiency, low power harvesting and scavenging at the global positioning system L1 frequency," *Appl. Phys. Lett.*, vol. 99, no. 11, 114101, 2011.

269 N. Zhu, R. W. Ziolkowski, and H. Xin, "Electrically small GPS L1 rectennas," *IEEE Antennas Wireless Propag. Lett.*, vol. 10, pp. 935–938, 2011.

270 W. Lin, R. W. Ziolkowski, and J. Huang, "Electrically small, low profile, highly efficient, Huygens dipole rectennas for wirelessly powering internet-of-things (IoT) devices," *IEEE Trans. Antennas Propag.*, vol. 67, no. 6, pp. 3670–3679, Jun 2019.

271 W. Lin and R. W. Ziolkowski, "Electrically small Huygens CP rectenna with a driven loop element maximizes its wireless power transfer efficiency," *IEEE Trans. Antennas Propag.*, vol. 68, no. 1, pp. 540–545, Jan 2020.

272 C. Song, Y. Huang, J. Zhou, P. Carter, S. Yuan, Q. Xu, and Z. Fei, "Matching network elimination in broadband rectennas for high-efficiency wireless power transfer and energy harvesting," *IEEE Trans. Ind. Electron.*, vol. 64, no. 5, pp. 3950–3961, May 2017.

273 W. Lin and R. W. Ziolkowski, "Electrically small Huygens antenna-based fully-integrated wireless power transfer and communication system," *IEEE Access*, vol. 7, pp. 39762–39769, Apr 2019.

274 F. J. Faisal, M. A. A. Haider, and A. A. Butt, "A technical review of simultaneous wireless information and power transfer (SWIPT)," in *Proceedings of 2017 International Symposium on Recent Advances in Electrical Engineering (RAEE)*, pp. 1–6, Islamabad, Pakistan, 2017.

275 H.-S. Nguyen, D.-T. Do, and M. Voznak, "Two-way relaying networks in green communications for 5G: Optimal throughput and tradeoff between relay distance on power splitting-based and time switching-based relaying SWIPT," *Int. J. Electron. Commun.*, vol. 70, no. 12, pp. 1637–1644, Dec 2016.

276 Z. Popović, E. A. Falkenstein, D. Costinett, and R. Zane, "Low-power farfield wireless powering for wireless sensors," *Proc. IEEE*, vol. 101, no. 6, pp. 1397–1409, Jun 2013.

277 H. J. Visser and R. J. M. Vullers, "RF energy harvesting and transport for wireless sensor network applications: Principles and requirements," *Proc. IEEE*, vol. 101, no. 6, pp. 1410–1423, Jun 2013.

278 S. Kim, R. Vyas, J. Bito, K. Niotaki, A. Collado, A. Georgiadis, and M. M. Tentzeris, "Ambient RF energy-harvesting technologies for self-sustainable standalone wireless sensor platforms," *Proc. IEEE*, vol. 102, no. 11, pp. 1649–1666, Nov 2014.

279 H. Khodr, N. Kouzayha, M. Abdallah, J. Costantine, and Z. Dawy, "Energy efficient IoT sensor with RF wake-up and addressing capability," *IEEE Sensors Lett.*, vol. 1, no. 6, pp. 1–4, Dec 2017.

280 O. Bjorkqvist, O. Dahlberg, G. Silver, C. Kolitsidas, O. Quevedo-Teruel, and B. L. G. Jonsson, "Wireless sensor network utilizing radio-frequency energy harvesting for smart building applications," *IEEE Antennas Propag. Mag.*, vol. 60, no. 5, pp. 124–136, Oct 2018.

281 W. Lin and R. W. Ziolkowski, "Wirelessly powered light and temperature sensors facilitated by electrically small omnidirectional and Huygens dipole antennas," *Sensors*, vol. 19, no. 9, 1998, Apr 2019.

282 W. Lin and R. W. Ziolkowski, "Electrically small, single-substrate Huygens dipole rectenna for ultra-compact wireless power transfer applications," *IEEE Trans. Antennas Propag.*, vol. 69, no. 2, pp. 1130–1134, Feb 2021.

283 W. Lin and R. W. Ziolkowski, "Wireless power transfer (WPT) enabled IoT sensors based on ultra-thin electrically small antennas," in *Proceedings of 2021 15th European Conference on Antennas and Propagation (EUCAP)*, Düsseldorf, Germany, 22–26 March 2021. IEEE.

284 T. C. Baum, R. W. Ziolkowski, K. Ghorbani, and K. J. Nicholson, "Embroidered active microwave composite preimpregnated electronics-Pregtronics," *IEEE Trans. Microw. Theory Tech.*, vol. 64, no. 10, pp. 3175–3186, Oct 2016.

285 T. C. Baum, R. W. Ziolkowski, K. Ghorbani, and K. J. Nicholson, "Investigations of a load-bearing composite electrically small Egyptian axe dipole antenna," *IEEE Trans. Antennas Propag.*, vol. 65, no. 8, pp. 3827–3837, Aug 2017.

286 M.-C. Tang, T. Shi, and R. W. Ziolkowski, "A study of 28 GHz, planar, multilayered, electrically small, broadside radiating, Huygens source antennas," *IEEE Trans. Antennas Propag.*, vol. 65, no. 12, pp. 6345–6354, Dec 2017.

287 W. Lin, R. W. Ziolkowski, and T. C. Baum, "28 GHz compact omnidirectional circularly polarized antenna for device-to-device communications in the future 5G systems," *IEEE Trans. Antennas Propag.*, vol. 65, no. 12, pp. 6904–6914, Dec 2017.

288 N. Ashraf, A. R. Sebak, and A. A. Kishk, "Packaged microstrip line feed network on a single surface for dual-polarized $2^N \times 2^M$ ME-dipole antenna array," *IEEE Antennas Wireless Propag. Lett.*, vol. 19, no. 4, pp. 596–600, 2020.

289 Y. Li and K. Luk, "A multibeam end-fire magnetoelectric dipole antenna array for millimeter-wave applications," *IEEE Trans. Antennas Propag.*, vol. 64, no. 7, pp. 2894–2904, Jul 2016.

290 W. Hong, Z. H. Jiang, C. Yu, J. Zhou, and P. Chen et al., "Multibeam antenna technologies for 5G wireless communications," *IEEE Trans. Antennas Propag.*, vol. 65, no. 12, pp. 6231–6249, Dec 2017.

291 R. A. Alhalabi and G. M. Rebeiz, "High-efficiency angled-dipole antennas for millimeter-wave phased array applications," *IEEE Trans. Antennas Propag.*, vol. 56, no. 10, pp. 3136–3142, Oct 2008.

292 J. Wang, Y. Li, L. Ge, J. Wang, and K.-M. Luk, "A 60 GHz horizontally polarized magneto-electric dipole antenna array with 2-D multibeam endfire radiation," *IEEE Trans. Antennas Propag.*, vol. 65, no. 11, pp. 5837–5845, Nov 2017.

293 J. Zeng and K.-M. Luk, "Wideband millimeter-wave end-fire magnetoelectric dipole antenna with microstrip-line feed," *IEEE Trans. Antennas Propag.*, vol. 68, no. 4, pp. 2658–2665, Apr 2020.

294 R. Li, T. Wu, B. Pan, K. Lim, J. Laskar, and M. M. Tentzeris, "Equivalent-circuit analysis of a broadband printed dipole with adjusted integrated balun and an array for base station applications," *IEEE Trans. Antennas Propag.*, vol. 57, no. 7, pp. 2180–2184, Jul 2009.

295 N. Nikolic and A. R. Weily, "Compact E-band planar quasi-Yagi antenna with folded dipole driver," *IET Microw. Antennas Propag.*, vol. 4, no. 11, pp. 1728–1734, Nov 2010.

296 S. X. Ta and I. Park, "Two-element folded-dipole-driven quasi-Yagi array with low mutual coupling," in *Proceedings of 2013 International Workshop on Antenna Technology (iWAT)*, pp. 236–239, Karlsruhe, Germany, 4–6 Mar 2013.

297 L. Lu, K. Ma, F. Meng, and K. S. Yeo, "Design of a 60-GHz quasi-Yagi antenna with novel ladder-like directors for gain and bandwidth enhancements," *IEEE Antennas Wireless Propag. Lett.*, vol. 15, pp. 682–685, 2016.

298 S. Zhu, H. Liu, Z. Chen, and P. Wen, "A compact gain-enhanced Vivaldi antenna array with suppressed mutual coupling for 5G mmWave application," *IEEE Antennas Wireless Propag. Lett.*, vol. 17, no. 5, pp. 776–779, May 2018.

299 N. Ashraf, A.-R. Sebak, and A. A. Kishk, "PMC packaged single-substrate 4 × 4 Butler matrix and double-ridge gap waveguide horn antenna array for multibeam applications," *IEEE Trans. Microw. Theory Tech.*, vol. 69, no. 1, pp. 248–261, Jan 2021.

300 Q. Sun, Y.-L. Ban, J.-W. Lian, Y. Liu, and Z. Nie, "Millimeter-wave multibeam antenna based on folded C-type SIW," *IEEE Trans. Antennas Propag.*, vol. 68, no. 5, pp. 3465–3476, May 2020.

301 P. Chen, W. Hong, Z. Kuai, J. Xu, and H. Wang et al., "A multibeam antenna based on substrate integrated waveguide technology for MIMO wireless communications," *IEEE Trans. Antennas Propag.*, vol. 57, no. 6, pp. 1813–1821, Jun 2009.

302 L.-H. Zhong, Y.-L. Ban, J.-W. Lian, Q.-L. Yang, J. Guo, and Z.-F. Yu, "Miniaturized SIW multi-beam antenna array fed by dual-layer 8 × 8 Butler matrix," *IEEE Antennas Wireless Propag. Lett.*, vol. 16, pp. 3018–3021, 2017.

303 J.-W. Lian, Y.-L. Ban, C. Xiao, and Z.-F. Yu, "Compact substrate-integrated 4 × 8 Butler matrix with sidelobe suppression for millimeter-wave multibeam application," *IEEE Antennas Wireless Propag. Lett.*, vol. 17, no. 5, pp. 928–932, May 2018.

304 Y. Hu, J. Zhan, Z. H. Jiang, C. Yu, and W. Hong, "An orthogonal hybrid analog-digital multibeam antenna array for millimeter-wave massive MIMO systems," *IEEE Trans. Antennas Propag.*, vol. 69, no. 3, pp. 1393–1403, Mar 2021.

305 Q. Wu, J. Hirokawa, J. Yin, C. Yu, H. Wang, and W. Hong, "Millimeter-wave multibeam endfire dual-circularly polarized antenna array for 5G wireless applications," *IEEE Trans. Antennas Propag.*, vol. 66, no. 9, pp. 4930–4935, Sep 2018.

306 K.-R. Xiang, F.-C. Chen, Q.-X. Chu, and M. J. Lancaster, "A broadband 3 × 4 Butler matrix and its application in multibeam antenna arrays," *IEEE Trans. Antennas Propag.*, vol. 67, no. 12, pp. 7622–7627, Dec 2019.

307 Y. Cao, K.-S. Chin, W. Che, W. Yang, and E. S. Li, "A compact 38 GHz multibeam antenna array with multifolded Butler matrix for 5G applications," *IEEE Antennas Wireless Propag. Lett.*, vol. 16, pp. 2996–2999, 2017.

308 C. Liu, S. Xiao, Y. Guo, M. Tang, Y. Bai, and B. Wang, "Circularly polarized beam-steering antenna array with Butler matrix network," *IEEE Antennas Wireless Propag. Lett.*, vol. 10, pp. 1278–1281, 2011.

309 Y. Zhang and Y. Li, "A dimension-reduction multibeam antenna scheme with dual integrated Butler matrix networks for low-complex massive MIMO systems," *IEEE Antennas Wireless Propag. Lett.*, vol. 19, no. 11, pp. 1938–1942, Nov 2020.

310 F. Y. Xia, Y. J. Cheng, Y. F. Wu, and Y. Fan, "V-band wideband circularly polarized endfire multibeam antenna with wide beam coverage," *IEEE Antennas Wireless Propag. Lett.*, vol. 18, no. 8, pp. 1616–1620, Aug 2019.

311 M. Ansari, H. Zhu, N. Shariati, and Y. J. Guo, "Compact planar beamforming array with end-fire radiating elements for 5G applications," *IEEE Trans. Antennas Propag.*, vol. 67, no. 11, pp. 6859–6869, Nov 2019.

312 B. Cetinoneri, Y. A. Atesal, and G. M. Rebeiz, "An 8 × 8 Butler matrix in 0.13-μm CMOS for 5-6-GHz multibeam applications," *IEEE Trans. Microw. Theory Tech.*, vol. 59, no. 2, pp. 295–301, Feb 2011.

313 Y. Li and K.-M. Luk, "A multibeam end-fire magnetoelectric dipole antenna array for millimeter-wave applications," *IEEE Trans. Antennas Propag.*, vol. 64, no. 7, pp. 2894–2904, Jul 2016.

314 W. Lin and R. W. Ziolkowski, "Theoretical analysis of beam-steerable, broadside-radiating Huygens dipole antenna arrays and experimental verification of an ultrathin prototype for wirelessly powered IoT applications," *IEEE Open J. Antennas Propag.*, vol. 2, pp. 954–967, 2021.

315 A. D. Yaghjian, "Increasing the supergain of electrically small antennas using metamaterials," in *2009 3rd European Conference on Antennas and Propagation*, pp. 858–860, Berlin, Germany, 23–27 March 2009. IEEE.

11

Overview of Rydberg Atom-Based Sensors/Receivers for the Measurement of Electric Fields, Power, Voltage, and Modulated Signals

Christopher L. Holloway, Matthew T. Simons, Alexandra B. Artusio-Glimpse, and Joshua A. Gordon

U.S. Department of Commerce, Boulder Laboratories, National Institute of Standards and Technology (NIST), Boulder, CO, USA

11.1 Introduction

Guglielmo Marconi performed the first wireless radio transmission measurements in 1895 when he sent and received the first radio signals in Italy. His efforts culminated in 1902 with the first transatlantic transmission. This work was due in part to the efforts of Nikola Tesla and others. While Marconi is widely credited for these first experiments, some authors also give credit to others [1–3].

Prior to radio, we can look even earlier to give tribute to the "truly" first propagation experiments. In 1887, Heinrich R. Hertz was the first to perform experiments to confirm the hypothesis laid out by Maxwell's equations [4]; that is, the equivalence of light and electromagnetic propagation. The research of Maxwell and Hertz is the single most influential work that started the field of electromagnetic theory, leading to today's numerous applications of wireless electromagnetic wave propagation ranging from biomedical imaging to the telecommunication revolution.

Since these first radio wave propagation experiments, individuals have needed to predict and measure an electric (E) field. As such, a relentless pursuit to acquire accurate and direct measurements of E-field began. At the core of all electromagnetic measurements are calibrated probes and antennas. The conventional probe or antenna relies on currents flowing (via conducting electrons) on some type of metallic structure, see Figure 11.1a. To make accurate measurements, a calibration of the probe or antenna is required. Calibrating E-field probes requires a complex and indirect traceability path and presents a chicken-or-egg dilemma. To calibrate a probe, one must place the probe (sensor) in a "known" field. However, to know the field we need a calibrated probe.

One of the keys to developing new science and technologies is to have sound metrological tools and techniques. Whenever possible, we would like these metrological techniques to make absolute measurements of the physical quantity of interest. A stated goal of international metrology organizations is to make all measurements directly traceable to the International System of Units (SI) [5–7]. Measurements based on atoms provide one direct SI traceability path and enable absolute measurements of physical quantities. Measurement standards based on atoms have been used for a number of years for a wide range of measurements: most notable are time (s), frequency (Hz), and length (m). There is a need to extend these atom-based techniques to other physical quantities, such as E-fields. E-field measurements based on the atom allow a simple and direct traceability path and eliminate the chicken-and-egg dilemma by separating the calibration of the probe from the calibration of the field.

Antenna and Array Technologies for Future Wireless Ecosystems, First Edition.
Edited by Y. Jay Guo and Richard W. Ziolkowski.

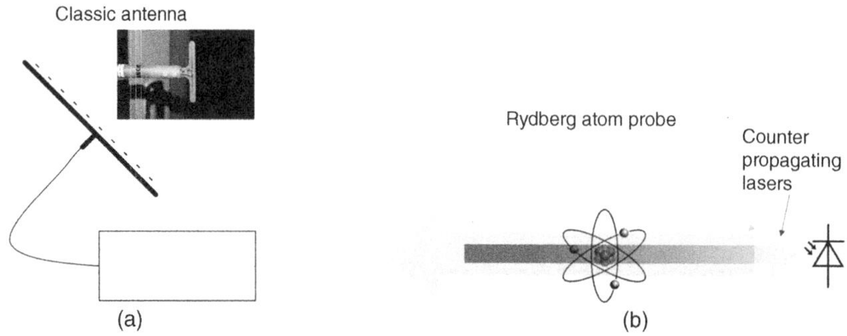

Figure 11.1 Sensor types. (a) Conventional sensors utilize free electrons flowing on metallic structure. (b) Rydberg atom-based sensors utilize atom-bound electrons.

Rydberg atoms provide a pathway to E-field measurements that are traceable to the SI. Rydberg atoms are atoms with one or more electrons excited to a very high principal quantum number n [8]. These atoms have several useful properties that scale as n. In particular, their dipole moments are exceptionally large, scaling with n^2. These large dipole moments make Rydberg atoms sensitive to E-fields and useful for field sensing. In effect, the Rydberg atoms act as the probe and/or antenna as illustrated in Figure 11.1. In this approach, the conventional probe shown in Figure 11.1a is replaced with a glass cell containing Rydberg atoms, and by probing the response of the atoms with light, the incident field strength of a modulated signal can be detected. If alkali atoms are used, the interaction between the Rydberg state and the radio frequency (RF) field can be reliably calculated.

Over the last few years, great progress has been made in the development of Rydberg atom-based RF E-field sensors [9–39]. The Rydberg atom-based sensors have been demonstrated to be capable of measuring amplitude, polarization, and phase of the RF field. As such, various applications are beginning to emerge. These include E-field probes traceable to the SI [11, 12, 15], power-sensors [23], spectrum analyzers [35], angle-of-arrival (AoA) sensors [36], ac and dc voltage measurements [40], receivers for modulated communication signals (AM/FM modulated and digital phase modulation signals) [26–31], and even receivers for recording musical instruments [37]. This new atom-based technology has allowed for many interesting and unforeseen applications. Furthermore, these new Rydberg atom-based sensors will be beneficial for 5G and beyond. In fact, they will allow for traceable calibrations of both field strength and power for frequencies above 100 GHz, which is currently not available. In this chapter we give an overview of this fundamentally new approach for the detection of the E-field and modulated signals.

11.2 Electric-Field Strength: EIT (On Resonant and Stark Shift)

In this section, we give the basic measurement approach. The approach utilizes electromagnetically induced transparency (EIT) [19, 41, 42] to determine the change in the atomic spectra of the Rydberg atoms for an applied E-field, where EIT is used to optically read-out the response of the Rydberg atoms to the applied field. These measurements are performed either when the RF field is on resonance with a Rydberg transition (using Autler–Townes (AT) splitting) or off resonance (using AC Stark shifts). One sensor can measure a field from tens of MHz to above 1 THz. Using standard EIT/AT techniques, E-field strength can be routinely measured down to a few mV/cm [10–12, 14]. With optical homodyne [43] or RF heterodyne (a Rydberg atom-based mixer) [33, 34] techniques with EIT, E-field strengths down to 55–700 nV/cm$\sqrt{\text{Hz}}$ have been measured. At the other extreme, measured field levels of 10 kV/m and higher have been demonstrated. For such strong field level measurements, the AC stark shift approach is used [16–18].

The basic concept uses a vapor of alkali atoms (placed in a glass cell, referred to as a "vapor" cell, see Figure 11.2a) as the active medium for the RF E-field measurement. Rubidium (^{85}Rb) and cesium (^{133}Cs) are the two atomic species that are typically used in the approach. The classic EIT technique involves using two lasers, one laser called the "probe" laser is used to monitor the optical response of the medium in the vapor cell and a second laser called the "coupling" laser is

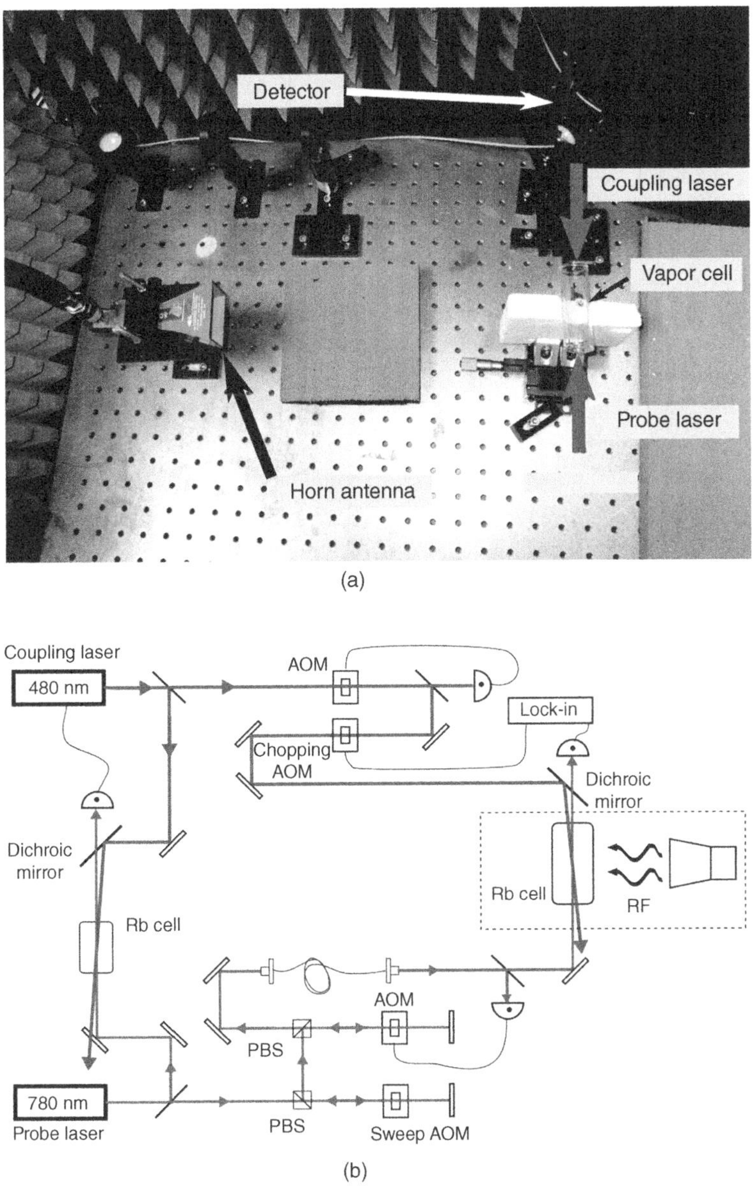

Figure 11.2 Experimental setup for E-field measurements using EIT. (a) Photo of the system. (b) Block diagram of the system indicating its polarizing beam splitter (PBS) and acousto-optic modulator (AOM). Source: From [12] / with permission from IEEE.

used to establish a coherence in the atomic system. When the RF E-field is applied, it alters the susceptibility of the atomic vapor seen by the probe laser as it propagates through the vapor cell. By detecting the probe light propagating through the cell, i.e. by the change in the susceptibility of the atom population, the RF E-field strength can be determined. We note that three lasers can be used, and the concept is the same with slightly increased complexity [44, 45].

To explain how the three fields (probe, coupling, and RF) interact and influence the atoms, we discuss how the measurement system, diagrammed in Figure 11.2, causes light to be transmitted through the vapor cell when it otherwise would be absorbed by the atoms in the cell. The system nominally consists of an RF source, a lock-in amplifier, a probe laser (780 nm laser), a coupling laser (480 nm laser), a photodetector, and a vapor cell. The alkali atoms ^{85}Rb or ^{133}Cs or sometimes both species are in the vapor cell [22, 28]. In Section 11.8, we look at a compact version of this system where the optical beams are brought to the vapor cell through optical fibers, which allows the vapor cell (the E-field probe) to be more mobile and taken off the optical table.

If we turn on only the probe laser, we find that when it is tuned to a ground state transition of the atoms, they absorb the laser light and little power is detected at the photodetector. For ^{85}Rb, this corresponds to a probe laser wavelength of 780.24 nm, its frequency being 384.231 THz. For ^{133}Cs, this corresponds to a probe laser wavelength of 852.35 nm, its frequency being 351.725 THz. As we scan the probe laser wavelength, the power on the detector changes and shows the so-called Doppler background, which corresponds to the absorption line of the alkali atoms. This spectral response is given in the bottom curve in Figure 11.3a. Here, $\Delta p = \omega_o - \omega_p$ is the frequency detuning of the probe laser, where ω_o is the on-resonance angular frequency of the atom ground state transition and ω_p is the angular frequency of the probe laser. The inverted bell-curve shape of the Doppler background is the typical signal obtained when performing atomic spectroscopy experiments, e.g. studies of absorption and emission properties of atoms. When $\omega_p = \omega_o$, the probe laser is on resonance with the atom ground state transition and the transmitted power through the vapor cell is at a minimum.

Next, we turn on the coupling laser. Typically the coupling laser is aligned to be counter-propagating with respect to the probe laser through the vapor cell. Such a configuration minimizes excessive Doppler broadening of the spectral signatures that arises from the large frequency

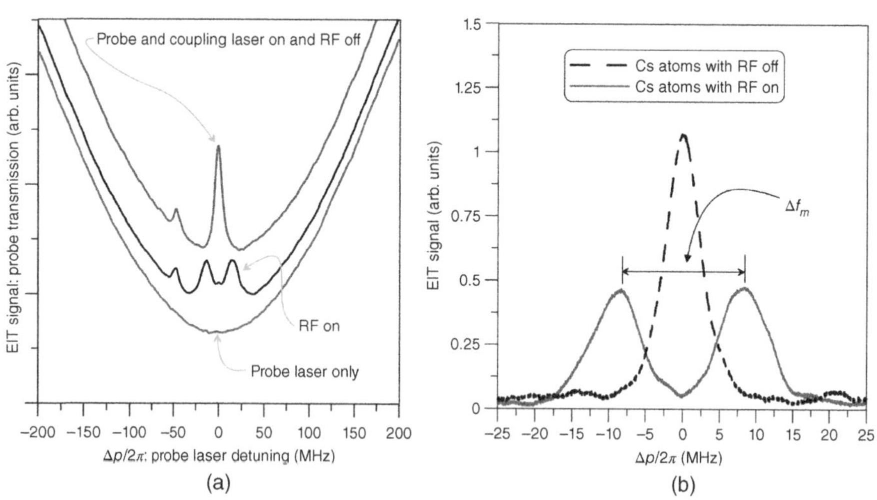

Figure 11.3 EIT signal. (a) With the Doppler background. (b) After the lock-in amplifier is used [12]. Source: From [12] / with permission from IEEE.

difference between the two lasers. The frequency of the coupling laser is tunable according to the measurement requirements, but is typically in the range of 480 nm (624.568 THz) or 510 nm (587.828 THz) for ^{85}Rb or ^{133}Cs, respectively. When the atoms are exposed to the coupling lasers, they are excited to a higher atomic state, i.e. to a Rydberg state. An interference of the two atomic states occurs due to the presence of both the probe and coupling lasers. It induces transparency of the atoms to the probe laser. This effect is known as EIT, where a medium that is normally absorbing becomes transparent in the presence of an electromagnetic field (in this case the coupling laser). The resulting spectrum of this effect is shown in the top curve of Figure 11.3a. Note that the three curves in this figure lay on top each other, but have been arbitrarily shifted them along the vertical axis for ease of viewing. A clear peak in this top curve indicates transmission near $\Delta_p = 0$; it appears as a result of exposing the atoms to the coupling laser. The additional peak to the left of the main EIT peak is related to the atomic structure of the atoms. Many additional peaks may be present for some configurations (see [14] for details).

If the wavelength of the coupling laser is judiciously chosen in order to excite the atoms to a high enough Rydberg state, the RF source will be resonant with a transition to a nearby Rydberg state, i.e. an RF atomic transition. In other words, when the atoms are preconditioned with both the probe and coupling lasers, the next atomic transition can be accessed with an RF source. This RF source, when resonant with the RF atomic transition, causes the EIT transmission line to split in two. This splitting is known as AT splitting, and an example of the effect on the EIT spectral curve is given in the middle (black) curve of Figure 11.3a.

The EIT spectral signal and the AT splitting can at times be weak signatures on the Doppler background (the probe absorption line). To increase the signal-to-noise ratio (SNR) of the EIT, we often modulate the coupling laser amplitude with a 50% duty cycle, 30 kHz square wave and detect any resulting modulation of the probe transmission with a lock-in amplifier. This removes the Doppler background and isolates the EIT signal. The black, dashed-line curve in Figure 11.3b shows a typical EIT signal from the lock-in amplifier. When the RF field is turned on, this EIT peak splits and the splitting frequency is indicated by Δf_m.

Frequency splitting of the laser transmission spectrum through the alkali atoms is easily measured and is directly proportional to the applied RF E-field amplitude. The optical frequency difference between the two new peaks, Δf_m, is directly related to the RF E-field strength through the following expression [10, 11]

$$|E| = 2\pi \frac{\hbar}{\wp} \frac{\lambda_p}{\lambda_c} \Delta f_m = 2\pi \frac{\hbar}{\wp} \Delta f_o \tag{11.1}$$

where $\hbar$ is Planck's constant, $\wp$ is the atomic dipole moment of the RF atomic transition (see the following), $\Delta f_o = \frac{\lambda_p}{\lambda_c} \Delta f_m$, and λ_p and λ_c are the wavelengths of the probe and coupling laser, respectively. The λ_p/λ_c ratio is needed to account for the Doppler mismatch of the probe and coupling lasers [19]. The coupling laser can also be scanned as opposed to the probe laser. In this case, the wavelength ratio in Eq. (11.1) is set to 1.0 (see Eq. (1) in [28]).

Changes in the power or strength of the RF E-field result in linear changes in the measured splitting frequency difference Δf_m. Thus, it is only necessary to know the dipole moment of the RF atomic transition, $\wp$, and the wavelengths of the two optical fields along with the universal constant, Planck's constant (which is known by definition [5–7]) in order to obtain the E-field strength of the RF field at the atoms. It should be emphasized that this simple measurement only detects the magnitude of the RF E-field. Section 11.5 will discuss extensions to the measurement scheme that allow for detection of the RF E-field phase in addition to the magnitude. Because the atoms only respond to the RF field where the probe and coupling lasers overlap, the spatial resolution of this field measurement can be quite small. This feature opens up the opportunity to perform imaging (even sub-wavelength imaging [14, 46]) or beam profiling measurements with arrays of such atom probes. These applications will also be discussed in Section 11.8.

This RF field measurement with the Rydberg atoms is a direct, SI-traceable measurement where the only unknown is the dipole moment, $\wp$, which can be calculated accurately to 0.1% [11, 22, 47]. Figure 11.4 shows the results of such a calculation for a particular RF atomic transition for both rubidium (^{85}Rb) and cesium (^{133}Cs) atoms. From these results we see that ^{85}Rb and ^{133}Cs have similar values for the radial dipole moment $\mathcal{R}_{RF}$ or the normalized dipole moment Q_{RF}, i.e. $\mathcal{R}_{RF}/a_0$, where $a_0 = 0.529\ 177 \times 10^{-10}$ m is the Bohr radius. These large values of $\mathcal{R}_{RF}$ for these Rydberg states is what makes this technique beneficial for measuring RF E-fields. While either of these two atoms can be used, ^{85}Rb nevertheless has some advantages when performing measurements at the high RF end of the spectrum. On the other hand, ^{133}Cs has some advantages at the low RF end of the spectrum. The other advantage of ^{133}Cs that may be useful in some situations is that ^{133}Cs has a higher vapor pressure than ^{85}Rb [48, 49]. Both ^{133}Cs and ^{85}Rb were used in the various experiments presented in this chapter.

Figure 11.5 shows a comparison of measurements of the E-field strength with the Rydberg atom sensor to results obtained from calculations based on Maxwell's equations (based on Eq. (3) in [38]). These results illustrate the validity of the Rydberg atom-based technique. They also illustrate that a single probe has the ability to measure fields ranging from hundreds of megahertz to the lower terahertz spectrum.

The broadband nature of this technique is due to the large number of possible Rydberg states that can exhibit a large response to an RF source. These states form a discrete set of frequencies that can be measured. The RF source must be tuned to the precise frequency (within a given bandwidth) that corresponds to an atomic transition. The wavelength of the coupling laser selects the transition that will respond to the RF source, which defines the frequency that will be measured. While there are a large number of possible atomic states with RF transition frequencies, several of these have

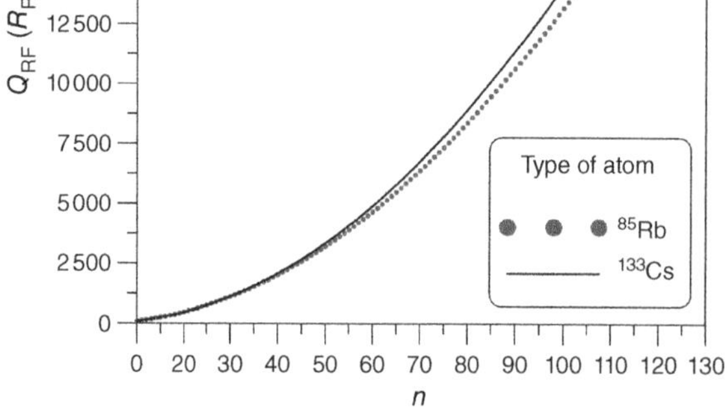

Figure 11.4 Calculations of Q_{RF} ($\mathcal{R}_{RF}/a_0$) for a particular RF atomic transition for ^{85}Rb and ^{133}Cs. The inset shows the atomic levels for ^{85}Rb and ^{133}Cs, where n is the principal quantum number and "D" and "P" indicate the angular momentum state of the atoms. Source: From [12] / with permission from IEEE.

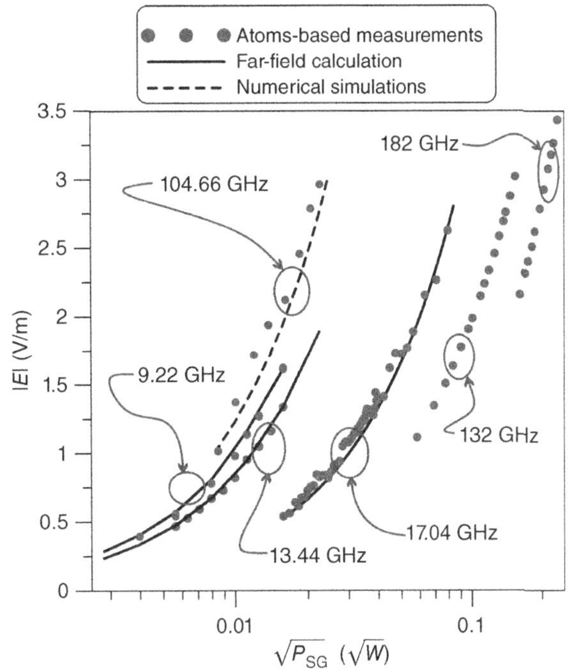

Figure 11.5 Comparison of the results obtained with Rydberg atom-based techniques to those calculated with Maxwell's equations. Source: From [50] / with permission from IEEE.

small atomic dipole moments. Since the measurement splitting (Δf_m) is directly proportional to the atomic dipole moment, as shown in Eq. (11.1), we want to use RF transitions with large dipole moments. One typically uses the four RF transitions corresponding to $nD_{5/2} - (n+1)P_{3/2}$, $nD_{5/2} - (n-1)F_{5/2}$, $nS_{1/2} - nP_{3/2}$, and $nS_{1/2} - (n-1)P_{3/2}$, where "*S*," "*P*," "*D*," and "*F*" indicate the angular momentum state of the atoms [51].

We note that a scheme has recently been developed that allows for continuous frequency detection of weak RF fields [52]. This technique allows for the detection of an RF field signal over a frequency range that covers from one Rydberg state with principal quantum number n to the next $n+1$ state. Such an approach allows for the development of a spectrum analyzer and waveform detector. Continuous frequency detection of strong field strengths can also be done with Stark shifts approaches [53].

While the on-resonant EIT/AT scheme discussion earlier has been very successful in detecting field levels in the range of 5 μV/m [10–12, 33, 34] to tens of V/m, the EIT/AT poses two problems. First, once the field strength becomes too large, AT peaks become difficult to resolve (e.g. see the +6 dBm case in Figure 18 in [11]). Secondly, the EIT/AT scheme requires the RF signal of interest to be resonant with the Rydberg transition as the scheme only detects frequencies within the narrow bandwidth of a Rydberg transition. For strong and off-resonance fields, AC Stark shifts [16–18] or, more generally, a Floquet method [18] are used. A strong RF field – either on or off resonance – can cause the EIT signal to shift in frequency, e.g. see Figure 11.6. The shift is related to the square of the magnitude of the applied RF field [54] and, hence, provides a means to detect strong field strengths.

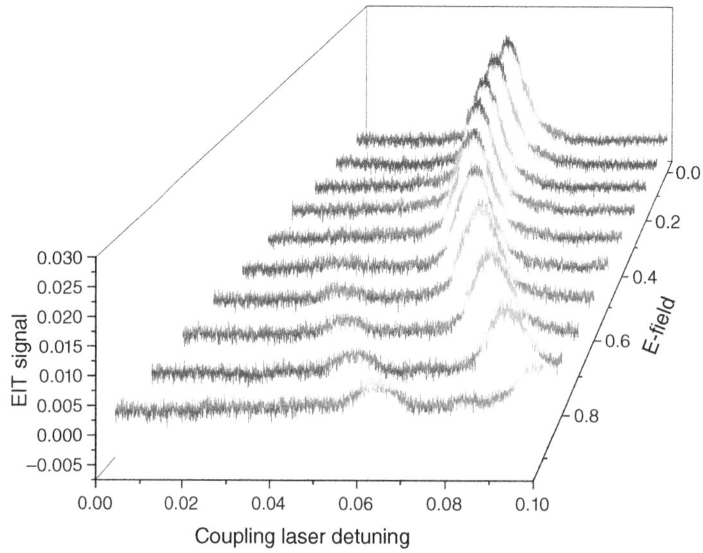

Figure 11.6 AC Stark shift due to a high intensity RF field.

11.3 Uncertainties

As introduced earlier, this method for RF E-field measurement provides a potential for absolute, SI-traceable calibrations. In order for these measurements to serve SI-traceable calibrations, it is important to understand the sources of uncertainty. Initial investigation into the uncertainty contributions have been done in [10, 20], and [55]. These studies found the most significant sources of uncertainty to be laser intensity noise, electrical detection noise, and the effect of the vapor cell on the microwave field. Laser and electronic noise can be minimized through various schemes. The vapor cell, typically constructed from dielectric glass, has a relatively low permittivity in the microwave range. However, the cell can still cause standing waves to arise inside where the atoms are probed. This causes a difference between the field seen by the atoms and the field outside the vapor cell, dependent on the microwave frequency, geometry of the vapor cell, and the location of the measurement inside the cell [56]. This discrepancy can be large. This can be calibrated out by measuring the whole standing wave pattern inside the cell [14, 55, 56].

While the cell can perturb the E-field measurement, vapor cell designs can be pursued to minimize and even eliminate the vapor cell perturbations. Various groups are investigating different approaches to modify the vapor cell used for these Rydberg atom-based sensors. Two examples include the use of vapor cells with honeycomb sides [57] or the use of metamaterials on the sides of the vapor cells [58].

11.4 Detection of AM and FM Modulated Signals

The Rydberg EIT system can respond to time-varying RF signals. This includes both amplitude modulated (AM) and frequency modulated (FM) signals. Modulating either the amplitude or frequency of a resonant RF field (or "carrier") causes the AT peak separation to vary in time (see [28] for details). The Rabi frequency is defined as:

$$\Omega = \frac{\wp}{\hbar}\,|E| \tag{11.2}$$

If the modulation amplitude effective Rabi frequency is large compared with the EIT or AT linewidth, the peak location in laser frequency detuning will vary proportionally to the modulation. If the modulation amplitude effective Rabi frequency is small compared with the EIT or AT linewidth, this AT peak shift appears as a variation in the probe transmission for fixed laser detunings.

For the case where the mean RF field amplitude Rabi frequency is less than the EIT linewidth, it is important to note the behavior of the EIT line. Below this threshold for AT splitting, the EIT line amplitude reduces with increased RF field amplitude. In this regime, an AM RF signal will cause a direct modulation of the EIT amplitude.

Figure 11.7a shows a simulation of the AT splitting for three different RF field amplitudes. By setting the laser detuning to zero, the probe transmission directly follows the amplitude modulation. Therefore, no demodulation circuitry is needed to retrieve the modulated signal from the carrier. The transmission value is highlighted by the circle for each case.

For higher field strengths, the AM can still be detected, as shown in Figure 11.7b. The transmission value at the laser frequency is again highlighted by the dots. In this case the laser must be detuned and locked near one of the AT peaks, which no longer is near $\Delta = 0$. As the field amplitude increases and decreases, the peak location shifts. This shift again causes the probe transmission to follow the baseband modulation.

Frequency modulation creates a very similar result [28] for relatively weak fields. The laser detunings can then be set to zero, and the probe transmission will directly follow the frequency modulation. For stronger fields, the lasers must again be detuned to find the shifted AT peak. The probe transmission will then continue to the follow the frequency modulation. This behavior is illustrated in Figure 11.7c. It effectively converts an RF FM signal into an optical AM signal directly, without demodulation electronics. More discussion on how the atoms actually perform the demodulation is given in [28].

Various groups have demonstrated AM/FM reception using Rydberg-atom sensors [26–31]. Figure 11.8 shows a comparison of an original waveform to the received waveform from an AM and a FM modulated signal. Figure 11.9 shows the results for a received digital AM waveform. In this figure, we also illustrate the effect of noise on the reception. Groups are currently investigating the effect of noise on these types of Rydberg atom-based receivers and sensors in order to understand if the atom-based systems have better noise rejection. For example, [25] investigates how white Gaussian noise effects these types of measurements, where it is shown that detection is possible even for very low SNRs.

11.5 Phase Detection and Phase Modulated Signal Detection

The Rydberg EIT signal is not directly sensitive to the phase of an RF field. The response rates that we have observed are limited to tens of MHz. They are sensitive to changes in the field amplitude due to constructive or destructive interference from the addition of other resonant RF fields. This sensitivity has been used to map out standing waves inside of the dielectric vapor cells [12, 14, 56], as well as inside of metallic waveguides (Section 11.6). In this section, we show how Rydberg EIT can detect the phase of an unknown RF signal by applying a second, well-characterized RF field of a similar frequency.

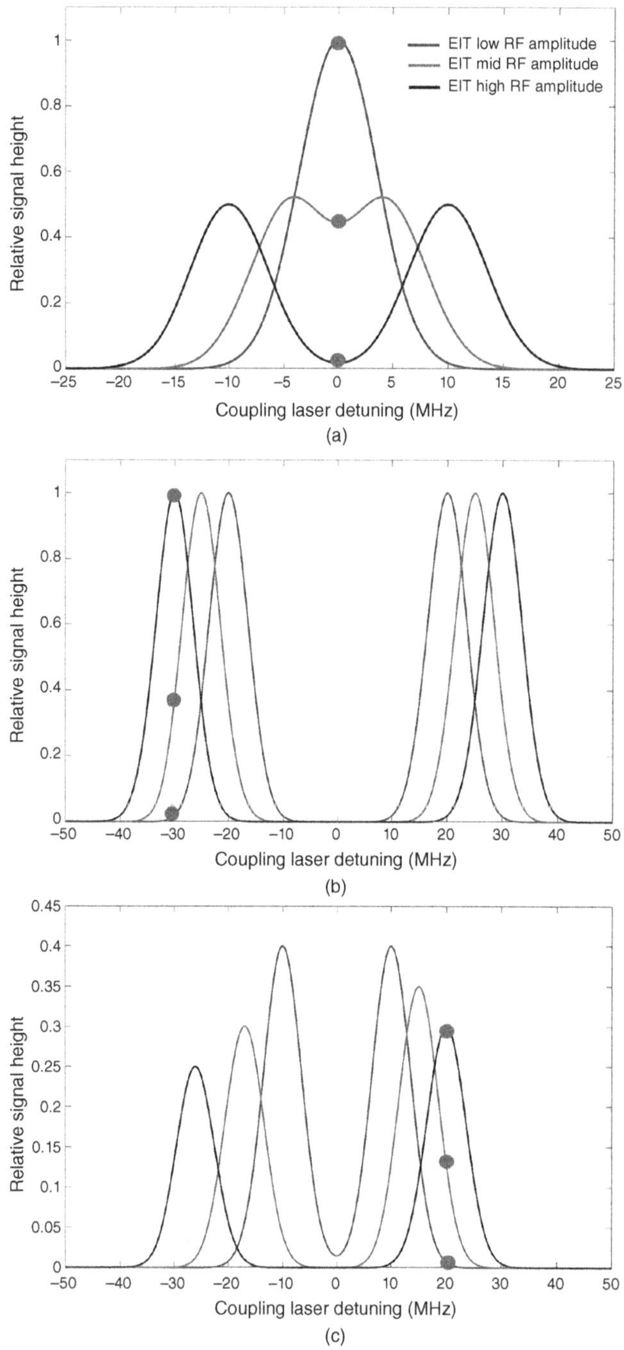

Figure 11.7 Simulations of AT spectra for different RF field amplitudes, demonstrating the effect of a modulated RF signal on the probe field transmission. The probe transmission is shown relative to a maximum EIT signal. The dot illustrates the probe transmission for fixed probe and coupling laser frequencies where there is significant modulation. Each trace is for a different RF field strength at different times within the modulation period. Three interesting cases are: (a) Amplitude Modulated RF signal for a relatively weak RF field, where maximum modulation is near zero laser detuning; (b) strong RF field carrier with an AM signal, where the lasers are detuned to an AT peak; (c) a Frequency Modulated RF signal, where the detection results in an amplitude modulated probe transmission.

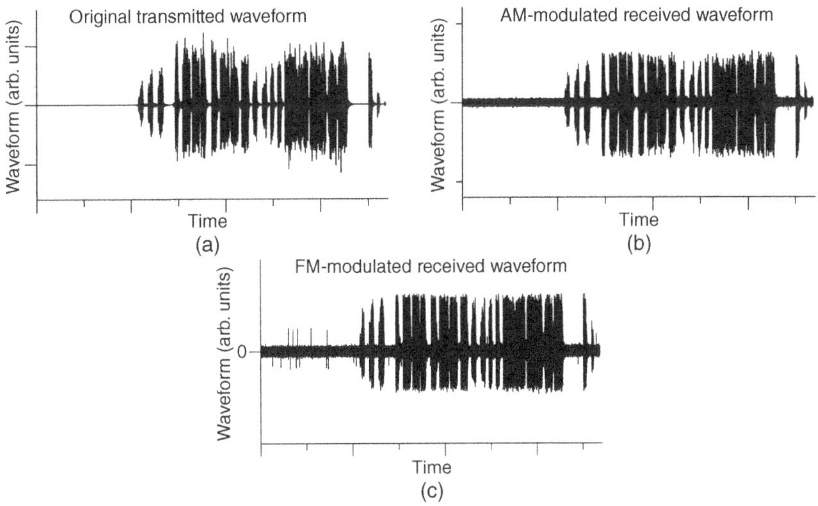

Figure 11.8 Waveforms associated with the atom-based stereo receiver when the signal is either amplitude or frequency modulated. (a) Transmitted waveform. (b) Received waveform from the AM modulation scheme. (c) Received waveform from the FM modulation scheme.

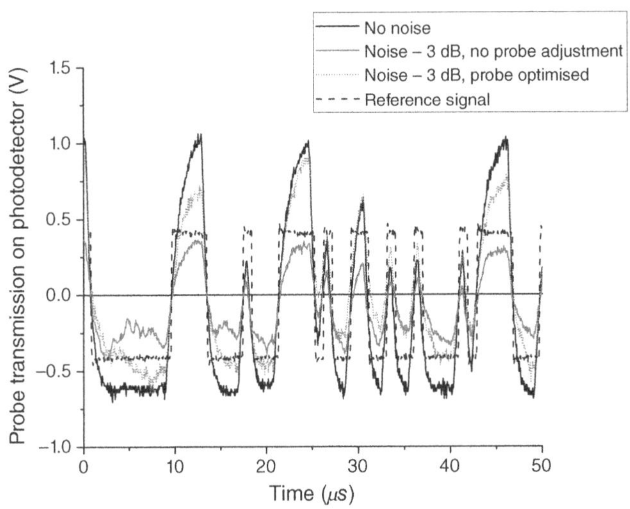

Figure 11.9 Examples of a digital AM waveform received by the atoms under different measurement conditions.

Details on how phase detection is achieved are given in [32, 39]. A single RF field near resonance, denoted as local oscillator or LO field, will cause a Rydberg peak to reduce in amplitude. It will then undergo AT splitting (as discussed in Section 11.2) as the LO field strength is increased. Fixing the probe laser frequency to the center of the EIT signal, the output will appear as a decrease from a maximum value (with zero LO field strength) to zero (corresponding to an LO field strength where the AT peaks are well-separated and beyond). Consider the case where the LO field strength is fixed such that the EIT signal is midway between its maximum and minimum values. Any constructive or destructive interference from another source will cause the total field strength to change, and thus cause the EIT signal to rise or fall. If a second RF signal (defined as

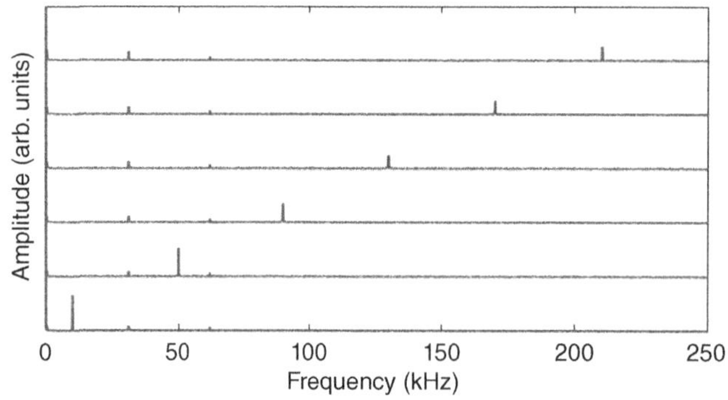

Figure 11.10 Example of the transmitted probe spectrum for various difference frequencies between the SIG and LO fields.

SIG) field is applied with the same frequency as the LO, the two field strengths sum according to their phase difference. In the case where the amplitudes are equal, the EIT signal will reduce to zero if they are in phase. On the other hand, if they are perfectly out of phase, the EIT signal increases to its maximum value as if there was no RF field present. This results in a phase sensitive homodyne detection of the RF SIG field [32, 39].

If a SIG field is applied that is detuned in frequency from the LO, the EIT signal will oscillate in time at the difference frequency (or intermediate frequency [IF]) between the SIG and LO. In this atom mixer or superheterodyne configuration, the phase of the IF signal is directly related to the difference between the SIG and LO phases. Figure 11.10 shows examples of the spectrum of the transmitted probe laser for various IF (Δf) frequencies. The power spectrum of the transmitted probe field shows the IF signals present; their amplitudes can then be directly related to the SIG field strength [32, 34].

The heterodyne technique allows for a much more sensitive detection of the RF signals. The AT splitting method is limited to measuring RF field amplitudes equivalent to Rabi frequencies close to the EIT linewidth [20], and the EIT amplitude-only approach is limited to 80 μV/m [10]. In contrast, the atom mixer approach can provide sensitivities down to 55–700 nV/cm$\sqrt{\text{Hz}}$ [33, 34]. As a low-frequency signal is created in the presence of the RF signal of interest, this signal can be detected through lock-in amplification of the photodetector signal. This allows for sub-Hz frequency selection [33]. Heterodyne reception can also be observed through the AC Stark effect with higher RF field strengths. Such a configuration was used to create an atom-based spectrum analyzer from DC - 20 GHz [35].

With phase-sensitive detection, the Rydberg atom sensor can now be used to receive traditional phase modulated (PM) signals. While AM and FM have been used for simple analog transmission, most digital communications schemes rely on phase modulation. Typical schemes include binary phase-shift keying (BPSK), quadrature phase-shift keying (QPSK), and quadrature and amplitude modulation (QAM). Rydberg atoms have been used to down-convert the carrier to the baseband modulation for these types of signals, where the resulting photodetector signal was sent to a vector signal analyzer for demodulation [30]. Figure 11.11 shows the IQ-diagram for QPSK and 16-QAM received signals obtained from the Rydberg atom-based receiver [30]. The atom-based mixer has also been used to measure the propagation constant of free space [32, 39] and for measuring the phase of a microwave device [39, 59].

While applying a second RF field (the LO) to create an atom-based mixer has been shown to be successful, this approach does have some disadvantages. Foremost is the need to apply a second

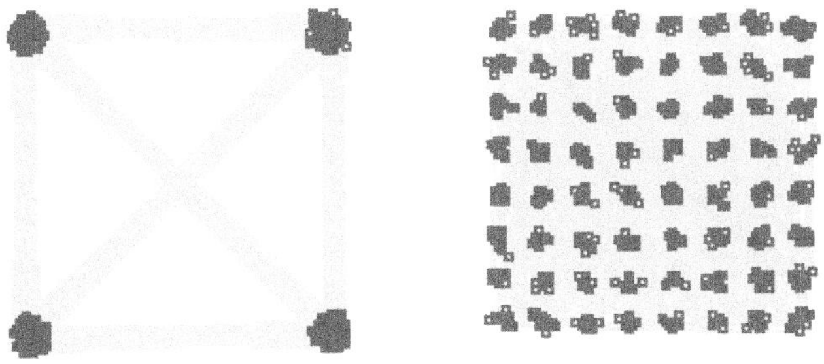

Figure 11.11 Example of the IQ-diagram from received phase-modulation signals: (a) QPSK, and (b) 64-QAM. Source: From [30] / with permission from IEEE.

RF field. To overcome this, an all-optical phase detection approach is desired. Various all-optical approaches are currently being investigated and one approach is given in [60].

11.6 RF Power Measurements

In addition to RF E-field strength measurements, RF power measurements and calibrations are very important in all electromagnetic systems, including radar and advanced communications applications. A Rydberg atom-based sensor placed in rectangular waveguide can be used for this purpose [23], see Figure 11.12. The power carried by the fundamental mode of a rectangular wave guide is given by

$$P = E_0^2 \, \frac{ab}{4} \sqrt{\frac{\epsilon_o}{\mu_o}} \sqrt{1 - \left(\frac{c}{2af}\right)^2} \tag{11.3}$$

where E_0 is the amplitude of the E-field at the center of the waveguide and a and b are the cross-sectional dimensions of the rectangular waveguide where a is the larger dimension as indicated in the inset in Figure 11.12. The frequency is f; ϵ_0 and μ_0 are the permittivity and

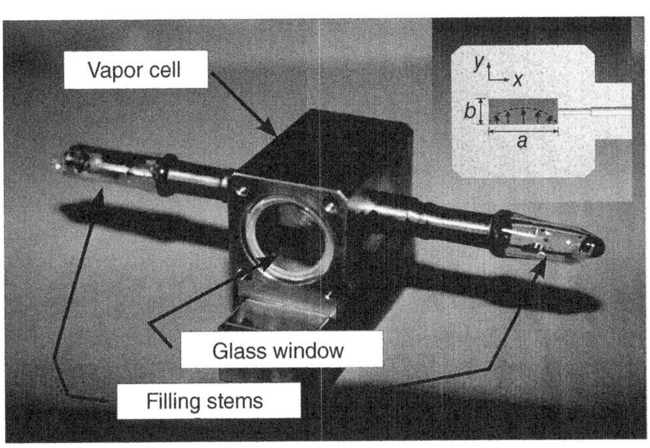

Figure 11.12 WR-42 rectangular waveguide vapor cell with waveguide dimensions. The vapor cell consists of a 34-mm section of waveguide with glass windows attached to each end that is filled with the atomic vapor. Source: From [23] / with permission from AIP Publishing LLC.

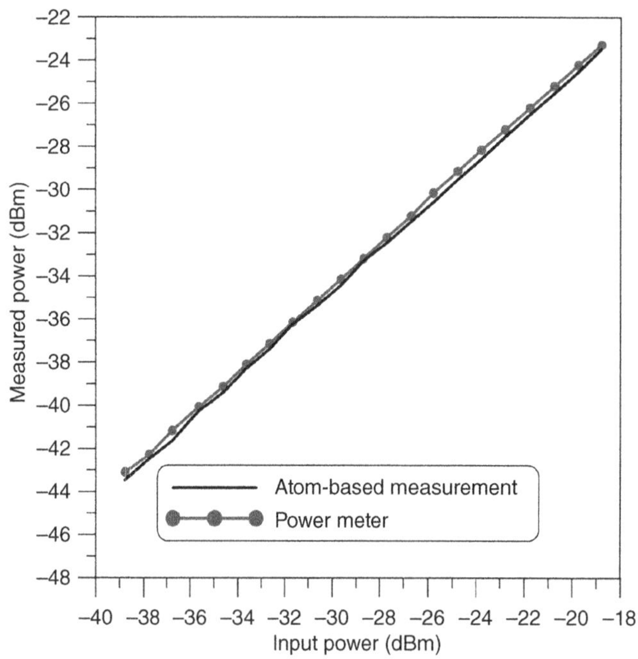

Figure 11.13 Measurements of the power in the waveguide versus the input power at 19.629 GHz. Source: From [23] / with permission from AIP Publishing LLC.

permeability of free space; and c is the speed of light in *vacuo*. As discussed in [23], the Rydberg atoms are used to measure E_0 for a given input power. The power propagating through the waveguide is determined with Eq. (11.3).

Figure 11.13 shows a comparison of the power measured with the Rydberg atom-based approach to that obtained with a conventional power meter. The comparison shows very good correlation between the two methods. When compared with conventional power metrology approaches, the Rydberg atom-based approach: (i) is a more direct SI-traceable approach, (ii) has the possibility of having much lower uncertainty, (iii) exhibits a large frequency range, and (iv) has much better dynamic range (i.e. power-level ranges). Furthermore, the Rydberg atom-based approach can lead to "real-time" in situ calibrated sources.

11.7 Voltage Measurements

EIT and Stark shifts in Rydberg atoms can be used for the measurement of direct current (dc) and 60 Hz alternating current (ac) voltages [40]. This section describes the atom-based measurements of the dc and ac voltages with an emphasis on how they may lead to the potential calibration of voltage instrumentation. A Rydberg atom-based dc (or 60 Hz) voltage measurement is performed by observing the dc (quasi-dc) differential Stark shift in the Rydberg states, which is given by the expression [54]:

$$\Delta = \frac{\alpha}{2} E^2 \tag{11.4}$$

where Δ is the Stark shift (in units of Hz), E is the applied E-field (in units of V/m) at the frequency f (in units of Hz), and α is the polarizability of the atom (in units of Hz/(V^2/m^2)).

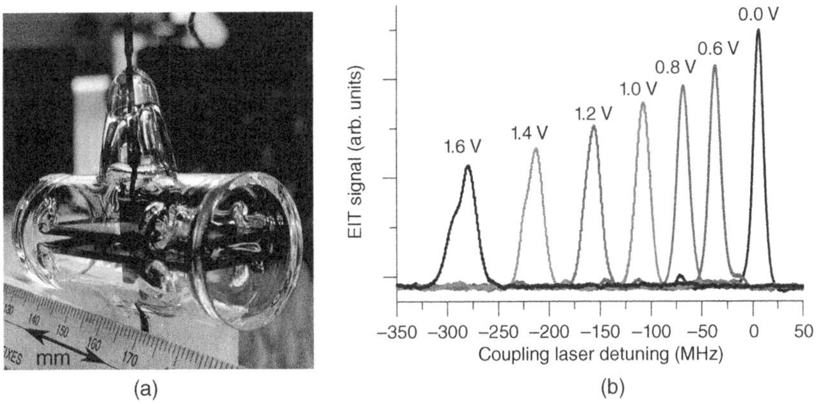

Figure 11.14 Rydberg-atom based voltage measurement. (a) Cylindrical vapor cell with stainless-steal parallel plates. The vapor cell is 50 mm in length and has an outside diameter of 25 mm. The plates are rectangular in shape with width of 18 mm, length of 45 mm, and separation of $\approx$ 2mm. (b) dc Stark shift due to an applied voltage across the parallel plates in the cell shown in (a). Source: (a) From [40] / arXivLabs / CC BY 4.0.

We measure the voltage induced between two parallel plates embedded in an atomic vapor cell as shown in Figure 11.14a by measuring the Stark shifts in the atomic spectra of the Rydberg atoms. If a voltage is applied across two parallel plates separated by the distance d, then the E-field between the plates is $E = V/d$. By substituting this result into Eq. (11.4), the voltage can be found by a measurement of the Stark shift as:

$$V = d\sqrt{\frac{2\,\Delta}{\alpha}} \tag{11.5}$$

Figure 11.14b illustrates the Stark shift of the EIT signal as a function of an applied voltage. By measuring the Δ and using Eq. (11.5), the applied voltage can be determined. These types of Rydberg atom-based voltage measurements are currently being investigated as an alternative for Josephson-junction based standards and Zener-diode voltage references [40].

11.8 Other Applications: RF Camera, Angle-of-Arrival, Waveform Analyzer, Plasma Measurements, and Thermometry

The atom-probe systems that we have described up to this point have been bound to an optical table due to the alignment requirements of the two (or many) lasers within the atomic vapor cell. Such constraints, though, are not absolute. The optical fields needed for the atom RF probe measurements can be transferred to the vapor cell through well-aligned optical fibers. Even the returning probe laser, which carries the susceptibility information linked to the RF field, is passed back through the optical fibers. This separates the laser generation and detection from the vapor cell where the measurement occurs, creating a movable RF probe head consists of only the dielectric vapor cell with bonded optical fibers. A photograph of such a sensor head is given in Figure 11.15.

Fiber connections afford greater mobility. Hence, they expand the range of E-field measurements and calibrations that can be completed with the atom probes because a probe can be brought to the RF source or into an environment of interest. In addition, the all-dielectric probe head is especially valuable as it minimally scatters the RF field(s). Traditional, metallic probes and antennas have a

Figure 11.15 Photograph of a fiber-coupled atom probe. Source: From [12] / with permission from IEEE.

relatively large coefficient of reflection. A significant fraction of the incident RF field will scatter off the probe and may return, after subsequent scattering events, back to the probe. Such scattering off the probe is problematic for accurate measurements of phase, AoA, and channel sounding measurements. The all-dielectric fiber-coupled (FC) atom probe, on the other hand, has a significantly lower coefficient of reflection at radio frequencies and offers an attractive alternative to traditional metallic probes, in addition to the SI-traceability and absolute accuracy benefits discussed in Section 11.2 of this chapter.

With the movable, all-dielectric FC atom probe, researchers can measure spatial variations of a static RF field by translating the probe or moving a horn antenna with respect to the probe. For example, a horn gain pattern was measured [24], showing the spatial distribution of the power emitted by the horn. The measurements confirmed that the fiber probe resolves differences in the side lobe level of the antenna pattern between the H plane and E plane of propagation from the horn and showed good agreement with measurements made with a traditional device. Furthering the demonstration, the authors traced out the field pattern above a co-planar waveguide and in a transverse electromagnetic (TEM) test cell [50]. Measuring across the waveguide axis, they showed an M-shaped field distribution in the E-field strength that decays as the probe is moved away from the co-planar waveguide.

By placing the FC probe on a translation stage, it is possible to map out the relative change in magnitude and phase of an RF signal using the superheterodyne method described in Section 11.5. For example, horns for both the SIG (19.626 GHz) and LO RF fields are placed inside an anechoic chamber with the FC probe and a two dimensional (2D) translation stage. A system diagram is given in Figure 11.16a. At each (X, Y) position of the stage, a new beat note at the IF of 1 kHz is collected and the magnitude and relative phase of that beat note is extracted. The 2D maps of the magnitude and phase in Figure 11.16b,c give the relative change in each measurement either over the spatial area with respect to the measurement at $(X = 0\ \text{mm},\ Y = 0\ \text{mm})$ or on-axis with the SIG horn.

This type of spatial acquisition of the RF field magnitude and phase through scanning can be used for channel sounding measurements. Channel sounding is a technique of evaluating the RF environment that a wireless signal interacts with between a transmitter and a receiver. Due to obstacles and scatterers in the environment, the wireless signal can follow multiple propagation paths. To optimize the signal at the receiver, engineers can use beamforming techniques and adaptive networks that overcome environmental losses experienced by the wireless

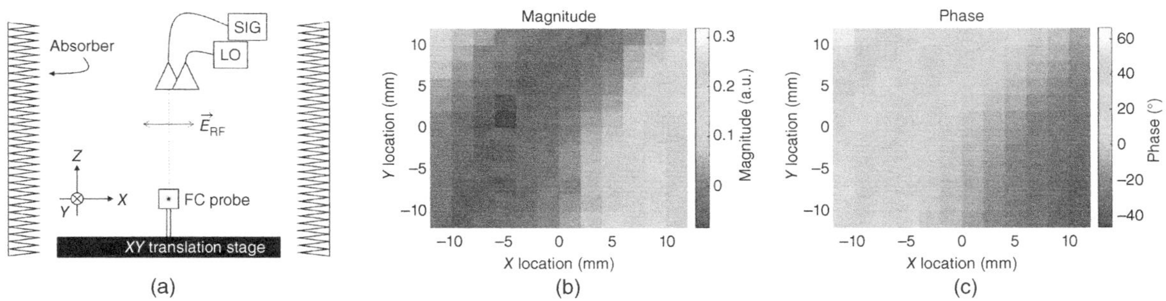

Figure 11.16 All-dielectric fiber-coupled (FC) Rydberg atom probe is placed inside an anechoic chamber with the RF signal (SIG, 19.626 GHz) and local oscillator (LO, 19.626 GHz + 1 kHz) horns. (a) Translating the probe in the *XY* plane results in measured spatial distributions of the (b) relative RF field magnitude and (c) phase.

signal. However, a well-characterized environment, or channel, is required to design such signal improvement techniques. The all-dielectric, FC atom probe with its minimized perturbations to the RF fields can be used to measure and characterize such a channel. We can demonstrate this by returning to the anechoic chamber example in Figure 11.16, but this time with a scatterer placed in the $+X$ direction to modify the relative magnitude and phase measurements. Differences between the measurements without the scatterer and with the scatter, shown in Figure 11.17, confirm that the probe does resolve the subtle changes to the RF field in the plane of the translation stage when the scatterer is added to the environment.

One drawback of the single-point FC probe, like that photographed in Figure 11.15, is that the probe must be translated in order to obtain spatially variant information from a scene. This means that the RF fields cannot change over the time it takes to complete a full 2D scan. An alternative to this technique is to build an atom-based sensor that consists of an array of probes.

The active area of the Rydberg atom probe is defined by the overlapping region of the probe and coupling lasers. This small, well defined area of the RF field sensitivity can, therefore, be leveraged to build RF field cameras. The design of such a camera is a simple extension to the FC probe discussed earlier. Within the vapor cell, the propagation paths of the two lasers are defined in such a manner that the vapor cell area is segmented into pixels. Each pixel is then addressed by either modulating or switching the lasers in each row and column. Optical fibers are again used to transfer the laser fields to the vapor cell. A photonic integrated circuit may be used to streamline the transfer of the laser fields to each pixel within a compact overall package. To date, such an RF field camera has yet to be demonstrated. Nevertheless, our group is currently developing such architectures for an RF camera and RF beam profiler. Rydberg atom probe arrays of this kind will have many benefits over the single-point probes currently in development. For one, acquisition of 2D field information can be captured within the time needed to read out each pixel of the array. If each pixel has its own detector for reading the transmitted probe power, then the read out time could be as fast as the integration time needed to obtain an acceptable SNR at a single pixel. RF E-field beam profiling and fast acquisition of channel sounding measurements would be possible with these arrays. Furthermore, the entire array can be constructed from purely dielectric materials. This removes the translation stage from the acquisition process, which can otherwise contribute to undesirable scattering of the RF field.

The ability to measure AoA is of great importance to radar and advanced communications applications. Measuring the phase of an RF field in different locations in space affords a determination AoA of an incident RF field. Recent work has shown that this can be accomplished by determining the phase of an incident RF field in various locations inside an atomic vapor cell [36]. Figure 11.18 shows a comparison of the measured AoA of an incident RF field to the actual incidence angle. These results show that compact atom-based sensors can be used for determining the AoA. One can envision (i) developing arrays of these Rydberg atom sensors or (ii) sampling the phase at numerous locations inside one vapor cell in order to detect the AoA from more general incidence angles. This measurement may also be used for the purpose of simultaneously detecting the AoA of several sources at once. Other work has shown that by sampling the phase of an incident RF field at various locations in a vapor cell, one can characterize other aspects of an RF source and increase its data capacity [61].

The phase measurement of an RF signal also allows for waveform analysis and even a Rydberg atom-based spectrum analyzer [35, 62]. As an example, Figure 11.19 shows the received waveform obtained with a Rydberg atom-based sensor in response to an incident chirp signal. The frequency of the RF source was linearly increased to generate the chirp. The atoms in the heterodyne mixer

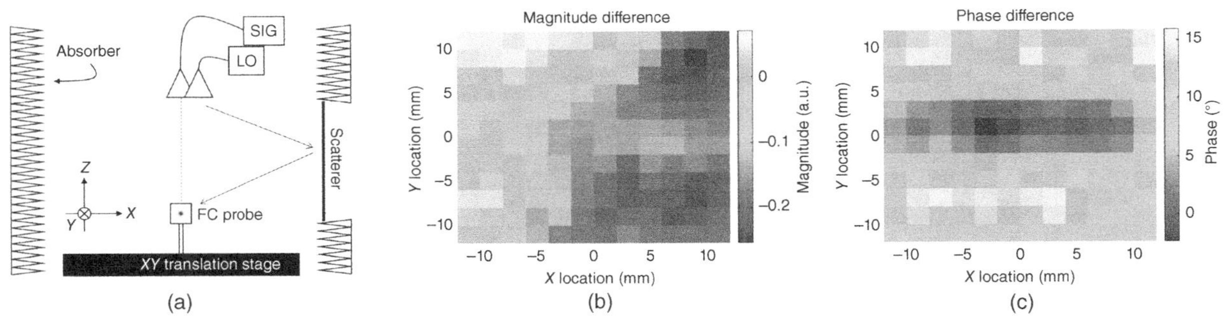

Figure 11.17 The setup (a) is the same as in Figure 11.16, but now a scatterer located in the $+X$ direction is added to the chamber resulting in measurable differences in the measured E-field (b) magnitude and (c) phase.

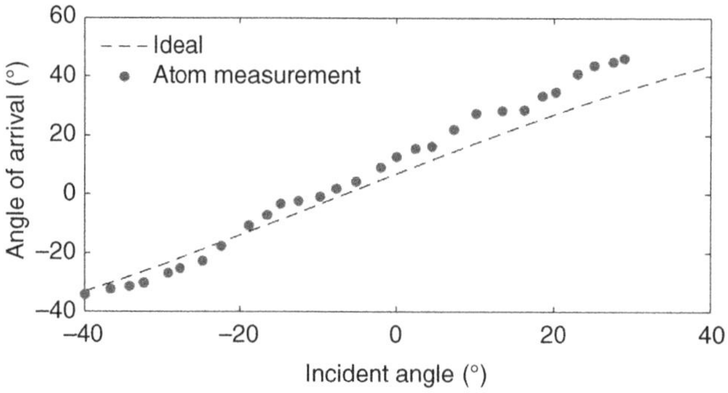

Figure 11.18 An RF SIG horn is rotated with respect to an LO horn such that the angle of incidence of the SIG RF field on a Rydberg atom vapor cell is changed. The difference in the phase measured by two optical probes within the vapor cell is used to determine a measured value of the angle-of-arrival (AoA) of the SIG RF field. These data demonstrate the ability to measure the AoA. Nonetheless, some refinement of the technique is required to improve its accuracy.

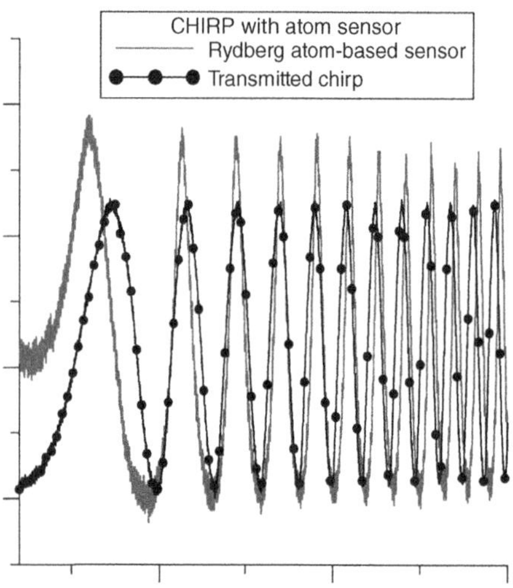

Figure 11.19 Received chirp signal.

configuration show the deviation of the RF source from the frequency of the applied LO by registering the beat note between the two. The atom signal obtained in this manner is compared with a classical mixer between the LO and RF source.

In addition to the aforementioned discussed applications, Rydberg atom sensors have been shown to be useful as a non-invasive method of measuring and characterising plasmas [63] and for possible blackbody thermometry [64]. In fact, current work is being performed to investigate the possibility of using Rydberg atom-based sensors for the detection of blackbody radiation (BBR). If successful, this could lead to a new SI-traceable method of calibrating BBR sources and instrumentation.

11.9 Conclusion and Discussion

Rydberg atoms show much promise as a paradigm shift in electromagnetic sensors wherein sensor elements are no longer realized through lumped circuit elements or semiconductor devices, but rather as a naturally occurring and fundamental physical system: the atom. As has been summarized in Sections 11.4 and 11.5, Rydberg atoms have the ability to act as atomic sensors and receivers in a variety of modalities. Thus far, they have been demonstrated for measuring a number of electromagnetic quantities including field amplitude and phase, modulated signals, power, and voltage. Fundamental aspects of Rydberg atoms should be pointed out for which future applications and discoveries will undoubtedly leverage the use of Rydberg atoms.

The ability of a Rydberg atom to act as a transducer between the RF amplitude and optical frequency via the Rabi frequency is at the core of many of the sensor applications. Through this property, the atoms are able to link two vastly differing size scales – some nine orders of magnitude – between the optical domain ($\lambda \approx 10^{-9}$ m) and the RF domain ($\lambda \approx 1$ m). This feature facilitates a unique and advantageous scenario wherein the Rydberg atoms are able to resonantly operate and detect electromagnetic fields well below the cutoff wavelength of RF systems. Applied in creative ways, one can realize new measurement opportunities. For instance, as discussed in Sections 11.1 and 11.6, the probe can act as an electrically small RF antenna or be applied to waveguide power measurements, respectively.

Another consequence of the transduction property between the RF and optical domains is that the required optical tuning range to access the atomic transitions provided by the Rydberg atoms for sensing applications are easily attainable by available laser systems. As such, a single laser system can now replace a whole host of RF equipment that would be needed in order to span the operating range of Rydberg atom species, e.g. from hundreds of MHz to 1 THz with Cs and Rb. This reduction in equipment requirements adds to the possibility for compact Rydberg atoms based sensors and arrays of sensors. Integration with other atomic devices, such as atomic clocks, also becomes possible as the laser systems are common for alkali atoms in general.

Being a natural system, Rydberg atoms provide a fundamental standard measurement of electromagnetic quantities. Another consequence of the transduction property via the Rabi Frequency is that measurements with Rydberg atoms become traceable to the SI via Planck's constant. This traceability is advantageous as oftentimes its path in more conventional electromagnetic measurements relies heavily on measuring the shape and/or position of an antenna within an anechoic environment. This can become time consuming and can result in a chicken-and-egg traceability path dilemma. The body of work related Rydberg atoms studies is not new per se, but the comparatively recent increased interest and advances in the application of Rydberg atoms as electromagnetic sensors points to an exciting future of new ideas and metrology methods using the ever so humble atom.

The techniques discussed in this chapter allow for the development of an E-field probe that does not require a calibration because an absolute value of the field is determined by the atomic properties of the Rydberg atom itself. Moreover, they provide self-calibrating, direct SI-traceable methods for RF E-field metrology. Furthermore, we have also discussed the ability to measure RF power. As a consequence, a Rydberg probe can lead to a direct SI-traceable approach for power metrology. The relationships and measurement demonstrations discussed herein are the first step toward the realization of quantum-based RF power and RF E-field strength measurement techniques and toward the realization of a more direct link to the newly redefined SI. Other applications include: waveform and spectrum analyzers, AoA sensors, voltage measurements, RF cameras and sensor arrays, and receivers for communication signals (AM/FM modulated and digital phase modulation signals). In the past ten years of the development of the Rydberg atom-based sensors, we have learned to control ensembles of room-temperature atoms in such a manner that we are able to develop interesting and unique applications. As a result, unforeseen applications are emerging each day.

References

1 J. S. Belrose, "Reginald Aubrey Fessenden and the birth of wireless telephony," *IEEE Antenna Propag. Mag.*, vol. 44, no. 2, pp. 33–47, 2002.

2 G. Pelosi, "Foreword from the associate editor," *IEEE Antenna Propag. Mag.*, vol. 54, no. 4, p. 283, 2012.

3 J. S. Belrose, "A Fessenden Christmas Eve broadcast retrospective," *IEEE Antenna Propag. Mag.*, vol. 54, no. 4, pp. 285–287, 2012.

4 M. Born, *Einstein's Theory of Relativity*. New York: Dover Publications, Inc., 1965.

5 NIST. Metric-Program. Si units. https://www.nist.gov/pml/weights-and-measures/metric-si/si-units, June 2021.

6 D. B. Newell, F. Cabiati, J. Fischer, S. G. Fujii, K. Karshenboim, H. S. Margolis, E. de Mirandés, P. J. Mohr, F. Nez, K. Pachucki, T. J. Quinn, B. N. Taylor, M. Wang, B. M. Wood, and Z. Zhang, "The CODATA 2017 values of h, e, k, and n_a for the revision of the SI," *Metrologia*, vol. 55, no. 1, pp. 13–16, 2018.

7 M. Stock, "The revision of the SI towards an international system of units based on defining constants," *Meas. Tech.*, vol. 60, no. 12, pp. 1169–1177, 2018.

8 T. F. Gallagher, *Rydberg Atoms*. Cambridge: Cambridge University Press, 1994.

9 J. A. Gordon, C. L. Holloway, S. Jefferts, and T. Heavner, "Quantum-based SI traceable electric-field probe," in *2010 IEEE International Symposium on Electromagnetic Compatibility, July 25–30*, pp. 321–324, 2010.

10 J. A. Sedlacek, A. Schwettmann, H. Kubler, R. Low, T. Pfau, and J. P. Shaffer, "Microwave electrometry with Rydberg atoms in a vapour cell using bright atomic resonances," *Nat. Phys.*, vol. 8, no. 11, pp. 816–824, 2012.

11 C. L. Holloway, J. A. Gordon, A. Schwarzkopf, D. A. Anderson, S. A. Miller, N. Thaicharoen, and G. Raithel, "Broadband Rydberg atom-based electric-field probe for SI-traceable, self-calibrated measurements," *IEEE Trans. Antenna Propag.*, vol. 62, no. 12, pp. 6169–6182, 2014.

12 C. L. Holloway, M. T. Simons, J. A. Gordon, P. F. Wilson, C. M. Cooke, D. A. Anderson, and G. Raithel, "Atom-based RF electric field metrology: From self-calibrated measurements to subwavelength and near-field imaging," *IEEE Trans. Electromagn. Compat.*, vol. 59, no. 2, pp. 717–728, 2017.

13 M. Tanasittikosol, J. D. Pritchard, D. Maxwell, A. Gauguet, K. J. Weatherill, R. M. Potvliege, and C. S. Adams, "Microwave dressing of Rydberg dark states," *J. Phys. B*, vol. 44, 184020, 2011.

14 C. L. Holloway, J. A. Gordon, A. Schwarzkopf, D. A. Anderson, S. A. Miller, N. Thaicharoen, and G. Raithel, "Sub-wavelength imaging and field mapping via electromagnetically induced transparency and Autler–Townes splitting in Rydberg atoms," *Appl. Phys. Lett.*, vol. 104, 244102, 2014.

15 J. A. Gordon, C. L. Holloway, A. Schwarzkopf, D. A. Anderson, S. A. Miller, N. Thaicharoen, and G. Raithel, "Millimeter wave detection via Autler–Townes splitting in rubidium Rydberg atoms," *Appl. Phys. Lett.*, vol. 105, 024104, 2014.

16 D. A. Anderson, S. A. Miller, A. Schwarzkopf, C. L. Holloway, J. A. Gordon, N. Thaicharoen, and G. Raithel, "Two-photon microwave transitions and strong-field effects in a room-temperature Rydberg-atom gas," *Phys. Rev. A*, vol. 90, 043419, 2014.

17 D. A. Anderson, E. Paradis, G. Raithel, R. E. Sapiro, and C. L. Holloway, "High-resolution antenna near-field imaging and sub-THz measurements with a small atomic vapor-cell

sensing element," *2018 11th Global Symposium Millimeter Waves (GSMM), Boulder, CO*, doi: 10.1109/GSMM.2018.8439437, 2018.

18 D. A. Anderson, S. A. Miller, G. Raithel, J. A. Gordon, M. L. Bulter, and C. L. Holloway, "Optical measurements of strong microwave fields with Rydberg atoms in a vapor cell," *Phys. Rev. Appl.*, vol. 5, 034003, 2016.

19 A. K. Mohapatra, T. R. Jackson, and C. S. Adams, "Coherent optical detection of highly excited Rydberg states using electromagnetically induced transparency," *Phys. Rev. Lett.*, vol. 98, 113003, 2007.

20 C. L. Holloway, M. T. Simons, J. A. Gordon, A. Dienstfrey, D. A. Anderson, and G. Raithel, "Electric field metrology for SI traceability: Systematic measurement uncertainties in electromagnetically induced transparency in atomic vapor," *J. Appl. Phys.*, vol. 121, no. 23, 233106, 2017.

21 M. T. Simons, J. A. Gordon, C. L. Holloway, D. A. Anderson, S. A. Miller, and G. Raithel, "Using frequency detuning to improve the sensitivity of electric field measurements via electromagnetically induced transparency and Autler–Townes splitting in Rydberg atoms," *Appl. Phys. Lett.*, vol. 108, 174101, 2016.

22 M. T. Simons, J. A. Gordon, and C. L. Holloway, "Simultaneous use of CS and RB Rydberg atoms for dipole moment assessment and RF electric field measurements via electromagnetically induced transparency," *J. Appl. Phys.*, vol. 120, no. 12, 123103, 2016.

23 C. L. Holloway, M. T. Simons, M. D. Kautz, A. H. Haddab, J. A. Gordon, and T. P. Crowley, "A quantum-based power standard: Using Rydberg atoms for a SI-traceable radio-frequency power measurement technique in rectangular waveguides," *Appl. Phys. Lett.*, vol. 113, 094101, 2018.

24 M. T. Simons, J. A. Gordon, and C. L. Holloway, "Fiber-coupled vapor cell for a portable Rydberg atom-based radio frequency electric field sensor," *Appl. Opt.*, vol. 57, no. 22, pp. 6456–6460, 2018.

25 M. T. Simons, M. D. Kautz, C. L. Holloway, D. A. Anderson, and G. Raithel, "Electromagnetically induced transparency (EIT) and Autler–Townes (AT) splitting in the presence of band-limited white Gaussian noise," *J. Appl. Phys.*, vol. 123, no. 20, 203105, 2018.

26 D. H. Meyer, K. C. Cox, F. K. Fatemi, and P. D. Kunz, "Digital communication with Rydberg atoms and amplitude-modulated microwave fields," *Appl. Phys. Lett.*, vol. 112, 211108, 2018.

27 K. C. Cox, D. H. Meyer, F. K. Fatemi, and P. D. Kunz, "Quantum-limited atomic receiver in the electrically small regime," *Phys. Rev. Lett.*, vol. 121, 110502, 2018.

28 C. L. Holloway, M. T. Simons, A. H. Haddab, J. A. Gordon, D. A. Anderson, and G. Raithel, "A multiple-band Rydberg atom-based receiver: AM/FM stereo reception," *IEEE Antenna Propag. Mag.*, vol. 63, no. 3, pp. 63–76, 2021.

29 D. A. Anderson, R. E. Sapiro, and G. Raithel, "An atomic receiver for AM and FM radio communication," *IEEE Trans. Antennas Propog.*, vol. 69, no. 5, pp. 2455–2462, 2021.

30 C. L. Holloway, M. T. Simons, A. H. Haddab, J. A. Gordon, and D. Novotny, "Detecting and receiving phase-modulated signals with a Rydberg atom-based receiver," *IEEE Antenna Wireless Propag. Lett.*, vol. 18, no. 9, pp. 1853–1857, 2019.

31 Z. Song, H. Liu, X. Liu, W. Zhang, H. Zou, J. Zhang, and J. Qu, "Rydberg-atom-based digital communication using a continuously tunable radio-frequency carrier," *Opt. Express*, vol. 27, no. 6, pp. 8848–8857, 2019.

32 M. T. Simons, A. H. Haddab, J. A. Gordon, and C. L. Holloway, "A Rydberg atom-based mixer: Measuring the phase of a radio frequency wave," *Appl. Phys. Lett.*, vol. 114, 114101, 2019.

33 J. A. Gordon, M. T. Simons, A. H. Haddab, and C. L. Holloway, "Weak electric-field detection with sub-1 Hz resolution at radio frequencies using a Rydberg atom-based mixer," *AIP Adv.*, vol. 9, no. 4, 045030, 2019.

34 M. Jing, Y. Hu, J. Ma, H. Zhang, L. Zhang, L. Xiao, and S. Jia, "Atomic superheterodyne receiver based on microwave-dressed Rydberg spectroscopy," *Nat. Phys.*, vol. 16, pp. 911–915, 2020.

35 D. H. Meyer, P. D. Kunz, and K. C. Cox, "Waveguide-coupled Rydberg spectrum analyzer from 0 to 20 GHz," *Phys. Rev. Appl.*, vol. 15, 014053, 2021.

36 A. K. Robinson, N. Prajapati, D. Senic, M. T. Simons, and C. L. Holloway, "Determining the angle-of-arrival of a radio-frequency source with a Rydberg atom-based sensor," *Appl. Phys. Lett.*, vol. 118, 114001, 2021.

37 C. L. Holloway, M. T. Simons, A. H. Haddab, C. J. Williams, and M. W. Holloway, "A "real-time" guitar recording using Rydberg atoms and electromagnetically induced transparency: Quantum physics meets music," *AIP Adv.*, vol. 9, no. 6, 065110, 2019.

38 A. K. Robinson, A. B. Artusio-Glimpse, M. T. Simons, and C. L. Holloway, "Atomic spectra in a six-level scheme for electromagnetically induced transparency and Autler–Townes splitting in Rydberg atoms," *Phys. Rev. A*, vol. 103, 023704, 2021.

39 M. T. Simons, A. H. Haddab, J. A. Gordon, D. Novotny, and C. L. Holloway, "Embedding a Rydberg atom-based sensor into an antenna for phase and amplitude detection of radio-frequency fields and modulated signals," *IEEE Access*, vol. 7, pp. 164975–164985, 2019.

40 C. L. Holloway, N. Prajapati, J. E. Kitching, J. A. Sherman, C. Teale, A. Rüfenacht, A. B. Artusio-Glimpse, M. T. Simons, A. K. Robinson, and E. B. Norrgard, "Electromagnetically induced transparency based Rydberg-atom sensor for quantum voltage measurements,"*arXiv:2110.02335, Oct 22*, 2021.

41 M. Fleischhauer, A. Imamoglu, and J. P. Marangos, "Electromagnetically induced transparency: Optics in coherent media," *Rev. Mod. Phys.*, vol. 77, pp. 633–673, 2005.

42 K. J. Boller, A. Imamolu, and S. E. Harris, "Observation of electromagnetically induced transparency," *Phys. Rev. Lett.*, vol. 66, no. 20, pp. 2593–2596, 1991.

43 S. Kumar, H. Fan, H. Kubler, J. Sheng, and J. P. Shaffer, "Atom-based sensing of weak radio frequency electric fields using homodyne readout," *Sci. Rep.*, vol. 7, 42981, 2017.

44 J. P. Shaffer and H. Kübler, "A read-out enhancement for microwave electric field sensing with Rydberg atoms," in *Quantum Technologies 2018*, vol. 10674, J. Stuhler, A. J. Shields, and M. J. Padgett, Ed. International Society for Optics and Photonics, SPIE, 2018, pp. 39–49. https://doi.org/10.1117/12.2309386.

45 H. M. Kwak, T. Jeong, Y.-S. Lee, and H. S. Moon, "Microwave-induced three-photon coherence of Rydberg atomic states," *Opt. Commun.*, vol. 380, pp. 168–173, 2016. ISSN 0030-4018. Doi: 10.1016/j.optcom.2016.06.004. https://www.sciencedirect.com/science/article/pii/S0030401816304989.

46 H. Q. Fan, S. Kumar, R. Daschner, H. Kübler, and J. P. Shaffer, "Subwavelength microwave electric-field imaging using Rydberg atoms inside atomic vapor cells," *Opt. Lett.*, vol. 39, no. 10, pp. 3030–3033, May 2014. Doi: 10.1364/OL.39.003030. http://ol.osa.org/abstract.cfm?URI=ol-39-10-3030.

47 I. I. Sobelman, *Atomic Spectra and Radiative Transitions*, 2nd ed. New York: Springer, 1992.

48 D. A. Steck. "Cesium D line data," *available online at* http://steck.us/alkalidata *(revision 2.1.4, 23 December September 2010)*, 2010.

49 D. A. Steck, "Rubidium 85 D line data," *available online at* http://steck.us/alkalidata *(revision 2.1.6, 20 September 2013)*, 2013.

50 C. L. Holloway, M. T. Simons, M. Kautz, P. F. Wilson, and J. A. Gordon, "Development and applications of a fiber-coupled atom-based electric field probe," in *2018 International Symposium on Electromagnetic Compatibility (EMC Europe 2018), Amsterdam, The Netherlands, Aug 27–30*, pp. 381–385, 2018.

51 W. Demtröder, *Atoms, Molecules and Photons: An Introduction to Atomic-, Molecular-, and Quantum Physics*, 2nd ed. New York: Springer, 2010.

52 M. T. Simons, A. B. Artusio-Glimpse, C. L. Holloway, E. Imhof, S. R. Jefferts, R. Wyllie, B. C. Sawyer, and T. G. Walker, "Continuous radio frequency electric-field detection through adjacent Rydberg resonance tuning," *Phys. Rev. A*, vol. 104, 032824, 2021.

53 D. A. Anderson and G. Raithel, "Continuous-frequency measurements of high-intensity microwave electric fields with atomic vapor cells," *Appl. Phys. Lett.*, vol. 111, 053504, 2017.

54 H. Friedrich, *Theoretical Atomic Physics*, 4th ed. New York: Springer.

55 M. T. Simons, M. D. Kautz, J. A. Gordon, and C. L. Holloway, "Uncertainties in Rydberg atom-based RF E-field measurements," in *2018 International Symposium on Electromagnetic Compatibility (EMC Europe 2018), Amsterdam, The Netherlands, Aug 27–30*, pp. 381–385, 2018.

56 H. Fan, S. Kumar, J. Sheng, J. P. Shaffer, J. A. Gordon, and C. L. Holloway, "Effect of vapor-cell geometry on Rydberg-atom-based measurements of radio-frequency electric fields," *Phys. Rev. Appl.*, vol. 4, 044015, 2015.

57 J. P. Shaffer, "Atom-based electromagnetic field sensing," *SPIE 11296, Optical, Opto-Atomic, and Entanglement-Enhanced Precision Metrology II, 27 March 2020*, https://doi.org/10.1117/12.2552626, 2020.

58 H. Mu, H. Zou, Q. Wang, and Z. Song, "A low permittivity metamaterial on a glass substrate for fabricating an atomic vapor cell," *2019 Photonics & Electromagnetics Research Symposium - Fall (PIERS - Fall), Xiamen, China, 2019*, pp. 344–350, 2019. doi: 10.1109/PIERS-Fall48861.2019.9021792.

59 M. T. Simons, A. H. Haddab, J. A. Gordon, and C. L. Holloway, "Applications with a Rydberg atom-based radio frequency antenna/receiver," *2019 International Symposium on Electromagnetic Compatibility (EMC Europe 2019), Barcelona, Spain, Sept 2–6*, pp. 885–889, 2019.

60 D. A. Anderson, R. E. Sapiro, and G. Raithel, "Rydberg atoms for radio-frequency communications and sensing: Atomic receivers for pulsed RF field and phase detection," *IEEE Aerospace Electron. Syst. Mag.*, vol. 35, no. 4, pp. 48–56, 2020.

61 S. Ottoa, M. K. Hunter, N. Kjærgaard, and A. B. Debb, "Data capacity scaling of a distributed Rydberg atomic receiver array," *J. Appl. Phys.*, vol. 129, 154503, 2021.

62 M. T. Simons, A. B. Artusio-Glimpse, A. K. Robinson, N. Prajapati, and C. L. Holloway, "Rydberg atom-based sensors for radio frequency electric field metrology, sensing, and communications," *XXIII IMEKO World Congress: Measurement: Sparking Tomorrow's Smart Revolution, Yokohama, Japan, Sept 2021*, 2021.

63 D. A. Anderson, G. Raithel, M. T. Simons, and C. L. Holloway, "Quantum-optical spectroscopy for plasma electric field measurements and diagnostics," *arXiv:1712.08717v1 [physics.atom-ph], Dec 23*, 2017.

64 E. B. Norrgard, S. P. Eckel, C. L. Holloway, and E. L. Shirley, "Quantum blackbody thermometry," *New J. Phys.*, vol. 23, 033037, 2021.

12

Quantum Antenna Arrays

Iñigo Liberal[1] and Richard W. Ziolkowski[2]

[1]Department of Electrical, Electronic and Communications Engineering, Institute of Smart Cities (ISC), Public University of Navarre (UPNA), Pamplona, Spain
[2]Global Big Data Technologies Centre, School of Electrical and Data Engineering, University of Technology Sydney, Ultimo, New South Wales, Australia

12.1 The Rise of Quantum Technologies

Quantum technologies are defined by their ability to harness quantum mechanical effects in order to achieve performances that cannot be attained within the boundaries of classical physics. In doing so, quantum technologies hold the promise to revolutionize communication, sensing, simulation and computing technologies, thus bringing transformative advances to science, industry and society [1]. Photonic quantum technologies, for which light acts a fundamental resource, are arguably the major driving force in promoting basic quantum technologies to commercial systems. For example, quantum key distribution (QKD) is the most promising example of a commercial quantum communication system [2]. QKD enables a secure channel with the guarantee that interceptions in the communication can be detected, an invaluable property to distribute cryptographic keys. By using optical links, several companies currently offer commercial QKD systems [3–5], and satellite QKD communications have been demonstrated by state-driven enterprises [6]. Similarly, quantum sensing takes advantage of nonclassical field correlations to enable metrology with an unprecedented precision. The use of nonclassical light at the Laser Interferometer Gravitational-wave Observatory (LIGO) is perhaps the most iconic example of a practical quantum metrology system [7]. However, sensing via the optical readout of solid state spins have enabled a myriad of applications, fascinating examples including temperature mapping within a neuron [8], intra-bacterial magnetic field mapping [9], nuclear magnetic resonance of single proteins [10], and nanoscale magnetic resonance imaging (MRI) [11], only to name a few. Quantum computing, while still a futuristic goal [12], is the most appealing quantum technology application for the general public. Despite the fact that many fundamental challenges must be addressed, several companies are investing in the development of a practical optical quantum computer [13, 14]. In fact, the first demonstration of quantum computation advantage was based on a photonic device [15]. These are only a few examples of how quantum photonic technologies are a growing reality that will only continue to increase its impact in the following years.

Nonclassical light states are the fundamental resource driving most of the aforementioned photonic quantum technologies. For example, single-photon states are the basis of QKD communications systems, squeezed light states and N00N states empower sensing below shot noise, and single-photon and cluster states give power to linear-optics quantum computers. Therefore, the

Antenna and Array Technologies for Future Wireless Ecosystems, First Edition.
Edited by Y. Jay Guo and Richard W. Ziolkowski.

design of light sources capable of generating prescribed nonclassical light states is one of the most important aspects in developing photonic quantum technologies.

In an entirely different field, the vast body of research nowadays known simply as antenna theory has led to a high degree of sophistication in the ability to tailor the radiation characteristics of electromagnetic waves emitted by innumerable types of sources. Previous chapters of this book have focused on classical antennas and antenna arrays; they exemplify how this is indeed the case. Without any doubts, much of this work can be applied to catalyze the design of nonclassical light sources. Based on this perspective, and following the general themes of this book, we will discuss in this chapter how photon statistics can be engineered by designing the geometry of quantum antenna arrays. The main goal is to summarize the basic theory behind quantum antenna arrays, providing antenna designers with the tools to explore how their know-how can be applied to the exciting field of quantum technologies. In doing so, we will explore intriguing nonclassical light sources with no classical counterparts, including directionally entangled photon bunches and perfectly isotropic single-photon sources.

To achieve our goal, we will begin with an overview of quantum antenna array theory. While arrays of electrically small classical antennas were described in Chapter 10, the radiating elements herein are physically extremely small being quantum emitters, i.e. atoms. We then follow with a relevant introduction to the concepts of photon statistics, emphasizing directional correlations. These two fields are then combined to reveal how the design of linear arrays of quantum emitters can enable beamforming of their photon statistics. We emphasize the effects that can and cannot be observed in classical systems. Examples of first, second and Nth order correlation outcomes demonstrate how their directivities can be tailored by a quantum array's configuration. We apply these results to understand how one could in fact achieve an isotropic single-photon source. We also address the difficulties associated with the very contrasting concept of quantum superdirectivity. A description of the technologies available today that are the most promising to realize these exciting quantum array results is given. Finally, we explore the many possible future directions of the field of quantum antenna arrays and the concurrent opportunities for electromagnetic scientists and engineers.

12.2 Quantum Antenna Array Theory

We start by introducing the basic quantum theory of radiation for an array of quantum emitters. We try throughout to establish the connections with classical antenna array theory that was the focus of most of the previous chapters of this book. In particular, the final expressions are written in forms that are readily connected to their classical counterparts.

12.2.1 Macroscopic Quantum Electrodynamics

As a first approach to quantum antenna arrays, we analyze the general scenario schematically depicted in Figure 10.1. An array of N quantum emitters is depicted that is coupled to a macroscopic photonic environment through the fields radiated by its individual elements. This complex environment is characterized by a relative permittivity distribution, $\varepsilon\left(\mathbf{r},\omega_f\right) = \varepsilon_R\left(\mathbf{r},\omega_f\right) + i\,\varepsilon_I\left(\mathbf{r},\omega_f\right)$. Quantum excitation in such macroscopic environment will be modeled as a bath of polaritonic modes (i.e. a polariton is a quasi-particle formed when a photon couples strongly with an exciton, e.g. a phonon, and, hence, it represents a joint state of light and matter) following the macroscopic quantum electrodynamics (QED) formalism (see, e.g. [16–18]). This approach allows for the analysis of lossy and dispersive media. We restrict the analysis in this chapter to the case of isotropic dielectric media, which is typical of most configurations at optical frequencies. However, generalizations of the same quantization procedure to account for magnetically polarizable, bi-anisotropic, and nonlocal media are available [19]. While most quantum optics textbooks treat the decay of quantum emitters from the Schrödinger and/or interaction pictures, here we will adopt the Heisenberg picture because fully time-evolving field operators bring a closer analogy to classical antenna theory.

The Hamiltonian of the system depicted in Figure 12.1 is given by the addition of the Hamiltonians for the array of quantum emitter's and the electromagnetic field plus polarizable matter (i.e. the polaritonic) system, as well as the interaction Hamiltonian between both systems:

$$\hat{\mathcal{H}} = \hat{\mathcal{H}}_{\text{emitters}} + \hat{\mathcal{H}}_{\text{fields}} + \hat{\mathcal{H}}_{\text{int}} \tag{12.1}$$

The emitters are modeled as two-level systems $\{|e_n\rangle, |g_n\rangle\}$ that are located at positions $\mathbf{r}_n$ and have the transition frequency ω_n and dipole moment $\mathbf{p}_n$. This choice leads to the Hamiltonian

$$\hat{\mathcal{H}}_{\text{emitters}} = \sum_{n=1}^{N} \frac{\hbar\omega_n}{2} \hat{\sigma}_n^z(t) \tag{12.2}$$

where $\hat{\sigma}_n^z$ is the Pauli operator with $\hat{\sigma}_n^z(t=0) = |e_n\rangle\langle e_n| - |g_n\rangle\langle g_n|$.

Excitations in the macroscopic photonic environment are modeled as a continuum of polariton modes for each position $\mathbf{r}$, frequency ω_f, and polarization $p = x,\ y,$ or z with annihilation operators $\hat{f}_p(\mathbf{r}, \omega_f;\ t)$. They can also be written in vector form as $\hat{\mathbf{f}}(\mathbf{r}, \omega_f;\ t) = \sum_p \mathbf{u}_p \hat{f}_p(\mathbf{r}, \omega_f;\ t)$ and obey equal-time bosonic commutation relations:

$$\left[\hat{\mathbf{f}}(\mathbf{r}, \omega_f;\ t), \hat{\mathbf{f}}^{\dagger}\left(\mathbf{r}', \omega_f';\ t\right)\right] = \hat{\mathbf{I}}\delta(\mathbf{r} - \mathbf{r}')\,\delta\left(\omega_f - \omega_f'\right) \tag{12.3}$$

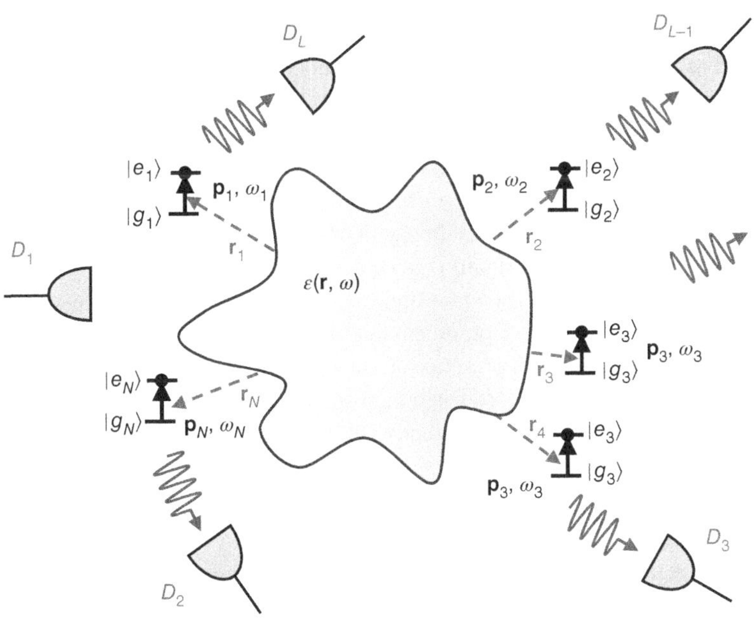

Figure 12.1 *Quantum antenna arrays in a complex environment.* Schematic depiction of a quantum antenna array composed of N two-level systems $\{|e_n\rangle, |g_n\rangle\}$ $n = 1, 2, \ldots, N$, with transition frequencies ω_n and transition dipole moments $\mathbf{p}_n$ located at positions $\mathbf{r}_n$ and immersed in a macroscopic photonic environment characterized by a lossy and dispersive permittivity distribution $\varepsilon(\mathbf{r}, \omega)$. A set of L detectors consisting of photon counters is situated around the array in the same environment.

In this manner, the Hamiltonian for the macroscopic photonic environment can be compactly written as

$$\hat{\mathcal{H}}_{\text{fields}} = \int d^3\mathbf{r} \int_0^\infty d\omega_f \, \hbar\omega_f \hat{\mathbf{f}}^\dagger \left(\mathbf{r}, \omega_f; t\right) \cdot \hat{\mathbf{f}}\left(\mathbf{r}, \omega_f; t\right) \tag{12.4}$$

Finally, we consider an interaction Hamiltonian within the electric dipole approximation

$$\hat{\mathcal{H}}_I = -\sum_{n=1}^{N} \hat{\mathbf{p}}_n(t) \cdot \hat{\mathbf{E}}\left(\mathbf{r}_n; t\right) \tag{12.5}$$

where $\hat{\mathbf{p}}_n(t) = \mathbf{p}_n \hat{\sigma}_n(t) + \mathbf{p}_n^* \hat{\sigma}_n^\dagger(t)$, with $\hat{\sigma}_n(t)$ and $\hat{\sigma}_n^\dagger(t)$ being the annihilation and creation operators associated with the nth emitter and with $\hat{\sigma}_n(t = 0) = \left|g_n\right\rangle \left\langle e_n\right|$. Note that because we are assuming point-like emitters with an electric dipole transition, the electric field operator is evaluated at the position of the emitters $\mathbf{r}_n$. This approximation is similar to assuming that the elements of a classical antenna array are electric Hertzian dipoles. However, similar to antenna theory, it is possible to further control the quantum radiated field by engineering the current distributions of the individual radiating emitters [20].

The electric field operator is given by

$$\hat{\mathbf{E}}(\mathbf{r}; t) = \int_0^\infty d\omega_f \int d^3\mathbf{r}' \, \mathbf{G}_E\left(\mathbf{r}, \mathbf{r}', \omega_f\right) \cdot \hat{\mathbf{f}}\left(\mathbf{r}', \omega_f; t\right) + h.c. \tag{12.6}$$

where we have introduced the response function

$$\mathbf{G}_E\left(\mathbf{r}, \mathbf{r}', \omega_f\right) = i\sqrt{\frac{\hbar}{\pi\varepsilon_0}\frac{\omega_f^2}{c^2}} \sqrt{\varepsilon_I\left(\mathbf{r}', \omega_f\right)} \mathbf{G}\left(\mathbf{r}, \mathbf{r}', \omega_f\right) \tag{12.7}$$

which is proportional to the classical dyadic Green's function $\mathbf{G}\left(\mathbf{r}, \mathbf{r}', \omega_f\right)$ [21]. Equation (12.6) intuitively conveys the idea that the electric field operator can be understood as the result of radiation by the polaritonic excitations in the macroscopic environment. Therefore, it is not surprising that the classical dyadic Green's functions play the role of a propagator in such a process. However, the antenna designer might be surprised that the propagation term also includes the square root of the imaginary part of the permittivity as a coupling term between fields and environmental excitations. Intuitively, it makes sense that the effective exchange of energy between fields and matter must be mediated by dissipation. From a more rigorous perspective, this term is exactly what is needed to reconstruct the macroscopic Maxwell's equations from the coupling of the free-space fields and matter when the quantum mechanical canonical quantization of the fields is required and employed in the presence of matter [17]. The presence of the square root of the imaginary part of the permittivity also ensures that the fluctuation dissipation theorem, as it is often invoked in the semiclassical treatment of thermal emission [22], directly emerges from this framework.

12.2.2 The Quantum Field in the Far-Zone

The Hamiltonian formulated in Section 12.2.1 is the standard formulation of macroscopic QED [16–18]. Next, we derive an expression for the radiated field in the far zone that has very clear analogies with antenna theory. To this end, we compute the Heisenberg equation of motion, $i\hbar\partial_t \hat{O}(t) = \left[\hat{O}, \hat{\mathcal{H}}\right]$, for the polaritonic operator $\hat{\mathbf{f}}\left(\mathbf{r}', \omega_f; t\right)$, and integrate it, leading to its decomposition into the free-evolving and source parts

$$\hat{\mathbf{f}}\left(\mathbf{r}', \omega_f; t\right) = \hat{\mathbf{f}}_0\left(\mathbf{r}', \omega_f; t\right) + \hat{\mathbf{f}}_S\left(\mathbf{r}', \omega_f; t\right) \tag{12.8}$$

Physically, the free-evolving part corresponds to the electromagnetic field and matter evolution that would be taking place in the absence of the interactions with the quantum emitters. It can be written as:

$$\hat{\mathbf{f}}_0\left(\mathbf{r}',\omega_f;\ t\right)=\hat{\mathbf{f}}\left(\mathbf{r}',\omega_f;\ t=0\right)e^{-i\omega_f t} \tag{12.9}$$

On the other hand, the source part corresponds to the excitations of the electromagnetic field and matter system induced by the interaction with the quantum emitter system. It is given by

$$\hat{\mathbf{f}}_S\left(\mathbf{r}',\omega_f;\ t\right)=\sum_{n=1}^{N}\hat{\mathbf{f}}_{Sn}\left(\mathbf{r}',\omega_f;\ t\right) \tag{12.10}$$

with

$$\hat{\mathbf{f}}_{Sn}\left(\mathbf{r}',\omega_f;\ t\right)=\frac{i}{\hbar}\int_0^t d\tau\ e^{-i\omega_f(t-\tau)}\ \hat{\mathbf{p}}_n(\tau)\cdot\mathbf{G}_E^*\left(\mathbf{r}_n,\mathbf{r}',\omega_f\right) \tag{12.11}$$

By introducing (12.8) into (12.6), we find that the electric field operator can be similarly decomposed into free-evolving and source driven parts

$$\hat{\mathbf{E}}(\mathbf{r};\ t)=\hat{\mathbf{E}}_0(\mathbf{r};\ t)+\hat{\mathbf{E}}_S(\mathbf{r};\ t) \tag{12.12}$$

with

$$\hat{\mathbf{E}}_S(\mathbf{r};\ t)=\sum_{n=1}^{N}\hat{\mathbf{E}}_{Sn}(\mathbf{r};\ t) \tag{12.13}$$

Introducing (12.11) into (12.6), using the completeness relation of the dyadic Green's function to compute the volume integral, and noting that $\mathrm{Im}\mathbf{G}\left(\mathbf{r},\mathbf{r}_n,\omega_f\right)$ is an odd function in ω_f, the contribution of each emitter to the source field can be written as a memory kernel acting on the electric dipole operator

$$\hat{\mathbf{E}}_{Sn}(\mathbf{r};\ t)=\int_0^t d\tau\ \mathbf{K}\left(\mathbf{r},\mathbf{r}_n,t-\tau\right)\cdot\hat{\mathbf{p}}_n(\tau) \tag{12.14}$$

with

$$\mathbf{K}\left(\mathbf{r},\mathbf{r}_n,t\right)=\frac{i}{\pi\varepsilon_0 c^2}\int_{-\infty}^{\infty}d\omega_f\ e^{-i\omega_f t}\omega_f^2\ \mathrm{Im}\mathbf{G}\left(\mathbf{r},\mathbf{r}_n,\omega_f\right) \tag{12.15}$$

Recalling that the electric dipole operator is $\hat{\mathbf{p}}_n(\tau)=\mathbf{p}_n\hat{\sigma}_n(t)+\mathbf{p}_n^*\hat{\sigma}_n^\dagger(t)$, the decomposition of the source electric field operator into positive and negative frequency components is a useful approach:

$$\hat{\mathbf{E}}_{Sn}(\mathbf{r};\ t)=\hat{\mathbf{E}}_{Sn}^{(+)}(\mathbf{r};\ t)+\hat{\mathbf{E}}_{Sn}^{(-)}(\mathbf{r};\ t) \tag{12.16}$$

with

$$\hat{\mathbf{E}}_{Sn}^{(+)}(\mathbf{r};\ t)=\int_0^t d\tau\ \mathbf{K}\left(\mathbf{r},\mathbf{r}_n,t-\tau\right)\cdot\mathbf{p}_n\ \hat{\sigma}_n(t) \tag{12.17}$$

and

$$\hat{\mathbf{E}}_{Sn}^{(-)}(\mathbf{r};\ t)=\left(\hat{\mathbf{E}}_{Sn}^{(+)}(\mathbf{r};\ t)\right)^\dagger \tag{12.18}$$

Physically, the positive frequency component is associated with the radiation of one photon by the emitter (as a transmitter), while the negative frequency component represents the absorption of one photon by the emitter (as a receiver). A very compact expression for the electric field operator can now be obtained by applying the Laplace transform to these operators:

$$\hat{O}(\omega)=\int_0^\infty d\omega\ \hat{O}(t)\,e^{i\omega t} \tag{12.19}$$

and then evaluating the integral over all frequencies via complex analysis techniques. The result is that Eq. (12.17) can be written as

$$\hat{\mathbf{E}}_{Sn}^{(+)}(\mathbf{r};\ \omega) = \omega^2 \mu_0\, \mathbf{G}\left(\mathbf{r}, \mathbf{r}_n, \omega\right) \cdot \mathbf{p}_n\, \hat{\sigma}_n(\omega) \tag{12.20}$$

The formulation up to this point has been developed for an arbitrary photonic environment. Particular cases can be addressed by introducing a specific form of the dyadic Green's function. Thus, we assume that the emitters are immersed in free-space and the observation point for the electric field is in the far-zone of this system to establish a closer connection with antenna theory.

The classical dyadic Green's function for this case is

$$\mathbf{G}\left(\mathbf{r}, \mathbf{r}_n, \omega\right) = \frac{e^{ikR_n}}{4\pi R_n}\left(\mathbf{I} - \mathbf{u}_{R_n}\mathbf{u}_{R_n}\right) \tag{12.21}$$

with $k = \omega/c$, $\mathbf{R}_n = \mathbf{r} - \mathbf{r}_n$, $R_n = \left|\mathbf{R}_n\right|$, and the unit vector $\mathbf{u}_{R_n} = \mathbf{R}_n/R_n$. The first-order approximation to R_n is given by the well-known expression: $R_n \simeq r - \mathbf{u}_r \cdot \mathbf{r}_n$. We keep this first-order approximation in the exponential and only the zero-order approximation in the other terms. The dyadic Greens function then becomes

$$\mathbf{G}\left(\mathbf{r}, \mathbf{r}_n, \omega\right) = \frac{1}{4\pi r}\, \mathbf{I}_T\, e^{i\omega\left[\frac{r - \mathbf{u}_r \cdot \mathbf{r}_n}{c}\right]} \tag{12.22}$$

with $\mathbf{I}_T = \mathbf{I} - \mathbf{u}_r\mathbf{u}_r$. In this manner, we can rewrite the electric field operator as

$$\hat{\mathbf{E}}_{S}^{(+)}(\mathbf{r};\ \omega) = \frac{\mu_0}{4\pi r}\sum_{n=1}^{N} \mathbf{I}_T \cdot \mathbf{p}_n\, \omega^2 \hat{\sigma}_n(\omega)\, e^{i\omega\left[\frac{r - \mathbf{u}_r \cdot \mathbf{r}_n}{c}\right]} \tag{12.23}$$

Finally, we apply the inverse Laplace transform and get the general form for the source electric field in the far zone

$$\hat{\mathbf{E}}_{S}^{(+)}(\mathbf{r};\ t) = -\frac{\mu_0}{4\pi r}\sum_{n=1}^{N} \mathbf{I}_T \cdot \mathbf{p}_n\, \partial_t^2 \hat{\sigma}_n\left(t_d + \frac{\mathbf{u}_r \cdot \mathbf{r}_n}{c}\right) \tag{12.24}$$

with $t_d = t - r/c$ being the delayed time with respect to the origin of the coordinates.

12.2.3 Decay Dynamics in the Noninteracting Markovian Regime

Equation (12.24) describes the electric field operator in the far-zone for an arbitrary time evolution of the quantum emitter operator. This formalism can be applied to a variety of configurations. We particularize it here to the case of the decay dynamics of noninteracting quantum emitters in the weak coupling regime. Thus, we will show that it exhibits a simple exponential decay. This case corresponds to the physical situation in which each emitter operates as an independent single-photon source.

First, we analyze the Heisenberg equations of motion, $i\hbar\partial_t\hat{O}(t) = \left[\hat{O}, \hat{\mathcal{H}}\right]$, for the quantum emitter annihilation operator $\hat{\sigma}(t)$. They lead to the relation for the nth emitter:

$$\partial_t \hat{\sigma}_n(t) = -i\omega_n \hat{\sigma}(t) + \frac{1}{i\hbar}\, \hat{\sigma}_n^z(t)\, \mathbf{p} \cdot \hat{\mathbf{E}}\left(\mathbf{r}_n; t\right) \tag{12.25}$$

Next, we clarify the approximations involved in assuming a simple exponential decay rate.

We begin by assuming that the macroscopic photonic environment is initially in the vacuum state. We can then substitute $\hat{\mathbf{E}}\left(\mathbf{r}_n; t\right) \simeq \hat{\mathbf{E}}_S\left(\mathbf{r}_n; t\right)$ for all relevant interactions. From an antenna theory perspective, this condition is similar to assuming that there are no external sources to the problem. In addition, we assume that the emitters are not interacting amongst themselves. This allows us to approximate $\hat{\mathbf{E}}_S\left(\mathbf{r}_n; t\right) \simeq \hat{\mathbf{E}}_{Sn}\left(\mathbf{r}_n; t\right)$. This is the typical assumption in antenna theory where each element of the array is driven independently, and the array geometry is designed so that the coupling between each of the antenna elements in the array is weak enough.

We also introduce the low-excitation approximation (also known as the one-photon correlation approximation). It consists of replacing $\hat{\sigma}^z(t) \to -1$. This approximation is similar to having started with a Hamiltonian within the rotating wave approximation. For the problem of initially-excited noninteracting quantum emitters, this means that each emitter effectively sees only a single excitation. It is a very accurate approximation.

Even with all of these approximations, the model still is capable of describing both the weak and strong coupling regimes. These are the ones in which most optical systems operate. However, the approximation will not be valid in the ultra-strong coupling regime where the number of excitations is not preserved.

With these approximations and the application of the Laplace transform, the equation of motion for $\hat{\sigma}(t)$ reduces to

$$-i\left(\omega-\omega_0\right)\hat{\sigma}_n(\omega)=\hat{\sigma}(t=0)+\frac{1}{i\hbar}\,\mathbf{p}_n^*\cdot\hat{\mathbf{E}}_{Sn}^{(+)}\left(\mathbf{r}_n;\omega\right) \tag{12.26}$$

Using (12.20) and rearranging its terms we find the following expression for the Laplace-transformed emitter operator:

$$\hat{\sigma}_n(\omega)=i\,\frac{\hat{\sigma}(t=0)}{\omega-\omega_0-\Sigma_n(\omega)} \tag{12.27}$$

where we have introduced the emitter's self-energy $\Sigma_n(\omega)$:

$$\Sigma_n(\omega)=\frac{\omega^2\mu_0}{\hbar}\,\mathbf{p}_n^*\cdot\mathbf{G}\left(\mathbf{r}_n,\mathbf{r}_n,\omega\right)\cdot\mathbf{p}_n \tag{12.28}$$

Finally, the standard Born–Markov approximation consists of neglecting the dispersion of the self-energy, i.e. $\Sigma_n(\omega)\simeq\Sigma_n\left(\omega_n\right)$. This approximation is accurate when the coupling between the emitter and the photonic environment is small. This situation is usually referred to as the weak-coupling regime. The approximation is also equivalent to assuming that the interaction between the emitter and the photonic environment has no memory effects.

Within this approximation, the imaginary part of the self-energy corresponds to the decay rate $\mathrm{Im}\left[\Sigma_n\left(\omega_n\right)\right]=-\gamma_n$ associated with the exponential decay. On the other hand, the real part of the self-energy results in a frequency shift, $\mathrm{Re}\left[\Sigma_n\left(\omega_n\right)\right]=\Delta\omega_n$.

Note that the real part of the dyadic Green's function diverges. As a consequence, the self-energy must be properly renormalized to properly compute this frequency shift which is well-known as the Lamb shift. However, since this small shift does not affect the overall decay dynamics, it is typically assumed that it was included in the transition frequency of the emitter, ω_n. Finally, within all of these approximations, we recover the usual exponential decay expression:

$$\hat{\sigma}_n(t)\simeq e^{-i\omega_n t}\,e^{-\gamma_n t}\hat{\sigma}(t=0) \tag{12.29}$$

After detailing all of the approximations leading to the simple exponential decay, it might seem that it is a crude approximation that can only be applied in very restricted circumstances. However, the truth is that for most weekly interacting single-photon sources, exponential decay is a very accurate approximation. On the other hand, observing non-exponential decay dynamics is a challenging experimental task in most configurations.

Finally, by introducing (12.29) into (12.24) we find the expression for the positive frequency electric field operator that we will use in the rest of the chapter:

$$\hat{\mathbf{E}}^{(+)}(\mathbf{r},t)=\sum_{p=\theta,\phi}\mathbf{u}_p\,\hat{E}_p^{(+)}(\mathbf{r},t) \tag{12.30}$$

The term

$$\hat{E}_p^{(+)}(\mathbf{r},t) = \sum_{n=1}^{N} \frac{B_{np}(\phi,\theta)}{r} \hat{\sigma}_n \, e^{-(i\omega_n+\gamma_n)t_d} \, e^{-i\frac{\omega_n}{c}\mathbf{u}_n\cdot\mathbf{r}_n} \tag{12.31}$$

where

$$B_{np}(\phi,\theta) = -\frac{\mu_0\omega_n^2}{4\pi}\mathbf{u}_p \cdot \mathbf{I}_T \cdot \mathbf{p}_n \tag{12.32}$$

12.3 Photon Statistics

The theory detailed in Section 12.2 shows that the temporal and spatial variations of the electric field operator are the same as those of a classical field. This outcome suggests that many aspects of the theory learned for the design of classical antenna arrays could be transferred to quantum antenna arrays. At the same time, it is expected that quantum antenna arrays will give access to effects that cannot be observed in classical systems. This is indeed the case.

Quantum effects fundamentally arise in the measurement of photon statistics, i.e. in coincidence photon measurements. Early experiments in the 1970s and 1980s confirmed that the measurements of photon statistics cannot be explained by a classical theory. They revealed numerous novel phenomena including antibunching [23], two-photon cascade emission [24] and violation of Bell's inequalities [25]. These effects provide the most direct evidence that a quantum theory of light is needed. Moreover, correlations that cannot be achieved by classical systems are the major driving force behind the currently evolving quantum technologies. Being the most direct experimental proof for quantum mechanics, early experiments have been very critically analyzed, motivating researchers to identify loopholes invalidating the conclusions extracted from them [26–28]. Nevertheless, the improved and loophole-free experimental setups continue to confirm the quantum theory [29–31].

Following the theme of antenna arrays, we outline in this section the basic theory needed to describe how the geometry of a quantum antenna array influences directional correlations in photon statistics. To this end, we assume that the quantum antenna array is surrounded by a number L of photon counters located at positions $\mathbf{r}_l$, $l = 1, \ldots, L$. We also assume that these devices operate via direct photon detection, though alternatives such as homodyne detection might be considered [32]. Therefore, the detectors can be phenomenologically modeled by using Glauber's theory of optical coherence [33, 34], which connects the probability of photon coincidence detections with field correlations.

Specifically, a measurement event can be modeled as a change in the quantum state of the detector from $|\psi_1\rangle$ to $|\psi_2\rangle$, triggered by the absorption of a photon. Since the annihilation of environmental excitations is described by the positive frequency electric field operator, the probability amplitude for a photon with polarization p to be detected at time t and location $\mathbf{r}$ is $\langle\psi_2|\hat{E}_p^{(+)}(\mathbf{r},t)|\psi_1\rangle$. Consequently, the total probability density of detecting a photon is proportional to the sum over all of the probabilities for each final state:

$$\begin{aligned} P_1(\mathbf{r},t,p) &\propto \sum_{\psi_2} \left|\langle\psi_2|\hat{E}_p^{(+)}(\mathbf{r},t)|\psi_1\rangle\right|^2 \\ &= \sum_{\psi_2} \langle\psi_1|\hat{E}_p^{(-)}(\mathbf{r},t)|\psi_2\rangle\langle\psi_2|\hat{E}_p^{(+)}(\mathbf{r},t)|\psi_1\rangle = \left\langle \hat{E}_p^{(-)}(\mathbf{r},t)\hat{E}_p^{(+)}(\mathbf{r},t)\right\rangle \end{aligned} \tag{12.33}$$

where we have made use of the identity operator $\sum_{\psi_2}|\psi_2\rangle\langle\psi_2| = \hat{I}$ to obtain the intermediate expression.

For a set of L detectors, the procedure outlined earlier can be generalized to describe the coincidence measurement of L photons with $p_1, \dots, p_L$ polarizations, at different times $t_1, \dots, t_L$, and for different positions $\mathbf{r}_1, \dots, \mathbf{r}_L$:

$$P_L\left(\mathbf{r}_1, t_1, p_1; \dots; \mathbf{r}_L, t_L, p_L\right) \propto \left\langle \widehat{E}_{p_1}^{(-)}\left(\mathbf{r}_1, t_1\right) \dots \widehat{E}_{p_L}^{(-)}\left(\mathbf{r}_L, t_L\right) \widehat{E}_{p_L}^{(+)}\left(\mathbf{r}_L, t_L\right) \dots \widehat{E}_{p_1}^{(+)}\left(\mathbf{r}_1, t_1\right)\right\rangle \tag{12.34}$$

If we assume that the detectors are in the far-zone of the array at positions $\mathbf{r}_n = \mathbf{u}_n r$ it is clear from Eq. (12.31) that the photon statistics are not affected by the separation distance r. Therefore, it is possible to define the probability density per unit (time)^L and (solid angle)^L as:

$$P_L\left(\mathbf{u}_1, t_1, p_1; \dots; \mathbf{u}_L, t_L, p_L\right) \propto \left\langle \widehat{E}_{p_1}^{(-)}\left(\mathbf{u}_1, t_1\right) \dots \widehat{E}_{p_L}^{(-)}\left(\mathbf{u}_L, t_L\right) \widehat{E}_{p_L}^{(+)}\left(\mathbf{u}_L, t_L\right) \dots \widehat{E}_{p_1}^{(+)}\left(\mathbf{u}_1, t_1\right)\right\rangle \tag{12.35}$$

The probability density $P_L\left(\mathbf{u}_1, t_1, p_1; \dots; \mathbf{u}_L, t_L, p_L\right)$ described by Eq. (12.35) enables the study of interesting spatio-temporal correlations. Importantly, it can be shown that any other photon statistic can be constructed from such a probability density [35, 36].

As classical antenna array theory and its practical applications emphasize, directivity is a major performance characteristic associated with any array. Consequently, we are most interested in the directional correlations that can be induced by the geometry of a quantum array. In order to emphasize the directional behavior of its emissions, independent of the times at which the measurements take place, we define the time integral of the probability density:

$$g_L\left(\mathbf{u}_1, p_1; \dots; \mathbf{u}_L, p_L\right) = \int_0^\infty dt_1 \dots dt_L \;\; P_L\left(\mathbf{u}_1, t_1, p_1; \dots; \mathbf{u}_L, t_L, p_L\right) \tag{12.36}$$

The integrated correlations g_L represent the average number of L-photon coincidence measurements for any time delay. They are a good figure of merit of how "bunched" the photons are in different directions after their emission.

The functions (12.36) include time integrals over rapidly oscillatory functions that are related to the transition frequency of the dipoles, ω_n. Unless the transition frequencies are very close to each other, these integrals will average to zero over time. As a consequence, interference phenomena might only be observable for short time intervals. In general, this property indicates that no quantum interference of any significance will take place unless the emitters have similar transition frequencies, i.e. $\omega_n \simeq \omega_0$, $\forall n$. In such a case, the time-integrated correlation (12.36) reduces to:

$$\begin{aligned} g\left(\mathbf{u}_1, p_1; \dots; \mathbf{u}_L, p_L\right) &\propto \sum_{n_1,\dots,n_L} \sum_{m_1,\dots,m_L} \prod_{l=1}^{L} \left|c_{pl}(\phi_l, \theta_l)\right|^2 \\ &\times \left\langle \widehat{\sigma}_{n_1}^\dagger \dots \widehat{\sigma}_{n_L}^\dagger \widehat{\sigma}_{m_L} \dots \widehat{\sigma}_{m_1}\right\rangle \prod_{l=1}^{L} e^{i\frac{\omega_0}{c}\mathbf{u}_l\cdot\mathbf{r}_{n_l}} e^{-i\frac{\omega_0}{c}\mathbf{u}_l\cdot\mathbf{r}_{m_l}} \end{aligned} \tag{12.37}$$

with $c_p(\phi, \theta) = \mathbf{u}_p \cdot \mathbf{I}_T \cdot \mathbf{u}_n$. With this general form, time-integrated correlation functions of arbitrary order can be evaluated for any array geometry and initial state of the emitters in the quantum array.

12.4 Linear Array of Quantum Emitters

We illustrate the applicability of the theory outlined earlier by studying the photon statistics of a uniform vertical linear array of quantum emitters, an archetypical geometry in textbooks

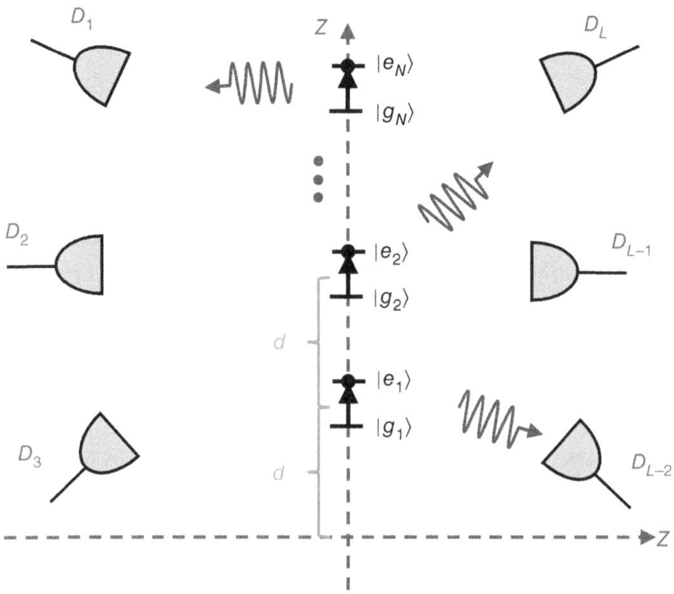

Figure 12.2 *Uniform linear vertical arrays.* Schematic depiction of a uniform vertical linear array of N quantum emitters. The emitters are located at positions $\mathbf{r}_n = \mathbf{u}_z d$ along the Z-axis and are separated uniformally by the distance d. Each emitter has a transition dipole moment, $\mathbf{p}_n = \mathbf{u}_z p_n$, that is also oriented along the Z-axis.

of antenna theory [37–40]. Specifically, we consider the geometry schematically depicted in Figure 12.2, where all of the emitters are located at the positions $\mathbf{r}_n = \mathbf{u}_z nd$ along the Z-axis with equal separation distance d, and each has its transition dipole moment also oriented along the Z-axis, i.e. $\mathbf{p}_n = \mathbf{u}_z p_n$.

For this particular configuration, the far-zone electric field operator has a single polarization component $\widehat{\mathbf{E}}^{(+)}(\mathbf{r}, t) = \mathbf{u}_\theta \widehat{E}_\theta^{(+)}(\mathbf{r}, t)$. Then, Eq. (12.24) reduces to

$$\widehat{E}_\theta^{(+)} = A\,\frac{e^{-(i\omega_0+\gamma)t_d}}{r}\,\sin\theta \sum_{n=1}^{N} \widehat{\sigma}_n\ e^{-ik_0\mathbf{u}_r\cdot\mathbf{r}_n} \tag{12.38}$$

where $A = p\ \mu_0\omega_n^2/(4\pi)$. In the same manner the time-integrated correlation function (12.36) can be compactly written as

$$g_L\left(\mathbf{u}_1, \dots, \mathbf{u}_L\right) \propto \left(\prod_{l=1}^{L}\sin^2\theta_l\right) f_L\left(\mathbf{u}_1, \dots, \mathbf{u}_L\right) \tag{12.39}$$

where we have defined a generalized L-order quantum array factor:

$$f_L\left(\mathbf{u}_1, \dots, \mathbf{u}_L\right) = \sum_{n_1,\dots,n_L=1}^{N}\ \sum_{m_1,\dots,m_L=1}^{N} \left\langle \widehat{\sigma}_{n_1}^\dagger \dots \widehat{\sigma}_{n_L}^\dagger \widehat{\sigma}_{m_L} \dots \widehat{\sigma}_{m_1}\right\rangle \prod_{p=1}^{L} e^{ik_0\mathbf{u}_p\cdot\mathbf{r}_{n_p}} \prod_{q=1}^{L} e^{-ik_0\mathbf{u}_q\cdot\mathbf{r}_{m_q}} \tag{12.40}$$

Similar to the classical array factor, the generalized array factor describes how the geometry of the array affects the photon statics of different order. For example, the $L = 1$ case corresponds to the average number of photons measured in a given direction. The $L = 2$ case provides a good figure of merit of how directionally bunched the emitted photons are. When each emitter acts as a single-photon source, $f_L = 0$ for $L > N$.

12.4.1 First-Order Correlations

The first-order correlation, $g_1(\mathbf{u}_1) \propto \sin\theta_1 f_1(\mathbf{u}_1)$, physically predicts the average number of photons measured by a detector placed along the direction $\mathbf{u}_1$. The generalized array factor for $L = 1$ reduces to

$$f_1(\mathbf{u}_1) = \sum_{n,m=1}^{N} \left\langle \hat{\sigma}_n^\dagger \hat{\sigma}_m \right\rangle e^{ik\mathbf{u}_1\cdot\mathbf{r}_n} e^{-ik\mathbf{u}_1\cdot\mathbf{r}_m} \tag{12.41}$$

Equation (12.41) reveals that the first-order generalized quantum array factor is very similar to the magnitude squared of the classical array factor [37–40]. This is not surprising since the classical power emission pattern should correspond to the average number of photons. At the same time, however, there is an important difference that results in nonclassical effects even in the average number of photons. Specifically, the initial-time correlation $\langle\hat{\sigma}_n^\dagger\hat{\sigma}_m\rangle$ plays the role of the signals feeding the nth and mth radiating elements. On the other hand, the importance difference is that the initial-time correlation is not always necessarily factored as the product of two c-numbers, e.g. as $\langle\hat{\sigma}_n^\dagger\hat{\sigma}_m\rangle = a_n^* a_m$. As a consequence, access is granted to configurations forbidden in the classical case.

In the same manner that classical antenna arrays control the emission pattern by changing the magnitude and phases of the signals driving their antenna elements, quantum antenna arrays control the emission pattern by controlling the initial state of the system, i.e. how the array is initially excited to start the emission process. There are several experimental techniques that enable the preparation of a given initial state of a quantum system. For example, initialization laser pulses and incoherent pumping can be employed for initial states corresponding to product states (see, e.g. [41]), where each emitter is individually excited. More complex initialization schemes are required for entangled states. Specific phase profiles for them could be implemented, for instance, if they are based on laser-assisted interactions [42, 43]).

We consider the single-excitation state:

$$|\psi_1\rangle = \sum_{r=1}^{N} a_r \hat{\sigma}_r^\dagger |0\rangle \tag{12.42}$$

as an example of an initial entangled state. This is actually a maximally entangled state in which a single excitation is shared by all the elements of the array. By placing the same phase on each emitter, the result is similar to that of a classical array with a uniform excitation. In fact, we recover a factorization analogous to the classical case $\langle\hat{\sigma}_n^\dagger\hat{\sigma}_m\rangle = a_n^* a_m$. The first-order generalized array factor then reduces to

$$f_{1,\psi_1}(\mathbf{u}_1) = \left| \sum_{m=1}^{N} a_m e^{-ik_0\mathbf{u}_1\cdot\mathbf{r}_m} \right|^2 \tag{12.43}$$

Despite the fact that we are using a nonclassical, maximally entangled state, it is found that the radiated field can be beamformed exactly as in the classical case. The probability amplitudes play exactly the same role as the currents driving the classical antenna elements. This is an exciting outcome because it means that all the machinery developed for the synthesis of classical antenna arrays can be directly applied to single-excitation state initiated quantum arrays.

Next, we consider the initial state in which all N emitters are initially excited, i.e.

$$|\psi_{SN}\rangle = \prod_{r=1}^{N} \hat{\sigma}_r^\dagger |0\rangle \tag{12.44}$$

Despite the fact that it is a factorized (not entangled) state, this configuration is very relevant from a practical standpoint. It represents the case in which all emitters are independently excited, which could be implemented in practice with electrically driven elements. Moreover, it leads to nonclassical photon statistics. Specifically, the correlator for this initial state reduces to $\langle \hat{\sigma}_n^\dagger \hat{\sigma}_m \rangle = \delta_{nm}$. Thus, the generalized array factor takes the very simple form:

$$f_{1,\psi_{SN}} = N \tag{12.45}$$

It can be concluded from this simple expression that the emitters do not interfere and, hence, do not modify the directional properties of the average number of photons. In other words, the emission pattern in terms of the average number of photons is identical to the individual emitter pattern, i.e. a $\sin^2\theta$ function, and is independent of the number and geometrical arrangement of the emitters. The fact that the emission pattern is independent of the number of emitters and the geometry of the array is a purely quantum effect with no counterpart in classical antenna theory.

In order to illustrate this point, Figure 12.3 depicts the first-order time-integrated correlation $g_1(\mathbf{u}_1)$ for a uniform linear vertical array composed by $N = 3$ identical emitters separated by one wavelength, i.e. for the emitter locations $\mathbf{r}_n = \mathbf{u}_z n \lambda_0$. The patterns are only depicted in the XZ-plane due to the fact that the geometry is rotationally symmetric around the Z-axis. Classical linear vertical arrays are known for their enhancement of the directivity, focusing the emitter radiation into the plane perpendicular to the array axis (i.e. the broadside direction $\theta = \pi/2$) [37]. Following the theory earlier, Figure 12.3 confirms that the emitted photons will be constrained to an narrower set of directions than in the single emitter case (shown as a dashed black line) when the array is initially driven into a single-excitation state. On the other hand, it is found that the emission pattern is identical to that of the individual elements when all three emitters are initially excited. This outcome confirms the peculiar quantum effect that the directionality on the average number of photons in the case in which all of the identical emitters are initially excited is not affected by their array configuration.

It can be concluded from the two examples earlier that it is possible to beamform photon statistics by selecting how the array is initially excited. While we have shown here the two main examples of

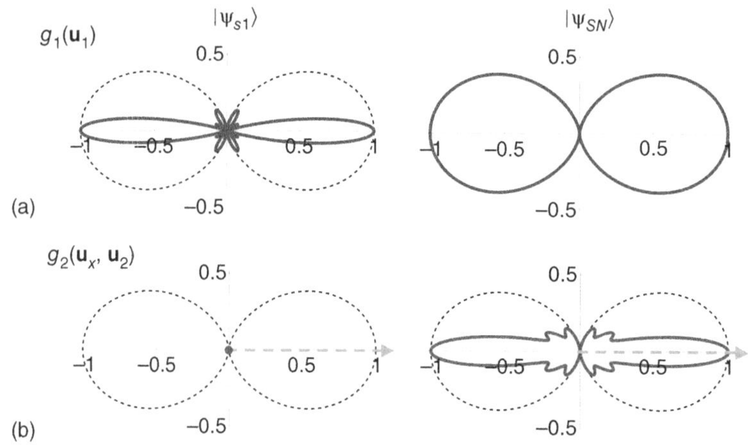

Figure 12.3 *Photon statistics from a linear vertical array.* Linear array of $N = 3$ identical quantum emitters located along the Z-axis with constant separation $d = \lambda_0$. (a) First-order correlation pattern, $g_1(\mathbf{u}_1)$, as a function of θ_1 for the (left) initial symmetrical single-excitation state $|\psi_{S1}\rangle$ and (right) the initial N-excitation state $|\psi_{SN}\rangle$. (b) Same as in (a) but for the second-order correlation function $g_1(\mathbf{u}_1, \mathbf{u}_2)$ as a function of $\mathbf{u}_2$ for $\mathbf{u}_1 = \mathbf{u}_x$ (indicated as the gray dashed arrows).

single-excitation and N-excitation states, many other options are available and could be explored. Based on similar considerations, the possibility of designing a perfectly isotropic antenna is discussed in Section 12.5. Similarly, the difficulties in harnessing these extra degrees of freedom to obtain quantum superdirectivity beyond classical limits are summarized in Section 12.6.

12.4.2 Second-Order Correlations

The second-order ($L = 2$) time-integrated correlation for the same linear array takes the form:

$$g_2(\mathbf{u}_1, \mathbf{u}_2) \propto \sin\theta_1 \sin\theta_2 f_2(\mathbf{u}_1, \mathbf{u}_2) \tag{12.46}$$

By definition, it represents the average number of two-photon coincidences for any time delay. In practice, it is a good figure of merit of how directionally bunched the emitted photons are. Specifically, since Glauber's probability densities are nonexclusive quantities [35], $g_2(\mathbf{u}_1, \mathbf{u}_2) = 0$ implies that no photon coincidences of any order will be recorded along the directions $\mathbf{u}_1$ and $\mathbf{u}_2$ for the same decay process. Consequently, if $g_2(\mathbf{u}_1, \mathbf{u}_2)$ has a highly directive pattern, it indicates that all of the emitted photons are bunched into a narrow set of directions.

For this reason, engineering a quantum antenna array with a highly-directive second-order correlation would lead to a nonclassical light source whose photons are emitted in directionally entangled bunches. Similar to the first-order array factor, the second-order array factor, $f_2(\mathbf{u}_1, \mathbf{u}_2)$, can be controlled by the initial state of the system, which affects the correlation function, $\left\langle \sigma_{n_1}^{\dagger} \sigma_{n_2}^{\dagger} \sigma_{m_2} \sigma_{m_1} \right\rangle$, as well as by the geometry of the array.

Importantly, the angular patterns of the second-order correlations do not have to match those obtained for the first-order correlation. They can have an entirely different behavior. In order to illustrate this point, Figure 12.3b depicts the time-integrated second order correlation $g_2(\mathbf{u}_1, \mathbf{u}_2)$ for the same initial configurations associated with Figure 12.3a. To obtain these results, we fixed the first evaluation direction to be $\mathbf{u}_1 = \mathbf{u}_x$ and then evaluated the correlation function as a function of $\mathbf{u}_2$.

First, for the single-excitation state the initial-time correlator trivially reduces to $\left\langle \sigma_{n_1}^{\dagger} \sigma_{n_2}^{\dagger} \sigma_{m_2} \sigma_{m_1} \right\rangle = 0$. Consequently, the second order correlation is identically zero for all evaluation angles, i.e. $g_2(\mathbf{u}_1, \mathbf{u}_2) = 0 \;\; \forall \mathbf{u}_1, \mathbf{u}_2$. This nonclassical effect is the directional counterpart to temporal antibunching. It describes the impossibility of measuring two excitations from a single-excitation state. As was noted, $g_L = 0$ for $L > N$ within our model. From a practical standpoint this condition implies that two different detectors would never detect a photon on the same decay run if the system was initialized with a single-excitation state.

In contrast, we obtain for the N-excitation state a nontrivial initial-time correlation

$$\left\langle \sigma_{n_1}^{\dagger} \sigma_{n_2}^{\dagger} \sigma_{m_2} \sigma_{m_1} \right\rangle = \left(1 - \delta_{m_1 m_2}\right)\left(\delta_{n_1 m_1}\delta_{n_1 m_2} + \delta_{n_1 m_2}\delta_{n_2 m_1}\right) \tag{12.47}$$

Then, the second-order generalized array factor can be written as:

$$f_{2,\psi_{SN}}(\mathbf{u}_1, \mathbf{u}_2) = N(N-2) + \left| \sum_{m=1}^{N} e^{-ik_0(\mathbf{u}_1 - \mathbf{u}_2)\cdot \mathbf{r}_m} \right|^2 \tag{12.48}$$

In this manner, we find that there are directional effects in the second order correlation factor for an array in which all emitters are independently excited. They occur despite the absence of any directionality in the first-order correlation. Therefore, it can be then concluded that the photons are probabilistically emitted with no directional preference other than the pattern of the single emitter,

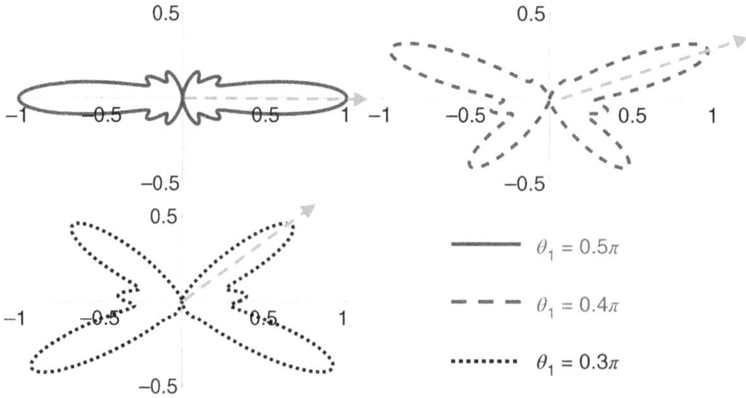

Figure 12.4 *Examples of second-order photon statistics.* Normalized second-order angular correlation function $g_2(\mathbf{u}_1, \mathbf{u}_2)$ evaluated for a vertical linear array of $N = 3$ emitters, with all emitters initially excited and $\mathbf{u}_1(\phi_1 = 0, \theta_1 = 0.5\pi)$, $\mathbf{u}_1(\phi_1 = 0, \theta_1 = 0.4\pi)$, and $\mathbf{u}_1(\phi_1 = 0, \theta_1 = 0.3\pi)$. Gray dashed arrow indicates the direction of $\mathbf{u}_1$ in each subpanel.

i.e. with no directionality for $g_1(\mathbf{u}_1)$. On the other hand, the photons will all be measured bunched around a set of directions for a given decay process, i.e. with the directionality of $g_2(\mathbf{u}_1, \mathbf{u}_2)$.

This effect is somewhat analogous to two-photon interference in beamsplitters. The two photons are equally probable to exit this system into any of the two output ports. Nevertheless, both are certain to exit into the same output port [32]. Here, a number N of photons are equally probable (up to the individual emitter pattern) to exit the system on a continuum of output channels (i.e. the directions), but they are certain to be measured within a narrow interval of directions. This effect can be more clearly appreciated in Figure 12.4. The second-order correlation $g_2(\mathbf{u}_1, \mathbf{u}_2)$ is presented for three different $\mathbf{u}_1$ given by $(\phi_1 = 0, \theta_1 = 0.3\pi)$, $(\phi_1 = 0, \theta_1 = 0.4\pi)$, and $(\phi_1 = 0, \theta_1 = 0.5\pi)$. The angular patterns consist of lobes centered around each $\mathbf{u}_1$, plus additional grating lobes.

These photon statistic outcomes can then be tailored by changing the geometry of the array. To this end, different design techniques could be developed. They are very likely to exhibit both similarities and differences with respect to the synthesis techniques developed for classical antenna arrays. In order to illustrate this point, Figure 12.5 depicts $g_2(\mathbf{u}_1, \mathbf{u}_2)$ as a function of the number of emitters and the distance of separation between them. In analogy with classical antenna arrays, we find that increasing the separation between the emitters results in the appearance of grating lobes, i.e. the emitted photons will be bunched around not one, but a set of multiple lobes. However, in contrast with classical antenna arrays, it is found that the directivity does not increase monotonically along with the number of emitters. Conversely, the $g_2(\mathbf{u}_1, \mathbf{u}_2)$ results in the superposition of the single emitter pattern with that of a highly directive lobe. These results suggest that nonclassical techniques will have to be developed for the synthesis of quantum antenna arrays and *each* of their different photon correlations.

12.4.3 *N*th-Order Correlations

Examining more closely the generalized array factor Eq. (12.40), it is clear that computing the photon statistics becomes a very challenging task as the order increases. Having addressed the first- and second-order correlations, here we address the Nth-order photon correlations for an antenna

(a)

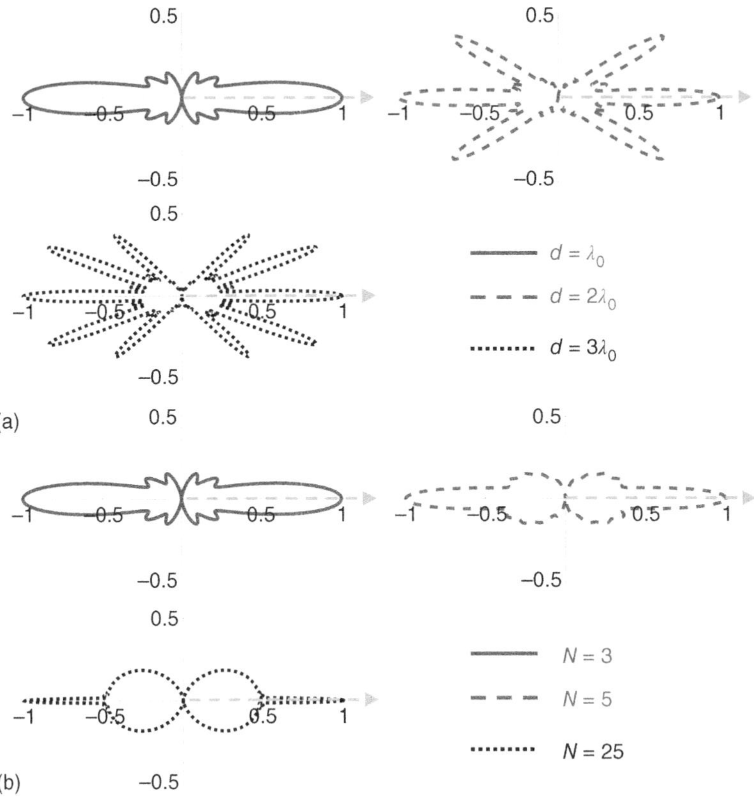

(b)

Figure 12.5 *Engineering the statistics of the emitted photon by tuning the array geometry.* Normalized second-order angular correlation function $g_2(\mathbf{u}_1, \mathbf{u}_2)$ evaluated with $\mathbf{u}_1\ (\phi_1 = 0, \theta_1 = 0.5\pi)$ for a linear vertical array. (a) $N = 3$ elements with separation distances $d = 1\lambda_0$, $d = 2\lambda_0$ and $d = 3\lambda_0$. (b) A separation distance $d = 1\lambda_0$ and arrays with $N = 3$, $N = 5$, and $N = 25$ emitters. Gray dashed arrow indicates the direction of $\mathbf{u}_1$ in each subpanel.

array in which all emitters are initially excited, i.e. for which

$$|\psi\rangle = \prod_{r=1}^{N} \hat{\sigma}_r^\dagger |0\rangle \tag{12.49}$$

It is the highest-order nonzero correlation for an array of N single photon sources. Interestingly, a tractable expression can be derived for this extreme case.

The main difficulty in computing the photon statistics stems from evaluating the Nth-order correlator:

$$\left\langle \hat{\sigma}_{n_1}^\dagger \dots \hat{\sigma}_{n_N}^\dagger \hat{\sigma}_{m_N} \dots \hat{\sigma}_{m_1} \right\rangle \tag{12.50}$$

In order to evaluate the correlator, let $\{|\zeta_n\rangle\}$ with $n = 1, \dots, 2^N$ be a basis for all possible states available to the array of N emitters. This basis can be arbitrary, but for convenience we define the first state of this basis as all emitter's being in the ground state, i.e. $|\zeta_1\rangle = |g_1, \dots, g_N\rangle$. Next, we find the following identity

$$\langle \zeta_n | \hat{\sigma}_{m_N} \dots \hat{\sigma}_{m_1} |\psi\rangle = \begin{cases} 1 & \text{if } |\zeta_n\rangle = |\zeta_1\rangle \quad \text{and} \quad m_i \neq m_j \quad \forall i,j \\ 0 & \text{else} \end{cases} \tag{12.51}$$

We then make use of the identity operator $\hat{I} = \sum_n |\zeta_n\rangle\langle\zeta_n|$ to evaluate Eq. (12.52) as follows

$$\langle\psi|\hat{\sigma}^{\dagger}_{n_1}\dots\hat{\sigma}^{\dagger}_{n_N}\hat{\sigma}_{m_N}\dots\hat{\sigma}_{m_1}|\psi\rangle = \sum_n \langle\psi|\hat{\sigma}^{\dagger}_{n_1}\dots\hat{\sigma}^{\dagger}_{n_N}|\zeta_n\rangle\langle\zeta_n|\hat{\sigma}_{m_N}\dots\hat{\sigma}_{m_1}|\psi\rangle = \xi_{m_1\cdots m_N} \tag{12.52}$$

where we define $\xi_{m_1\cdots m_N} = 1$ for $m_i \neq m_j$ $\forall i,j$ and $\xi_{m_1\cdots m_N} = 0$ in the rest of the cases.

With this result, the generalized quantum array factor can be written as

$$f_N(\mathbf{u}_1,\dots,\mathbf{u}_N) = \left|\sum_{m_1,\dots,m_N=1}^{N}\xi_{m_1\cdots m_N}\prod_{p=1}^{N}e^{ik_0\mathbf{u}_p\cdot\mathbf{r}_{m_p}}\right|^2 \tag{12.53}$$

Consequently, the definition of $\xi_{m_1\cdots m_N}$ implies that the sum in Eq. (12.53) runs over all possible permutations of the $\mathbf{r}_n$ position vectors of the following product of exponentials:

$$e^{ik_0\mathbf{u}_1\cdot\mathbf{r}_1}e^{ik_0\mathbf{u}_2\cdot\mathbf{r}_2}\ \dots\ e^{ik_0\mathbf{u}_N\cdot\mathbf{r}_N} \tag{12.54}$$

For the example of three emitters used in this chapter, the generalized array factor is explicitly:

$$\begin{aligned} f_{N=3}(\mathbf{u}_1,\mathbf{u}_2,\mathbf{u}_3) = \Big| & e^{ik_0\mathbf{u}_1\cdot\mathbf{r}_1}e^{ik_0\mathbf{u}_2\cdot\mathbf{r}_2}e^{ik_0\mathbf{u}_3\cdot\mathbf{r}_3} + e^{ik_0\mathbf{u}_1\cdot\mathbf{r}_1}e^{ik_0\mathbf{u}_2\cdot\mathbf{r}_3}e^{ik_0\mathbf{u}_3\cdot\mathbf{r}_2} \\ & + e^{ik_0\mathbf{u}_1\cdot\mathbf{r}_2}e^{ik_0\mathbf{u}_2\cdot\mathbf{r}_1}e^{ik_0\mathbf{u}_3\cdot\mathbf{r}_3} + e^{ik_0\mathbf{u}_1\cdot\mathbf{r}_2}e^{ik_0\mathbf{u}_2\cdot\mathbf{r}_3}e^{ik_0\mathbf{u}_3\cdot\mathbf{r}_1} \\ & + e^{ik_0\mathbf{u}_1\cdot\mathbf{r}_3}e^{ik_0\mathbf{u}_2\cdot\mathbf{r}_2}e^{ik_0\mathbf{u}_3\cdot\mathbf{r}_1} + e^{ik_0\mathbf{u}_1\cdot\mathbf{r}_3}e^{ik_0\mathbf{u}_2\cdot\mathbf{r}_1}e^{ik_0\mathbf{u}_3\cdot\mathbf{r}_2}\Big|^2 \end{aligned} \tag{12.55}$$

This result can be understood intuitively in terms of the "which-path" information that is available in the state of the quantum emitter subsystem. After measuring a N-photon coincidence event, the emitter's subsystem must necessarily be in the ground state, $|e_1 \cdots e_N\rangle \to |g_1 \cdots g_N\rangle$. Therefore, the state of the emitters contain no information in terms of what emitter radiated the photon measured at a given detector. In other words, there is no which-path information and, hence, quantum interference can occur. In this manner, each of the permutations in (12.40) can be understood as a different path that connects the detection in each $\mathbf{u}_p$ direction to the emitter in the $\mathbf{r}_{m_p}$ location. For these reasons, computing the photon statistics become "simpler." This result also shows that the directivity of the photon statistics is expected to increase along with their order since less and less which-path information is available as more photons are measured (see Figure 12.3).

12.5 Isotropic Single-Photon Sources

We provide in this section yet another example of a quantum antenna device with functionalities that cannot be obtained with a classical antenna system. In particular, we demonstrate that it is possible to have a perfectly isotropic and temporally coherent single-photon source with full-polarization coverage [44]. Isotropic antennas are used as a reference in the very definition of directivity. However, it is a well-known result from classical antenna theory that perfectly isotropic and temporally coherent radiators are not possible [45–48]. This conclusion is best visualized by the hairy ball theorem [49, 50], an algebraic topology result stating that there is no nonvanishing continuous tangent vector field over the entire surface of a sphere. In other words, "you cannot comb a hairy ball flat without creating a cowlick." Therefore, perfect isotropy with a well-defined polarization is impossible.

Despite this impossibility, antenna designers continue to seek efficient quasi-isotropic antennas that approximate the properties of an isotropic radiator [51–54]. The motivation behind

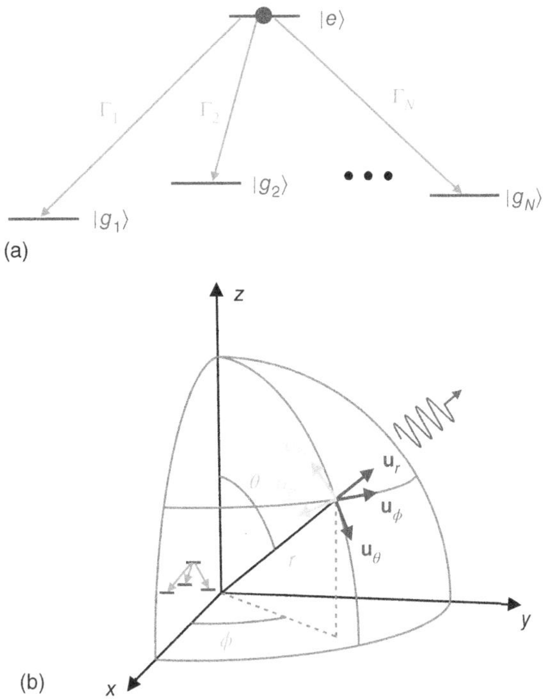

Figure 12.6 *Isotropic single photon sources.* (a) Emitter level structure showing a single excited state connected via radiative transitions to a set of degenerate ground states. (b) Coordinate system and polarization basis. Emission of a photon into the direction $\mathbf{u}_r$, whose polarization is described by the transverse zenith $\mathbf{u}_\theta$ and azimuthal $\mathbf{u}_\phi$ vectors, or, equivalently, by an arbitrary polarization basis composed of two orthogonal vectors $\mathbf{u}_a$ and $\mathbf{u}_b$.

quasi-isotropic antennas is that they provide full-coverage, an invaluable property for any wireless technology in which transmitting and receiving devices cannot be aligned [55–57]. As a separate note, isotropic sources do exist in the form of thermal radiators [22], and multifrequency systems [58], at the cost of being temporally incoherent. Beyond electromagnetic systems, isotropic sources do exist in other physical systems such as in acoustics [59].

Nevertheless, an isotropic single-photon source can be designed by using the emitter schematically depicted in Figure 12.6. It consists of a multi-level system with a single excited state $|e\rangle$, and N different ground states $|g_1\rangle \dots |g_N\rangle$. The transition from the excited state to each ground place is characterized by the transition frequency ω_n and the dipole moment $\mathbf{p}_n$. The electric field operator takes the same form as that derived for an array of quantum emitters, albeit with all the emitters being located at the same position. Consequently, Eq. (12.38) reduces to

$$\hat{E}_p^{(+)}(\mathbf{r},t) = \frac{\mu_0}{4\pi r}\sum_{n=1}^{N}\omega_n^2\left[\mathbf{u}_p\cdot\mathbf{I}_T\cdot\mathbf{p}_n\right]\hat{\sigma}_n\, e^{-(i\omega_n+\gamma_n)t_d} \tag{12.56}$$

The probability density of measuring a photon with polarization p at the position $\mathbf{r}$ and time t is given by Eq. (12.33). For the multi-level emitter under consideration it can be explicitly written as

$$\left\langle \hat{E}_p^-(\mathbf{r},t)\,\hat{E}_p^+(\mathbf{r},t)\right\rangle = \left(\frac{\mu_0}{4\pi r}\right)^2\sum_{n,m}\omega_n^2\omega_m^2\left\langle\hat{\sigma}_n^\dagger\hat{\sigma}_m\right\rangle$$
$$\times e^{-i(\omega_m-\omega_n)t_d}\, e^{-(\gamma_m+\gamma_n)t_d}\left(\mathbf{u}_p\cdot\mathbf{I}_T\cdot\mathbf{p}_m\right)\left(\mathbf{u}_p\cdot\mathbf{I}_T\cdot\mathbf{p}_n^*\right) \tag{12.57}$$

Again, the nonclassical properties of the emission pattern in the proposed configuration arise from the $\left\langle\hat{\sigma}_n^\dagger\hat{\sigma}_m\right\rangle$ correlators in Eq. (12.57). In contrast with the classical case, the correlation does not need to be decomposed into the product of two complex numbers, opening up additional degrees of freedom. For a multi-level emitter with a single excited state and multiple ground states, we have that $\left\langle \hat{\sigma}_n^{\dagger}\hat{\sigma}_m \right\rangle = \delta_{nm}$. Interestingly, the result of such a correlator is that there is no interference between the field operators associated with each transition. The absence of interference can be intuitively understood as the result of having distinguishable ground states and thus access to perfect which-path information. In this sense, the distinguishable ground states play the same role as passive observers similar to popular effects such as two-slit experiments [60] and quantum erasers [61, 62].

The lack of interference is a nonclassical property arising from the superposition of distinguishable decay paths. It can be harnessed to design an isotropic single-photon source. To this end, we particularize the multi-level emitter to the case with three degenerate ground-states with the same transition frequency ($\omega_n = \omega_0 \ \forall n$) and with transition dipole moments of the same magnitude ($|\mathbf{p}_n| = |\mathbf{p}| \ \forall n$), but oriented along each of the Cartesian axis, i.e. $\mathbf{p}_1 = p\,\mathbf{u}_z$, $\mathbf{p}_2 = p\,\mathbf{u}_x$ and $\mathbf{p}_3 = p\,\mathbf{u}_y$. For this configuration, the field correlation reduces to

$$\left\langle \hat{E}_p^-(\mathbf{r},t)\,\hat{E}_p^+(\mathbf{r},t) \right\rangle = \left(\frac{\mu_0\omega_0^2}{4\pi r}\right)^2 |p|^2 e^{-2\gamma t_d} \sum_{n=x,y,z} \left|\mathbf{u}_p \cdot \mathbf{I}_T \cdot \mathbf{u}_n\right|^2 \tag{12.58}$$

In order to show that the emission will be isotropic for all arbitrary polarizations, we recall that an arbitrary polarization basis $(\mathbf{u}_p, \mathbf{u}_q)$ can be obtained from the usual $(\mathbf{u}_\theta, \mathbf{u}_\phi)$ basis via rotation by an angle φ and the introduction a phase shift ζ (see Figure 12.6):

$$\mathbf{u}_p = \cos\varphi\, e^{i\frac{\zeta}{2}}\,\mathbf{u}_\theta - \sin\varphi\, e^{i\frac{\zeta}{2}}\,\mathbf{u}_\phi \tag{12.59}$$

$$\mathbf{u}_q = \sin\varphi\, e^{-i\frac{\zeta}{2}}\,\mathbf{u}_\theta + \cos\varphi\, e^{-i\frac{\zeta}{2}}\,\mathbf{u}_\phi \tag{12.60}$$

By using this decomposition, the probability density of measuring a photon with arbitrary polarization p at time t and location $\mathbf{r}$ is found to be:

$$\begin{aligned}\left\langle \hat{E}_p^-(\mathbf{r},t)\,\hat{E}_p^+(\mathbf{r},t) \right\rangle &= \left(\frac{\mu_0\omega_0^2}{4\pi r}\right)^2 |p|^2 e^{-2\gamma t_d} \left\{ \left|\cos\varphi e^{i\frac{\zeta}{2}}\ \sin\theta\right|^2 \right.\\ &\quad + \left|\cos\varphi e^{i\frac{\zeta}{2}}\ \cos\phi\ \cos\theta + \sin\varphi e^{i\frac{\zeta}{2}}\ \sin\phi\right|^2 \\ &\quad \left. + \left|\cos\varphi e^{i\frac{\zeta}{2}}\ \sin\phi\ \cos\theta - \sin\varphi e^{i\frac{\zeta}{2}}\ \cos\phi\right|^2 \right\} \\ &= \left(\frac{\mu_0\omega_0^2}{4\pi r}\right)^2 |p|^2 e^{-2\gamma t_d}\end{aligned} \tag{12.61}$$

Equation (12.61) reveals that the probability density of detecting a photon is independent of the measurement direction, which is described by the angles θ and ϕ. Therefore, it can be concluded that the emission is isotropic, i.e. the same probability is obtained for all directions. Moreover, since the probability density is independent of the choice of the polarization vector $\mathbf{u}_p$, it can also be concluded that the same result holds for all polarizations. In this manner, it is demonstrated that a perfectly isotropic and with full polarization coverage single-photon source does exist.

The emission from an isotropic single-photon source resembles black-body radiation from a heated object, i.e. the emitted radiation is both isotropic and unpolarized. The main difference is that isotropic single-photon sources are coherent radiators. Thus, the emitted light corresponds to a single photon with a well-defined wavepacket. For this reason, it is a coherent source for which interference phenomena can be observed. For example, interference fringes will appear if the emitted photon is sent through a Mach–Zehnder interferometer. In conclusion, isotropic single-photon sources are a quantum light source with properties that cannot be obtained with classical antennas.

12.6 Quantum Superdirectivity?

It was shown in earlier sections of this chapter that quantum antenna systems provide access to radiation patterns as the average number of photons per direction that cannot be obtained with classical antenna systems. This poses the question of whether or not quantum sources can be used to increase the directivity of antenna systems beyond their classical limits, leading to quantum superdirectivity. A simple proof seems to indicate that this is not possible [63]. However, researchers are known for relentlessly searching for ways to circumvent fundamental limits, and the situation might change in the future. Since the proof provides an additional insight into the mode of operation of quantum antenna arrays, we quickly reproduce it here.

For the sake of simplicity let us consider here an array of all identical emitters, even though the proof can be generalized to more complex scenarios. The starting point is then the first-order time-integrated correlation function – the average number of photons measured along the direction $\mathbf{u}_1$ – discussed previously:

$$g_1\left(\mathbf{u}_1\right) \propto \sin^2\theta_1\, f_1\left(\mathbf{u}_1\right) \tag{12.62}$$

where $f_1\left(\mathbf{u}_1\right)$ is the first-order generalized array factor:

$$f_1\left(\mathbf{u}_1\right) = \sum_{n,m=1}^{N} \left\langle \hat{\sigma}_n^{\dagger}\hat{\sigma}_m \right\rangle e^{ik_0\mathbf{u}_1\cdot\mathbf{r}_n} e^{-ik_0\mathbf{u}_1\cdot\mathbf{r}_m} \tag{12.63}$$

The classical case is recovered for $\left\langle \hat{\sigma}_n^{\dagger}\hat{\sigma}_m \right\rangle \to a_n^* a_m$. On the other hand, $\left\langle \sigma_n^{\dagger}\sigma_m \right\rangle$ does not need to be factorized as the product of two complex numbers for a quantum antenna array. Again, as a consequence, it provides additional degrees of freedom.

It was shown in Section 12.5 that such freedom can be used to design an isotropic single-photon source. At the same time, the initial-time correlation $\left\langle \hat{\sigma}_n^{\dagger}\hat{\sigma}_m \right\rangle$ has some restrictions that limit the achievable radiation patterns. To this end, we define a matrix $\mathbf{M}$ with elements $M_{nm} = \left\langle \hat{\sigma}_n^{\dagger}\hat{\sigma}_m \right\rangle$ and a vector $\mathbf{e}$ with elements $e_n = e^{-ik_0\mathbf{u}_1\cdot\mathbf{r}_n}$. We use them to express $f_1\left(\mathbf{u}_1\right)$ in matrix form as:

$$f_1\left(\mathbf{u}_1\right) = \mathbf{e}^{\dagger}\cdot\mathbf{M}\cdot\mathbf{e} \tag{12.64}$$

In contrast, the radiation pattern of a classical antenna array would be restricted to the more constrained form

$$f_{1,\text{classical}}\left(\mathbf{u}_1\right) = \mathbf{e}^{\dagger}\cdot\left(\mathbf{a}\cdot\mathbf{a}^{\dagger}\right)\cdot\mathbf{e} \tag{12.65}$$

Nevertheless, quantum antenna arrays also have some restrictions of their own. First, we note that $\left\langle \hat{\sigma}_n^{\dagger}\hat{\sigma}_m \right\rangle = \left\langle \hat{\sigma}_m^{\dagger}\hat{\sigma}_n \right\rangle^*$ and, hence, $\mathbf{M}$ is a Hermitian matrix. In addition, since $f_1\left(\mathbf{u}_1\right)$ is a positive real number, we then have that $\mathbf{M}$ is a positive semidefinite matrix. Importantly, this implies that all its eigenvalues are positive. Next, let λ_j and $\mathbf{m}_j$, $j = 1, \ldots, N$, be the eigenvalues and associated eigenvectors of $\mathbf{M}$. Then the spectral decomposition of $\mathbf{M}$ is given by

$$\mathbf{M} = \sum_{j=1}^{N} \lambda_j\ \mathbf{m}_j\cdot\mathbf{m}_j^{\dagger} \tag{12.66}$$

Subsequently, $f_1\left(\mathbf{u}_1\right)$ can be rewritten as:

$$f_1\left(\mathbf{u}_1\right) = \sum_j \lambda_j\, \mathbf{e}^{\dagger}\cdot\left(\mathbf{m}_j\cdot\mathbf{m}_j^{\dagger}\right)\cdot\mathbf{e} = \sum_j \lambda_j \left|\sum_{n=1}^{N} m_{n(j)}\, e^{-ik_0\mathbf{u}_1\cdot\mathbf{r}_n}\right|^2 \tag{12.67}$$

Equation (12.67) reveals that the quantum generalized array factor describing the average number of photons measured as a function of the direction can be written as a linear combination of classical power patterns for the same array, but with different driving coefficients. From this perspective, the power behind quantum antenna arrays is that they enable the superposition of classical power radiation patterns. However, since **M** is a positive definite matrix, its eigenvalues λ_j must be positive. Therefore, there cannot be destructive interference between the classical radiation patterns. This fact suggests that the directivity cannot be increased by the superposition. In many senses, the situation is very similar to polarization considerations in the maximization of the directivity of classical antennas [64]. The polarization degrees of freedom provide orthogonal channels in which the power can be distributed. However, they cannot be used to increase the directivity.

To demonstrate this explicitly, we first define the quantum directivity as follows:

$$\begin{aligned} D_1\left(\mathbf{u}_1\right) &= 4\pi \frac{g_1\left(\mathbf{u}_1\right)}{\int d\Omega_1\, g_1\left(\mathbf{u}_1\right)} \\ &= 4\pi \frac{\sin^2\theta_1\, \sum_j \lambda_j \left|\sum_{n=1}^{N} m_{n(j)} e^{-ik_0 \mathbf{u}_1 \cdot \mathbf{r}_n}\right|^2}{\sum_j \lambda_j \int d\Omega_1\, \sin^2\theta_1 \left|\sum_{n=1}^{N} m_{n(j)} e^{-ik_0 \mathbf{u}_1 \cdot \mathbf{r}_n}\right|^2} \end{aligned} \tag{12.68}$$

Next, if we optimize the $m_{n(j)}$ coefficients, we would obtain the same optimal coefficients for any j, and therefore the sums in the numerator and denominator would compensate in obtaining the maximal directivity. Consequently, the directivity of a quantum antenna array in terms of the average number of photons cannot exceed the upper limit for a classical antenna array.

12.7 Quantum Antenna Array Technologies

Having described the basic theory and its particularly interesting outcomes, we discuss in this section those material platforms that are currently the most promising candidates for implementing quantum antenna arrays. First, the basis of a quantum antenna array is the single-photon sources that act as their individual elements. The history of single-photon sources dates back to the 1970s with the first demonstration of a single-photon source based on a beam of sodium atoms [23]. This effort was followed a decade later by experiments with trapped ions [65]. The first solid-state single-photon source was demonstrated in the 1990s using a single molecule dye trapped in a solid [66]. Since then, research on single-photon sources has exploded, leading to their rapid development. Today, there is a variety of solid-state single-photon sources in the form of quantum dots (QDs) [67–70]; color centers in semiconductors, such as diamond, silicon carbide, and silicon [71–76]; and defects in two-dimensional (2D) materials [77–80]. In general, solid-state sources suffer from their coupling to a complex environment, e.g. phonons in a crystal lattice, charge accumulations, and interactions with nearby spins. This coupling limits their coherence times. Moreover, it poses difficulties in the fabrication of identical emitters. However, the quality of solid-state single-photon sources has been steadily improving and is rapidly approaching the required standards in terms of purity, indistinguishability and on-demand operation.

Free from a disturbing environment, ultra-cold atoms [81, 82] and trapped ions [83] are excellent single-photon sources. Atomic-based sources provide better single-photon performance, longer coherence times and a finer control over the photon's wavepacket than their solid-state counterparts do. Nevertheless, the integration of atomic-based devices into commercial systems is likely to be restricted to very high-end applications. Superconducting circuits [84, 85] exhibit the best performance in terms of the coupling strength to the environment and their coherence times. However, the low energy of microwave photons forces these circuits to operate at very low temperatures. The entire system must be contained within a cryogenic setup. Furthermore, nonlinear processes such as spontaneous parametric down conversion were identified early as very high quality nonclassical light sources [86–88], albeit at the cost of being bulky and nondeterministic in nature.

It must be noted that single-photon sources seldom consist of a quantum emitter alone. The pioneering work of Purcell [89] showed that emission processes are affected by their photonic environments. Subsequently, the use of nanostructures to tailor light-matter interactions and to enhance the performance of single-photon sources has been aggressively pursued [69, 90]. Different strategies include the use of resonant cavities [91], a field usually referred to as cavity QED [90]. The resonant interaction with a cavity mode increases the brightness and efficiency of the source, locks the emitted photons into a well-defined mode, and empowers the observation of reversible decay dynamics in the strong coupling regime. Commonly used resonators include nanopillars [92], nanodisks [93], photonic crystal [94, 95], superconducting [96], and plasmonic [97, 98] cavities. More recently, attention has been partially shifted to the integration of quantum emitters into waveguides [99–101]. Waveguides exhibit advantages similar to cavities in terms of the coupling efficiency to a specific mode and the Purcell enhancement associated with slow light, while at the same time providing intrinsic long-range interactions that enable the exploration of collective effects [102, 103]. Light emission can not only be enhanced, but also suppressed by the environment. Examples include photonic crystals that exhibit a band gap [104, 105], closed cavities far from resonance [106, 107] and near-zero-index (NZI) media [108–111] which lead to the excitation of bound states [112–114] and the associated many-body physics [115].

Beyond the design of individual emitters, the deterministic fabrication of arrays of quantum emitters with control over their position, orientation, and emission frequency still poses important technological challenges. Again, the most advanced technology is that of atomic-based emitters. Intensive research on the trapping and cooling of atoms in atomic physics has enabled the deterministic generation of arrays of quantum emitters with hundreds of elements, including one-dimensional (1D) [116], two-dimensional (2D) [117], and three-dimensional (3D) [118, 119] topologies with quite arbitrary geometries. Additional advantages of atomic-based systems are that all emitters are ensured to be identical, and the initialization and coherent control of the quantum state of the array is much more advanced. We believe that all of the theory presented in earlier sections of this chapter is within the reach of atomic-based technologies.

Unfortunately, atomic physics setups are not a viable option for most practical applications, and research is needed in the fabrication of quantum antenna arrays in solid-state platforms. However, the fabrication of arrays of solid-state emitters is much more challenging. As solid-state emitters mostly consist of defects in a solid, the interaction with the complex solid-state environment makes it very difficult to obtain identical emitters with the same orientation, emission wavelength, linewidth, etc. The accurate positioning of the emitters is also a challenging task. Nevertheless, there have been some promising demonstrations of arrays of color centers based on selectively growing on arrays of diamond nanopillars and nanopyramids [120, 121]. Moreover, emitters based on atomically-thin semiconductors, e.g. tungsten diselenide (WSe_2) and hexagonal boron nitride (hBN), are currently showing the most promising advances toward the deterministic fabrication of arrays of solid-state quantum emitters. In an atomically-thin semiconductor, an emitter at a specific location can be induced by using nanostructures producing local strains [78, 122, 123], as well as by laser [124] and electron beam [125] irradiation. With these fabrication techniques, we believe that planar quantum antenna arrays based on atomically-thin semiconductors should be feasible in the short term.

12.8 Forward-Looking Directions

This chapter has outlined the basic theory of quantum antenna arrays, establishing connections with the antenna theory field whenever possible. While multiple opportunities can be identified at the fundamental and technological levels, quantum antenna arrays are still a nascent field, and it is difficult to predict how it will evolve in the following years. At the fundamental level, it has been demonstrated that quantum antenna arrays enable directional correlations that cannot be achieved with classical systems. However, the complete set of available correlations in the quantum regime is not known, and their impact on physical observables is yet to be discovered. Note that we have used a simple two-level system model for the emitters. We have not discussed how information would be encoded in the photons emitted by them. Research on more complex emitters, e.g. emitters with a non-dipolar current distributions and/or multi-particle transitions, is likely to facilitate even better performance characteristics and other nontrivial phenomena. Information on the emitted photons might be encoded by using polarization and/or temporal control over them. At the same time, the directional correlations induced by the initial state of the systems are a form of embedded information that could be investigated for some practical information exchanges, i.e. communication-enabled applications. In general, we anticipate that in the next years there will be a large number of theoretical efforts both at the very basic level, as well as in the development of design and synthesis procedures for quantum antenna arrays.

From a practical standpoint, technological applications of quantum antenna arrays are yet to be explored. Classical antenna arrays radiate into free-space, being the basis of wireless technologies. However, current photonic quantum technologies are mostly based on guided platforms such as optical fibers and photonic integrated circuits. The main reason for the predominant use of guided setups is that these quantum optical systems utilize single or a very reduced number of photons. Therefore, propagation photon losses can be prohibitively high for most applications. Nevertheless, aggressive investments on quantum experiments at the space scale are being funded by state agencies, including the recent demonstrations of satellite QKD distribution, seem to indicate that there will be a future for wireless quantum communications.

We expect that the benefits of wireless connections will also be exploited at small scales and within integrated systems. For example, quantum sensing technologies based on solid-state spins rely on their optical readout. Currently, optical readout tasks are carried out with bulky far field optics, an approach that is incompatible with a practical implementation outside the laboratory. They need to be addressed in a wireless fashion. Optical antenna arrays might solve this problem by enabling compact and efficient interfaces between solid-state spins and integrated photonics. Moreover, with the size of solid-state spins being at the atomic scale, dense arrays of spins cannot be individually resolved with far-field optics. Having the need to address a sample with many of these emitters, quantum antenna array theory is likely to be needed for the design of advanced optical interfaces, where quantum correlations might be harnessed to improve the performance of their readout.

These are only a few examples of how future directions might look like for quantum antenna arrays. We re-emphasize that quantum antenna arrays are an emerging field of science and engineering. Many aspects and applications are yet to be explored both from the fundamental and technology perspectives. There are numerous opportunities for electromagnetic researchers with quantum analysis, simulation, and measurement experiences to contribute their skills to this nascent field.

References

1 M. F. Riedel, D. Binosi, R. Thew, and T. Calarco, "The European quantum technologies flagship programme," *Quantum Sci. Technol.*, vol. 2, no. 3, 030501, 2017.

2 V. Scarani, H. Bechmann-Pasquinucci, N. J. Cerf, M. Dušek, N. Lütkenhaus, and M. Peev, "The security of practical quantum key distribution," *Rev. Mod. Phys.*, vol. 81, no. 3, pp. 1301–1350, 2009.

3 Id Quanqique, https://www.idquantique.com.

4 Magiq, https://www.magiqtech.com.

5 Quintessence labs, https://www.quintessencelabs.com/.

6 S.-K. Liao, W.-Q. Cai, W.-Y. Liu, L. Zhang, Y. Li, J.-G. Ren, J. Yin, Q. Shen, Y. Cao, Z.-P. Li *et al.*, "Satellite-to-ground quantum key distribution," *Nature*, vol. 549, no. 7670, pp. 43–47, 2017.

7 J. Aasi, J. Abadie, B. P. Abbott, R. Abbott, T. D. Abbott, M. R. A., C. Adams, T. Adams, P. Addesso, R. X. Adhikari *et al.*, "Enhanced sensitivity of the LIGO gravitational wave detector by using squeezed states of light," *Nat. Photonics*, vol. 7, no. 8, pp. 613–619, 2013.

8 D. A. Simpson, E. Morrisroe, J. M. McCoey, A. H. Lombard, D. C. Mendis, F. Treussart, L. T. Hall, S. Petrou, and L. C. L. Hollenberg, "Non-neurotoxic nanodiamond probes for intraneuronal temperature mapping," *ACS Nano*, vol. 11, no. 12, pp. 12077–12086, 2017.

9 D. Le Sage, K. Arai, D. R. Glenn, S. J. DeVience, L. M. Pham, L. Rahn-Lee, M. D. Lukin, A. Yacoby, A. Komeili, and R. L. Walsworth, "Optical magnetic imaging of living cells," *Nature*, vol. 496, no. 7446, pp. 486–489, 2013.

10 I. Lovchinsky, A. O. Sushkov, E. Urbach, N. P. de Leon, S. Choi, K. De Greve, R. Evans, R. Gertner, E. Bersin, C. Müller *et al.*, "Nuclear magnetic resonance detection and spectroscopy of single proteins using quantum logic," *Science*, vol. 351, no. 6275, pp. 836–841, 2016.

11 F. Ziem, M. Garsi, H. Fedder, and J. Wrachtrup, "Quantitative nanoscale MRI with a wide field of view," *Sci. Rep.*, vol. 9, no. 1, pp. 1–9, 2019.

12 J. I. Cirac, "Quantum computing and simulation: Where we stand and what awaits US," *Nanophotonics*, vol. 10, no. 1, pp. 453–456, 2020.

13 Psiquantum, https://psiquantum.com/.

14 Xanadu, https://www.xanadu.ai/.

15 H.-S. Zhong, H. Wang, Y.-H. Deng, M.-C. Chen, L.-C. Peng, Y.-H. Luo, J. Qin, D. Wu, X. Ding, Y. Hu *et al.*, "Quantum computational advantage using photons," *Science*, vol. 370, no. 6523, pp. 1460–1463, 2020.

16 W. Vogel and D.-G. Welsch, *Quantum Optics*. John Wiley & Sons, 2006.

17 T. G. Philbin, "Canonical quantization of macroscopic electromagnetism," *New J. Phys.*, vol. 12, no. 12, 123008, 2010.

18 S. Scheel and S. Y. Buhmann, "Macroscopic QED-concepts and applications," *arXiv preprint arXiv:0902.3586*, 2009.

19 S. Y. Buhmann, D. T. Butcher, and S. Scheel, "Macroscopic quantum electrodynamics in nonlocal and nonreciprocal media," *New J. Phys.*, vol. 14, no. 8, 083034, 2012.

20 I. Liberal, I. Ederra, and R. W. Ziolkowski, "Control of a quantum Emitter's bandwidth by managing its reactive power," *Phys. Rev. A*, vol. 100, no. 2, 023830, 2019.

21 C.-T. Tai, *Dyadic Green Functions in Electromagnetic Theory*. IEEE Press, 1994.

22 K. Joulain, J.-P. Mulet, F. Marquier, R. Carminati, and J.-J. Greffet, "Surface electromagnetic waves thermally excited: Radiative heat transfer, coherence properties and Casimir forces revisited in the near field," *Surf. Sci. Rep.*, vol. 57, no. 3–4, pp. 59–112, 2005.

23 H. J. Kimble, M. Dagenais, and L. Mandel, "Photon antibunching in resonance fluorescence," *Phys. Rev. Lett.*, vol. 39, no. 11, pp. 691–695, 1977.
24 J. F. Clauser, "Experimental distinction between the quantum and classical field-theoretic predictions for the photoelectric effect," *Phys. Rev. D*, vol. 9, no. 4, pp. 853–860, 1974.
25 A. Aspect, J. Dalibard, and G. Roger, "Experimental test of Bell's inequalities using time-varying analyzers," *Phys. Rev. Lett.*, vol. 49, no. 25, pp. 1804–1807, 1982.
26 A. Garg and N. D. Mermin, "Detector inefficiencies in the Einstein-Podolsky-Rosen experiment," *Phys. Rev. D*, vol. 35, no. 12, pp. 3831–3835, 1987.
27 P. H. Eberhard, "Background level and counter efficiencies required for a loophole-free Einstein-Podolsky-Rosen experiment," *Phys. Rev. A*, vol. 47, no. 2, pp. R747–R750, 1993.
28 J. Barrett, D. Collins, L. Hardy, A. Kent, and S. Popescu, "Quantum nonlocality, bell inequalities, and the memory loophole," *Phys. Rev. A*, vol. 66, no. 4, 042111, 2002.
29 B. Hensen, H. Bernien, A. E. Dréau, A. Reiserer, N. Kalb, M. S. Blok, J. Ruitenberg, R. F. L. Vermeulen, R. N. Schouten, C. Abellán *et al.*, "Loophole-free bell inequality violation using electron spins separated by 1.3 kilometres," *Nature*, vol. 526, no. 7575, pp. 682–686, 2015.
30 L. K. Shalm, E. Meyer-Scott, B. G. Christensen, P. Bierhorst, M. A. Wayne, M. J. Stevens, T. Gerrits, S. Glancy, D. R. Hamel, M. S. Allman *et al.*, "Strong loophole-free test of local realism," *Phys. Rev. Lett.*, vol. 115, no. 25, 250402, 2015.
31 M. Giustina, M. A. M. Versteegh, S. Wengerowsky, J. Handsteiner, A. Hochrainer, K. Phelan, F. Steinlechner, J. Kofler, J.-Å. Larsson, C. Abellán *et al.*, "Significant-loophole-free test of Bell's theorem with entangled photons," *Phys. Rev. Lett.*, vol. 115, no. 25, 250401, 2015.
32 R. Loudon, *The Quantum Theory of Light*. OUP Oxford, 2000.
33 R. J. Glauber, "Photon correlations," *Phys. Rev. Lett.*, vol. 10, no. 3, pp. 84–86, 1963.
34 R. J. Glauber, "The quantum theory of optical coherence," *Phys. Rev.*, vol. 130, no. 6, pp. 2529–2539, 1963.
35 P. L. Kelley and W. H. Kleiner, "Theory of electromagnetic field measurement and photoelectron counting," *Phys. Rev.*, vol. 136, no. 2A, pp. A316–A334, 1964.
36 H. J. Carmichael, *Statistical Methods in Quantum Optics 2: Non-Classical Fields*. Springer Science & Business Media, 2009.
37 C. A. Balanis, *Antenna Theory: Analysis and Design*. John Wiley & Sons, 2005.
38 R. C. Hansen, *Phased Array Antennas*, vol. 213. John Wiley & Sons, 2009.
39 R. J. Mailloux, *Phased Array Antenna Handbook*, vol. 2. Boston, MA: Artech House, 2005.
40 P.-S. Kildal, *Foundations of Antenna Engineering: A Unified Approach for Line-of-sight and Multipath*. Artech House, 2015.
41 A. Predojević and M. W. Mitchell, *Engineering the Atom-Photon Interaction*. Springer, 2015.
42 M. Saffman, T. G. Walker, and K. Mølmer, "Quantum information with Rydberg atoms," *Rev. Mod. Phys.*, vol. 82, no. 3, pp. 2313–2363, 2010.
43 A. Grankin, D. V. Vasilyev, P. O. Guimond, B. Vermersch, and P. Zoller, A 'few-atom' quantum optical antenna. *arXiv preprint arXiv:1802.05592*, 2018.
44 I. Liberal, I. Ederra, and Y. Li, "Isotropic single-photon sources," *Opt. Lett.*, vol. 43, no. 12, pp. 2736–2739, 2018.
45 H. F. Mathis, "A short proof that an isotropic antenna is impossible," *Proc. IRE*, vol. 39, p. 970, 1951.
46 W. K. Saunders, "On the Unity Gain Antenna," in *Electromagnetic Theory and Antennas*, vol. 2, E. C. Jordan, Ed. Pergamon Press, 1963.
47 W. Scott and K. Soo Hoo, "A theorem on the polarization of null-free antennas," *IEEE Trans. Antennas Propag.*, vol. 14, no. 5, pp. 587–590, 1966.
48 H. Matzner and K. T. McDonald, "Isotropic radiators,". *arXiv preprint physics/0312023*, 2003.

49 P. Renteln, *Manifolds, Tensors, and Forms: An Introduction for Mathematicians and Physicists.* Cambridge University Press, 2013.
50 R. E. Schwartz, *Mostly Surfaces*, vol. 60. American Mathematical Society, 2011.
51 Y. Pan and S. Zheng, "A compact quasi-isotropic shorted patch antenna," *IEEE Access*, vol. 5, pp. 2771–2778, 2017.
52 P. Liu, Y. Li, Z. Zhang, and Z. Feng, "Omnidirectional dual-polarized antenna with sabre-like structure," *IEEE Trans. Antennas Propag.*, vol. 65, no. 6, pp. 3221–3225, 2017.
53 Q. Li, W.-J. Lu, S.-G. Wang, and L. Zhu, "Planar quasi-isotropic magnetic dipole antenna using fractional-order circular sector cavity resonant mode," *IEEE Access*, vol. 5, pp. 8515–8525, 2017.
54 J.-H. Kim and S. Nam, "A compact quasi-isotropic antenna based on folded split-ring resonators," *IEEE Antennas Wireless Propag. Lett.*, vol. 16, pp. 294–297, 2017.
55 Z. N. Chen, X. Qing, T. S. P. See, and W. K. Toh, "Antennas for WiFi connectivity," *Proc. IEEE*, vol. 100, no. 7, pp. 2322–2329, 2012.
56 S. X. Ta, I. Park, and R. W. Ziolkowski, "Crossed dipole antennas: A review," *IEEE Antennas Propag. Mag.*, vol. 57, no. 5, pp. 107–122, 2015.
57 M. Arrawatia, M. S. Baghini, and G. Kumar, "Broadband bent triangular omnidirectional antenna for RF energy harvesting," *IEEE Antennas Wireless Propag. Lett.*, vol. 15, pp. 36–39, 2016.
58 F. Fulton, "The combined radiation pattern of three orthogonal dipoles," *IEEE Trans. Antennas Propag.*, vol. 13, no. 2, pp. 323–324, 1965.
59 A. D. Yaghjian, "Scattering-matrix analysis of linear periodic arrays," *IEEE Trans. Antennas Propag.*, vol. 50, no. 8, pp. 1050–1064, Aug 2002.
60 M. O. Scully and M. S. Zubairy, *Quantum Optics*. Cambridge University Press, 1997.
61 M. O. Scully and K. Drühl, "Quantum eraser: A proposed photon correlation experiment concerning observation and "delayed choice" in quantum mechanics," *Phys. Rev. A*, vol. 25, no. 4, pp. 2208–2213, 1982.
62 Y. Aharonov and M. S. Zubairy, "Time and the quantum: Erasing the past and impacting the future," *Science*, vol. 307, no. 5711, pp. 875–879, 2005.
63 I. Liberal, I. Ederra, and R. W. Ziolkowski, "Quantum antenna arrays: The role of quantum interference on direction-dependent photon statistics," *Phys. Rev. A*, vol. 97, no. 5, 053847, 2018.
64 R. Harrington, "On the gain and beamwidth of directional antennas," *IRE Trans. Antennas Propag.*, vol. 6, no. 3, pp. 219–225, 1958.
65 F. Diedrich and H. Walther, "Nonclassical radiation of a single stored ion," *Phys. Rev. Lett.*, vol. 58, no. 3, pp. 203–206, 1987.
66 Th. Basché, W. E. Moerner, M. Orrit, and H. Talon, "Photon antibunching in the fluorescence of a single dye molecule trapped in a solid," *Phys. Rev. Lett.*, vol. 69, no. 10, pp. 1516–1519, 1992.
67 N. Tomm, A. Javadi, N. O. Antoniadis, D. Najer, M. C. Löbl, A. R. Korsch, R. Schott, S. R. Valentin, A. D. Wieck, A. Ludwig *et al.*, "A bright and fast source of coherent single photons," *Nat. Nanotechnol.*, vol. 16, no. 4, pp. 399–403, 2021.
68 P. Senellart, G. Solomon, and A. White, "High-performance semiconductor quantum-dot single-photon sources," *Nat. Nanotechnol.*, vol. 12, no. 11, pp. 1026–1039, 2017.
69 P. Lodahl, S. Mahmoodian, and S. Stobbe, "Interfacing single photons and single quantum dots with photonic nanostructures," *Rev. Mod. Phys.*, vol. 87, no. 2, pp. 347–400, 2015.
70 S. Buckley, K. Rivoire, and J. Vučković, "Engineered quantum dot single-photon sources," *Rep. Prog. Phys.*, vol. 75, no. 12, 126503, 2012.

71 M. Atatüre, D. Englund, N. Vamivakas, S.-Y. Lee, and J. Wrachtrup, "Material platforms for spin-based photonic quantum technologies," *Nat. Rev. Mater.*, vol. 3, no. 5, pp. 38–51, 2018.

72 D. D. Awschalom, R. Hanson, J. Wrachtrup, and B. B. Zhou, "Quantum technologies with optically interfaced solid-state spins," *Nat. Photonics*, vol. 12, no. 9, pp. 516–527, 2018.

73 I. Aharonovich, D. Englund, and M. Toth, "Solid-state single-photon emitters," *Nat. Photonics*, vol. 10, no. 10, pp. 631–641, 2016.

74 S. Castelletto, B. C. Johnson, V. Ivády, N. Stavrias, T. Umeda, A. Gali, and T. Ohshima, "A silicon carbide room-temperature single-photon source," *Nat. Mater.*, vol. 13, no. 2, pp. 151–156, 2014.

75 A. Sipahigil, R. E. Evans, D. D. Sukachev, M. J. Burek, J. Borregaard, M. K. Bhaskar, C. T. Nguyen, J. L. Pacheco, H. A. Atikian, C. Meuwly *et al.*, "An integrated diamond nanophotonics platform for quantum-optical networks," *Science*, vol. 354, no. 6314, pp. 847–850, 2016.

76 I. Aharonovich and E. Neu, "Diamond nanophotonics," *Adv. Opt. Mater.*, vol. 2, no. 10, pp. 911–928, 2014.

77 M. Toth and I. Aharonovich, "Single photon sources in atomically thin materials," *Annu. Rev. Phys. Chem.*, vol. 70, pp. 123–142, 2019.

78 T. T. Tran, K. Bray, M. J. Ford, M. Toth, and I. Aharonovich, "Quantum emission from hexagonal boron nitride monolayers," *Nat. Nanotechnol.*, vol. 11, no. 1, pp. 37–41, 2016.

79 Y.-M. He, G. Clark, J. R. Schaibley, Y. He, M.-C. Chen, Y.-J. Wei, X. Ding, Q. Zhang, W. Yao, X. Xu *et al.*, "Single quantum emitters in monolayer semiconductors," *Nat. Nanotechnol.*, vol. 10, no. 6, pp. 497–502, 2015.

80 A. Srivastava, M. Sidler, A. V. Allain, D. S. Lembke, A. Kis, and A. Imamoğlu, "Optically active quantum dots in monolayer WSE 2," *Nat. Nanotechnol.*, vol. 10, no. 6, pp. 491–496, 2015.

81 A. Kuhn, M. Hennrich, and G. Rempe, "Deterministic single-photon source for distributed quantum networking," *Phys. Rev. Lett.*, vol. 89, no. 6, 067901, 2002.

82 J. McKeever, A. Boca, A. D. Boozer, R. Miller, J. R. Buck, A. Kuzmich, and H. J. Kimble, "Deterministic generation of single photons from one atom trapped in a cavity," *Science*, vol. 303, no. 5666, pp. 1992–1994, 2004.

83 M. Keller, B. Lange, K. Hayasaka, W. Lange, and H. Walther, "Continuous generation of single photons with controlled waveform in an ion-trap cavity system," *Nature*, vol. 431, no. 7012, pp. 1075–1078, 2004.

84 A. Wallraff, D. I. Schuster, A. Blais, L. Frunzio, R.-S. Huang, J. Majer, S. Kumar, S. M. Girvin, and R. J. Schoelkopf, "Strong coupling of a single photon to a superconducting qubit using circuit quantum electrodynamics," *Nature*, vol. 431, no. 7005, pp. 162–167, 2004.

85 A. A. Houck, D. I. Schuster, J. M. Gambetta, J. A. Schreier, B. R. Johnson, J. M. Chow, L. Frunzio, J. Majer, M. H. Devoret, S. M. Girvin *et al.*, "Generating single microwave photons in a circuit," *Nature*, vol. 449, no. 7160, pp. 328–331, 2007.

86 F. Kaneda and P. G. Kwiat, "High-efficiency single-photon generation via large-scale active time multiplexing," *Sci. Adv.*, vol. 5, no. 10, eaaw8586, 2019.

87 P. G. Kwiat, K. Mattle, H. Weinfurter, A. Zeilinger, A. V. Sergienko, and Y. Shih, "New high-intensity source of polarization-entangled photon pairs," *Phys. Rev. Lett.*, vol. 75, no. 24, pp. 4337–4341, 1995.

88 C. K. Hong and L. Mandel, "Experimental realization of a localized one-photon state," *Phys. Rev. Lett.*, vol. 56, no. 1, pp. 58–60, 1986.

89 E. M. Purcell, " Spontaneous emission probabilities at radio frequencies," in *Confined Electrons and Photons*. Springer, 1995, p. 839.

90 S. Haroche and J.-M. Raimond, *Exploring the Quantum: Atoms, Cavities, and Photons*. Oxford University Press, 2006.

91 K. J. Vahala, "Optical microcavities," *Nature*, vol. 424, no. 6950, pp. 839–846, 2003.

92 J. P. Reithmaier, G. Sek, A. Löffler, C. Hofmann, S. Kuhn, S. Reitzenstein, L. V. Keldysh, V. D. Kulakovskii, T. L. Reinecke, and A. Forchel, "Strong coupling in a single quantum dot–semiconductor microcavity system," *Nature*, vol. 432, no. 7014, pp. 197–200, 2004.

93 K. Srinivasan and O. Painter, "Linear and nonlinear optical spectroscopy of a strongly coupled microdisk–quantum dot system," *Nature*, vol. 450, no. 7171, pp. 862–865, 2007.

94 T. Yoshie, A. Scherer, J. Hendrickson, G. Khitrova, H. M. Gibbs, G. Rupper, C. Ell, O. B. Shchekin, and D. G. Deppe, "Vacuum Rabi splitting with a single quantum dot in a photonic crystal nanocavity," *Nature*, vol. 432, no. 7014, pp. 200–203, 2004.

95 K. Hennessy, A. Badolato, M. Winger, D. Gerace, M. Atatüre, S. Gulde, S. Fält, E. L. Hu, and A. Imamoğlu, "Quantum nature of a strongly coupled single quantum dot–cavity system," *Nature*, vol. 445, no. 7130, pp. 896–899, 2007.

96 S. Kuhr, S. Gleyzes, C. Guerlin, J. Bernu, U. B. Hoff, S. Deléglise, S. Osnaghi, M. Brune, J.-M. Raimond, S. Haroche *et al.*, "Ultrahigh finesse Fabry-Pérot superconducting resonator," *Appl. Phys. Lett.*, vol. 90, no. 16, 164101, 2007.

97 G. M. Akselrod, C. Argyropoulos, T. B. Hoang, C. Ciracì, C. Fang, J. Huang, D. R. Smith, and M. H. Mikkelsen, "Probing the mechanisms of large Purcell enhancement in plasmonic nanoantennas," *Nat. Photonics*, vol. 8, no. 11, pp. 835–840, 2014.

98 R. Chikkaraddy, B. D. Nijs, F. Benz, S. J. Barrow, O. A. Scherman, E. Rosta, A. Demetriadou, P. Fox, O. Hess, and J. J. Baumberg, "Single-molecule strong coupling at room temperature in plasmonic nanocavities," *Nature*, vol. 535, no. 7610, pp. 127–130, 2016.

99 M. Arcari, I. Söllner, A. Javadi, S. L. Hansen, S. Mahmoodian, J. Liu, H. Thyrrestrup, E. H. Lee, J. D. Song, S. Stobbe *et al.*, "Near-unity coupling efficiency of a quantum emitter to a photonic crystal waveguide," *Phys. Rev. Lett.*, vol. 113, no. 9, 093603, 2014.

100 A. V. Akimov, A. Mukherjee, C. L. Yu, D. E. Chang, A. S. Zibrov, P. R. Hemmer, H. Park, and M. D. Lukin, "Generation of single optical plasmons in metallic nanowires coupled to quantum dots," *Nature*, vol. 450, no. 7168, pp. 402–406, 2007.

101 A. F. Van Loo, A. Fedorov, K. Lalumiere, B. C. Sanders, A. Blais, and A. Wallraff, "Photon-mediated interactions between distant artificial atoms," *Science*, vol. 342, no. 6165, pp. 1494–1496, 2013.

102 A. Goban, C.-L. Hung, J. D. Hood, S.-P. Yu, J. A. Muniz, O. Painter, and H. J. Kimble, "Superradiance for atoms trapped along a photonic crystal waveguide," *Phys. Rev. Lett.*, vol. 115, no. 6, 063601, 2015.

103 A. Goban, C.-L. Hung, S.-P. Yu, J. D. Hood, J. A. Muniz, J. H. Lee, M. J. Martin, A. C. McClung, K. S. Choi, D. E. Chang *et al.*, "Atom–light interactions in photonic crystals," *Nat. Commun.*, vol. 5, no. 1, pp. 1–9, 2014.

104 V. P. Bykov, "Spontaneous emission from a medium with a band spectrum," *Sov. J. Quantum Electron.*, vol. 4, no. 7, pp. 861–871, 1975.

105 E. Yablonovitch, "Inhibited spontaneous emission in solid-state physics and electronics," *Phys. Rev. Lett.*, vol. 58, no. 20, pp. 2059–2062, 1987.

106 D. Kleppner, "Inhibited spontaneous emission," *Phys. Rev. Lett.*, vol. 47, no. 4, pp. 233–236, 1981.

107 R. G. Hulet, E. S. Hilfer, and D. Kleppner, "Inhibited spontaneous emission by a Rydberg atom," *Phys. Rev. Lett.*, vol. 55, no. 20, pp. 2137–2140, 1985.

108 M. Lobet, I. Liberal, E. N. Knall, M. Z. Alam, O. Reshef, R. W. Boyd, N. Engheta, and E. Mazur, "Fundamental radiative processes in near-zero-index media of various dimensionalities," *ACS Photonics*, vol. 7, no. 8, pp. 1965–1970, 2020.

109 I. Liberal and N. Engheta, "Zero-index structures as an alternative platform for quantum optics," *Proc. Natl. Acad. Sci. U.S.A.*, vol. 114, no. 5, pp. 822–827, 2017.

110 I. Liberal and N. Engheta, "Nonradiating and radiating modes excited by quantum emitters in open epsilon-near-zero cavities," *Sci. Adv.*, vol. 2, no. 10, e1600987, 2016.

111 A. Erentok and R. W. Ziolkowski, "A hybrid optimization method to analyze metamaterial-based electrically small antennas," *IEEE Trans. Antennas Propag.*, vol. 55, no. 3, pp. 731–741, 2007.

112 S. John and J. Wang, "Quantum electrodynamics near a photonic band gap: Photon bound states and dressed atoms," *Phys. Rev. Lett.*, vol. 64, no. 20, pp. 2418–2421, 1990.

113 S. John and T. Quang, "Spontaneous emission near the edge of a photonic band gap," *Phys. Rev. A*, vol. 50, no. 2, pp. 1764–1769, 1994.

114 T. Shi, Y.-H. Wu, A. González-Tudela, and J. I. Cirac, "Bound states in boson impurity models," *Phys. Rev. X*, vol. 6, no. 2, 021027, 2016.

115 D. E. Chang, J. S. Douglas, A. González-Tudela, C.-L. Hung, and H. J. Kimble, "Colloquium: Quantum matter built from nanoscopic lattices of atoms and photons," *Rev. Mod. Phys.*, vol. 90, no. 3, 031002, 2018.

116 M. Endres, H. Bernien, A. Keesling, H. Levine, E. R. Anschuetz, A. Krajenbrink, C. Senko, V. Vuletic, M. Greiner, and M. D. Lukin, "Atom-by-atom assembly of defect-free one-dimensional cold atom arrays," *Science*, vol. 354, no. 6315, pp. 1024–1027, 2016.

117 H. Labuhn, D. Barredo, S. Ravets, S. De Léséleuc, T. Macrì, T. Lahaye, and A. Browaeys, "Tunable two-dimensional arrays of single Rydberg atoms for realizing quantum Ising models," *Nature*, vol. 534, no. 7609, pp. 667–670, 2016.

118 D. Barredo, V. Lienhard, S. De Leseleuc, T. Lahaye, and A. Browaeys, "Synthetic three-dimensional atomic structures assembled atom by atom," *Nature*, vol. 561, no. 7721, pp. 79–82, 2018.

119 A. Kumar, T.-Y. Wu, F. Giraldo, and D. S. Weiss, "Sorting ultracold atoms in a three-dimensional optical lattice in a realization of Maxwell's demon," *Nature*, vol. 561, no. 7721, pp. 83–87, 2018.

120 T. Jaffe, N. Felgen, L. Gal, L. Kornblum, J. P. Reithmaier, C. Popov, and M. Orenstein, "Deterministic arrays of epitaxially grown diamond nanopyramids with embedded silicon-vacancy centers," *Adv. Opt. Mater.*, vol. 7, no. 2, 1800715, 2019.

121 J. L. Zhang, K. G. Lagoudakis, Y.-K. Tzeng, C. Dory, M. Radulaski, Y. Kelaita, K. A. Fischer, S. Sun, Z.-X. Shen, N. A. Melosh, S. Chu, and J. Vučković, "Complete coherent control of silicon vacancies in diamond nanopillars containing single defect centers," *Optica*, vol. 4, no. 11, pp. 1317–1321, Nov 2017.

122 C. Palacios-Berraquero, D. M. Kara, A. R.-P. Montblanch, M. Barbone, P. Latawiec, D. Yoon, A. K. Ott, M. Loncar, A. C. Ferrari, and A. Mete, "Large-scale quantum-emitter arrays in atomically thin semiconductors," *Nat. Commun.*, vol. 8, 15093, 2017.

123 C. Li, N. Mendelson, R. Ritika, Y. L. Chen, Z.-Q. Xu, M. Toth, and I. Aharonovich, "Scalable and deterministic fabrication of quantum emitter arrays from hexagonal boron nitride," *Nano Lett.*, vol. 21, no. 8, pp. 3626–3632, 2021.

124 S. Hou, M. D. Birowosuto, S. Umar, M. A. Anicet, R. Y. Tay, P. Coquet, B. K. Tay, H. Wang, and E. H. T. Teo, "Localized emission from laser-irradiated defects in two-dimensional hexagonal boron nitride," *2D Materials*, vol. 5, pp. 2053–1583, 2017.

125 C. Fournier, A. Plaud, S. Roux, A. Pierret, M. Rosticher, K. Watanabe, T. Taniguchi, S. Buil, X. Quélin, J. Barjon *et al.*, "Position-controlled quantum emitters with reproducible emission wavelength in hexagonal boron nitride," *Nat. Commun.*, vol. 12, 3779, 2021.

Index

3D magnetic EZ antenna 344
3D printed launcher 235
3D printing 217
3-dB fractional bandwidth (FBW) 337
5G 43, 204, 305
5G antenna requirements 285
6G 214, 305

a

Acousto-optic modulator (AOM) 395
Active
  impedances 287
  patterns 288
  rectangular slot-based FSS 85
Adaptive smart antennas 119, 131
Adjacent power leakage ratio (APLR) 201
Amplitude
  control 26
  modulated RoF signal 263
  modulation (AM) 400, 402
Analog subarrays 305
Analog-to-digital converters (ADC) 305
Angle-of-arrival (AoA) 305, 410
  AoA estimation 305, 308, 310, 314, 323
    narrowband 314
    wideband 318
Angular scanning range 122, 137
Anisotropic
  patches 2
  pixels 3
Antenna arrays 253, 254, 269, 285, 309
Antenna-in-package (AiP) 183
  Fan-out AiP (FO-AiP) 185, 207
  FC-AiP layout 207
  Flip-Chip AiP (FC-AiP) 204
  module 185
  testing 200
  wire-bond 201
Antenna-on-chip (AoC) 183
Antenna site (AS) 253, 256
Anytime-wireless-everywhere (AWE) 335
Aperture sharing 11, 29, 30
Arbitrary waveform generator (AWG) 264
Array
  analysis 297
  elements reduction 83
  factor 287, 428, 438
  feeding 255
  limitation 294
  receivers 297, 298, 299
Arrayed modules 270
Arrays of phased subarrays 314
Artificial intelligence (AI) 140
Artificial magnetic conductor (AMC) 197, 340
Automated test equipment (ATE) 200

b

Ball grid array (BGA) 183
Band pass compensation 194
Band-pass filter (BPF) 264
Bandwidth 22, 336
Bandwidth-gain product 22
Base station 253, 307
  base station array 295
Beamforming 83, 110, 131, 253, 261, 269, 271, 273, 276, 294
  low complexity 83
  low-cost 110
  Mm-wave 253
  multi-user beam-forming (MU-BF) 131

Antenna and Array Technologies for Future Wireless Ecosystems, First Edition.
Edited by Y. Jay Guo and Richard W. Ziolkowski.

Beamforming (*contd.*)
photonics 253, 254
remote antenna 261
RoF-based 270
Beam scanning 57, 119, 131
2D 66, 68
longitudinal 66
response 120
transversal 68
Beam shaping 15
Beam steering 108, 294, 363
1D 109
2D 110
mechanical 106
Beam switching 98, 99
Benzocyclobutene (BCB) 186
Bessel beam 232, 235
Beyond 5G (B5G)
antenna requirements 43
communications systems 305
Bismaleimide-triazine (BT) resin 186
Boundary conditions (BCs) 1
continuous BC synthesis 9
Broadside radiating 354
Bumps 194
Butler matrix 110, 309, 324
Butler matrix-based transceiver (BMT) 324

c

Canopy antenna 346
Cardioid 364
Channel slicing 131
Chirp signal 412
Chromatic dispersion (CD) 253, 256, 263
Circular polarized antenna (CP) 2
CP Lens 222, 228
left hand CP (LHCP) 11
omnidirectional CP (OCP) 362
right hand CP (RHCP) 11
Circulator 280
Classical array 285
classical array factor 286
Coherent signal-subspace method (CSM) 311
Coincidence measurements 426
Compact antenna 267, 268
Compartment shields (CPS) 198
Complementary metal oxide semiconductor (CMOS) 183
Composite right-/left-handed (CRLH) 106
Compression point 201
Conformal shields (CFS) 198
Contemporary array analysis 285
Control site (CS) 253, 256
Cooling
active 200
mechanism 199
passive 199
Cost reduction 83
Cyber-physical systems 335

d

Data transmission 271
Decay dynamics 424
Defected ground plane 190
Defected ground structure (DGS) 356
Delay control 253
Desired permittivity 45
Device under test (DUT) 200
DFT beam difference (DBD) 326
Dielectric polarizer 222
Dielectric superstrate 67
Die micrograph 184
Digital signal
beamforming 273
transmission 264
Digital-to-analog converters (DAC) 305
Direct amplitude modulation (DAM) 345
Direct delay control 271
Directional antennas 83
Direction of arrival (DoA) 140, 141
Directivity 130, 338, 351
Discrete antenna arrays (DAAs) 308
Discrete dielectric lens (DDL) 220
Discrete Fourier transform (DFT) 308
Dispersion curve 121
Dispersion engineering 121
Doppler broadening 396
Double-feed 11
Double negative (DNG) 339
Double positive (DPS) 339
Drone radar 208
Dual-beam 29

e

Effective isotropic radiated power (EIRP) 201
Efficiency 15, 338
  antenna 15
  aperture 15, 17
  aperture tapering 15, 17
  conversion 16
  feed 19
  high aperture 20
  maximum aperture 17
  ohmic 15
  overall 338
  radiation 338
  spill over 15
Efficient communications 131
Efficient wireless systems 119, 130
Egyptian axe dipole (EAD) 342, 343
Electrically small antennas (ESA) 335, 336, 343, 345, 354
Electrical to optical (E/O) conversion 259
Electric-field
  measurements 393
  strength 394
Electric Hertzian dipole (EHD) 338
Electro-absorption modulator-integrated with laser diode (EML) 276
Electromagnetically induced transparency (EIT) 394
Electromagnetic bandgap (EBG) 356
Electronically steerable parasitic array (ESPA) 88
Electronically tunable component 89
Elemental radiators 338
Element pattern overlap matrix 290
Element synthesis 9
Elevation pattern shaping 15
Ellipsoidal lens 100
Embedded element patterns (EEPs) 288
  accuracy of EEPs 302
Embedded wafer level ball (eWLB) 187
Endfire radiation 354
Epoxy molding compound (EMC) 185
Epsilon-negative (ENG) 339
Erbium-doped fiber amplifier (EDFA) 259
Error vector magnitude (EVM) 201, 265
Estimation error 321
Exact design methods 292
Excitation 10
  double-mode 10
  peripheral 12
Exponential modulation 24
Extended hemispherical lens 98

f

Fabry–Perot (FP)
  cavity antenna 108, 109, 110
  resonator antenna 108
Fan-out wafer level packaging (FOWLP) 186
Far-field focusing Lenses 225, 228, 232
Far-zone 422
Fast multipole method (FMM) 10
Fiber-coupled (FC) atom probe 408
Field programmable gate array (FPGA) 94
Figures of merit 299, 336
Fine-phase tuning 73
Fine structures printing 218
Fingerprinting (FP) 140
Finite array 287
Finite-difference time-domain (FDTD) 292
Finite element method (FEM) 293
Fired ceramic materials 185
First-order correlations 429
  first-order correlation pattern 430
Fixed-frequency beam-scanning 105, 107
Flame retardant 4 (FR4) 186
Flat optics (FO) 1
Flexible array pattern 267
Flip-chip interconnect 194
Formlabs 217
Foster's theorem 345
Frequency agility 345
Frequency division multiplexing (FDM) 257
Frequency modulation (FM) 400, 402
Frequency selective surfaces (FSSs) 83
  active FSS (AFSS) 83, 84, 85
  beam-switching antenna 85
  cross type 84
Front-to-back ratio (FTBR) 351

g

Gain enhancement 57, 62, 64
Gain-to-noise temperature ratio (G/T) 306
Gaussian rings basis functions (GRBFs) 34
Geosynchronous Earth orbit (GEO) 311

Ground-signal-ground (GSG) 191
Ground-signal-ground-signal-ground (GSGSG) 191

h

Hamiltonian
  interaction 422
  macroscopic 422
  photonic 422
High-density interconnect (HDI) 186
High-directivity antenna 105
High frequency structure simulator (HFSS) 258
High gain 15, 21, 43, 61
High-impedance surface (HIS) 340
High power microwave (HPM) 344
High-speed trains (HSTs) 307
High-volume manufacturing (HVM) 185
Holey dielectric sheets 46
Huygens dipole antenna (HDA) 338, 353, 354
  array (HDAA) 363
Huygens source 351
Hybrid antenna arrays 305

i

Impedance
  reactance 342
  resistance 342
Improved gain ESPAR 90
Incoherent signal sub-space method (ISSM) 311
Infinite array 287
Instantaneous bandwidth 345
Integrated photonic antenna module 267
Integrated photonic array-antenna (IPA) 258
  beam control 266
Intensity modulator (IM) 255
Intermediate frequency (IF) 264
Internet of things (IoT) 119, 363
Isolated element impedances 287
Isotropic 2, 3
  patches 2
  pixels 3
  radiators 292
Isotropic single-photon sources 434, 435

j

Jig 260

l

Large-scale
  antenna arrays 307, 308, 314
  hybrid arrays 305, 308, 309
Laser tuning 402
Leakage rates 121
Leaky-mode dispersion 125
Leaky-wave antenna (LWA) 105, 106, 119
  array (LAA) 309, 312
  fixed-frequency electronically-scanned 143
  forward-scanning 121
  frequency-scanning LWA (FS-LWAs) 120, 123
    FS-LWAs for DoA 144
    smart passive FS-LWA 143
  half-width MLWA 123, 137
  homogeneous LWAs 121
  low-cost LWAs 123
  microstrip LWA (MLWA) 123
  smart 119
  traditional 105
Leaky-wave mode 106
Lens antennas measurements 218
Light source 276
Linear quantum arrays 427
Line-of-sight (LoS) 201
Liquid crystal polymer (LCP) 186
Liquid metal actuation 74
Loaded microstrip patch 94
Localization 140
Local oscillator (LO) 264
Lossless, resonant, minimum scattering approximation (LRMSA) 290, 291
  LRMSA *versus* MoM methods 292, 300
Low Earth orbit (LEO) 15
Low pass compensation 193
Low power wide area networks (LPWAN) 344
Low SNR regions 327
Low temperature co-fired ceramic (LTCC) 185
LP lenses 225, 232, 235
Luneburg lens 44, 100
  cylindrical 102
  dual LP 50, 51
  flat 47, 53
  hemispherical 55
  parallel-plate 102

patch-fed 50
waveguide-fed 47

m

Machine learning (ML) 140
Macroscopic quantum electrodynamics 420
Magnetoelectric (ME) 363
Markovian regime 424
Massive antenna arrays 305
Maximum directivity beamformer 295
Mean squared error (MSE) 321
Mean squared error lower bound (MSELB) 321
Measurements
AM signals 400, 403
atom-based 399
electric fields 393
FM signals 400
modulated signals 393
plasmas 412
platforms 218
power 393
splitting 399
uncertainties 400
voltage 393, 405
Meshed ground plane 186
Mesoscale 8
Metal Fill 196
Metamaterial (MTM) 196, 339, 341
metamaterial-based 339
metamaterial-inspired 341, 342
Metasurfaces (MTSs) 1, 196
antennas 14
currents flowing 1
dual-band 35
dual polarized 11, 13, 14
excitation 18
requirements 9
surface wave-based 1
Method of moments (MoM) 8, 290, 293
Microstrip patch array 53
Microwave power transmission (MPT) 357
Millimeter-wave (mmWave) 305, 362
NFRP ESAs 362
WPT (mmWPT) 363
MIMO 183
multi-user MIMO (MU-MIMO) 131
Minimum scattering 290
Mm-wave antenna array 307
Modulation and coding scheme (MCS) 131
Modulation parameters 5
Module integrated photodiodes antenna 267
Mueller
matrix imaging 215
polarimetry 216
Multibeam 29, 30, 306, 309
lens 98
terabit satellite communications 306
Multi-channel single radio (MCSR) 132
Multi-core fiber (MCF) 257, 272
Multi-frequency 29, 34
Multifunctional 348
Multiple input multiple output (MIMO) 285
Multi-point sources 30
Multiscale 7
analysis 7
synthesis 7
Mu-negative (MNG) 339
Mutual coupling 285

n

Narrowband hybrid array 310
Near field 108
focusing lenses 225, 227, 228
resonant parasitic (NFRP) 340
Needle-like antenna 305
Network Model 297
Nomadic wireless devices 183
Non-Foster (NF)
circuits 344
elements 345
Nth-order correlations 432

o

Omnidirectional circularly polarized (OCP) 362
On-die spiral inductor 197
Operating principles 342
Optical
circulator (CIR) 280
domain 413
fiber 256
signal feed 260
spectrum 277
Optical to electrical (O/E) conversion 259
Over-the-air (OTA) 184

p

Parasitic
 elements 88
 patch 187
Partially-connected hybrid arrays 309
Partially reflective surface (PRS) 108
Patch array with wing 60
Patch inverted-F antenna (PIFA) 94
Patterned ground plane 191
Pencil-beam frequency scanning 131
Phase center (PHC) 28
 stability 28
Phase constant 106
Phased array 83, 131
 phased array transceivers 184
 phased subarrays 310, 311, 314
Phase detection 401
Phase modulation 22, 401, 404
 non-uniform 22
 uniform 22
Phase shift calculation 315
Phase shifter 70
 liquid metal-based 72
 reactively loaded 73
 reconfigurable 70
 switched-line 71
 U-shaped 67
Phase-shift keying (BPSK) 404
Photodiodes 253, 254, 257
Photonic 253
 technologies 419
Photon statistics 426, 430
Pillbox 110, 111, 112
Pixels 1
 design 20
 topology 2
Plasma measurements 412
Polaritonic modes 420
Polarization 10
 circular polarization (CP) 2
 control 4, 6
 dual linear polarization (LP) 10, 53
 left hand circular polarization (LHCP) 11
 linear polarization (LP) 10
 right hand circular polarization (RHCP) 11
Polarizing beam splitter (PBS) 395
Polytetrafluoroethylene (PTFE) 186
Positive intrinsic negative (PIN)
 diodes 83
 varactors 83
Power added efficiency (PAE) 201
Power amplifier 274
Power photodiodes (PDs) 253
 PD-integrated antenna substrate 258
Praseodymium-doped fiber amplifier (PDFA) 276
Pre-preg electronics (Pregtronics) 360
Printing
 horizontal 217
 slanted 217
 vertical 217
Probe calibration 393

q

Quadrature amplitude modulation (QAM) 404
Quadrature phase-shift keying (QPSK) 404
Quantum
 antenna arrays 419, 420, 438
 antennas 437
 computing 419
 electrodynamics (QED) 420
 emitters 427
 fields 422
 key distribution (QKD) 419
 phase shift keying (QPSK) 264
 superdirectivity 437
 technologies 419
Quasi-Yagi 351, 352

r

Rabi frequency 401
Radio frequency (RF) 305
Radio-over-fiber (RoF) 253
 signals 255, 256
 technique 254
Rao–Wilton–Glisson (RWG) 10
Real-time localization systems (RTLSs) 140
 indoor RTLSs solutions 140
Received signal strength indicator (RSSI) 138
 RSSI-based DoA estimation 142, 144
Receive-only systems 297
Receiver 201, 393
Receiver temperature calculation 300
Reciprocity 298

Reconfigurable antennas 94, 349
Reconfiguring surface impedance 106
Rectangular waveguide 405
Rectennas 357, 358
Rectifiers 358
Redistribution layers (RDLs) 186
Reflectarray (AR) 91
 1-bit 91, 92, 96
 classic 94
 element 91
 reconfigurable 94
Resonance 394
Resonant cavity antenna 108
RF
 beam profiler 410
 camera 410
 chains 305
 domain 413
 field carrier 402
 heterodyne 394
 microelectromechanical systems (MEMS) 92
 phases 255, 256
 PIN diodes 91
 power measurement 405
Rotman lens 110
Rydberg
 atom 393
 transition 394

s

Satellite communications 306
Scan loss 48
Scanned monopulse antenna 142
Scanning bandwidths 137
Scanning electron microscope (SEM) 196
Scan patterns 288
Second-order correlation 431
Second-order photon statistics 432
Self-diplex 36
Sensors 393
Shielding 197
Sidelobe level (SLL) 56
Signal-to-noise ratio (SNR) 205, 265, 308, 348
Simultaneous wireless information and power transfer (SWIPT) 306, 358
Single-input-single-output (SISO) 214
Single mode fiber 255
Single-photon sources 419
Single point source 11, 29, 35
SIW (substrate integrated waveguide) 70, 73, 111, 112
SLL reduction 67
Slotted waveguide antenna array (SWAA) 61, 62, 66, 67
 V-shaped 63, 65
Small cells 253
Smart modulated apertures 1
Smart railway communication system 307
Spatial sectorization 131
Split-ring resonators (SRRs) 256
SPnT switches 99
Spurious free dynamic range (SFDR) 201
Stacked microstrip patch antenna 187
Stark shift 394
Stepped-index dielectric lens 99
Superdirective 339
Surface waves (SWs) 1
 inward 12
 MTS antennas 3
 outward 12
 perturbation 1
Switched beam antenna 142
System-in-package (SiP) 183
System-on-chip (SoC) 183
System temperature 300

t

Terahertz (THz) 1, 4, 214, 306
 design 224
 fabrication 224
 lens antennas 214, 224
 testing 224
Testing strategy 201
Test system 200
Thermometry 412
Thin wires 293, 294
Tightly packed arrays 285
Time delay equalizer metasurface 108
Time of arrival (ToA) 140, 141
Transmitarray (TA) 91
 1-bit 91, 92, 94, 96
 aperture phase distribution 91
 FSS-based 98
Transverse equivalent network 124

Transverse plane 130
Transverse resonance equation (TRE) 125
Tunable dielectrics 103
Tunable high impedance surface (HIS) 109
Tunable light source (TLS) 255
Tuning vias 75

u

Ultra-high frequency (UHF) 343
Uniform linear array (ULA) 286
Unit cells
  CRLH 107
  electric 340
  magnetic 340
Uni-traveling carrier photodiode 259
User terminal direction 279

v

Vacuum states 424
Vapor cell 395, 405
Variable delay line (VDL) 255
Variable optical attenuator (VOA) 259
Variable optical delay 271
Vertical electric dipole (VED) 30
Very high frequency (VHF) 144, 343
Via 194
Virtual Prototyping 10
Voltage controlled oscillator (VCO) 201

w

Waveform analyzer 407
Wavelength division multiplexing (WDM) 257
  WDM RoF transmission 261
Wheeler cap 336
Wideband hybrid array 311
Wide-beam synthesis 323
Wi-Fi 144
Wire-bonding interconnect 193
Wireless energy harvesting (WEH) 358
Wireless information transfer (WIT) 307
Wirelessly powered sensors 359
Wireless power transfer (WPT) 307, 335

x

X-band 11

y

Yagi-Uda 187, 285

z

Z-antenna 343

Printed and bound by CPI Group (UK) Ltd, Croydon, CR0 4YY
14/07/2022
03136109-0001